Ergebnisse der Mathematik und ihrer Grenzgebiete

Volume 56

3. Folge

A Series of Modern Surveys in Mathematics

For further volumes:
www.springer.com/series/728

Franc Forstnerič

Stein Manifolds and Holomorphic Mappings

The Homotopy Principle in Complex Analysis

Second Edition

Franc Forstnerič
Faculty of Mathematics and Physics
University of Ljubljana
Ljubljana
Slovenia

ISSN 0071-1136 ISSN 2197-5655 (electronic)
Ergebnisse der Mathematik und ihrer Grenzgebiete. 3. Folge / A Series of Modern Surveys in Mathematics
ISBN 978-3-319-61057-3 ISBN 978-3-319-61058-0 (eBook)
DOI 10.1007/978-3-319-61058-0

Library of Congress Control Number: 2017953831

Mathematics Subject Classification: Primary 32E10, 32H02, 32L05, 32M12, 32M17, Secondary 14M17, 58D15

Printed on acid-free paper

This Springer imprint is published by Springer Nature
The registered company is Springer International Publishing AG
The registered company address is: Gewerbestrasse 11, 6330 Cham, Switzerland

Preface to the Second Edition

The main stimulus for preparing this new edition was the lively development of the subject since the publication of the first edition in 2011. I have tried to incorporate the new results as much as possible, although some of them could only be mentioned or sketched due to their complexity and space limitations. For the sake of consistency and completeness, I also added some material predating 2011 that was not included in the first edition due to a self-imposed limitation on the size. On the other hand, some inessential material has been removed, so neither edition is a subset of the other one. Needless to say, I eliminated the errors, gaps and inconsistencies which had been found.

Let me now describe the main changes in this edition.

Chapter 2, which summarizes the basic Stein manifold theory, has been slightly expanded. In particular, I added the statement and proof of the parametric version of the Cartan-Oka-Weil theorem in Sect. 2.8. Although these extensions follow easily from the classical theory, it seems impossible to find them in the literature. Since they are very important in Oka theory, I decided to include them here.

Chapter 3 is basically unchanged, except for Sect. 3.12 where some important recent works are mentioned.

In the theory of holomorphic automorphisms, discussed in Chap. 4, major new developments occurred in the study of the density property of Stein manifolds and affine algebraic manifolds, and of geometric structures on them; Sect. 4.10 has been rewritten accordingly. The class of complex manifolds called long $\mathbb{C}^n$'s is now somewhat better understood, and these new developments are presented in Sect. 4.21. There are several other improvements and new results in Chap. 4 which are too numerous to mention here. This beautiful and very useful subject would clearly deserve a book in its own right.

Chapters 5 and 6, which constitute the core of the book, received some topical improvements, minor rearrangements, and additions of newly discovered examples of Oka manifolds. The presentation of the relative Oka principle on 1-convex manifolds in Sect. 6.13 reflects new developments which complete the outline of proof given in the first edition. I added Sect. 6.14 where a relative Oka principle for sections of branched holomorphic maps is presented. This result was originally proved in 2003, but was not included in the first edition. In the meantime new applications have been found, two of which are included here. This result may have the potential of extending the Oka principle for Oka pairs of sheaves due to Forster and Ramspott; however, further investigations are needed.

Chapter 7 is new. Sections 7.1–7.2 contain a discussion of the known relationships between the Oka property and several other holomorphic flexibility properties (by which I mean anti-hyperbolicity properties) of complex manifolds. Section 7.3 brings a summary of what we know about which compact complex surfaces are Oka manifolds. There is an expanded version of Sect. 7.4 on Oka maps. In Sect. 7.5, which was contributed by Finnur Lárusson, the reader can find a homotopy theoretic point of view on Oka theory. The final Sect. 7.6 contains miscellaneous new results and a collection of open problems.

Chapters 8–10 roughly correspond to Chaps. 7–9 in the first edition. Sections 8.10–8.13 are new, although the material predates 2011. Section 9.1 has been expanded and moved here from the last chapter. This enabled me to simplify the topological aspects of proofs of some of the subsequent results. Section 9.7 offers a new and considerably simpler proof of the splitting lemma for biholomorphic maps close to the identity on Cartan pairs, also in complex spaces with singularities. Besides its original use in the construction of holomorphic submersions from Stein manifolds, many new applications of this gluing technique have been found, especially to the problem of exposing boundary points of domains. These results play a major role in the study of the so-called squeezing function and of the boundary behavior of invariant metrics. I have rewritten Sect. 9.8 in light of the new results concerning the existence of proper holomorphic immersions and embeddings of Stein manifolds into Stein manifolds with the (volume) density property. Also, there is a new application to the Hodge theory of q-complete manifolds. Section 9.11 contains several new recent results on the existence of proper holomorphic embeddings of open Riemann surfaces into the affine plane $\mathbb{C}^2$. In particular, it has been proved by Wold and the author than any circular planar domain (possibly infinitely connected) with at most finitely many punctures embeds properly holomorphically into $\mathbb{C}^2$.

The changes in Chap. 10 pertain mainly to the *soft Oka principle* in Sects. 10.9–10.11. A more complete presentation of this subject is now available in the book by Cieliebak and Eliashberg, *From Stein to Weinstein and back*, published in 2012. The improvements which concern this text are described at the relevant places, but the overall presentation and the proofs have not been changed since the interested reader can consult their book. These improvements mainly pertain to complex dimension 2 where there is an abundance of exotic Stein structures on surfaces of suitable topological type.

The Oka principle for sections of holomorphic maps over Stein spaces, presented in Chaps. 5 and 6, has reached a mature stage and has essentially been axiomatized. For stratified fibre bundles, considered in Chap. 5, the relevant condition implying the Oka principle is that the fibres satisfy the *Convex Approximation Property*, CAP, characterizing the class of *Oka manifolds* which became one of the central notions of the theory. For holomorphic submersions, the relevant condition is the *Homotopy Approximation Property*, HAP, for sections over small open sets in the base. The gist of the proofs in these two chapters, which follow ideas of Grauert, Henkin and Leiterer, and Gromov, is that these local conditions imply the Oka principle for global sections. This reduces the problem to finding geometric conditions implying CAP or HAP. The notions of ellipticity and subellipticity continue to play an important role, although some finer conditions such as holomorphic flexibility and the density property have become major new sources of examples. A possible next step would be to axiomatize the relevant geometric conditions on the source side; here we are talking of Cartan pairs and $\mathcal{C}$-strings in Stein spaces. This has the potential of extending the Oka principle way beyond its current scope. I leave this task for another day and another person.

I wish to thank the numerous colleagues who have contributed to this book by collaboration, discussions, or simply by pointing out the deficiencies and proposing improvements. First and foremost, I thank Finnur Lárusson for having contributed Sect. 7.5 and for his lively interest and collaboration on this subject, Frank Kutzschebauch for having explained to me the developments concerning the flexibility properties of affine algebraic and Stein manifolds and for numerous discussions of other topics, and the collaborators and colleagues whose ideas and communications contributed in an essential way to this text: Rafael Andrist, Luka Boc Thaler, Gregery Buzzard, Barbara Drinovec Drnovšek, Josip Globevnik, Jürgen Leiterer, Erik Løw, Takeo Ohsawa, Jasna Prezelj, Tyson Ritter, Jean-Pierre Rosay, Marko Slapar, Dror Varolin, Jörg Winkelmann, and Erlend F. Wold. I sincerely thank Peter Landweber who carefully read the draft and proposed numerous improvements.

A major part of the preparation of this edition was made during my visit at the Centre for Advanced Study in Oslo, Norway, as a Fellow of the program *Several Complex Variables and Complex Dynamics* during Fall 2016. I wish to thank Erlend F. Wold and Berit Stensønes for their kind invitation and the institution for excellent working conditions.

Finally, my sincere thanks to the staff of Springer-Verlag for their professional work in the technical preparation of the book.

Ljubljana
May 1, 2017

Franc Forstnerič

Preface to the First Edition

This book is an attempt to present a coherent account of Oka theory, from the classical Oka-Grauert theory originating in the works of Kiyoshi Oka and Hans Grauert to the contemporary developments initiated by Mikhael Gromov.

At the core of Oka theory lies the heuristic *Oka principle*, a term coined by Jean-Pierre Serre in 1951: *Analytic problems on Stein manifolds admit analytic solutions if there are no topological obstructions.* The Cartan-Serre Theorems A and B are primary examples. The main exponent of the classical Oka-Grauert theory is the equivalence between topological and holomorphic classification of principal fiber bundles over Stein spaces. On the interface with affine algebraic geometry the Oka principle holds only rarely, while in projective geometry we have Serre's GAGA principle, the equivalence of analytic and algebraic coherent sheaves on compact projective algebraic varieties. In smooth geometry there is the analogous *homotopy principle* originating in the Smale-Hirsch homotopy classification of smooth immersions.

Modern Oka theory focuses on those properties of a complex manifold Y which ensure that any continuous map $X \to Y$ from a Stein source space X can be deformed to a holomorphic map; the same property is considered for sections of a holomorphic submersion $Y \to X$. By including the Runge approximation and the Cartan extension condition one obtains several ostensibly different Oka properties. Gromov's main result is that a geometric condition called ellipticity—the existence of a dominating holomorphic spray on Y—implies all forms of the Oka principle for maps or sections $X \to Y$. Subsequent research culminated in the result that all Oka properties of a complex manifold Y are equivalent to the following Runge approximation property:

A complex manifold Y is said to be an Oka manifold *if every holomorphic map $f\colon K \to Y$ from a neighborhood of a compact convex set $K \subset \mathbb{C}^n$ to Y can be approximated uniformly on K by entire maps $\mathbb{C}^n \to Y$.*

The related concept of an *Oka map* pertains to the Oka principle for lifting holomorphic maps from Stein sources. The class of Oka manifolds is dual to the class of Stein manifolds in a sense that can be made precise by means of abstract homotopy theory. Finnur Lárusson constructed a model category containing all complex manifolds in which Stein manifolds are cofibrant, Oka manifolds are fibrant, and Oka maps are fibrations. This means that

Stein manifolds are the natural sources of holomorphic maps, while Oka manifolds are the natural targets.

Oka manifolds seem to be few and special; in particular, no compact complex manifold of Kodaira general type is Oka. However, special and highly symmetric objects are often more interesting than average generic ones.

A few words about the content. Chapter 1 contains some preparatory material, and Chap. 2 is a brief survey of Stein space theory. In Chap. 3 we construct open Stein neighborhoods of certain types of sets in complex spaces that are used in Oka theory. Chapter 4 contains an exposition of the theory of holomorphic automorphisms of Euclidean spaces and of the density property, a subject closely intertwined with our main theme. In Chap. 5 we develop Oka theory for stratified fiber bundles with Oka fibers (this includes the classical Oka-Grauert theory), and in Chap. 6 we treat Oka-Gromov theory for stratified subelliptic submersions over Stein spaces. Chapters 7 and 8 contain applications ranging from classical to the recent ones. In Chap. 8 we present results on regular holomorphic maps of Stein manifolds; highlights include the optimal embedding theorems for Stein manifolds and Stein spaces, proper holomorphic embeddings of some bordered Riemann surfaces into $\mathbb{C}^2$, and the construction of noncritical holomorphic functions, submersions and foliations on Stein manifolds. In Chap. 9 we explore implications of Seiberg-Witten theory for the geometry of Stein surface, and we present the Eliashberg-Gompf construction of Stein structures on manifolds with suitable handlebody decomposition. A part of this story is the *Soft Oka principle*.

This book would not have existed without my collaboration with Jasna Prezelj who explained parts of Gromov's work on the Oka principle in her dissertation (University of Ljubljana, 2000). Josip Globevnik suggested that we look into this subject, while many years earlier Edgar Lee Stout proposed that I study the Oka-Grauert principle. My very special thanks go to the colleagues who read parts of the text and offered suggestions for improvements: Barbara Drinovec-Drnovšek, Frank Kutzschebauch, Finnur Lárusson, Takeo Ohsawa, Marko Slapar, and Erlend Fornæss Wold. I am grateful to Reinhold Remmert for his invitation to write a volume for the Ergebnisse series, and to the staff of Springer-Verlag for their professional help.

Finally, I thank Angela Gheorghiu for all those incomparably beautiful arias, and my family for their patience.

Ljubljana
May 1, 2011

Franc Forstnerič

Contents

'Say at last—who art thou?'
'That Power I serve which wills forever evil yet does forever good.'

J.W. Goethe, *Faust*

Part I
Stein Manifolds

'Forgive me, but I don't believe you,' said Woland. 'That cannot be. Manuscripts don't burn.'

M.A. Bulgakov, *The Master and Margarita*

Chapter 1
Preliminaries

In this chapter we review the basic notions and constructions that are indispensable for reading this book. A comprehensive account is available in a number of sources; for smooth manifolds see e.g. [4, 563]; for complex and algebraic manifolds, see [261, 264, 274, 427, 568], among others; and for the theory of Stein manifolds and Stein spaces, see the monographs [260, 274, 300]. A comprehensive introduction to complex differential and analytic geometry is Demailly's unpublished monograph [110]. For the theory of Riemann surfaces, see the texts by Donaldson [130] and Forster [173].

1.1 Complex Manifolds and Holomorphic Maps

We denote by $\mathbb{R}$ the field of real numbers and by $\mathbb{C}$ the field of complex numbers. Let $n \in \mathbb{N} = \{1, 2, 3, \ldots\}$ be a positive integer. The model n-dimensional complex manifold is the complex Euclidean space $\mathbb{C}^n$, the Cartesian product of n copies of $\mathbb{C}$. Let $z = (z_1, \ldots, z_n)$ denote the complex coordinates on $\mathbb{C}^n$. Write $z_j = x_j + \mathrm{i}\, y_j$, where $x_j, y_j \in \mathbb{R}$ and $\mathrm{i} = \sqrt{-1}$. Given a differentiable complex valued function $f\colon D \to \mathbb{C}$ on a domain $D \subset \mathbb{C}^n$, the differential $\mathrm{d}f$ splits as the sum of the $\mathbb{C}$-linear part ∂f and the $\mathbb{C}$-antilinear part $\overline{\partial} f$:

$$\mathrm{d}f = \partial f + \overline{\partial} f = \sum_{j=1}^{n} \frac{\partial f}{\partial z_j}\, \mathrm{d}z_j + \sum_{j=1}^{n} \frac{\partial f}{\partial \bar{z}_j}\, \mathrm{d}\bar{z}_j. \tag{1.1}$$

Here $\mathrm{d}z_j = \mathrm{d}x_j + \mathrm{i}\, \mathrm{d}y_j$, $\mathrm{d}\bar{z}_j = \mathrm{d}x_j - \mathrm{i}\, \mathrm{d}y_j$, and

$$\frac{\partial f}{\partial z_j} = \frac{1}{2}\left(\frac{\partial f}{\partial x_j} - \mathrm{i} \frac{\partial f}{\partial y_j} \right), \qquad \frac{\partial f}{\partial \bar{z}_j} = \frac{1}{2}\left(\frac{\partial f}{\partial x_j} + \mathrm{i} \frac{\partial f}{\partial y_j} \right). \tag{1.2}$$

F. Forstnerič, *Stein Manifolds and Holomorphic Mappings*, Ergebnisse der Mathematik und ihrer Grenzgebiete. 3. Folge / A Series of Modern Surveys in Mathematics 56, DOI 10.1007/978-3-319-61058-0_1

The function f is *holomorphic* if $\mathrm{d}f = \partial f$ on D; that is, if the differential $\mathrm{d}f_z$ is $\mathbb{C}$-linear at every point $z \in D$. Equivalently, f is holomorphic if and only if $\overline{\partial} f = 0$, and this is equivalent to the system of n equations

$$\frac{\partial f}{\partial \bar{z}_j} = 0, \quad j = 1, \ldots, n.$$

Writing $f = u + \mathrm{i}v$ with u and v real, the equation $\partial f/\partial \bar{z}_j = 0$ is equivalent to the system of Cauchy-Riemann equations

$$\frac{\partial u}{\partial x_j} = \frac{\partial v}{\partial y_j}, \qquad \frac{\partial u}{\partial y_j} = -\frac{\partial v}{\partial x_j}. \tag{1.3}$$

A mapping $f = (f_1, f_2, \ldots, f_m) \colon D \to \mathbb{C}^m$ is holomorphic if each component function f_j is such. When $m = n$, f is *biholomorphic* onto its image $D' = f(D) \subset \mathbb{C}^n$ if it is bijective and its inverse $f^{-1} \colon D' \to D$ is also holomorphic. An injective holomorphic map of a domain $D \subset \mathbb{C}^n$ to $\mathbb{C}^n$ is always biholomorphic onto its image [264, p. 19].

A *topological manifold of dimension n* is a second countable Hausdorff topological space which is locally Euclidean, in the sense that each point has an open neighborhood homeomorphic to an open set in $\mathbb{R}^n$. Such a space is metrizable, countably compact, and paracompact.

Assume that X is a topological manifold of even dimension $2n$. A *complex atlas* on X is a collection $\mathcal{U} = \{(U_\alpha, \phi_\alpha)\}_{\alpha \in A}$, where $\{U_\alpha\}_{\alpha \in A}$ is an open cover of X and ϕ_α is a homeomorphism of U_α onto an open subset U'_α of $\mathbb{R}^{2n} = \mathbb{C}^n$ such that for every pair of indexes $\alpha, \beta \in A$ the *transition map*

$$\phi_{\alpha,\beta} = \phi_\alpha \circ \phi_\beta^{-1} : \phi_\beta(U_{\alpha,\beta}) \to \phi_\alpha(U_{\alpha,\beta}) \tag{1.4}$$

is biholomorphic. Here $U_{\alpha,\beta} = U_\alpha \cap U_\beta$. An element (U_α, ϕ_α) of a complex atlas is called a *complex chart*, or a *local holomorphic coordinate system* on X. We also say that charts in a complex atlas are *holomorphically compatible*. For any three indexes $\alpha, \beta, \gamma \in A$ we have

$$\phi_{\alpha,\alpha} = \mathrm{Id}, \qquad \phi_{\alpha,\beta} = \phi_{\beta,\alpha}^{-1}, \qquad \phi_{\alpha,\beta} \circ \phi_{\beta,\gamma} = \phi_{\alpha,\gamma} \tag{1.5}$$

on the respective domains of these maps. Two complex atlases $\mathcal{U}$, $\mathcal{V}$ on a topological manifold X are said to be *holomorphically compatible* if their union $\mathcal{U} \cup \mathcal{V}$ is also a complex atlas. This is an equivalence relation on the set of all complex atlases on X. Each equivalence class contains a unique *maximal complex atlas*—the union of all complex atlases in the given class.

A *complex manifold of complex dimension n* is a topological manifold X of real dimension $2n$ equipped with a complex atlas. Two complex atlases determine the same complex structure on X if and only if they are holomorphically compatible. We write $n = \dim_{\mathbb{C}} X$. A complex manifold of dimension one is called a *Riemann surface*, or a *complex curve* when it is seen as a complex submanifold in another complex manifold. A *complex surface* is a complex manifold of dimension $n = 2$.

A function $f \colon X \to \mathbb{C}$ on a complex manifold is said to be holomorphic if for any chart (U, ϕ) from the maximal atlas on X the function $f \circ \phi^{-1} \colon \phi(U) \to \mathbb{C}$ is holomorphic on the open set $\phi(U) \subset \mathbb{C}^n$. We denote by $\mathcal{O}(X)$ the Fréchet algebra of all holomorphic functions on X with the compact-open topology.

Let X and Y be complex manifolds of dimensions n and m, respectively. A continuous map $f: X \to Y$ is said to be *holomorphic* if for any point $p \in X$ there are complex charts (U, ϕ) on X and (V, ψ) on Y such that $p \in U$, $f(U) \subset V$, and the map $\widetilde{f} = \psi \circ f \circ \phi^{-1}: \phi(U) \longrightarrow \psi(V) \subset \mathbb{C}^m$ is holomorphic on the open set $\phi(U) \subset \mathbb{C}^n$. Since the charts in a complex atlas are holomorphically compatible, the choice of charts is unimportant in this definition.

We denote by $\mathcal{O}(X, Y)$ the space of all holomorphic maps $X \to Y$ endowed with the compact-open topology. The basis of this topology is determined by sets of the form $\{f \in \mathcal{O}(X, Y) : f(K) \subset U\}$ for all pairs of a compact set $K \subset X$ and an open set $U \subset Y$. The same topology is used on the space $\mathcal{C}(X, Y)$ of all continuous maps $X \to Y$, and $\mathcal{O}(X, Y)$ is then a closed subspace of $\mathcal{C}(X, Y)$. A sequence $f_j \in \mathcal{C}(X, Y)$ converges in this topology to a map $f \in \mathcal{C}(X, Y)$ if and only if the sequence of restriction $f_j|_K$ to any compact set $K \subset X$ converges uniformly on K to $f|_K$. The compact-open topology is metrizable which can be seen as follows. Pick a complete distance function dist on Y which induces the manifold topology. For any compact set K in X and continuous maps $f, g: K \to Y$ let $d_K(f, g) = \sup_{x \in K} \mathrm{dist}(f(x), g(x))$. Choose an exhaustion of X by an increasing family of compact sets $K_1 \subset K_2 \subset \cdots \subset \bigcup_j K_j = X$ such that $K_j \subset \mathring{K}_{j+1}$ for all j. The function

$$d(f, g) = \sum_{j=1}^{\infty} 2^{-j} \frac{d_{K_j}(f, g)}{1 + d_{K_j}(f, g)}, \qquad f, g \in \mathcal{C}(X, Y)$$

is then a complete distance function on $\mathcal{C}(X, Y)$.

A *Baire space* is a topological space in which the union of every countable collection of closed sets with empty interior has empty interior. Equivalently, the intersection of a countable family of dense open sets is dense; any such set is said to be *residual*, or a set of the *second category*. A property of elements in a Baire space is said to be *generic* if it holds for all elements in a residual set. Every complete metric space is a Baire space [332, Theorem 34]. In particular, for any pair of complex manifolds X and Y, the space $\mathcal{O}(X, Y)$ is a separable Baire space, a closed subspace of the Baire space $\mathcal{C}(X, Y)$.

If K is a compact set in a complex manifold X then a map $K \to Y$ is said to be *holomorphic on* K if it is holomorphic in an open neighborhood of K in X; two such maps are considered the same if they agree in some neighborhood of K. A continuous family (for example, a homotopy) of maps $K \to Y$ is holomorphic on K if all maps in the family are holomorphic in the same open neighborhood of K.

The notation $D \Subset X$ will mean that the closure of the set D in X is compact; we shall say that D is a *relatively compact subset* of X.

If D is a relatively compact domain with piecewise $\mathcal{C}^1$ boundary in a complex manifold X and $r \in \{0, 1, \ldots, \infty\}$, then $\mathcal{A}^r(D)$ denotes the Banach algebra (Fréchet if $r = \infty$) consisting of all functions $\bar{D} \to \mathbb{C}$ of class $\mathcal{C}^r(\bar{D})$ (i.e., whose derivatives up to order r exist and are continuous on $\bar{D}$) that are holomorphic on D. We use the analogous notation $\mathcal{A}^r(D,Y)$ for maps $\bar{D} \to Y$ of class $\mathcal{A}^r(D)$ to another complex manifold Y.

A map $f: X \to Y$ is *biholomorphic* if it is bijective and both f and its inverse $f^{-1}: Y \to X$ are holomorphic. (This requires that $\dim X = \dim Y$.) As before, the latter condition is superfluous—a bijective holomorphic map between complex manifolds is biholomorphic. Note that every local chart $\phi: U \to \mathbb{C}^n$ on X is a biholomorphic map of U onto $\phi(U) \subset \mathbb{C}^n$.

A biholomorphic self-map $f: X \to X$ is called a *holomorphic automorphism* of X; the collection of all automorphisms is the *holomorphic automorphism group* $\mathrm{Aut}(X)$. We denote by $\mathrm{Aut}_{\mathrm{alg}}(X)$ the group of all algebraic automorphisms of an algebraic manifold. In many cases, $\mathrm{Aut}(X)$ has the structure of a real or complex *Lie group* (see Example 1.2.4). For instance, $\mathrm{Aut}(\mathbb{C})$ consists of all affine linear maps $z \mapsto \alpha z + \beta$ ($\alpha \in \mathbb{C}^* = \mathbb{C} \setminus \{0\}$, $\beta \in \mathbb{C}$) and is a complex two dimensional Lie group. The automorphism group of any bounded domain $D \subset \mathbb{C}^n$ is a finite dimensional real Lie group; the maximal dimension is obtained when D is the ball $\mathbb{B}^n = \{z \in \mathbb{C}^n : |z|^2 = \sum_{j=1}^n |z_j|^2 < 1\}$. The group $\mathrm{Aut}(\mathbb{B}^n)$ acts transitively on $\mathbb{B}^n$, and the isotropy group of the origin $0 \in \mathbb{B}^n$ is the unitary group $U(n)$ (see [483]). Most bounded domains $D \subset \mathbb{C}^n$ have no automorphisms other than the identity. On the other hand, for $n > 1$ the group $\mathrm{Aut}(\mathbb{C}^n)$ is infinite dimensional (see Chap. 4).

Given a holomorphic map $f = (f_1, \dots, f_m): D \to \mathbb{C}^m$ on a domain $D \subset \mathbb{C}^n$, we denote by $\mathrm{rank}_p f$ the complex rank of f at a point $p \in D$; that is, the rank of the complex $m \times n$ Jacobian matrix

$$f'(p) = \left(\frac{\partial f_j}{\partial z_k}(p)\right). \tag{1.6}$$

This matrix represents the differential $df_p = \partial f_p: \mathrm{T}_p\mathbb{C}^n \to \mathrm{T}_{f(p)}\mathbb{C}^m$ in standard bases on the tangent spaces $\mathrm{T}_p\mathbb{C}^n$, $\mathrm{T}_{f(p)}\mathbb{C}^m$, respectively. (See Sect. 1.6.) Clearly $\mathrm{rank}_p f \le \min\{m,n\}$. The map f is an *immersion* at p if $\mathrm{rank}_p f = n$, and is a *submersion* at p if $\mathrm{rank}_p f = m$. These notions coincide when $n = m$, and in this case f is said to be *locally biholomorphic* at p. These notions, being local, extend to holomorphic maps between complex manifolds.

Let X be a complex manifold of dimension n. A subset M of X is a *complex submanifold* of dimension $m \in \{0, 1, \dots, n\}$ (and codimension $d = n - m$) if every point $p \in M$ admits an open neighborhood $U \subset X$ and a holomorphic chart $\phi: U \to U' \subset \mathbb{C}^n$ such that $\phi(U \cap M) = U' \cap (\mathbb{C}^m \times \{0\}^{n-m})$. Any such chart (U, ϕ) on X is said to be *adapted* to M. Let $\pi: \mathbb{C}^n \to \mathbb{C}^m$ denote the coordinate projection $\pi(z_1, \dots, z_m, \dots, z_n) = (z_1, \dots, z_m)$. For each adapted complex chart (U, ϕ) on X we get a complex chart $(U \cap M, \pi \circ \phi|_{U \cap M})$ on M with values in $\mathbb{C}^m$. The collection of all such charts is a complex atlas on M, and the corresponding complex structure on M is the *complex submanifold structure* induced by the inclusion map $\iota: M \hookrightarrow X$. Considering M with this submanifold structure as a complex manifold in its own right, the inclusion ι is a *holomorphic embedding* of M in X, that is, an injective holomorphic immersion of M onto the complex submanifold $\iota(M)$ of X.

The image of an injective holomorphic immersion $f: M \to X$ need not be a submanifold of X, not even a topological one, due to possible accumulation of the image on itself. The following important property prevents this behavior.

Definition 1.1.1 A continuous map $f: X \to Y$ of topological spaces is *proper* if for any compact set $K \subset Y$ the preimage $f^{-1}(K)$ is compact.

A map $f: X \to Y$ between manifolds is proper if and only if it maps any discrete sequence in X to a discrete sequence in Y. If X and Y are complex manifolds and $f: X \to Y$ is a proper injective holomorphic immersion, then $f(X)$ is a closed complex submanifold of Y; such f is called a *proper holomorphic embedding*. More generally, if X and Y are complex spaces (see Sect. 1.3) and $f: X \to Y$ is a proper holomorphic map, then $f(X)$ is a closed complex subvariety of Y according to a theorem of Remmert [471].

1.2 Examples of Complex Manifolds

Example 1.2.1 (Riemann surfaces) A 1-dimensional complex manifold is called a *Riemann surface*; it can be either open or compact.

By the Riemann-Koebe-Poincaré uniformization theorem [344, 456] the only connected and simply connected Riemann surfaces up to a biholomorphism are the complex plane $\mathbb{C}$, the *Riemann sphere* $\mathbb{C} \cup \{\infty\} = \mathbb{P}^1$, and the disc $\mathbb{D} = \{z \in \mathbb{C} : |z| < 1\}$. If R is a connected Riemann surface then its universal covering space X is one of the surfaces $\mathbb{P}^1, \mathbb{C}, \mathbb{D}$, and R is biholomorphic to the quotient X/Γ for some group $\Gamma \subset \mathrm{Aut}(X)$ acting without fixed points and properly discontinuously. The automorphism group $\mathrm{Aut}(\mathbb{P}^1) = \{z \mapsto \frac{az+b}{cz+d} : ad - bc = 1\}$ does not contain any nontrivial subgroups with these properties; hence $\mathbb{P}^1$ has no nontrivial holomorphic quotients. The subgroups $\Gamma \subset \mathrm{Aut}(\mathbb{C})$ with these properties are *lattices*, i.e., discrete $\mathbb{Z}$-submodules of $\mathbb{C}$ acting on $\mathbb{C}$ by translations. Such Γ has either one or two generators: $\Gamma = \mathbb{Z}a$ $(a \neq 0)$, or $\Gamma = \mathbb{Z}a + \mathbb{Z}b$ where $a, b \in \mathbb{C}$ are nonzero numbers with $ab^{-1} \notin \mathbb{R}$. The quotient $\mathbb{C}/\Gamma$ is $\mathbb{C}^* = \mathbb{C} \setminus \{0\}$ in the case of a single generator, and a complex one dimensional torus in the case of two generators. All other Riemann surfaces are quotients of the disc $\mathbb{D}$.

Excellent introductions to Riemann surfaces are the texts by Donaldson [130] and Forster [173]. □

Example 1.2.2 (Affine algebraic manifolds) An *affine algebraic variety* in $\mathbb{C}^n$ is the common zero set of finitely many holomorphic polynomials in n complex variables. An affine algebraic variety without singular points is called an *affine algebraic manifold*. □

Example 1.2.3 (Stein manifolds) The class of Stein manifolds was introduced by Karl Stein in 1951 [524] (under the name *holomorphically complete manifolds*) by a system of three axioms, postulating the existence of many global holomorphic functions, in analogy to the properties of domains of holomorphy (see Definition 2.2.1 on p. 49). The simplest characterization of this class is given by the Remmert embedding theorem [470]: A complex manifold is Stein if and only if it is biholomorphic to a closed complex submanifold of a Euclidean space $\mathbb{C}^N$ (see Theorem 2.4.1). Stein manifolds are holomorphic analogues of affine algebraic manifolds, a fact that is made precise by algebraic approximation theorems (see p. 52). □

Example 1.2.4 (Lie groups and homogeneous manifolds) A complex manifold G that is also a group, with holomorphic group operations, is called a *complex Lie group*. The main examples include the *general linear group* $GL_n(\mathbb{C})$ (the group of invertible complex $n \times n$ matrices) and its subgroups such as $SL_n(\mathbb{C}) = \{A \in GL_n(\mathbb{C}) : \det A = 1\}$, the symplectic subgroup $Sp_n(\mathbb{C}) \subset GL_{2n}(\mathbb{C})$, certain quotients such as the *projective linear group* $PGL_n(\mathbb{C}) = GL_{n+1}(\mathbb{C})/\mathbb{C}^* = \mathrm{Aut}(\mathbb{P}^n)$ (the holomorphic automorphism group of the complex projective space $\mathbb{P}^n$), universal coverings, etc. A complex manifold X is said to be *G-homogeneous* if there exists a transitive holomorphic action $G \times X \to X$ of G on X by holomorphic automorphisms. Fixing a point $p \in X$, we see that X is biholomorphic to the quotient G/H where $H = \{g \in G : g(p) = p\}$ is the isotropy subgroup of the point p. For results on this subject see e.g. [6, 68]. □

Example 1.2.5 (Projective spaces) The complex projective spaces $\mathbb{P}^n$ (also denoted $\mathbb{CP}^n$ in some texts) play the analogous role in algebraic geometry as the Euclidean spaces play in affine and Stein geometry. As a set, $\mathbb{P}^n$ consists of all complex lines through the origin in $\mathbb{C}^{n+1}$. A complex line $\lambda \subset \mathbb{C}^{n+1}$ is determined by any point $0 \neq z = (z_0, \ldots, z_n) \in \lambda$; we denote this line by $[z] = [z_0 : z_1 : \cdots : z_n]$ and call these the *homogeneous coordinates* on $\mathbb{P}^n$. Clearly, $[z] = [w]$ if and only if $w = tz$ for some $t \in \mathbb{C}^*$. There is a unique complex manifold structure on $\mathbb{P}^n$ in which the projection $\pi : \mathbb{C}^{n+1}_* = \mathbb{C}^{n+1} \setminus \{0\} \to \mathbb{P}^n$, $\pi(z) = [z] \in \mathbb{P}^n$ is holomorphic. A complex atlas is given by the collection (U_j, ϕ_j) $(j = 0, 1, \ldots, n)$ where $U_j = \{[z_0 : z_1 : \cdots : z_n] \in \mathbb{P}^n : z_j \neq 0\}$ and

$$\phi_j\big([z_0 : z_1 : \cdots : z_n]\big) = \left(\frac{z_0}{z_j}, \ldots, \frac{z_{j-1}}{z_j}, \frac{z_{j+1}}{z_j}, \ldots, \frac{z_n}{z_j}\right) \in \mathbb{C}^n.$$

It is immediate that ϕ_j maps U_j bijectively onto $\mathbb{C}^n$ and that the transition maps $\phi_i \circ \phi_j^{-1}$ are linear fractional.

The real projective spaces shall be denoted by $\mathbb{RP}^n$. □

Example 1.2.6 (Projective manifolds) A nonzero holomorphic polynomial $P(z_0, \ldots, z_n)$ is *homogeneous* of degree d if $P(tz_0, \ldots, tz_n) = t^d P(z_0, \ldots, z_n)$ for all $t \in \mathbb{C}$. It determines a complex hypersurface

$$V = V(P) = \big\{[z_0 : z_1 : \cdots : z_n] \in \mathbb{P}^n : P(z_0, \ldots, z_n) = 0\big\}.$$

More generally, homogeneous polynomials $P_1, \ldots, P_m$ on $\mathbb{C}^{n+1}$ determine a complex subvariety $V(P_1, \ldots, P_m) = V(P_1) \cap \cdots \cap V(P_m) \subset \mathbb{P}^n$. Subvarieties of this type in $\mathbb{P}^n$ are called *projective varieties*, or *projective manifolds* when they are nonsingular. A *quasi-projective variety* is a variety of the form $V = X \setminus Y$, where X and Y are closed complex subvarieties of $\mathbb{P}^n$. By Chow's theorem [98, 264, 274], every closed complex subvariety of $\mathbb{P}^n$ equals $V(P_1, \ldots, P_m)$ for some homogeneous polynomials in $n+1$ variables. A compact complex manifold (resp. a complex space) is said to be *projective algebraic* if it is biholomorphic to a projective manifold (resp. to a projective subvariety) in some $\mathbb{P}^n$. A major extension of Chow's theorem is the *GAGA principle* of J-P. Serre [498] concerning the equivalence between analytic and algebraic coherent sheaves over projective algebraic varieties. □

Example 1.2.7 (Stiefel manifolds) Pick integers $1 \le k \le n$. The complex *Stiefel manifold* $V_{k,n}$ consists of all complex $k \times n$ matrices $A \in M_{k,n}(\mathbb{C}) \cong \mathbb{C}^{kn}$ with $\operatorname{rank} A = k$. Clearly $V_{k,n}$ is an open subset of $M_{k,n} = M_{k,n}(\mathbb{C})$.

The group $GL_n(\mathbb{C})$ acts transitively on $V_{k,n}$ by right multiplication, so $V_{k,n}$ is a complex homogeneous manifold. We have $V_{k,n} = M_{k,n} \setminus \Sigma_{k,n}$ where $\Sigma_{k,n}$ consists of all complex $k \times n$ matrices of less than maximal rank. Note that $\Sigma_{k,n}$ is an algebraic subvariety of $M_{k,n}(\mathbb{C}) \cong \mathbb{C}^{kn}$ defined by the vanishing of all maximal $k \times k$ minors; these are homogeneous polynomial equations of order k (so $\Sigma_{k,n}$ is a complex cone in $\mathbb{C}^{kn}$), and at every point of $\Sigma_{k,n}$ at least $n-k+1$ of these equations are independent. In fact we have a stratification $\Sigma_{k,n} = \Sigma^1_{k,n} \supset \Sigma^2_{k,n} \supset \cdots$ where for every $i = 1, \ldots, k$ the set

$$\Sigma^i_{k,n} = \{A \in M_{k,n} : \operatorname{rank} A = k - i\}$$

is an algebraic subvariety of complex codimension $\operatorname{codim} \Sigma^i_{k,n} = i(n-k+i)$. (See e.g. [244, Proposition 5.3, p. 60].) In particular, $\operatorname{codim} \Sigma_{k,n} = n-k+1$. It follows from the transversality theorem that the homotopy groups of $V_{k,n}$ vanish in the range up to $2(n-k)$:

$$\pi_q(V_{k,n}) = 0, \quad q = 1, 2, \ldots, 2n-2k. \tag{1.7}$$

Example 1.2.8 (Grassmann manifolds) The complex *Grassmann manifold* $G_{k,n} = G_k(\mathbb{C}^n)$ is the set of all k-dimensional complex linear subspaces of $\mathbb{C}^n$. (Thus $G_{1,n} = \mathbb{P}^{n-1}$.) Let $V_{k,n}$ be the Stiefel manifold (see Example 1.2.7 above). We have a surjective map $\pi \colon V_{k,n} \to G_{k,n}$ which sends $A \in V_{k,n}$ to the $\mathbb{C}$-linear span of the row vectors in A. There is a unique complex structure on $G_{k,n}$ which makes this projection holomorphic. The group $GL_k(\mathbb{C})$ acts on $V_{k,n}$ by left multiplication, and we have $\pi(A) = \pi(B)$ for $A, B \in V_{k,n}$ if and only if $A = GB$ for some $G \in GL_k(\mathbb{C})$, so the Grassmann manifold $G_{k,n}$ is the orbit space of this action. Grassmann manifolds are projective algebraic; the *Plücker embedding* $G_{k,n}(\mathbb{C}) \to \mathbb{P}(\wedge^k \mathbb{C}^n)$ is induced by the map $V_{k,n} \to \wedge^k \mathbb{C}^n$ sending a matrix $A \in V_{k,n}$ with rows $a_1, \ldots, a_k$ to $a_1 \wedge \cdots \wedge a_k \in \wedge^k \mathbb{C}^n$ [568, p. 11]. An important property for us is that every point in $G_{k,n}$ contains a Zariski open neighborhood isomorphic to $\mathbb{C}^{k(n-k)}$. □

Example 1.2.9 (Complexification) For every real analytic manifold M there exists a complex manifold X obtained by complexifying the transition maps defining M [575]. Such X contains M as a maximal totally real submanifold, and it can be chosen Stein according to Grauert [255, Sect. 3]. (See Sect. 3.5.) □

Example 1.2.10 (Hyperbolic manifolds) A complex manifold X is said to be

- *Brody hyperbolic* [69] if every holomorphic map $\mathbb{C} \to X$ is constant;
- *Brody-Eisenman volume hyperbolic* if every holomorphic map $\mathbb{C}^n \to X$ ($n = \dim X$) has rank $< n$ at each point (a degenerate map);
- *Kobayashi (complete) hyperbolic* [339, 340] if the *Kobayashi pseudometric* k_X is a (complete) metric on X.

Recall that k_X is the integrated form of the infinitesimal pseudometric

$$|v| = \inf\left\{\frac{1}{|\lambda|} : f\colon \mathbb{D} \to X \text{ holomorphic}, \ f(0) = x, \ f'(0) = \lambda v\right\}, \qquad v \in \mathrm{T}_x X.$$

The Kobayashi pseudometric is the biggest pseudometric which equals the Poincaré metric on the unit disc and is such that holomorphic maps are distance decreasing. A Kobayashi hyperbolic manifold is also Brody hyperbolic; the converse holds for compact manifolds (Brody [69]).

Furthermore, for every integer $k \in \{1, \dots, \dim X\}$ there is the notion of *Eisenman k-hyperbolicity* [141, 320]; for $k = 1$ it coincides with Kobayashi hyperbolicity.

Most complex manifolds have at least some amount of rigidity. In particular, every compact complex manifold X of Kodaira general type (i.e., of Kodaira dimension $\kappa_X = \dim X$) is volume hyperbolic according to Kobayashi and Ochiai [342]. Such a manifold X is believed to be almost Kobayashi hyperbolic, in the sense that it admits a proper complex subvariety X' which contains the image of any nonconstant entire map $\mathbb{C} \to X$. $\square$

1.3 Subvarieties and Complex Spaces

Let X be a complex manifold. We denote by $\mathcal{O}_x = \mathcal{O}_{X,x}$ the ring of germs of holomorphic functions at a point $x \in X$. A germ $[f]_x \in \mathcal{O}_x$ is represented by a holomorphic function in an open neighborhood of x; two such functions determine the same germ at x if and only if they agree in some neighborhood of x. The ring $\mathcal{O}_{X,x}$ is isomorphic to the ring $\mathcal{O}_{\mathbb{C}^n,0}$ via any holomorphic coordinate map sending x to 0. We can identify $\mathcal{O}_{\mathbb{C}^n,0}$ with the ring of convergent power series in n complex variables $(z_1, \dots, z_n)$. This ring is Noetherian and a unique factorization domain. Its units are precisely the germs that do not vanish at 0. The set of germs vanishing at 0 is the unique maximal ideal $\mathfrak{m}_0 \subset \mathcal{O}_{\mathbb{C}^n,0}$ and $\mathcal{O}_{\mathbb{C}^n,0}/\mathfrak{m}_0 = \mathbb{C}$. The disjoint union $\mathcal{O}_X = \bigcup_{x \in X} \mathcal{O}_{X,x}$ is equipped with the topology whose basis is given by sets $\{[f]_x : x \in U\}$, where $f\colon U \to \mathbb{C}$ is a holomorphic function on an open set $U \subset X$. This makes $\mathcal{O}_X$ into a sheaf of commutative rings, called the *sheaf of germs of holomorphic functions* or *the structure sheaf* of X. The identity principle shows that the sheaf $\mathcal{O}_X$ is Hausdorff. We denote by $\mathcal{C}_X$ the sheaf of germs of continuous functions on X.

Since the ring $\mathcal{O}_x$ has no zero divisors, we can form its quotient field $\mathcal{M}_x$, called the field of germs of *meromorphic functions* on X at the point x. Thus a meromorphic function on X is locally at every point $x \in X$ given as the quotient f/g of two holomorphic functions whose germs at x are coprime. Such a function is holomorphic off the zero locus of g, also called the *polar set* of f/g, and its *indeterminacy set* is $\{f=0,\ g=0\}$.

A subset A of a complex manifold X is a *complex (analytic) subvariety* of X if for every point $p \in A$ there exist a neighborhood $U \subset X$ of p and functions $f_1, \ldots, f_d \in \mathcal{O}(U)$ such that

$$A \cap U = \{x \in U : f_1(x) = 0, \ldots, f_d(x) = 0\}.$$

If such A is topologically closed in X then A is a *closed complex subvariety* of X. Since the local ring $\mathcal{O}_x$ is Noetherian, a subset of X that is locally defined by infinitely many holomorphic equations is still a subvariety and can be locally defined by finitely many equations.

A point p in a subvariety A is a *regular* (or *smooth*) point if A is a complex submanifold near p; the set of all regular points is denoted A_{reg}. A point $p \in A \setminus A_{\mathrm{reg}} = A_{\mathrm{sing}}$ is a *singular point* of A.

Let A be a closed complex subvariety of X. For every point $x \in X$ we denote by $\mathcal{J}_{A,x}$ the ideal in $\mathcal{O}_x$ consisting of all holomorphic function germs at x whose restriction to A vanishes. In particular, $\mathcal{J}_{A,x} = \mathcal{O}_x$ for every $x \in X \setminus A$. The corresponding sheaf $\mathcal{J}_A = \bigcup_{x\in X} \mathcal{J}_{A,x}$ is the *sheaf of ideals* (or the *ideal sheaf*, or simply the *ideal*) of A in X. The restriction of the quotient sheaf $\mathcal{O}_X/\mathcal{J}_A = \bigcup_{x\in X} \mathcal{O}_{X,x}/\mathcal{J}_{A,x}$ to A is the sheaf of germs of holomorphic functions on A, denoted $\mathcal{O}_A$ and called the structure sheaf of A.

The notion of a *complex space* was first introduced in 1951 by Behnke and Stein [52] and Cartan; their definitions correspond to what is now called a *normal complex space* (see Grauert and Remmert [258]). The standard definition which is used in this book was given by J-P. Serre in [498] (see also Grauert and Remmert [251, 258] and Cartan [89]):

A *reduced complex space* is a pair $(X, \mathcal{O}_X)$, where X is a paracompact Hausdorff topological space and $\mathcal{O}_X$ is a sheaf of rings of continuous functions on X (a subsheaf of the sheaf $\mathcal{C}_X$ of germs of continuous functions) such that for every point $x \in X$ there is a neighborhood $U \subset X$ and a homeomorphism $\phi : U \to A \subset \mathbb{C}^n$ onto a locally closed complex subvariety of $\mathbb{C}^n$ so that the homomorphism $\phi^* : \mathcal{C}_A \to \mathcal{C}_X$, $f \mapsto f \circ \phi$, induces an isomorphism of $\mathcal{O}_A$ onto $\mathcal{O}_U = \mathcal{O}_X|_U$. Intuitively speaking, X is obtained by gluing pieces of subvarieties in Euclidean spaces using biholomorphic transition maps. Similarly one defines an *algebraic space* [498]. We get a *nonreduced complex space* by allowing local models $(A, \mathcal{F})$, where A is a closed complex subvariety in an open set $\Omega \subset \mathbb{C}^n$ and $\mathcal{F} = (\mathcal{O}_\Omega/\mathcal{I})|_A$ for some sheaf of ideals $\mathcal{I} \subset \mathcal{J}_A$ supported on A (i.e., $\mathcal{I}_x = \mathcal{O}_x$ for $x \notin A$). The ring $\mathcal{F}_x$ may have nilpotent elements. By the Nullstellensatz the radical $\sqrt{\mathcal{I}_x}$ of any such ideal equals $\mathcal{J}_{A,x}$.

Let $(X, \mathcal{O}_X)$ and $(Y, \mathcal{O}_Y)$ be complex spaces. A continuous map $f : X \to Y$ is said to be holomorphic if for every $x \in X$ the composition $\mathcal{C}_{Y,f(x)} \ni g \mapsto g \circ f \in \mathcal{C}_{X,x}$ defines a homomorphism $f_x^* : \mathcal{O}_{Y,f(x)} \to \mathcal{O}_{X,x}$. For such map we can define the differential $\mathrm{d}f_x : \mathrm{T}_x X \to \mathrm{T}_{f(x)} Y$ as a $\mathbb{C}$-linear map on the *Zariski tangent space* (see (1.28) on p. 24). This is the usual differential at smooth points, while at singular points we locally embed the two spaces as complex subvarieties of Euclidean space of minimal dimension $n_x = \mathrm{embdim}_x X$, $m_y = \mathrm{embdim}_y Y$, respectively, and take the differential dF_x of the local holomorphic extension F of f (a $\mathbb{C}$-linear map $\mathbb{C}^{n_x} \to \mathbb{C}^{m_y}$).

Among the most fundamental results in the theory of complex spaces is Remmert's theorem [471] saying that the image $f(X)$ of a proper holomorphic map $f: X \to Y$ is a closed analytic subvariety of Y. A more general result of Grauert (see [261]) gives the coherence of the direct image $f_*\mathcal{F}$ of any coherent analytic sheaf $\mathcal{F}$ under a proper holomorphic map. (See also p. 55.)

Definition 1.3.1 Let Z and X be reduced complex spaces. A holomorphic map $\pi: Z \to X$ is a *holomorphic submersion* if for every point $z_0 \in Z$ there exist an open neighborhood $V \subset Z$ of z_0, an open neighborhood $U \subset X$ of $x_0 = \pi(z_0)$, an open set W in $\mathbb{C}^p$, and a biholomorphic map $\phi: V \to U \times W$ such that $pr_1 \circ \phi = \pi$. (Here $pr_1: U \times W \to U$ is the projection on the first factor.) Each such local chart ϕ will be called *adapted to* π.

Note that each fibre $Z_x = \pi^{-1}(x)$ $(x \in X)$ of a holomorphic submersion is a closed complex submanifold of Z, and the dimension $\dim Z_x$ is constant on every connected component of Z.

Definition 1.3.2 Assume that $h: Z \to X$ is a holomorphic submersion onto a complex space X, X' is a closed complex subvariety of X, and $\mathcal{S} \subset \mathcal{O}_X$ is a sheaf of ideals with support X', i.e., $\mathcal{S}_x = \mathcal{O}_{X,x}$ when $x \in X \setminus X'$. Local holomorphic sections f_0, f_1 of $h: Z \to X$ in a neighborhood of a point $x \in X'$ are *$\mathcal{S}$-tangent at x* if there is a neighborhood $V \subset Z$ of the point $z = f_0(x) = f_1(x) \in Z$ and a holomorphic embedding $\phi: V \hookrightarrow \mathbb{C}^N$ such that the germ at x of any component of the map $\phi \circ f_0 - \phi \circ f_1 : U \to \mathbb{C}^N$ belongs to $\mathcal{S}_x$. If f_0 and f_1 are holomorphic in a neighborhood of X' and $\mathcal{S}$-tangent at each point $x \in X'$, then we say that f_0 and f_1 are $\mathcal{S}$-tangent and write $\delta(f_0, f_1) \in \mathcal{S}$. If this holds for the r-th power of the ideal sheaf $\mathcal{J}_{X'}$ of the subvariety X' then f_0 and f_1 are said to be *tangent to order r along X'*.

Definition 1.3.3 A *stratification* of a finite dimensional complex space X is a finite descending sequence $X = X_0 \supset X_1 \supset \cdots \supset X_m = \varnothing$ of closed complex subvarieties X_k such that each connected component S (stratum) of a difference $X_k \setminus X_{k+1}$ is a complex manifold and $\overline{S} \setminus S \subset X_{k+1}$.

Every finite dimensional complex space admits a stratification [573, p. 227]: Take X_1 to be the union of the singular locus of $X = X_0$ and of all irreducible components of X_0 of less than maximal dimension; define X_2 in the same way with respect to X_1, etc. By considering substratifications we can ask for many additional properties. For example, a finite dimensional Stein space admits a stratification whose strata are Stein manifolds. Whitney's condition (a) is used in transversality theorems proved in Sect. 8.8.

1.4 Holomorphic Fibre Bundles

Fibre bundles represent one of the most important constructions of new manifolds from existing ones.

Definition 1.4.1 A *holomorphic fibre bundle* is a triple (Z, π, X), where X and Z are complex spaces and $\pi: Z \to X$ is a holomorphic map of Z onto X such that there exist a complex manifold Y, an open cover $\mathcal{U} = \{U_\alpha\}$ of X, and for every α a biholomorphic map

$$\theta_\alpha: Z|_{U_\alpha} = \pi^{-1}(U_\alpha) \to U_\alpha \times Y, \quad \theta_\alpha(z) = \big(\pi(z), \vartheta_\alpha(z)\big). \tag{1.8}$$

The manifold Z is the *total space*, X is the *base space*, and Y is the *fibre*.

A holomorphic submersion $\pi: Z \to X$ is a *stratified holomorphic fibre bundle* if there is a stratification $X = X_0 \supset X_1 \supset \cdots \supset X_m = \varnothing$ of X such that the restriction of Z to every stratum $S \subset X_k \setminus X_{k+1}$ is a holomorphic fibre bundle over S. (Fibres over different strata may be different.)

The simplest example is a *product bundle* $\pi: Z = X \times Y \to X$, $(x, y) \mapsto x$. By definition, every fibre bundle is isomorphic to the product bundle over small open sets in X. The fibre $Z_x = \pi^{-1}(x)$ over any point $x \in X$ is biholomorphic to Y. A map θ_α (1.8) is a *fibre bundle chart* on Z, and the collection $\{(U_\alpha, \theta_\alpha)\}$ is a *holomorphic fibre bundle atlas* on Z. The *transition maps* $\theta_{\alpha,\beta} = \theta_\alpha \circ \theta_\beta^{-1}: U_{\alpha,\beta} \times Y \to U_{\alpha,\beta} \times Y$ are of the form

$$\theta_{\alpha,\beta}(x, y) = \big(x, \vartheta_{\alpha,\beta}(x, y)\big), \quad x \in U_{\alpha,\beta}, \ y \in Y, \tag{1.9}$$

and they satisfy the *cocycle condition*

$$\theta_{\alpha,\alpha} = \mathrm{Id}, \qquad \theta_{\alpha,\beta} \circ \theta_{\beta,\gamma} \circ \theta_{\gamma,\alpha} = \mathrm{Id} \quad \text{on } U_{\alpha,\beta,\gamma} \times Y. \tag{1.10}$$

For every fixed $x \in U_{\alpha,\beta}$ we have $\vartheta_{\alpha,\beta}(x, \cdot) \in \mathrm{Aut}(Y)$. Conversely, given an open cover $\mathcal{U} = \{U_\alpha\}$ of X and a collection of biholomorphic self-maps (1.9) satisfying the cocycle condition (1.10), we get a holomorphic fibre bundle $Z \to X$ with these transition maps by taking Z to be the disjoint union of all $U_\alpha \times Y$, modulo the identifications provided by the transition maps.

A *section* of $\pi: Z \to X$ is a map $f: X \to Z$ such that $\pi \circ f$ is the identity on X; that is, $f(x) \in Z_x$ for every $x \in X$. Any section of the product bundle $X \times Y \to X$ is of the form $f(x) = (x, g(x))$, where $g: X \to Y$ is a map to the fibre. If $\{(U_\alpha, \theta_\alpha)\}$ is a holomorphic fibre bundle atlas on $Z \to X$ with the transition maps $\theta_{\alpha,\beta}$ (1.9), then a holomorphic section $f: X \to Z$ is given by a collection of holomorphic maps $f_\alpha: U_\alpha \to Y$ into the fibre satisfying the compatibility conditions

$$f_\alpha(x) = \vartheta_{\alpha,\beta}\big(x, f_\beta(x)\big), \quad x \in U_{\alpha,\beta}.$$

Definition 1.4.2 A *holomorphic isomorphism* of holomorphic fibre bundles $\pi: Z \to X$, $\pi': Z' \to X$ is a biholomorphic map $\Phi: Z \to Z'$ such that $\pi' \circ \Phi = \pi$; if such Φ exists then the bundles are *holomorphically isomorphic*. A fibre bundle is *trivial* if it is isomorphic to the product bundle.

Isomorphisms of a fibre bundle onto itself are *fibre bundle automorphisms*. Holomorphic automorphisms of a product bundle $X \times Y \to X$ are biholomorphic self-maps of $X \times Y$ of the form $(x, y) \mapsto (x, \varphi(x, y))$, with $\varphi(x, \cdot) \in \mathrm{Aut}(Y)$ for every $x \in X$. In general we choose an open cover $\mathcal{U} = \{U_\alpha\}$ of X and fibre bundle atlases $\{(U_\alpha, \theta_\alpha)\}$ for (Z, π, X), and $\{(U_\alpha, \theta'_\alpha)\}$ for (Z', π', X). A fibre bundle isomorphism Φ of (Z, π, X) to (Z', π', X) is then given by a collection of fibre preserving biholomorphic self-maps $\phi_\alpha: U_\alpha \times Y \to U_\alpha \times Y$ of the form $\phi_\alpha(x, y) = (x, \varphi_\alpha(x, y))$ so that the following diagrams commute:

$$\begin{array}{ccc} U_{\alpha,\beta} \times Y & \xleftarrow{\theta_{\alpha,\beta}} & U_{\alpha,\beta} \times Y \\ \downarrow{\scriptstyle \phi_\alpha} & & \downarrow{\scriptstyle \phi_\beta} \\ U_{\alpha,\beta} \times Y & \xleftarrow{\theta'_{\alpha,\beta}} & U_{\alpha,\beta} \times Y \end{array}$$

If there exists a fibre bundle atlas on $Z \to X$ (in the given isomorphism class) such that all transition maps belong to a certain subgroup G of $\mathrm{Aut}(Y)$, then we say that the bundle has *structure group* G.

If $\pi: Z \to X$ is a holomorphic fibre bundle and X' is a complex subvariety of X then the restriction $\pi: Z' = Z|_{X'} \to X'$ is a holomorphic fibre bundle over X', called the *restricted bundle*.

Given a holomorphic fibre bundle $\pi: Z \to X$ with fibre Y and a holomorphic map $f: W \to X$, the *pullback bundle* $\pi': f^*Z \to W$ and the map $F: f^*Z \to Z$ are defined as follows:

$$f^*Z = \big\{(w, z) \in W \times Z : f(w) = \pi(z)\big\}, \quad \pi'(w, z) = w, \quad F(w, z) = z. \tag{1.11}$$

Let $\{(U_\alpha, \theta_\alpha)\}$ be a fibre bundle atlas for $\pi: Z \to X$ with

$$\theta_\alpha(z) = \big(\pi(z), \vartheta_\alpha(z)\big) \in U_\alpha \times Y.$$

Set $V_\alpha = f^{-1}(U_\alpha) \subset W$ and define a map $\theta'_\alpha: f^*Z|_{V_\alpha} \to V_\alpha \times Y$ by $\theta'_\alpha(w, z) = (w, \vartheta_\alpha(z))$. (The map f appears implicitly in the above definition by the condition $\pi(z) = f(w)$.) This is a holomorphic fibre bundle atlas on $\pi': f^*Z \to W$ with transition maps $\vartheta'_{\alpha,\beta}(w, y) = \vartheta_{\alpha,\beta}(f(w), y)$; hence $f^*Z \to W$ is indeed a holomorphic fibre bundle with fibre Y.

Example 1.4.3 (Vector bundles) These are fibre bundles with fibre $\mathbb{C}^n$ and structure group $GL_n(\mathbb{C})$. They are considered in the following section. □

Example 1.4.4 (Affine bundles) These are fibre bundles with fibre $\mathbb{C}^n$ and structure group consisting of affine linear maps:

$$v \mapsto a + Bv, \quad a \in \mathbb{C}^n, \ B \in GL_n(\mathbb{C}).$$

By [316] every projective analytic variety X carries an affine bundle $E \to X$ whose total space E is a Stein space. (This will be of interest in Sect. 7.6.) We recall the construction in the basic case $X = \mathbb{P}^n$. Consider the *Segre embedding* $\rho: \mathbb{P}^n \times \mathbb{P}^n \hookrightarrow \mathbb{P}^{n^2+2n}$,

$$\big([z_0 : \cdots : z_n], [w_0 : \cdots : w_n]\big) \longmapsto [z_0 w_0 : \cdots : z_i w_j : \cdots : z_n w_n].$$

Let Σ denote the quadratic hypersurface in $\mathbb{P}^n \times \mathbb{P}^n$ defined by the equation $z_0 w_0 + \cdots + z_n w_n = 0$ and set $E = \mathbb{P}^n \times \mathbb{P}^n \setminus \Sigma$. Choose homogeneous coordinates $[\zeta_0 : \zeta_1 : \cdots : \zeta_{n^2+2n}]$ on $\mathbb{P}^{n^2+2n}$ such that $\zeta_j = z_j w_j$ for $j = 0, \ldots, n$ under the embedding ρ. Let H denote the hyperplane in $\mathbb{P}^{n^2+2n}$ given by the equation $\sum_{j=0}^{n} \zeta_j = 0$. Then ρ embeds E properly into $\mathbb{P}^{n^2+2n} \setminus H = \mathbb{C}^{n^2+2n}$, so E is Stein. The projection $\pi : E \to \mathbb{P}^n$ onto the first component is an affine bundle with fibre $\mathbb{C}^n$. The restriction $E|_X$ to any closed complex subvariety $X \subset \mathbb{P}^n$ is still an affine bundle whose total space (being a closed analytic subvariety of E) is Stein. For quasi-projective varieties there is some work to see that one can get an affine bundle with Stein total space. □

Example 1.4.5 (Principal bundles) Let G be a finite dimensional complex Lie group. For every $g \in G$ let $\sigma_g \in \mathrm{Aut}(G)$ be the left multiplication on G by g: $\sigma_g(g') = gg'$ $(g' \in G)$. Set $\Gamma = \{\sigma_g : g \in G\} \subset \mathrm{Aut}(G)$. A *holomorphic principal G-bundle* over X is a holomorphic fibre bundle $\pi : Z \to X$ with fibre G and structure group Γ. Such a bundle is determined by a 1-cocycle $g_{\alpha,\beta} : U_{\alpha,\beta} \to G$ over an open cover $\mathcal{U} = \{U_\alpha\}$ of X; the corresponding transition maps are $\phi_{\alpha,\beta}(x, g) = (x, g_{\alpha,\beta}(x)g)$. The group G acts holomorphically on Z by the right multiplication on the fibres Z_x, and these fibres are precisely the orbits of the action. See Sects. 8.1–8.2 for further results. □

Example 1.4.6 (Fibre bundles associated to principal bundles) Assume that a complex Lie group G acts holomorphically on a complex manifold Y. Every holomorphic 1-cocycle $g_{\alpha,\beta} : U_{\alpha,\beta} \to G$ on an open cover $\mathcal{U} = \{U_\alpha\}$ of X determines a holomorphic fibre bundle with fibre Y, structure group G and transition maps $\theta_{\alpha,\beta}(x, v) = (x, g_{\alpha,\beta}(x)v)$. In this way we associate to a principal G-bundle $Z \to X$ a fibre bundle $E \to X$ with fibre Y, structure group G and the same structure cocycle that determines the bundle Z. In particular, to a principal $GL_n(\mathbb{C})$-bundle we associate a holomorphic vector bundle of rank n. Conversely, to a holomorphic vector bundle $E \to X$ of rank n we associate the principal $GL_n(\mathbb{C})$-bundle $Z = F(E) \to X$, called *the frame bundle* of the vector bundle $E \to X$. The elements of $Z_x = F(E_x)$ are *frames* (complex bases) on the vector space $E_x \cong \mathbb{C}^n$. □

Example 1.4.7 (Bundles with Euclidean fibres) These are fibre bundles with fibre $\mathbb{C}^n$ and structure group $\mathrm{Aut}(\mathbb{C}^n)$, the holomorphic automorphism group of $\mathbb{C}^n$. For $n = 1$, every such bundle is affine since $\mathrm{Aut}(\mathbb{C})$ consists of affine linear maps $z \mapsto az + b$, $a \neq 0$. However, for $n > 1$ the group $\mathrm{Aut}(\mathbb{C}^n)$ is very big (see Chap. 4), and there exist fibre bundles with fibre $\mathbb{C}^n$ over the disc or the plane whose total space is not Stein; see Theorems 8.3.12 and 8.3.13 due to Demailly and Rosay, respectively. □

Example 1.4.8 (Flat bundles) A holomorphic fibre bundle $Z \to X$ with fibre Y is said to be *flat* if it admits a holomorphic fibre bundle atlas whose transition maps $\vartheta_{\alpha,\beta} : U_{\alpha,\beta} \to \mathrm{Aut}(Y)$ are locally constant, and hence constant on every connected component of $U_{\alpha,\beta}$. The trivial horizontal foliations of $U_\alpha \times Y$ with leaves $U_\alpha \times \{y\}$ ($y \in Y$) patch together to a horizontal holomorphic foliation of the total space Z. A flat holomorphic fibre bundle is trivial over every simply connected open set in the base X, and every holomorphic isomorphism class of flat bundles is determined by a representation of the fundamental group $\pi_1(X, p)$ in the automorphism group $\mathrm{Aut}(Y)$. □

We conclude by mentioning the following important result.

Theorem 1.4.9 (Fischer and Grauert [159]) *If X and Z are compact manifolds and $\pi : Z \to X$ is a holomorphic submersion whose fibres $Z_x = \pi^{-1}(x)$ $(x \in X)$ are compact and biholomorphic to each other, then $Z \to X$ is a holomorphic fibre bundle.*

1.5 Holomorphic Vector Bundles

Vector bundles are a principal tool used to linearize problems in analysis and geometry. They are also a subject of intrinsic investigation with a profound impact on modern mathematics. We focus on holomorphic vector bundles, recalling those constructions that will be important to us. Similar constructions apply to other classes of vector bundles (topological, smooth, and with $\mathbb{C}$ replaced by another field such as $\mathbb{R}$).

Definition 1.5.1 A *holomorphic vector bundle of rank n* over a complex space X is a holomorphic fibre bundle $E \to X$ (Definition 1.4.1) with fibre $Y = \mathbb{C}^n$ and structure group $GL_n(\mathbb{C})$. A vector bundle of rank $n = 1$ is also called a (holomorphic) *line bundle*.

This means that we have an open cover $\mathcal{U} = \{U_\alpha\}$ of X and vector bundle charts $\theta_\alpha : E|_{U_\alpha} \xrightarrow{\cong} U_\alpha \times \mathbb{C}^n$ with transition maps of the form

$$\theta_\alpha \circ \theta_\beta^{-1}(x, v) = \theta_{\alpha,\beta}(x, v) = \big(x, g_{\alpha,\beta}(x)\, v\big), \quad x \in U_{\alpha,\beta},\ v \in \mathbb{C}^n, \tag{1.12}$$

where $g_{\alpha,\beta} : U_{\alpha,\beta} \to GL_n(\mathbb{C})$ is a holomorphic multiplicative 1-cocycle:

$$g_{\alpha,\alpha} = 1, \qquad g_{\alpha,\beta}\, g_{\beta,\alpha} = 1, \qquad g_{\alpha,\beta}\, g_{\beta,\gamma}\, g_{\gamma,\alpha} = 1. \tag{1.13}$$

Every fibre E_x is a complex vector space such that the fibre bundle charts $\theta_\alpha : Z|_{U_\alpha} \to U_\alpha \times \mathbb{C}^n$ are $\mathbb{C}$-vector space isomorphisms on each fibre.

If X is a real manifold of class $\mathcal{C}^r$ ($r \in \{0, 1, 2, \ldots, \infty\}$) and $E \to X$ is a complex vector bundle whose transition maps $g_{\alpha,\beta} : U_{\alpha,\beta} \to GL_n(\mathbb{C})$ are of class $\mathcal{C}^r$, then we have a *complex vector bundle of class $\mathcal{C}^r$* over X. For $r = 0$ we have a *topological complex vector bundle*. Replacing $\mathbb{C}$ by any field F we get topological F-vector bundles over topological spaces, or smooth F-vector bundles over smooth manifolds. For $F = \mathbb{R}$ we speak of *real vector bundles*.

Every vector bundle has the *zero section* sending each point $x \in X$ to the origin $0_x \in E_x$. Given a holomorphic vector bundle atlas $\{(U_\alpha, \theta_\alpha)\}$ on E with transition maps $g_{\alpha,\beta}$ (1.12), a section $f: X \to E$ is determined by a collection of maps $f_\alpha: U_\alpha \to \mathbb{C}^n$ satisfying the compatibility conditions

$$f_\alpha = g_{\alpha,\beta} f_\beta \quad \text{on } U_{\alpha,\beta}. \tag{1.14}$$

Example 1.5.2 (The tangent bundle) Let X be a manifold of class $\mathscr{C}^r$ and of dimension n. Given an atlas $\{(U_\alpha, \phi_\alpha)\}$ on X with transition maps $\phi_{\alpha,\beta}$, the tangent bundle $\mathrm{T}X \to X$ is a real vector bundle of rank n and of class $\mathscr{C}^{r-1}$ with vector bundle charts $\theta_\alpha: \mathrm{T}X|_{U_\alpha} \to U_\alpha \times \mathbb{R}^n$ and the transition cocycle $g_{\alpha,\beta} = (\phi_{\alpha,\beta})' \circ \phi_\beta$, where $(\phi_{\alpha,\beta})'$ is the Jacobian matrix of $\phi_{\alpha,\beta}$. The cotangent bundle T^*X is the dual bundle of $\mathrm{T}X$. For the tangent and cotangent bundles of a complex manifold, see Sects. 1.6–1.7. □

Example 1.5.3 (The universal bundle) Let $G_{k,n}$ be the Grassmann manifold whose points are k-dimensional subspaces of $\mathbb{C}^n$ (see Example 1.2.8), and set

$$U_{k,n} = \{(\lambda, z) \in G_{k,n} \times \mathbb{C}^n : z \in \lambda\}.$$

The projection $\pi: U_{k,n} \to G_{k,n}$, $\pi(\lambda, z) = \lambda$, admits the structure of a holomorphic vector bundle (a holomorphic vector subbundle of the trivial bundle $G_{k,n} \times \mathbb{C}^n$), called the universal bundle over $G_{k,n}$.

In particular, $U_{1,n+1} \to G_{1,n+1} = \mathbb{P}^n$ is a holomorphic line bundle over the projective space $\mathbb{P}^n$. This bundle is trivial over every coordinate neighborhood $V_j = \{[z_0: \cdots : z_n] : z_j \neq 0\} \cong \mathbb{C}^n$ in $\mathbb{P}^n$; a vector bundle chart is given by

$$\theta_j\big([z_0: \cdots : z_n], (v_0, \ldots, v_n)\big) = \big([z_0: \cdots : z_n], v_j\big) \in V_j \times \mathbb{C}.$$

The colinearity condition $v \in [z_0: \cdots : z_n]$ defining $U_{1,n+1}$ implies $v_i z_j = v_j z_i$, which shows that the transition maps equal $g_{i,j}([z_0: \cdots : z_n]) = \frac{z_i}{z_j}$. □

Definition 1.5.4 Let $\pi: E \to X$ and $\pi': E' \to X$ be holomorphic vector bundles. A *holomorphic vector bundle morphism* of (E, π, X) to (E', π', X) is a holomorphic map $\Phi: E \to E'$ such that $\pi' \circ \Phi = \pi$ and $\Phi_x: E_x \to E'_x$ is $\mathbb{C}$-linear for every $x \in X$. Such Φ is a *isomorphism* if $\Phi_x: E_x \to E'_x$ is an isomorphism of $\mathbb{C}$-vector spaces for every $x \in X$. A $\mathscr{C}^r$ *morphism* is a $\mathscr{C}^r$ map $\Phi: E \to E'$ that is $\mathbb{C}$-linear on every fibre.

The *kernel* and *cokernel* of a morphism $\Phi: E \to E'$ are defined by

$$\ker \Phi = \bigcup_{x \in X} \ker \Phi_x \subset E, \qquad \operatorname{im} \Phi = \bigcup_{x \in X} \operatorname{im} \Phi_x \subset E'.$$

Definition 1.5.5 Let $\pi: E \to X$ be a holomorphic vector bundle of rank n. A *holomorphic vector subbundle* of rank $m \in \{0, 1, \ldots, n\}$ of (E, π, X) is a complex submanifold $E' \subset E$, with the restricted projection $\pi' = \pi|_{E'}: E' \to X$ onto X, such that every point $x_0 \in X$ admits an open neighborhood $U \subset X$ and a holomorphic vector bundle chart $\theta\colon E|_U \xrightarrow{\cong} U \times \mathbb{C}^n$ satisfying the condition

$$\theta\big(E'|_U\big) = U \times \big(\mathbb{C}^m \times \{0\}^{n-m}\big). \tag{1.15}$$

Any such chart θ is said to be *adapted* to E'. Denote by

$$pr_1\colon \mathbb{C}^n = \mathbb{C}^m \times \mathbb{C}^{n-m} \to \mathbb{C}^m, \qquad pr_2\colon \mathbb{C}^n = \mathbb{C}^m \times \mathbb{C}^{n-m} \to \mathbb{C}^{n-m}$$

the projections onto the first and the second factor, respectively. For every θ as above the map $pr_1 \circ \theta\colon E'|_U \to U \times \mathbb{C}^m$ is a vector bundle chart on E', and the collection of all such charts is a holomorphic vector bundle atlas on E'. In this structure the inclusion map $E' \hookrightarrow E$ is a holomorphic vector bundle map. We have the following elementary result.

Proposition 1.5.6 *Let $\Phi\colon E \to E'$ be a holomorphic morphism of holomorphic vector bundles $E \to X$, $E' \to X$. If* $\dim\ker\Phi_x$ *is independent of the point $x \in X$, then the kernel* $\ker\Phi$ *is a holomorphic vector subbundle of E and the image* $\operatorname{im}\Phi$ *is a holomorphic vector subbundle of E'.*

We give a description of morphisms in local charts, beginning with the simplest case of product bundles. Let $\mathrm{Hom}(\mathbb{C}^n, \mathbb{C}^m)$ denote the set of all $\mathbb{C}$-linear maps $\mathbb{C}^n \to \mathbb{C}^m$. With respect to any pair of complex bases on the two space this equals $M_{m,n}(\mathbb{C}) \cong \mathbb{C}^{mn}$, the set of all complex $m \times n$ matrices. A morphism $\Phi\colon X \times \mathbb{C}^n \to X \times \mathbb{C}^m$ of product bundles is of the form $(x, v) \to (x, \varphi(x)v)$ for a holomorphic map $\varphi\colon X \to M_{m,n}(\mathbb{C})$. In particular, an automorphism of $X \times \mathbb{C}^n$ is given by a map $X \to GL_n(\mathbb{C}) \subset M_{n,n}(\mathbb{C})$.

Assume now that $E \to X$ and $E' \to X$ are holomorphic vector bundles of rank n, m, respectively. Choose holomorphic vector bundle atlases $\{(U_\alpha, \theta_\alpha)\}$, $\{(U_\alpha, \theta'_\alpha)\}$ for E, E', with the transition maps $g_{\alpha,\beta}$, $g'_{\alpha,\beta}$, respectively. A morphism $\Phi\colon E \to E'$ is given by a collection of maps $\varphi_\alpha\colon U_\alpha \to M_{m,n}(\mathbb{C})$ satisfying the compatibility conditions

$$\varphi_\alpha\, g_{\alpha,\beta} = g'_{\alpha,\beta}\, \varphi_\beta \quad \text{on } U_{\alpha,\beta}.$$

If Φ is an isomorphism, then $\varphi_\alpha\colon U_\alpha \to GL_n(\mathbb{C})$ and we can write

$$g'_{\alpha,\beta} = \varphi_\alpha\, g_{\alpha,\beta}\, \varphi_\beta^{-1}. \tag{1.16}$$

We say that the 1-cocycle $g' = (g'_{\alpha,\beta})$ is obtained by *twisting the 1-cocycle* $g = (g_{\alpha,\beta})$ *by the* 0-*cochain* $\varphi = (\varphi_\alpha)$, and we write $g' = \varphi \,\square\, g$. This leads to the observation that the isomorphism classes of holomorphic vector bundles of rank n over X are given by elements of the cohomology group $H^1(X; \mathcal{O}^{GL_n(\mathbb{C})})$ with coefficients in the multiplicative sheaf of germs of holomorphic maps $X \to GL_n(\mathbb{C})$. (See Sect. 8.1 for a further discussion of this topic.)

The group $H^1(X; \mathcal{O}^{GL_n(\mathbb{C})})$ is Abelian only for $n = 1$ when it equals $H^1(X; \mathcal{O}^*)$. The multiplicative group $H^1(X; \mathcal{O}^*) = \mathrm{Pic}(X)$ of equivalence classes of holomorphic line bundles on X is called the *Picard group* of X. The product on $\mathrm{Pic}(X)$ corresponds to the tensor product of line bundles.

Example 1.5.7 (Line bundles and divisors) A *divisor* D on a complex manifold X is determined by an open cover $\mathcal{U} = \{U_\alpha\}_{\alpha\in A}$ of X and a collection of meromorphic functions $f_\alpha \in \mathcal{M}(U_\alpha)$ that are not identically zero on any connected component of U_α such that for any pair of indexes $\alpha, \beta \in A$ there exists a nowhere vanishing holomorphic function $f_{\alpha,\beta} \in \mathcal{O}^*(U_{\alpha,\beta})$ satisfying

$$f_\alpha = f_{\alpha,\beta} f_\beta \quad \text{on } U_{\alpha,\beta}. \tag{1.17}$$

The 1-cocycle $(f_{\alpha,\beta})$ determines a holomorphic line bundle $E = [D]$ over X, and the collection (f_α) is a meromorphic section of $[D]$ in view of (1.17). In particular, a meromorphic function $f \in \mathcal{M}(X)$ that is not identically zero on any connected component determines a trivial line bundle on X. Conversely, if a line bundle E is presented over an open cover $\mathcal{U} = \{U_\alpha\}_{\alpha\in A}$ of X by a 1-cocycle $(f_{\alpha,\beta})$ with coefficients in the sheaf $\mathcal{O}_X^*$, then each meromorphic section f of E is given in the respective holomorphic trivializations of $E|_{U_\alpha} \cong U_\alpha \times \mathbb{C}$ by a collection of meromorphic functions $f_\alpha \in \mathcal{M}(U_\alpha)$ satisfying (1.17). If D is the divisor determined by (f_α) then clearly $E \cong [D]$.

A complex hypersurface $V \subset X$ determines a divisor D given by a collection of local defining functions for V. Conversely, every divisor D on X can be represented by a locally finite formal combination $D = \sum_i a_i V_i$ of irreducible complex hypersurfaces $V_i \subset X$ with integer coefficients $a_i \in \mathbb{Z}$ [264, p. 130]. The divisors on X form an Abelian group, $\mathrm{Div}(X)$, and by the above discussion we have a natural homomorphism $\mathrm{Div}(X) \to \mathrm{Pic}(X)$. This homomorphism is surjective on any quasi-projective manifold. The line bundle $[D]$ determined by a divisor D is trivial if and only if D is given by a global meromorphic function on X. (In this connection see the discussion in Sect. 5.2 concerning the *second Cousin problem*.) Two divisors D, D' are *linearly equivalent* if $D = D' + (f)$ for some $f \in \mathcal{M}(X)$; thus linear equivalence of a pair of divisors corresponds to holomorphic equivalence of the corresponding line bundles $[D]$ and $[D']$. For a sheaf theoretic interpretation of divisors and linear equivalence, see [264]. □

Definition 1.5.8 Given holomorphic vector bundles $\pi\colon E \to X$, $\pi'\colon E' \to X'$, a morphism of the first to the second bundle is a pair of holomorphic maps $f\colon X \to X'$, $F\colon E \to E'$ such that $\pi' \circ F = f \circ \pi$ and F is $\mathbb{C}$-linear on fibres.

An example is the tangent map $F = \mathrm{T}f\colon \mathrm{T}X \to \mathrm{T}X'$ of a holomorphic map $f\colon X \to X'$; in this case, $F_x\colon \mathrm{T}_x X \to \mathrm{T}_{f(x)}X'$ is the differential $\mathrm{d}f_x$ of f at x. The analogous definition applies to $\mathcal{C}^r$ vector bundles for $r \ge 1$.

Given a holomorphic vector bundle $\pi\colon E \to X$ and a holomorphic map $f\colon W \to X$, the *pullback bundle* $f^*E \to W$ (see (1.11) on p. 14) is a holomorphic vector bundle over W. We have a natural morphism $f^*E \to E$ over f which maps each fibre $(f^*E)_x$ isomorphically onto the fibre $E_{f(x)}$.

Assume that $E \to X$ and $F \to X$ are complex (or holomorphic) vector bundles. Using standard functors on complex vector spaces we obtain the following derived complex (resp. holomorphic) vector bundles over X:

(a) $E \oplus F = \bigcup_{x\in X} E_x \oplus F_x$, the *direct sum* or the *Whitney sum*,

(b) $E \otimes F = \bigcup_{x \in X} E_x \otimes F_x$, the *tensor product*,
(c) $E^* = \bigcup_{x \in X} E_x^*$, the *dual bundle* of E,
(d) $\mathrm{Hom}(E, F) = \bigcup_{x \in X} \mathrm{Hom}(E_x, F_x) = E^* \otimes F$,
(e) $\wedge^k E = \bigcup_{x \in X} \wedge^k E_x$, the *k-th exterior power* of E,
(f) $S^k(E) = \bigcup_{x \in X} S^k(E_x)$, the *k-th symmetric power* of E.

The transition maps in these bundles are obtained by applying the respective functor fibrewise to the transition maps of the original bundles. For example, if E and E' are given by cocycles $g_{\alpha,\beta}$, $g'_{\alpha,\beta}$ over the same open cover $\mathcal{U} = \{U_\alpha\}$ of X then the direct sum $E \oplus E'$ is given by the cocycle

$$\begin{pmatrix} g_{\alpha,\beta} & 0 \\ 0 & g'_{\alpha,\beta} \end{pmatrix}.$$

Given a subbundle E' of E, the *quotient bundle* $E/E' \to X$ is defined by $E/E' = \bigcup_{x \in X} E_x/E'_x$. For any vector bundle chart θ on E satisfying (1.15) the map $pr_2 \circ \theta \colon E|_U \to U \times \mathbb{C}^{n-m}$ factors through $(E/E')|_U$ and induces a bijective map $\widetilde{\theta} \colon (E/E')|_U \to U \times \mathbb{C}^{n-m}$. The collection of all such maps is a complex (resp. holomorphic) vector bundle atlas on E/E'. If $E = E' \oplus E''$ is a direct sum of its subbundles $E', E'' \subset E$ then the projection $\tau \colon E \to E''$ with the kernel $\ker \tau = E'$ induces an isomorphism of E/E' onto E''.

A sequence of vector bundle maps over X,

$$\cdots \longrightarrow E_{k-1} \xrightarrow{\sigma_{k-1}} E_k \xrightarrow{\sigma_k} E_{k+1} \longrightarrow \cdots$$

is a *complex* if $\sigma_k \circ \sigma_{k-1} = 0$ (equivalently, $\mathrm{im}\,\sigma_{k-1} \subset \ker \sigma_k$) for every k. The sequence is *exact* at E_k if $\mathrm{im}\,\sigma_{k-1} = \ker \sigma_k$. A *short exact sequence* is an exact sequence of the form

$$0 \longrightarrow E' \xrightarrow{\sigma} E \xrightarrow{\tau} E'' \longrightarrow 0. \tag{1.18}$$

This means that σ is injective, τ is surjective, and $\mathrm{im}\,\sigma = \ker \tau$. Hence τ induces an isomorphism of the quotient bundle $E/\sigma(E')$ onto E''.

A short exact sequence (1.18) *splits* if there exists a vector bundle homomorphism $\rho \colon E'' \to E$ such that $\tau \circ \rho$ is the identity on E''. Such ρ is called a *splitting map* for the sequence. In this case E is isomorphic to the Whitney sum $E = \sigma(E') \oplus \rho(E'')$ of its subbundles $\sigma(E')$ and $\rho(E'')$. Note that every short exact sequence splits locally over small open subsets of the base, and any convex linear combination of splittings is again a splitting. By patching local splittings with a partition of unity one gets the following.

Proposition 1.5.9 *Every short exact sequence* (1.18) *of complex vector bundle maps of class* $\mathscr{C}^r$ $(r \in \{0, 1, \ldots, \infty\})$ *admits a* $\mathscr{C}^r$ *splitting. In particular, we have* $E \cong E' \oplus E''$ *as complex vector bundles of class* $\mathscr{C}^r$.

The analogous result for holomorphic vector bundles over Stein spaces follows from Cartan's Theorem B; see Corollary 2.6.5 on p. 57.

1.6 The Tangent Bundle

We assume that the reader is familiar with the construction of the real tangent bundle TX of a smooth manifold X (see Example 1.5.2). A tangent vector $V_x \in T_x X$ is viewed as a derivation $\mathcal{C}^\infty_x \ni f \mapsto V_x(f) \in \mathbb{R}$ on the algebra of germs of smooth functions at x. Sections $X \to TX$ are called *vector fields* on X. The complexification $\mathbb{C}TX = TX \otimes_{\mathbb{R}} \mathbb{C}$ of TX is the *complexified tangent bundle* of X; its sections are called *complex vector fields* on X.

Assume now that X is a complex manifold. There is a unique real linear endomorphism $J \in \operatorname{End}_{\mathbb{R}} TX$, called the *almost complex structure operator*, which is given in any local holomorphic coordinate system $z = (z_1, \dots, z_n)$ $(z_j = x_j + iy_j)$ on X by

$$J \frac{\partial}{\partial x_j} = \frac{\partial}{\partial y_j}, \qquad J \frac{\partial}{\partial y_j} = -\frac{\partial}{\partial x_j}. \tag{1.19}$$

The operator J clearly satisfies $J^2 = -\mathrm{Id}$. It extends to $\mathbb{C}TX$ by $J(v \otimes \alpha) = J(v) \otimes \alpha$ for $v \in TX$ and $\alpha \in \mathbb{C}$. From $J^2 = -\mathrm{Id}$ we infer that the eigenvalues of J are $+i$ and $-i$. Hence we have a decomposition

$$\mathbb{C}TX = T^{1,0}X \oplus T^{0,1}X \tag{1.20}$$

into the $+i$ eigenspace $T^{1,0}X$ and the $-i$ eigenspace $T^{0,1}X$ of J. In holomorphic coordinates $z = (z_1, \dots, z_n)$ on an open subset $U \subset X$ we have

$$T^{1,0}X|_U = \operatorname{Span}_{\mathbb{C}}\left\{\frac{\partial}{\partial z_1}, \dots, \frac{\partial}{\partial z_n}\right\}, \qquad T^{0,1}X|_U = \operatorname{Span}_{\mathbb{C}}\left\{\frac{\partial}{\partial \bar{z}_1}, \dots, \frac{\partial}{\partial \bar{z}_n}\right\},$$

where

$$\frac{\partial}{\partial z_j} = \frac{1}{2}\left(\frac{\partial}{\partial x_j} - i\frac{\partial}{\partial y_j}\right), \qquad \frac{\partial}{\partial \bar{z}_j} = \frac{1}{2}\left(\frac{\partial}{\partial x_j} + i\frac{\partial}{\partial y_j}\right).$$

Note that $T^{1,0}X$ is a holomorphic vector bundle whose transition functions are the complex Jacobians (1.6) of the holomorphic transition maps (1.4) between complex charts on X. The conjugation $v \otimes \alpha \to v \otimes \bar{\alpha}$ induces a $\mathbb{C}$-antilinear isomorphism of $T^{1,0}X$ onto $T^{0,1}X$, and hence $T^{0,1}X = \overline{T^{1,0}X}$ is an antiholomorphic complex vector bundle.

Remark 1.6.1 A smooth real manifold X of even dimension $2n$ endowed with a smooth endomorphism $J \in \operatorname{End}_{\mathbb{R}} TX$ satisfying $J^2 = -\mathrm{Id}$ is called an *almost complex manifold*. If $n > 1$ then such (X, J) need not admit any local J-holomorphic functions f, i.e., smooth complex valued functions f satisfying $df \circ J = i \cdot df$. Such J induces a complex manifold structure on X in the sense of (1.19) if and only if the Nijenhuis tensor of J vanishes identically (this always holds when $n = 1$). We refer the reader to the paper by Newlander and Nirenberg [432] and the monograph [412] by McDuff and Salamon. □

We have an $\mathbb{R}$-linear isomorphism $\Phi\colon \mathrm{T}X \xrightarrow{\cong} \mathrm{T}^{1,0}X$ given by

$$\mathrm{T}X \ni V \longmapsto \Phi(V) = \frac{1}{2}(V - \mathrm{i}JV) \in \mathrm{T}^{1,0}X \tag{1.21}$$

with the inverse $\Phi^{-1}(W) = 2\Re W$ $(W \in \mathrm{T}^{1,0}X)$. We have $\Phi(JV) = \mathrm{i}\,\Phi(V)$ for $V \in \mathrm{T}X$; i.e., the following diagram commutes:

$$\begin{array}{ccc} \mathrm{T}X & \xrightarrow{J} & \mathrm{T}X \\ \Phi\downarrow & & \downarrow\Phi \\ \mathrm{T}^{1,0}X & \xrightarrow{\mathrm{i}} & \mathrm{T}^{1,0}X \end{array}$$

In local holomorphic coordinates the isomorphism Φ is given by

$$\sum_{j=1}^{n}\left(a_j\frac{\partial}{\partial x_j} + b_j\frac{\partial}{\partial y_j}\right) \overset{\Phi}{\longmapsto} \sum_{j=1}^{n}(a_j + \mathrm{i}b_j)\frac{\partial}{\partial z_j}. \tag{1.22}$$

Definition 1.6.2 A real vector field V on X is said to be *holomorphic* if $\Phi(V) = \frac{1}{2}(V - \mathrm{i}JV)$ is a holomorphic section of $\mathrm{T}^{1,0}X$.

We see from (1.22) that the vector field $V = \sum_{j=1}^{n} a_j\frac{\partial}{\partial x_j} + b_j\frac{\partial}{\partial y_j}$ is holomorphic if and only if the functions $a_j + \mathrm{i}b_j$ $(j = 1, \ldots, n)$ are holomorphic.

Denote by $\aleph_r(X)$ the space of all real vector fields of class $\mathcal{C}^r$ on a (real or complex) manifold X $(r = 0, 1, \ldots, \infty, \omega)$, by $\aleph_{\mathcal{O}}(X)$ the space of all holomorphic vector fields on a complex manifold X, and by $\aleph_{\mathcal{A}}(X)$ the space of all algebraic vector fields on a complex algebraic manifold X (algebraic sections of the tangent bundle $\mathrm{T}X$).

The *commutator* or *Lie bracket* of vector fields V, W is defined by

$$[V, W](f) = V\big(W(f)\big) - W\big(V(f)\big), \quad f \in \mathcal{C}^\infty(X). \tag{1.23}$$

If V, W are of class $\mathcal{C}^r$ for some $r \ge 1$ then $[V, W]$ is a vector field of class $\mathcal{C}^{r-1}$. If

$$V = \sum_{j=1}^{n} a_j(x)\frac{\partial}{\partial x_j}, \qquad W = \sum_{j=1}^{n} b_j(x)\frac{\partial}{\partial x_j} \tag{1.24}$$

then

$$[V, W] = \sum_{j=1}^{n}\big(V(b_j) - W(a_j)\big)\frac{\partial}{\partial x_j}. \tag{1.25}$$

This operation is linear in both factors, anticommutative ($[V, W] = -[W, V]$, hence $[V, V] = 0$), and it satisfies the *Jacobi identity*:

$$\big[U, [V, W]\big] + \big[V, [W, U]\big] + \big[W, [U, V]\big] = 0.$$

The $\mathbb{R}$-vector spaces $\aleph_\infty(X)$ and $\aleph_\omega(X)$, endowed with the commutator bracket, are real Lie algebras. The commutator operation extends to sections of $\mathbb{C}\mathrm{T}X$, $\mathrm{T}^{1,0}X$ and $\mathrm{T}^{0,1}X$, with the same properties.

Proposition 1.6.3 *The space $\aleph_{\mathcal{O}}(X)$ of all holomorphic vector fields is a complex Lie subalgebra of $\aleph_\infty(X)$. The map Φ (1.21) induces a complex Lie algebra isomorphism of $\aleph_{\mathcal{O}}(X)$ onto the space of all holomorphic sections of $\mathrm{T}^{1,0}X$.*

Proof The Cauchy-Riemann equations (1.3) (p. 4) show that for a real vector field V and a holomorphic function f we have $\Phi(V)(f) = V(f)$. Hence V is holomorphic if and only if, as a derivation $f \to V_x(f)$ on the algebra $\mathcal{C}_x^\infty$ of germs of smooth functions at any point x, it maps the subalgebra $\mathcal{O}_x \subset \mathcal{C}_x^\infty$ consisting of holomorphic germs to itself. The proposition now follows by applying (1.23) to holomorphic test functions f on small open sets in X. □

In the sequel we tacitly use the identification $\mathrm{T}X \cong \mathrm{T}^{1,0}X$, and the corresponding identification of $\aleph_{\mathcal{O}}(X)$ with the space of holomorphic sections of $\mathrm{T}^{1,0}X$.

To a smooth map $f: X \to Y$ we associate its *tangent map* $\mathrm{T}f: \mathrm{T}X \to \mathrm{T}Y$; this is a vector bundle morphism over f that is given on any fibre $\mathrm{T}_x X$ by the differential $\mathrm{d}f_x: \mathrm{T}_x X \to \mathrm{T}_{f(x)}Y$. If (X, J_X) and (Y, J_Y) are complex manifolds then a map $f: X \to Y$ is holomorphic precisely when the differential $\mathrm{d}f$ commutes with the almost complex structure operators on X and Y:

$$\mathrm{d}f \circ J_X = J_Y \circ \mathrm{d}f.$$

In this case $\mathrm{d}f$ respects the decomposition $\mathbb{C}\mathrm{T}X = \mathrm{T}^{1,0}X \oplus \mathrm{T}^{0,1}X$ (1.20) and $\mathrm{d}f_x: \mathrm{T}_x^{1,0}X \to \mathrm{T}_{f(x)}^{1,0}Y$ is $\mathbb{C}$-linear for every $x \in X$.

If $f: X \to Y$ is biholomorphic then every holomorphic vector field V on X is pushed forward by f to a holomorphic vector field $(f_* V)_{f(x)} = \mathrm{d}f_x \cdot V_x$ on Y. Conversely, a holomorphic vector field W on Y pulls back to a holomorphic vector field $f^* W = (f^{-1})_* W$.

If M is a complex submanifold of a complex manifold X then its tangent bundle $\mathrm{T}M$ is a holomorphic vector subbundle of the restricted bundle $\mathrm{T}X|_M$. The *normal bundle* of M in X is the quotient bundle $\mathrm{N}_{M,X} = \mathrm{T}X|_M / \mathrm{T}M$. We thus have a short exact sequence

$$0 \longrightarrow \mathrm{T}M \longrightarrow \mathrm{T}X|_M \xrightarrow{\beta} \mathrm{N}_{M/X} \longrightarrow 0. \tag{1.26}$$

By Proposition 1.5.9 we can realize $\mathrm{N}_{M/X}$ as a smooth complex vector subbundle of $\mathrm{T}X|_M$ such that

$$\mathrm{T}X|_M = \mathrm{T}M \oplus \mathrm{N}_{M/X}. \tag{1.27}$$

If the submanifold M is Stein, it admits an open Stein neighborhood $\Omega \subset X$ (Theorem 3.1.1 on p. 66); hence Cartan's Theorem B (Theorem 2.6.1 on p. 54) implies that the sequence (1.26) also splits holomorphically and we get a holomorphic direct sum (1.27).

The notion of tangent space is also defined for a complex space X with singularities. We get a *linear space* $\pi: \mathrm{T}X \to X$, that is, a (not necessarily reduced) complex space $L = \mathrm{T}X$ and a holomorphic projection π onto X such that the fibres $L_x = \mathrm{T}_x X$ are vector spaces whose dimension depends on x. Here is a local description; for further details see [158]. Fix a point $x_0 \in X$ and represent an open neighborhood $W \subset X$ of x_0 as a closed complex subvariety in an open set $U \subset \mathbb{C}^n$. After shrinking U we find holomorphic functions $f_1, \ldots, f_k \in \mathcal{O}(U)$ which generate the ideal sheaf $\mathcal{J}_W$ at each point. Let $(z_1, \ldots, z_n, w_1, \ldots, w_n)$ be complex coordinates on $\mathbb{C}^n \times \mathbb{C}^n$. Then $\mathrm{T}W = \mathrm{T}X|_W$ is the complex subspace of $U \times \mathbb{C}^n$ determined by the functions

$$f_1, \ldots, f_k \quad \text{and} \quad \frac{\partial f_i}{\partial z_1} w_1 + \cdots + \frac{\partial f_i}{\partial z_n} w_n \quad \text{for } i = 1, \ldots, k. \tag{1.28}$$

The projection $\mathrm{T}W \to W$ is the coordinate projection $W \times \mathbb{C}^m \to W$ restricted to $\mathrm{T}W$. Different local representations of X as subvarieties of $\mathbb{C}^n$ give isomorphic representations of the tangent space. The restriction of $\mathrm{T}X$ to the regular locus X_{reg} is the usual tangent bundle of X_{reg}. The fibre $\mathrm{T}_x X$ is called the *Zariski tangent space* of X at the point $x \in X$. A holomorphic map $f: X \to Y$ of complex spaces induces the Zariski differential $\mathrm{d}f_x: \mathrm{T}_x X \to \mathrm{T}_{f(x)} Y$.

If $h: Z \to X$ is a holomorphic submersion of a complex space Z onto a complex space X (Definition 1.3.1), then the fibres $Z_x = h^{-1}(x)$ are smooth and the tangent spaces $\mathrm{T}_z Z_{h(z)} = \ker \mathrm{d}h_z$ define a holomorphic vector bundle $\mathrm{VT}Z \to Z$, called the *vertical tangent bundle* of Z, or the *relative tangent bundle*, sometimes denoted $\mathrm{T}Z/X$.

Assume now that M is a real submanifold of a complex manifold (X, J). Let $m = \dim_{\mathbb{R}} M$, $n = \dim_{\mathbb{C}} X$ and $d = 2n - m$ (the real codimension of M in X). For every point $x \in M$ we define subspaces $\mathrm{T}_x^{\mathbb{C}} M, L_x \subset \mathrm{T}_x M$ by

$$\mathrm{T}_x^{\mathbb{C}} M = \mathrm{T}_x M \cap J(\mathrm{T}_x M), \quad \mathrm{T}_x M = \mathrm{T}_x^{\mathbb{C}} M \oplus L_x.$$

Clearly $\mathrm{T}_x^{\mathbb{C}} M$ is the unique maximal complex subspace of $\mathrm{T}_x M$. The choice of the complementary real subspace L_x can be specified by requiring that the direct sum is orthogonal with respect to a chosen Hermitian metric on $\mathrm{T}X$. If $k = \dim_{\mathbb{C}} \mathrm{T}_x^{\mathbb{C}} M$ is independent of the point $x \in M$ then M is called a *CR submanifold* of X of *CR dimension* k. If $k + d = n$ (hence $\dim_{\mathbb{R}} M = 2k + d$) then M is said to be *generic*. If $\dim M = 2k + 1$ then M is *of hypersurface type*. (For analysis on CR manifolds see [42].)

1.7 The Cotangent Bundle and Differential Forms

Let X be a complex manifold of dimension n. We denote by $\mathrm{T}^* X$ its cotangent bundle, the real dual of the tangent bundle $\mathrm{T}X$. The $\mathbb{R}$-linear endomorphism J of $\mathbb{C}\mathrm{T}X$ induces the dual endomorphism J^* of the complexified cotangent bundle $\mathbb{C}\mathrm{T}^* X = \mathrm{T}^* X \otimes_{\mathbb{R}} \mathbb{C}$ with $(J^*)^2 = -\mathrm{Id}$. We have a splitting

$$\mathbb{C}\mathrm{T}^* X = \mathrm{T}^{*1,0} X \oplus \mathrm{T}^{*0,1} X$$

into the $+\mathrm{i}$ and $-\mathrm{i}$ eigenspaces of J^*. Given holomorphic coordinates $z = (z_1, \ldots, z_n)$ on an open subset $U \subset X$ and a point $p \in U$, the 1-forms $\mathrm{d}z_1, \ldots, \mathrm{d}z_n$ at p are a complex basis of $\mathrm{T}_p^{*1,0} X$, while $\mathrm{d}\bar z_1, \ldots, \mathrm{d}\bar z_n$ are a complex basis of

$\mathrm{T}^{*0,1}_p X$. Note that $\mathrm{T}^{*1,0}X$ and $\mathrm{T}^{*0,1}X$ are the complex dual bundles of $\mathrm{T}^{1,0}X$ and $\mathrm{T}^{0,1}X$, respectively.

The complex exterior algebra bundles $\wedge\mathrm{T}^{*1,0}X$, $\wedge\mathrm{T}^{*0,1}X$ admit natural bundle injections into the exterior algebra bundle $\wedge\mathbb{C}\mathrm{T}^*X$. We have $\wedge\mathbb{C}\mathrm{T}^*X = \bigoplus_{k=0}^{2n}\wedge^k\mathbb{C}\mathrm{T}^*X$, $\wedge\mathrm{T}^{*1,0}X = \bigoplus_{k=0}^{n}\wedge^k\mathrm{T}^{*1,0}X$, and $\wedge\mathrm{T}^{*0,1}X = \bigoplus_{k=0}^{n}\wedge^k\mathrm{T}^{*0,1}X$. For every $k = 0, 1\ldots, 2n$ we have a decomposition

$$\wedge^k\mathbb{C}\mathrm{T}^*X = \bigoplus_{p+q=k} (\wedge^p\mathrm{T}^{*1,0}X)\wedge(\wedge^q\mathrm{T}^{*0,1}X).$$

Let $\mathcal{E}_k(X)$ denote the space of smooth sections of the bundle $\wedge^k\mathbb{C}\mathrm{T}^*X$; these are smooth complex valued differential forms of degree k on X. In particular, $\mathcal{E}_0(X) = \mathcal{C}^\infty(X)$ is the space of smooth functions on X. We have $\mathcal{E}_k(X) = \bigoplus_{p+q=k}\mathcal{E}_{p,q}(X)$ where $\mathcal{E}_{p,q}(X)$ is the space of smooth sections of $(\wedge^p\mathrm{T}^{*1,0}X)\wedge(\wedge^q\mathrm{T}^{*0,1}X)$; these are called differential (p,q)-forms. In complex coordinates $z = (z_1, \ldots, z_n)$ every (p,q)-form has the expression

$$\alpha = \sum_{|I|=p,\ |J|=q} a_{I,J}(z)\,\mathrm{d}z_I\wedge\mathrm{d}\bar{z}_J \tag{1.29}$$

where $I = (i_1, \ldots, i_p)$, $J = (j_1, \ldots, j_q)$ $(1\le i_1 < \cdots < i_p \le n,\ 1\le j_1 < \cdots < j_q \le n)$, $\mathrm{d}z_I = \mathrm{d}z_{i_1}\wedge\cdots\wedge\mathrm{d}z_{i_p}$, and $\mathrm{d}\bar{z}_J = \mathrm{d}\bar{z}_{j_1}\wedge\cdots\wedge\mathrm{d}\bar{z}_{j_q}$.

A *holomorphic p-form* is a holomorphic section of the holomorphic vector bundle $\wedge^p\mathrm{T}^{*1,0}X$. In local holomorphic coordinates,

$$\alpha = \sum_{|I|=p} a_I(z)\,\mathrm{d}z_I = \sum_{|I|=p} a_I(z)\,\mathrm{d}z_{i_1}\wedge\cdots\wedge\mathrm{d}z_{i_p}$$

where the coefficients a_I are holomorphic functions. The space of all holomorphic p-forms on X is denoted $\Omega_p(X)$; hence $\Omega_0(X) = \mathcal{O}(X)$.

The *pullback* of a form (1.29) by a smooth map $f = (f_1, \ldots, f_n)$ equals

$$f^*\alpha = \sum_{|I|=p,\ |J|=q} (a_{I,J}\circ f)\,\mathrm{d}f_I\wedge\mathrm{d}\bar{f}_J.$$

We have $\mathrm{d}(f^*\alpha) = f^*(\mathrm{d}\alpha)$, where $\mathrm{d}\colon\mathcal{E}_k(X)\to\mathcal{E}_{k+1}(X)$ is the exterior derivative. If $z = f(w)$ is holomorphic in w then

$$\mathrm{d}f_j = \sum_k \frac{\partial f_j}{\partial w_k}\mathrm{d}w_k, \qquad \mathrm{d}\bar{f}_j = \sum_k \frac{\partial\bar{f}_j}{\partial\overline{w}_k}\mathrm{d}\overline{w}_k$$

for every j, and we see that $f^*\alpha$ is again a (p,q)-form. Thus, the pullback by a holomorphic map respects the decomposition (1.29) and takes holomorphic forms to holomorphic forms. The exterior derivative splits as $\mathrm{d} = \partial + \overline{\partial}$, where

$$\partial\colon\mathcal{E}_{p,q}(X)\to\mathcal{E}_{p+1,q}(X), \qquad \overline{\partial}\colon\mathcal{E}_{p,q}(X)\to\mathcal{E}_{p,q+1}(X)$$

are defined in local holomorphic coordinates by applying the corresponding operators (1.1) to the coefficient functions of α (1.29):

$$\partial\alpha = \sum_{|I|=p,\ |J|=q} \partial a_{I,J}\wedge\mathrm{d}z_I\wedge\mathrm{d}\bar{z}_J, \qquad \overline{\partial}\alpha = \sum_{|I|=p,\ |J|=q} \overline{\partial}a_{I,J}\wedge\mathrm{d}z_I\wedge\mathrm{d}\bar{z}_J.$$

The definition is independent of the choice of holomorphic coordinates since the pullback by a holomorphic map f commutes with ∂ and $\overline{\partial}$:

$$f^*(\partial\alpha) = \partial(f^*\alpha), \qquad f^*(\overline{\partial}\alpha) = \overline{\partial}(f^*\alpha).$$

The operators d, ∂ and $\overline{\partial}$ are local and satisfy the properties

$$\mathrm{d}^2 = \mathrm{d}\circ\mathrm{d} = 0, \qquad \partial^2 = 0, \qquad \overline{\partial}^2 = 0, \qquad \partial\overline{\partial} = -\overline{\partial}\partial.$$

The *conjugate differential* is defined by

$$\mathrm{d}^c = -J^*\mathrm{d}. \tag{1.30}$$

Explicitly, on functions we have $(\mathrm{d}^c f)_x \cdot v = -\mathrm{d}f_x \cdot Jv$ for $v \in \mathrm{T}_x X$. In local holomorphic coordinates $z = (z_1, \ldots, z_n)$, $z_j = x_j + \mathrm{i}y_j$, we have

$$\mathrm{d}^c = \sum_{j=1}^{n} -\frac{\partial}{\partial y_j}\,\mathrm{d}x_j + \frac{\partial}{\partial x_j}\,\mathrm{d}y_j. \tag{1.31}$$

Clearly, d^c is a real operator and we have the following relations:

$$\mathrm{d} = \partial + \overline{\partial}, \qquad \mathrm{d}^c = \mathrm{i}(\overline{\partial} - \partial), \qquad \mathrm{d}\mathrm{d}^c = 2\mathrm{i}\,\partial\overline{\partial}.$$

A differential form α is *closed* if $\mathrm{d}\alpha = 0$, and is $\overline{\partial}$-*closed* if $\overline{\partial}\alpha = 0$. We denote by $\mathcal{Z}_k(X)$ the space of all closed k-forms, and by $\mathcal{Z}_{p,q}(X)$ the space of all $\overline{\partial}$-closed (p,q)-forms. Note that $\mathcal{Z}_{0,0}(X) = \mathcal{O}(X)$ is the space of holomorphic functions, and $\mathcal{Z}_{p,0}(X) = \Omega_p(X)$ is the space of holomorphic p-forms.

A form α of degree $k \geq 1$ is *exact* if $\alpha = \mathrm{d}\beta$ for some form β (this implies $\mathrm{d}\alpha = 0$); α is $\overline{\partial}$-*exact* if $\alpha = \overline{\partial}\beta$ for some form β (this implies $\overline{\partial}\alpha = 0$).

Each of the operators $\mathrm{d}, \partial, \overline{\partial}$ is *locally exact*. For d this is the *Poincaré lemma* which states that for every closed k-form α the equation $\mathrm{d}\beta = \alpha$ has a solution on any contractible open set. For $\overline{\partial}$ we have the following lemma of Grothendieck, also called the $\overline{\partial}$-*Poincaré lemma* [264, p. 25].

Lemma 1.7.1 *For every smooth (p,q)-form α on a polydisc $\mathbb{D}^n \subset \mathbb{C}^n$ $(q \geq 1)$ such that $\overline{\partial}\alpha = 0$ there exists a smooth $(p, q-1)$-form β on $\mathbb{D}^n$ satisfying $\overline{\partial}\beta = \alpha$. The analogous result holds for the operator ∂.*

The complex vector spaces

$$H^p_{\mathrm{d}}(X) = \frac{\mathcal{Z}_p(X)}{\mathrm{d}(\mathcal{E}_{p-1}(X))}, \qquad p = 0, 1, \ldots$$

are called the *de Rham cohomology groups* of X. (For $p < 0$ set $\mathcal{E}_p(X) = \{0\}$.) In view of Stokes' theorem, the integration of closed p-forms over singular p-cycles in X defines a map $H^p_{\mathrm{d}}(X) \to H^p_{\mathrm{sing}}(X; \mathbb{C})$ to the singular cohomology groups of X with complex coefficients; this map is an isomorphism by de Rham's theorem ([264, p. 43], [563]). Hence, the de Rham cohomology ring $H^*_{\mathrm{d}}(X) = \bigoplus_{p\geq 0} H^p_{\mathrm{d}}(X)$ is a topological invariant of X. For $p > \dim_{\mathbb{R}} X$ the group $H^p_{\mathrm{d}}(X)$ vanishes since there are no nonvanishing p-forms. The vector spaces

$$H^{p,q}_{\overline{\partial}}(X) = \frac{\mathcal{Z}_{p,q}(X)}{\overline{\partial}(\mathcal{E}_{p,q-1}(X))}, \quad p,q \geq 0 \tag{1.32}$$

are called the *Dolbeault cohomology groups* of X. They depend on the holomorphic structure of X. Clearly, $H^{p,0}_{\overline{\partial}}(X) = \Omega_p(X)$ is the space of holomorphic p-forms on X, and $H^{p,q}_{\overline{\partial}}(X) = 0$ if $\max\{p,q\} > \dim_{\mathbb{C}} X$. Lemma 1.7.1 says that $H^{p,q}_{\overline{\partial}}(P) = 0$ for all $p \geq 0$, $q \geq 1$ when P is a polydisc in $\mathbb{C}^n$. It is a much deeper result that

$$X \text{ Stein} \Longrightarrow H^{p,q}_{\overline{\partial}}(X) = 0, \quad p \geq 0,\ q \geq 1$$

(see Theorem 2.6.7). If X is a compact complex manifold then the cohomology groups $H^{p,q}_{\overline{\partial}}(X)$ are finite dimensional vector spaces.

1.8 Plurisubharmonic Functions and the Levi Form

Plurisubharmonic functions were introduced in 1942 by Oka [444, Chap. VI] and Lelong [388]. This is undoubtedly one of the most important classes of functions in complex analysis and geometry. A complex manifold with a strongly plurisubharmonic exhaustion function is a Stein manifold (see Theorem 2.5.2 on p. 53). Plurisubharmonic functions play a major role in the L^2-theory for solving the $\overline{\partial}$-problem (see Sect. 2.7). Pluripotential theory provides an important link with Kähler geometry via the complex Monge-Ampère equation. For more on this subject, see [301, 337, 389].

An upper semicontinuous function u on a complex manifold X which is not identically $-\infty$ on any connected component is said to be *plurisubharmonic* if for every continuous map $f\colon \overline{\mathbb{D}} \to X$ which is holomorphic in the disc $\mathbb{D} = \{z \in \mathbb{C} : |z| < 1\}$ (such f is called an *analytic disc* in X) we have the submeanvalue property

$$u\big(f(0)\big) \leq \int_0^{2\pi} u\big(f\big(e^{i\theta}\big)\big)\, \frac{d\theta}{2\pi}.$$

This is equivalent to asking that $u \circ f$ be subharmonic on $\mathbb{D}$ for every such f. For a domain $X \subset \mathbb{C}^n$ it suffices to assume that this holds for small linear complex discs in X. The definition extends to functions on a complex space X by local plurisubharmonic extendibility from local embeddings of X in $\mathbb{C}^N$. We denote by $\mathrm{Psh}(X)$ the set of all plurisubharmonic functions on X.

Here are some basic examples of plurisubharmonic functions:

- $\log|f|$ where f is holomorphic and not identically zero.
- $|f|^p$ for any $p > 0$, where f is holomorphic.
- $\log(\sum_{j=1}^m |f_j|^2)$ where the f_j are holomorphic and not all zero.

We collect the main properties of the class $\mathrm{Psh}(X)$:

- Plurisubharmonicity is a local property: A function is plurisubharmonic if and only if it is such in an open neighborhood of any point.
- *The maximum principle:* If $u \in \mathrm{Psh}(X)$ has a local maximum at some point, then u is constant on the respective connected component of X.
- If $u_j \in \mathrm{Psh}(X)$ and $c_j \geq 0$ for $j = 1, \dots, k$ then $\sum_{j=1}^{k} c_j u_j \in \mathrm{Psh}(X)$.
- If $u_j \in \mathrm{Psh}(X)$ for $j = 1, \dots, k$ then $\max\{u_1, \dots, u_k\} \in \mathrm{Psh}(X)$.
- If $u_1 \leq u_2 \leq \cdots \leq u = \lim_{j \to \infty} u_j < +\infty$ and $u_j \in \mathrm{Psh}(X)$ for every $j = 1, 2, \dots$, then u is plurisubharmonic provided that it is upper semicontinuous. If u is locally bounded from above then its upper regularization $u^*(z) = \limsup_{\zeta \to z} u(\zeta)$ is plurisubharmonic.
- If $u_1 \geq u_2 \geq \cdots \geq u = \lim_{j \to \infty} u_j$ and $u_j \in \mathrm{Psh}(X)$ for every $j = 1, 2, \dots$, then u is either plurisubharmonic or identically $-\infty$ on any connected component of X.
- Every $u \in \mathrm{Psh}(X)$ is the pointwise limit of a decreasing sequence of smooth plurisubharmonic functions on any relatively compact subset $\Omega \Subset X$.
- If $u \in \mathrm{Psh}(X)$ and $h: [-\infty, \infty) \to [-\infty, \infty)$ is an increasing convex function then $h \circ \rho \in \mathrm{Psh}(X)$.
- If u is plurisubharmonic and f is a holomorphic map then $u \circ f$ is plurisubharmonic or $\equiv -\infty$ on each connected component of its domain.

In order to place plurisubharmonicity in its proper geometric context we recall the notion of the *Levi form*. Let $z = (z_1, \dots, z_n)$ be complex coordinates on $\mathbb{C}^n$. Using the isomorphism $\mathrm{T}_z\mathbb{C}^n \cong \mathrm{T}_z^{1,0}\mathbb{C}^n \cong \mathbb{C}^n$ (1.21) we identify $v = 2\Re \sum_{j=1}^{n} v_j \frac{\partial}{\partial z_j} \in \mathrm{T}_z\mathbb{C}^n$ with $(v_1, \dots, v_n) \in \mathbb{C}^n$. Given a $\mathcal{C}^2$ function ρ in a neighborhood of a point $z \in \mathbb{C}^n$, we have

$$\rho(z+v) = \rho(z) + 2\Re \sum_{j=1}^{n} \frac{\partial \rho}{\partial z_j}(z)\, v_j + \Re \sum_{j,k=1}^{n} \frac{\partial^2 \rho}{\partial z_j \partial z_k}(z)\, v_j v_k$$
$$+ \frac{1}{2} \sum_{j,k=1}^{n} \frac{\partial^2 \rho}{\partial z_j \partial \bar{z}_k}(z)\, v_j \overline{v}_k + o\big(|v|^2\big).$$

The *complex Hessian* of ρ at z is the Hermitian bilinear form

$$\mathrm{T}_z\mathbb{C}^n \times \mathrm{T}_z\mathbb{C}^n \ni (v, w) \mapsto H_{\rho,z}(v, w) = \sum_{j,k=1}^{n} \frac{\partial^2 \rho}{\partial z_j \partial \bar{z}_k}(z)\, v_j \overline{w}_k. \tag{1.33}$$

The associated Hermitian quadratic form is called the *Levi form* of ρ at z:

$$\mathcal{L}_{\rho,z}(v) = H_{\rho,z}(v, v) = \sum_{j,k=1}^{n} \frac{\partial^2 \rho}{\partial z_j \partial \bar{z}_k}(z)\, v_j \bar{v}_k = \frac{\partial^2}{\partial \zeta\, \partial \bar{\zeta}}\bigg|_{\zeta=0} \rho(z + \zeta v). \tag{1.34}$$

Clearly this equals one quarter of the Laplacian $\Delta = \frac{\partial^2}{\partial x^2} + \frac{\partial^2}{\partial y^2}$ of the function $\zeta = x + \mathrm{i}y \mapsto \rho_{z,v}(\zeta) = \rho(z + \zeta v)$ at $\zeta = 0$:

$$\mathcal{L}_{\rho,z}(v) = \frac{1}{4}\Delta\rho_{z,v}(0). \tag{1.35}$$

Hence, *a $\mathcal{C}^2$ function $\rho\colon \Omega \to \mathbb{R}$ on a domain $\Omega \subset \mathbb{C}^n$ is plurisubharmonic if and only if $\mathcal{L}_{\rho,z} \ge 0$ is nonnegative definite on $\mathrm{T}_z\mathbb{C}^n$ for every $z \in \Omega$.*

Definition 1.8.1 A $\mathcal{C}^2$ function $\rho\colon \Omega \to \mathbb{R}$ is *strongly plurisubharmonic* if its Levi form $\mathcal{L}_{\rho,z}$ is positive definite for every point $z \in \Omega$.

Clearly this condition is stable under small $\mathcal{C}^2$ deformations.

Example 1.8.2 If $f_1, \ldots, f_k$ are holomorphic functions then $\rho = \sum_{j=1}^k |f_j|^2$ is plurisubharmonic, and is strongly plurisubharmonic at each point z where the complex differentials $(\mathrm{d}f_j)_z$ span the cotangent space. □

Consider now the effect of coordinate changes. Assume that $f\colon \Omega \to \Omega'$ is a holomorphic map between domains $\Omega \subset \mathbb{C}^n$, $\Omega' \subset \mathbb{C}^m$. Let $\rho'\colon \Omega' \to \mathbb{R}$ be a $\mathcal{C}^2$ function and set $\rho = \rho' \circ f\colon \Omega \to \mathbb{R}$. A computation gives

$$H_{\rho,z}(v,w) = H_{\rho',f(z)}(\mathrm{d}f_z\cdot v, \mathrm{d}f_z\cdot w), \quad v, w \in \mathrm{T}_z\mathbb{C}^n.$$

The Levi form transforms accordingly:

$$\mathcal{L}_{\rho,z}(v) = \mathcal{L}_{\rho',f(z)}(\mathrm{d}f_z\cdot v), \quad v \in \mathrm{T}_z\mathbb{C}^n.$$

In particular, if ρ' is plurisubharmonic on Ω' then $\rho = \rho' \circ f$ is plurisubharmonic on Ω. If f is a holomorphic immersion and ρ' is strongly plurisubharmonic on Ω', then $\rho' \circ f$ is strongly plurisubharmonic on Ω.

We now give a coordinate-free expression for the Levi form. It is natural to look at the more general case of the *canonical form* associated to a Hermitian metric. Let h be a contravariant Hermitian tensor field of type $(1,1)$ on a complex manifold X. In local holomorphic coordinates, h is of the form

$$h = \sum_{j,k=1}^n h_{jk}(z)\,\mathrm{d}z_j \otimes \mathrm{d}\bar{z}_k, \qquad \overline{h}_{jk} = h_{kj}. \tag{1.36}$$

It defines a Hermitian bilinear form $\langle h_z, v \otimes w\rangle = \sum_{j,k=1}^n h_{jk}(z) v_j \overline{w}_k$ on each tangent space T_zX. Write $h = S - \mathrm{i}\omega$ with S, ω real. Then

$$2\langle S, v \otimes w\rangle = \sum_{j,k=1}^n h_{jk} v_j \overline{w}_k + \sum_{j,k=1}^n h_{kj} w_k \overline{v}_j = \sum_{j,k=1}^n h_{jk}(v_j\overline{w}_k + \overline{v}_k w_j).$$

Thus S is a symmetric tensor field given by

$$S = \Re h = \frac{1}{2}\sum_{j,k=1}^n h_{jk}(\mathrm{d}z_j \otimes \mathrm{d}\bar{z}_k + \mathrm{d}\bar{z}_k \otimes \mathrm{d}z_j).$$

Similarly, $2\mathrm{i}\,\omega = \bar{h} - h$, and hence

$$-2\mathrm{i}\langle\omega, v \otimes w\rangle = \sum_{j,k=1}^{n} (h_{jk} v_j \overline{w}_k - h_{kj} w_k \overline{v}_j) = \sum_{j,k=1}^{n} h_{jk}(v_j \overline{w}_k - w_j \overline{v}_k).$$

Thus ω is antisymmetric, whence a 2-form. Since

$$\langle \mathrm{d}z_j \wedge \mathrm{d}\bar{z}_k, v \wedge w\rangle = \big\langle \mathrm{d}z_j \wedge \mathrm{d}\bar{z}_k, (V + \overline{V}) \wedge (W + \overline{W})\big\rangle = v_j \overline{w}_k - w_j \overline{v}_k,$$

we have

$$\omega = -\Im h = \frac{\mathrm{i}}{2} \sum_{j,k=1}^{n} h_{jk} \mathrm{d}z_j \wedge \mathrm{d}\bar{z}_k. \tag{1.37}$$

We can express h and S in terms of ω as follows. Note that

$$\langle h, v \otimes w\rangle = \langle S, v \otimes w\rangle - \mathrm{i}\langle\omega, v \wedge w\rangle.$$

Replacing the vector $w = W + \overline{W} \in \mathrm{T}X$ by $Jw = \mathrm{i}W - \mathrm{i}\overline{W}$ gives

$$\langle h, v \otimes Jw\rangle = \langle S, v \otimes Jw\rangle - \mathrm{i}\langle\omega, v \wedge Jw\rangle.$$

The Hermitian property of h, together with the previous identity, implies

$$\langle h, v \otimes Jw\rangle = -\mathrm{i}\langle h, v \otimes w\rangle = \langle\omega, v \wedge w\rangle - \mathrm{i}\langle S, v \otimes w\rangle.$$

Comparing the imaginary parts in the last two equations gives $\langle S, v \otimes w\rangle = \langle\omega, v \wedge Jw\rangle$. If $v = w$ then $\langle h, v \otimes v\rangle$ is real, and hence

$$\langle h, v \otimes v\rangle = \langle S, v \otimes v\rangle = \langle\omega, v \wedge Jv\rangle.$$

When $h = H_\rho$ is the complex Hessian (1.33) of a $\mathcal{C}^2$ function ρ in some local holomorphic coordinates $z = (z_1, \ldots, z_n)$, we have

$$\omega = \frac{\mathrm{i}}{2} \sum_{j,k=1}^{n} \frac{\partial^2 \rho}{\partial z_j \partial \bar{z}_k} \mathrm{d}z_j \wedge \mathrm{d}\bar{z}_k = \frac{\mathrm{i}}{2} \partial\overline{\partial}\rho = \frac{1}{4} \mathrm{dd}^c \rho. \tag{1.38}$$

Comparing (1.34) and (1.38) gives the following coordinate free definition of the Levi form on any complex manifold (X, J):

$$\mathcal{L}_{\rho,x}(v) = \frac{1}{4} \big\langle \mathrm{dd}^c \rho|_x, v \wedge Jv \big\rangle, \quad v \in \mathrm{T}_x X. \tag{1.39}$$

This is the restriction to the diagonal $v = w$ of the symmetric bilinear form

$$\mathcal{L}_{\rho,x}(v, w) = \frac{1}{4} \big\langle \mathrm{dd}^c \rho|_x, v \wedge Jw \big\rangle, \quad v, w \in \mathrm{T}_x X,$$

which is the coordinate free expression for the complex Hessian $H_{\rho,x}(v, w)$. Clearly $\mathcal{L}_{\rho,x}(Jv) = \mathcal{L}_{\rho,x}(v)$ and $\mathcal{L}_{\rho,x}(Jv, Jw) = \mathcal{L}_{\rho,x}(v, w)$.

It follows that a $\mathcal{C}^2$ function $\rho\colon X \to \mathbb{R}$ on a complex manifold is plurisubharmonic if the Levi form $\mathcal{L}_{\rho,x} \ge 0$ is nonnegative for every $x \in X$, and is strongly plurisubharmonic if $\mathcal{L}_{\rho,x}$ is positive definite for every $x \in X$. These conditions are commonly written as $\mathrm{dd}^c \rho \ge 0$, resp. $\mathrm{dd}^c \rho > 0$.

Assume now that h is a Hermitian metric on TX. Then any matrix (h_{jk}), representing h in a local holomorphic coordinate system on X, is positive definite. The associated form ω (1.37) is called the *canonical form* of h. Since $\langle \omega, v \wedge Jv \rangle > 0$ for every $0 \neq v \in T_x X$, ω is said to be a *positive* $(1,1)$*-form*. We say that h is a *Kähler metric*, and that $\omega = -\Im h$ is a *Kähler form*, if $d\omega = 0$; in such case (X, h) is a *Kähler manifold*. Note that $\omega = dd^c \rho$ is a Kähler form if and only if the function ρ is strongly plurisubharmonic. If $\dim X = n$ and ω is a Kähler form on X then its n-th power ω^n is a *volume form* on X, that is, a nonvanishing (n,n)-form.

Example 1.8.3 The standard Kähler form on $\mathbb{C}^n$, associated to the Euclidean metric $h = \sum_{j=1}^n dz_j \otimes d\bar{z}_j$, is $\omega = \frac{1}{4} dd^c |z|^2 = \frac{i}{2} \sum_{j=1}^n dz_j \wedge d\bar{z}_j$. The *Fubini-Study metric* on $\mathbb{P}^n$ is given in any affine coordinates $z \in \mathbb{C}^n \subset \mathbb{P}^n$ by the Kähler form $\omega = \frac{1}{4} dd^c \log(1 + |z|^2)$. □

For a general plurisubharmonic function ρ we consider $dd^c \rho$ as a positive current of bidimension $(1,1)$. This generalizes the fact that the Laplacian of a subharmonic function of one complex variable is a positive measure.

Definition 1.8.4 A set E in a complex manifold X is *pluripolar* if for each point $p \in E$ there exist a neighborhood $U \subset X$ of p and a plurisubharmonic function $u: U \to [-\infty, +\infty)$ such that $E \cap U \subset \{x \in U : u(x) = -\infty\}$. The set E is *complete pluripolar* if it is locally near each point $p \in E$ the exact $-\infty$ level set of a plurisubharmonic function.

By a theorem of Josefson [315], if $E \subset \mathbb{C}^n$ is pluripolar then there is a plurisubharmonic function u on $\mathbb{C}^n$ such that $E \subset \{u = -\infty\}$. That is, a set $E \subset \mathbb{C}^n$ is locally pluripolar if and only if it is globally pluripolar.

Example 1.8.5 Every closed complex analytic set A in a complex manifold X is complete pluripolar. If A is defined locally in an open set $U \subset X$ by holomorphic functions $f_1, \ldots, f_k \in \mathcal{O}(U)$ then $u = \log(|f_1|^2 + \cdots + |f_k|^2)$ is plurisubharmonic in U and $\{u = -\infty\} = A \cap U$. □

Pluripolar sets are removable singularities for plurisubharmonic functions:

Theorem 1.8.6 *If E is a closed pluripolar set in a complex manifold X then every plurisubharmonic function on $X \setminus E$ which is locally bounded at every point of E extends to a plurisubharmonic function on X.*

Functions satisfying the equation $dd^c u = 0$ are also of great interest.

Lemma 1.8.7 *A real $\mathcal{C}^2$ function u satisfies $dd^c u = 0$ if and only if u is locally the real part of a holomorphic function. Such u is called* pluriharmonic.

Proof If $f = u + \mathrm{i}v$ is holomorphic then $\mathrm{dd}^c f = 2\mathrm{i}\,\partial\overline{\partial} f = 0$. Since dd^c is a real operator, we get $\mathrm{dd}^c u = \mathrm{dd}^c v = 0$. Conversely, if $\mathrm{dd}^c u = 0$ then $\mathrm{d}^c u$ is a closed 1-form, and hence on any simply connected domain we have $\mathrm{d}^c u = \mathrm{d}v$ for some $\mathcal{C}^1$ function v. Hence $\langle \mathrm{d}u, -J\xi\rangle = \langle \mathrm{d}^c u, \xi\rangle = \langle \mathrm{d}v, \xi\rangle$ for every tangent vector ξ. Replacing ξ by $-J\xi$ in this identity gives $\mathrm{d}u = -\mathrm{d}^c v$. It follows that $2\overline{\partial}(u + \mathrm{i}v) = (\mathrm{d} - \mathrm{i}\mathrm{d}^c)(u + \mathrm{i}v) = 0$, so the function $u + \mathrm{i}v$ is holomorphic. □

Note that a real $\mathcal{C}^2$ function u is pluriharmonic if and only if both u and $-u$ are plurisubharmonic. This definition also applies of u is merely continuous, but the conclusion of the lemma remains the same.

The following lemma gives an expression for the Levi form of a real hypersurface in terms of commutators of complex tangent vector fields.

Lemma 1.8.8 *Let $M = \{\rho = 0\}$ be a real hypersurface in a complex manifold X with a $\mathcal{C}^2$ defining function ρ satisfying $\mathrm{d}\rho \neq 0$ on M. For every pair of vector field V, W on X that are tangent to $T^{\mathbb{C}}M = TM \cap JTM$ we have*

$$4\mathcal{L}_\rho(V, W) = \big\langle \mathrm{dd}^c\rho, V \wedge JW\big\rangle = -\big\langle \mathrm{d}^c\rho, [V, JW]\big\rangle. \tag{1.40}$$

Proof Since V and JW are tangent to M, so is their commutator $[V, JW]$. For every 1-form η and vector fields V, V' we have [4, p. 432]

$$\big\langle \mathrm{d}\eta, V \wedge V'\big\rangle = V\big\langle \eta, V'\big\rangle - V'\langle \eta, V\rangle - \big\langle \eta, \big[V, V'\big]\big\rangle.$$

Taking $\eta = \mathrm{d}^c\rho$, $V' = JW$, and observing that $\ker \mathrm{d}^c\rho|_{TM} = T^{\mathbb{C}}M$ (hence $\langle \eta, V\rangle = 0 = \langle \eta, JW\rangle$) gives (1.40). □

1.9 Vector Fields, Flows and Foliations

We recall the basic notions on vector fields that are used in this book. Let X be a smooth manifold. Given a differentiable path $\phi: I \to X$ from an interval I in the t-axis $\mathbb{R}$, we denote its velocity vector at time t by $\dot\phi(t) = (\mathrm{d}\phi)_t \cdot \frac{\partial}{\partial t} \in T_{\phi(t)}X$. An *integral curve* of a vector field V on X is a path $\phi: I \to X$ such that $\dot\phi(t) = V_{\phi(t)}$ for all $t \in I$. We also write $V_x = V(x)$ when there is no danger of confusing the latter notation with the value of the vector field V on a function x. The initial value problem

$$\dot x = V_x, \qquad x(0) = x^0 \tag{1.41}$$

asks for an integral curve which passes through the point x^0 at time $t = 0$. In local coordinates $x = (x_1, \dots, x_n)$ on X, with $V_x = \sum_{j=1}^n a_j(x)\frac{\partial}{\partial x_j}$ and $x^0 = (x_1^0, \dots, x_n^0)$, the flow equation (1.41) is equivalent to the system of autonomous ordinary differential equations

$$\dot x_j = a_j(x_1, \dots, x_n), \qquad x_j(0) = x_j^0, \quad j = 1, \dots, n.$$

If V is Lipschitz continuous then for every point $p \in X$ there exist a neighborhood $U \subset X$ of p and a number $t_0 > 0$ such that (1.41) has a unique solution $x(t, x^0) = \phi_t(x^0)$ for every $x^0 \in U$ and every $|t| < t_0$. This solution and its t-derivative are continuous in (t, x^0). The map $t \to \phi_t(x)$ is called the (local) *flow* of V. If V is of class $\mathcal{C}^k$ for some $k \geq 1$ then $\phi_t(x)$ and $\dot{\phi}_t(x)$ are also of class $\mathcal{C}^k$ in (t, x). (See e.g. [4, Sect. 4.1].) For every fixed $t \in \mathbb{R}$ the map ϕ_t is a diffeomorphism of its domain $\Omega_t \subset X$ onto $\phi_t(\Omega_t) \subset X$, called the *time-t map*. These maps satisfy the group law

$$\phi_t \circ \phi_s = \phi_{t+s}, \quad t, s \in \mathbb{R}$$

on the set in X where both sides are defined.

By uniqueness of trajectories the flow $t \mapsto \phi_t(x)$ extends to the maximal interval $I_x = (\alpha(x), \omega(x)) \subset \mathbb{R}$ around the origin. The functions $-\alpha, \omega: X \to (0, +\infty]$ are lower semicontinuous. If $\omega(x) < +\infty$ for some $x \in X$, then the point $\phi_t(x)$ leaves every compact set in X as $t \nearrow \omega(x)$. Similarly, if $\alpha(x) > -\infty$ then the point $\phi_t(x)$ leaves every compact set in X as $t \searrow \alpha(x)$. The open set

$$\Omega = \big\{(t, x) \in \mathbb{R} \times X : \alpha(x) < t < \omega(x)\big\} \subset \mathbb{R} \times X \tag{1.42}$$

is called the *fundamental domain* of the vector field V.

Definition 1.9.1 A vector field V is $\mathbb{R}$-*complete* (or *complete in real time*) if $\Omega = \mathbb{R} \times X$; that is, if the flow $\phi_t(x)$ exists for all $t \in \mathbb{R}$ and $x \in X$.

Every vector field on a compact manifold without boundary is complete. The flow $\{\phi_t\}_{t\in\mathbb{R}}$ of an $\mathbb{R}$-complete vector field is a real 1-parameter subgroup of the diffeomorphism group Diff(X) of X, representing an action of the group $(\mathbb{R}, +)$ on X by diffeomorphisms. Conversely, every action $\phi: \mathbb{R} \times X \to X$ of $(\mathbb{R}, +)$ by diffeomorphisms of X is the flow of the vector field $V_x = \frac{\partial}{\partial t}\phi(t, x)|_{t=0}$, called the *infinitesimal generator* of ϕ.

A diffeomorphism $f: X \to Y$ pushes a vector field V on X to the vector field $W = f_* V$ on Y (explicitly, $W_{f(x)} = \mathrm{d}f_x V_x$); a vector field W on Y pulls back to the vector field $V = f^* W = (f^{-1})_* W$ on X. The chain rule shows that if $\phi_t(x)$ is the flow of V then $\psi_t(y) = f \circ \phi_t(f^{-1}(y))$ is the flow of W.

More generally, if $f: X \to Y$ is a smooth map, V is a vector field on X, and W is a vector fields on Y such that $\mathrm{d}f_x V_x = W_{f(x)}$ holds for all $x \in X$ (in such case we write $V = f^* W$), then f conjugates the flow ϕ_t of V to the flow ψ_t of W: $f(\phi_t(x)) = \psi_t(f(x))$. In particular, if V is tangent to a submanifold $M \subset X$, then for each $x \in M$ the flow $\phi_t(x)$ remains in M.

Lemma 1.9.2 *Let V be a $\mathcal{C}^1$ vector field on an n-dimensional manifold X. If $V_p \neq 0$ for some $p \in X$ then there exist local coordinates $x = (x_1, \ldots, x_n)$ in a neighborhood of p such that $V = \frac{\partial}{\partial x_1}$.*

Proof A suitable choice of local coordinates ensures that $p = 0 \in \mathbb{R}^n$ and that $V_0 = \frac{\partial}{\partial x_1}|_0$. Let ϕ_t denote the flow of V. Consider the map

$$g(x_1, x_2, \ldots, x_n) = \phi_{x_1}(0, x_2, \ldots, x_n).$$

It is immediate that $g_* \frac{\partial}{\partial x_1} = V$ and that $\mathrm{d}g_0$ is the identity map. Hence g is a diffeomorphism near the origin. □

Let $\mathbb{R} \times X$ be the extended phase space with the projection $\pi: \mathbb{R} \times X \to X$. The tangent bundle $\mathrm{T}X$ pulls back to a bundle on $\mathbb{R} \times X$, still denoted $\mathrm{T}X$. A *time-dependent* (also called *non-autonomous*) vector field on a domain $\Omega \subset \mathbb{R} \times X$ is a section V of $\mathrm{T}X|_\Omega$. For a fixed t, V_t is a vector field on the domain

$$\Omega_t = \{x \in X : (t,x) \in \Omega\}. \tag{1.43}$$

In local coordinates $x = (x_1, \ldots, x_n)$ on an open set $U \subset X$ we have $V_t(x) = \sum_{j=1}^n v_j(t,x) \frac{\partial}{\partial x_j}$, where the coefficient functions $v_j(t,x)$ are defined for $(t,x) \in \Omega \cap \pi^{-1}(U)$. Fix a point $x \in \Omega_s$ for some $s \in \mathbb{R}$. If V is continuous in (t,x) and Lipschitz continuous in x, then the equation

$$\frac{\partial}{\partial t} \phi_{t,s}(x) = V_t\big(\phi_{t,s}(x)\big), \qquad \phi_{s,s}(x) = x \tag{1.44}$$

has a unique local solution. The flow $\phi_{t,s}(x)$ is defined as long as the trajectory remains in the domain of the vector field. We have the semigroup property

$$\phi_{t,u} \circ \phi_{u,s} = \phi_{t,s}, \qquad \phi_{s,t} = \phi_{t,s}^{-1}.$$

The *Grönwall inequality* says that if nonnegative continuous functions $f, g: [a,b) \to \mathbb{R}_+$ satisfy the inequality $f(t) \le A + \int_a^t f(\tau) g(\tau) \mathrm{d}\tau$ for some $A \ge 0$, then $f(t) \le A \exp(\int_a^t g(\tau) \mathrm{d}\tau)$ for all $t \in [a,b)$ [4, Lemma 4.1.8]. This inequality gives the following estimate on the distance between trajectories.

Lemma 1.9.3 *Let V be a time-dependent continuous vector field on a domain $\Omega \subset \mathbb{R}^{1+n}$ satisfying a uniform Lipschitz estimate*

$$\big|V_t(x) - V_t(y)\big| \le B|x-y|. \tag{1.45}$$

Then for any $s \in \mathbb{R}$ and any pair of points $x, y \in \Omega_s$ (1.43) we have

$$\big|\phi_{t,s}(x) - \phi_{t,s}(y)\big| \le \mathrm{e}^{B|t-s|}|x-y| \tag{1.46}$$

for all t such that the trajectories exist and remain in the domain Ω_t (1.43).

Proof Let $f(t) = |\phi_{t,s}(x) - \phi_{t,s}(y)|$. Assume first that $t \ge s$. By integrating Eq. (1.44) we get

$$\begin{aligned} f(t) &= \left| x + \int_s^t V_\tau\big(\phi_{\tau,s}(x)\big)\,\mathrm{d}\tau - y - \int_s^t V_\tau\big(\phi_{\tau,s}(y)\big)\,\mathrm{d}\tau \right| \\ &\le |x-y| + \int_s^t \big|V_\tau\big(\phi_{\tau,s}(x)\big) - V_\tau\big(\phi_{\tau,s}(y)\big)\big|\,\mathrm{d}\tau \\ &\le |x-y| + B\int_s^t \big|\phi_{\tau,s}(x) - \phi_{\tau,s}(y)\big|\,\mathrm{d}\tau \\ &= |x-y| + B\int_s^t f(\tau)\,\mathrm{d}\tau. \end{aligned}$$

Since $f(s) = |x - y|$, Grönwall's inequality implies $f(t) \leq e^{B|t-s|}|x - y|$ which gives (1.46). The estimate for $t < s$ is obtained by replacing V by $-V$. □

Assume now that $\Omega_0 \neq \varnothing$. Fix a compact subset $K \subset \Omega_0$ and let $t_0 > 0$ be such that the flow $\phi_t(x) = \phi_{t,0}(x)$ exists and remains in Ω_t when $x \in K$ and $t \in [0, t_0]$. Set $K_t = \phi_t(K) \subset \Omega_t$. For any $\epsilon > 0$ we let

$$K(\epsilon) = \Big\{x \in \mathbb{R}^n : \mathrm{dist}(x, K) = \inf_{y\in K} |x - y| < \epsilon\Big\},$$

$$S(\epsilon) = \big\{(t, x) \in \mathbb{R} \times \mathbb{R}^n : 0 \leq t \leq t_0,\ \mathrm{dist}(x, K_t) < \epsilon\big\}.$$

Set $\eta_0 = (1 + t_0)e^{Bt_0} > 1$, where B is the Lipschitz constant (1.45). Choose $\epsilon_0 > 0$ sufficiently small such that $S(\epsilon_0\eta_0) \Subset \Omega$.

Lemma 1.9.4 (Notation as above) *Assume that for some $\epsilon \in (0, \epsilon_0)$ we have a continuous map $V^\epsilon : \Omega \to \mathbb{R}^n$ (a time-dependent vector field) satisfying*

$$\big\| V - V^\epsilon \big\|_{L^\infty(S(\epsilon\eta_0))} \leq \epsilon.$$

Then the flow $\phi_t^\epsilon(x)$ of V_t^ϵ with $\phi_0^\epsilon(x) = x$ exists for all $x \in K(\epsilon)$ and for all $t \in [0, t_0]$, and it satisfies the estimate

$$\big\|\phi_t - \phi_t^\epsilon\big\|_{L^\infty(K(\epsilon))} \leq t_0 e^{Bt} \big\| V - V^\epsilon \big\|_{L^\infty(S(\epsilon\eta_0))}, \quad 0 \leq t \leq t_0. \tag{1.47}$$

Proof Let $A(\epsilon) = \|V - V^\epsilon\|_{L^\infty(S(\epsilon\eta_0))}$. Fix a point $x \in K(\epsilon)$ and set $f(t) = |\phi_t(x) - \phi_t^\epsilon(x)|$ for $t \geq 0$. We have $f(0) = 0$ and

$$\begin{aligned} f(t) &= \left| \int_0^t \Big(V_s(\phi_s(x)) - V_s^\epsilon(\phi_s^\epsilon(x))\Big) \mathrm{d}s \right| \leq \int_0^t \big|V_s(\phi_s(x)) - V_s^\epsilon(\phi_s^\epsilon(x))\big| \mathrm{d}s \\ &\leq \int_0^t \big|V_s(\phi_s(x)) - V_s(\phi_s^\epsilon(x))\big| \,\mathrm{d}s + \int_0^t \big|V_s(\phi_s^\epsilon(x)) - V_s^\epsilon(\phi_s^\epsilon(x))\big| \,\mathrm{d}s \\ &\leq B \int_0^t f(s)\,\mathrm{d}s + t_0 A(\epsilon). \end{aligned}$$

The last inequality holds for all $t \in [0, t_0]$ such that $\phi_t^\epsilon(x) \in K_t(\epsilon\eta_0)$. For such values of t, Grönwall's inequality implies $f(t) \leq A(\epsilon)t_0 e^{Bt}$ which gives (1.47). For $x \in K(\epsilon)$, Lemma 1.9.3 also gives $\phi_t(x) \in K_t(\epsilon e^{Bt})$. Hence,

$$\begin{aligned} \mathrm{dist}\big(\phi_t^\epsilon(x), K_t\big) &\leq \big|\phi_t^\epsilon(x) - \phi_t(x)\big| + \mathrm{dist}\big(\phi_t(x), K_t\big) \\ &\leq A(\epsilon)t_0 e^{Bt} + \epsilon e^{Bt} \leq \epsilon(t_0 + 1)e^{Bt_0} = \epsilon\eta_0. \end{aligned}$$

We used the assumption $A(\epsilon) \leq \epsilon$ and the definition of η_0. This shows that $\phi_t^\epsilon(x) \in K_t(\epsilon\eta_0)$ for all $t \in [0, t_0]$, so the estimate (1.47) holds. □

The k-th power of a smooth vector field V is a differential operator of order k defined inductively by $V^k(f) = V(V^{k-1}(f))$, where f is a smooth function. Let $\phi_t(x)$ be the flow of V. By differentiating the function $t \mapsto f(\phi_t(x))$ for a fixed

$x \in X$ we get $\frac{\mathrm{d}^k}{\mathrm{d}t^k} f(\phi_t(x)) = V^k(f)(\phi_t(x))$. This gives the formal Taylor expansion of $t \mapsto f(\phi_t(x))$, also called the *Lie series*:

$$f\big(\phi_t(x)\big) = \sum_{k \geq 0} \frac{1}{k!} t^k V^k(f)(x) = f(x) + tV(f)(x) + \frac{1}{2} t^2 V^2(f)(x) + \cdots. \tag{1.48}$$

Let $x = (x_1, \ldots, x_n)$ be a coordinate map and $V_x = \sum_{j=1}^n a_j(x) \frac{\partial}{\partial x_j}$. By applying V to the j-th component x_j of x we get $V(x_j) = a_j$ (the j-th coefficient function of V). Thus $V(x_1, \ldots, x_n) := (V(x_1), \ldots, V(x_n)) = (a_1, \ldots, a_n)$. Using (1.48) with the identity map $f(x) = x$ we obtain the Lie series of the flow:

$$\phi_t(x) = x + tV(x) + \frac{1}{2} t^2 V^2(x) + \cdots + \frac{1}{k!} t^k V^k(x) + \cdots. \tag{1.49}$$

Note that the j-th component function of $V^k(x)$ equals $V^{k-1}(a_j)$.

Definition 1.9.5 Let ϕ_t be the flow of a vector field V. The *Lie derivative* $L_V W$ of a tensor field W is defined by

$$L_V W = \frac{\mathrm{d}}{\mathrm{d}t}\bigg|_{t=0} \phi_t^* W. \tag{1.50}$$

The Lie derivative of a function f equals $L_V f = V(f) = \mathrm{d}f \cdot V$. Suppose now that W is a vector field. Writing $s = t + u$ we have $\phi_s^* W = \phi_t^*(\phi_u^* W)$ by the group property of the flow. Differentiating on s at $s = t$ gives

$$\frac{\mathrm{d}}{\mathrm{d}t} \phi_t^* W = \frac{\mathrm{d}}{\mathrm{d}u}\bigg|_{u=0} \phi_t^*(\phi_u^* W) = \phi_t^* \left(\frac{\mathrm{d}}{\mathrm{d}u}\bigg|_{u=0} \phi_u^* W \right) = \phi_t^*(L_V W). \tag{1.51}$$

Hence $L_V W = 0$ implies $\phi_t^* W = W$ for all t. If $f \colon X \to Y$ is a smooth map and $\widetilde{V}$, $\widetilde{W}$ are vector fields on Y such that $V = f^* \widetilde{V}$ and $W = f^* \widetilde{W}$, then

$$f^*(L_{\widetilde{V}} \widetilde{W}) = L_V W. \tag{1.52}$$

To see this, let $\widetilde{\phi}_t$ denote the flow of $\widetilde{V}$. Then $f \circ \phi_t = \widetilde{\phi}_t \circ f$ and hence

$$\phi_t^* W = \phi_t^* f^* \widetilde{W} = (f \circ \phi_t)^* \widetilde{W} = (\widetilde{\phi}_t \circ f)^* \widetilde{W} = f^* \big(\widetilde{\phi}_t^* \widetilde{W}\big).$$

Differentiation at $t = 0$ gives (1.52). If f is a diffeomorphism then (1.52) implies $f_*(L_V W) = L_{f_* V} f_* W$, so the Lie derivative can be calculated in any coordinate system. It is now easy to show that

$$L_V W = [V, W] = VW - WV,$$

where the commutator $[V, W]$ is defined by (1.23) on p. 22. Indeed, at a point $p \in X$ where $V_p \neq 0$ Lemma 1.9.2 gives local coordinates $x = (x_1, \ldots, x_n)$ such that $V = \frac{\partial}{\partial x_1}$. Its flow is $\phi_t(x_1, \ldots, x_n) = (x_1 + t, x_2, \ldots, x_n)$. If $W = \sum_{j=1}^n b_j(x) \frac{\partial}{\partial x_j}$, a trivial calculation gives

$$L_V W = \sum_{j=1}^n \frac{\partial b_j}{\partial x_1}(x) \frac{\partial}{\partial x_j} = [V, W].$$

If V equals zero in some neighborhood of p then $L_V W = 0 = [V, W]$ near p. The equality at all points then follows by the continuity of both fields.

Lemma 1.9.6 *If V, W are vector fields with flows ϕ_t, ψ_t, then $[V, W] = 0$ if and only if $\phi_t \circ \psi_s = \psi_s \circ \phi_t$ holds on the domain of the compositions.*

Proof Suppose that $\phi_t \circ \psi_s = \psi_s \circ \phi_t$. Differentiating on s at $s = 0$ gives $(\phi_t)_* W = W \circ \phi_t$ which is equivalent to $\phi_t^* W = W$. Differentiation at $t = 0$ gives $L_V W = 0$. Conversely, if $L_V W = 0$ then $(\phi_t)_* W = W \circ \phi_t$ by (1.51). Consider the path $\mathbb{R} \ni s \mapsto \gamma(s) = \phi_t(\psi_s(x))$. Then

$$\frac{d\gamma}{ds}(s) = (\phi_t)_* \frac{d}{ds}\psi_s(x) = (\phi_t)_* W\big(\psi_s(x)\big) = W\big(\phi_t \circ \psi_s(x)\big) = W\big(\gamma(s)\big),$$

so γ is an integral curve of W. Also, $\sigma(s) = \psi_s \circ \phi_t(x)$ is obviously an integral curve of W. Since $\gamma(0) = \phi_t(x) = \sigma(0)$, the uniqueness of integral curves implies $\gamma(s) = \sigma(s)$. This means that $\phi_t \circ \psi_s = \psi_s \circ \phi_t$. □

Lemma 1.9.6 can also be seen as follows. Denote by $a(x), b(x) \in \mathbb{R}^n$ the coefficient vectors of V, W, respectively. Using the Lie series (1.49) we get

$$\begin{aligned}
\psi_s\big(\phi_t(x)\big) &= x + ta(x) + sb(x) + stV(b)(x) + O\big(s^2, t^2\big), \\
\phi_t\big(\psi_s(x)\big) &= x + ta(x) + sb(x) + stW(a)(x) + O\big(s^2, t^2\big), \\
\frac{\partial^2}{\partial s\, \partial t}\bigg|_{s=t=0} &\big[\psi_s\big(\phi_t(x)\big) - \phi_t\big(\psi_s(x)\big)\big] = V(b)(x) - W(a)(x) = [V, W](x).
\end{aligned}$$

If the flows ϕ_t and ψ_s commute then clearly $[V, W] = 0$. The converse is shown in the same way as above by using uniqueness of integral curves.

A similar calculation, using the Lie series (1.49) of the flow, implies

Lemma 1.9.7 *If V and W are vector fields with flows ψ_t, ψ_t, resp., then*

$$[V, W]_x = \frac{d}{dt}\bigg|_{t=0} \psi_{-\sqrt{t}} \circ \phi_{-\sqrt{t}} \circ \psi_{\sqrt{t}} \circ \phi_{\sqrt{t}}(x). \tag{1.53}$$

Lemma 1.9.6 gives a simple proof of the classical Frobenius theorem. Let X be a smooth manifold of dimension n, and let E be a smooth vector subbundle of TX of rank m and corank $d = n - m$. A vector field V on an open set $U \subset X$ is *tangent to* E if $V_x \in E_x$ for all $x \in U$. The subbundle E of TX is said to be *involutive* if for any pair of local vector fields V, W that are tangent to E, their commutator $[V, W]$ is also tangent to E.

A *nonsingular smooth foliation* $\mathcal{F}$ of dimension $m \in \{1, \dots, n\}$ on X is a subdivision of X into a disjoint union $X = \bigcup_{\alpha \in A} F_\alpha$ of m dimensional, locally closed, connected smooth submanifolds $F_\alpha \subset X$, called the *leaves* of $\mathcal{F}$, such that every point $x^0 \in X$ has an open neighborhood $U \subset X$ and a diffeomorphism $f = (f', f''): U \to D^m \times D^{n-m} \subset \mathbb{R}^n$ such that for every $\alpha \in A$, the intersection $F_\alpha \cap U$ is a union of plaques $f'' = \text{const}$. It is evident that the tangent bundle to a foliation is an involutive subbundle of TX.

Theorem 1.9.8 (Frobenius) *Every involutive smooth subbundle E of $\mathrm{T}X$ is the tangent bundle of a smooth foliation of X.*

Proof We may work in local coordinates around $0 \in \mathbb{R}^n$. Assume that $E_0 = \mathbb{R}^m \times \{0\}^d$. Choose vector fields $V_j = \sum_{k=1}^n a_{j,k}(x)\frac{\partial}{\partial x_k}$ $(j = 1, \ldots, m)$ spanning E near 0. We may assume that the $m \times m$ matrix $A(x) = (a_{j,k}(x))_{j,k=1}^m$ is invertible for x near 0. Replacing the V_j's by their linear combinations obtained by applying the inverse matrix $A(x)^{-1}$ we get a new system of spanning vector fields of the simpler form

$$V_j = \frac{\partial}{\partial x_j} + \sum_{k=m+1}^{n} a_{j,k}(x)\frac{\partial}{\partial x_k}, \qquad j = 1, \ldots, m.$$

Assuming that E is involutive, each commutator $[V_j, V_l]$ is a linear combination of the vector fields $V_1, \ldots, V_m$. On the other hand, a computation shows that $[V_j, V_l]$ is a linear combination of $\frac{\partial}{\partial x_{m+1}}, \ldots, \frac{\partial}{\partial x_n}$. It follows that $[V_j, V_l] = 0$ for all $j, l = 1, \ldots, m$. Hence, the flows $\phi_{j,t}$ of V_j commute by Lemma 1.9.6. The map

$$g(x_1, \ldots, x_n) = \phi_{1,x_1} \circ \cdots \circ \phi_{m,x_m}(\underbrace{0, \ldots, 0}_{m}, x_{m+1}, \ldots, x_n)$$

is then a diffeomorphism near $0 \in \mathbb{R}^n$ satisfying $g_* \frac{\partial}{\partial x_j} = V_j$ for $j = 1, \ldots, m$. For $j = 1$ this is immediate from the definition of g and of the flow. For other values of j we use commutativity of the flows to bring ϕ_{j,x_j} to the left-most place in the definition of g, and hence the same argument applies. This shows that the foliation with leaves $x_{m+1} = c_{m+1}, \ldots, x_n = c_n$ is mapped by g to a foliation whose leaves are tangent to E. □

Given a vector field V and a differential r-form α on a manifold X, we denote by $V\rfloor\alpha$ the *contraction* (or the *interior product*) of α by V. This is an $(r-1)$-form defined at a point $x \in X$ by

$$\langle V\rfloor\alpha, V_1 \wedge \cdots V_{r-1}\rangle = \langle \alpha, V_x \wedge V_1 \wedge \cdots V_{r-1}\rangle, \qquad V_1, \ldots, V_{r-1} \in \mathrm{T}_x X. \quad (1.54)$$

(Some texts use the notation $V\rfloor\alpha = i_V\alpha$.) It is easily seen that $i_V = V\rfloor\cdot$ is an antiderivation with respect to $\wedge$. The properties relating the interior product i_V, the Lie derivative L_V, and the exterior derivative d, are summarized by [4, Theorem 6.4.8.]. In particular, we have the Cartan formulas:

$$V\rfloor\mathrm{d}f = V(f), \quad L_V\alpha = V\rfloor\mathrm{d}\alpha + \mathrm{d}(V\rfloor\alpha), \quad \mathrm{d}(L_V\alpha) = L_V\mathrm{d}\alpha. \qquad (1.55)$$

Assume now that V is a holomorphic vector field on a complex manifold X (see Definition 1.6.2 on p. 22). Let $W = \frac{1}{2}(V - \mathrm{i}JV)$ be the associated holomorphic section of $\mathrm{T}^{1,0}X$. The flow equation (1.41) (p. 32) can now be considered for a complex time variable $\zeta = t + \mathrm{i}s$. In local holomorphic coordinates $z = (z_1, \ldots, z_n)$ with $z_j = x_j + \mathrm{i}y_j$ we have $V = \sum_{j=1}^n a_j\frac{\partial}{\partial x_j} + b_j\frac{\partial}{\partial y_j}$ and $W = \sum_{j=1}^n \alpha_j\frac{\partial}{\partial z_j}$ with

$\alpha_j = a_j + \mathrm{i}b_j$. We are looking for holomorphic functions $z_j = z_j(\zeta)$ of a complex variable $\zeta = t + \mathrm{i}s \in \mathbb{C}$ satisfying

$$\frac{\mathrm{d}z_j}{\mathrm{d}\zeta}(\zeta) = \alpha_j\big(z_1(\zeta), \ldots, z_n(\zeta)\big), \quad z_j(0) = z_j^0, \quad j = 1, \ldots, n. \tag{1.56}$$

This system has a unique local solution $\zeta \mapsto \phi_\zeta(z^0)$ that is holomorphic in all variables (ζ, z^0). Write $\zeta = t + \mathrm{i}s$ with $t, s \in \mathbb{R}$. By the Cauchy-Riemann equations, (1.56) is equivalent to the system of $2n$ real equations

$$\frac{\partial x}{\partial t} = V(x), \quad \frac{\partial x}{\partial s} = JV(x), \quad x(0) = x^0.$$

In fact, $[V, JV] = [W + \overline{W}, \mathrm{i}W - \mathrm{i}\overline{W}] = -2\mathrm{i}[W, \overline{W}] = 0$ by (1.25) (p. 22); hence the flows ϕ_t, ψ_t commute, and the flow in complex time is given by

$$\phi_{t+\mathrm{i}s}(x) = \phi_t \circ \psi_s(x) = \psi_s \circ \phi_t(x). \tag{1.57}$$

Definition 1.9.9 A holomorphic vector field V on a complex manifold X is $\mathbb{C}$-*complete* (or *completely integrable*) if its flow $\phi_{t+\mathrm{i}s}(x)$ exists for all $x \in X$ and $t + \mathrm{i}s \in \mathbb{C}$.

A holomorphic vector field V is $\mathbb{C}$-complete if and only if V and JV are both $\mathbb{R}$-complete. The flow of an $\mathbb{R}$-complete (resp. $\mathbb{C}$-complete) holomorphic vector field is an action of the additive group $(\mathbb{R}, +)$ (resp. of $(\mathbb{C}, +)$) on X by holomorphic automorphisms, and every action is of this form. Two such actions $\phi, \psi : \mathbb{R} \times X \to X$ are *holomorphically equivalent* if there exists a holomorphic automorphism $f \in \mathrm{Aut}(X)$ such that $f(\phi(t, x)) = \psi(t, f(x))$ for all $t \in \mathbb{R}$ (resp. $t \in \mathbb{C}$). Letting V, W denote the infinitesimal generator of ϕ, ψ, respectively, this holds if and only if $\mathrm{d}f_x \cdot V_x = W_{f(x)}$ for every $x \in X$.

Remark 1.9.10 If X is a connected Stein manifold without nonconstant bounded plurisubharmonic functions (a complex manifold with the latter property is called *Liouville*) then every $\mathbb{R}$-complete holomorphic vector field on X is also $\mathbb{C}$-complete [187, Corollary 2.2]. This holds in particular when $X = \mathbb{C}^n$, or more generally when X is a Stein Oka manifold (see Definition 5.4.1 on p. 215). Furthermore, if X is as above, then a holomorphic vector field on X that is complete in positive real time is also $\mathbb{C}$-complete [8]. □

1.10 What is the H-Principle?

In a broad sense, we say that the *h-principle* (an abbreviation for the *homotopy principle*) holds in a certain analytic problem if solutions exists in the absence of topological obstructions. In complex analysis, instances of the h-principle are called the *Oka principle*. The term h-principle started being widely used after Gromov's work [268].

The purpose of this section is to recall a simple case of Gromov's h-principle that is used at several places in the book. For a more complete introduction to this subject, see the monographs [152, 265, 521].

Let $f = (f_1, \ldots, f_m): U \to \mathbb{R}^m$ be a smooth map in an open neighborhood $U \subset \mathbb{R}^n$ of a point $x \in \mathbb{R}^n$. The k-jet $\mathrm{j}^k_x f$ of f at x is determined by the Taylor polynomial of f of order k at x, that is, by the partial derivatives

$$c_{j,\alpha} = \frac{\partial^{|\alpha|} f_j}{\partial x_1^{\alpha_1} \cdots \partial x_n^{\alpha_n}}(x), \quad 0 \le |\alpha| \le k, \ j = 1, \ldots, m.$$

The manifold of all k-jets $\mathrm{J}^k(\mathbb{R}^n, \mathbb{R}^m)$ is a Euclidean space with the coordinates $x = (x_1, \ldots, x_n)$ and $c_{j,\alpha} \in \mathbb{R}$ for all $|\alpha| \le k$, $j = 1, \ldots, m$. In particular, the 0-jet is $\mathrm{j}^0_x f = (x, f(x))$, and $\mathrm{J}^0(\mathbb{R}^n, \mathbb{R}^m) = \mathbb{R}^n \times \mathbb{R}^m$. The 1-jet of f at x is $\mathrm{j}^k_x f = (x, f(x), \mathrm{d}f_x)$, and the space of 1-jets

$$\mathrm{J}^1(\mathbb{R}^n, \mathbb{R}^m) = \{(x, y, \lambda) : x \in \mathbb{R}^n, \ y \in \mathbb{R}^m, \ \lambda \in \mathrm{Hom}(\mathrm{T}_x\mathbb{R}^n, \mathrm{T}_y\mathbb{R}^m)\}$$

is isomorphic to $\mathbb{R}^{n+m+nm}$.

Assume now that X and Y are smooth manifolds. For a fixed pair of points $x \in X$, $y \in Y$ we consider smooth maps $f: U \to Y$ in neighborhoods of x such that $f(x) = y$. Choosing local coordinates around x and y brings us back to the Euclidean case. Clearly the maps f and g have the same k-jet at $x \in X$ if and only if their difference $f - g$ vanishes to order k at x in some (and hence in any) pair of local coordinates. We denote the space of k-jets of maps $X \to Y$ by $\mathrm{J}^k(X, Y)$; thus $\mathrm{J}^0(X, Y) = X \times Y$ and

$$\mathrm{J}^1(X, Y) = \{(x, y, \lambda) : x \in X, \ y \in Y, \ \lambda \in \mathrm{Hom}(\mathrm{T}_x X, \mathrm{T}_y Y)\}.$$

If X and Y are complex manifolds of dimension n, m, respectively, then we consider the complex (holomorphic) derivatives with respect to the local holomorphic coordinates $z = (z_1, \ldots, z_n)$ on X:

$$c_{j,\alpha} = \frac{\partial^{|\alpha|} f_j}{\partial z_1^{\alpha_1} \cdots \partial z_n^{\alpha_n}}(z), \quad |\alpha| \le k, \ j = 1, \ldots, m.$$

In this case we denote by $\mathrm{J}^k(X, Y) = \mathrm{J}^k_{\mathbb{C}}(X, Y)$ the manifold of complex k-jets and by $\mathrm{J}^k_{\mathbb{R}}(X, Y)$ the manifold of real k-jets.

The chain rule shows that $\mathrm{J}^k(X, Y)$ is a smooth manifold, resp. a complex manifold in the holomorphic case. Every smooth (resp. holomorphic) map $f: X \to Y$ determines the k-jet map $\mathrm{j}^k f: X \to \mathrm{J}^k(X, Y)$, $x \mapsto \mathrm{j}^k_x f$. We have natural source point and image point projections

$$p^k : \mathrm{J}^k(X, Y) \to X, \qquad q^k : \mathrm{J}^k(X, Y) \to Y.$$

We denote by $\mathrm{J}^k_x(X, Y) = (p^k)^{-1}(x)$ the set of all k-jets with the source point x, and by $\mathrm{J}^k_{x,y}(X, Y)$ the set of all k-jets with the source x and the target y.

For every $k \in \mathbb{N}$ we have the projection

$$\tau_k : \mathrm{J}^k(X, Y) \to \mathrm{J}^{k-1}(X, Y) \tag{1.58}$$

which forgets the partial derivatives of pure order k. The chain rule shows that (1.58) carries a natural structure of an affine bundle; this no longer holds for the composition of two or more of these projections.

Similarly, if $\pi: Z \to X$ is a smooth or a holomorphic submersion, we let $Z^{(k)}$ denote the manifold of all k-jets of smooth (resp. holomorphic) sections $f: X \to Z$, $\pi \circ f = \mathrm{Id}_X$. These have similar properties as the jet manifolds $\mathrm{J}^k(X, Y)$. In particular, the natural 'forgetful' projection $\tau_k: Z^{(k)} \to Z^{(k-1)}$ is a smooth (resp. holomorphic) affine bundle.

Definition 1.10.1 A map $g: X \to \mathrm{J}^k(X, Y)$ such that $g(x) \in \mathrm{J}^k_x(X, Y)$ (i.e., a section of the projection $p^k: \mathrm{J}^k_x(X, Y) \to X$) is *holonomic* if there exists a smooth (resp. holomorphic) map $f: X \to Y$ such that $g(x) = \mathrm{j}^k_x f$ for all $x \in X$. The analogous definition applies to sections $X \to Z^{(k)}$ of the jet bundle projection $Z^{(k)} \to X$ associated to a submersion $\pi: Z \to X$.

Every section $g: X \to \mathrm{J}^0(X, Y)$ is of the form $g(x) = (x, f(x))$ for some map $f: X \to Y$, and hence g is holonomic. However, being holonomic is a nontrivial condition on a section $X \to \mathrm{J}^k(X, Y)$ for every $k \geq 1$. A section $g: X \to \mathrm{J}^1(X, Y)$ is of the form $g(x) = (x, f(x), \lambda(x))$ with $\lambda(x): \mathrm{T}_x X \to \mathrm{T}_{f(x)} Y$ an arbitrary linear map; g is holonomic if and only if $\lambda(x) = \mathrm{d} f_x$.

The following notions due to Gromov [265] apply both in the smooth and the holomorphic category.

Definition 1.10.2 A *differential relation of order* $k \in \mathbb{N}$ for maps $X \to Y$ (resp. for section $X \to Z$ of a submersion $\pi: Z \to X$) is a subset $\Omega \subset \mathrm{J}^k(X, Y)$ (resp. $\Omega \subset Z^{(k)}$). The differential relation Ω is *open* (or *closed*) if Ω is an open (or a closed) subset of $\mathrm{J}^k(X, Y)$. A map $f: X \to Y$ is *solution* of Ω over a subset $V \subset X$ if the k-jet extension $\mathrm{j}^k f: X \to \mathrm{J}^k(X, Y)$ maps V to Ω.

A continuous map $\phi: X \to Y$ is said to be a *weak homotopy equivalence* if it induces a bijection of the path connected components $\pi_0(X) \to \pi_0(Y)$ and for each $x \in X$ an isomorphism of the homotopy groups

$$\pi_k(\phi) : \pi_k(X, x) \xrightarrow{\cong} \pi_k\big(Y, \phi(x)\big), \quad k = 1, 2, \ldots.$$

Definition 1.10.3 Let $\Omega \subset \mathrm{J}^k(X, Y)$ be a differential relation of order k.

(i) Ω satisfies the *basic h-principle* if every section $X \to \Omega$ of the projection $p^k: \mathrm{J}^k(X, Y) \to X$ is homotopic to a holonomic section by a continuous homotopy of sections $X \to \Omega$.

(ii) Ω satisfies the 1*-parametric h-principle* if, in addition to (i), any homotopy $g_t: X \to \Omega$ ($t \in [0, 1]$) between a pair of holonomic sections $g_0 = \mathrm{j}^k f_0$, $g_1 = \mathrm{j}^k f_1$ can be deformed, with fixed ends at $t = 0, 1$, to a homotopy $\mathrm{j}^k f_t$ consisting of holonomic sections of Ω.

(iii) Ω satisfies the *weak homotopy equivalence principle* if the inclusion

$$\big\{\mathrm{j}^k f: X \to \Omega \;\big|\; f: X \to Y\big\} \hookrightarrow \{g: X \to \Omega\}$$

of the space of holonomic sections of Ω into the space of all sections of Ω is a weak homotopy equivalence. (We use $\mathcal{C}^k$ topology on the space of maps $X \to Y$ and $\mathcal{C}^0$ topology on the space of maps $X \to \Omega$.)

The analogous definitions apply to differential relations $\Omega \subset Z^{(k)}$ for sections of a submersion $Z \to X$.

The validity of the h-principle for a certain differential relation means that the corresponding analytic or geometric problem has a solution provided that there are no topological obstructions (these would amount to absence of a nonholonomic section). There are many natural examples when the h-principle applies. For example, the Smale-Hirsch principle [295, 518] says that regular homotopy classes of smooth immersions $X \to Y$ of a manifold X to a higher dimensional manifold Y are in one-to-one correspondence with the homotopy classes of fibrewise injective vector bundle maps $TX \hookrightarrow TY$ of their tangent bundles. In particular, immersions $X \to \mathbb{R}^m$ for $m > n = \dim X$ are classified by the homotopy classes of vector bundle injections $TX \hookrightarrow X \times \mathbb{R}^m$. The same holds when $m = n = \dim X$ provided that X is an open manifold.

Similarly, if X is a smooth open manifold then smooth submersions $X \to Y$ satisfy the h-principle which is expressed in terms of the existence of fibrewise surjective vector bundle maps $TX \to TY$ [267, 451]. For a more complete discussion, see the monographs [152, 265] and Sect. 9.14 below.

The differential relations describing immersion and submersions, as well as many other natural relations arising from geometric problems, are particular examples in a certain class of first order open differential relations that we now describe. The condition is local and can be explained for sections of a trivial bundle $Z = \mathbb{R}^n \times \mathbb{R}^m \to \mathbb{R}^n$. Every local section is of the form $x \mapsto (x, f(x))$ for a map $f: U \to \mathbb{R}^m$ on an open set $U \subset \mathbb{R}^n$. A 1-jet $z^{(1)} \in Z^{(1)}$ is of the form $z^{(1)} = (z, v_1, \ldots, v_n)$ where $z = (x, y)$, $x = (x_1, \ldots, x_n) \in \mathbb{R}^n$, $y = (y_1, \ldots, y_m) \in \mathbb{R}^m$, and $v_j \in \mathbb{R}^m$ for $j = 1, \ldots, n$. In a holonomic section we have $v_j = \frac{\partial f}{\partial x_j}(x)$. The restriction of a jet to a coordinate hyperplane $\{x_k = \text{const}\} \subset \mathbb{R}^n$ is determined by omitting the k-th vector v_k of the jet:

$$z_k^{(1)} = (z, v_1, \ldots, v_{k-1}, v_{k+1}, \ldots, v_m).$$

Definition 1.10.4 ([268]) An open differential relation $\Omega \subset Z^{(1)}$ is *ample* (in coordinate directions) if, for every restricted 1-jet $z_k^{(1)}$ as above, the subset

$$\left\{v \in \mathbb{R}^m : (z, v_1, \ldots, v_{k-1}, v, v_{k+1}, \ldots) \in \Omega\right\} \subset \mathbb{R}^m$$

is either empty or else the convex hull of each of its connected components equals $\mathbb{R}^m$.

The following result is due to Gromov [268].

Theorem 1.10.5 (h-Principle for ample differential relations) *Assume that π: $Z \to X$ is a smooth submersion and $\Omega \subset Z^{(1)}$ is an open subset of the manifold* *of 1-jets of sections $X \to Z$. Assume that for any point $z^0 \in Z$ we can find local coordinates (x, y) near z^0 such that, in these coordinates, $\pi(x, y) = x$ and Ω is ample (see Definition 1.10.4). Then solutions of Ω satisfy the weak homotopy equivalence principle.*

For a proof see [268], [265, Sect. 2.4], or [152, Sect. 18.2] (in particular Corollary 18.2.2). One proceeds by triangulating the manifold X and extending a solution inductively from a neighborhood of a skeleton of certain dimension $q < \dim X$ to the skeleton of dimension $q + 1$. The main point is that a solution of an ample open differential relation can be extended from the boundary of a cube in $\mathbb{R}^n$ to the interior of the cube provided there are no topological obstructions. Here is the precise statement.

Lemma 1.10.6 ([268, p. 339, Lemma 3.1.3]) *Let $P = [0, 1]^n \subset \mathbb{R}^n$, let $Z = P \times \mathbb{R}^m \to P$ be the trivial bundle, and let $\Omega \subset Z^{(1)}$ be an ample open differential relation. Suppose that $f_0 \colon P \to Z$ and $\phi_0 \colon P \to \Omega$ are smooth sections such that $\tau_1(\phi_0) = f_0$ on P (see (1.58)) and $j^1_x f_0 = \phi_0(x)$ holds for every $x \in bP$. Then for any $\epsilon > 0$ there exists a smooth section $f \colon P \to Z$ satisfying the following conditions:*

(i) *$j^1_x f = \phi_0(x)$ for every $x \in bP$,*
(ii) *$j^1_x f \in \Omega$ for every $x \in P$,*
(iii) *$|f(x) - f_0(x)| < \epsilon$ for every $x \in P$, and*
(iv) *there is a homotopy of sections $\phi_t \colon P \to \Omega$ $(t \in [0, 1])$ that is fixed on bP such that ϕ_0 is the given section and $\phi_1 = j^1 f$.*

This lemma is proved by Gromov's *convex integration method*. Theorem 1.10.5 and Lemma 1.10.6 are used in several places in the text.

Chapter 2
Stein Manifolds

This chapter is a brief survey of the theory of Stein manifolds and Stein spaces, with emphasis on the results that are frequently used in this book. After the initial developments by Karl Weierstrass, Bernhard Riemann, Fritz Hartogs, Eugenio E. Levi, Karl Reinhardt, Hellmuth Kneser, Henri Cartan, Peter Thullen and many others, the main contributions were made in the period 1942–1965 by Kiyoshi Oka, by the French school around Henri Cartan including Pierre Dolbeault, Alexander Grothendieck and Jean-Pierre Serre, and by the Münster school founded by Heinrich Behnke and including Karl Stein, Hans Grauert, Reinhold Remmert and Friedrich Hirzebruch. In 1942, Oka [444, Chap. VI] published the first solution to the Levi problem on two dimensional domains, while the year 1965 marks the publication of Lars Hörmander's fundamental paper [299] in which the $\overline{\partial}$-equation was solved by L^2-methods. Another contemporary work using the L^2-approach on q-convex manifolds is due to Aldo Andreotti and Edoardo Vesentini [28]. Together with the works of Joseph J. Kohn [345, 346], these provide the basis for quantitative methods in complex analysis. Comprehensive accounts of the theory of Stein manifolds and Stein spaces are available in [260, 274, 300], while the article of Schumacher [490] provides a historical survey. An introduction to topics in L^2-theory can be found in Ohsawa's book [441], while his recent book [442] presents an L^2 approach to problems in several complex variables and differential and algebraic geometry.

2.1 Domains of Holomorphy

A basic notion in complex analysis is that of analytic continuation. Karl Weierstrass knew already in 1841 that a holomorphic function in an annulus in the complex plane $\mathbb{C}$ admits a development into what is now called a Laurent series. By estimating the coefficients in this series, Bernhard Riemann showed in his dissertation in 1851 that a function which is analytic in a punctured neighborhood of a point $p \in \mathbb{C}$ and is bounded near p extends to a holomorphic function in a neighborhood of p. It was known early on that on any open relatively compact set $D \Subset \mathbb{C}$ in $\mathbb{C}$ there exist holomorphic functions that do not extend holomorphically across any boundary

F. Forstnerič, *Stein Manifolds and Holomorphic Mappings*, Ergebnisse der Mathematik und ihrer Grenzgebiete. 3. Folge / A Series of Modern Surveys in Mathematics 56, DOI 10.1007/978-3-319-61058-0_2

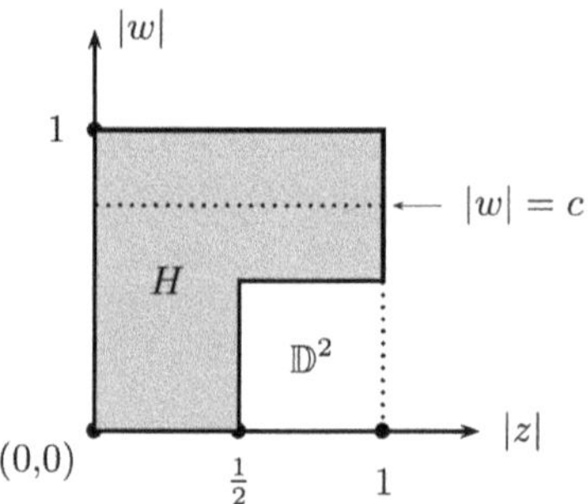

Fig. 2.1 A Hartogs figure in the bidisc

point of D. An explicit example on the disc $\mathbb{D} = \{|z| < 1\}$ is Kronecker's function $f(z) = \sum_{n=1}^{\infty} z^{n^2}$; further examples were given by Weierstrass.

A fundamental discovery was the phenomenon of *simultaneous analytic continuation*. In 1897 Adolph Hurwitz showed in his lecture at the first International Congress of Mathematicians that a holomorphic function of two or more variables does not have isolated singularities. More interesting examples of analytic continuation were found by Friedrich Hartogs in 1906 [280]. The simplest *Hartogs figure* is the domain H in the bidisc $\mathbb{D}^2 \subset \mathbb{C}^2$ defined by

$$H = \left\{ (z, w) \in \mathbb{D}^2 : |z| < \frac{1}{2} \text{ or } |w| > \frac{1}{2} \right\}. \tag{2.1}$$

(See Fig. 2.1.) Every function $f \in \mathcal{O}(H)$ extends to a holomorphic function on the bidisc $\mathbb{D}^2$. Indeed, pick a number $\frac{1}{2} < c < 1$ and consider the Cauchy integral

$$F(z, w) = \frac{1}{2\pi \mathrm{i}} \int_{|\zeta|=c} \frac{f(z,\zeta)}{\zeta - w} \, \mathrm{d}\zeta, \quad |z| < 1, \ |w| < c.$$

Then F is a holomorphic function on $D = \mathbb{D} \times c\mathbb{D}$ which agrees with f on $H \cap D$. (Since the disc $\{z\} \times c\mathbb{D}$ is contained in H when $|z| < \frac{1}{2}$, we have $f = F$ there by the Cauchy integral formula; the equality elsewhere follows by the identity principle.) This extends f to a holomorphic function on $H \cup D = \mathbb{D}^2$.

Fifteen years later, Karl Reinhardt [469] studied domains of convergence of power series $\sum_{\alpha \in \mathbb{Z}_+^n} c_\alpha z^\alpha$ in several variables $z = (z_1, \ldots, z_n)$. It is immediate that the domain of convergence is a union of open polydiscs centered at the origin. By introducing the map $\phi \colon \mathbb{C}^n \to (\{-\infty\} \cup \mathbb{R})^n$,

$$\phi(z_1, \ldots, z_n) = \big(\log|z_1|, \ldots, \log|z_n|\big)$$

we see that each union of polydiscs is of the form $\Omega = \phi^{-1}(D)$ where D is a domain in $(\{-\infty\} \cup \mathbb{R})^n$ such that $(x_1, \ldots, x_n) \in D$ and $y_j \le x_j$ for $j = 1, \ldots, n$ implies that $(y_1, \ldots, y_n) \in D$. Reinhardt showed that Ω is the domain of convergence of a power series if and only if the corresponding domain $D \subset (\{-\infty\} \cup \mathbb{R})^n$ is *convex*. This gives analytic continuation of holomorphic functions from a complete Reinhardt domain $\Omega \subset \mathbb{C}^n$ to the smallest logarithmically convex complete Reinhardt domain $\widetilde{\Omega} \subset \mathbb{C}^n$ containing Ω.

In 1932 Hellmuth Kneser reformulated Hartogs' result into a more useful form known as the *Kontinuitätssatz*: Given an embedded family of closed analytic discs $D_t \subset \mathbb{C}^n$ ($t \in [0, 1]$) such that D_0 and all the boundaries bD_t belong to a domain $\Omega \subset \mathbb{C}^n$, every holomorphic function on Ω admits an analytic continuation along this family to a neighborhood of the disc D_1.

Hartogs' discovery initiated research on 'natural domains' of holomorphic functions. Analytic continuation in general yields a multi-valued function. Following an idea of Riemann, multi-valued functions are considered as single-valued functions on *Riemann domains* over $\mathbb{C}^n$: a complex manifold X together with a locally biholomorphic map $\pi: X \to \mathbb{C}^n$. The central concept became that of a *domain of holomorphy*—a domain in $\mathbb{C}^n$, or over $\mathbb{C}^n$, with a holomorphic function that does not extend holomorphically to any bigger domain, not even as a multi-valued function. (See Oka [444, Chap. II].) Much of the classical theory developed around the problem of characterizing domains of holomorphy, and of constructing the *envelope of holomorphy* $\widetilde{\Omega}$ of a given domain $\Omega \subset \mathbb{C}^n$—the largest domain such that every holomorphic function on Ω extends to a holomorphic function on $\widetilde{\Omega}$.

Another important discovery was made by Eugenio E. Levi in 1911 [393]. He investigated domains $D \Subset \mathbb{C}^n$ with $\mathcal{C}^2$ boundaries. Let $\rho: \mathbb{C}^n \to \mathbb{R}$ be a $\mathcal{C}^2$ defining function for D, i.e., $D = \{z \in \mathbb{C}^n : \rho(z) < 0\}$ and $d\rho_z \neq 0$ for every $z \in bD = \{\rho = 0\}$. Levi noticed that, if for some boundary point $p \in bD$ and some vector $v \in T^{\mathbb{C}}_p bD$ that is complex tangent to the boundary (i.e., such that $\sum_{j=1}^n \frac{\partial \rho}{\partial z_j}(p) v_j = 0$) the Levi form $\mathcal{L}_{\rho,p}(v) < 0$ is negative, then holomorphic functions on D continue to a neighborhood of p in $\mathbb{C}^n$. The condition $\mathcal{L}_{\rho,p}(v) < 0$ implies that we can holomorphically embed a Hartogs pair $(H, \mathbb{D}^n)$ in $\mathbb{C}^n$ such that H is mapped into D but the image of $\mathbb{D}^n$ contains a neighborhood of p. Levi conjectured that any domain $D \subset \mathbb{C}^n$ as in the following definition is a domain of holomorphy; this became known as the *Levi problem*.

Definition 2.1.1 A domain $D = \{\rho < 0\}$ with a $\mathcal{C}^2$ defining function ρ such that $d\rho \neq 0$ on $bD = \{\rho = 0\}$ is *Levi pseudoconvex* if $\mathcal{L}_{\rho,p}(v) \geq 0$ for every $p \in bD$ and $v \in T^{\mathbb{C}}_z bD$. The domain D is *strongly pseudoconvex* if $\mathcal{L}_{\rho,p}(v) > 0$ for every $p \in bD$ and $0 \neq v \in T^{\mathbb{C}}_p bD$.

It is easily seen that the definition is independent of the choice of a defining function. A strongly pseudoconvex domain is locally at each boundary point biholomorphic to a piece of a strongly convex domain, and is osculated by a ball in suitable coordinates. This is commonly known as *Narasimhan's lemma*, although it was already known to Kneser in 1936 [338].

An important characterization of domains of holomorphy was obtained by Henri Cartan and Peter Thullen in 1932. To a compact set K in a complex space X we associate its $\mathcal{O}(X)$*-convex hull*, also called $\mathcal{O}(X)$*-hull*:

$$\widehat{K}_{\mathcal{O}(X)} = \Big\{ p \in X : |f(p)| \leq \max_{x \in K} |f(x)|, \ \forall f \in \mathcal{O}(X) \Big\}. \tag{2.2}$$

If K is a compact set in $\mathbb{C}^n$ then $\widehat{K} = \widehat{K}_{\mathcal{O}(\mathbb{C}^n)}$ is the *polynomial hull* of K.

Definition 2.1.2 A compact set K in a complex space X is *$\mathcal{O}(X)$-convex* if $K = \widehat{K}_{\mathcal{O}(X)}$. If $X = \mathbb{C}^n$ then such a set K is *polynomially convex*. A complex space X is *holomorphically convex* if for every compact set $K \subset X$ its $\mathcal{O}(X)$-hull $\widehat{K}_{\mathcal{O}(X)}$ is also compact.

Theorem 2.1.3 (Cartan and Thullen [91]) *A Riemann domain over $\mathbb{C}^n$ is a domain of holomorphy if and only if it is holomorphically convex.*

The hull of any compact set in a domain $\Omega \subset \mathbb{C}^n$ is a bounded closed subset of Ω, but it may fail to be compact as is seen in the Hartogs figure (2.1): Since every $f \in \mathcal{O}(H)$ extends to a function in $\mathcal{O}(\mathbb{D}^2)$, the maximum principle shows that the $\mathcal{O}(H)$-hull of the circle $\{(z_0, w) : |w| = \frac{3}{4}\}$ is the intersection of the disc $\{(z_0, w) : |w| \le \frac{3}{4}\}$ with Ω; this set is not compact if $\frac{1}{2} < |z_0| < 1$.

Theorem 2.1.3 is not difficult to prove. On the one hand, the derivatives of a holomorphic function $f \in \mathcal{O}(\Omega)$ satisfy the same bounds on $\widehat{K}_{\mathcal{O}(\Omega)}$ as on K, and hence the Taylor series of f centered around a point $p \in \widehat{K}_{\mathcal{O}(\Omega)}$ has the same domain of convergence as for points in K. If Ω is a domain of holomorphy, it follows that for any compact set $K \subset \Omega$ we have

$$\mathrm{dist}(\widehat{K}_{\mathcal{O}(\Omega)}, b\Omega) = \mathrm{dist}(K, b\Omega), \tag{2.3}$$

so $\widehat{K}_{\mathcal{O}(\Omega)}$ is compact. Conversely, using holomorphic convexity one can easily construct holomorphic functions tending to infinity along a given discrete sequence, so Ω is a domain of holomorphy.

A more challenging problem was to find a geometric characterization of domains of holomorphy. It follows from (2.3) that any closed holomorphic disc D in a domain of holomorphy Ω satisfies $\mathrm{dist}(D, b\Omega) = \mathrm{dist}(bD, b\Omega)$. This condition, which can be formulated in terms of *Hartogs pairs* (biholomorphic images of a standard pair $H \subset \mathbb{D}^n$, where H is a Hartogs figure in the polydisc $\mathbb{D}^n$), is known as *Hartogs pseudoconvexity* of Ω. Essentially it means that an analytic disc in $\overline{\Omega}$ with boundary in Ω must be contained in Ω. Oka showed that in such a case the function $\Omega \ni z \mapsto -\log \mathrm{dist}(z, b\Omega)$ is plurisubharmonic on Ω. Clearly this function blows up at $b\Omega$, so by adding the term $|z|^2$ we get a strongly plurisubharmonic exhaustion function on Ω. Similarly, Levi pseudoconvexity of a domain $\Omega \Subset \mathbb{C}^n$ easily implies that the function $-\log \mathrm{dist}(\cdot\,, b\Omega)$ is plurisubharmonic on Ω.

Could this be a characterization of domains of holomorphy?

This *Levi problem* was solved affirmatively by Oka in 1942 for domains in $\mathbb{C}^2$ [444, Chap. VI]; the higher dimensional case followed ten years later by Oka [444, Chap. IX], Bremermann [67], and Norguet [437]. In summary, we have the following result [300, Theorem 2.6.7].

Theorem 2.1.4 *The following conditions are equivalent for a domain Ω in $\mathbb{C}^n$, or a domain over $\mathbb{C}^n$:*

(a) *Ω is a domain of holomorphy.*
(b) *Ω is Hartogs pseudoconvex.*

(c) *The function* $-\log \mathrm{dist}(\cdot\,, b\Omega)$ *is plurisubharmonic.*
(d) *There exists a (strongly) plurisubharmonic exhaustion function on* Ω.

A domain $\Omega \subset \mathbb{C}^n$ *with* $\mathscr{C}^2$ *boundary is a domain of holomorphy if and only if it is Levi pseudoconvex.*

Every domain in (or over) $\mathbb{C}^n$ admits an envelope of holomorphy which can be constructed by 'pushing analytic discs' countably many times. A construction of the envelope in one step for domains in $\mathbb{C}^2$, and also in any two dimensional Stein manifold, was given by Jöricke in 2009 [314]. For results in this direction see also Merker and Porten [414].

2.2 Stein Manifolds and Stein Spaces

The class of Stein manifolds was introduced by Karl Stein in 1951 [524].

Definition 2.2.1 A complex manifold X is a *Stein manifold* if the following conditions hold:

(a) For every pair of distinct points $x \neq y$ in X there is a holomorphic function $f \in \mathcal{O}(X)$ such that $f(x) \neq f(y)$.
(b) For every point $p \in X$ there exist functions $f_1, \ldots, f_n \in \mathcal{O}(X)$, $n = \dim X$, whose differentials $df_1, \ldots, df_n$ are $\mathbb{C}$-linearly independent at p.
(c) X is holomorphically convex (see Definition 2.1.2).

Property (b) means that global holomorphic functions provide local charts at each point. Property (c) implies that a Stein manifold X admits an exhaustion $K_1 \subset K_2 \subset \cdots \subset \bigcup_{j=1}^{\infty} K_j = X$ by compact $\mathcal{O}(X)$-convex subsets such that $K_j \subset \mathring{K}_{j+1}$ holds for every $j = 1, 2, \ldots$.

Here are some basic properties and examples of Stein manifolds:

- An open set in $\mathbb{C}^n$ is Stein if and only if it is a domain of holomorphy. (This follows from the Cartan-Thullen theorem; see Theorem 2.1.3.)
- A Stein manifold does not contain any compact complex subvariety of positive dimension. (Apply axiom (a) and the maximum principle.)
- The Cartesian product $X \times Y$ of a pair of Stein manifolds is Stein.
- A closed complex submanifold X of $\mathbb{C}^N$ is Stein. (Use coordinate functions restricted to X. For the converse, see Theorem 2.4.1.)
- More generally, a closed complex submanifold of a Stein manifold is Stein.
- Every open Riemann surface is a Stein manifold (Behnke and Stein [50, 51], [260, p. 134]).
- If $X \to Y$ is a holomorphic covering space and Y is Stein, then X is Stein. (This is due to Stein [525].)
- If $X \to Y$ is a finite branched holomorphic covering, then X is Stein if and only if Y is Stein (Gunning [272, p. 151]).

- If $E \to X$ is a holomorphic vector bundle over a Stein base X, then the total space E is also Stein.
- (The Serre problem.) There exist holomorphic fibre bundles over the disc or the plane, with fibre $\mathbb{C}^2$ and with transitions maps given by polynomial automorphisms of $\mathbb{C}^2$, whose total space is not Stein; see Theorems 8.3.12 and 8.3.13.

The notion of a Stein space was introduced by Grauert in 1955 [251]. The standard definition is the following one.

Definition 2.2.2 A second countable complex space X is said to be a *Stein space* if it satisfies properties (a), (c) in Definition 2.2.1 and also

(b′) Every local ring $\mathcal{O}_{X,x}$ is generated by functions in $\mathcal{O}(X)$.

Condition (b′) means that there is a holomorphic map $X \to \mathbb{C}^N$ which embeds a neighborhood of x as a local complex subvariety of $\mathbb{C}^N$. Grauert showed in [251] that one gets an equivalent definition by keeping (c) and replacing (a) and (b) (resp. (b′)) by the following property.

Definition 2.2.3 A complex space X is called *K-complete* if for every point $x \in X$ there is a holomorphic map $f: X \to \mathbb{C}^N$ (with $N = N_x$) such that x is an isolated point of the fibre $f^{-1}(f(x))$.

It is immediate that axiom (a) implies K-completeness. In summary:

Theorem 2.2.4 ([251]) *A complex space X is a Stein space if and only if it is holomorphically convex and it satisfies one of the following two properties:*

(i) *Holomorphic functions separate points on X (axiom* (a) *in Definition* 2.2.1).
(ii) *X is K-complete in the sense of Definition* 2.2.3.

For further characterizations of Stein spaces see [260, p. 152].

2.3 Holomorphic Convexity and the Oka-Weil Theorem

The following *Oka-Weil theorem* generalizes Runge's theorem. See Theorem 2.6.8 for an analogous result concerning sections of coherent analytic sheaves.

Theorem 2.3.1 *If X is a Stein space and K is a compact $\mathcal{O}(X)$-convex subset of X, then every holomorphic function in an open neighborhood of K can be approximated uniformly on K by functions in $\mathcal{O}(X)$.*

Theorem 2.3.1 was proved for domains of holomorphy by Oka [444, Chap. I] using his *Oka lemma*; see [300, Lemma 2.7.5]. It is an immediate consequence of the definition that an $\mathcal{O}(X)$-convex set K can be approximated from the outside by *analytic polyhedra*, i.e., by Stein open sets of the form

$$U = \{x \in X : |h_j(x)| < 1,\ j = 1, \ldots, m\}, \quad h_1, \ldots, h_m \in \mathcal{O}(X).$$

By adding more functions, we can ensure that $h = (h_1, \ldots, h_m): X \to \mathbb{C}^m$ embeds U properly into the polydisc $\mathbb{D}^m \subset \mathbb{C}^m$. The key point proved by Oka is that for any function $f \in \mathcal{O}(U)$ there is a function $g \in \mathcal{O}(\mathbb{D}^m)$ such that $g \circ h = f$. (This is a special case of the Oka-Cartan extension theorem, see Corollary 2.6.3.) By expanding g in power series and approximating it by Taylor polynomials $P \in \mathbb{C}[z_1, \ldots, z_m]$ we get functions $P \circ h \in \mathcal{O}(X)$ approximating f on K.

Another proof of the Oka-Weil theorem can be given by the L^2-methods for solving the nonhomogeneous $\overline{\partial}$-equation. We follows this approach in Sect. 2.8 to prove a stronger parametric version of the Oka-Weil theorem, combined with the Oka-Cartan extension theorem; see Theorem 2.8.4.

Definition 2.3.2 A domain Ω in a complex space X is *Runge in* X if every holomorphic function $f \in \mathcal{O}(\Omega)$ can be approximated uniformly on compacts in Ω by functions in $\mathcal{O}(X)$; equivalently, if the subalgebra $\{f|_\Omega : f \in \mathcal{O}(X)\}$ of $\mathcal{O}(\Omega)$ is dense in $\mathcal{O}(\Omega)$ in the compact-open topology.

Theorem 2.3.3 ([300, p. 91]) *A Stein domain Ω in a Stein space X is Runge in X if and only if for every compact set $K \subset \Omega$ we have $\widehat{K}_{\mathcal{O}(\Omega)} = \widehat{K}_{\mathcal{O}(X)}$.*

There exist several notions of *ambient holomorphic convexity* of a compact set (see [534]); we shall use the following ones.

Definition 2.3.4 Assume that K is a compact set in a complex space X.

(i) K is a *Stein compact* if it admits a basis of Stein neighborhoods in X.
(ii) K is *holomorphically convex* if it admits an open Stein neighborhood Ω in X such that K is $\mathcal{O}(\Omega)$-convex.

2.4 Embedding Stein Manifolds into Euclidean Spaces

An important characterization of Stein manifolds is that they are embeddable as closed complex submanifolds of complex Euclidean spaces. It is an immediate consequence of Definition 2.2.1 that for every relatively compact domain Ω in a Stein manifold X there is a holomorphic map $f: X \to \mathbb{C}^N$ for a big enough N such that $f|_\Omega: \Omega \to \mathbb{C}^N$ is an injective holomorphic immersion. In 1956, Remmert proved a substantially stronger result that every Stein manifold admits a *proper* holomorphic embedding into some Euclidean space $\mathbb{C}^N$ [470]. In 1960–1961, Bishop and Narasimhan independently proved that Remmert's theorem holds with $N = 2\dim X + 1$. We now summarize these classical results. (For smooth manifolds, part (a) is due to Whitney [571].)

Theorem 2.4.1 ([58], [423, Theorem 5], [424])

(a) *If* X *is a Stein manifold of dimension* n, *then the set of proper holomorphic maps* $X \to \mathbb{C}^{n+1}$ *is dense in* $\mathcal{O}(X)^{n+1}$, *the set of proper holomorphic immersions* $X \to \mathbb{C}^{2n}$ *is dense in* $\mathcal{O}(X)^{2n}$, *and the set of proper holomorphic embeddings* $X \to \mathbb{C}^{2n+1}$ *is dense in* $\mathcal{O}(X)^{2n+1}$.
(b) *If* X *is a Stein space of dimension* n, *then the set of holomorphic maps* $X \to \mathbb{C}^{2n+1}$ *which are proper, injective, and regular (immersions) on the regular part* X_{reg} *is dense in* $\mathcal{O}(X)^{2n+1}$.
(c) *If* X *is a Stein space of dimension* n *and of finite embedding dimension* m, *then for* $N = \max\{n+m, 2n+1\}$ *the set of proper holomorphic embeddings* $X \hookrightarrow \mathbb{C}^N$ *is dense in* $\mathcal{O}(X)^N$.

More precise embedding theorems for Stein manifolds and Stein spaces are proved in Sects. 9.3–9.5, and for Riemann surfaces in Sects. 9.10–9.11. Unlike Theorem 2.4.1, those results depend on the Oka theory developed in Chaps. 5 and 6.

Since every real analytic manifold admits a Stein complexification [255], we get the following consequence of Theorem 2.4.1 which answers a question of Whitney [571, p. 645].

Corollary 2.4.2 ([255, Theorem 3]) *Every real analytic manifold admits a proper real analytic embedding into a Euclidean space* $\mathbb{R}^N$.

Since Stein manifolds are complex submanifolds of Euclidean spaces, it is not surprising that they can be approximated by affine algebraic manifolds. It was proved by Stout [533] that any relatively compact domain in a Stein manifold is biholomorphically equivalent to a domain in an affine algebraic manifold. (For the real algebraic case, see Nash [428].) More precise algebraic approximation results were obtained by Demailly, Lempert and Schiffman [117, 390] and by Lisca and Matič [398] (see Theorem 10.7.1 on p. 506).

2.5 Characterization by Plurisubharmonic Functions

It is a fundamental fact that Stein manifolds and Stein spaces are characterized by plurisubharmonicity (see Theorem 2.5.2). Quite often, the most efficient way to show that a complex space is Stein is to find a strongly plurisubharmonic exhaustion function on it. This is how Siu proved in 1976 [504] that a Stein subvariety of any complex space has a basis of open Stein neighborhoods (see Theorem 3.1.1 on p. 66). Stein neighborhoods often allow us to transfer a problem on a complex space to a more tractable problem on an ambient Euclidean space; Chap. 3 focuses on such methods.

It follows from holomorphic convexity that every Stein space X is exhausted by an increasing sequence of compacts $K_1 \subset K_2 \subset \cdots \subset \bigcup_{j=1}^{\infty} K_j = X$ such that

$K_j = \widehat{K}_j$ and $K_j \subset \mathring{K}_{j+1}$ for every $j \in \mathbb{N}$. Using such exhaustions and axioms (a), (b′), one can easily find strongly plurisubharmonic exhaustion functions of the form

$$\rho = \sum_{j=1}^{\infty} |f_j|^2 : X \to \mathbb{R}_+, \quad f_j \in \mathcal{O}(X), \ j = 1, 2, \ldots .$$

By a more precise argument one obtains the following result on approximation of compact $\mathcal{O}(X)$-convex subsets of a Stein space by sublevel sets of strongly plurisubharmonic functions (see [300, Theorem 5.1.6, p. 117]).

Proposition 2.5.1 *If K is a compact $\mathcal{O}(X)$-convex set in a Stein space X, then for every open set $U \subset X$ containing K there exists a smooth strongly plurisubharmonic function $\rho: X \to \mathbb{R}$ such that $\rho < 0$ on K and $\rho > 1$ on $X \setminus U$. Furthermore, there exists a plurisubharmonic exhaustion function $\rho: X \to \mathbb{R}_+$ such that $\rho^{-1}(0) = K$ and ρ is strongly plurisubharmonic on $X \setminus K = \{\rho > 0\}$.*

Note that the function $\rho_a: \mathbb{C}^N \to \mathbb{R}_+$ given by $\rho_a = |z - a|^2$ is strongly plurisubharmonic on any complex subvariety $X \subset \mathbb{C}^N$; if X is closed, then this is an exhaustion function on X. Furthermore, if X is smooth, then $\rho_a|_X$ is a Morse function on X for most choices of the point $a \in \mathbb{C}^N$.

These observations show that a Stein space admits plenty of smooth strongly plurisubharmonic exhaustion functions. The following converse is the most useful characterization of Stein manifolds and Stein spaces.

Theorem 2.5.2

(a) [128, 255] *A complex manifold which admits a strongly plurisubharmonic exhaustion function is a Stein manifold.*
(b) [163, 425] *A complex space which admits a strongly plurisubharmonic exhaustion function is a Stein space.*

Furthermore, if $\rho: X \to \mathbb{R}$ is a strongly plurisubharmonic exhaustion function, then each sublevel set $\{x \in X : \rho(x) \le c\}$ is $\mathcal{O}(X)$-convex.

Corollary 2.5.3 *For every compact set K in a Stein space X, the $\mathcal{O}(X)$-hull of K coincides with its plurisubharmonic hull:*

$$\widehat{K}_{\mathcal{O}(X)} = \widehat{K}_{\mathrm{Psh}(X)}.$$

Hence, every holomorphic function in a neighborhood of a compact $\mathrm{Psh}(X)$*-convex set $K = \widehat{K}_{\mathrm{Psh}(X)}$ is a uniform limit on K of functions in $\mathcal{O}(X)$.*

The most efficient proof of Theorem 2.5.2 and Corollary 2.5.3 can be given by the L^2-method for solving nonhomogeneous $\overline{\partial}$-equations with weights of the form $e^{-\rho}$ with $\rho \in \mathrm{Psh}(X)$ (see e.g. [299, 300, 442]; see also Sect. 2.8 where we prove a parametric version of the Oka-Weil approximation theorem).

Theorem 2.5.2 implies the following solution of the Levi problem.

Corollary 2.5.4 *Let X be a Stein space. If a domain $\Omega \subset X$ admits a plurisubharmonic exhaustion function $\rho: \Omega \to \mathbb{R}$, then Ω is Stein. In particular, every Levi (or Hartogs) pseudoconvex domain in a Stein space is Stein.*

The notion of a holomorphically convex set and of a Stein compact was introduced in Definition 2.3.4. Proposition 2.5.1 and Theorem 2.5.2 imply the following characterization of these notions by plurisubharmonicity.

Proposition 2.5.5 *A compact set K in a Stein space X is holomorphically convex if and only if there exists a plurisubharmonic function $\rho: U \to \mathbb{R}_+$ in an open neighborhood U of K such that $\rho^{-1}(0) = K$ and ρ is strongly plurisubharmonic on $U \setminus K = \{\rho > 0\}$.*

The sets $\Omega_c = \{x \in U : \rho(x) < c\}$ for small $c > 0$ then form a basis of Stein neighborhoods of K such that K is $\mathcal{O}(\Omega_c)$-convex.

Here is another useful sufficient condition for a set to be Stein compact.

Proposition 2.5.6 *Let K be a compact set in a complex space X. Assume that there exist a neighborhood $U \subset X$ of K, a strongly plurisubharmonic function $\rho: U \to \mathbb{R}$, and a weakly plurisubharmonic function $\tau: U \to \mathbb{R}_+$ such that $K = \{\tau = 0\}$. Then, K is a Stein compact.*

Proof Fix an open neighborhood $V \Subset U$ of K. It is easy to find a fast growing convex increasing function $\chi: \mathbb{R} \to \mathbb{R}$ such that the strongly plurisubharmonic function $\phi = \rho + \chi \circ \tau: U \to \mathbb{R}$ satisfies $\phi|_K < 0$ and $K \subset V_c = \{\phi < c\} \Subset V$ for some $c > 0$. The domain V_c is then Stein by Theorem 2.5.2. □

The closure of a smooth weakly pseudoconvex domain $D \Subset \mathbb{C}^n$ need not be a Stein compact; an example is the *worm domain* [121]. For the existence of bounded strongly plurisubharmonic exhaustion functions on weakly Levi pseudoconvex domains see [120].

2.6 Cartan-Serre Theorems A & B

The famous Theorems A and B were proved in Cartan's seminar in 1951–1954; see [87, 90, 260]. It would be impossible to overstate the importance of these results for the development of analytic and algebraic geometry.

Theorem 2.6.1 *Let $\mathcal{F}$ be a coherent analytic sheaf on a Stein space X. Then:*

(A) *The stalk of $\mathcal{F}_x$ of $\mathcal{F}$ at any point $x \in X$ is generated as an $\mathcal{O}_{X,x}$-module by global sections of the sheaf $\mathcal{F}$.*
(B) *$H^p(X; \mathcal{F}) = 0$ for all $p = 1, 2, \ldots$.*

The corresponding results hold for every coherent algebraic sheaf over an affine algebraic variety $X \subset \mathbb{C}^N$ (Serre [497, p. 237, Théorème 2]).

An analogue of Theorems A and B for coherent analytic sheaves with continuous boundary values on strongly pseudoconvex domains was proved by Heunemann [293] and Leiterer [385].

We recall the relevant notions; a comprehensive account is available in [261]. Let X be a complex space. An *analytic sheaf* (or $\mathcal{O}_X$-sheaf) on X is a sheaf $\mathcal{F}$ of $\mathcal{O}_X$-modules; that is, a sheaf whose stalk $\mathcal{F}_x$ over any point $x \in X$ is a module over the local ring $\mathcal{O}_{X,x}$. The sheaf $\mathcal{F}$ is *locally finitely generated* if for every point $x_0 \in X$ there exist an open neighborhood $U \subset X$ and finitely many sections $f_1, \ldots, f_k \in \mathcal{F}(U) = \Gamma(U, \mathcal{F})$ whose germs at any point $x \in U$ generate $\mathcal{F}_x$ as an $\mathcal{O}_{X,x}$-module. The simplest example is $\mathcal{O}_X^k$, the direct sum of k copies of the structure sheaf $\mathcal{O}_X$ for any $k \in \mathbb{N}$; this is the sheaf of holomorphic sections of the trivial bundle $X \times \mathbb{C}^k \to X$.

An analytic sheaf is *coherent* if it is locally finitely generated and if for any set of local sections $f_1, \ldots, f_k \in \mathcal{F}(U)$ the corresponding *sheaf of relations* $\mathcal{R} = \mathcal{R}(f_1, \ldots, f_k)$ is also locally finitely generated. The latter sheaf has stalks

$$\mathcal{R}_x = \left\{ (g_{1,x}, \ldots, g_{k,x}) \in \mathcal{O}_{X,x}^k : \sum_{j=1}^k g_{j,x} f_{j,x} = 0 \right\}, \quad x \in U. \tag{2.4}$$

From the above description, we see that an analytic sheaf $\mathcal{F}$ over X is coherent if and only if each point $x \in X$ has an open neighborhood $U \subset X$ and a short exact sequence of analytic sheaf homomorphisms

$$\mathcal{O}_U^m \stackrel{\alpha}{\longrightarrow} \mathcal{O}_U^k \stackrel{\beta}{\longrightarrow} \mathcal{F}|_U \longrightarrow 0 \tag{2.5}$$

where

$$\beta(g_{1,x}, \ldots, g_{k,x}) - \sum_{j=1}^k g_{j,x} f_{j,x}.$$

Hence, β maps the standard basis sections $e_j = (0, \ldots, 1, \ldots, 0)$ of $\mathcal{O}_U^k$ onto the generators f_j of $\mathcal{F}|_U$ and $\mathcal{R} = \ker \beta = \operatorname{im} \alpha$ is the sheaf of relations (2.4). If X is a Stein space then a resolution (2.5) exists over any relatively compact open subset $U \Subset X$.

Here are the main examples of coherent sheaves on a complex space X:

- The structure sheaf $\mathcal{O}_X$ (Oka [444, Chap. VII]; see also [261, p. 59] and [435]).
- The sheaf of ideals $\mathcal{O}_A$ of a complex subvariety A in X (the Oka-Cartan coherence theorem; [90, p. 631], [261, p. 84], [435]).
- A locally free analytic sheaf, i.e., a sheaf of holomorphic sections of a holomorphic vector bundle. In particular, we have the *tangent sheaf* $\mathcal{T}_X$ and the *cotangent sheaf* $\mathcal{T}_X^*$ on a complex manifold X.
- The Whitney sum $\mathcal{E} \oplus \mathcal{F}$ and the tensor product $\mathcal{E} \otimes \mathcal{F}$ of coherent sheaves.

- The sheaf $\mathcal{H}om(\mathcal{E}, \mathcal{F})$ of $\mathcal{O}_X$-homomorphisms $\mathcal{E} \to \mathcal{F}$ between a pair of coherent analytic sheaves. In particular, the dual $\mathcal{E}^*$ of a coherent analytic sheaf.
- If $\beta: \mathcal{F} \to \mathcal{G}$ is a homomorphism of coherent analytic sheaves, then the kernel $\ker\beta$ and the image $\mathrm{im}\beta$ are coherent analytic sheaves. In summary, given a short exact sequence of homomorphisms of $\mathcal{O}_X$-analytic sheaves

$$0 \longrightarrow \mathcal{E} \xrightarrow{\alpha} \mathcal{F} \xrightarrow{\beta} \mathcal{G} \longrightarrow 0,$$

  if two sheaves are coherent then so is the third one [261, p. 236].
- The direct image of an $\mathcal{O}_X$-coherent sheaf by a proper holomorphic map $X \to Y$ of complex spaces is a coherent $\mathcal{O}_Y$-sheaf (Grauert's coherence theorem; see [261, p. 207]).

Each coherent analytic sheaf $\mathcal{F}$ can be represented as the sheaf of germs of fibrewise linear holomorphic functions on a linear space $\pi: L \to X$ [158]. More precisely, there is a contravariant equivalence between the category of coherent analytic sheaves and the category of linear spaces such that locally free sheaves correspond to vector bundles. The sheaf of germs of holomorphic sections $X \to L$ of any linear space is also coherent [158, p. 53, Corollary].

Given a coherent analytic sheaf $\mathcal{F}$ on a complex space X, the $\mathcal{O}(X)$-module $\mathcal{F}(X) = \Gamma(X, \mathcal{F})$ of all global sections is endowed with a Fréchet space topology (the topology of uniform convergence on compacts in X) such that for every point $x \in X$ the natural restriction map $\mathcal{F}(X) \mapsto \mathcal{F}_x$ is continuous (see [260, Theorem 5, p. 167]). The topology on the stalks $\mathcal{F}_x$ is the *sequence topology* (cf. [259, p. 86ff]). In particular, $\mathcal{F}(X)$ is a Baire space.

We now mention some applications of Theorems A and B; for more on this subject, see [496] and [260, Chap. V].

Corollary 2.6.2 *Let $\beta: \mathcal{F} \to \mathcal{G}$ be an epimorphism of analytic sheaves over a Stein space X. If the kernel $\mathcal{E} = \ker\beta$ is coherent, then the induced map on sections $\mathcal{F}(X) \to \mathcal{G}(X)$, $f \mapsto \beta(f)$ is surjective.*

Proof Since $H^1(X; \mathcal{E}) = 0$ by Theorem 2.6.1, the conclusion follows from the exact cohomology sequence $\mathcal{F}(X) \to \mathcal{G}(X) \to H^1(X; \mathcal{E}) = 0$. □

Applying Corollary 2.6.2 to the exact sequence

$$0 \longrightarrow \mathcal{J}_A \longrightarrow \mathcal{O}_X \longrightarrow \mathcal{O}_X/\mathcal{J}_A \longrightarrow 0$$

where A is a closed complex subvariety of X, we obtain

Corollary 2.6.3 (Oka-Cartan extension theorem) *Every holomorphic function on a closed complex subvariety of a Stein space X extends to a holomorphic function on X.*

Corollary 2.6.4 (Cartan's division theorem) *If $\mathcal{F}$ is a coherent analytic sheaf on a Stein space X and if $f_1, \ldots, f_k \in \mathcal{F}(X)$ generate each stalk $\mathcal{F}_x$ ($x \in X$), then every section $f \in \mathcal{F}(X)$ is of the form $f = \sum_{j=1}^k g_j f_j$ for some $g_j \in \mathcal{O}(X)$.*

Proof Consider the exact sequence $0 \to \mathcal{R} \to \mathcal{O}^k \xrightarrow{\beta} \mathcal{F} \to 0$ as in (2.5). Since $\mathcal{R} = \ker\beta$ is coherent, the conclusion follows from Corollary 2.6.2. □

Corollary 2.6.5 *Given a short exact sequence $0 \to \mathcal{E} \to \mathcal{F} \xrightarrow{\beta} \mathcal{G} \to 0$ of coherent analytic sheaves on a Stein space such that $\mathcal{G}$ is locally free, there exists a sheaf homomorphism $\phi: \mathcal{G} \to \mathcal{F}$ such that $\beta \circ \phi = \mathrm{Id}_{\mathcal{G}}$.*

Proof Consider the induced exact sequence

$$0 \longrightarrow \mathcal{H}om(\mathcal{G},\mathcal{E}) \longrightarrow \mathcal{H}om(\mathcal{G},\mathcal{F}) \xrightarrow{\beta} \mathcal{H}om(\mathcal{G},\mathcal{G}) \longrightarrow 0.$$

Surjectivity of β is due to $\mathcal{G}$ being locally free. By Theorem B we have $H^1(X; \mathcal{H}om(\mathcal{G},\mathcal{E})) = 0$, and hence β is surjective also on the level of sections. Hence, $\mathrm{Id}_{\mathcal{G}}$ lifts to a homomorphism $\phi: \mathcal{G} \to \mathcal{F}$ with $\beta \circ \phi = \mathrm{Id}_{\mathcal{G}}$. □

The following is a special case of Corollary 2.6.5.

Corollary 2.6.6 *If E' is a holomorphic vector subbundle of a holomorphic vector bundle E over a Stein space X, then there exists a holomorphic vector subbundle E'' of E such that $E = E' \oplus E''$ is a holomorphic direct sum.*

Theorem 2.6.7 *On any Stein manifold X, the Dolbeault cohomology groups vanish: $H^{p,q}_{\overline{\partial}}(X) = 0$ for all $p \ge 0$, $q \ge 1$.*

Proof The sheaf Ω_p of holomorphic p-forms on X admits a resolution

$$0 \to \Omega_p \hookrightarrow \mathcal{E}_{p,0} \xrightarrow{\overline{\partial}} \mathcal{E}_{p,1} \xrightarrow{\overline{\partial}} \mathcal{E}_{p,2} \cdots \xrightarrow{\overline{\partial}} \mathcal{E}_{p,n} \longrightarrow 0.$$

Since the sheaves $\mathcal{E}_{p,q}$ of smooth (p,q)-forms on X are fine, their cohomology vanishes. Leray's theorem implies that $H^{p,q}_{\overline{\partial}}(X) \cong H^q(X; \Omega_p)$. Since the sheaf Ω_p is coherent analytic, these groups are zero by Cartan's Theorem B. □

Another proof of Theorem 2.6.7 is obtained by Hörmander's L^2 theory.

Serre proved that each element of a de Rham cohomology group $H^p(X;\mathbb{C})$ ($p = 1, 2, \ldots, \dim X$) of a Stein manifold is represented by a closed holomorphic p-form on X (see [496, Theorem 1], [260, p. 155]). The de Rham cohomology of an affine algebraic manifold is represented by algebraic forms (see Grothendieck [270]).

We have the following approximation theorem for sections of coherent analytic sheaves over Stein spaces (see e.g. [260, p. 170]).

Theorem 2.6.8 (Oka-Weil theorem for coherent analytic sheaves) *Let $\mathcal{F}$ be a coherent analytic sheaf on a Stein space X. If K is a compact $\mathcal{O}(X)$-convex set in X, then any section of $\mathcal{F}$ over an open neighborhood of K can be approximated uniformly on K by sections in $\mathcal{F}(X)$. More precisely, if sections $f_1, \ldots, f_m \in \mathcal{F}(X)$ generate every stalk $\mathcal{F}_x$, $x \in K$, then every section of $\mathcal{F}$ over an open neighborhood of K can be approximated uniformly on K by sections of the form $\sum_{j=1}^m g_j f_j$ for some $g_j \in \mathcal{O}(X)$.*

Proof Assume that f is a section of $\mathcal{F}$ over an open neighborhood $\Omega \subset X$ of K. We may assume that Ω is Stein and relatively compact in X. Since a coherent analytic sheaf $\mathcal{F}$ is locally finitely generated, there exist sections $f_1, \ldots f_m \in \mathcal{F}(X)$ which generate every stalk $\mathcal{F}_x$ for $x \in \Omega$. By Corollary 2.6.4 we have $f = \sum_{j=1}^m h_j f_j$ for some functions $h_1, \ldots, h_m \in \mathcal{O}(\Omega)$. By Theorem 2.3.1 we can approximate every h_j uniformly on K by a function $g_j \in \mathcal{O}(X)$. The section $F = \sum_{j=1}^m g_j f_j \in \mathcal{F}(X)$ then approximates f on K. □

2.7 The $\overline{\partial}$-Problem

The $\overline{\partial}$-problem asks for a solution of the equation $\overline{\partial} u = f$ for a given $\overline{\partial}$-closed form f. By Theorem 2.6.7, this problem is always solvable on a Stein manifold. A more direct approach, which also gives L^2 estimates of solutions, is provided by Hörmander's theory [299, 300]; see also Andreotti and Vesentini [28], Kohn [345, 346], and the monographs [95, 442].

We shall frequently use the following result for (0, 1)-forms. Let $\mathrm{d}\lambda$ denote the Lebesgue measure on $\mathbb{C}^n = \mathbb{R}^{2n}$.

Theorem 2.7.1 ([300, Theorem 4.4.2, p. 94]) *Let Ω be a pseudoconvex (Stein) domain in $\mathbb{C}^n$ and ϕ be a plurisubharmonic function in Ω. For every (0, 1)-form $f = \sum f_j \, \mathrm{d}\bar{z}_j$ such that $f_j \in L^2_{loc}(\Omega)$ and $\overline{\partial} f = 0$ (in the weak sense) there exists $u \in L^2_{loc}(\Omega)$ such that*

$$\overline{\partial} u = f \quad \text{and} \quad \int_\Omega \frac{|u|^2}{(1+|z|^2)^2} \mathrm{e}^{-\phi} \mathrm{d}\lambda \le \int_\Omega \sum_{j=1}^n |f_j|^2 \mathrm{e}^{-\phi} \mathrm{d}\lambda.$$

If f is smooth then so is u.

By taking Ω bounded and $\phi = 0$ we get the estimate

$$\overline{\partial} u = f \quad \text{and} \quad \int_\Omega |u|^2 \mathrm{d}\lambda \le C \int_\Omega \sum_j |f_j|^2 \, \mathrm{d}\lambda, \tag{2.6}$$

where the constant C depends on the radius of Ω and on the dimension n. The analogous results hold on relatively compact domains in Stein manifolds.

To pass from L^2 to $\mathcal{C}^k$ estimates, one needs the following well-known lemma which follows from the Bochner-Martinelli formula [215, Lemma 3.2].

Lemma 2.7.2 (Interior elliptic regularity estimates) *Let $\mathbb{B}^n$ denote the open unit ball in $\mathbb{C}^n$. For each $s \in \mathbb{Z}_+$ there is a constant $c_s > 0$ such that if $f \in \mathcal{C}^{s+1}(r\mathbb{B})$ for some $r > 0$ and $\alpha \in \mathbb{Z}_+^{2n}$ is a multi-index with $|\alpha| = s$ then*

$$c_s \left| \partial^\alpha f(0) \right| \le r^{-n-s} \| f \|_{L^2(r\mathbb{B})} + \sum_{|\beta| \le s} r^{|\beta|+1-s} \left\| \partial^\beta (\overline{\partial} f) \right\|_{L^\infty(r\mathbb{B})}.$$

In particular, we have the sup-norm estimate

$$c_0 \left| f(0) \right| \le r^{-n} \|f\|_{L^2(r\mathbb{B})} + r \|\overline{\partial} f\|_{L^\infty(r\mathbb{B})}. \tag{2.7}$$

On bounded strongly pseudoconvex domains in Stein manifolds, the $\overline{\partial}$-equation can also be solved by means of integral formulas with holomorphic kernels. This kernel method gives optimal Hölder estimates. The first results of this type were obtained by Henkin [288] and R. de Arellano (see [289]). We shall use the following result due to Range and Siu [467] and Lieb and Range [395, Theorem 1]; see also [396, 397], [416, Theorem 1′], and [394].

If D is a domain in a complex manifold X, we denote by $\mathcal{C}^l_{p,q}(D)$ the space of (p,q)-forms whose coefficients (in any local chart on X) are of class $\mathcal{C}^l(D)$, i.e., l times continuously differentiable. If D has piecewise $\mathcal{C}^1$ boundary then $\mathcal{C}^l_{p,q}(\bar{D})$ stands for the space of (p,q)-forms on $\bar{D}$ of class $\mathcal{C}^l(\bar{D})$. If $l = k + \alpha$ with $k \in \mathbb{Z}_+$ and $0 < \alpha \le 1$, then $\mathcal{C}^l = \mathcal{C}^{k,\alpha}$ denotes the Hölder space.

Theorem 2.7.3 *Given a relatively compact strongly pseudoconvex domain D in a Stein manifold, there exists a linear operator $T: \mathcal{C}^0_{0,1}(D) \to \mathcal{C}^{1/2}(D)$ such that, if $f \in \mathcal{C}^0_{0,1}(\bar{D}) \cap \mathcal{C}^1_{0,1}(D)$ and $\overline{\partial} f = 0$ in D then*

$$\overline{\partial}(Tf) = f \quad \text{and} \quad \|Tf\|_{\mathcal{C}^{1/2}(\bar{D})} \le c_D \|f\|_{\mathcal{C}^0_{0,1}(\bar{D})}.$$

The constant c_D can be chosen uniform for all domains sufficiently $\mathcal{C}^2$-close to D. If D has boundary of class $\mathcal{C}^\ell$ for some $\ell \in \{2,3,\ldots\}$ then there exists a linear operator $T: \mathcal{C}^0_{0,1}(D) \to \mathcal{C}^0(D)$ satisfying the following properties:

(i) *If $f \in \mathcal{C}^0_{0,1}(\bar{D}) \cap \mathcal{C}^1_{0,1}(D)$ and $\overline{\partial} f = 0$ then $\overline{\partial}(Tf) = f$.*
(ii) *If $f \in \mathcal{C}^0_{0,1}(\bar{D}) \cap \mathcal{C}^r_{0,1}(D)$ for some $r \in \{1,\ldots,\ell)$ then*

$$\|Tf\|_{\mathcal{C}^{l,1/2}(\bar{D})} \le C_{l,D} \|f\|_{\mathcal{C}^l_{0,1}(\bar{D})}, \quad l = 0,1,\ldots,r. \tag{2.8}$$

Although these results are stated in the original papers for domains with $\mathcal{C}^\infty$ boundaries, one only needs $\mathcal{C}^\ell$ boundary to get estimates up to order ℓ; this is implicitly contained in the paper by Michel and Perotti [416].

2.8 Cartan-Oka-Weil Theorem with Parameters

In this section we prove a parametric version of the classical Oka-Cartan extension theorem combined with the Oka-Weil approximation theorem; see Theorem 2.8.4. We begin with the following simple version on $\mathbb{C}^n$.

Proposition 2.8.1 (The Oka-Weil theorem with parameters on $\mathbb{C}^n$) *Let K be a compact polynomially convex set in $\mathbb{C}^n$ and let $U \subset \mathbb{C}^n$ be an open set containing K. Assume that P is a compact Hausdorff space and $f: P \times U \to \mathbb{C}$ is* *a continuous function such that* $f_p = f(p,\cdot): U \to \mathbb{C}$ *is holomorphic for every* $p \in P$. *Given* $\epsilon > 0$ *there exists a continuous function* $F: P \times \mathbb{C}^n \to \mathbb{C}$ *such that* $F_p = F(p,\cdot): \mathbb{C}^n \to \mathbb{C}$ *is holomorphic for every* $p \in P$ *and*

$$\sup_{z\in K,\, p\in P} \left|F_p(z) - f_p(z)\right| < \epsilon.$$

Proof Since the set K is polynomially convex, Proposition 2.5.1 gives a smooth strongly plurisubharmonic function $\phi: \mathbb{C}^n \to \mathbb{R}$ such that $\phi < 0$ on K and $\phi > 0$ on $\mathbb{C}^n \setminus U$. Let $h: \mathbb{R} \to \mathbb{R}_+$ be a smooth convex increasing function such that $h(t) = 0$ for $t \le 0$ and h is positive and strictly increasing on $t > 0$. Then, the function $\psi = h \circ \phi: \mathbb{C}^n \to [0,\infty)$ is smooth plurisubharmonic, it vanishes in an open neighborhood U_0 of K, and is positive on $\mathbb{C}^n \setminus U$. Pick a neighborhood $U_1 \Subset U$ of K such that

$$\psi \ge c > 0 \quad \text{on } \overline{U} \setminus U_1 \tag{2.9}$$

for some positive constant $c > 0$. Choose a smooth function $\chi: \mathbb{C}^n \to [0,1]$ such that $\chi = 1$ on $\overline{U}_1$ and $\operatorname{supp}\chi \subset U$. For every $p \in P$ set

$$\alpha_p = \overline{\partial}(\chi f_p) = f_p\,\overline{\partial}\chi = \sum_{i=1}^{n} \alpha_{i,p} d\bar{z}_i.$$

Note that α_p is a smooth $(0,1)$-form on $\mathbb{C}^n$ with compact support contained in $U \setminus U_1$ and depending continuously on $p \in P$. By Theorem 2.7.1 there exists for every $t > 0$ a smooth solution $u_{p,t}: \mathbb{C}^n \to \mathbb{C}$ of the equation $\overline{\partial} u_{p,t} = \alpha_p$ satisfying the estimate

$$\int_{\mathbb{C}^n} \frac{|u_{p,t}|^2}{(1+|z|^2)^2} e^{-t\psi} d\lambda \le \int_U \sum_{i=1}^{n} |\alpha_{i,p}|^2 e^{-t\psi} d\lambda.$$

Moreover, since Hörmander's solution to the $\overline{\partial}$-equation is given by a linear solution operator, we can choose solutions depending continuously on the parameters $p \in P$ and $t \in (0,+\infty)$. In view of (2.9), the right hand side of the above estimate approaches 0 when $t \to +\infty$. Since the weight ψ vanishes in U_0, it follows that

$$\int_{U_0} \frac{|u_{p,t}|^2}{(1+|z|^2)^2} d\lambda \to 0 \quad \text{as } t \to +\infty$$

and the convergence is uniform in $p \in P$. Note that $u_{p,t}$ is a holomorphic function on U_0 since α_p vanishes there. By the interior elliptic estimate (see Lemma 2.7.2 and in particular the estimate (2.7)) it follows that

$$\lim_{t\to+\infty} \sup_{z\in K,\, p\in P} \left|u_{p,t}(z)\right| = 0.$$

Hence, for a sufficiently big $t > 0$ the function

$$F_p = \chi f_p - u_{p,t} : \mathbb{C}^n \to \mathbb{C}, \quad p \in P$$

satisfies the proposition. □

Before proving the general result, we need some preparations. The first one is Michael's *Convex Selection Theorem* [415] which we now present.

Given topological spaces P and B and a set-valued map $\phi: P \to 2^B$ whose values are subsets of B, we say that a map $f: P \to B$ is a *selection* of ϕ if $f(p) \in \phi(p)$ for every $p \in P$. Such ϕ is said to be lower semicontinuous if for every open set $V \subset B$ the set $\{p \in P: \phi(p) \cap V \neq \varnothing\}$ is open in P. The following is a special case of Michael's theorem; a similar result was proved by Cartan [88, Appendix]. We do not prove it here.

Theorem 2.8.2 ([415]) *Assume that B is a Banach space, P is a paracompact Hausdorff space, and $\phi: P \to 2^B$ is a lower semicontinuous set-valued map such that $\phi(p)$ is a nonempty closed convex subset of B for every $p \in P$. For every closed subset P_0 of P and every continuous selection $f: P_0 \to B$ of $\phi|_{P_0}$ there exists a continuous selection $F: P \to B$ of ϕ extending f.*

Denote by $H^\infty(D)$ the Banach space of all bounded holomorphic functions on a complex space D. We need the following lemma [218, Lemma 3.1] on the existence of a linear bounded extension operator.

Lemma 2.8.3 *Assume that X is a reduced Stein space, X' is a closed complex subvariety of X and $\Omega \Subset \Omega'$ are relatively compact Stein domains in X. There exists a bounded linear extension operator*

$$S: H^\infty\big(X' \cap \Omega'\big) \to H^\infty(\Omega)$$

such that $(Sf)(x) = f(x)$ for every $f \in H^\infty(X' \cap \Omega')$ and $x \in X' \cap \Omega$.

Proof We replace X by a relatively compact Stein subdomain containing $\overline{\Omega}'$ and embed it as a closed complex subvariety in a Euclidean space $\mathbb{C}^n$. Since every Stein domain $D \Subset X$ is the intersection $D = X \cap \widetilde{D}$ of X with an open Stein domain $\widetilde{D} \Subset \mathbb{C}^n$, it suffices to prove the lemma for the case $X = \mathbb{C}^n$.

Since Ω' is Stein, the restriction operator $R: \mathcal{O}(\Omega') \to \mathcal{O}(X' \cap \Omega')$ is surjective by Cartan's extension theorem (Corollary 2.6.3). Choose a domain $\Omega_1 \subset \mathbb{C}^n$ such that $\Omega \Subset \Omega_1 \Subset \Omega'$. By the open mapping theorem for Fréchet spaces, the image by R of the set $\{f \in \mathcal{O}(\Omega'): \|f\|_{L^\infty(\Omega_1)} < 1\}$ contains a neighborhood of the origin in $\mathcal{O}(X' \cap \Omega')$. This means that there are a relatively compact subset $Y \Subset X' \cap \Omega'$ and a constant $M < +\infty$ such that every $h \in \mathcal{O}(X' \cap \Omega')$ extends to a function $h' \in \mathcal{O}(\Omega')$ satisfying the estimate

$$\big\|h'\big\|_{L^\infty(\Omega_1)} \le M\|h\|_{L^\infty(Y)}.$$

We may assume that $\Omega_1 \cap X' \subset Y$. The restriction $h'|_{\Omega_1}$, being bounded, belongs to the Bergman space $H = L^2(\Omega_1) \cap \mathcal{O}(\Omega_1)$. Note that H is a Hilbert space containing the closed subspace $H_0 = \{f \in H: f|_{X'} = 0\}$.

Let H_1 be the orthogonal complement of H_0 in H. Projecting h' orthogonally to H_1 gives a function $\tilde{h} \in H_1$ such that $\tilde{h}|_{X' \cap \Omega_1} = h|_{X' \cap \Omega_1}$ and $\tilde{h}$ has the minimal $L^2(\Omega_1)$-norm among all L^2-holomorphic extensions of h to Ω_1. Clearly, such $\tilde{h}$ is unique, and $S: h \to \tilde{h}$ gives a bounded linear operator $S: H^\infty(X' \cap \Omega') \to L^2(\Omega_1)$. By restricting $\tilde{h}$ to $\Omega \Subset \Omega_1$ we get a bounded linear extension operator $S: H^\infty(X' \cap \Omega') \to H^\infty(\Omega)$. □

Theorem 2.8.4 (Cartan-Oka-Weil theorem with parameters) *Let X be a reduced Stein space. Assume that K is an $\mathcal{O}(X)$-convex subset of X, X' is a closed complex subvariety of X, and $P_0 \subset P$ are compact Hausdorff spaces. Let $f: P \times X \to \mathbb{C}$ be a continuous function such that*

(a) *for every $p \in P$, $f(p, \cdot): X \to \mathbb{C}$ is holomorphic on a neighborhood of K (independent of p) and $f(p, \cdot)|_{X'}$ is holomorphic, and*
(b) *$f(p, \cdot)$ is holomorphic on X for every $p \in P_0$.*

Then there exists for every $\epsilon > 0$ a continuous function $F: P \times X \to \mathbb{C}$ satisfying the following conditions:

(i) *$F_p = F(p, \cdot)$ is holomorphic on X for all $p \in P$,*
(ii) *$|F - f| < \epsilon$ on $P \times K$, and*
(iii) *$F = f$ on $(P_0 \times X) \cup (P \times X')$.*

The same result holds for sections of any holomorphic vector bundle over X.

Proof It suffices to show that a function F with the stated properties exists on $P \times D$, where $D \Subset X$ is any given Stein Runge domain in X containing K; the result then follows by an obvious induction over an exhaustion of X. Fix such a domain D and replace X by a relatively compact Stein neighborhood of $\bar{D}$. By Theorem 2.4.1 we can embed this new X as a closed complex subvariety of a Euclidean space $\mathbb{C}^n$. Choose bounded pseudoconvex Runge domains $\Omega \Subset \Omega' \Subset \mathbb{C}^n$ such that $\bar{D} \subset \Omega \cap X$. Lemma 2.8.3 furnishes bounded linear extension operators

$$S: H^\infty\big(X \cap \Omega'\big) \longrightarrow H^\infty(\Omega), \qquad S': H^\infty\big(X' \cap \Omega'\big) \longrightarrow H^\infty(\Omega)$$

such that

$$S(g)|_{X\cap\Omega} = g|_{X\cap\Omega}, \qquad S'(g)|_{X'\cap\Omega} = g|_{X'\cap\Omega}$$

holds for every g is the respective space. With f_p as in the theorem we set

$$h_p = S(f_p|_{X\cap\Omega'}) - S'(f_p|_{X'\cap\Omega'}) \in H^\infty(\Omega), \quad p \in P_0.$$

Then, h_p belongs to the closed subspace of $H^\infty(\Omega)$ defined by

$$H^\infty_{X'}(\Omega) = \big\{h \in H^\infty(\Omega) : h = 0 \text{ on } X' \cap \Omega\big\}.$$

Since these are Banach spaces, Theorem 2.8.2 furnishes a continuous extension of the map $P_0 \to H^\infty_{X'}(\Omega)$, $p \to h_p$, to a map $P \ni p \to \tilde{h}_p \in H^\infty_{X'}(\Omega)$. Set

$$G_p = \tilde{h}_p + S'(f_p|_{X'\cap\Omega'}) \in H^\infty(\Omega), \quad p \in P.$$

We then clearly have

$$G_p|_{X'\cap\Omega} = f_p|_{X'\cap\Omega} \ (\forall p \in P), \qquad G_p|_{X\cap\Omega} = f_p|_{X\cap\Omega} \ (\forall p \in P_0).$$

Thus, the family of holomorphic functions $G_p|_{X\cap\Omega}: X \cap \Omega \to \mathbb{C}$ $(p \in P)$ satisfies conditions (i) and (iii) in the theorem, but not necessarily the approximation condition (ii). However, by continuity there is a small open neighborhood $P_0' \subset P$ of P_0 such that condition (ii) does hold for $p \in P_0'$.

To achieve condition (ii) for all $p \in P$, we proceed as follows. Choose functions $\xi_1, \ldots, \xi_m \in \mathcal{O}(\mathbb{C}^n)$ which generate the sheaf of ideals of the subvariety $X' \subset \mathbb{C}^n$ on the subset $\Omega' \Subset \mathbb{C}^n$. By using Lemma 2.8.3 in exactly the same way as above, we can extend the family of holomorphic functions $\{f_p\}_{p\in P}$ from an open neighborhood of the set K in X to an open Stein Runge domain $\Omega_0 \subset \mathbb{C}^n$ such that $K \subset \Omega_0 \Subset \Omega$. As before, we keep their values on the subvariety $X' \cap \Omega_0$, so we have $G_p = f_p$ on $X' \cap \Omega_0$. Cartan's division theorem (see Corollary 2.6.4) gives

$$G_p = f_p + \sum_{i=1}^{m} g_{i,p}\xi_i \tag{2.10}$$

where $g_{i,p} \in \mathcal{O}(\Omega_0)$ for $p \in P$ and $i = 1, \ldots m$.

We now show that, after shrinking their domain slightly, the families $g_{i,p}$ can be chosen to depend continuously on $p \in P$.

The Oka-Cartan extension theorem (Corollary 2.6.3) shows that the map

$$\begin{aligned} \Phi: \mathcal{O}(\Omega_0)^m \to \mathcal{O}_{X'}(\Omega_0) = \big\{h \in \mathcal{O}(\Omega_0) : h|_{X'} = 0\big\}, \\ \Phi(g_1, \ldots, g_m) = \sum_{i=1}^{m} g_i\xi_i \end{aligned} \tag{2.11}$$

is surjective. Choose a Stein domain $\Omega_1 \subset \mathbb{C}^n$ such that $K \subset \Omega_1 \Subset \Omega_0$. Consider the Hilbert spaces

$$H = L^2(\Omega_1) \cap \mathcal{O}(\Omega_1), \qquad H' = \{h \in H : h|_{X'} = 0\}.$$

Note that (2.11) defines a linear Hilbert space map $\Phi: H \to H'$. Clearly, the functions $(G_p - f_p)|_{\Omega_1}$ $(p \in P)$ belong to H'. Let $g_p = (g_{1,p}, \ldots, g_{m,p}) \in H$ be the unique preimage of $(G_p - f_p)|_{\Omega_1}$ which is orthogonal to $\ker\Phi$; this family is continuous in $p \in P$. Now, apply Proposition 2.8.1 to approximate g_p by a continuous family of holomorphic maps $\tilde g_p = (\tilde g_{1,p}, \ldots, \tilde g_{m,p}) \in \mathcal{O}(\Omega)^m$ $(p \in P)$ and set

$$\widetilde{G}_p = G_p - \sum_{i=1}^{m} \tilde g_{i,p}\xi_i \in \mathcal{O}(\Omega), \quad p \in P.$$

Comparing with (2.10) we see that, if the approximation of g by $\tilde g$ is close enough, the family $\widetilde{G}_p$ satisfies conditions (i)–(iii), except that $\widetilde{G}_p$ need not agree with f_p for $p \in P_0$. This is corrected by choosing a continuous function $\chi: P \to [0,1]$ which equals 1 on P_0 and has support contained in P_0', and setting

$$F_p = \chi(p)G_p + \big(1 - \chi(p)\big)\widetilde{G}_p \in \mathcal{O}(\Omega), \quad p \in P.$$

The family $F_p|_{X\cap\Omega}$ then satisfies Theorem 2.8.4 on the domain $D \subset X$. This completes the proof for functions.

Suppose now that $E \to X$ is a holomorphic vector bundle over X and f_p $(p \in P)$ is a continuous family of sections of E. As before, it suffices to find a family of sections $\{F_p\}_{p \in P}$ satisfying the stated conditions on any given relatively compact open Stein domain $U \Subset X$; the proof is then completed by an induction over an exhaustion of X. Replacing X by such a subset, the bundle $E \to X$ is finitely generated, and hence there exists a surjective holomorphic vector bundle map $\Psi : X \times \mathbb{C}^N \to E$ for some $N \in \mathbb{N}$. Let $E' = \ker \Psi$. By Corollary 2.6.6 we can embed E as a holomorphic vector subbundle of $X \times \mathbb{C}^N$ such that $X \times \mathbb{C}^N = E \oplus E'$, and there is a holomorphic vector bundle projection $\phi : X \times \mathbb{C}^N \to E$ with $\ker \phi = E'$. Hence, sections of E can be seen as maps $X \to \mathbb{C}^N$. Applying the already proved result for functions componentwise and projecting the resulting map $F : P \times X \to \mathbb{C}^N$ back to E by using ϕ completes the proof. □

Chapter 3
Stein Neighborhoods and Approximation

In this chapter we find Stein neighborhoods of certain types of sets in complex manifolds and complex spaces, and we prove several extension and approximation theorems for holomorphic maps. We focus on results that are essential for our subsequent constructions in Oka theory.

We begin in Sect. 3.1 with Siu's theorem that every Stein subvariety in a complex space admits an open Stein neighborhood. The analogous result for q-complete spaces (Theorem 3.1.5), due to Colţoiu and Demailly, is obtained without much additional effort. In Sect. 3.2 we give some extensions of Siu's theorem. An important application in Sect. 3.3 is the existence of local holomorphic retractions onto Stein submanifolds, or a family of such retractions in the foliated case. In Sect. 3.4 we obtain a semiglobal extension theorem for holomorphic sections of holomorphic submersion onto Stein spaces; this result is very important in our proof of the Oka principle over singular base spaces.

In the second part of the chapter we consider Stein compacts with attached totally real submanifolds. We begin in Sect. 3.5 by considering tubular Stein neighborhoods of totally real submanifolds. In Sect. 3.6 we find Stein neighborhoods of certain compact sets in laminated manifolds. In Sect. 3.7 we construct Stein neighborhoods of holomorphically convex compact sets with suitably attached totally real manifolds, and in Sect. 3.8 we prove a Mergelyan type approximation theorem on such objects. In Sect. 3.9 we construct thin strongly pseudoconvex handlebodies around the union of a strongly pseudoconvex domain with a totally real submanifold. This result is used in the Eliashberg-Gompf construction of Stein manifold structures on smooth manifolds with the correct handlebody decomposition (see Chap. 10). In Sects. 3.10–3.11 we analyze the geometry of a q-convex function at a Morse critical point. Finally, in Sect. 3.12 we describe the topological structure of Stein spaces and of q-complete complex spaces.

3.1 Q-Complete Neighborhoods

This and the following section are devoted to the proof and some generalizations of the following seminal result of Y.-T. Siu from 1976.

F. Forstnerič, *Stein Manifolds and Holomorphic Mappings*, Ergebnisse der Mathematik und ihrer Grenzgebiete. 3. Folge / A Series of Modern Surveys in Mathematics 56, DOI 10.1007/978-3-319-61058-0_3

Theorem 3.1.1 (Siu [504]) *Every locally closed Stein subvariety Y of any complex space X admits a basis of open Stein neighborhoods in X.*

The complex space X in Siu's theorem is not necessarily reduced. After a simple reduction to the case in which X is reduced and finite dimensional, the first step consists in showing that every compact subset K of the subvariety Y admits an open Stein neighborhood in X; this was proved independently by Schneider [487]. Here, we present a proof, due to Colţoiu [102] and Demailly [113, Theorem 1], which also applies to q-complete subvarieties for any integer $q \geq 1$; Siu's theorem corresponds to the case $q = 1$.

We begin by recalling the notion of a q-convex function and of a q-convex and q-complete complex space. For more information, see Andreotti and Grauert [27] and the surveys by Colţoiu [103] and Grauert [256].

Definition 3.1.2 Let X be a complex manifold. A function $\rho\colon X \to \mathbb{R}$ of class $\mathcal{C}^2$ is *q-convex* if its Levi form $\mathcal{L}_\rho$ (1.39) has at most $q-1$ negative or zero eigenvalues at each point of X.

Equivalently, ρ is q-convex if its Levi form has at least $\dim X - q + 1$ positive eigenvalues at every point.

Definition 3.1.3 Let X be a possibly nonreduced complex space.

(a) A function $\phi\colon X \to \mathbb{R}$ is *q-convex* if there exists a cover of X by open sets A_λ, isomorphic to closed analytic sets in domains $\Omega_\lambda \subset \mathbb{C}^{N_\lambda}$, such that for each λ the restriction $\phi|_{A_\lambda}$ admits a q-convex extension to Ω_λ.
(b) The space X is *q-convex* if it admits an exhaustion function $\rho\colon X \to \mathbb{R}_+$ that is q-convex on $\{\rho > c\}$ for some $c > 0$; if ρ can be chosen q-convex on all of X then X is *q-complete.*

Remark 3.1.4 By a theorem of Narasimhan [425] (see Theorem 2.5.2 on p. 53), a complex space X is 1-complete if and only if it is Stein. (For complex manifolds, this was proved by Grauert [255].) Furthermore, a 1-convex complex space X contains a maximal compact complex subvariety of positive dimension, S, called the *exceptional variety* of X. There is a proper holomorphic map $\pi : X \to \widetilde{X}$ onto a Stein space $\widetilde{X}$ that maps each connected component of S to a point in $\widetilde{X}$ such that $\pi : X \setminus \pi^{-1}(\pi(S)) \to \widetilde{X} \setminus \pi(S)$ is biholomorphic and the composition operator $\mathcal{O}(\widetilde{X}) \ni f \mapsto f \circ \pi \in \mathcal{O}(X)$ induces an isomorphism of $\mathcal{O}(\widetilde{X})$ onto $\mathcal{O}(X)$. The Stein space $\widetilde{X}$ together with the map $\pi : X \to \widetilde{X}$ is the *Remmert reduction* of X.

Similarly, Vâjâitu proved [557] that the following conditions are equivalent for a finite dimensional holomorphically convex complex space X:

1. X is q-complete;
2. X is cohomologically q-complete;
3. X does not contain compact analytic subsets of dimension $\geq q$.

Every purely n-dimensional complex space without compact irreducible components is n-convex (see Ohsawa [439] and Greene and Wu [263]). □

Theorem 3.1.1 corresponds to the case $q = 1$ of the following result due to Colţoiu [102] and Demailly [113].

Theorem 3.1.5 *A q-complete complex subvariety Y of a complex space X has a basis of open q-complete neighborhoods in X.*

By essentially the same proof, Colţoiu [102, Theorem 2] obtained Theorem 3.1.5 under the weaker condition that Y is a complete locally pluripolar set in X which is *ambiently q-complete*, in the sense that there exist an open neighborhood U of Y in X and a q-convex function ϕ on U such that $\phi|_Y$ is an exhaustion function on Y. We begin by showing that a q-complete complex subvariety is ambiently q-complete.

Lemma 3.1.6 ([113, Theorem 4]) *Let Y be an analytic subvariety of a complex space X and let ψ be a smooth q-convex function on Y. For every continuous function $\delta > 0$ on Y there exists a smooth q-convex function ϕ on a neighborhood V of Y in X such that $\psi \le \phi|_Y < \psi + \delta$.*

Proof Replacing X by a suitable open neighborhood of Y, we may assume that Y is closed in X. Choose a stratification

$$\varnothing = Y_{-1} \subset Y_0 \subset Y_1 \subset \cdots \subset \bigcup_k Y_k = Y$$

by closed complex subvarieties such that $S_k = Y_k \setminus Y_{k-1}$ is a complex k-dimensional submanifold of Y for every k. (We may let Y_k be the union of all irreducible components of dimension $\le k$ of one of the varieties Y, Y_{sing}, $(Y_{\mathrm{sing}})_{\mathrm{sing}}$, etc.) It suffices to prove the following statement by induction on k:

(*) *There exists a smooth function ϕ_k on X such that $\phi_k|_Y$ is q-convex, ϕ_k is q-convex on a neighborhood $\overline{V}_k$ of Y_k in X, and $\psi \le \phi_k|_Y < \psi + \delta$.*

Any smooth extension ϕ_{-1} of ψ to X satisfies this condition with $Y_{-1} = V_{-1} = \varnothing$. Assume inductively that V_{k-1} and ϕ_{k-1} have been chosen. The closed set $Y_k \setminus V_{k-1} \subset S_k = Y_k \setminus Y_{k-1}$ admits a locally finite cover by open coordinate patches $A_\lambda \subset \Omega_\lambda \subset \mathbb{C}^{N_\lambda}$ as in Definition 3.1.3 (p. 66) such that $Y_k \cap A_\lambda$ is given by the equations $z'_\lambda = (z_{\lambda,k+1}, \ldots, z_{\lambda,N_\lambda}) = 0$. (Here, $z_\lambda = (z_{\lambda,1}, \ldots, z_{\lambda,N_\lambda})$ are the complex coordinates on $\mathbb{C}^{N_\lambda}$.) Let $\theta_\lambda \colon A_\lambda \to [0, 1]$ be smooth functions with compact support in A_λ such that $\sum_\lambda \theta_\lambda = 1$ on $Y_k \setminus V_{k-1}$. For $\epsilon_\lambda > 0$ chosen small enough, the function

$$\phi_k = \phi_{k-1} + \sum_\lambda \theta_\lambda \, \epsilon_\lambda^3 \log\big(1 + \epsilon_\lambda^{-4} |z'_\lambda|^2\big)$$

satisfies $\psi \le \phi_{k-1}|_Y \le \phi_k|_Y < \psi + \delta$. We now check that ϕ_k is q-convex along Y and on $\overline{V}_{k-1}$ (where ϕ_{k-1} was such), and that ϕ_k is also q-convex near every point

$x_0 \in Y_k \setminus V_{k-1}$. If x_0 does not belong to the support of θ_λ for any λ, then $\phi_k = \phi_{k-1}$ in a neighborhood of x_0 and this holds. Assume now that $x_0 \in \operatorname{Supp} \theta_\mu$ for some μ. Pick a neighborhood $W \Subset \Omega_\mu$ of x_0 such that the following conditions hold:

(a) If $x_0 \subset Y_k \setminus V_{k-1} \subset S_k$, then $\theta_\mu(x_0) > 0$ and $A_\mu \cap W \Subset \{\theta_\mu > 0\}$.
(b) If $x_0 \in A_\lambda$ for some λ (there are finitely many such λ's), then $A_\mu \cap W \Subset A_\lambda$ and $z_\lambda|_{A_\mu \cap W}$ admits a holomorphic extension $\widetilde{z}_\lambda$ to $\overline{W}$.
(c) If $x_0 \in \overline{V}_{k-1}$, then $\phi_{k-1}|_{A_\mu \cap W}$ admits a q-convex extension $\widetilde{\phi}_{k-1}$ to $\overline{W}$.
(d) If $x_0 \in Y \setminus \overline{V}_{k-1}$, then $\phi_{k-1}|_{Y \cap W}$ admits a q-convex extension $\widetilde{\phi}_{k-1}$ to $\overline{W}$.

If $x \notin Y \cup \overline{V}_{k-1}$, we take an arbitrary smooth extension $\widetilde{\phi}_{k-1}$ of $\phi_{k-1}|_{A_\mu \cap W}$ to $\overline{W}$. Let $\widetilde{\theta}_\lambda$ denote a smooth extension of $\theta_\lambda|_{A_\mu \cap W}$ to $\overline{W}$. The function

$$\widetilde{\phi}_k = \widetilde{\phi}_{k-1} + \sum_\lambda \widetilde{\theta}_\lambda \, \epsilon_\lambda^3 \log\big(1 + \epsilon_\lambda^{-4} |\widetilde{z}'_\lambda|^2\big)$$

is then a smooth extension of $\phi_k|_{A_\mu \cap W}$ to $\overline{W}$, resp. of $\phi_k|_{Y \cap W}$ to $\overline{W}$ in case (d). Since the function $\log(1 + \epsilon_\lambda^{-4}|\widetilde{z}'_\lambda|^2)$ is plurisubharmonic and

$$\partial \log\big(1 + \epsilon_\lambda^{-4} |\widetilde{z}'_\lambda|^2\big) = \big(\epsilon_\lambda^4 + |\widetilde{z}'_\lambda|^2\big)^{-1} \sum_{j=k+1}^{N_\lambda} \overline{\widetilde{z}}_{\lambda,j} \, \partial \widetilde{z}_{\lambda,j}$$

is bounded by $O(\epsilon_\lambda^{-2})$, we obtain

$$\mathrm{i}\partial\overline{\partial}\widetilde{\phi}_k \ge \mathrm{i}\partial\overline{\partial}\widetilde{\phi}_{k-1} - O\Big(\sum \epsilon_\lambda\Big).$$

Therefore, for sufficiently small numbers $\epsilon_\lambda > 0$ the function $\widetilde{\phi}_k$ is q-convex on $\overline{W}$ in cases (c) and (d). Since all functions $\widetilde{z}'_\lambda$ vanish along $Y_k \cap W$, the following estimates hold at every point of $Y_k \cap W$:

$$\mathrm{i}\partial\overline{\partial}\widetilde{\phi}_k \ge \mathrm{i}\partial\overline{\partial}\widetilde{\phi}_{k-1} + \sum_\lambda \theta_\lambda \epsilon_\lambda^{-1} \, \mathrm{i}\partial\overline{\partial}|\widetilde{z}'_\lambda|^2 \ge \mathrm{i}\partial\overline{\partial}\widetilde{\phi}_{k-1} + \theta_\mu \epsilon_\mu^{-1} \, \mathrm{i}\partial\overline{\partial}|\widetilde{z}'_\mu|^2.$$

Since $S_k \subset Y$, the $(1,1)$-form $\mathrm{i}\partial\overline{\partial}\widetilde{\phi}_{k-1}$ has at most $q-1$ nonpositive eigenvalues on TS_k, whereas $\mathrm{i}\partial\overline{\partial}|\widetilde{z}'_\lambda|^2$ is positive definite in the normal directions to S_k; hence in case (a) the function $\widetilde{\phi}_k$ is q-convex on $\overline{W}$ for small enough $\epsilon_\mu > 0$. Note that only finitely many conditions are required on each number ϵ_λ if we choose a locally finite cover of $\bigcup_\lambda \operatorname{Supp} \theta_\lambda$ by neighborhoods W as above. Hence, for numbers $\epsilon_\lambda > 0$ chosen small enough, ϕ_k is q-convex on a neighborhood $\overline{V}'_k$ of $Y_k \setminus V_{k-1}$. The function ϕ_k and the set $V_k = V_{k-1} \cup V'_k$ then satisfy (*) and the induction step is complete. □

The second main ingredient are *almost plurisubharmonic functions* with a logarithmic pole along a closed complex subvariety.

Definition 3.1.7 A function $v\colon X \to \mathbb{R} \cup \{-\infty\}$ on a complex manifold X is *almost plurisubharmonic* if every point $x_0 \in X$ admits an open neighborhood $U \subset X$ such that $v|_U = w_U + h_U$, where w_U is plurisubharmonic on U and h_U is smooth on U. A function on a complex space is almost plurisubharmonic if it can be locally extended to an almost plurisubharmonic function on the ambient manifold of a local holomorphic embedding.

An equivalent definition is that v is almost plurisubharmonic if it is upper semicontinuous and its Levi form has a bounded negative part on each compact subset of X (i.e., its eigenvalues are bounded away from $-\infty$).

Lemma 3.1.8 ([102, Theorem 1], [113, Lemma 5]) *For every closed analytic subvariety Y in a complex space X there exists an almost plurisubharmonic function v on X which is smooth on $X \setminus Y$ and has a logarithmic pole on $Y = \{v = -\infty\}$.*

Proof Let $\mathcal{J}_Y$ denote the sheaf of ideals of Y in X. There is a locally finite cover of X by patches A_λ, isomorphic to closed analytic subsets $A_\lambda \subset \mathbb{B}_\lambda$ in unit balls $\mathbb{B}_\lambda \subset \mathbb{C}^{N_\lambda}$, such that $\mathcal{J}_Y$ admits a finite system of generators $g_\lambda = (g_{\lambda,j})$ on an open neighborhood of each set $\bar{A}_\lambda$ in X. Let z denote the complex coordinate on $\mathbb{B}_\lambda$ and $z(x)$ its value at $x \in A_\lambda$. The function

$$v_\lambda(x) = \log\left|g_\lambda(x)\right|^2 - \frac{1}{1 - |z(x)|^2}, \qquad x \in A_\lambda$$

is almost plurisubharmonic on A_λ since the first summand is plurisubharmonic and the second one is smooth. Note that $v_\lambda = -\infty$ on $(A_\lambda \cap Y) \cup (\bar{A}_\lambda \cap bB_\lambda)$. We patch the functions v_λ into an almost plurisubharmonic function v on X by using a *regularized maximum function* defined as follows. Select a nonnegative smooth function ρ on $\mathbb{R}$, with support in $[-\frac{1}{2}, \frac{1}{2}]$, such that $\int \rho(u)\mathrm{d}u = 1$ and $\int_{\mathbb{R}} u\rho(u)\mathrm{d}u = 0$, and set

$$\operatorname{rmax}\{t_1, \ldots, t_p\} = \int_{\mathbb{R}^p} \max\{t_1 + u_1, \ldots, t_p + u_p\} \prod_{j=1,\ldots,p} \rho(u_j)\,\mathrm{d}u_j. \tag{3.1}$$

Since rmax is clearly increasing and convex in all variables, the composition with rmax preserves plurisubharmonicity. Moreover, we have

$$\operatorname{rmax}\{t_1, \ldots, t_j, \ldots, t_p\} = \operatorname{rmax}\{t_1, \ldots, t_{j-1}, t_{j+1}, \ldots, t_p\}$$

as soon as $t_j < \max\{t_1, \ldots, t_{j-1}, t_{j+1}, \ldots, t_p\} - 1$. Set

$$v(x) = \operatorname{rmax}\left\{v_\lambda(x) : x \in A_\lambda\right\}, \qquad x \in X.$$

Near every point x there are only finitely many terms under rmax as the cover A_λ is locally finite. Since the generators g_λ and g_μ for $\mathcal{J}_Y$ can be expressed in terms of each other in a neighborhood of $\bar{A}_\lambda \cap \bar{A}_\mu$, the quotient $|g_\lambda|/|g_\mu|$ is bounded on this set, and hence none of the values $v_\lambda(x)$ for $x \in A_\lambda$ sufficiently near bA_λ (where v_λ is large negative) contributes to the value of $v(x)$. It follows that v is smooth on $X \setminus Y$, has a logarithmic pole on $Y = \{v = -\infty\}$, and is almost plurisubharmonic on X since it is locally the regularized maximum of finitely many almost plurisubharmonic functions v_λ. □

Proof of Theorem 3.1.5 By Lemma 3.1.6, applied to a q-convex exhaustion of Y and $\delta = 1$, there exists a q-convex function ρ on a neighborhood W_0 of Y in X such that $\rho|_Y$ is an exhaustion of Y. Choose a neighborhood W of Y such that $\overline{W} \subset W_0$ and $\rho|_{\overline{W}}$ is an exhaustion (every sufficiently thin neighborhood of Y satisfies these requirements). Let v be an almost plurisubharmonic function on X with a logarithmic pole along $Y = \{v = -\infty\}$, furnished by Lemma 3.1.8. Choose a smooth convex increasing function $g\colon \mathbb{R} \to \mathbb{R}$ and set

$$\widetilde{\rho} = v + g \circ \rho, \qquad V = \big\{x \in W : \widetilde{\rho}(x) < 0\big\}.$$

Note that $Y \subset V$ since $v|_Y = -\infty$. Choosing g to grow fast enough, we get $\widetilde{\rho} > 0$ on bW and hence $\overline{V} \subset W$. Fast growth of g also ensures that the positive Levi form of $g \circ \rho$ on a suitable family of $(q-1)$-codimensional subspaces of the tangent bundle (on which $dd^c \rho > 0$) compensates the locally bounded negative part of the Levi form of v. For such g, the function $\widetilde{\rho}$ is q-convex on W. Let $\tau\colon (-\infty, 0) \to \mathbb{R}_+$ be a smooth convex increasing function such that $\tau(t) = 0$ for $t \le -1$ and $\lim_{t\to 0} \tau(t) = +\infty$. Then, the function

$$\psi = \rho + \tau \circ \widetilde{\rho}\colon V \to \mathbb{R}$$

is q-convex exhaustion of V, and hence V is q-complete. □

3.2 Stein Neighborhoods of Stein Subvarieties

In this section we generalize Siu's theorem (Theorem 3.1.1 on p. 66) to the form that will be used in Oka theory in Chap. 5. Recall that a compact set K in a complex space X is holomorphically convex if K is $\mathcal{O}(\Omega)$-convex in an open Stein neighborhood $\Omega \subset X$ (see Definition 2.3.4 on p. 51).

Theorem 3.2.1 ([195, Theorem 1.2]) *Let Y be a closed Stein subvariety of a complex space X. Assume that K is a compact holomorphically convex set in X such that $K \cap Y$ is $\mathcal{O}(Y)$-convex. Then $K \cup Y$ has a basis of open Stein neighborhoods $V \subset X$ such that K is $\mathcal{O}(V)$-convex.*

Remark 3.2.2 The necessity of $\mathcal{O}(Y)$-convexity of $K \cap Y$ is seen by taking $X = \mathbb{C}^2$, $Y = \mathbb{C} \times \{0\}$, and $K = \{(z,w) \in \mathbb{C}^2 : 1 \le |z| \le 2,\ |w| \le 1\}$. It is immediate that every Stein neighborhood of $K \cup Y$ in $\mathbb{C}^2$ also contains the bidisc $\{(z,w) \in \mathbb{C}^2 : |z| \le 2,\ |w| \le 1\}$. □

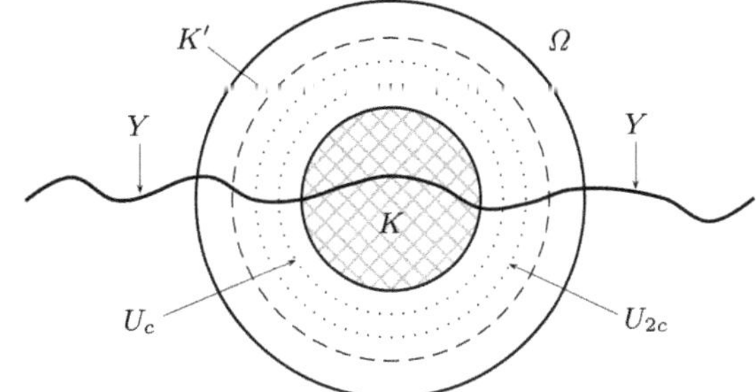

Fig. 3.1 Sets in the proof of Theorem 3.2.1

Proof Let U be an open set in X containing $K \cup Y$. We shall find an open Stein set V in X such that $K \cup Y \subset V \subset U$ and K is $\mathcal{O}(V)$-convex.

By the assumption, K has an open Stein neighborhood $\Omega \subset X$ such that K is $\mathcal{O}(\Omega)$-convex. Hence, K has a basis of open Stein neighborhoods; replacing Ω by one of them, we may assume that $\Omega \subset U$.

Since the set $K_0 = K \cap X_0$ is $\mathcal{O}(Y)$-convex, it has a compact $\mathcal{O}(Y)$-convex neighborhood K_0' in Y that is contained in Ω. Choose a compact neighborhood K' of K such that $K' \subset \Omega$ and $K' \cap Y = K_0'$. As K is $\mathcal{O}(\Omega)$-convex, Proposition 2.5.1 furnishes a smooth strongly plurisubharmonic function ρ_0 on Ω such that $\rho_0 < 0$ on K and $\rho_0 > 1$ on $\Omega \setminus K'$. Set $U_c = \{x \in \Omega : \rho_0(x) < c\}$. Fix $c \in (0, \frac{1}{2})$; then $K \subset U_c \subset U_{2c} \subset K'$ (see Fig. 3.1).

The restriction $\rho_0|_{Y \cap \Omega}$ is smooth and strongly plurisubharmonic. Since $K_0' = K' \cap Y$ is $\mathcal{O}(Y)$-convex, there is a smooth strongly plurisubharmonic exhaustion function $\rho_0' \colon Y \to \mathbb{R}$ that agrees with ρ_0 on K_0' and satisfies $\rho_0' > c$ on $Y \setminus \overline{U}_c$. (To obtain such ρ_0', take a smooth strongly plurisubharmonic exhaustion function $\tau \colon Y \to \mathbb{R}$ such that $\tau < 0$ on K_0' and $\tau > 1$ on $Y \setminus \Omega$; also, choose a smooth convex increasing function $\xi \colon \mathbb{R} \to \mathbb{R}_+$ with $\xi(t) = 0$ for $t \le 0$, and a smooth function $\chi \colon X \to [0, 1]$ such that $\chi = 1$ on $\{x \in Y : \tau(x) \le 1/2\}$ and $\chi = 0$ on $\{x \in Y : \tau(x) \ge 1\}$. The function $\rho_0' = \chi\rho_0 + \xi \circ \tau$ then satisfies the stated properties, provided that $\xi(t)$ is chosen to grow sufficiently fast for $t > 0$.) Let $\widetilde{\rho}_0 \colon K' \cup Y \to \mathbb{R}$ be defined by $\widetilde{\rho}_0|_{K'} = \rho_0|_{K'}$ and $\widetilde{\rho}_0|_Y = \rho_0'$.

Choose a smooth convex increasing function $h \colon \mathbb{R} \to \mathbb{R}$ satisfying $h(t) \ge t$ for all $t \in \mathbb{R}$, $h(t) = t$ for $t \le c$, and $h(t) > t + 1$ for $t \ge 2c$. The function $\rho_1 := h \circ \widetilde{\rho}_0$ is smooth strongly plurisubharmonic on $K' \cup Y$; on the set $\overline{U}_c = \{\rho_0 \le c\}$ we have $\rho_1 = \widetilde{\rho}_0 = \rho_0$, while outside of U_{2c} we have $\rho_1 > \widetilde{\rho}_0 + 1$.

By Lemma 3.1.6, there exists a smooth strongly plurisubharmonic function ρ_2 in an open neighborhood of Y in X satisfying

$$\rho_1(x) - 1 < \rho_2(x) < \rho_1(x), \quad x \in Y.$$

On a neighborhood of $\overline{U}_c \cap Y = \{x \in Y : \widetilde{\rho}_0(x) \le c\}$ we have $\rho_2 < \rho_1 = \widetilde{\rho}_0$, while on $Y \setminus U_{2c}$ we have $\rho_2 > \rho_1 - 1 > \widetilde{\rho}_0$. It follows that the function

$$\rho = \max\{\widetilde{\rho}_0, \rho_2\}$$

is well defined and strongly plurisubharmonic in an open set $W \subset X$ satisfying $\overline{U}_c \cup Y \subset W \subset U$. (To see this, note that the union of the domains of $\widetilde{\rho}_0$ and ρ_2 contains a neighborhood of $\overline{U}_c \cup Y$, and before running out of the domain of one of these two functions, the second function is the larger one and hence takes over.) After shrinking W around the set $\overline{U}_c \cup Y$, the function $\rho\colon W \to \mathbb{R}$ satisfies the following conditions:

(i) $\rho = \widetilde{\rho}_0 = \rho_0$ on $\overline{U}_c$ (hence $\rho < 0$ on K),
(ii) $\rho > c$ on $W \setminus \overline{U}_c$, and
(iii) $\rho = \rho_2$ on $W \setminus U_{2c}$.

Using rmax instead of max (see the proof of Theorem 3.1.5), we can also ensure that ρ is smooth. After shrinking W we may assume that ρ is a strongly plurisubharmonic exhaustion of $\overline{W} \supset \overline{U}_c \cup Y$ satisfying $\rho > c$ on bW. The compact set $L = \{x \in W : \rho(x) \le 0\}$ contains K in its interior.

Let v be an almost plurisubharmonic function on X with a logarithmic pole on $Y = \{v = -\infty\}$, furnished by Lemma 3.1.8. By subtracting a constant from v, we may assume that $v|_K < 0$. Let $g\colon \mathbb{R} \to \mathbb{R}$ be a convex increasing function with $g(t) = t$ for $t \le 0$. For a small $\epsilon > 0$ we set

$$\widetilde{\rho} = \epsilon v + g \circ \rho, \qquad V = \big\{x \in W : \widetilde{\rho}(x) < 0\big\}.$$

Clearly, $\widetilde{\rho}|_Y = -\infty$ and hence $Y \subset V$. As $v|_K < 0$ and $g \circ \rho|_K = \rho|_K < 0$, we have $K \subset V$. To complete the proof, we show that for suitable choices of ϵ and g the set V is Stein and K is $\mathcal{O}(V)$-convex.

On the set $L = \{\rho \le 0\}$ we have $g \circ \rho = \rho$ which is strongly plurisubharmonic. Since the Levi form of v is bounded on L, $\widetilde{\rho}$ is strongly plurisubharmonic on L for a sufficiently small $\epsilon > 0$; fix such ϵ. By choosing g to grow sufficiently fast on $(0, +\infty)$, we can ensure that the Levi form of $g \circ \rho$ compensates the bounded negative part of the Levi form of ϵv on each compact in $\overline{W}$, and hence $\widetilde{\rho}$ is strongly plurisubharmonic on $\overline{W}$. Furthermore, rapid growth of g ensures that $\widetilde{\rho}|_{bW} > 0$, and hence $\overline{V} \subset W$.

Let $\tau\colon (-\infty, 0) \to \mathbb{R}_+$ be a smooth convex increasing function such that $\tau(t) = 0$ for $t \le -1$ and $\lim_{t\to 0} \tau(t) = +\infty$. Then, $\psi = \rho + \tau \circ \widetilde{\rho}$ is a strongly plurisubharmonic exhaustion function on V, and hence V is Stein according to Theorem 2.5.2 on p. 53.

It remains to see that V can be chosen such that K is $\mathcal{O}(V)$-convex. Since K is $\mathcal{O}(\Omega)$-convex, there is a plurisubharmonic function $\theta\colon \Omega \to \mathbb{R}_+$ that vanishes precisely on K and is strongly plurisubharmonic on $\Omega \setminus K$ (Proposition 2.5.1 on p. 53). As before, we find an open set $W \subset X$ containing $K \cup Y$ and a smooth plurisubharmonic function $\widetilde{\theta}\colon W \to \mathbb{R}_+$ such that, for some $c > 0$, we have $\theta = \widetilde{\theta}$ in $O_c = \{x \in \Omega : \theta(x) < c\} \supset K$ and $\widetilde{\theta} > c$ on $W \setminus \overline{O}_c$. (See (i) and (ii) above.) The first part of the proof gives an open Stein domain V with $K \cup Y \subset V \subset W$. The restriction $\widetilde{\theta}|_V \ge 0$ is then a plurisubharmonic function on V that vanishes precisely on K, so K is $\mathcal{O}(V)$-convex by Corollary 2.5.3. □

A related result was obtained by J. Prezelj [461]; it concerns the following situation. Let X be a 1-convex complex space with the exceptional subvariety S (the maximal compact complex subvariety of positive dimension in X). Assume that $h\colon Z \to X$ is a holomorphic submersion of a complex space Z onto X and $f\colon X \to Z$ is a holomorphic section. The subvariety $f(X) \subset Z$ does not have a Stein neighborhood in Z since it contains the compact subvariety $f(S)$. However, if $g \in \mathcal{O}(X)$ is a holomorphic functions that vanishes on S, then $X_g = X \setminus \{g = 0\}$ is an open Stein subset of X, and hence $f(X_g)$ admits an open Stein neighborhood $\Omega_g \subset Z$ by Siu's theorem (Theorem 3.1.1 on p. 66). Call such a neighborhood Ω_g *conic along* $g^{-1}(0)$ if its diameter in the fibre direction decays at most as a power of $|g(x)|$ as the point $x \in X$ approaches the zero locus $g^{-1}(0)$. Over any compact subset of X this notion is independent of the choice of a distance function on Z.

Theorem 3.2.3 ([461, Theorem 3.2] (Hypotheses as above)) *For every compact $\mathcal{O}(X)$-convex subset K of X there exists an open Stein neighborhood Ω_g of $f(K \setminus g^{-1}(0))$ in Z which is conic along $g^{-1}(0)$.*

Note that finitely many functions $g_1, \ldots, g_k \in \mathcal{O}(X)$ define the exceptional set S; the corresponding conic Stein neighborhoods Ω_{g_j} then cover $f(X \setminus S)$.

3.3 Holomorphic Retractions onto Stein Submanifolds

In this section we prove the Docquier-Grauert tubular neighborhood theorem for Stein submanifolds in arbitrary complex manifolds (Theorem 3.3.3).

Proposition 3.3.1 *Let X be a Stein manifold. There exist an open Stein neighborhood Ω of the zero section in the tangent bundle $\mathrm{T}X$ and a holomorphic map $s\colon \Omega \to X$ that maps the fibre Ω_x over any point $x \subset X$ biholomorphically onto a neighborhood of x in X, with $s(0_x) = x$. It is possible to choose Ω to be Runge in $\mathrm{T}X$ and to have convex fibres.*

Proof By Cartan's Theorem B (see Theorem 2.6.1) there are finitely many holomorphic vector fields $V_1, \ldots, V_N$ on X spanning the tangent space at every point. Let ϕ_t^j denote the local flow of V_j for complex values of t (1.56). The map

$$F(x, t_1, \ldots, t_N) = \phi_{t_1}^1 \circ \cdots \circ \phi_{t_N}^N(x), \quad x \in X,\ t_j \in \mathbb{C} \tag{3.2}$$

is defined and holomorphic in an open neighborhood of $X \times \{0\}^N$ in $X \times \mathbb{C}^N$, and it satisfies $F(x, 0) = x$ and $\frac{\partial}{\partial t_j} F(x, t)|_{t=0} = V_j(x)$ for every $x \in X$ and $j = 1, \ldots, N$. As the V_j's span the tangent space $\mathrm{T}_x X$ at every point, the partial differential

$$\Theta_x = \partial_t|_{t=0} F(x, t) : \mathbb{C}^N \to \mathrm{T}_x X \tag{3.3}$$

is surjective for every $x \in X$. Hence, the subset $E' \subset X \times \mathbb{C}^N$ with fibres $E'_x = \ker \Theta_x$ is a holomorphic vector subbundle of $X \times \mathbb{C}^N$. Since X is Stein, we have $X \times \mathbb{C}^N = E \oplus E'$ for a holomorphic vector subbundle E of $X \times \mathbb{C}^N$. Then $\Theta_x \colon E_x \to \mathrm{T}_x X$ is bijective for every $x \in X$, and hence $E \cong \mathrm{T}X$; we identify E with $\mathrm{T}X$. The implicit function theorem gives an open neighborhood $\Omega \subset \mathrm{T}X$ of the zero section such that the restricted map $s = F|_\Omega$ satisfies the conclusion of the proposition. It is possible to choose Ω to be Stein, Runge in $\mathrm{T}X$, and to have convex fibres by taking $\Omega_x = \{v \in E_x : \mathrm{e}^{\phi(x)}|v|^2 < 1\}$, where $\phi \colon X \to \mathbb{R}_+$ is a fast growing strongly plurisubharmonic function. □

Essentially the same proof gives the following result.

Proposition 3.3.2 *Assume that $\pi : Z \to X$ is a holomorphic submersion of a complex space Z onto a Stein space X. There is a Stein neighborhood Ω of the zero section in the vertical tangent bundle* $\mathrm{VT}Z = \ker \mathrm{d}\pi$ *(see p. 24) and a holomorphic map $s \colon \Omega \to Z$ such that the following hold for every $z \in Z$:*

(i) *s maps the fibre Ω_z of Ω over z biholomorphically onto a neighborhood of z in the fibre $Z_{\pi(z)}$, and*
(ii) *$s(0_z) = z$. (Here, 0_z is the origin in* $\mathrm{VT}_z Z$.)

We may choose Ω to be Runge in $\mathrm{VT}Z$ *and to have convex fibres. (A map s with these properties is a* local fibre dominating spray; *see Definition* 5.9.1.)

As a consequence of Proposition 3.3.1 and of Siu's theorem (see Theorem 3.1.1) we obtain the following *tubular neighborhood theorem* of Docquier and Grauert ([128], [274, p. 257, Theorem 8]).

Theorem 3.3.3 (Tubular neighborhood theorem) *Let S be a Stein submanifold of a complex manifold X. Denote by $\mathrm{N}_{S/X}$ the normal bundle of S in X. There exist an open Stein neighborhood U of S in X, biholomorphic to an open neighborhood Ω of the zero section in $\mathrm{N}_{S/X}$, and a homotopy of holomorphic maps $\iota_t \colon U \to U$ ($t \in [0,1]$) such that ι_0 is the identity map on U, $\iota_t|_S$ is the identity map on S for all $t \in [0,1]$, and $\iota_1(U) = S$.*

The family $\{\iota_t\}_{t\in[0,1]}$ is a deformation retraction of U onto S consisting of holomorphic mappings.

Remark 3.3.4 There is no analogue of this result when S has singularities. More precisely, if Theorem 3.3.3 holds locally for a normal Stein subvariety S of a complex manifold X, then S must be smooth according to a result of Rossi [479, Theorem 7.1]. □

Proof By Corollary 3.1.1 we may assume that X is Stein. Choose vector fields $V_1, \ldots, V_N$ as in the proof of Proposition 3.3.1 and let F and Θ_x be defined by

(3.2), (3.3), respectively. The subset $E' \subset S \times \mathbb{C}^N$ with the fibres

$$E'_x = \Theta_x^{-1}(\mathrm{T}_x S), \quad x \in S$$

is a holomorphic vector subbundle of $S \times \mathbb{C}^N$. Since S is Stein, there is holomorphic vector subbundle $E \subset S \times \mathbb{C}^N$ such that $S \times \mathbb{C}^N = E \oplus E'$. Then, $\Theta\colon E \to \mathrm{T}X|_S$ is an injective holomorphic vector bundle map such that $\mathrm{T}X|_S = \mathrm{T}S \oplus \Theta(E)$. Thus, $\Theta\colon E \xrightarrow{\cong} \mathrm{N}_{S/X}$ is an isomorphism of E onto the normal bundle of S in X. By the inverse function theorem, F (3.2) maps an open neighborhood Ω of the zero section in E biholomorphically onto an open neighborhood $U = F(\Omega)$ of S in X. Hence, F conjugates the family of radial dilations

$$\iota'_t(v) = (1-t)v, \quad v \in \Omega,\ t \in [0,1]$$

to a family of holomorphic maps $\iota_t\colon U \to U$ satisfying the stated properties. As in the proof of Proposition 3.3.1, we can choose the domain Ω to be Stein, Runge in $\mathrm{N}_{S/X}$, and with convex fibres. □

Similarly one proves the following foliated version of Theorem 3.3.3.

Theorem 3.3.5 ([202, Proposition 3.2]) *Assume that $\pi\colon Z \to X$ is a holomorphic submersion of a complex space Z onto a complex space X. Let S be a locally closed Stein subvariety of Z whose fibres $S_x = S \cap Z_x$ are complex manifolds of dimension independent of $x \in X$. Then, there exist an open Stein neighborhood U of S in Z and a homotopy of holomorphic maps $\iota_t\colon U \to U$ satisfying conditions* (a)–(c) *in Theorem* 3.3.3 *and $\pi \circ \iota_t = \pi$ $(t \in [0,1])$. In particular, $\iota_1\colon U \to S$ is a fibre preserving holomorphic retraction of U onto S.*

3.4 A Semiglobal Holomorphic Extension Theorem

In this section we prove the following semiglobal extension and approximation theorem that will be used in Oka theory in Chap. 5.

Theorem 3.4.1 ([195, Theorem 3.1], [202, Proposition 4.1]) *Assume that X is a complex space, X' is a closed Stein subvariety of X, and K is a compact holomorphically convex set in X (see Definition 2.3.4) such that $K \cap X'$ is $\mathcal{O}(X')$-convex. Let $\pi\colon Z \to X$ be a holomorphic submersion of a complex space Z onto X. Given an open set $U \subset X$ containing K and a section $f\colon U \cup X' \to Z|_{U \cup X'}$ whose restrictions to U and to X' are holomorphic, there exist for every $\epsilon > 0$ an open Stein neighborhood $V = V^\epsilon$ of $K \cup X'$ in X and a holomorphic section $f^\epsilon\colon V \to Z|_V$ such that*

$$f^\epsilon|_{X'} = f|_{X'} \quad \textit{and} \quad \sup_{x \in K} \mathrm{dist}\big(f^\epsilon(x), f(x)\big) < \epsilon.$$

Theorem 3.4.1 fails in general if $\pi: Z \to X$ is a holomorphic map that is not a submersion. For example, the map $\mathbb{C}^2 \ni (z,w) \mapsto z^p w^q \in \mathbb{C}$ for coprime integers $p, q \in \mathbb{Z}$ admits a Hölder continuous section at $(0,0)$, but there are no local holomorphic sections.

The following corollary follows immediately by replacing the submersion $\pi: Z \to X$ by $\widetilde{\pi}: Z \times S \to X \times S$, where $\widetilde{\pi}(z,s) = (\pi(z), s)$.

Corollary 3.4.2 *Assume that $\pi: Z \to X$ is a holomorphic submersion of a complex space Z onto a complex space X. Fix a point $x_0 \in X$. Let S be a Stein space and $f: S \to Z_{x_0} = \pi^{-1}(x_0)$ a holomorphic map into the fibre over x_0. For every relatively compact domain $U \Subset S$ there exist an open neighborhood $B \subset X$ of x_0 and a holomorphic map $F: B \times U \to Z$ such that $F(x,s) \in Z_x$ for all $s \in U$ and $F(x_0, \cdot) = f|_U$.*

Proof of Theorem 3.4.1 By Theorem 3.2.1 (p. 70) there is an open Stein neighborhood $V \subset X$ of $K \cup X'$ such that K is $\mathcal{O}(V)$-convex. We replace X by V and assume in the sequel that K is $\mathcal{O}(X)$-convex. By shrinking $U \supset K$ we may also assume that U is Stein and $\overline{U}$ is compact.

Consider first the case when $Z = X \times \mathbb{C}^p$ and $\pi: Z \to X$ is the projection $\pi(x, \zeta) = x$. We identify sections $X \to Z$ with maps $f: X \to \mathbb{C}^p$ to the fibre. By the Cartan extension theorem (Corollary 2.6.3) there is a holomorphic map $\phi: X \to \mathbb{C}^p$ such that $\phi|_{X'} = f|_{X'}$. By Theorem A (p. 54) there exist finitely many functions $h_1, \ldots, h_m \in \mathcal{O}(X)$ that vanish on X' and generate the ideal sheaf $\mathcal{J}_{X'}$ of X' at every point of U. Since $f - \phi$ vanishes on $X' \cap U$, Theorem B gives holomorphic maps $g_j: U \to \mathbb{C}^p$ $(j = 1, \ldots, m)$ such that

$$f = \phi + \sum_{j=1}^{m} g_j h_j \quad \text{on } U.$$

By the Oka-Weil theorem (see Theorem 2.3.1) we can approximate each g_j, uniformly on K, by a holomorphic map $\widetilde{g}_j: X \to \mathbb{C}^p$. The map

$$\widetilde{f} = \phi + \sum_{j=1}^{m} \widetilde{g}_j h_j : X \to \mathbb{C}^p \tag{3.4}$$

then approximates f uniformly on K and it agrees with f on X'.

With the aid of Theorems 3.2.1 (p. 70) and 3.3.5 (p. 75) we now reduce the general case to this special case. Consider the following subsets of Z:

$$\widetilde{K} = f(K), \qquad \widetilde{X}' = f\big(X'\big), \qquad \widetilde{U} = f(U).$$

Since U is Stein and $f: U \to \widetilde{U}$ is biholomorphic, $\widetilde{U}$ is a closed Stein subvariety of $Z|_U$. Theorem 3.2.1 furnishes an open Stein set $\widetilde{\Omega} \subset Z|_U$ containing $\widetilde{U}$ as a closed subvariety. As K is $\mathcal{O}(X)$-convex and hence $\mathcal{O}(U)$-convex, we infer that $\widetilde{K}$ is $\mathcal{O}(\widetilde{U})$-convex, and hence also $\mathcal{O}(\widetilde{\Omega})$-convex. Since X' is Stein and $f: X' \to \widetilde{X}'$

is biholomorphic, $\widetilde{X}'$ is also Stein. Finally, since $K \cap X'$ is $\mathcal{O}(X')$-convex, the set $\widetilde{K} \cap \widetilde{X}' = f(K \cap X')$ is $\mathcal{O}(\widetilde{X}')$-convex. By Theorem 3.2.1, the set $\widetilde{K} \cup \widetilde{X}'$ admits a basis of open Stein neighborhoods in Z.

Lemma 3.4.3 *There exist an open Stein neighborhood W of $\widetilde{K} \cup \widetilde{X}'$ in Z and a holomorphic embedding $G\colon W \hookrightarrow X \times \mathbb{C}^N$ for some $N \in \mathbb{N}$ such that for every $x \in X$, G embeds the fibre $W_x = W \cap \pi^{-1}(x)$ onto a locally closed submanifold $G(W_x)$ of $\{x\} \times \mathbb{C}^N$. ($W_x$ may be empty for some $x \in X$.)*

Proof Let $\mathrm{VT}Z = \mathrm{T}Z/X$ denote the vertical tangent bundle, a holomorphic vector bundle of rank p over Z (p. 24). For each holomorphic function g on an open subset of Z we denote by $\mathrm{V}dg$ the differential of g restricted to $\mathrm{VT}Z$; thus $\mathrm{V}dg$ is a holomorphic section of the vertical cotangent bundle VT^*Z. Choose an open Stein neighborhood $W_0 \subset Z$ of $\widetilde{K} \cup \widetilde{X}'$. By Cartan's Theorem A there exist functions $g_1, \ldots, g_N \in \mathcal{O}(W_0)$ whose vertical differentials $\mathrm{V}dg_j$ span $\mathrm{VT}^*_z Z$ at each point $z \in W_0$. Consider the holomorphic map

$$G\colon W_0 \to X \times \mathbb{C}^N, \quad G(z) = \big(\pi(z), g_1(z), \ldots, g_N(z)\big). \tag{3.5}$$

Our choice of the g_j's implies that G embeds an open Stein neighborhood $W \subset W_0$ of $\widetilde{K} \cup \widetilde{X}'$ biholomorphically onto a locally closed subvariety $G(W)$ of $X \times \mathbb{C}^N$. Clearly, G respects the fibres, i.e., $pr_1 \circ G = \pi$ where $pr_1\colon X \times \mathbb{C}^N \to X$ is the projection $pr_1(x, \zeta) = x$, and $G(W_x)$ is a locally closed Stein submanifold of $\{x\} \times \mathbb{C}^N$ for every $x \in \pi(W)$. □

Let $G\colon W \hookrightarrow X \times \mathbb{C}^N$ be a holomorphic embedding furnished by Lemma 3.4.3. Theorem 3.3.5 (p. 75), applied to the Stein subvariety $S = G(W)$ of $X \times \mathbb{C}^N$, gives an open Stein neighborhood $\Omega \subset X \times \mathbb{C}^N$ of $G(W)$ and a fibre preserving holomorphic retraction $\iota\colon \Omega \to S$ which retracts the fibre $\Omega_x = \Omega \cap (\{x\} \times \mathbb{C}^N)$ onto $G(W_x)$ for each $x \in \pi(W)$. After shrinking $U \supset K$ we may assume that $f(U) \subset W$. Consider the composed section

$$G \circ f : U \cup X' \to \big(U \cup X'\big) \times \mathbb{C}^N \subset X \times \mathbb{C}^N.$$

By the special case proved above, we can approximate $G \circ f$ uniformly on K by a section $F\colon V \to V \times \mathbb{C}^N$ which is holomorphic in an open neighborhood V of $K \cup X'$ such that $F = G \circ f$ on X'. If the approximation is sufficiently close and V is chosen small enough, then $F(V) \subset \Omega$, and the holomorphic section $f^\epsilon = G^{-1} \circ \iota \circ F\colon V \to Z|_V$ fulfills Theorem 3.4.1. □

Remark 3.4.4 Our proof also shows the following. Assume that f satisfies the hypotheses of Theorem 3.4.1. Let $\phi\colon W \to Z|_W$ be a holomorphic section in an open neighborhood W of X' such that $f = \phi$ on X' and ϕ is tangent to f to order $\ell \in \mathbb{N}$ along $K \cap X'$. Then, the maps f^ϵ in Theorem 3.4.1 can be chosen tangent to ϕ to order ℓ along X'. This follows from the setup (3.4) by choosing $h_1, \ldots, h_m \in \mathcal{O}(X)$ that vanish to order ℓ on the subvariety X' and that generate the sheaf $\mathcal{J}_{X'}^{\ell}$ (the ℓ-th power of the ideal sheaf $\mathcal{J}_{X'}$) at every point of the compact set K. □

We have the analogous result for a family of sections depending continuously on a parameter in a compact Hausdorff space.

Theorem 3.4.5 *Assume the setup of Theorem* 3.4.1. *Let L be a compact set in X with $K \subset L$. Given an open neighborhood U of K and a family of holomorphic section $f_p\colon U \cup X' \to Z|_{U\cup X'}$, depending continuously on a parameter p in a compact Hausdorff space P, there exist for every $\epsilon > 0$ an open Stein neighborhood $V \subset X$ of $K' = K \cup (L \cap X')$ and holomorphic sections $f_p^{\epsilon}\colon V \to Z|_V$, depending continuously on $p \in P$, such that*

$$f_p^{\epsilon}\big|_{X'} = f_p|_{X'} \ (p \in P), \qquad \sup_{x\in K,\ p\in P} \operatorname{dist}\big(f_p^{\epsilon}(x), f_p(x)\big) < \epsilon.$$

Proof For every fixed value of the parameter $p \in P$ we set

$$\widetilde{K}_p = f_p(K), \qquad \widetilde{X}'_p = f_p\big(X'\big), \qquad \widetilde{U}_p = f_p(U).$$

Lemma 3.4.3 (p. 77) gives a holomorphic embedding $G_p\colon W_p \hookrightarrow X \times \mathbb{C}^N$ of an open Stein neighborhood $W_p \subset Z$ of $\widetilde{K}_p \cup \widetilde{X}'_p$ (with $N = N_p \in \mathbb{N}$) and a fibre preserving holomorphic retraction $\iota_p\colon \Omega_p \to G_p(W_p)$. Set $K' = K \cup (L \cap X')$. Every point $p \in P$ admits an open neighborhood $P'_p \subset P$ such that for every $q \in P'_p$ we have $f_q(K') \subset W_p$. By the compactness of P we obtain a finite open cover $\{P'_j\}$ of P and points $p_j \in P'_j$ such that $f_q(K') \subset W_{p_j}$ for all $q \in \bar{P}'_j$. Write $N_j = N_{p_j}$, $G_j = G_{p_j}$, $W_j = W_{p_j}$, $\Omega_j = \Omega_{p_j}$ and $\iota_j = \iota_{p_j}$. As in the proof of Theorem 3.4.1 we obtain for every j an open neighborhood $V_j \supset K'$ and a continuous family of holomorphic sections $F_{j,p}\colon V_j \to V_j \times \mathbb{C}^{N_j}$ $(p \in \bar{P}'_j)$ such that $F_{j,p}$ approximates f_p uniformly on K and it agrees with f_p on $V_j \cap X'$. If the approximation is sufficiently close on K and the neighborhood $V_j \supset K'$ is chosen sufficiently small, we also have $F_{j,p}(V_j) \subset \Omega_j$. Set

$$g_{j,p} = \iota_j \circ F_{j,p}, \qquad f_{j,p} = G_j^{-1} \circ g_{j,p}, \quad p \in \bar{P}'_j.$$

Each of the families $\{f_{j,p} : p \in \bar{P}'_j\}$ satisfies the conclusion of Theorem 3.4.5 for the respective values of the parameter p. To complete the proof we patch these families of sections into a global family depending continuously on $p \in P$. The following method will be used in many subsequent places (e.g. in Sect. 6.5).

The method of successive patching. Choose compact sets $P_j \subset P'_j$ such that $\bigcup_{j=1}^m P_j = P$. Pick a continuous function $\chi\colon P \to [0,1]$ with support in P'_1 which equals 1 in a neighborhood $\widetilde{P}_1 \subset P'_1$ of P_1. For $p \in \widetilde{P}_1 \cup P'_2$ we set

$$f_p = G_2^{-1} \circ \iota_2 \circ \big(\chi(p)\, G_2 \circ f_{1,p} + \big(1 - \chi(p)\big) g_{2,p}\big). \tag{3.6}$$

The expression is well defined on a neighborhood of K' in X provided that all approximations were sufficiently close on K. For $p \in \widetilde{P}_1$ we have $\chi(p) = 1$ and hence $f_p = f_{1,p}$; for $p \in P_2' \setminus P_1'$ we have $\chi(p) = 0$ and hence $f_p = f_{2,p}$. A nontrivial convex linear combination of the sections $G_2 \circ f_{1,p}$ and $g_{2,p}$ (with values in $\mathbb{C}^{N_2}$) only occurs for $p \in P_2' \setminus \widetilde{P}_1 \subset P_2'$, and our choices ensure that these sections lie in the domain of the retraction ι_2. This gives a continuous family of holomorphic sections for p in a neighborhood of $P_1 \cup P_2$ in P. We continue in the same way to patch the family (3.6) with $\{f_{3,p} : p \in P_3'\}$ into a new family defined for p in a neighborhood of $P_1 \cup P_2 \cup P_3$. After finitely many steps we obtain a desired solution. □

3.5 Approximation on Totally Real Submanifolds

Assume that M is a real submanifold of a complex manifold (X, J). The question whether a compact set $K \subset M$ admits an open Stein neighborhood in X, or a basis of such neighborhoods, is rather intricate. The simplest case are totally real submanifolds. Recall that a submanifold M of X is said to be *totally real* (or J-real) if for every $p \in M$ the tangent space T_pM contains no complex line: $T_pM \cap J(T_pM) = \{0\}$.

Let dist be a distance function on X induced by a Riemannian metric.

The following result and its corollary are well known (see [468] or [534, p. 281, Theorem 6.1.6]).

Proposition 3.5.1 *Let M be a totally real submanifold of class $\mathscr{C}^1$ in a complex manifold X. Assume that $\rho \ge 0$ is a nonnegative $\mathscr{C}^2$ function in a neighborhood of M such that $\rho^{-1}(0) = M$. If for every point $p \in M$ there are a neighborhood $U_p \subset X$ and a number $c = c_p > 0$ such that $\rho(x) \ge c\,\mathrm{dist}(x, M)^2$ for $x \in U_p$, then ρ is strongly plurisubharmonic in a neighborhood of M.*

Proof Let $m = \dim_{\mathbb{R}} M \le n = \dim_{\mathbb{C}} X$. Assume first that $m = n$. Fix a point $p \in M$ and choose local holomorphic coordinates $z = (z_1, \ldots, z_n) = x + \mathrm{i}y$ near p such that $z(p) = 0$ and $T_0M = \mathbb{R}^n \oplus \{\mathrm{i}0\}^n = \{y = 0\}$. We shall calculate the Levi form $\mathscr{L}_{\rho,0}$ of ρ at 0. We have

$$4\frac{\partial^2 \rho}{\partial z_j \partial \bar z_k} = \frac{\partial^2 \rho}{\partial x_j \partial x_k} + \frac{\partial^2 \rho}{\partial y_j \partial y_k} + \mathrm{i}\left(\frac{\partial^2 \rho}{\partial x_j \partial y_k} - \frac{\partial^2 \rho}{\partial x_k \partial y_j}\right).$$

The conditions imply that all first order partial derivatives of ρ vanish identically on M; hence each second order derivative of ρ that includes a differentiation with respect to one of the variables $x_1, \ldots, x_n$ vanishes at 0 (since these directions are tangential to M at 0). It follows that $4\frac{\partial^2 \rho}{\partial z_j \partial \bar z_k}(0) = \frac{\partial^2 \rho}{\partial y_j \partial y_k}(0)$. Thus, $\mathscr{L}_{\rho,0}$ equals a quarter of the real Hessian of $y \to \rho(\mathrm{i}y)$ at $y = 0$. Since $\rho(\mathrm{i}y) \ge c|y|^2$, its Hessian is positive definite, and hence $\mathscr{L}_{\rho,0} > 0$.

If $m < n$, set $d = n - m$ and choose local coordinates at $p = 0$ with $T_0M = \mathbb{R}^m \oplus \{\mathrm{i}0\}^m \oplus \{0\}^d$. The previous argument shows that $\mathscr{L}_{\rho,0}$ is positive on $\mathbb{C}^m \oplus \{0\}^d$.

Consider now a vector $v \in \mathbb{C}^n \setminus (\mathbb{C}^m \oplus \{0\}^d)$. The complex line $\mathbb{C}v$ intersects M transversely at 0, and hence the function $\zeta \to \lambda_v(\zeta) = \rho(\zeta v) \ge 0$ vanishes precisely to order two at $\zeta = 0$; thus its Hessian at 0 is positive definite. The Laplacian of λ_v at 0 equals the trace of the Hessian and hence is positive. But the Laplacian of λ_v at 0 equals $4\mathcal{L}_{\rho,0}(v)$. □

Corollary 3.5.2 *For every totally real submanifold M of class $\mathcal{C}^1$ in a complex manifold X there exists a strongly plurisubharmonic function $\rho \ge 0$ in a neighborhood V of M satisfying*

$$\frac{1}{2}\operatorname{dist}(x, M)^2 \le \rho(x) \le \operatorname{dist}(x, M)^2, \quad x \in V.$$

Further, every compact set $K \subset M$ admits a neighborhood $U \subset X$ and a strongly plurisubharmonic function $\tau \colon U \to \mathbb{R}_+$ such that $\tau^{-1}(0) = K$. The sets $\{\rho < c\}$ for small $c > 0$ form a basis of open Stein neighborhoods of K.

Proof If M is of class $\mathcal{C}^2$ then the squared distance function $\rho = \operatorname{dist}(\cdot, M)^2$ satisfies the hypothesis in Lemma 3.5.1, and hence is strongly plurisubharmonic near M. If M is merely $\mathcal{C}^1$, we can still find a $\mathcal{C}^2$ function $\rho \ge 0$ that is comparable to $\operatorname{dist}(\cdot, M)^2$ near M by using Whitney's extension theorem [534, Theorem 6.1.6.]. For the second statement, choose a smooth function $\chi \ge 0$ on X such that $\chi^{-1}(0) = K$ and χ vanishes to at least third order at every point of K. Then $\tau = \rho + \chi$ satisfies the stated properties near K. □

By a refinement of this proof it can be shown that even a noncompact $\mathcal{C}^1$ totally real submanifold M in a complex manifold X admits a basis of open tubular Stein neighborhoods (see [255, Sect. 3]). Since a real analytic manifold M embeds as a totally real submanifold of maximal dimension in its complexification, we have the following corollary.

Corollary 3.5.3 ([255, Sect. 3], [260, p. 157]) *Every real analytic manifold is real analytically isomorphic to a real analytic totally real submanifold M of maximal dimension in a Stein manifold X. Furthermore, X can be chosen diffeomorphic to the normal bundle $\mathrm{N}_{M/X}$.*

Continuous or smooth functions on totally real submanifolds can be approximated by functions that are holomorphic in open neighborhoods. A classical case is the Weierstrass theorem on approximation of continuous functions on the real axis $\mathbb{R} \subset \mathbb{C}$ by holomorphic polynomials. Alexander proved that every continuous function on a rectifiable arc in $\mathbb{C}^n$ can be approximated uniformly by functions that are holomorphic in open neighborhoods of the arc [20, 534]. The first general results on totally real submanifolds of higher dimension go back to the papers [302] and [282]. The following optimal result is due to Range and Siu [468, Theorem 1]; see also [534, Sect. 6.3].

Theorem 3.5.4 *If M is a totally real submanifold of class $\mathcal{C}^k$ ($k \in \mathbb{N} \cup \{\infty\}$) in a complex manifold X, then there exists a Stein open neighborhood U of M in X such that the set of restrictions to M of all holomorphic functions on U is dense in the Fréchet space $\mathcal{C}^k(M)$.*

Another approach was given by Manne, Øvrelid and Wold [407, 408] who used convolution with the Gauss kernel to obtain Carleman approximation on totally real submanifolds. Theorem 3.5.4 also follows from the more precise result [215, Theorem 3.1] on solutions of the $\overline{\partial}$-equation in tubes around totally real submanifolds and the following standard lemma on the existence of $\overline{\partial}$-flat extensions. We omit the proof.

Lemma 3.5.5 ([302, Lemma 4.3]) *Let M be a totally real submanifold of class $\mathcal{C}^k$ ($k \in \{1, 2, \ldots, \infty\}$) in a complex manifold X. Every function $f \in \mathcal{C}^k(M)$ extends to a function $F \in \mathcal{C}^k(X)$ that is smooth on $X \setminus M$ and $\overline{\partial}$-flat to order $k-1$ along M, in the sense that all partial derivatives of order $\le k-1$ of the coefficients of $\overline{\partial} F$ (in any local holomorphic coordinates) vanish on M.*

Theorem 3.5.4 extends to maps with values in a complex manifold, except that the neighborhood may depend on the map. Here is the precise result. For a more general approximation result see Theorem 3.8.1 on p. 88.

Corollary 3.5.6 *If M is a compact totally real submanifold of class $\mathcal{C}^r$ ($r \ge 1$) in a complex manifold X and Y is a complex manifold, then every map $f\colon M \to Y$ of class $\mathcal{C}^s$ for some $s \le r$ can be approximated in $\mathcal{C}^s(M)$ by holomorphic maps $F\colon U \to Y$ in open Stein neighborhoods $U = U_F$ of M.*

Proof By an initial approximation we may assume that f is of class $\mathcal{C}^r$. The graph $\widetilde{M} = \{(x, f(x)) : x \in M\}$ is a totally real submanifold of $X \times Y$, and hence it has an open Stein neighborhood $\Omega \subset X \times Y$. The domain Ω embeds as a closed Stein submanifold of some $\mathbb{C}^N$, so we can view $M \ni x \mapsto (x, f(x))$ as a $\mathbb{C}^N$-valued map. We approximate it by a holomorphic map, compose with a holomorphic retraction onto Ω and project to Y. □

In the same direction we mention the following result of Stout.

Theorem 3.5.7 ([535]) *Assume that K is a compact set in a complex space X such that every continuous function $f\colon K \to \mathbb{C}$ can be approximated uniformly on K by holomorphic functions on open neighborhoods of K. Then every continuous map $f\colon K \to Y$ to an arbitrary complex manifold Y can be uniformly approximated by holomorphic maps $F\colon U_F \to Y$ in open neighborhoods $U_F \supset K$.*

Proof Choose a smooth embedding $\phi\colon Y \hookrightarrow \mathbb{R}^m$ for some $m \in \mathbb{N}$. Considering $\mathbb{R}^m$ as the real subspace of $\mathbb{C}^m$, the graph

$$Z = \big\{(y, \phi(y)) : y \in Y\big\} \subset Y \times \mathbb{C}^m$$

is a totally real submanifold of $Y \times \mathbb{C}^m$, and hence it has an open Stein neighborhood Ω in $Y \times \mathbb{C}^m$ by Corollary 3.5.2. Embed Ω as a complex submanifold of some $\mathbb{C}^N$. Given a continuous map $f\colon K \to Y$, the continuous map $K \ni x \mapsto (f(x), \phi(f(x))) \in \Omega \subset \mathbb{C}^N$ can be approximated by holomorphic maps $U \to \mathbb{C}^N$ in open neighborhoods $U \subset X$ of K. The proof is now concluded exactly as in Corollary 3.5.6. □

Theorem 3.5.7 gives the following corollary to a theorem of Alexander [20] concerning the approximation of continuous function on rectifiable arcs in $\mathbb{C}^n$ by holomorphic functions. For results on approximation of maps to almost complex manifolds, see Chakrabarti [94].

Corollary 3.5.8 *If C is a compact rectifiable arc in $\mathbb{C}^n$ and $f : C \to Y$ is a continuous map to a complex manifold Y, then f can be approximated uniformly on C by maps to Y that are holomorphic on neighborhoods of C.*

The next simplest case where things are fairly well understood are real two dimensional surfaces in a complex surface; we consider these in Sects. 10.1–10.7.

Generic CR submanifolds of positive CR dimension (see p. 24) are only rarely locally holomorphically convex. If M is *minimal* at a point $p \in M$, in the sense that there is no local smooth CR submanifold $N \subset M$ passing through p such that $\dim_p N < \dim_p M$ but $\mathrm{CRdim}_p N = \mathrm{CRdim}_p M$, then there exist analytic discs in X with boundaries in M that sweep out a wedge-like domain W with edge M, and any CR function on M near p extends to a holomorphic function on W (see [42] for a survey of results on this subject). Clearly M is not locally holomorphically convex at such a point. In spite of this, CR functions on any CR submanifold of $\mathbb{C}^n$ are locally approximable by holomorphic polynomials [43].

3.6 Stein Neighborhoods of Laminated Sets

The main result of this section is used in the proof of Theorem 5.13.4 (p. 254) on the equivalence of the basic and the parametric Oka property.

Let Z be a complex space. Consider the product $\mathbb{C}^n \times Z$ with the projection $\pi : \mathbb{C}^n \times Z \to \mathbb{C}^n$. Let S be a compact subset of $\mathbb{C}^n \times Z$. Under what conditions on the projection $\pi(S) \subset \mathbb{C}^n$ and on the fibres $S_\zeta = \{z \in Z : (\zeta, z) \in S\}$ is S a Stein compact in $\mathbb{C}^n \times Z$? We have the following precise answer when the projection $\pi(S)$ is contained in $\mathbb{R}^n$, the real subspace of $\mathbb{C}^n$.

Theorem 3.6.1 ([227]) *Let S be a compact set in $\mathbb{C}^n \times Z$ such that $P = \pi(S) \subset \mathbb{R}^n$. Then S is a Stein compact in $\mathbb{C}^n \times Z$ if and only if for any open neighborhood $U \subset \mathbb{R}^n \times Z$ of S there exist open sets $V, \Omega \subset \mathbb{R}^n \times Z$, with $S \subset V \Subset \Omega \subset U$, such that for any $u \in P$ the fibre Ω_u is Stein and $\widehat{(S_u)}_{\mathcal{O}(\Omega_u)} \subset V$. The same holds if $\pi(S)$ belongs to a totally real subset of $\mathbb{C}^n$.*

The following example illustrates why it is not enough to assume that each fibre of S is a Stein compact.

Example 3.6.2 Let $Z = \mathbb{C}$ and $P = \{u \in \mathbb{R} : 0 \le u \le 1\} \subset \mathbb{C}$. For $0 \le u \le \frac{1}{2}$ let $S_u = \{z \in \mathbb{C} : |z| \le 1\}$, and for $\frac{1}{2} < u \le 1$ let $S_u = \{z \in \mathbb{C} : \frac{1}{2} \le |z| \le 1\}$. Clearly, each fibre is a Stein compact, but any Stein domain in $\mathbb{C}^2$ which contains S also contains $P \times \bar{\mathbb{D}}$. □

Proof Assume that S is a Stein compact. Then, for every open neighborhood U of S in $\mathbb{R}^n \times Z$ there exists a Stein neighborhood $\widetilde{\Omega}$ of S in $\mathbb{C}^n \times Z$ such that $\Omega := \widetilde{\Omega} \cap (\mathbb{R}^n \times Z) \subset U$. The hull $\widehat{S}_{\mathcal{O}(\widetilde{\Omega})} \subset \widetilde{\Omega}$ is compact, and we have that $\pi(\widehat{S}_{\mathcal{O}(\widetilde{\Omega})}) = \pi(S) \subset \mathbb{R}^n$ (since every compact subset of $\mathbb{R}^n$ is polynomially convex in $\mathbb{C}^n$). Taking $V \Subset \Omega$ to be an open neighborhood of $\widehat{S}_{\mathcal{O}(\widetilde{\Omega})}$ in $\mathbb{R}^n \times Z$, we see that the conditions in the theorem are satisfied.

Conversely, assume that the conditions hold. Fix an open set $\widetilde{U} \subset \mathbb{C}^n \times Z$ containing S and let $U = \widetilde{U} \cap (\mathbb{R}^n \times Z)$. Assume that the open sets $V \Subset \Omega \subset U$ satisfy the hypotheses of the theorem. Choose an open set $V' \subset \mathbb{R}^n \times Z$ such that $V \Subset V' \Subset \Omega$. Given $\epsilon > 0$, set

$$V(\epsilon) = \{(u + \mathrm{i}v, z) \in \mathbb{C}^n \times Z : (u, z) \in V,\ |v| < \epsilon\}.$$

Choose $\epsilon_0 > 0$ sufficiently small such that $V'(\epsilon_0) \Subset \widetilde{U}$. We shall construct a (Stein) plurisubharmonic polyhedral neighborhood of S contained in $V(\epsilon_0)$.

For $\delta > 0$ and $u \in \mathbb{R}^n$ let $\mathbb{B}(u, \delta)$ denote the open ball in $\mathbb{R}^n$ of radius δ centered at u.

Lemma 3.6.3 *There exists a positive strongly plurisubharmonic function ρ_0 in an open neighborhood of S in $\mathbb{C}^n \times Z$.*

Proof Fix a point $s_0 = (u_0, z_0) \in S$. By compactness it suffices to show that there exists a plurisubharmonic function in a neighborhood of S that is strongly plurisubharmonic near s_0. Since $\overline{V'} \subset \Omega$ and Ω is open in $\mathbb{R}^n \times Z$, there exists a $\delta > 0$ such that $V'_u \Subset \Omega_{u_0}$ for all u in the ball $\mathbb{B}(u_0, 2\delta) = \{|u - u_0| < 2\delta\}$. As $W = (\mathbb{B}(u_0, 2\delta) + \mathrm{i}\mathbb{R}^n) \times \Omega_{u_0}$ is Stein, there exists a positive strongly plurisubharmonic function ρ_1 on W. We may assume that $\rho_1(s_0) = 1$. Set

$$M = \sup\{\rho_1(u, z) : |u - u_0| = \delta,\ z \in V'_u\}.$$

Let ρ_2 be a plurisubharmonic function on $\mathbb{C}^n$ such that $\rho_2(u_0) = 0$ and $\rho_2(u) > M$ for all $u \in b\mathbb{B}(u_0, \delta)$. If $\epsilon > 0$ is small enough, then $\max\{\rho_1, \rho_2\}$ is well defined and plurisubharmonic on $V'(\epsilon)$, and it is strongly plurisubharmonic near the point $s_0 = (u_0, z_0)$. □

Shrinking the initial neighborhood $\widetilde{U} \subset Z$ of S, we may assume by Lemma 3.6.3 that there exists a positive strongly plurisubharmonic function ρ_0 on $\widetilde{U}$. We adjust the sets V, V' and Ω to the new $U = \widetilde{U} \cap (\mathbb{R}^n \times Z)$.

Lemma 3.6.4 *For any point q belonging to the boundary of V in $\mathbb{R}^n \times Z$ there exist an $\epsilon > 0$ and a continuous plurisubharmonic function ρ on $V'(\epsilon)$ such that $\rho(q) = 2$ and $\rho < \frac{1}{2}$ on S.*

Proof If $\pi(q) \notin \pi(S) = P$, we may find a holomorphic function $g \in \mathcal{O}(\mathbb{C}^n)$ such that $\rho = |g \circ \pi|$ works.

Assume now that $q = (u, z)$ with $u \in P$. Since $\widehat{(S_u)}_{\mathcal{O}(\Omega_u)} \subset V_u$ by the assumption, there is a holomorphic function $f \in \mathcal{O}(\Omega_u)$ such that $f(z) = 2$ and $|f| < \frac{1}{2}$ on S_u. By continuity of f and since $\overline{V'} \subset \Omega$, there is a $\delta > 0$ such that for all $u' \in \overline{\mathbb{B}(u, \delta)}$ we have $V'_{u'} \Subset \Omega_u$ and $|f| < \frac{1}{2}$ on $S_{u'}$. Let

$$M = \sup\{\left|f(z)\right| : z \in \overline{V}'_{u'},\ \left|u' - u\right| \le \delta\}.$$

We consider f as a holomorphic function on $\mathbb{C}^n \times \Omega_u$ which is independent of the first variable. Let $\chi \in \mathcal{C}_0^\infty(\mathbb{B}(u, \delta))$ be a smooth function with compact support such that $0 \le \chi \le 1$ and $\chi(u) = 1$. Let $g \in \mathcal{O}(\mathbb{C}^n)$ approximate χ close enough on $\overline{\mathbb{B}(u, \delta)}$ such that $|g|_{b\mathbb{B}(u,\delta)} < 1/3M$ and $g(u) = 1$. Then, $\rho_1 = |fg|$ is plurisubharmonic on $[\overline{\mathbb{B}(u, \delta)} \oplus \mathrm{i}\mathbb{R}^n] \times \Omega_u$, $\rho_1(q) = |f(z)g(u)| = 2$, and $\rho_1(w) < \frac{1}{3}$ for all $w = (u', z)$ such that $u' \in b\mathbb{B}(u, \delta)$ and $z \in V'_{u'}$.

Let $h \in \mathcal{O}(\mathbb{C}^n)$ be such that $h(u) = 0$, $0 \le |h| \le \frac{1}{2}$ on $\mathbb{B}(u, \delta)$, and $\frac{1}{3} < |h| < \frac{1}{2}$ on $\mathbb{R}^n \setminus \mathbb{B}(u, \delta)$. The function $\rho_2 = |h \circ \pi|$ is plurisubharmonic on $\mathbb{C}^n \times Z$. If $\epsilon > 0$ is small enough, then $\rho = \max\{\rho_1, \rho_2\}$ is well defined on $V'(\epsilon)$ and satisfies the conclusion of the lemma. □

We now complete the proof of Theorem 3.6.1. By compactness and Lemma 3.6.4 there exist an $\epsilon > 0$ with $2\epsilon < \epsilon_0$ and plurisubharmonic functions $\rho_1, \ldots, \rho_m$ on $V'(2\epsilon)$ such that $\rho_j < \frac{1}{2}$ on S for $j = 1, \ldots, m$, and such that for every $q = (u, z) \in bV \subset \mathbb{R}^n \times Z$ we have $\rho_j(q) > \frac{3}{2}$ for at least one $j \in \{1, \ldots, m\}$. Denote the variables on $\mathbb{C}^n$ by $\zeta = u + \mathrm{i}v$. If we replace ρ_j by $\rho_j + C|v|^2$ for a sufficiently big $C > 0$, then we have

$$b\big[V(\epsilon)\big] \subset \bigcup_{j=1}^{m} \left\{w \in V'(2\epsilon) : \rho_j(w) > \frac{3}{2}\right\}.$$

Let

$$W_0 = \{w \in V'(2\epsilon) : \rho_j(w) < 1,\ j = 1, \ldots, m\},$$

and let W be the union of all connected components of W_0 which intersect S; then $S \subset W \Subset V(\epsilon)$. Choose a convex increasing function $\varphi \in \mathcal{C}^\infty((-\infty, 1))$ with $\lim_{t\to 1} \varphi(t) = +\infty$. Recall that ρ_0 is a strongly plurisubharmonic function on $\widetilde{U} \supset S$. The function ρ defined by

$$\rho = \rho_0 + \max_{1\le j\le m} \varphi \circ \rho_j$$

is a strongly plurisubharmonic exhaustion of W, and hence W is Stein by Theorem 2.5.2 (p. 53). Since $\overline{W} \subset V(\epsilon_0)$, this concludes the proof. □

Corollary 3.6.5 *Assume that S is a compact set in $\mathbb{R}^n \times Z$ with the following property. Given an open neighborhood U of S, there exist for each point $u \in \pi(S) \subset \mathbb{R}^n$ a Stein neighborhood $\Omega_u \Subset U_u \subset Z$ of the fibre S_u and a number $\delta = \delta_u > 0$ such that for every point $u' \in \mathbb{B}(u, \delta)$ the fibre $S_{u'}$ is holomorphically convex in Ω_u. Then S is a Stein compact in $\mathbb{C}^n \times Z$.*

Proof We verify the assumptions in Theorem 3.6.1. By compactness there is a finite number of u_j's, Ω_{u_j}'s and δ_j's such that $\{\mathbb{B}(u_j, \delta_j)\}$ is an open cover of $\pi(S)$ in $\mathbb{C}^n$. If all δ_j's are small enough then $\bigcup_j \mathbb{B}(u_j, \delta_j) \times \Omega_{u_j}$ is contained in U. Let Ω be a neighborhood of S in $\mathbb{C}^n \times Z$ with the fibres

$$\Omega_u = \bigcap \bigl\{\Omega_{u_k} : u \in \overline{\mathbb{B}(u_k, \delta_k)}\bigr\}.$$

Then Ω is contained in U and S_u is $\mathcal{O}(\Omega_u)$-convex for every u. Hence, the assumptions in Theorem 3.6.1 hold for any open set $V \Subset \Omega$ containing S. □

Corollary 3.6.6 *Assume that $h \colon Z \to X$ is a holomorphic submersion of a complex space Z onto a Stein space X, K is a compact $\mathcal{O}(X)$-convex subset of X, $U \subset X$ is an open set containing K, and P is a compact set in $\mathbb{R}^n$ (the real subspace of $\mathbb{C}^n$). Let $f \colon P \times U \to Z$ be a continuous map such that $h \circ f(p, x) = x$ for all $(p, x) \in P \times U$, and $f_p = f(p, \cdot) \colon U \to Z$ is holomorphic for every fixed $p \in P$. Then, the set*

$$\Sigma = \bigl\{(p, f(p, x)) : p \in P,\ x \in K\bigr\} \subset \mathbb{C}^n \times Z$$

admits an open Stein neighborhood Θ in $\mathbb{C}^n \times Z$ such that Σ is $\mathcal{O}(\Theta)$-convex.

Proof By shrinking U around K, we may assume that U is Stein. For every $p \in P$ the set $V_p = f_p(U)$ is then a closed Stein subvariety of the complex space $Z|_U = h^{-1}(U)$, and the set $K_p = f_p(K)$ is $\mathcal{O}(V_p)$-convex.

Fix a point $p \in P$. By Siu's theorem (Theorem 3.1.1 on p. 66) there exists an open Stein neighborhood $\Omega_p \subset Z$ of K_p such that $\overline{V_p \cap \Omega_p} \subset V_p$. If $q \in P$ is sufficiently near p, then $\Sigma_q \subset V_q \cap \Omega_p$ due to the continuity of f. Since Σ_q is $\mathcal{O}(V_q)$-convex, it is also $\mathcal{O}(V_q \cap \Omega_p)$-convex. As $V_q \cap \Omega_p$ is a closed subvariety of the Stein domain Ω_p, it follows that Σ_q is also $\mathcal{O}(\Omega_p)$-convex. Indeed, for any point $z \in \Omega_p \setminus V_q$ there exists by Cartan's theorem a holomorphic function on Ω_p that equals one at z and that vanishes on $V_q \cap \Omega_p$; hence no such point can belong to the $\mathcal{O}(\Omega_p)$-hull of Σ_q.

This shows that the set Σ satisfies the assumptions of Corollary 3.6.5, and hence it admits a basis of open Stein neighborhoods in $\mathbb{C}^n \times Z$.

Since P is contained in $\mathbb{R}^n$, $\mathcal{O}(\mathbb{C}^n)|_P$ is dense in the space of continuous complex-valued functions on P by the Weierstrass approximation theorem. By using cut-off functions in the Euclidean variable and approximating them by entire

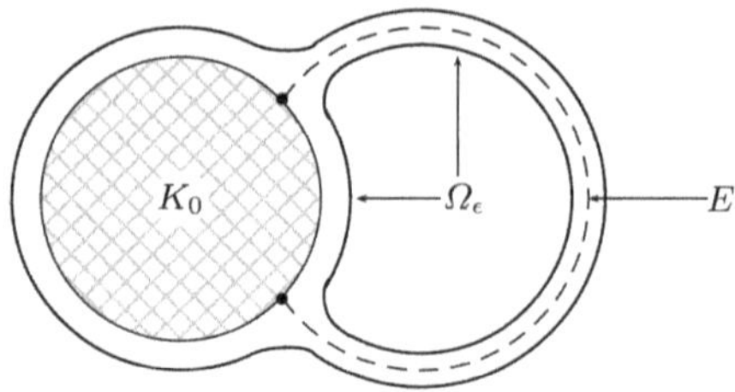

Fig. 3.2 A Stein neighborhood Ω_ϵ of $S = K_0 \cup E$. (Modified from [223, p. 632, Fig. 7] and [222, p. 348, Fig. 3])

functions we see that Σ is $\mathcal{O}(\Theta)$-convex in every open Stein set $\Theta \subset \mathbb{C}^n \times Z$ containing Σ. □

3.7 Stein Compacts with Totally Real Handles

We construct Stein neighborhoods of certain Stein compacts with attached totally real submanifolds. The following result was proved by Hörmander and Wermer when $X = \mathbb{C}^n$ [302]; the general case is [194, Theorem 3.1].

Theorem 3.7.1 *Let $S = K_0 \cup E$ be a closed subset of a complex space X, where K_0 is a compact holomorphically convex set* (*Definition* 2.3.4) *and E is a $\mathcal{C}^1$ totally real submanifold contained in X_{reg}. Assume that there exists a compact holomorphically convex set $K_1 \subset S$ that is a relative neighborhood of K_0 in S. Then, every compact set K with $K_1 \subset K \subset S$ is holomorphically convex. For any such K and any neighborhood $N \supset K_0$ there exists for every sufficiently small $\epsilon > 0$ an open Stein domain $\Omega_\epsilon \subset X$ (see Fig.* 3.2) *such that*

(i) *Ω_ϵ contains all points with distance $< \epsilon$ from K, and*
(ii) *all points $x \in \Omega_\epsilon \setminus N$ have distance $< 2\epsilon$ from S.*

Here, dist is any distance function on X compatible with the complex space topology and whose restriction to X_{reg} is induced by a smooth Riemannian metric. For example, we can embed a Stein neighborhood of K into a Euclidean space and use the induced distance function.

Proof By Proposition 2.5.5 there exist for $j = 0, 1$ an open set $U_j \supset K_j$ and a plurisubharmonic exhaustion function $\rho_j : U_j \to \mathbb{R}_+$ such that $K_j = \{\rho_j = 0\}$. We may choose $U_0 \subset U_1$. Pick $c_2 > 0$ small enough such that

$$\{x \in U_0 : \rho_0(x) \le c_2\} \subset N, \qquad K_1' := S \cap \{\rho_0 \le c_2\} \subset K_1.$$

Choose a convex increasing function $h : \mathbb{R}_+ \to \mathbb{R}_+$ which vanishes on $[0, c_2]$, is positive on $(c_2, +\infty)$, and satisfies $\lim_{t\to\infty} h(t) = +\infty$. The function

$$\rho_1' = \rho_1 + h \circ \rho_0 : U_0 \to \mathbb{R}_+$$

is a plurisubharmonic exhaustion with $\{\rho_1' = 0\} = K_1'$, and hence K_1' is $\mathcal{O}(U_0)$-convex. We now replace K_1 by K_1' and ρ_1 by ρ_1'. Choose constants $0 < c_0 < c_0' < c_1 < c_1' < c_2$. The function $\rho_1 \ge 0$ has a local minimum at each point $x \in K_1$. After multiplying ρ_1 by a suitably chosen positive constant we have $\rho_1(x) < \frac{1}{2}\operatorname{dist}(x, K_1)^2$ for $x \in U_0$. Fix a compact subset K of S with $K_1 \subset K$. Since E is a totally real submanifold of X_{reg}, Corollary 3.5.2 (p. 80) gives a $\mathcal{C}^2$ strongly plurisubharmonic function $\tau \ge 0$ in an open set $V_0 \supset E \cap K$ that vanishes precisely on $E \cap K$ and satisfies

$$\frac{1}{2}\operatorname{dist}(x, E)^2 \le \tau(x) \le \operatorname{dist}(x, E \cap K)^2, \quad x \in V_0.$$

If E is of class $\mathcal{C}^r$ for some $r > 1$, we can take $\tau(x) = \frac{1}{2}\operatorname{dist}(x, E)^2 + \tau'(x)$ where $\tau' \ge 0$ is a smooth function that vanishes to third order on $E \cap K$ and is small positive on $E \setminus K$. Note that $K \subset U_0 \cup V_0$.

Choose a smooth function $\chi \ge 0$ on X such that $\chi = 1$ on $\{\rho_0 \le c_0\}$ and $\operatorname{supp}\chi \Subset \{\rho_0 < c_0'\}$. For small $\delta > 0$ the function $\tau_\delta = \tau - \delta\chi$ is strongly plurisubharmonic on V_0; fix such δ. Choose an open set V satisfying $K \subset V \subset U_0 \cup V_0$ and define a function $\rho\colon V \to \mathbb{R}_+$ by

$$\rho = \begin{cases} \rho_1 & \text{on } V \cap \{\rho_0 < c_0\}, \\ \max\{\rho_1, \tau_\delta\} & \text{on } V \cap \left\{c_0 \le \rho_0 \le c_0'\right\}, \\ \tau & \text{on } V \cap \left\{\rho_0 > c_0'\right\}. \end{cases}$$

These choices are compatible if the neighborhood $V \supset K$ is sufficiently small:

- near $E \cap \{\rho_0 = c_0\}$ we have $\rho_1 \ge 0$ and $\tau_\delta < 0$, hence $\rho = \rho_1$;
- for x near $E \cap \{\rho_0 = c_0'\}$ we have $\rho_1(x) < \frac{1}{2}\operatorname{dist}(x, E)^2 \le \tau(x) = \tau_\delta(x)$, hence $\rho = \tau$.

By the construction, $\rho \ge 0$ is a continuous plurisubharmonic function on V that vanishes precisely on K and satisfies the following conditions:

- $\rho(x) \le \operatorname{dist}(x, K)^2$ for all $x \in V$, and
- $\rho(x) = \tau(x) \ge \frac{1}{2}\operatorname{dist}(x, E \cap K)^2$ for $x \in V \cap \{\rho_0 \ge c_0'\} \supset V \setminus N$.

Hence, for every sufficiently small $\epsilon > 0$ the set $\Omega_\epsilon = \{x \in V : \rho(x) < \epsilon^2\}$ is a pseudoconvex open neighborhood of K satisfying the following conditions:

$$\left\{x \in X : \operatorname{dist}(x, K) < \epsilon\right\} \subset \Omega_\epsilon, \qquad \Omega_\epsilon \setminus N \subset \left\{x \in X : \operatorname{dist}(x, E \cap K) < \sqrt{2}\epsilon\right\}.$$

It remains to show that the domains Ω_ϵ are Stein. Choose an increasing convex function $h_\epsilon\colon (-\infty, \epsilon^2) \to \mathbb{R}$ with $\lim_{t\to\epsilon^2} h(t) = +\infty$; then, $h_\epsilon \circ \rho\colon \Omega_\epsilon \to \mathbb{R}$ is a plurisubharmonic exhaustion. Choose a smooth strongly plurisubharmonic function ξ on the Stein domain U_0. Let $\chi \ge 0$ be a smooth cut-off function on X such that $\chi = 1$ on $\{\rho_0 \le c_1\}$ and $\operatorname{supp}\chi \subset \{\rho_0 < c_1'\}$. We claim that, for sufficiently small $\delta > 0$, the function $\widetilde{\rho} = \rho + \delta\chi\xi$ is strongly plurisubharmonic on V. Indeed,

- on $V \cap \{\rho_0 \le c_1\}$ the function $\widetilde{\rho} = \rho + \delta\xi$ is strongly plurisubharmonic,
- on $V \cap \{\rho_0 > c_1\}$ the function ρ is strongly plurisubharmonic, and hence so is $\widetilde{\rho}$ provided that $\delta > 0$ is chosen sufficiently small.

For such δ, the function $h_\epsilon \circ \rho + \widetilde{\rho} : \Omega_\epsilon \to \mathbb{R}$ is a strongly plurisubharmonic exhaustion and hence Ω_ϵ is Stein by Theorem 2.5.2. Since ρ is plurisubharmonic on $\Omega_{\epsilon_0} = \{\rho < \epsilon_0^2\}$ for a small $\epsilon_0 > 0$, the domain Ω_ϵ is Runge in $\Omega_{\epsilon'}$ for all $0 < \epsilon < \epsilon' \le \epsilon_0$. □

3.8 A Mergelyan Approximation Theorem

In this section we prove the following Mergelyan type approximation theorem in the geometric situation of Theorem 3.7.1. The original sources are [194, Theorem 3.2] and [302, Theorem 4.1].

Theorem 3.8.1 *Assume that X and Z are complex spaces, $\pi : Z \to X$ is a holomorphic submersion, and X' is a closed complex subvariety of X containing X_{sing}. Let K_0 and $S = K_0 \cup E$ be compact holomorphically convex subsets of X, where $E \subset X \setminus X'$ is a compact totally real m-dimensional submanifold of class $\mathcal{C}^r$ for some $r \ge \frac{m}{2} + 1$. Assume that k is an integer satisfying $0 \le k \le r - \frac{m}{2} - 1$. Given an open set $U \supset K_0$ and a section $f : U \cup E \to Z|_{U \cup E}$ such that $f|_U$ is holomorphic and $f|_E \in \mathcal{C}^k(E)$, there exist for every $s \in \mathbb{N}$ a sequence of open sets $V_j \supset S$ and holomorphic sections $f_j : V_j \to Z|_{V_j}$ $(j \in \mathbb{N})$ such that f_j agrees with f to order s along $X' \cap V_j$ for each $j \in \mathbb{N}$, and $\lim_{j\to\infty} f_j|_S = f|_S$ in the $\mathcal{C}^k(S)$-topology.*

Remark 3.8.2 Theorem 3.8.1 is not optimal due to loss of derivatives, but it suffices for our applications, and the proof by the L^2-method is simple. To prove the same result without loss of derivatives (i.e., with approximation in the $\mathcal{C}^r$ norm on a totally real submanifold E of class $\mathcal{C}^r$ attached to K_0), one can use the optimal approximation theorem in tubular neighborhoods of totally real submanifolds, due to Løw, Øvrelid and the author [215, Theorem 1.1], and solve a Cousin problem with bounds to glue this approximation with the original map f on a neighborhood of K_0. Another approach is due to Manne, Øvrelid and Wold [407, 408] who used convolution with the Gauss kernel to obtain Carleman approximation on totally real submanifolds. □

Proof We begin with the special case when X is a manifold and $Z = X \times \mathbb{C}$, identifying sections with functions $X \to \mathbb{C}$. Let dist be a distance function on X induced by a smooth Riemannian metric. By a $\mathcal{C}^k$-small deformation of $f|_E$ away from K_0, we may assume that $f|_E \in \mathcal{C}^r(E)$. Lemma 3.5.5 (p. 81) shows that f extends to a $\mathcal{C}^r$ function in an open neighborhood of S that agrees with the original function near K_0 and satisfies

$$\left|\overline{\partial} f(x)\right| = o\big(\mathrm{dist}(x, E)^{r-1}\big), \quad x \to E.$$

Note that $\overline{\partial} f = 0$ in a neighborhood of K_0. Let Ω_ϵ be a Stein neighborhood basis of S furnished by Theorem 3.7.1, so we have

$$\{x \in X : \mathrm{dist}(x, S) < \epsilon\} \subset \Omega_\epsilon \subset \{x \in X : \mathrm{dist}(x, S) < 2\epsilon\}.$$

Thus,

$$\|\overline{\partial} f\|_{L^\infty(\Omega_\epsilon)} = o(\epsilon^{r-1}).$$

Since the volume of the ϵ-tube around E is proportional to ϵ^{2n-m}, we get

$$\|\overline{\partial} f\|_{L^2(\Omega_\epsilon)} \le \|\overline{\partial} f\|_{L^\infty(\Omega_\epsilon)} \sqrt{\mathrm{Vol}(U_{2\epsilon})} = o(\epsilon^{r-1+n-m/2}).$$

By [299] (see Theorem 2.7.1 for $X = \mathbb{C}^n$) there exists a solution u_ϵ of

$$\begin{cases} \overline{\partial} u_\epsilon = \overline{\partial} f & \text{on } \Omega_\epsilon, \\ \|u_\epsilon\|_{L^2(\Omega_\epsilon)} \le \mathrm{const} \cdot \|\overline{\partial} f\|_{L^2(\Omega_\epsilon)} = o(\epsilon^{r-1+n-m/2}). \end{cases}$$

Recall that every $\mathcal{C}^1$ function on the ball $\epsilon\mathbb{B} = \{\zeta \in \mathbb{C}^n : |\zeta| < \epsilon\}$ satisfies the following estimate (see Lemma 2.7.2 on p. 58):

$$|u(0)| \le \mathrm{const} \cdot \left(\epsilon^{-n} \|u\|_{L^2(\epsilon\mathbb{B})} + \epsilon \|\overline{\partial} u\|_{L^\infty(\epsilon\mathbb{B})}\right).$$

Applying this estimate to u_ϵ at points $z \in S$, observing also that $\|\overline{\partial} u_\epsilon\|_{L^\infty(\Omega_\epsilon)} = \|\overline{\partial} f\|_{L^\infty(\Omega_\epsilon)} = o(\epsilon^{r-1})$, we obtain

$$\|u_\epsilon\|_{L^\infty(S)} \le o(\epsilon^{r-1-m/2}) + \epsilon\, o(\epsilon^{r-1}) = o(\epsilon^{r-1-m/2}).$$

The function $f_\epsilon = f - u_\epsilon$ is holomorphic on Ω_ϵ, and assuming that $r \ge 1 + \frac{m}{2}$ we get $\lim_{\epsilon \to 0} \|f - f_\epsilon\|_{L^\infty(S)} = 0$. By repeated differentiation we also get $\lim_{\epsilon \to 0} \|\partial^\alpha (f - f_\epsilon)\|_{L^\infty(S)} = 0$ for $|\alpha| \le r - 1 - \frac{m}{2}$.

Interpolation along a subvariety X' is easily built into the construction as follows. By the proof of Theorem 3.4.1 (p. 75) we have

$$f = \phi + \sum_\nu g_\nu h_\nu \quad \text{on } U,$$

where ϕ is a holomorphic function in a Stein neighborhood $V \subset X$ of S, h_ν is a finite collection of holomorphic functions in V that vanish to order s along X' and whose common zero set is $X' \cap V$, and the coefficients g_ν are holomorphic on U and of class $\mathcal{C}^r$ on E. By the special case proved above, g_ν is the limit of a sequence of holomorphic functions $g_{\nu,j}$ in open neighborhoods V_j of S. The sequence of functions

$$f_j = \phi + \sum_\nu g_{\nu,j} h_\nu, \quad j \in \mathbb{N}$$

then fulfills Theorem 3.8.1.

For $Z = X \times \mathbb{C}^N$ the result follows immediately. Suppose now that X is a complex space and $\pi : Z \to X$ is a holomorphic submersion.

Lemma 3.8.3 *Let X and Z be complex spaces and $\pi : Z \to X$ be a holomorphic submersion. Assume that $S = K_0 \cup E$ and $f : U \cup E \to Z$ satisfy the hypotheses of Theorem* 3.8.1. *Then $f(S)$ has a basis of Stein neighborhoods in Z.*

Proof We may assume that U is a Stein domain in X and K_0 is $\mathcal{O}(U)$-convex. Since the real submanifold $f(E)$ of Z is projected by π onto the totally real submanifold E of X, $f(E)$ is totally real in Z. Since $f|_U$ is holomorphic, $f(U)$ is a closed Stein subvariety of $Z|_U = \pi^{-1}(U)$, and hence it has an open Stein neighborhood $\widetilde{U} \subset Z|_U$ by Corollary 3.1.1 (p. 66). For any compact $\mathcal{O}(U)$-convex subset K of U, the set $f(K)$ is holomorphically convex in $f(U)$, and hence also in $\widetilde{U}$ (since $f(U)$ is a closed complex submanifold of $\widetilde{U}$). Applying this to $K = K_0$, and also to $K = S \cap N$ for some compact holomorphically convex neighborhood $N \subset U$ of K_0, we see that $f(S)$ satisfies the hypothesis of Theorem 3.7.1, and hence it has a basis of Stein neighborhoods in Z. □

We now complete the proof of Theorem 3.8.1. By Lemma 3.4.3 (p. 77) there exist a Stein neighborhood W_0 of $f(S)$ in Z and a fibre preserving holomorphic embedding $G \colon W_0 \hookrightarrow X \times \mathbb{C}^N$ (see (3.5) on p. 77). By the special case proved above, the section $G \circ f$ (with values in $X \times \mathbb{C}^N$) is a limit of a sequence of holomorphic sections F_j in open neighborhoods of S. Let $\iota \colon \Omega \to G(W_0)$ be a fibre preserving holomorphic retraction from an open neighborhood $\Omega \subset X \times \mathbb{C}^N$ onto $G(W_0)$, furnished by Theorem 3.3.5 on p. 75. The sequence of holomorphic sections $f_j = G^{-1} \circ \iota \circ F_j$ in open neighborhoods V_j of S then fulfills Theorem 3.8.1. □

3.9 Strongly Pseudoconvex Handlebodies

Let W be a relatively compact, smoothly bounded domain in a complex manifold X. Assume that M is a smooth submanifold of X that is contained in $X \setminus W$ and intersects the boundary $\Sigma = bW$ transversely along its boundary $bM \subset \Sigma$. By smoothly connecting the boundary of a tube around M with Σ, we obtain a smoothly bounded domain $\Omega \subset X$ that contains W and agrees with W outside a small neighborhood of M, and such that $\Omega \setminus W$ is diffeomorphic to the normal bundle of M in X. Such Ω (or its closure) is called a *handlebody with core* $W \cup M$. Replacing W in this construction with a slightly bigger set $W' \supset W$ diffeomorphic to W, we still call the resulting set, obtained by thickening $W' \cup M$, a handlebody with the core $\overline{W} \cup M$. We address the following problem which is important in applications.

Problem 3.9.1 Let $W \cup M \subset X$ be as above, with M totally real and W strongly pseudoconvex. Given a neighborhood $V \subset X$ of M, does there exist a strongly pseudoconvex handlebody $\Omega \subset W \cup V$ with core $W \cup M$?

Theorem 3.7.1 (p. 86) shows that the existence of a Stein neighborhood basis of $\overline{W} \cup M$ depends only on the attachment of M to $\overline{W}$ along the boundary $bM \subset bW$. At least for handles of maximal dimension $\dim_{\mathbb{R}} M = \dim_{\mathbb{C}} X$, a necessary condition is that bM is a complex tangential (Legendrian) submanifold of $\Sigma = bW$, that is,

$$\mathrm{T}_p bM \subset \mathrm{T}_p^{\mathbb{C}} \Sigma = \mathrm{T}_p(\Sigma) \cap J(\mathrm{T}_p \Sigma), \quad p \in bM. \tag{3.7}$$

In addition, we assume that M is *J-orthogonal to Σ* along bM:

$$J(\mathrm{T}_p M) \subset \mathrm{T}_p \Sigma, \quad p \in bM. \tag{3.8}$$

In view of (3.7) this orthogonality condition is equivalent to $Jw \in \mathrm{T}\Sigma$, where w is a tangential vector field to M along bM that is normal to bM.

The following result is used in the Eliashberg-Gompf construction of Stein structures on an almost complex manifold (X, J) with suitable handlebody decomposition (see Sects. 10.8–10.9). Results in this direction were obtained in [148, Lemmas 3.3.1 and 3.4.3], [475, Lemma 2], [209]; see also Cieliebak and Eliashberg [99, Theorem 8.4].

Theorem 3.9.2 *Let W be a relatively compact strongly pseudoconvex domain with smooth boundary $\Sigma = bW$ in a complex manifold X, and let $M \subset X \setminus W$ be a smoothly embedded totally real disc (a diffeomorphic image of the standard disc $D^k \subset \mathbb{R}^k$ for some $1 \le k \le n = \dim X$) with boundary in Σ. If conditions* (3.7) *and* (3.8) *hold, then for every open set $U \supset \overline{W} \cup M$ there is a smooth strongly pseudoconvex handlebody $\widetilde{W}$ with core $\overline{W} \cup M$ such that $\overline{W} \cup M \subset \widetilde{W} \Subset U$. If W is Stein, then $\widetilde{W}$ can also be chosen Stein and such that the sets $\overline{W}$ and $\overline{W} \cup M$ are $\mathcal{O}(\widetilde{W})$-convex.*

A construction of well attached totally real discs satisfying conditions (3.7) and (3.8) is given in Sect. 10.8. These two results together give a strong tool for constructing Stein manifold structures (see Sect. 10.9).

A complete proof of Theorem 3.9.2 can be found in [99, Theorem 8.4]. The first step is a reduction to a certain model situation in $\mathbb{C}^n$. While we shall not give the details of this reduction, we will consider the model case in detail due to its independent interest. In particular, this model situation arises in Sect. 3.11 where we find strongly pseudoconvex handlebodies associated to critical points of strongly plurisubharmonic functions. We wish to point out that, in the Oka theory in Chaps. 5 and 6, only this model situation is used. The general case of Theorem 3.9.2 is applied in Chap. 10.

Let J_{st} denote the standard complex structure on $\mathbb{C}^n$. For a fixed $k \in \{1, \ldots, n\}$ let $z = (z_1, \ldots, z_n) = x + \mathrm{i}y = (x' + \mathrm{i}y', x'' + \mathrm{i}y'')$, with $z_j = x_j + \mathrm{i}y_j$, denote the coordinates on $\mathbb{C}^n$ corresponding to the decomposition

$$\mathbb{C}^n = \mathbb{C}^k \oplus \mathbb{C}^{n-k} = \mathbb{R}^k \oplus \mathrm{i}\mathbb{R}^k \oplus \mathbb{R}^{n-k} \oplus \mathrm{i}\mathbb{R}^{n-k}.$$

Let $D = D^k \subset \mathbb{R}^k$ be the closed unit ball in $\mathbb{R}^k$ and $S = S^{k-1} = bD$ its boundary $(k-1)$-sphere. Identifying D^k with its image in $\mathbb{R}^k \oplus \{0\}^{2n-k} \subset \mathbb{C}^n$ we obtain the core of the standard index k handle

$$H_\delta = (1+\delta) D^k \times \delta D^{2n-k} \subset \mathbb{C}^n, \quad \delta > 0. \tag{3.9}$$

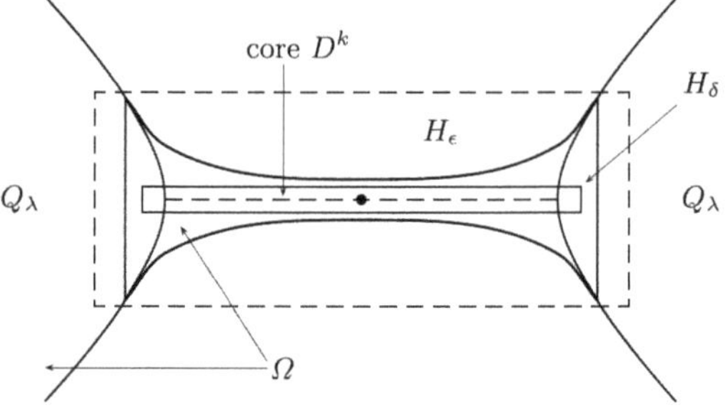

Fig. 3.3 The handlebody Ω. (Modified from [223, p. 620, Fig. 1])

A *standard handlebody of index k* in $\mathbb{C}^n$ is a set $K_{\lambda,\delta} = Q_\lambda \cup H_\delta$ for some $0 < \lambda < 1$ and $0 < \delta < \frac{2\lambda}{1-\lambda}$ (see Fig. 3.3), where

$$Q_\lambda = \{ z = (x' + \mathrm{i}y', z'') \in \mathbb{C}^k \oplus \mathbb{C}^{n-k} : |y'|^2 + |z''|^2 < \lambda(|x'|^2 - 1) \}. \tag{3.10}$$

The condition $\lambda < 1$ ensures that Q_λ is strongly pseudoconvex, and the bound on δ implies that $(1+\delta)bD^k \times \delta D^{2n-k} \subset Q_\lambda$.

Lemma 3.9.3 (Eliashberg [148]) *Let Q_λ be the domain* (3.10) *for some* $0 < \lambda < 1$. *For every $\epsilon > 0$ there exist $\delta \in (0, \epsilon)$ and a smooth strongly pseudoconvex handlebody $\Omega \subset \mathbb{C}^n$ with core $Q_\lambda \cup D^k$ such that $K_{\lambda,\delta} \subset \Omega \subset K_{\lambda,\epsilon}$ (see Fig.* 3.3).

Proof The main case to consider is $k = n$. In this case, we have

$$Q_\lambda = \{ x + \mathrm{i}y \in \mathbb{C}^n : |y|^2 \le \lambda(|x|^2 - 1) \}, \qquad D^n = \{ x + \mathrm{i}0 : x \in \mathbb{R}^n, \ |x| \le 1 \}.$$

Given $\epsilon > 0$ we shall find a number $\sigma = \sigma(\epsilon) \in (0, \epsilon)$ and a smooth increasing function $f : \mathbb{R}_+ \to (0, \infty)$ such that the domain

$$\Omega = \{ x + \mathrm{i}y \in \mathbb{C}^n : |y|^2 \le f(|x|^2) \} \tag{3.11}$$

is strongly pseudoconvex and we have

$$f(t) = \sigma \quad \text{for } t \le 1 - \epsilon, \qquad f(t) = \lambda(t - 1) \quad \text{for } t \ge 1 + \epsilon. \tag{3.12}$$

Hence, for every neighborhood $V \subset \mathbb{C}^n$ of the disc D^n we can choose a handlebody Ω as above such that $Q_\lambda \cup D^n \subset \Omega \subset Q_\lambda \cup V$ and $\Omega \setminus V = Q_\lambda \setminus V$.

The construction proceeds in two steps. First, we identify conditions on a function $f : \mathbb{R}_+ \to (0, \infty)$ that characterize strong pseudoconvexity of the tubular domain (3.11). Let U be a nonempty open set in $\mathbb{R}^n$ ($n > 1$) invariant under the action of the orthogonal group $O(n)$. Set $I = \{|x|^2 : x \in U\} \subset \mathbb{R}_+$. Assume that $f : I \to (0, +\infty)$ is a positive function of class $\mathcal{C}^2$.

Proposition 3.9.4 ([209, Proposition 2.1]) *The domain*

$$\Omega = \{ x + \mathrm{i}y \in \mathbb{C}^n : x \in U, \ |y|^2 < f(|x|^2) \} \tag{3.13}$$

is strongly pseudoconvex along the hypersurface $\Sigma = \{|y|^2 = f(|x|^2)\} \subset b\Omega$ *if and only if* f *satisfies the following differential inequalities for* $t \in I$:

$$f'(t) < 1 \quad \text{and} \quad 2tf(t)f''(t) < \big[1 - f'(t)\big]\cdot\big[tf'(t)^2 + f(t)\big]. \tag{3.14}$$

Proof Set $\rho(x + \mathrm{i}y) = |y|^2 - f(|x|^2)$. A calculation gives for $1 \le j \ne k \le n$

$$-\rho_{z_k} = x_k f' + \mathrm{i}y_k, \qquad -2\rho_{z_k\bar z_k} = 2x_k^2 f'' + f' - 1, \qquad -2\rho_{z_j\bar z_k} = 2x_j x_k f''.$$

Here, f and its derivatives are evaluated at $t = |x|^2$. The calculation of the Levi form of $\Sigma = \{\rho = 0\}$ can be simplified by observing that ρ is invariant under the action of $O(n)$ on $\mathbb{C}^n$ given by $A(x + \mathrm{i}y) = Ax + \mathrm{i}Ay$ $(A \in O(n))$. Fix a point $p \in \Sigma$. After an orthogonal rotation we have $p = (x_1 + \mathrm{i}y_1, \mathrm{i}\widetilde{y}_2, \ldots, \mathrm{i}\widetilde{y}_n) \in \mathbb{C} \times \mathrm{i}\mathbb{R}^{n-1}$. Applying another orthogonal map fixing $\mathbb{C} \times \{0\}^{n-1}$, we may further assume that $p = (x_1 + \mathrm{i}y_1, \mathrm{i}y_2, 0, \ldots, 0)$. At this point we have

$$\rho_{z_1}(p) = -x_1 f' - \mathrm{i}y_1, \qquad \rho_{z_2}(p) = -\mathrm{i}y_2, \qquad \rho_{z_k}(p) = 0 \quad \text{for } k = 3, \ldots, n.$$

Hence, the complex tangent space $T_p^{\mathbb{C}}\Sigma = \{v \in \mathbb{C}^n : \sum_{k=1}^n \frac{\partial\rho}{\partial z_k}(p)v_k = 0\}$ consists of all $v \in \mathbb{C}^n$ satisfying $v_1 = -\lambda\mathrm{i}y_2$, $v_2 = \lambda(x_1 f' + \mathrm{i}y_1)$ for arbitrary choices of $\lambda \in \mathbb{C}$ and $v'' = (v_3, \ldots, v_n) \in \mathbb{C}^{n-2}$. We also have

$$\begin{aligned} &2\rho_{z_1\bar z_1}(p) = 1 - f' - 2x_1^2 f'', && \\ &2\rho_{z_k\bar z_k}(p) = 1 - f', && k = 2, \ldots, n, \\ &2\rho_{z_j\bar z_k}(p) = 0, && 1 \le j \ne k \le n. \end{aligned}$$

For $v \in T_p^{\mathbb{C}}\Sigma$ we thus get (noting that $y_1^2 + y_2^2 = f(x_1^2)$) the following:

$$\begin{aligned} 2\mathcal{L}_{\rho,p}(v) &= \big(1 - f' - 2x_1^2 f''\big)|\lambda|^2 y_2^2 + \big(1 - f'\big)|\lambda|^2\big(x_1^2 f'^2 + y_1^2\big) + \big(1 - f'\big)|v''|^2 \\ &= |\lambda|^2\big(-2x_1^2 y_2^2 f'' + (1 - f')\big(x_1^2 f'^2 + f\big)\big) + \big(1 - f'\big)|v''|^2. \end{aligned} \tag{3.15}$$

The function f and its derivatives are evaluated at x_1^2. Thus, $\mathcal{L}_{\rho,p}(v) > 0$ for all choices of $\lambda \in \mathbb{C}$ and $v'' \in \mathbb{C}^{n-2}$ with $|\lambda|^2 + |v''|^2 > 0$ if and only if

$$f' < 1 \quad \text{and} \quad 2x_1^2 y_2^2 f'' < \big(1 - f'\big)\big(x_1^2 f'^2 + f\big).$$

Note that $0 \le y_2^2 \le |s|^2 = f(x_1^2)$, and y_2^2 assumes both extreme values 0 and $f(x_1^2)$ when (y_1, y_2) traces the circle $y_1^2 + y_2^2 = f(x_1^2)$. Thus, the second inequality above holds at all points of this circle precisely when it holds at the point $y_1 = 0$, $y_2 = \sqrt{f(x_1^2)}$. This shows that the condition

$$f' < 1 \quad \text{and} \quad 2x_1^2 f f'' < \big(1 - f'\big)\big(x_1^2 f'^2 + f\big)$$

characterizes strong pseudoconvexity of Ω along the mentioned circle in Σ. Since $x_1^2 = |x(p)|^2$, this is equivalent to (3.14). Similarly, we see that the domain $\Omega' = \{|y|^2 > f(|x|^2)\}$ is strongly pseudoconvex along the hypersurface Σ if and only if the reverse inequalities hold in (3.14). □

The second step in the proof of Lemma 3.9.3 amounts to finding functions $f\colon \mathbb{R} \to (0,\infty)$ satisfying Proposition 3.9.4 and the boundary conditions (3.12). A detailed construction of such functions is given in [209, Sect. 3]; we do not reproduce it here since it only involves elementary calculus. □

For handles of lower dimension, we obtain strongly pseudoconvex handlebodies satisfying Lemma 3.9.3 by taking

$$\Omega = \{(x' + \mathrm{i}y', z'') : |y'|^2 + |z''|^2 \le f(|x'|^2)\},$$

where f has the same properties as above.

3.10 Morse Critical Points of q-Convex Functions

Recall that a smooth function is a *Morse function* if all its critical points are *Morse points*, i.e., have nondegenerate Hessian. In this section we describe the quadratic normal form of a q-convex Morse function at a critical point.

We begin with the case when ρ is strongly plurisubharmonic ($q = 1$). The following lemma appeared in a 1924 paper of Takagi [542, Theorem II]; it was rediscovered in [491] and [282, p. 166].

Lemma 3.10.1 *Let ρ be a strongly plurisubharmonic function of class $\mathcal{C}^2$ in a neighborhood of the origin in $\mathbb{C}^n$ with a Morse critical point of index k at $0 \in \mathbb{C}^n$. Then $k \in \{0, 1, \ldots, n\}$. Write $z = (z', z'') = (x' + \mathrm{i}y', x'' + \mathrm{i}y'') \in \mathbb{C}^k \oplus \mathbb{C}^{n-k}$. After a $\mathbb{C}$-linear change of coordinates on $\mathbb{C}^n$ we have*

$$\rho(z) = \rho(0) - |x'|^2 + |x''|^2 + \sum_{j=1}^{n} \lambda_j y_j^2 + o(|z|^2) \tag{3.16}$$

where $\lambda_j > 1$ for $j \in \{1, \ldots, k\}$ and $\lambda_j \ge 1$ for $j \in \{k+1, \ldots, n\}$.

Proof We follow the proof in [282]. We may assume that $\rho(0) = 0$. Since the Levi form of ρ at 0 is positive definite, a $\mathbb{C}$-linear change of coordinates normalizes it to $|z|^2$. In these coordinates we have

$$\rho(z) = |z|^2 + \Re(z^t S z) + o(|z|^2),$$

where $S = A + \mathrm{i}B$ is a symmetric $n \times n$ matrix with entries $S_{j,l} = \frac{\partial^2 \rho(0)}{\partial z_j \partial z_l}$. The expression for ρ in terms of the real coordinates $s = (x, y)^t \in \mathbb{R}^{2n}$ is

$$\rho(s) = |s|^2 + s^t T s + o(|s|^2), \qquad T = \begin{pmatrix} A & -B \\ -B & -A \end{pmatrix}.$$

Let J denote the standard complex structure on $\mathbb{C}^n$. For every tangent vector $v = (a, b)^t \in T_0\mathbb{R}^{2n} \cong \mathbb{R}^{2n}$ we have $Jv = J(a, b)^t = (-b, a)^t$. If $Tv = \lambda v$ then a calculation gives $T(Jv) = -\lambda Jv$. Thus, for each eigenvector v of T with the eigenvalue λ, Jv is also an eigenvector of T with the eigenvalue $-\lambda$. Since T is symmetric, it follows that there exists an orthonormal basis $v_1, \ldots, v_n, Jv_1, \ldots, Jv_n$ of $\mathbb{R}^{2n}$ consisting of eigenvectors of T. Interchanging v_j and Jv_j if necessary, we may assume that the eigenvalue of v_j is $-\mu_j \le 0$.

Let $e_1, \ldots, e_{2n}$ denote the standard basis of $\mathbb{R}^{2n}$, chosen so that $Je_j = e_{n+j}$ for $j = 1, \ldots, n$. Let C denote the orthogonal linear transformation on $\mathbb{R}^{2n}$ determined by $Ce_j = v_j$ and $Ce_{n+j} = Jv_j$ for $j = 1, \ldots, n$. Then $CJ = JC$, and hence C determines a unitary transformation $\widetilde{C}$ of $\mathbb{C}^n$. We introduce new coordinates $\sigma = (x, y)^t \in \mathbb{R}^{2n}$ by $s = C\sigma$. In these coordinates,

$$\begin{aligned} \rho(C\sigma) &= (C\sigma)^t (I^{2n} + T) C\sigma + o(|\sigma|^2) \\ &= \sigma^t (I + C^t T C)\sigma + o(|\sigma|^2) \\ &= |\sigma|^2 + \sigma^t D\sigma + o(|\sigma|^2), \end{aligned}$$

where D is a diagonal matrix with entries $-\mu_1, \ldots, -\mu_n, \mu_1, \ldots, \mu_n$ on its main diagonal. Denoting again $z = x + \mathrm{i}y \in \mathbb{C}^n$, the above reads

$$\rho(\widetilde{C}z) = \sum_{j=1}^{n} \big((1 - \mu_j)x_j^2 + (1 + \mu_j)y_j^2\big) + o(|z|^2).$$

Since 0 is a Morse critical point of ρ, we have $\mu_j \ne 1$ or all j. Let $k \in \{0, 1, \ldots, n\}$ denote the number of indexes $j \in \{1, \ldots, n\}$ for which $\mu_j > 1$. Thus, the Hessian of ρ at 0 has precisely k negative eigenvalues, so k is the *Morse index* of ρ at 0. After a permutation of the variables, we may assume that $\mu_j > 1$ precisely for $j = 1, \ldots, k$. Setting

$$\lambda_j = \frac{1 + \mu_j}{|1 - \mu_j|} \ge 1, \quad j = 1, \ldots, n$$

and replacing $\sqrt{|1 - \mu_j|}\, z_j$ by z_j we get the normal form (3.16). □

Definition 3.10.2 A critical point p of a strongly plurisubharmonic function ρ is *nice* if, in some local holomorphic coordinates in a neighborhood of p, ρ is of the form (3.16) without the remainder term.

Lemma 3.10.3 *Every strongly plurisubharmonic function $\rho\colon X \to \mathbb{R}$ can be approximated in the fine $\mathcal{C}^2$ Whitney topology by a Morse strongly plurisubharmonic function all of whose critical points are nice.*

Proof By the Morse lemma, we can approximate ρ in the fine $\mathcal{C}^2$ topology on X by a strongly plurisubharmonic Morse function. Let $p \in X$ be a (Morse) critical point of ρ. In local holomorphic coordinates z in an open set $p \in U \subset X$, with $z(p) = 0$, ρ is of the form (3.16). Choose a smooth function $\chi : \mathbb{C}^n \to [0, 1]$ which equals zero on the unit ball $\mathbb{B} \subset \mathbb{C}^n$ and equals one on $\mathbb{C}^n \setminus 2\mathbb{B}$. For sufficiently small $\epsilon > 0$, the function

$$\widetilde{\rho}(z) = \rho(0) - |x'|^2 + |x''|^2 + \sum_{j=1}^{n} \lambda_j y_j^2 + \chi(\epsilon^{-1}z)\, o(|z|^2)$$

is strongly plurisubharmonic, has a nice critical point at 0 and no other critical points nearby, and it satisfies $\widetilde{\rho} = \rho$ outside of a small neighborhood of p. Performing such local change at every critical point we obtain a new strongly plurisubharmonic function with nice critical points. □

We now consider the general case when ρ is q-convex near $0 \in \mathbb{C}^n$ for some $q \in \{1, 2, \ldots, n+1\}$, i.e., its Levi form $\mathcal{L}_{\rho,0}$ has at least $r = n - q + 1$ positive eigenvalues; the remaining $s = q - 1$ eigenvalues can be of any sign. By a linear change of coordinates on $\mathbb{C}^n$ we can achieve that the subspace $\mathbb{C}^r \times \{0\}^s$ is spanned by some of the eigenvectors corresponding to the positive eigenvalues of $\mathcal{L}_{\rho,0}$ and 0 is a Morse critical point of $\rho(\cdot, 0)$. We denote the coordinates on $\mathbb{C}^n = \mathbb{C}^r \times \mathbb{C}^s = \mathbb{C}^r \times \mathbb{R}^{2s}$ by $z = (\zeta, u)$, where $\zeta = x + \mathrm{i}y \in \mathbb{C}^r$ $(x, y \in \mathbb{R}^r)$ and $u \in \mathbb{R}^{2s}$. By shrinking the domain of ρ to a small polydisc $P = P^r \times P^s \subset \mathbb{C}^n$ around 0, we can assume that the function $\zeta \to \rho(\zeta, u)$ is strongly plurisubharmonic on P^r for each $u \in P^s$.

Lemma 3.10.1 applied to the strongly plurisubharmonic function $\zeta \mapsto \rho(\zeta, 0)$ gives a complex linear change of coordinates on $\mathbb{C}^r$ and an integer $k \in \{0, 1, \ldots, r\}$ such that, in the new coordinates,

$$\rho(\zeta, 0) = \sum_{j=1}^{r} (\delta_j x_j^2 + \lambda_j y_j^2) + o(|\zeta|^2)$$

where $\lambda_j > 1$, $\delta_j = -1$ for $j = 1, \ldots, k$, and $\lambda_j \geq 1$, $\delta_j = +1$ for $j = k+1, \ldots, r$. Note that k is the Morse index of $\rho(\zeta, 0)$ at $\zeta = 0$. Writing $x' = (x_1, \ldots, x_k)$ and $x'' = (x_{k+1}, \ldots, x_r)$ we obtain

$$\rho(\zeta, 0) = -|x'|^2 + |x''|^2 + \sum_{j=1}^{r} \lambda_j y_j^2 + o(|\zeta|^2).$$

We now consider the full second order Taylor expansion of ρ at $0 \in \mathbb{C}^n$:

$$\rho(z) = \rho(\zeta, u) = \rho(\zeta, 0) + \sum_{j=1}^{2s} u_j a_j(x, y) + \sum_{i,j=1}^{2s} c_{ij}\, u_i u_j + o\big(|z|^2\big).$$

Here, $a_j(x, y) = \sum_{l=1}^{r}(\alpha_{jl}x_l + \beta_{jl}y_l)$ are real-valued linear functions on $\mathbb{C}^r = \mathbb{R}^{2r}$ and $c_{ij} = c_{ji}$ are real constants. Our next aim is to remove the mixed terms $u_j a_j(x, y)$ by a shear of the form $(\zeta, u) \mapsto (\zeta + h(u), u)$ for a suitable $\mathbb{R}$-linear map $h\colon \mathbb{R}^{2s} \to \mathbb{C}^r$; such transformation clearly preserves plurisubharmonicity in the ζ variable. Consider the critical point equation $\partial_\zeta \rho^{(2)}(\zeta, u) = 0$, where $\rho^{(2)}$ is the 2nd order homogeneous polynomial of ρ:

$$\frac{\partial \rho^{(2)}}{\partial x_i}(\zeta, u) = 2\delta_i x_i + \sum_{j=1}^{2s} u_j \alpha_{ji} = 0; \qquad \frac{\partial \rho^{(2)}}{\partial y_i}(\zeta, u) = 2\lambda_i y_i + \sum_{j=1}^{2s} u_j \beta_{ji} = 0.$$

This system has a unique (linear) solution $\zeta = x + \mathrm{i}y = h(u)$, and the quadratic map $\zeta \to \rho^{(2)}(\zeta + h(u), u)$ has a unique critical point at $\zeta = 0$ for every u. Writing $\rho(\zeta, u) = \widetilde{\rho}(\zeta + h(u), u)$, the function $\widetilde{\rho}$ is of the same form as ρ but with $a_j(x, y) = 0$ for all $j = 1, \ldots, 2s$. We drop the tilde and denote the new function again by ρ. Sylvester's theorem furnishes an $\mathbb{R}$-linear transformation of the u-coordinates which puts $\sum_{i,j=1}^{2s} c_{ij}\, u_i u_j$ into a normal form $-|u'|^2 + |u''|^2$ where $u' = (u_1, \ldots, u_m)$ and $u'' = (u_{m+1}, \ldots, u_{2s})$ for some $m \in \{0, 1, \ldots, 2s\}$. This gives $\rho(\zeta, u) = \widetilde{\rho}(\zeta, u) + o(|\zeta|^2 + |u|^2)$ where

$$\widetilde{\rho}(\zeta, u) = -\left|x'\right|^2 - \left|u'\right|^2 + \left|x''\right|^2 + \left|u''\right|^2 + \sum_{j=1}^{r} \lambda_j y_j^2 \tag{3.17}$$

and

$$\lambda_j > 1 \quad \text{for } j = 1, \ldots, k, \qquad \lambda \geq 1 \quad \text{for } j = k+1, \ldots, r. \tag{3.18}$$

We say that a function (3.17) satisfying conditions (3.18) is a *q-convex quadratic normal form* at a critical point. Note that $k + m$ is the Morse index at 0. We summarize the above discussion in the following lemma.

Lemma 3.10.4 *Let X be an n-dimensional complex manifold and $\rho\colon X \to \mathbb{R}$ be a $\mathcal{C}^2$ function with a Morse critical point at $p_0 \in X$. If ρ is q-convex at p_0 for some $q \in \{1, \ldots, n+1\}$, then there exist*

- *a local holomorphic coordinate map $z = (\zeta, w)\colon U \to \mathbb{C}^r \times \mathbb{C}^s$ on an open neighborhood $U \subset X$ of p_0, with $z(p_0) = 0$, $r = n - q + 1$ and $s = q - 1$,*
- *a change of coordinates $\psi(z) = \psi(\zeta, w) = (\zeta + h(w), g(w))$ on $\mathbb{C}^n$ which is $\mathbb{R}$-linear in $w \in \mathbb{C}^s = \mathbb{R}^{2s}$, and*
- *a quadratic normal form $\widetilde{\rho}(\zeta, u)$ of type* (3.17), (3.18)

such that, setting $\phi(p) = \psi(z(p)) \in \mathbb{C}^n$ *for* $p \in U$, *we have*

$$\rho(p) = \rho(p_0) + \widetilde{\rho}(\phi(p)) + o(|\phi(p)|^2), \quad p \in U.$$

Furthermore, we can approximate ρ *as closely as desired in the* $\mathcal{C}^2$ *topology by a* q*-convex function* ρ' *that agrees with* ρ *outside of* U *and has a* nice critical point *at* p_0, *i.e.*, $\rho'(p) = \rho(p_0) + \widetilde{\rho}(\phi(p))$ *holds near* p_0 (*see Definition* 3.10.2).

Proof Everything except the claim in the last sentence has already been proved. The latter is seen by taking

$$\rho'(p) = \rho(p_0) + \widetilde{\rho}(\phi(p)) + \chi(\epsilon^{-1}\phi(p))o(|\phi(p)|^2),$$

where $\chi\colon \mathbb{C}^n \to [0,1]$ is a smooth function which equals zero in the unit ball $\mathbb{B} \subset \mathbb{C}^n$ and equals one on $\mathbb{C}^n \setminus 2\mathbb{B}$. When $\epsilon > 0$ decreases to zero, the $\mathcal{C}^2$-norm of the last summand tends to zero uniformly on U. □

3.11 Critical Levels of a q-Convex Function

The results of this section serve to pass a critical level of a strongly plurisubharmonic function or, more generally, of a q-convex function, in many subsequent analytic constructions. We begin with the case $q = 1$. Let $\rho\colon \mathbb{C}^n \to \mathbb{R}$ be a strongly plurisubharmonic quadratic normal form

$$\rho(z) = -|x'|^2 + |x''|^2 + \sum_{j=1}^{n} \lambda_j y_j^2 \tag{3.19}$$

where $z = (z', z'') = (x' + \mathrm{i}y', x'' + \mathrm{i}y'') \in \mathbb{C}^k \oplus \mathbb{C}^{n-k}$, $k \in \{1, \ldots, n\}$, $\lambda_j > 1$ for $j \in \{1, \ldots, k\}$, and $\lambda_j \ge 1$ for $j \in \{k+1, \ldots, n\}$ (see Lemma 3.10.1). This function has a unique (Morse) critical point with Morse index k at the origin. Set $\lambda = \min\{\lambda_1, \ldots, \lambda_k\} > 1$. Choose a number $c_0 > 0$ and let

$$E = \{(x' + \mathrm{i}y', z'') \in \mathbb{C}^n : y' = 0,\ z'' = 0,\ |x'|^2 \le c_0\}. \tag{3.20}$$

Choose a number $\mu \in (1, \lambda)$ and set

$$t_0 = t_0(c_0, \mu) = c_0\left(1 - \frac{1}{\mu}\right)^2 \in (0, c_0). \tag{3.21}$$

Lemma 3.11.1 *There exists a smooth strongly plurisubharmonic function* τ *on* $\mathbb{C}^n$ *satisfying the following conditions* (*see Fig.* 3.5):

(a) $\{\rho \le -c_0\} \cup E \subset \{\tau \le 0\} \subset \{\rho \le -t_0\} \cup E$,
(b) $\{\rho \le c_0\} \subset \{\tau \le 2c_0\} \subset \{\rho < 3c_0\}$,

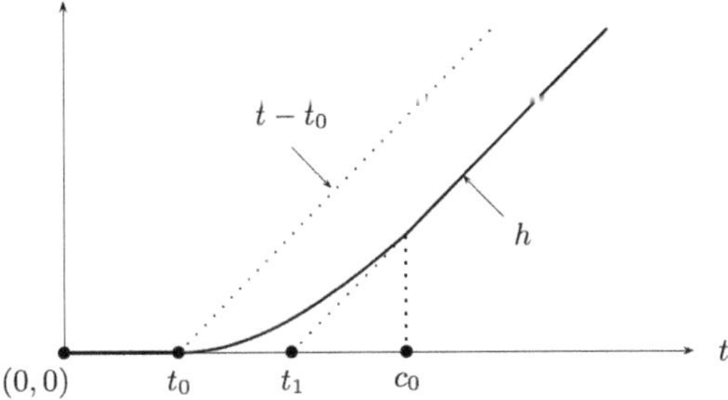

Fig. 3.4 The function h

(c) $\tau = \rho + t_1$ *on* $\{|x'|^2 \geq c_0\}$ *for some* $t_1 \in (t_0, c_0)$, *and*
(d) τ *has no critical values in* $(0, +\infty)$.

Proof We shall find τ of the form

$$\tau(z) = -h\big(|x'|^2\big) + |x''|^2 + \sum_{j=1}^{n} \lambda_j y_j^2 \tag{3.22}$$

where $h\colon \mathbb{R} \to [0, +\infty)$ is a smooth convex increasing function satisfying

(i) $h(t) = 0$ for $t \leq t_0$,
(ii) for $t \geq c_0$ we have $h(t) = t - t_1$ with $t_1 = c_0 - h(c_0) \in (t_0, c_0)$,
(iii) for $t_0 \leq t \leq c_0$ we have $t - t_1 \leq h(t) \leq t - t_0$, and
(iv) for all $t \in \mathbb{R}$ we have $0 \leq \dot{h}(t) \leq 1$ and $2t\ddot{h}(t) + \dot{h}(t) < \lambda$.

(See Fig. 3.4.) To construct h, we first consider the function

$$\xi(t) = \begin{cases} 0 & \text{if } t \leq t_0; \\ \mu(\sqrt{t} - \sqrt{t_0}\,)^2 & \text{if } t_0 \leq t \leq c_0, \\ t - c_0\big(1 - \frac{1}{\mu}\big) & \text{if } c_0 \leq t. \end{cases}$$

On $[t_0, c_0]$, ξ solves the initial value problem $2t\ddot{\xi} + \dot{\xi} = \mu$, $\xi(t_0) = \dot{\xi}(t_0) = 0$. It is $\mathcal{C}^1$ and piecewise $\mathcal{C}^2$, with discontinuities of the second derivative $\ddot{\xi}$ at t_0 and c_0. The value of t_0 is chosen such that $\dot{\xi}(c_0) = 1$. We have $\ddot{\xi}(t) = \mu\sqrt{t_0}/2\sqrt{t}^{\,3} > 0$ for $t \in [t_0, c_0]$, $\ddot{\xi}(t) = 0$ for t outside this interval, and $\int_{t_0}^{c_0} \ddot{\xi}(t)\mathrm{d}t = 1$.

Choose a smooth function $\chi \geq 0$ which vanishes outside $[t_0, c_0]$, equals $\ddot{\xi} + \epsilon$ on $[t_0 + \delta, c_0 - \delta]$ for small $\epsilon, \delta > 0$, and interpolates between 0 and $\ddot{\xi}$ on the intervals $[t_0, t_0 + \delta]$ and $[c_0 - \delta, c_0]$. We can find $\delta, \epsilon > 0$ arbitrarily small such that $\int_{t_0}^{c_0} \chi(t)dt = \int_{t_0}^{c_0} \ddot{\xi}(t)\mathrm{d}t = 1$. The function $h\colon \mathbb{R}_+ \to \mathbb{R}_+$ obtained by integrating χ twice with the initial conditions $h(t_0) = \dot{h}(t_0) = 0$ will satisfy the properties (i)–(iii) provided that ϵ and δ were chosen sufficiently small (since h is then $\mathcal{C}^1$-close to ξ and $\ddot{h} \leq \ddot{\xi} + \epsilon$). In particular, $t_1 = c_0 - h(c_0) \approx c_0 - \xi(c_0) = (1 - \frac{1}{\mu})c_0$ and hence $t_0 < t_1 < c_0$.

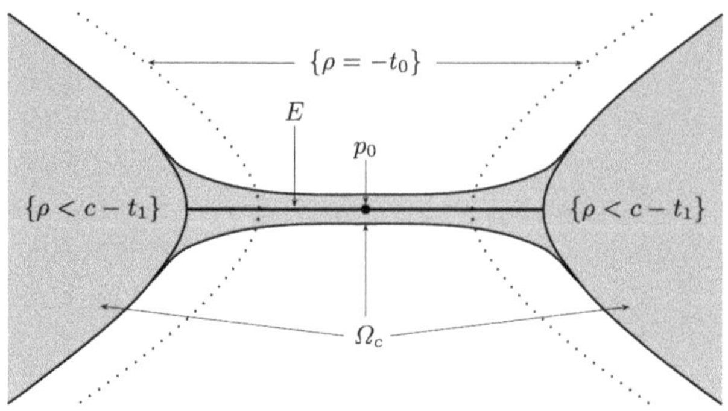

Fig. 3.5 The set $\Omega_c = \{\tau < c\}$. (Modified from [135, p. 246, Fig. 5])

Lemma 3.11.2 *Let $A = (a_{jl}) > 0$ be a symmetric real $n \times n$ matrix with the smallest eigenvalue $\lambda > 0$. If a $\mathcal{C}^2$ function $h\colon I \subset \mathbb{R}_+ \to \mathbb{R}$ satisfies*

$$\dot{h} < \lambda \quad \text{and} \quad 2t\ddot{h} + \dot{h} < \lambda, \quad t \in I, \tag{3.23}$$

then the function $\tau(z) = \tau(x + \mathrm{i}y) = \langle Ay, y\rangle - h(|x|^2)$ is strongly plurisubharmonic on the set $\{z = x + \mathrm{i}y \in \mathbb{C}^n : |x|^2 \in I\}$. (Here $\langle \cdot, \cdot \rangle$ denotes the Euclidean inner product on $\mathbb{R}^n$.)

Proof Let $A = (a_{jl})$. A calculation gives $-\tau_{z_j} = x_j\dot{h} + \mathrm{i}\sum_{s=1}^n a_{js}y_s$ and

$$-2\tau_{z_j\bar{z}_l} = \begin{cases} 2x_j^2\ddot{h} + \dot{h} - a_{jj} & \text{if } j = l; \\ 2x_jx_l\ddot{h} - a_{jl} & \text{if } j \neq l. \end{cases}$$

Thus the complex Hessian $H_\tau = (\frac{\partial^2\tau}{\partial z_j \partial\bar{z}_l})$ of τ satisfies

$$-2H_\tau = 2\ddot{h}\cdot xx^t + \dot{h}I - A,$$

where xx^t is the matrix product of the column $x \in \mathbb{R}^n$ with the row x^t and I denotes the identity matrix. For any $v \in \mathbb{R}^n$ we have $\langle (xx^t)v, v\rangle = v^txx^tv = |\langle x, v\rangle|^2$ which lies between 0 and $|x|^2|v|^2$. Hence $0 \le xx^t \le |x|^2 I$. (Here we write $A \le B$ if $B - A$ is nonnegative definite.) At points $|x|^2 = t$ where $\ddot{h}(t) \ge 0$ we thus get

$$-2H_\tau \le (2t\ddot{h} + \dot{h})I - A < \lambda I - A \le 0$$

and hence $H_\tau > 0$ (we used the second inequality in (3.23)). At points where $\ddot{h} < 0$ we can omit $2\ddot{h}xx^t \le 0$ to get $-2H_\tau \le \dot{h}I - A \le (\dot{h} - \lambda)I < 0$, so $H_\tau > 0$ as before. Thus, τ is strongly plurisubharmonic. □

We continue with the proof of Lemma 3.11.1. Let h be the function constructed above. By Lemma 3.11.2, the associated function τ (3.22) is strongly plurisubharmonic on $\mathbb{C}^n$. The properties (i)–(iii) of h imply

(α) $\rho \le \tau \le \rho + t_1$ (since $t - t_1 \le h(t) \le t$ for all $t \ge 0$),
(β) $\rho + t_0 \le \tau$ on the set $\{|x'|^2 \ge t_0\}$ (from (ii) and (iii)), and
(γ) $\tau = \rho + t_1$ on $\{|x'|^2 \ge c_0\}$ (from (ii)).

Conditions (a)–(c) in Lemma 3.11.1 now follow immediately. To see that τ has no critical values in $(0, +\infty)$ (condition (d)), note that $\tau(z) > 0$ implies (in view of $h \ge 0$) that $|x''|^2 + \sum_{j=1}^n \lambda_j y_j^2 > 0$, and this function has nonvanishing differential except where it equals zero. □

We obtain the analogous result for the q-convex normal form (3.17) on $\mathbb{C}^n = \mathbb{C}^r \oplus \mathbb{R}^{2s}$ ($r = n - q + 1$, $s = n - r = q - 1$). Using the coordinates $\zeta = (\zeta', \zeta'') = (x' + \mathrm{i}y', x'' + \mathrm{i}y'') \in \mathbb{C}^k \oplus \mathbb{C}^{r-k}$, $u = (u', u'') \in \mathbb{R}^{2s}$, we have

$$\rho(\zeta, u) = -|x'|^2 - |u'|^2 + Q(y, x'', u'') \tag{3.24}$$

where

$$Q(y, x'', u'') = |x''|^2 + |u''|^2 + \sum_{j=1}^r \lambda_j y_j^2,$$

with $\lambda_j > 1$ for $j = 1, \ldots, k$, and $\lambda \ge 1$ for $j = k + 1, \ldots, r$. Set $\lambda = \min\{\lambda_1, \ldots, \lambda_k\} > 1$; if $k = 0$ we take $\lambda = 2$. Choose $c_0 > 0$ and set

$$E = \{(\zeta, u) : y' = 0,\ \zeta'' = 0,\ u'' = 0,\ |x'|^2 + |u'|^2 \le c_0\}. \tag{3.25}$$

Choose a number $\mu \in (1, \lambda)$ and let $t_0 \in (0, c_0)$ be given by (3.21). With $h : \mathbb{R} \to \mathbb{R}_+$ chosen as in the proof of Lemma 3.11.1, we set

$$\tau(\zeta, u) = -h(|x'|^2 + |u'|^2) + |x''|^2 + |u''|^2 + \sum_{j=1}^r \lambda_j y_j^2. \tag{3.26}$$

Lemma 3.11.3 *For every fixed $u \in \mathbb{R}^{2s}$ the function $\tau(\cdot, u)$ (3.26) is strongly plurisubharmonic on $\mathbb{C}^r$ (hence τ is q-convex on $\mathbb{C}^n$), and τ satisfies conditions* (a)–(d) *in Lemma* 3.11.1 *with respect to the set E given by* (3.25), *except that condition* (c) *must be replaced by*

(c′) $\tau = \rho + t_1$ *on* $\{|x'|^2 + |u'|^2 \ge c_0\}$ *for some number* $t_1 \in (t_0, c_0)$.

Proof For a fixed $u \in \mathbb{R}^{2s}$ we set $c = |u'|^2 \ge 0$ and $h_c(t) = h(t + c)$ ($t \in \mathbb{R}$). We have $\dot{h}_c(t) = \dot{h}(t + c) \le 1 < \lambda$ and

$$2t\ddot{h}_c(t) + \dot{h}_c(t) \le 2(t + c)\ddot{h}(t + c) + \dot{h}(t + c) < \lambda$$

(we have used $\ddot{h} \ge 0$ and the property (iv) of h). Lemma 3.11.2 now shows that the function

$$\mathbb{C}^r \ni \zeta \mapsto -h_c(|x'|^2) + Q(y, x'', u'') = \tau(\zeta, u)$$

is strongly plurisubharmonic. The other properties are verified as before. □

Although Lemmas 3.11.1 and 3.11.3 only apply locally near a nice q-convex critical point, the property (c) (resp. (c′)) of the resulting function τ gives the same construction in the following global setting. Assume that ρ is a q-convex function in an open subset Ω of an n-dimensional complex manifold X, with a nice critical point at the point $p_0 \in \Omega$. Without loss of generality we may assume that $\rho(p_0)=0$. Assume in addition that for some $c_0>0$ the set

$$\Omega_{c_0}=\{p\in\Omega:-c_0<\rho(p)<3c_0\}$$

does not contain any critical point of ρ other than p_0. Choose an open neighborhood $U\subset\Omega$ of p_0 and a coordinate map $\phi\colon U\to P$ onto a polydisc P in $\mathbb{C}^n$ such that the function $\widetilde{\rho}=\rho\circ\phi^{-1}\colon P\to\mathbb{R}$ is a q-convex normal form (3.24). By decreasing $c_0>0$ if necessary we may assume that

$$\{(x+\mathrm{i}y,u)\in\mathbb{C}^r\times\mathbb{R}^{2s}:|x'|^2+|u'|^2\le c_0,\ Q(y,x'',u'')\le 4c_0\}\subset P.$$

Let $\widetilde{E}\subset\mathbb{C}^n$ be given by (3.25). Its preimage $E=\phi^{-1}(\widetilde{E})\subset U$ is an embedded disc of dimension $k+m$ (the Morse index of ρ at p_0) that is attached from the outside to the sublevel set $\{\rho\le -c_0\}$ along the sphere $bE\subset\{\rho=-c_0\}$. (In the metric on U inherited by ϕ from the standard metric in $\mathbb{C}^n$, E is the local stable manifold of p_0 for the gradient flow of ρ.) Let $t_0\in(0,c_0)$ be given by (3.21). Also, set

$$U'=\{p\in U:|x'(p)|^2+|u'(p)|^2<c_0\}.$$

Proposition 3.11.4 *There is a q-convex function $\tau:\{\rho<3c_0\}\to\mathbb{R}$ satisfying the following conditions:*

(a) $\{\rho\le -c_0\}\cup E\subset\{\tau\le 0\}\subset\{\rho\le -t_0\}\cup E$,
(b) $\{\rho\le c_0\}\subset\{\tau\le 2c_0\}\subset\{\rho<3c_0\}$,
(c) $\tau=\rho+t_1$ *holds outside of U' for some $t_1\in(t_0,c_0)$, and*
(d) τ *has no critical values in the open interval $(0,3c_0)\subset\mathbb{R}$.*

Proof Set $V=\{\rho<3c_0\}$. Let $\widetilde{\tau}$ be a q-convex function on $\mathbb{C}^n$ given by Lemma 3.11.3. We define a function $\tau: V\to\mathbb{R}$ by

$$\tau=\begin{cases}\widetilde{\tau}\circ\phi & \text{on } U\cap V,\\ \rho+t_1 & \text{on } V\setminus U.\end{cases}$$

By Lemma 3.11.3(c) these two definitions agree on $(U\setminus U')\cap V$; hence τ is well defined, and the stated properties follow immediately. □

3.12 Topological Structure of a Stein Space

A complex manifold X of dimension n, being triangulable, has the homotopy type of a CW complex of real dimension $\le 2n$. Since a Stein manifold admits a strongly plurisubharmonic Morse exhaustion function and each critical point of such function has Morse index at most $\dim_{\mathbb{C}} X$ by Proposition 3.10.1, Morse theory implies that an n dimensional Stein manifold is homotopy equivalent to a CW complex of real dimension $\leq n$. More precisely, assume that $\rho\colon X \to \mathbb{R}$ is a smooth Morse exhaustion function. Given real numbers $r' < r$, set

$$X_r = \{x \in X : \rho(x) \leq r\}, \qquad X_{r',r} = \{x \in X : r' \leq \rho(x) \leq r\} = \overline{X_r \setminus X_{r'}}.$$

If ρ is strongly plurisubharmonic on $X_{r',r}$ then the pair $(X_r, X_{r'})$ is a relative CW complex of dimension $\leq n$, which means that X_r is obtained from $X_{r'}$ by attaching cells of index $\leq n$. (See [520, p. 401] for the precise definition.) This was first observed by Lefschetz in the proof of his theorem on hyperplane sections of projective algebraic varieties (see [26] or [417, Sect. 7]). By the same token, Lemma 3.10.4 shows that a q-complete complex manifold of dimension n has the homotopy type of a CW complex of dimension at most $n+q-1$. These results were generalized by Hamm [276, 277] to complex spaces, both in the absolute and in the relative form as in the following theorem.

Theorem 3.12.1 ([276, 277]) *Let X be a complex space of dimension n, $\rho\colon X \to \mathbb{R}$ a real analytic q-convex exhaustion function on X, and $A \subset X$ a closed complex subvariety. Then, for every $r \in \mathbb{R}$ the pair $(X, A \cup X_r)$ is a relative CW complex of dimension at most $n+q-1$. In particular, if ρ is strongly plurisubharmonic (and hence X is Stein), then $(X, A \cup X_r)$ is a relative CW complex of dimension at most n.*

The following corollary is immediate.

Corollary 3.12.2 *A q-complete complex space X of complex dimension n has the homotopy type of a CW complex of dimension at most $n+q-1$. In particular, an n-dimensional Stein space is homotopy equivalent to a CW complex of dimension at most n.*

Proof of Theorem 3.12.1 We describe the main idea and refer to [277] for the details.

Let us consider the case when X is Stein. By a finite induction on strata it suffices to prove the result when the difference $X \setminus A$ is smooth (without singularities) and of pure dimension n. Pick functions $f_1, \ldots, f_k \in \mathcal{O}(X)$ such that $A = \{x \in X : f_1(x) = 0, \ldots, f_k(x) = 0\}$ and set $\varphi = \sum_{j=1}^k |f_j|^2$. (By a substratification one can even assume that $A = \{f = 0\}$ is a hypersurface and that $X \setminus A$ is a Stein manifold.) It suffices to show that for any given number $r' > r$ such that r' is a regular values of $\rho|_{X\setminus A}$ the set $X_{r'} = \{\rho \leq r'\}$ is obtained from $K = (A \cap X_{r'}) \cup X_r$ by attaching cells of index $\leq n$. Choose $M > 0$ big enough such that $X_{r'} \subset U = \{x \in X : \varphi(x) < M\}$. Consider the function $\chi\colon U \to \mathbb{R}_+$ given by $\chi = 0$ on $A \cup X_r$ any by

$$\chi(x) = \frac{\rho(x) - r}{\rho(x) - r + \log \frac{M}{\varphi(x)}} \qquad \text{for } x \in U' = U \setminus (A \cup X_r).$$

Clearly, χ is continuous on U and smooth on U'. A calculation shows that the Levi form $\mathcal{L}_\chi$ on U' is positive definite in the directions tangent to the level set of χ [277, Lemma 1]. Furthermore, there is a number $\alpha > 0$ such that χ and its restriction to the hypersurface $\{\rho = r'\} \setminus A$ have no critical points x with $0 < \chi(x) \leq \alpha$ [277, Lemma 2]. It follows that the set $\{x \in X_{r'} : \chi(x) \leq \alpha\}$ admits a strong deformation retraction onto K. (That K is a strong neighborhood deformation retract also follows from the fact that it is a semianalytic set.) The proof is completed in [277] by applying Morse theory on manifolds with smooth boundaries; in our case, $X_{r'} \setminus A$.

Here is an alternative argument for the last step. (Compare with the proof of Proposition 5.12.1 on p. 247.) After decreasing $\alpha > 0$ we may assume that α is a regular value of χ and the hypersurfaces $\{\rho = r'\}$ and $\{\chi = \alpha\}$ intersect transversely along the codimension two submanifold $\Sigma = \{\rho = r'\} \cap \{\chi = \alpha\}$ contained in $X \setminus A$. Hence, $D_0 = \{\chi \leq \alpha\} \cap X_{r'}$ is a piecewise smooth compact strongly pseudoconvex domain. Choosing $\lambda > 0$ big enough, the function $\tau = \mathrm{e}^{\lambda\chi}$ is strongly plurisubharmonic on the set $\{\tau \geq \alpha' = \mathrm{e}^{\lambda\alpha}\}$. For each number $s \in [0, 1]$ let

$$\rho_s = (1-s)\big(\tau - \alpha'\big) + s\big(\rho - r'\big), \qquad D_s = \{\rho_s \leq 0\} \cap X_{r'}.$$

(The sets D_s are similar to those shown in Fig. 5.4 on p. 250.) As the parameter s increases from 0 to 1, the compact sets D_s increase from D_0 to $D_1 = X_{r'}$. For any $0 \leq s < s' \leq 1$ we have $D_{s'} \setminus D_s \subset D_1 \setminus D_0 \Subset X \setminus A$. There are at most finitely many values of the parameter s at which the topology of D_s changes, and this change is described by attaching a handle of index $\leq n = \dim X$. (See the proof of Proposition 5.12.1.) The conclusion follows. A similar proof applies in the case of a q-complete complex space. □

Theorem 3.12.1, together with standard topological arguments, implies the following corollary which generalizes results of Narasimhan [426, Theorem 3] and Kaup [331, Satz 1].

Corollary 3.12.3 ([276, Korollar]) *If X is a q-complete complex space and A is a closed complex subvariety of X, then for any abelian group G we have*

$$H^k(X, A; G) = 0 = H_k(X, A; G), \quad k \geq n + q,$$

and $H_{n+q-1}(X; G)$ is free. The same holds if A is replaced by $A \cup X_r$, where $X_r = \{\rho \leq r\}$ for some real analytic q-convex exhaustion function $\rho \colon X \to \mathbb{R}$.

In the special case $q = 1$ (when X is Stein), another proof of these vanishing theorems was given by Colţoiu and Michalache [105].

Morse-theoretic methods also give information on the topology of complements of q-convex sets. The following case is of particular interest.

Theorem 3.12.4 ([184], [534, p. 93]) *If K is a compact polynomially convex set in $\mathbb{C}^n$, then its complement $\mathbb{C}^n \setminus K$ is $(n-1)$-connected:*

$$\pi_k\big(\mathbb{C}^n \setminus K\big) = 0, \quad k = 0, 1, \ldots, n-1.$$

Proof Pick an open set $U \subset \mathbb{C}^n$ with $K \subset U$. Using standard techniques we find a strongly plurisubharmonic Morse exhaustion function $\rho\colon \mathbb{C}^n \to \mathbb{R}$ that equal $|z|^2$ near infinity and satisfies $K \subset \{\rho \le 0\} \subset U$. We may assume that 0 is a regular value of ρ. For $r > 0$ big enough we have $\{\rho \ge r\} = \{|z| \ge \sqrt{r}\}$, and this set is clearly homotopy equivalent to the sphere S^{2n-1}. Morse theory, applied with the function $-\rho$ (which has Morse indexes $\ge n$), shows that the set $\{\rho \ge 0\}$ is obtained from $\{\rho \ge r\} \simeq S^{2n-1}$ by attaching cells of indexes $\ge n$. Such cells do not add any nontrivial elements to the homotopy groups in the range up to $n-1$, and hence the set $\{\rho \ge 0\}$ is $(n-1)$-connected. By exhaustion we get the same conclusion for $\mathbb{C}^n \setminus K$. □

Theorem 3.12.4, coupled with the Alexander duality theorem, gives the following cohomology vanishing theorem of A. Browder [70] (see also [534, p. 96]): For any polynomially convex set $K \subset \mathbb{C}^n$ and abelian group G we have

$$H^k(K; G) = 0, \quad k \ge n.$$

Furthermore, Zeron [592] proved that for any compact set $K \subset \mathbb{C}^n$ with a basis of open Stein neighborhoods, in particular for any compact rationally convex set, we have $\pi_k(\mathbb{C}^n \setminus K) = 0$ for $k = 0, \ldots, n-2$. (Recall that K is rationally convex if for any point $p \in \mathbb{C}^n \setminus K$ there exists a holomorphic polynomial f on $\mathbb{C}^n$ such that $f(p) = 0$ and $f \ne 0$ on K. See Stout [534] for more information.) By using the Alexander duality theorem it follows that $H^k(K; G) = 0$ for $k \ge n+1$.

Interesting converse results were obtained by Eliashberg [148], Gompf [246], and Cieliebak and Eliashberg [100]. For the Eliashberg-Gompf theorem, see Theorem 10.9.1 on p. 517. Here we present without proof the main result of [100].

Theorem 3.12.5 (Cieliebak and Eliashberg [100]) *Let W be a compact domain with smooth boundary in $\mathbb{C}^n$ for some $n \ge 3$.*

(a) *W is smoothly isotopic to a rationally convex domain if and only if it has a defining Morse function without critical points of index $> n$.*
(b) *W is smoothly isotopic to a polynomially convex domain if and only if it has a defining Morse function without critical points of index $> n$ and $H_n(W; G) = 0$ for any abelian group G.*

In view of the universal coefficient theorem, we have $H_n(W; G) = 0$ for every abelian group G if and only if $H_n(W; \mathbb{Z}) = 0$ and $H_{n-1}(W; \mathbb{Z})$ has no torsion. When W is simply connected, a third equivalent property is that W has a Morse defining function all of whose critical points have index less than n. In [100], the authors proved more precise results under an additional assumption, called *flexibility*, on the contact structure on the boundary bW of a strongly pseudoconvex domain. In particular, any two flexible strongly pseudoconvex domains in $\mathbb{C}^n$, $n \ge 3$, that are smoothly isotopic are isotopic through strongly pseudoconvex domains; this fails without the flexibility assumption. Furthermore, every flexible strongly pseudoconvex domain W in $\mathbb{C}^n$, $n \ge 3$, is isotopic through strongly pseudoconvex domains to a rationally convex domain, and, if $H_n(W;\mathbb{Z}) = 0$ and $H_{n-1}(W;\mathbb{Z})$ has no torsion, even to a polynomially convex domain. The proofs require a considerable amount of symplectic and contact topology and are beyond the scope of this book. A general background to this subject is available in the monograph [99] by Cieliebak and Eliashberg.

Another interesting work on this topic is due to Nemirovski and Siegel [430] who gave a complete characterization of disc bundles over smooth surfaces whose total spaces embed as rationally convex strongly pseudoconvex domains in $\mathbb{C}^2$.

Chapter 4
Automorphisms of Complex Euclidean Spaces

In this chapter we develop the theory of holomorphic automorphisms of complex Euclidean spaces and of other complex manifolds with big automorphism groups. These results play an important role in Oka theory, and they provide a rich family of examples and counterexamples which delineate the borderline between holomorphic flexibility and holomorphic rigidity.

The holomorphic automorphism group $\mathrm{Aut}(\mathbb{C}^n)$ is big and complicated when $n > 1$. Following pioneering works of Dixon and Esterle and Rosay and Rudin, Andersén and Lempert made a crucial discovery which marks the beginning of a new theory. They showed that the group generated by simple automorphisms called *shears* is dense in $\mathrm{Aut}(\mathbb{C}^n)$; see Theorem 4.9.1. The key underlying observation is that every polynomial holomorphic vector field on $\mathbb{C}^n$ is a finite sum of complete polynomial fields whose flows consist of shears. It follows that every isotopy of biholomorphic maps $F_t\colon \Omega \to \Omega_t = F_t(\Omega)$ between Runge domains in $\mathbb{C}^n$, beginning at $t = 0$ with the identity map, can be approximated by holomorphic automorphisms of $\mathbb{C}^n$ (see Theorem 4.9.2). The Andersén-Lempert theory applies to any Stein manifold whose Lie algebra of holomorphic vector fields is densely generated by the complete vector fields; this *holomorphic density property* was introduced by Varolin. Every Stein manifold with this property is an Oka manifold.

These methods are used in many interesting constructions: of nonstraightenable embedded complex lines in $\mathbb{C}^n$ (Theorem 4.18.7), of proper holomorphic embeddings with interpolation along discrete sets (Theorem 4.17.1), of proper holomorphic embeddings $f\colon \mathbb{C}^k \hookrightarrow \mathbb{C}^n$ whose complement $\mathbb{C}^n \setminus f(\mathbb{C}^k)$ is $(n-k)$-hyperbolic (Theorem 4.18.9), of nonlinearizable periodic holomorphic automorphisms (Theorem 4.19.1), of non-Runge Fatou-Bieberbach domains (Corollary 4.20.3), of long $\mathbb{C}^n$'s without nonconstant holomorphic or plurisubharmonic functions (Theorem 4.21.1), to mention just a few.

4.1 Shears

We denote by $\mathrm{Aut}(\mathbb{C}^n)$ the group of all holomorphic automorphisms of $\mathbb{C}^n$, endowed with the topology of uniform convergence on compact sets. This topology is

F. Forstnerič, *Stein Manifolds and Holomorphic Mappings*, Ergebnisse der Mathematik und ihrer Grenzgebiete. 3. Folge / A Series of Modern Surveys in Mathematics 56,
DOI 10.1007/978-3-319-61058-0_4

metrizable such that $\mathrm{Aut}(\mathbb{C}^n)$ becomes a complete metric space. For example, we can take the distance function

$$d(\Phi,\Psi)=\sum_{k=1}^{\infty}2^{-k}\frac{d_k(\Phi,\Psi)}{1+d_k(\Phi,\Psi)}$$

for any $\Phi,\Psi\in\mathrm{Aut}(\mathbb{C}^n)$, where

$$d_k(\Phi,\Psi)=\sup_{|z|\le k}\big(\big|\Phi(z)-\Psi(z)\big|+\big|\Phi^{-1}(z)-\Psi^{-1}(z)\big|\big).$$

Similarly, for every Stein manifold X the group $\mathrm{Aut}(X)$ carries a complete metric obtained by embedding X into a Euclidean space.

Every automorphism of $\mathbb{C}$ is affine linear: $z\mapsto az+b$ $(a,b,\in\mathbb{C},\ a\neq 0)$. From now on we assume that $n>1$. Let $z=(z_1,\dots,z_n)=(z',z_n)$ denote the coordinates on $\mathbb{C}^n$. Among the simplest automorphisms of $\mathbb{C}^n$ are maps

$$z\mapsto\big(z',z_n+f(z')\big),\tag{4.1}$$

$$z\mapsto\big(z',\mathrm{e}^{f(z')}z_n\big)\tag{4.2}$$

and their conjugates by the elements of $GL_n(\mathbb{C})$, where f is an entire function on $\mathbb{C}^{n-1}$. Their inverses are of the same form, with f replaced by $-f$. Automorphisms of type (4.1) and their $SL_n(\mathbb{C})$-conjugates are called *additive shears*, or *volume preserving shears* since they preserve the *standard holomorphic volume form* on $\mathbb{C}^n$,

$$\omega=\mathrm{d}z_1\wedge\mathrm{d}z_2\wedge\cdots\wedge\mathrm{d}z_n.\tag{4.3}$$

Automorphisms of type (4.2) and their $GL_n(\mathbb{C})$-conjugates are called *multiplicative shears* (also called *overshears* in [25]). In this book, the term *shear* refers to maps of both types. By composing one shear of each type we get a *generalized shear*:

$$z\mapsto\big(z',\mathrm{e}^{f(z')}z_n+g(z')\big).\tag{4.4}$$

Every additive shear (4.1) is a limit of polynomial additive shears obtained by approximating the entire function f by polynomials. A nontrivial multiplicative shear (4.2) is never polynomial.

Each shear is the time-one map of a one-parameter group of automorphisms, with the complex parameter $t\in\mathbb{C}$, consisting entirely of shears:

$$\Phi_t(z)=\big(z',z_n+tf(z')\big),\tag{4.5}$$

$$\Psi_t(z)=\big(z',\mathrm{e}^{tf(z')}z_n\big).\tag{4.6}$$

This means that $\Phi_0=\mathrm{Id}$, $\Phi_t\circ\Phi_s=\Phi_{t+s}$ for $t,s\in\mathbb{C}$, and similarly for Ψ_t.

Let us give a coordinate-free expression of shears. Recall that $\langle z,w\rangle=\sum_{i=1}^n z_i\overline{w}_i$ is the standard Hermitian form on $\mathbb{C}^n$. Given a $\mathbb{C}$-linear map $\lambda\colon\mathbb{C}^n\to\mathbb{C}^k$ for some

$k < n$, a vector $v = (v_1, \dots, v_n) \in \ker \lambda$, and an entire function $f \in \mathcal{O}(\mathbb{C}^k)$, we set for every $t \in \mathbb{C}$ and $z \in \mathbb{C}^n$:

$$\Phi_t(z) = z + tf(\lambda z)\, v, \tag{4.7}$$

$$\Psi_t(z) = z + \frac{1}{|v|^2}\big(\mathrm{e}^{t|v|^2 f(\lambda z)} - 1\big)\langle z, v\rangle v. \tag{4.8}$$

These are one-parameter groups of holomorphic automorphisms of $\mathbb{C}^n$ whose infinitesimal generators are the holomorphic vector fields

$$V_z = \frac{\mathrm{d}}{\mathrm{d}t}\Phi_t(z)|_{t=0} = f(\lambda z)v, \tag{4.9}$$

$$W_z = \frac{\mathrm{d}}{\mathrm{d}t}\Psi_t(z)|_{t=0} = f(\lambda z)\langle z, v\rangle v. \tag{4.10}$$

We have identified the vector $v = (v_1, \dots, v_n)$ with the constant vector field $\sum_{j=1}^n v_j \frac{\partial}{\partial z_j}$. When $\lambda_j(z_1, \dots, z_n) = (z_1, \dots, \hat{z}_j, \dots, z_n) \in \mathbb{C}^{n-1}$ and $v = e_j$ for some $j = 1, \dots, n$, we have

$$V_z = f(\lambda_j z)\frac{\partial}{\partial z_j}, \qquad W_z = f(\lambda_j z) z_j \frac{\partial}{\partial z_j}. \tag{4.11}$$

The following lemma is seen by a simple calculation [183, Appendix].

Lemma 4.1.1 *Let Φ_t and Ψ_t be given by* (4.7) *and* (4.8), *respectively.*

(i) *If $A \in GL_n(\mathbb{C})$ is such that $Av = e_n$, then $A \circ \Phi_t(z) = \widetilde{\Phi}_t(Az)$, where $\widetilde{\Phi}_t$ is given by* (4.5) *with f replaced by $f \circ \lambda \circ A^{-1}$.*
(ii) *If $A \in U(n)$ is such that $Av = |v|e_n$, then $A \circ \Psi_t(z) = \widetilde{\Psi}_t(Az)$, where $\widetilde{\Psi}_t$ is given by* (4.6) *with f replaced by $|v|^2 f \circ \lambda \circ A^{-1}$.*

In particular, every shear group (4.7), (4.8) *is $SU(n)$-conjugate to a shear group of the form* (4.5), (4.6), *respectively.*

Given a holomorphic map $F = (F_1, \dots, F_n)$ from a domain in $\mathbb{C}^n$ to $\mathbb{C}^n$, we denote by $F' = (\partial F_j / \partial z_k)$ its complex Jacobian matrix and by $\mathrm{J}F = \det F'$ its Jacobian determinant. Then, $F^*\omega = (\mathrm{J}F)\,\omega$ where ω is given by (4.3). We say that F is *volume preserving* if $F^*\omega = \omega$; equivalently, if $\mathrm{J}F = 1$. Let

$$\mathrm{Aut}_1\big(\mathbb{C}^n\big) = \big\{F \in \mathrm{Aut}\big(\mathbb{C}^n\big) : \mathrm{J}F = 1\big\}.$$

All additive shears (4.1) and their $SL_n(\mathbb{C})$-conjugates are volume preserving; in particular, the family Φ_t (4.7) is a one-parameter subgroup of $\mathrm{Aut}_1(\mathbb{C}^n)$.

More generally, a *holomorphic volume form* on an n-dimensional complex manifold X is a nowhere vanishing holomorphic n-form, i.e., a nonzero section of the canonical bundle $K_X = \wedge^n \mathrm{T}^{*1,0}X$. In local holomorphic coordinates on X, such a form equals $\omega = a(z)\mathrm{d}z_1 \wedge \cdots \wedge \mathrm{d}z_n$ where a is a nowhere vanishing holomorphic function. A holomorphic map $F\colon D \to X$ on a domain $D \subset X$ is *volume preserving* if $F^*\omega = \omega$. The *divergence* of a holomorphic vector field V with respect to a holomorphic volume form ω is the holomorphic function $\mathrm{div}_\omega V$ satisfying the equation

$$L_V\omega = \mathrm{d}(V\rfloor\omega) = \mathrm{div}_\omega V \cdot \omega. \tag{4.12}$$

Here, $L_V\omega$ is the Lie derivative of ω (p. 36) and $V\rfloor\omega$ is the contraction of ω by V (p. 38). The first equality follows from Cartan's formula (1.55). For the standard volume form (4.3) on $\mathbb{C}^n$ we get

$$\mathrm{div}_\omega\left(\sum_{j=1}^n a_j(z)\frac{\partial}{\partial z_j}\right) = \sum_{j=1}^n \frac{\partial a_j}{\partial z_j}(z).$$

Let ϕ_t denote the local flow of a holomorphic vector field V. Cartan's formula (1.55) and (4.12) give

$$\frac{\mathrm{d}}{\mathrm{d}t}\,\phi_t^*\omega = \phi_t^* L_V\omega = \phi_t^*\mathrm{d}(V\rfloor\omega) = \phi_t^*(\mathrm{div}_\omega V \cdot \omega).$$

Since $\phi_t^*\omega = (\mathrm{J}\phi_t)\,\omega$ and $\phi_0 = \mathrm{Id}$, we conclude that

$$\mathrm{J}\phi_t \equiv 1 \quad\Longleftrightarrow\quad \mathrm{div}_\omega V = 0; \tag{4.13}$$

that is, divergence zero vector fields correspond to volume preserving flows. The analogous result holds for time-dependent holomorphic vector fields.

For the shear vector fields (4.9) and (4.10) with $|v| = 1$ on $\mathbb{C}^n$, and ω the standard volume form (4.3), we have

$$\mathrm{div}_\omega\big(f(\lambda z)v\big) = 0, \qquad \mathrm{div}_\omega\big(f(\lambda z)\langle z, v\rangle v\big) = f(\lambda z). \tag{4.14}$$

This is obvious when $v = e_n$, and the general case follows from Lemma 4.1.1 since conjugation by maps in $SL_n(\mathbb{C})$ preserves the divergence.

We also consider *symplectic holomorphic maps*. Let $x = (z, w)$ be complex coordinates on $\mathbb{C}^{2n}$, with $z = (z_1, \ldots, z_n) \in \mathbb{C}^n$ and $w = (w_1, \ldots, w_n) \in \mathbb{C}^n$. Let $\{e^j : j = 1, \ldots, 2n\}$ denote the standard basis of $\mathbb{C}^{2n}$, and let $u = (u_1, \ldots, u_{2n})$, $v = (v_1, \ldots, v_{2n}) \in \mathbb{C}^{2n}$. The alternating bilinear form

$$\omega(u, v) = \sum_{j=1}^n u_j v_{n+j} - u_{n+j} v_j \tag{4.15}$$

is the *standard symplectic form* on $\mathbb{C}^{2n}$. The *linear complex symplectic group* $Sp_n(\mathbb{C})$ consists of all $A \in GL_{2n}(\mathbb{C})$ satisfying

$$\omega(Au, Av) = \omega(u, v), \quad u, v \in \mathbb{C}^{2n}.$$

Introducing the linear operator J on $\mathbb{C}^{2n}$ by

$$Je^k = -e^{n+k}, \qquad Je^{n+k} = e^k, \quad k = 1, \ldots, n$$

we have that $\omega(u,v)=u^t Jv$ and hence

$$Sp_n(\mathbb{C})=\{A\subset GL_{2n}(\mathbb{C}):A^tJA=J\}. \tag{4.16}$$

Recall [308, p. 373] that the linear symplectic group $Sp_n(\mathbb{C})$ is generated by linear *symplectic transvections*

$$A_{\lambda,v}(x)=x+\lambda\,\omega(x,v)\,v,\quad \lambda\in\mathbb{C}^*,\ v\in\mathbb{C}^{2n}. \tag{4.17}$$

The symplectic differential form on $\mathbb{C}^{2n}$ corresponding to (4.15) is

$$\omega=\sum_{j=1}^{n}\mathrm{d}z_j\wedge\mathrm{d}w_j=\sum_{j=1}^{n}\mathrm{d}x_j\wedge\mathrm{d}x_{n+j}. \tag{4.18}$$

We shall use the same letter ω for both (4.15) and (4.18). Note that

$$\frac{(-1)^{[n/2]}}{n!}\omega^n=\mathrm{d}x_1\wedge\mathrm{d}x_2\wedge\cdots\wedge\mathrm{d}x_{2n}$$

is the standard volume form on $\mathbb{C}^{2n}$.

A holomorphic mapping $F\colon D\to\mathbb{C}^{2n}$ on a domain $D\subset\mathbb{C}^{2n}$ is said to be *symplectic holomorphic* if $F^*\omega=\omega$. Such a map also satisfies $F^*\omega^n=\omega^n$ and hence $JF=1$. Clearly, the composition of symplectic holomorphic maps is again symplectic holomorphic. Denote by $\mathrm{Aut}_{\mathrm{sp}}(\mathbb{C}^{2n})$ the group of all symplectic holomorphic automorphisms of $\mathbb{C}^{2n}$. We have

$$Sp_n(\mathbb{C})\subset\mathrm{Aut}_{\mathrm{sp}}(\mathbb{C}^{2n})\subset\mathrm{Aut}_1(\mathbb{C}^{2n}),\quad \mathrm{Aut}_{\mathrm{sp}}\mathbb{C}^2=\mathrm{Aut}_1(\mathbb{C}^2).$$

If $v\in\mathbb{C}^{2n}$ and f is an entire function on $\mathbb{C}$, then the map

$$F(x)=x+f\big(\omega(x,v)\big)v,\quad x\in\mathbb{C}^{2n} \tag{4.19}$$

is an additive shear automorphism of $\mathbb{C}^{2n}$ which is easily seen to be symplectic; we call such a map a *symplectic shear*. Indeed, its differential equals

$$\mathrm{d}F_x\cdot u=u+f'\big(\omega(x,v)\big)\omega(u,v)\,v$$

and hence $\omega(\mathrm{d}F_x\cdot u,\mathrm{d}F_x\cdot u')=\omega(u,u')$, so $F^*\omega=\omega$. (Every additive shear on $\mathbb{C}^2$ is symplectic.) Note that F (4.19) is the time-one map of the flow

$$\phi_t(x)=x+tf\big(\omega(x,v)\big)v,\quad t\in\mathbb{C}$$

belonging to $\mathrm{Aut}_{\mathrm{sp}}(\mathbb{C}^{2n})$. Its infinitesimal generator

$$V_x=f\big(\omega(x,v)\big)v,\quad x\in\mathbb{C}^{2n} \tag{4.20}$$

is a complete *holomorphic Hamiltonian vector field* on $\mathbb{C}^{2n}$.

We mention a few other types of automorphisms. Choose nonnegative integers $k_1, \ldots, k_n$ and complex numbers $c_1, \ldots, c_n$ such that $\sum_{j=1}^n c_j k_j = 0$, and let $f \colon \mathbb{C} \to \mathbb{C}$ be an entire function. The map $w = F(z)$ given by

$$w_j = z_j \, \mathrm{e}^{c_j f(z_1^{k_1} \cdots z_n^{k_n})}, \quad j = 1, \ldots, n$$

is an automorphism satisfying $w_1^{k_1} \cdots w_n^{k_n} = z_1^{k_1} \cdots z_n^{k_n}$, and its Jacobian is $w_1 \cdots w_n / z_1 \cdots z_n$. In the special case $n = 2$, $k_1 = k_2 = 1$, $c_1 = 1$ and $c_2 = -1$ we get a one-parameter group $\{\Theta_t\}_{t \in \mathbb{C}} \in \mathrm{Aut}_1(\mathbb{C}^2)$ given by

$$\Theta_t(z_1, z_2) = \big(z_1 \mathrm{e}^{tf(z_1 z_2)}, z_2 \mathrm{e}^{-tf(z_1 z_2)}\big).$$

According to Nishimura [434], every injective holomorphic map $\mathbb{C}^2 \to \mathbb{C}^2$ with Jacobian one and sending the coordinate axes to themselves is of this form.

Let $n = 2$. By composing a shear (4.1) with an affine linear map $(z_1, z_2) \to (z_2, \alpha z_1 + \beta)$ $(\alpha, \beta \in \mathbb{C},\ \alpha \neq 0)$ we get a *Hénon map*

$$(z_1, z_2) \mapsto \big(z_2 + f(z_1), \alpha z_1 + \beta\big). \tag{4.21}$$

4.2 Automorphisms of $\mathbb{C}^2$

In this section we review some classical results on the polynomial automorphism group $\mathrm{Aut}_{\mathrm{alg}}(\mathbb{C}^2)$ of the plane, and we mention generalizations to the subgroups of $\mathrm{Aut}(\mathbb{C}^2)$ generated by entire shears.

Let $z = (x, y)$ be complex coordinates on $\mathbb{C}^2$. We denote by $\mathcal{A}$ the group of all affine automorphisms of $\mathbb{C}^2$:

$$a(x, y) = (\alpha x + \beta y + \xi, \gamma x + \delta y + \eta), \quad \alpha\delta - \beta\gamma \neq 0. \tag{4.22}$$

Let $\mathcal{E} \subset \mathrm{Aut}_{\mathrm{alg}}(\mathbb{C}^2)$ consist of all *elementary automorphisms*:

$$e(x, y) = \big(\alpha x + p(y), \beta y + \gamma\big), \tag{4.23}$$

where $\alpha, \beta \in \mathbb{C}^*$, $\gamma \in \mathbb{C}$, and $p(y)$ is a polynomial in y. Note that every affine linear map is a composition of elementary linear maps (in fact, of linear shears and dilations in the coordinate directions).

More generally, automorphisms of $\mathbb{C}^n$ whose k-th component only depends on the first k variables are called *elementary* or *de Jonquières*; they form a group under composition, called the *de Jonquières group*. These notions extend to fields other than $\mathbb{C}$, and (in the case of $\mathbb{C}$) to entire automorphisms.

By a theorem of Jung [317], for any field k, the group $\mathrm{Aut}(k^2)$ of polynomial automorphisms of the plane k^2 is a free product of the affine group $\mathcal{A}$ and the elementary group $\mathcal{E}$, amalgamated over $\mathcal{A} \cap \mathcal{E}$. (See also Friedland and Milnor [233].) In particular, the following holds.

Theorem 4.2.1 *The polynomial automorphism group* $\mathrm{Aut}_{\mathrm{alg}}(\mathbb{C}^2)$ *is an amalgamated free product* $\mathcal{A} * \mathcal{E}$ *of its subgroups* $\mathcal{A}$ *(affine) and* $\mathcal{E}$ *(elementary).*

The amalgamated free product structure $\mathrm{Aut}_{\mathrm{alg}}(\mathbb{C}^2) = \mathcal{A} * \mathcal{E}$ means that every automorphism $g \in \mathrm{Aut}_{\mathrm{alg}}(\mathbb{C}^2) \setminus (\mathcal{A} \cap \mathcal{E})$ is a represented by a reduced word $g = g_m \circ g_{m-1} \circ \cdots \circ g_1$, where $g_j \in \mathcal{A} \cup \mathcal{E} \setminus (\mathcal{A} \cap \mathcal{E})$ for every j and every two consecutive elements g_j, g_{j+1} belong to different groups $\mathcal{A}, \mathcal{E}$. Such a representation is unique up to elements in $\mathcal{A} \cap \mathcal{E}$; in particular, no reduced word of length $m > 0$ gives the identity element.

We do not give a detailed proof of Theorem 4.2.1, but only outline the main steps. The first main point is that the subgroups $\mathcal{A}$ and $\mathcal{E}$ generate $\mathrm{Aut}_{\mathrm{alg}}(\mathbb{C}^2)$. This is accomplished by showing that, if P and Q are holomorphic polynomials in two complex variables (x, y) such that the map $(x, y) \mapsto F(x, y) = (P(x, y), Q(x, y))$ is an automorphism of $\mathbb{C}^2$, then the degree of one of the polynomials divides the degree of the other one. One can proceed via the Abhyankar-Moh-Suzuki theorem [2, 540] as follows. (See also Sect. 4.18.) Consider the polynomial embedding $\mathbb{C} \ni x \mapsto (P(x, 0), Q(x, 0)) \in \mathbb{C}^2$ of $\mathbb{C}$ into $\mathbb{C}^2$. Let p, q be the degrees of the two components, considered as polynomials in a single complex variable x. We may assume that $p \le q$. The key step is to show that p divides q, $q = dp$. There exist several algebraic proofs of this result, but perhaps the simplest proof, due to L. Rudolph [484], uses elementary knot theory. Once this is known, we can decrease the degree by composing the given embedding with a polynomial shear $(P, Q) \mapsto (P, Q - \alpha P^d)$ for a suitable choice of $\alpha \in \mathbb{C}^*$. Continuing inductively, we see that by composing F with a finite composition of polynomial shears we obtain an automorphism of $\mathbb{C}^2$ which preserves the line $\mathbb{C} \times \{0\}$. Clearly the other lines $\mathbb{C} \times \{y\}$ for $y \neq 0$ get mapped onto lines $\mathbb{C} \times \{h(y)\}$ with $h(y) \neq 0$ (since they get mapped to $\mathbb{C} \times \mathbb{C}^*$ and the projection to the second coordinate is a nowhere vanishing polynomial on $\mathbb{C}$, hence a constant). So h is a linear function of y, and the new map $(x, y) \mapsto (\alpha x + g(y), by)$ is elementary.

To prove uniqueness of the decomposition into reduced words, one shows that the degree of a reduced word equals the product of the degrees of its characters g_j. (See [233] for this. The degree of $g \in \mathrm{Aut}_{\mathrm{alg}}(\mathbb{C}^2)$ is the larger of the degrees of its polynomial components.) It follows that every $g \in \mathrm{Aut}_{\mathrm{alg}}(\mathbb{C}^2)$ is conjugate either to an element of $\mathcal{E}$, or to a *cyclically reduced word* $a_r \circ e_r \circ \cdots \circ a_1 \circ e_1$ of even length, with $a_j \in \mathcal{A}$ and $e_j \in \mathcal{E}$ for every j. In the latter case we can arrange that every a_j equals the map $\sigma(x, y) = (y, x)$. The length of a cyclically reduced word is invariant under conjugation, and the degrees of its characters are invariant up to cyclic permutations. The simplest examples are $g = a \circ e$ with $a \in \mathcal{A} \setminus \mathcal{E}$ and $e \in \mathcal{E} \setminus \mathcal{A}$. These include Hénon mappings $h(x, y) = (y, p(y) - \delta x)$ where $\delta \in \mathbb{C}^*$ and p is a polynomial (take $a = \sigma$ and $e(x, y) = (-\delta x + p(y), y)$).

While maps in $\mathcal{E} \cup \mathcal{A}$ have simple dynamics, automorphisms represented by cyclically reduced words of even length exhibit nontrivial dynamical properties; see e.g. [48, 49, 165, 166, 233, 290].

If a group $\mathcal{G}$ is an amalgamated free product $\mathcal{A} * \mathcal{E}$ of its subgroups $\mathcal{A}$ and $\mathcal{E}$, then certain types of subgroups $G \subset \mathcal{G}$ are conjugate to a subgroup of $\mathcal{A}$ or to a subgroup of $\mathcal{E}$. This holds if G has bounded length with respect to the amalgamated product decomposition of $\mathcal{G}$ [499, Chap. I, No. 4.3, Théorème 8]. If $\mathcal{G}$ is a topological group and $\mathcal{A}, \mathcal{E}$ are closed subgroups then the same holds for every connected Lie group G that is a subgroup of $\mathcal{G}$ (see [33]). In particular, we have the following result [588].

Theorem 4.2.2 *Every subgroup of* $\mathrm{Aut}_{\mathrm{alg}}(\mathbb{C}^2)$ *isomorphic to* $\mathbb{R}$ *or to* $\mathbb{C}$ *is conjugate to a subgroup in the affine group* $\mathcal{A}$ *or in the elementary group* $\mathcal{E}$.

There is a simple reason why Hénon maps (4.21) and their compositions do not belong to any flow. Denote by h^k the k-th iterate of an automorphism (or a diffeomorphism) h, defined inductively by

$$h^1 = h, \qquad h^k = h \circ h^{k-1} \text{ for } k \in \mathbb{N}, \qquad h^{-k} = \left(h^{-1}\right)^k. \tag{4.24}$$

A point p is a *periodic point* of h is $h^k(p) = p$ for some $k \in \mathbb{N}$; the smallest such k is the *period* of p. If this holds for $k = 1$ then p is a *fixed point* of h.

Lemma 4.2.3 *Assume that X is a smooth manifold and that h is a diffeomorphism of X with the following two properties*:

(i) *the set of periodic points of h does not contain any nontrivial curves, and*
(ii) *there exists a periodic point that is not a fixed point.*

Then h is not the time-one map of any complete flow on X.

Proof Assume that $h = \phi_1$ where $\{\phi_t\}_{t\in\mathbb{R}}$ is a complete flow. Let p be a periodic point of h of period $k > 1$. Using the group property of ϕ_t and the assumption $h = \phi_1$ (hence $h^k = \phi_k$), we have

$$h^k\left(\phi_t(p)\right) = \phi_k\left(\phi_t(p)\right) = \phi_t\left(\phi_k(p)\right) = \phi_t\left(h^k(p)\right) = \phi_t(p).$$

Thus, the orbit $\{\phi_t(p) : t \in \mathbb{R}\}$ (which is a nontrivial curve since $\phi_1(p) \neq p$) consists of periodic points of period k, a contradiction to (i). □

Corollary 4.2.4 ([75]) *If* $\{\phi_t\}_{t\in\mathbb{C}} \subset \mathrm{Aut}(\mathbb{C}^2)$ *is a flow such that* $\phi_1 \in \mathrm{Aut}_{\mathrm{alg}}\mathbb{C}^2$, *then* ϕ_1 *is conjugate in* $\mathrm{Aut}_{\mathrm{alg}}\mathbb{C}^2$ *to an elementary map in* $\mathcal{E}$. *In particular, a nontrivial Hénon map* (4.21) *does not belong to any flow.*

Proof (sketch) Using the explicit form of polynomial Hénon maps, one can see by Bézout's theorem that every composition of Hénon maps has finitely many periodic points of any fixed period k, and this number grows as $k \to +\infty$. Hence we find a periodic point that is not a fixed point; Lemma 4.2.3 implies that such a map does not belong to any flow. Hence, if $h = \phi_1 \in \mathrm{Aut}_{\mathrm{alg}}\mathbb{C}^2$ then h is conjugate in $\mathrm{Aut}_{\mathrm{alg}}\mathbb{C}^2$ to an element of $\mathcal{A}$ or to an element of $\mathcal{E}$. Further, every affine map $z \mapsto Az + b$ is linearly conjugate to a map in $\mathcal{E}$ (by conjugating the matrix A to its lower triangular Jordan form). □

Complete flows on $\mathbb{C}^2$ consisting of polynomial maps were classified by Suzuki [541]; another proof was given in [46]. The classification was extended by Ahern, Varolin and the author [9, 10] who proved that any holomorphic flow $\{\phi_t\}_{t\in\mathbb{C}} \subset \mathrm{Aut}(\mathbb{C}^2)$ such that $\phi_1 \in \mathrm{Aut}_{\mathrm{alg}}\mathbb{C}^2$ is conjugate to a flow in the polynomial group $\mathrm{Aut}_{\mathrm{alg}}\mathbb{C}^2$. Brunella [71] gave a full classification of complete polynomial vector fields on $\mathbb{C}^2$ with entire flows.

A result analogous to Theorem 4.2.1 holds for subgroups of $\mathrm{Aut}(\mathbb{C}^2)$ generated by entire shears. Denote by $\mathcal{E} \subset \mathrm{Aut}(\mathbb{C}^2)$ the set of all maps

$$e(x, y) = \left(e^{g(y)}x + h(y), \beta y + \gamma\right), \tag{4.25}$$

where g and h are entire functions on $\mathbb{C}$, $\beta \in \mathbb{C}^*$, and $\gamma \in \mathbb{C}$. In analogy to the polynomial case we call such maps elementary. Let $\mathcal{E}_1$ denote the set of maps (4.25) with Jacobian one; this requires that g is constant (so $\alpha = e^g \in \mathbb{C}^*$) and that $\alpha\beta = 1$. Let $\mathcal{E}_c$ denote the set of maps (4.25) with constant nonvanishing Jacobian; this requires that g is constant. Denote by $\mathcal{S}(2)$ the subgroup of $\mathrm{Aut}(\mathbb{C}^2)$ generated by $\mathcal{E}$ and by the affine group $\mathcal{A}$. Similarly, let $\mathcal{S}_1(2)$ be the subgroup of $\mathrm{Aut}(\mathbb{C}^2)$ generated by $\mathcal{E}_1$ and $\mathcal{A}_1$, where $\mathcal{A}_1$ contains all affine maps with Jacobian one. Finally, let $\mathcal{S}_c(2) = \mathcal{S}_1(2) \times \mathbb{C}^*$ be the group generated by $\mathcal{E}_c$ and $\mathcal{A}$; $\mathcal{S}_c(2)$ consists of maps in $\mathcal{S}(2)$ with constant Jacobian. It was proved by Andersén [23] that for every $n \geq 2$ the shear group $\mathcal{S}_1(n)$ is a proper subgroup of $\mathrm{Aut}_1(\mathbb{C}^n)$; similarly, $\mathcal{S}(n)$ is a proper subgroup of $\mathrm{Aut}(\mathbb{C}^n)$ for every $n > 1$ [25]. Ahern and Rudin proved [12] that $\mathcal{S}(2)$ is a free product of its subgroups $\mathcal{E}$ and $\mathcal{A}$, amalgamated over their intersection $\mathcal{A} \cap \mathcal{E}$. Earlier C. de Fabritiis showed [109] that the groups $\mathcal{S}_1(2)$, resp. $\mathcal{S}_c(2)$, are amalgamated free products of their subgroups $\mathcal{E}_1$ and $\mathcal{A}_1$, resp. of $\mathcal{E}_c$ and $\mathcal{A}$. So in these cases we have an analogue of Theorem 4.2.1 of Jung and Van der Kulk.

The polynomial automorphism group $\mathrm{Aut}_{\mathrm{alg}}\mathbb{C}^n$ for $n \geq 3$ is much more complicated and contains automorphisms that are not a composition of elementary ones; such are called wild. In particular, Shestakov and Umirbaev proved in [500] that the following *Nagata automorphism* of $\mathbb{C}^3$ is wild:

$$\begin{aligned} x' &= x + \left(x^2 - yz\right)z, \\ y' &= y + 2\left(x^2 - yz\right)x + \left(x^2 - yz\right)^2 z, \\ z' &= z. \end{aligned}$$

However, this automorphism becomes tame after adding sufficiently many additional variables.

The main open problem in this area is the *Jacobian conjecture* which claims that every polynomial map $F\colon \mathbb{C}^n \to \mathbb{C}^n$ with nonvanishing (hence constant) Jacobian is an automorphism of $\mathbb{C}^n$. This holds if F is injective [482].

4.3 Attracting Basins and Fatou-Bieberbach Domains

By the Riemann mapping theorem and Liouville's theorem, the complex number field $\mathbb{C}$ does not contain any proper subsets biholomorphic to itself. The situation is entirely different in higher dimensional Euclidean spaces. In a couple of classical papers, Fatou [156] and Bieberbach [56] investigated basins of attraction of holomorphic automorphisms of $\mathbb{C}^2$ which were proper subdomains but nevertheless biholomorphic to the whole space.

Definition 4.3.1 A proper subdomain Ω of $\mathbb{C}^n$ that is biholomorphic to $\mathbb{C}^n$ is called a *Fatou-Bieberbach domain*. A biholomorphic map $F\colon \mathbb{C}^n \to \Omega$ onto such Ω (and its inverse map) is called a *Fatou-Bieberbach map*.

According to Rudin [482], all injective polynomial maps $\mathbb{C}^n \to \mathbb{C}^n$ are automorphisms, so each Fatou-Bieberbach map is non-algebraic.

Let $F \in \mathrm{Aut}(\mathbb{C}^n)$. We denote by F^k its k-th iterate (4.24). A point $p \in \mathbb{C}^n$ for which $F(p) = p$ is called a *fixed point* of F; a fixed point of an iterate F^k is a *periodic point*. A fixed point p is said to be *attracting* if all eigenvalues λ_j of the derivative $F'(p)$ satisfy $|\lambda_j| < 1$, and is said to be *repelling* if all eigenvalues satisfy $|\lambda_j| > 1$. In the attracting case the set

$$\Omega_{F,p} = \left\{ z \in \mathbb{C}^n : \lim_{k\to\infty} F^k(z) = p \right\} \tag{4.26}$$

is called the *basin* of p. In this section we prove the following result due to Sternberg [527] and Rosay and Rudin [477].

Theorem 4.3.2 *If $n > 1$ and $p \in \mathbb{C}^n$ is an attracting fixed point of a holomorphic automorphism $F \in \mathrm{Aut}(\mathbb{C}^n)$, then there exists a biholomorphic map ψ from the basin Ω (4.26) onto $\mathbb{C}^n$. If the Jacobian $\mathrm{J}F$ is constant, then ψ can be chosen such that $\mathrm{J}\psi \equiv 1$.*

This theorem is related to the problem of finding normal forms for germs of biholomorphic maps, a subject with a long history. The proof of Theorem 4.3.2 that we give here is due to Buchmann, Nærland and Wold (unpublished, private communication). The analogous result holds for an attracting periodic point. On the hand, this fails in general at a parabolic fixed point. In fact, Bracci, Raissy and Stensønes [66, Theorem 0.1] have recently constructed an example of a holomorphic automorphism F of $\mathbb{C}^2$ having an invariant, non-recurrent, attracting Fatou component biholomorphic to $\mathbb{C} \times \mathbb{C}^*$; its point of attraction is $(0,0)$ and the eigenvalues of $dF(0,0)$ are λ and $\bar\lambda$ where $|\lambda| = 1$ and λ is not a root of unity.

Remark 4.3.3 (A) By Corollary 4.9.5 on p. 136, a Fatou-Bieberbach domain $\Omega \subset \mathbb{C}^n$ is Runge in $\mathbb{C}^n$ if and only if the associated Fatou-Bieberbach map $F\colon \mathbb{C}^n \to \Omega$ is a locally uniform limit of holomorphic automorphisms of $\mathbb{C}^n$. Hence, the proof of Theorem 4.3.2 shows that an attracting basin of a holomorphic automorphism of $\mathbb{C}^n$ is Runge in $\mathbb{C}^n$. Examples of non-Runge Fatou-Bieberbach domains are given in Sect. 4.20. Wold also showed in [580] that there exist Fatou-Bieberbach domains which are not basins of a single automorphism, but rather of a sequence of automorphisms (see Sect. 4.4).

(B) Our proof of Theorem 4.3.2 only uses the local behavior of F at a fixed point; hence it applies to biholomorphic automorphisms $F: X \to X$ of any complex manifold and shows that the basin of any attracting fixed or periodic point of F is biholomorphic to $\mathbb{C}^n$ with $n = \dim X$. □

We begin with some preparations. We may assume that the fixed point p is the origin $0 \in \mathbb{C}^n$. By writing $F: \mathbb{C}^n_0 \to \mathbb{C}^n_0$ we mean that F is an injective holomorphic map near the origin in $\mathbb{C}^n$ satisfying $F(0) = 0$. Such F is *attracting* at 0 if the eigenvalues $\lambda_1, \ldots, \lambda_n$ of $F'(0)$ are all less than one in absolute value. (Injectivity implies that no eigenvalue equals zero.) Two such maps $F, G: \mathbb{C}^n_0 \to \mathbb{C}^n_0$ are said to be *conjugate* if there exists an injective holomorphic map $\psi: \mathbb{C}^n_0 \to \mathbb{C}^n_0$ such that $F = \psi^{-1} \circ G \circ \psi$ holds near the origin.

Lemma 4.3.4 *Let $F, G \in \mathrm{Aut}(\mathbb{C}^n)$ satisfy $F(0) = G(0) = 0$. If F and G are attracting at the origin and are locally conjugate at the origin, then they are globally conjugate on their basins of attraction. In particular, the basin $\Omega_{F,0}$ of F is biholomorphic to the basin $\Omega_{G,0}$ of G.*

Proof Let ψ denote the local conjugating map, i.e., $\psi = G^{-1} \circ \psi \circ F$ on $r\mathbb{B}$ for some $r > 0$. Clearly, ψ extends to the basin of F by observing that $\psi = G^{-j} \circ \psi \circ F^j$ and letting $j \to \infty$. □

In view of this lemma, it suffices to prove that any F in Theorem 4.3.2 is locally near an attracting fixed point conjugate to an automorphism whose basin of attraction is all of $\mathbb{C}^n$. We will actually show that any such F is locally conjugate to a *lower triangular polynomial automorphism* of the form

$$G(z) = Az + \big(0, g_2(z), \ldots, g_n(z)\big), \quad z \in \mathbb{C}^n,$$

where A is a lower triangular matrix with the same eigenvalues as $F'(0)$ and each g_j is a polynomial of degree at least two in the variables $z_1, \ldots, z_{j-1}$, with $g_j(0) = 0$. It is evident that a lower triangular polynomial map $G: \mathbb{C}^n_0 \to \mathbb{C}^n_0$ is an automorphism of $\mathbb{C}^n$, and if G is attracting at the origin then $\Omega_{G,0} = \mathbb{C}^n$. In view of Lemma 4.3.4 this will prove Theorem 4.3.2.

Let $F = (f_1, \ldots, f_n): \mathbb{C}^n_0 \to \mathbb{C}^n_0$ be attracting at the origin. After a linear change of coordinates we may assume that

(∗) the matrix $A = F'(0) = (a_{ij})$ is lower triangular, and the eigenvalues $\lambda_j = a_{jj}$ of A are ordered so that $1 > |\lambda_1| \ge |\lambda_2| \ge \cdots \ge |\lambda_n| > 0$.

Theorem 4.3.2 is a simple consequence of the following two results.

Theorem 4.3.5 *Let $F: \mathbb{C}^n_0 \to \mathbb{C}^n_0$ be attracting at the origin. If $k \in \mathbb{N}$ is chosen such that the eigenvalues λ_j of $F'(0)$ satisfy the condition*

$$|\lambda_l|^{k+1} < |\lambda_m|, \quad l, m = 1, \ldots, n, \tag{4.27}$$

then F is locally conjugate at $0 \in \mathbb{C}^n$ to its k-jet at the origin.

A monomial $a_\mu z^\mu$ in a component f_j of F is said to be *resonant* if $\lambda_1^{\mu_1}\cdots\lambda_n^{\mu_n}=\lambda_j$.

Theorem 4.3.6 *Let $F:\mathbb{C}^n_0\to\mathbb{C}^n_0$ be attracting at the origin. For any $k\in\mathbb{N}$, F is locally conjugate at the origin to a map $G:\mathbb{C}^n_0\to\mathbb{C}^n_0$ satisfying $(*)$ whose k-jet contains only resonant monomials.*

Let us first see how these results imply Theorem 4.3.2.

Proof of Theorem 4.3.2 We may assume that $(*)$ holds. Choose a sufficiently large integer $k\in\mathbb{N}$ such that (4.27) is satisfied. By Theorem 4.3.6, the map F is conjugate at the origin to a map $G:\mathbb{C}^n_0\to\mathbb{C}^n_0$ whose k-jet G_k contains only resonant terms. Theorem 4.3.5 implies that G is locally conjugate to its k-jet, and hence we may assume that G is a polynomial map.

Since the eigenvalues λ_j decrease in size according to $(*)$, any resonant monomial in a component g_j of G only depends on the variables $z_1,\ldots,z_{j-1}$. Hence G is a lower triangular polynomial automorphism whose basin equals $\mathbb{C}^n$. Since F is locally conjugate to G at 0, they are globally conjugate on their basins by Lemma 4.3.4. In particular, $\Omega_{F,0}$ is biholomorphic to $\mathbb{C}^n$.

Assume next that the Jacobian of F is constant, $\mathrm{J}F\equiv c\in\mathbb{C}^*$. Since G is polynomial, $\mathrm{J}G$ is constant, and clearly $\mathrm{J}G(0)=c$. On the basin $\Omega_{F,0}$ we have $G^{-j}\psi F^j=\psi$ for all $j\in\mathbb{N}$, so for $z\in\Omega_{F,0}$ we have

$$\mathrm{J}\psi(z)=c^{-j}\mathrm{J}\psi\big(F^j(z)\big)c^j=\mathrm{J}\psi\big(F^j(z)\big),\quad j=1,2,\ldots.$$

Since $F^j(z)\to 0$ as $j\to\infty$, we obtain $\mathrm{J}\psi(z)=\mathrm{J}\psi(0)$, and by rescaling ψ we may arrange that this constant equals one. □

Proof of Theorem 4.3.5 Let F satisfy $(*)$. Then (4.27) is equivalent to

$$|\lambda_1|^k<|\lambda_n|.$$

The matrix $A=F'(0)$ is linearly conjugate to a lower triangular matrix a_{ij} with $\tilde{a}_{ii}=\lambda_i$. Moreover, by conjugating with a diagonal matrix (ϵ_{ii}) where all ratios $\frac{\epsilon_{ii}}{\epsilon_{jj}}$ are small for $j>i$, we can arrange that all off-diagonal entries in A are arbitrary small. Therefore, by choosing numbers $\alpha,\beta\in\mathbb{R}$ such that

$$0<\alpha<|\lambda_n|,\quad |\lambda_1|<\beta<1,\quad \beta^{k+1}<\alpha$$

we may assume that for a sufficiently small $r>0$ we have

(a) $A=F'(0)=(a_{ij})$ is lower triangular, $|a_{ii}|=|\lambda_i|\ge|\lambda_j|=|a_{jj}|$ for all $j\ge i$, and
(b) $\alpha|z|\le|F(z)|\le\beta|z|$ for all $z\in r\mathbb{B}$.

Since the preliminary conjugacies performed so far were all linear, it is enough to prove the theorem under these assumptions. Let G denote the k-jet of F at the origin. Choosing $r > 0$ small enough and $C > 0$ big enough, the following hold for $z, z' \in r\mathbb{B}$:

(1) $|G^{-1} \circ F(z) - z| \leq C|z|^{k+1}$,
(2) $|F(z)| \leq \beta|z|$, and
(3) $|G^{-1}(z) - G^{-1}(z')| \leq \alpha^{-1}|z - z'|$.

To see (3), we write

$$G^{-1}(z') = G^{-1}(z) + (G^{-1})'(z)(z - z') + O(|z - z'|^2).$$

By continuity of partial derivatives and the fact that $(G^{-1})'(0) = F'(0)^{-1}$ we get (3) by decreasing α from (b) slightly and choosing $r > 0$ small enough.

We claim that the sequence $\psi_j = G^{-j} \circ F^j$ converges uniformly on some ball $\rho\mathbb{B}$ centered at the origin. We will show that if ρ is chosen such that $\rho + \rho^k C \sum_{j=1}^{\infty} (\beta^{k+1}\alpha^{-1})^j < r$, then

(4) $|\psi_j(z) - \psi_{j-1}(z)| \leq C(\beta^{k+1}\alpha^{-1})^j \, |z|^{k+1}$ for all $z \in \rho\mathbb{B}$.

We prove this by induction. Note that (4) holds for $j = 1$ in view of (1). Suppose that (4) holds for all $j = 1, \ldots, m$. Write $w_m = \psi_m(F(z))$, $w_{m-1} = \psi_{m-1}(F(z))$. By (4) and the choice of ρ we see that both w_m and w_{m-1} are contained in $r\mathbb{B}$, and we get

$$\begin{aligned} \left|G^{-m-1} \circ F^{m+1}(z) - G^{-m} \circ F^m(z)\right| &= \left|G^{-1}(w_m) - G^{-1}(w_{m-1})\right| \\ &\leq \alpha^{-1}\left|\psi_m\big(F(z)\big) - \psi_{m-1}\big(F(z)\big)\right| \\ &\leq \alpha^{-1} C\big(\beta^{k+1}\alpha^{-1}\big)^m \big(\beta|z|\big)^{k+1}. \end{aligned}$$

This is the desired estimate for $m + 1$. The induction may proceed.

The estimate (4) shows that ψ_j converges uniformly on $\rho\mathbb{B}$ to an injective holomorphic map $\psi: \rho\mathbb{B} \to \mathbb{C}^n_0$. Clearly, ψ satisfies $G^{-1} \circ \psi \circ F = \psi$. □

Proof of Theorem 4.3.6 Write

$$F(z) = Q_m(z) + \big(P_1^m(z), \ldots, P_n^m(z)\big) + O\big(|z|^{m+1}\big),$$

where Q_m is a polynomial map of degree less than m and P_j^m is homogeneous of degree m. For each component f_j of F, let $f_{j,k}$ denote its k-jet at the origin. For some $j \in \{1, , \ldots, n\}$, constant $c \in \mathbb{C}$ and multi-index $\mu = (\mu_1, \ldots, \mu_n)$ with $|\mu| = m$, let $\varphi_c: \mathbb{C}^n \to \mathbb{C}^n$ denote the polynomial map

$$\varphi_c(z) = \big(z_1, \ldots, z_{j-1}, z_j + cz^\mu, z_{j+1}, \ldots, z_n\big).$$

The key to the proof is to describe the m-jet of $\varphi_c^{-1} \circ F \circ \varphi_c$.

Write $F \circ \varphi_c = (\tilde{f}_1, \ldots, \tilde{f}_n)$. Since $A = F'(0)$ is lower triangular with the eigenvalues λ_j (see (*)), we have

$$\tilde{f}_{i,m} = f_{i,m} \quad \text{for } i < j, \qquad \tilde{f}_{j,m} = f_{j,m} + \lambda_j c z^\mu. \tag{4.28}$$

Write $\varphi_c^{-1} \circ F \circ \varphi_c = (\widehat{f}_1, \dots, \widehat{f}_n)$. Then

$$\widehat{f}_{i,m} = \tilde{f}_{i,m} = f_{i,m} \quad \text{for } i < j$$

and

$$\widehat{f}_{j,m}(z) = f_{j,m}(z) + c\big(\lambda_j - \lambda^\mu\big)z^\mu + \sum_{|\beta|=m} a_\beta z^\beta, \tag{4.29}$$

where we have

$$a_\beta = 0 \quad \text{unless} \quad \beta_s < \mu_s, \quad \text{where } s = \max\{j : \mu_j \neq 0\}. \tag{4.30}$$

If z^μ is not a resonant monomial, the constant c may be chosen such that $\widehat{f}_{j,m}$ does not contain any monomial of type z^μ.

The proof will consist of performing a suitable sequence of conjugacies of this type. To this end, we define an ordering of monomials of degree m in the variables $z_1, \dots, z_n$ by induction on n. For $n = 2$, we say that $z_1^{\mu_1} z_2^{\mu_2} > z_1^{\tilde{\mu}_1} z_2^{\tilde{\mu}_2}$ if $\mu_2 > \tilde{\mu}_2$. For $n \geq 3$, we say that $z^\mu > z^{\tilde{\mu}}$ if either $\mu_n > \tilde{\mu}_n$, or $\mu_n = \tilde{\mu}_n$ and $z_1^{\mu_1} \cdots z_{n-1}^{\mu_{n-1}} > z_1^{\tilde{\mu}_1} \cdots z_{n-1}^{\tilde{\mu}_{n-1}}$. Then, (4.30) amounts to saying that

$$a_\beta = 0 \quad \text{unless} \quad z^\beta < z^\mu. \tag{4.31}$$

For fixed j and m we define the following procedure $R_{j,m}$ for creating a conjugation. Consider the largest monomial az^μ in $P_j^m(z)$. If z^μ is not resonant, apply a map φ_{c_1} as above to cancel this term. If z^μ is resonant, then simply ignore it and proceed to the next largest monomial smaller than z^μ. Repeat and define conjugating maps φ_{c_l} for $l = 1, \dots, N$ until there are no more monomials of degree m left. Set $\phi_{j,m} = \varphi_{c_1} \circ \cdots \circ \varphi_{c_N}$ and write $\phi_{j,m}^{-1} F \phi_{j,m} = (f_1', \dots, f_n')$. It follows from (4.28) that $f_{i,m}' = f_{i,m}$ for all $i < j$, and from (4.29) that $f_{j,m-1}' = f_{j,m-1}$. Now, (4.31) implies that any monomial of degree m in f_j' is resonant.

For a fixed m we define a procedure S_m for creating a conjugation. Let $\Phi_m = \phi_{1,m} \circ \cdots \circ \phi_{n,m}$, where $\phi_{j,m}$ is obtained by following procedure $R_{j,m}$ with $(\phi_{1,m} \circ \cdots \circ \phi_{j-1,m})^{-1} \circ F \circ (\phi_{1,m} \circ \cdots \circ \phi_{j-1,m})$ as input. Then, the $(m-1)$-jet of $\Phi_m^{-1} \circ F \circ \Phi_m$ equals the $(m-1)$-jet of F, and its homogeneous part of degree m contains only resonant monomials.

We begin by following procedure S_2 to create a map Φ_2. We inductively follow procedure S_{m+1} with $(\Phi_2 \circ \cdots \circ \Phi_m)^{-1} \circ F \circ (\Phi_2 \circ \cdots \circ \Phi_m)$ as data to create a map Φ_{m+1}. Then, $\psi = \Phi_2 \circ \cdots \circ \Phi_k$ accomplishes the task. □

Example 4.3.7 ([477, Example 9.4]) We give an example of a Fatou-Bieberbach domain $\Omega \subset \mathbb{C}^2$ whose intersection with every affine complex line is bounded. Choose a number $\alpha \in \mathbb{C}$ with $0 < |\alpha| < 1$. The polynomial automorphism

$$F(z, w) = \big(\alpha w, \alpha z + w^2\big) = \big(z', w'\big)$$

fixes $(0,0)$, and the eigenvalues of $F'(0,0)$ are $\pm\alpha$. Let Ω denote its basin at $(0,0)$. Set $E = \{(z,w) : |w| > 1 + 2|\alpha| + |z|\}$. If $(z,w) \in E$, then

$$\begin{aligned} |w'| &\ge |w^2| - |\alpha z| > |w|^2 - |\alpha w| = |w|(|w| - |\alpha|) \\ &> |w|(1 + |\alpha|) > 1 + 2|\alpha| + |z'|, \end{aligned}$$

which means that $(z', w') = F(z,w) \in E$. Hence $E \cap \Omega = \varnothing$. It is a simple exercise to see that for every affine complex line L in $\mathbb{C}^2$ the set $F(L) \setminus E$ is bounded; hence $F(L) \cap \Omega \subset F(L) \setminus E$ is also bounded. Since $F(\Omega) \subset \Omega$, we also have $F(L \cap \Omega) \subset F(L) \cap \Omega$, and hence $L \cap \Omega$ is bounded. □

Bedford and Smillie showed that this is actually a general phenomenon.

Theorem 4.3.8 ([48, Theorem 1]) *Let Ω be a basin of a polynomial automorphism of $\mathbb{C}^2$. If $\Omega \neq \mathbb{C}^2$, then Ω intersects every algebraic curve $A \subset \mathbb{C}^2$ and the set $\overline{\Omega \cap A}$ is compact.*

We shall not prove this here, but we mention the following consequence.

Let Ω be as above, with a Fatou-Bieberbach map $\Phi \colon \Omega \to \mathbb{C}^2$. If L is an affine complex line in $\mathbb{C}^2$ then each connected component D of $L \cap \Omega$ is a bounded domain which is Runge in L, and hence is biholomorphic to the disc. The image $\Phi(D)$ of D is a closed embedded holomorphic disc in $\mathbb{C}^2$.

Example 4.3.9 ([477, Example 9.5]) We give an example of a Fatou-Bieberbach domain in $\mathbb{C}^2$ avoiding a complex line. Let $F(z,w) = (z', w')$ be given by

$$z' = z + w, \qquad w' = \frac{1}{2}\big(1 - w - \mathrm{e}^{z+w}\big). \tag{4.32}$$

Its fixed points are $p_m = (2m\pi\mathrm{i}, 0)$ for $m \in \mathbb{Z}$, and the eigenvalues of $F'(p_m)$ are $\pm\frac{1}{\sqrt{2}}$. Let Ω_m be the basin at p_m. Clearly these domains are pairwise disjoint. Furthermore, $F((z,w) + p_m) = F(z,w) + p_m$ and hence

$$\Omega_m = \Omega_0 + p_m, \quad m = 0, \pm 1, \pm 2, \ldots. \tag{4.33}$$

Thus, there exist countably many pairwise disjoint Fatou-Bieberbach domains $\Omega_m \subset \mathbb{C}^2$ $(m \in \mathbb{Z})$ which are translates of each other.

Consider now the map $E \colon \mathbb{C}^2 \to \mathbb{C}^2$ given by $E(z,w) = (\mathrm{e}^z, w\mathrm{e}^{-z})$. Note that $JE \equiv 1$ and that $E(z,w) = E(z',w')$ if and only if $w = w'$ and $z' = z + 2m\pi\mathrm{i}$ for some $m \in \mathbb{Z}$. Since the domains Ω_m are pairwise disjoint, it follows from (4.33) that E is injective on each Ω_m and the image $\Omega^* = E(\Omega_m)$ is independent of m. Note that Ω^* is Fatou-Bieberbach domain in $\mathbb{C}^2$ which does not intersect the complex line $\{z = 0\}$. □

It is easy to generalize the construction in Example 4.3.11 to find Fatou-Bieberbach domains in the complement of a hyperplane in $\mathbb{C}^n$. This is no longer the case if we replace a hyperplane by a more general analytic subvariety, even by a zero dimensional one (see Theorem 4.7.2), but it is true for tame subvarieties that do not contain any hypersurfaces (Proposition 4.11.9 on p. 155). Further results in this direction are given in Sects. 4.6 and 4.11.

Example 4.3.10 ([477, Examples 9.5 and 9.6]) We construct a Fatou-Bieberbach domain omitting countably many real 2-planes. Let F be the map (4.32). Set

$$\Pi_k = \{(x + (2k+1)\pi \mathrm{i}, y) : x, y \in \mathbb{R}\}, \quad k = 0, \pm 1, \pm 2, \ldots.$$

Note that $\Pi_k = \Pi_0 + ((2k+1)\pi \mathrm{i}, 0)$, and Π_0 is the standard $\mathbb{R}^2$ in $\mathbb{C}^2$. Then $F(\Pi_k) = \Pi_k$, and no p_m lies in any Π_k; therefore no point of any Π_k is attracted to any p_m by F. Hence, no Π_k intersects any Ω_m.

The same can be done with finitely many rotated copies of $\mathbb{R}^2$ in $\mathbb{C}^2$. Choose $N \in \mathbb{N}$, put $\alpha = \mathrm{e}^{\pi \mathrm{i}/2N}$ and $E_k = \alpha^k \mathbb{R}^2$ for $k = 0, 1, \ldots, 2N-1$. (Here we identified $\mathbb{R}^2$ with the subset of $\mathbb{C}^2$ consisting of points with real coordinates.) Define $F(z, w) = (z', w')$ by

$$z' = z + w, \qquad w' = \frac{1}{2N+1}\big(z + (z+w)^{2N+1}\big).$$

Then $F \in \mathrm{Aut}(\mathbb{C}^2)$, $F(E_k) = E_k$ for all k, and the fixed points of F are $(0,0)$ and $p_m = (\alpha^m, 0)$ for odd m. The eigenvalues of $F'(p_m)$ are $\pm\frac{1}{\sqrt{2N+1}}$. By Theorem 4.3.2 on p. 116, the basin Ω_m at p_m is a Fatou-Bieberbach domain. These N domains (one for each odd m between 1 and $2N-1$) are pairwise disjoint and are contained in $\mathbb{C}^2 \setminus (E_0 \cup E_2 \cup \cdots \cup E_{2N-2})$. Since $F(\alpha^2 z, \alpha^2 w) = \alpha^2 F(z, w)$, the rotation $(z, w) \to (\alpha^2 z, \alpha^2 w)$ permutes the domains Ω_m. □

Example 4.3.11 ([477, Example 9.7]) We give an example of a Fatou-Bieberbach domain in $\mathbb{C}^2$ whose closure misses a complex line. Choose a number $\alpha \in \mathbb{C}$, $0 < |\alpha| < 1$. Let f be an entire function on $\mathbb{C}$ such that

$$\mathrm{e}^{f(0)} = \frac{1}{\alpha}, \quad f'(0) = 0, \quad f(1) = 1, \quad f'(1) = \frac{1+\alpha^2}{1-\alpha^2}.$$

Define a map $F(z, w) = (z', w')$ by

$$z' = 1 - \alpha^2 + \alpha^2 z \mathrm{e}^{f(zw)}, \qquad w' = w\mathrm{e}^{-f(zw)}.$$

Then $F \in \mathrm{Aut}(\mathbb{C}^2)$, $JF \equiv \alpha^2$, $F(1,1) = (1,1)$, and the eigenvalues of $F'(1,1)$ are $\pm\alpha \mathrm{i}$. Let Ω_0 be the basin of attraction of $(1,1)$. F also fixes the point $(1+\alpha, 0)$ and $F'(1+\alpha, 0) = \alpha I$. Let Ω_1 denote the basin of $(1+\alpha, 0)$. Since $F(z, 0) = (1 - \alpha^2 + \alpha z, 0)$ for all $z \in \mathbb{C}$, we see that Ω_1 contains the complex line $\{w = 0\}$, and hence $\overline{\Omega_0}$ does not intersect this line. □

The image of a nondegenerate holomorphic self-map of $\mathbb{C}^2$ cannot omit a pair of parallel complex lines (by Picard's theorem), and it cannot omit three complex lines

(Green [262]). It is a long-standing open problem whether an *injective* self-map of $\mathbb{C}^2$ can omit a pair of intersecting complex lines:

Problem 4.3.12 Is there a Fatou-Bieberbach domain contained in $(\mathbb{C}^*)^2$? Is there a holomorphic automorphism of $\mathbb{C}^2$ fixing both coordinate axes and with an attracting fixed point at $(1, 1) \in (\mathbb{C}^*)^2$? In this connection, see the paper of Nishimura [434].

We now mention without proof some generalizations of Theorem 4.3.2.

Suppose that X is a complex manifold and $F\colon X \to X$ is a holomorphic automorphism which is uniformly hyperbolic on an invariant compact set $K \subset X$. (For the definitions, see e.g. Shub [502]. Note that F is hyperbolic at a fixed point $p \in X$ if the derivative $F'(p)$ does not have any eigenvalue of modulus one; for a periodic point, we consider the eigenvalues of a suitable iterate.) The *stable manifold* W^s_p of a point $p \in K$ is defined by

$$W^s_p = \Big\{q \in X : \lim_{n\to\infty} \operatorname{dist}\big(F^n(q), F^n(p)\big) = 0\Big\};$$

this is an immersed complex submanifold of X without self-intersections which is diffeomorphic to $\mathbb{R}^{2k}$, where $k = \dim W^s_p$. A very natural question, asked in particular by Bedford, is whether stable manifolds of points $p \in K$ are all biholomorphic to $\mathbb{C}^k$. At fixed and periodic points this holds by applying Theorem 4.3.2 and Remark 4.3.3 to the restricted map $F\colon W^s_p \to W^s_p$. Jonsson and Varolin proved in [312] that the stable manifold is biholomorphic to $\mathbb{C}^k$ for almost all points $p \in K$ with respect to any invariant measure. Although it is still an open question whether every stable manifold is a $\mathbb{C}^k$, the following result of Fornæss and Stensønes gives some information on their complex structure.

Theorem 4.3.13 ([167, Theorem 3.5]) *Let $F\colon X \to X$ be a holomorphic automorphism of a complex manifold X and $K \subset X$ be an invariant compact hyperbolic set of stable dimension k. Then, the stable manifold of F at any point $p \in K$ is biholomorphic to a domain in $\mathbb{C}^k$ which is a Stein increasing union of balls and has vanishing Kobayashi-Royden pseudometric.*

A complex manifold Ω of dimension k which is an increasing union of (biholomorphic images of) balls is called a *short* $\mathbb{C}^k$. If $k = 2$ and the Kobayashi-Royden pseudometric of Ω is not identically zero, then such Ω is biholomorphic to the ball or to $\mathbb{D} \times \mathbb{C}$, where $\mathbb{D}$ is the unit disc (see Fornæss and Sibony [164]). Fornæss gave an example of a short $\mathbb{C}^k$ different from $\mathbb{C}^k$, with vanishing Kobayashi pseudometric and supporting a nonconstant bounded plurisubharmonic function [162].

4.4 Random Iterations and the Push-Out Method

A sequences of compositions $\Psi_m \circ \cdots \circ \Psi_1$ $(m = 1, 2, \ldots)$ of holomorphic automorphisms $\Psi_k \in \operatorname{Aut}(\mathbb{C}^n)$ is called a *random sequence* of automorphisms. In this section we prove that, under rather mild assumptions on the maps Ψ_k, the domain of convergence of such a random sequence is a Fatou-Bieberbach domain; see Corollary 4.4.2. We begin with the following more general result; the analogous result holds in any complex manifold.

Proposition 4.4.1 *Let D be a connected open set in $\mathbb{C}^n$ that is exhausted by a sequence of compact sets $K_0 \subset K_1 \subset \cdots \subset \bigcup_{j=0}^{\infty} K_j = D$ such that $K_{j-1} \subset \mathring{K}_j$ for each $j \in \mathbb{N}$. Choose numbers ϵ_j $(j \in \mathbb{N})$ such that*

$$0 < \epsilon_j < \operatorname{dist}\big(K_{j-1}, \mathbb{C}^n \setminus K_j\big) \quad (\forall j \in \mathbb{N}), \qquad \sum_{j=1}^{\infty} \epsilon_j < \infty. \tag{4.34}$$

Suppose that Ψ_j $(j \in \mathbb{N})$ is a holomorphic automorphism of $\mathbb{C}^n$ satisfying

$$\big|\Psi_j(z) - z\big| < \epsilon_j, \qquad z \in K_j,\ j \in \mathbb{N}. \tag{4.35}$$

Set $\Phi_m = \Psi_m \circ \Psi_{m-1} \circ \cdots \circ \Psi_1$. Then there is an open set $\Omega \subset \mathbb{C}^n$ such that the limit $\lim_{m\to\infty} \Phi_m = \Phi$ exists uniformly on compacts in Ω, and Φ is a biholomorphic map of Ω onto D. In fact, $\Omega = \bigcup_{m=1}^{\infty} \Phi_m^{-1}(K_m)$.

Proof Set $\Phi_0(z) = z$ for $z \in \mathbb{C}^n$. Let

$$L_m = \Phi_m^{-1}(K_m) \quad (\forall m \in \mathbb{N}), \qquad \Omega = \bigcup_{m=1}^{\infty} L_m \subset \mathbb{C}^n. \tag{4.36}$$

From (4.34) and (4.35) we get $\Psi_{m+1}(K_m) \subset \mathring{K}_{m+1}$ for all $m \ge 0$, and hence

$$\Phi_{m+1}(L_m) = \Psi_{m+1}\big(\Phi_m(L_m)\big) = \Psi_{m+1}(K_m) \subset \mathring{K}_{m+1}.$$

Thus $L_m \subset \mathring{L}_{m+1}$ for all $m \in \mathbb{N}$, and hence Ω is an open set in $\mathbb{C}^n$. By induction on j we also get that $\Phi_j(L_m) \subset K_j$ for all $j \ge m$. Hence (4.35) gives the following for all $l > m \ge 1$ and $z \in L_m$:

$$\begin{aligned}\big|\Phi_l(z) - \Phi_m(z)\big| &\le \sum_{j=m+1}^{l} \big|\Phi_j(z) - \Phi_{j-1}(z)\big| \\ &= \sum_{j=m+1}^{l} \big|\Psi_j\big(\Phi_{j-1}(z)\big) - \Phi_{j-1}(z)\big| < \sum_{j=m+1}^{l} \epsilon_j.\end{aligned} \tag{4.37}$$

This shows that $\lim_{l\to\infty} \Phi_l = \Phi$ exists on L_m and satisfies

$$\big|\Phi(z) - \Phi_m(z)\big| \le \sum_{j=m+1}^{\infty} \epsilon_j < \operatorname{dist}\big(K_m, \mathbb{C}^n \setminus D\big), \quad z \in L_m. \tag{4.38}$$

The last inequality above follows from (4.34). Thus, $\Phi(L_m) \subset D$ holds for each $m \in \mathbb{N}$, and hence $\Phi(\Omega) \subset D$.

For $z \in L_m$ we write $\Phi_m(z) = w \in K_m$. Then, (4.38) implies

$$\left|\Phi \circ \Phi_m^{-1}(w) - w\right| \leq \sum_{j=m+1}^{\infty} \epsilon_j, \quad w \in K_m. \tag{4.39}$$

We claim that Φ maps Ω onto D. Fix an integer $m \in \mathbb{N}$ and choose $l > m$ big enough such that $\sum_{j=l+1}^{\infty} \epsilon_j < \operatorname{dist}(K_m, \mathbb{C}^n \setminus K_{m+1})$. The map $\widetilde{\Phi}_l = \Phi \circ \Phi_l^{-1}$ is holomorphic in a neighborhood of K_l. From (4.39) we get that

$$\left|\widetilde{\Phi}_l(w) - w\right| < \sum_{j=l+1}^{\infty} \epsilon_j < \operatorname{dist}\big(K_m, \mathbb{C}^n \setminus K_{m+1}\big), \quad w \in K_l.$$

This shows that the map $\widetilde{\Phi}_l|_{bK_l} \colon bK_l \to \mathbb{C}^n \setminus K_m$ has degree one with respect to any point of K_m. Rouché's theorem [98, p. 110] implies that $\widetilde{\Phi}_l(K_l) \supset K_m$. From $\widetilde{\Phi}_l(K_l) = \Phi(\Phi_l^{-1}(K_l)) = \Phi(L_l)$ we infer that $K_m \subset \Phi(L_l) \subset \Phi(\Omega)$. Since $\bigcup_{m\in\mathbb{N}} K_m = D$, it follows that $\Phi(\Omega) = D$. Since Φ is a locally uniform limit of injective holomorphic maps and is nondegenerate, it is also injective. □

Corollary 4.4.2 *Let $K_0 \subset K_1 \subset \cdots \subset \bigcup_{j=0}^{\infty} K_j = \mathbb{C}^n$ be compact sets such that $K_{j-1} \subset \mathring{K}_j$ for each $j \in \mathbb{N}$. Let $\epsilon_j > 0$, $\Psi_j \in \operatorname{Aut}(\mathbb{C}^n)$ and Φ_j be as in Proposition* 4.4.1, *satisfying conditions* (4.34) *and* (4.35). *Let $\Omega \subset \mathbb{C}^n$ consist of all points $z \in \mathbb{C}^n$ with bounded orbits $\{\Phi_m(z) \colon m \in \mathbb{N}\} \subset \mathbb{C}^n$. Then, the limit $\lim_{m\to\infty} \Phi_m = \Phi$ exists uniformly on compacts in Ω and maps Ω biholomorphically onto $\mathbb{C}^n$.*

Although Corollary 4.4.2 is analogous to the result of Fornæss and Stensønes [167] mentioned at the end of Sect. 4.3 (replace the maps by their inverses and consider the attracting basin of a random sequence), it will be more convenient for our purposes to use the result in this particular form.

Corollary 4.4.2 lies at the heart of the *push-out method* first used by Dixon and Esterle [125] to construct Fatou-Bieberbach domains with certain properties. By taking Ψ_k to be suitably chosen shears in coordinate directions, Globevnik [238] and Stensønes [526] constructed Fatou-Bieberbach domains with smooth boundaries by a method originating in their joint paper [242] on embedding certain plane domains into $\mathbb{C}^2$ (see Theorem 9.10.2 on p. 446).

Theorem 4.4.3 ([526]) *For every $n > 1$ there exists a Fatou-Bieberbach domain in $\mathbb{C}^n$ with $\mathcal{C}^\infty$-smooth boundary.*

It is unknown whether this phenomenon can occur also for basins of an attracting or semi-attracting fixed point. The existing information suggests to the contrary. Bedford and Smillie showed in [49] that for any basins of attraction $\Omega_1, \Omega_2, \ldots$ of a polynomial automorphism of $\mathbb{C}^2$, their boundaries coincide, and if such a mapping has at least two basins then the boundaries cannot be 3-manifolds. The interested reader should consult Bedford's very informative review MR1441879 of the paper [526].

Problem 4.4.4 Is there a Fatou-Bieberbach domain in $\mathbb{C}^n$ with real analytic boundary?

The boundary of a smooth Fatou-Bieberbach domain in $\mathbb{C}^2$ is easily seen to be pseudoconvex from both sides, and hence Levi-flat. Globevnik [238] constructed Fatou-Bieberbach domains with less regular boundary, but with a more precise behavior in a given compact set. For instance, he found Fatou-Bieberbach domains $\Omega \subset \mathbb{C}^2$ such that for a given $r > 1$ the intersection $\Omega \cap r\mathbb{D}^2$ is a $\mathcal{C}^2$-small perturbation of the tube $\mathbb{D} \times r\mathbb{D}$. Furthermore, Wold constructed a Fatou-Bieberbach domain in $\mathbb{C}^2$ such that a connected component of its intersection with the z-axis is precisely the unit disc [586].

4.5 Mittag-Leffler Theorem for Entire Maps

In this and the following two sections we present some of the results of Rosay and Rudin [477] concerning the action of holomorphic self-mappings of $\mathbb{C}^n$ on different types of countable subsets. The first result is motivated by the classical Mittag-Leffler interpolation theorem.

Theorem 4.5.1 ([477, Theorem 1.1]) *Let $n > 1$ and assume that $\{a_j\}_{j\in\mathbb{N}}$ is a discrete sequence without repetition in $\mathbb{C}^n$. For every sequence $\{b_j\}_{j\in\mathbb{N}}$ in $\mathbb{C}^n$ there exists an entire map $\Phi\colon \mathbb{C}^n \to \mathbb{C}^n$ satisfying*

(i) $\Phi(a_j) = b_j$ *for* $j = 1, 2, \ldots,$ *and*
(ii) $(\mathrm{J}\Phi)(z) = 1$ *for every* $z \in \mathbb{C}^n$.

The proof of Theorem 4.5.1 relies on the following simple lemma.

Lemma 4.5.2 *Assume that $n > 1$, $a_1, \ldots, a_m$ are points in a compact convex set $K \subset \mathbb{C}^n$, and p, q are points in a complex affine hyperplane $\Lambda \subset \mathbb{C}^n \setminus K$. For every $\epsilon > 0$ there exists a volume preserving shear τ on $\mathbb{C}^n$ that moves p to q, fixes every point a_j, and satisfies $|\tau(z) - z| < \epsilon$ for all $z \in K$.*

Proof Choose a $\mathbb{C}$-linear functional $\lambda\colon \mathbb{C}^n \to \mathbb{C}$ that is constant on Λ. The point $\lambda(p) = \lambda(q)$ then lies outside the compact convex set $\lambda(K) \subset \mathbb{C}$. Choose a unit vector $v \in \ker\lambda$; then $q = p + cv$ for some $c \in \mathbb{C}$. Choose a holomorphic polynomial g on $\mathbb{C}$ such that $g(\lambda(p)) = c$, $g(\lambda(a_j)) = 0$ for $j = 1, \ldots, m$, and $|g| < \epsilon$ on $\lambda(K)$. The shear $\tau(z) = z + g(\lambda(z))v$ does the job. □

Corollary 4.5.3 *If $a_1, \dots, a_m$, K and ϵ are as in Lemma 4.5.2, and if p, q are points in $\mathbb{C}^n \setminus K$, then some composition of two volume preserving shears moves p to q, fixes every point a_j, and moves no point of K by as much as ϵ.*

Proof There are affine complex hyperplanes $p \in \Lambda$, $q \in \Lambda'$ that do not intersect K and are not parallel. Pick $w \in \Lambda \cap \Lambda'$ and apply Lemma 4.5.2 twice to move p to w and then w to q, fixing the points a_j and moving no point of K by as much as $\frac{\epsilon}{2}$ at every step. □

Proof of Theorem 4.5.1 By an affine change of coordinates of $\mathbb{C}^n$ we may assume that $0 < |a_1| < |a_2| < \cdots$ and that the hyperplane $\{z_1 = 0\}$ does not contain any points of the b_j sequence. Let $\Phi = E \circ F$ where

$$E(z_1, z_2, \dots, z_n) = \left(\mathrm{e}^{z_1}, z_2\mathrm{e}^{-z_2}, z_3, \dots, z_n\right)$$

and F will be found as a limit of a certain sequence of compositions of additive shears (hence $JF = 1$). Note that every point $w = (w_1, \dots, w_n)$ with $\pi_1(w) = w_1 \neq 0$ belongs to the range of E, and we can choose $v \in \mathbb{C}^n$ such that $E(v) = w$ and $|\pi_1(v)|$ is arbitrary large.

Set $F_0(z) = z$. Assume inductively that $F_{k-1} \in \mathrm{Aut}(\mathbb{C}^n)$ has been chosen for some $k \in \mathbb{N}$. Choose $v_k \in \mathbb{C}^n$ such that $E(v_k) = b_k$ and $|\pi_1(v_k)|$ is so big that v_k lies outside of the compact set $F_{k-1}(|a_k|\overline{\mathbb{B}})$. Hence, there exists a point $q_k \in \mathbb{C}^n$ such that $F_{k-1}(q_k) = v_k$ and $|q_k| > |a_k|$. Choose a number δ_k, $0 < \delta_k < |a_k| - |a_{k-1}|$, such that for all $z, w \in |a_k|\overline{\mathbb{B}}$ with $|z - w| < \delta_k$ we have $|F_{k-1}(z) - F_{k-1}(w)| < 2^{-k}$. Corollary 4.5.3 furnishes $G_k \in \mathrm{Aut}(\mathbb{C}^n)$ (a composition of two additive shears) such that $G_k(a_j) = a_j$ for $j = 1, \dots, k-1$, $G_k(a_k) = b_k$, and $|G_k(z) - z| < \delta_k$ for $|z| \leq |a_{k-1}|$. The map $F_k = F_{k-1} \circ G_k$ then satisfies

$$F_k(a_k) = v_k, \qquad F_k(a_j) = F_{k-1}(a_j) \quad \text{for } j = 1, \dots, k-1,$$

and

$$\left|F_k(z) - F_{k-1}(z)\right| < 2^{-k}, \quad |z| \leq |a_{k-1}|.$$

The limit $F = \lim_{k\to\infty} F_k$ then exists uniformly on compacts in $\mathbb{C}^n$, F is injective holomorphic, $JF \equiv 1$, and $F(a_k) = v_k$ for all $k = 1, 2, \dots$. Since $E(v_k) = b_k$, we see that $\Phi(a_k) = b_k$ and the proof is complete. □

4.6 Tame Discrete Sets in $\mathbb{C}^n$

A discrete set is always assumed to be closed, and hence without limit points.

Rosay and Rudin [477] showed that, in general, one cannot map an infinite discrete set in $\mathbb{C}^n$ onto another discrete set by an automorphism of $\mathbb{C}^n$. Indeed, the infinite discrete sets in $\mathbb{C}^n$ form uncountably many different equivalence classes under $\mathrm{Aut}(\mathbb{C}^n)$-equivalence (see Corollary 4.7.7 on p. 132). They introduced the following notion.

Definition 4.6.1 A discrete set $E \subset \mathbb{C}^n$ is *tame* if there exists a holomorphic automorphism $F \in \mathrm{Aut}(\mathbb{C}^n)$ such that

$$F(E) = N = \{je_1 : j \in \mathbb{N}\} \subset \mathbb{C}^n. \tag{4.40}$$

The set E is *very tame* if the above holds for some $F \in \mathrm{Aut}_1(\mathbb{C}^n)$.

It is easily seen that any permutation of the standard tame set $N \subset \mathbb{C}^n$ (4.40) is achieved by a composition of three additive shears (4.1) (see [477, Proposition 3.1]). Hence, we can permute points in any tame set by holomorphic automorphisms, and for every tame sequence $\{a_j\}_{j\in\mathbb{N}} \subset \mathbb{C}^n$ there is an automorphism $F \in \mathrm{Aut}(\mathbb{C}^n)$ such that

$$F(a_j) = je_1 = (j, 0, \dots, 0), \quad j = 1, 2, \dots.$$

Let $\mathbb{C}^n = \mathbb{C}^k \oplus \mathbb{C}^m$ where $k, m \in \mathbb{N}$. Write the coordinates on $\mathbb{C}^n$ accordingly as $z = (z', z'') \in \mathbb{C}^k \oplus \mathbb{C}^m$ and set $\pi'(z', z'') = z'$, $\pi''(z', z'') = z''$. The following result combines Theorems 3.3 and 3.9 in [477].

Theorem 4.6.2 ([477]) *Suppose that E is an infinite discrete set in $\mathbb{C}^n$.*

(a) *If the set $\pi'(E)$ is discrete in $\mathbb{C}^k$, then E is tame.*
(b) *If in addition to (a) the set $(\{p\} \times \mathbb{C}^m) \cap E$ is finite for every $p \in \mathbb{C}^k$, then E is very tame.*

Proof Let $\pi'(E) = \{p_1, p_2, \dots\} \subset \mathbb{C}^k$. We first prove part (b). By induction we find $w_1, w_2, \dots \in \mathbb{C}^m$ such that for all points $(p_j, q) \in E$ and for all $(p_i, z'') \in E$ with $i < j$ we have $|q + w_j| > j + |z'' + w_i|$. By the Mittag-Leffler interpolation theorem there is a holomorphic map $f : \mathbb{C}^k \to \mathbb{C}^m$ such that $f(p_i) = w_i$ for $i = 1, 2, \dots$. Define

$$\sigma_1(z', z'') = (z', z'' + f(z')), \quad E_1 = \sigma_1(E).$$

Our choice of $\{w_j\}$ implies that π'' is one-to-one on E_1 and that $\pi''(E_1)$ is discrete in $\mathbb{C}^m$. Let $z = (z_1, \dots, z_n)$. The above implies that there is a function $\varphi : \pi''(E_1) \to \mathbb{C}$ such that $\{z_1 + \varphi(z'') : z \in E_1\} = \mathbb{N}$. Choose a holomorphic function g on $\mathbb{C}^m$ such that $g = \varphi$ on $\pi''(E_1)$ and set

$$\sigma_2(z) = (z_1 + g(z''), z_2, \dots, z_n), \quad E_2 = \sigma_2(E_1).$$

The first coordinate projection $\pi_1(z) = z_1$ then maps E_2 bijectively onto $\mathbb{N}$. Choose holomorphic functions $g_2, \dots, g_n$ on $\mathbb{C}$ such that

$$E_2 = \{(k, g_2(k), \dots, g_n(k)) : k \in \mathbb{N}\}.$$

The shear

$$\sigma_3(z) = (z_1, z_2 - g_2(z_1), \dots, z_n - g_n(z_1))$$

then takes E_2 onto the standard tame set N (4.40). Setting $\sigma = \sigma_3 \circ \sigma_2 \circ \sigma_1$ we have that $\sigma(E) = N$ and $\mathrm{J}\sigma = 1$. This proves (b).

To prove (a), we may suppose by a translation that $z' \neq 0$ and $z'' \neq 0$ for every point $z = (z', z'') \in E$. As before, let $\pi'(E) = \{p_1, p_2, \ldots\}$. Put

$$\delta_j = \min\{|z''| : (p_j, z'') \in E\} > 0, \quad j \in \mathbb{N}.$$

Choose a holomorphic function h on $\mathbb{C}^k$ such that $\Re h(p_j) > \log \frac{|p_j|}{\delta_j}$ for $j \in \mathbb{N}$. Define $F \in \mathrm{Aut}(\mathbb{C}^n)$ by $F(z', z'') = (z', \mathrm{e}^{h(z')} z'')$. If $(p_j, z'') \in E$ then

$$\left|\mathrm{e}^{h(p_j)} z''\right| \geq \mathrm{e}^{\Re h(p_j)} \delta_j > |p_j|.$$

This shows that each point of $F(E)$ has the form (p_j, w'') with $|w''| > |p_j|$. Hence, the projection π'' is finite-to-one on $F(E)$ and the set $\pi''(F(E))$ is discrete in $\mathbb{C}^m$. By part (b), $F(E)$ is tame in $\mathbb{C}^n$, and hence E is tame. □

Corollary 4.6.3 *Let $n > 1$ be an integer.*

(a) *Every infinite discrete set in $\mathbb{C}^{n-1}$ is very tame in $\mathbb{C}^n$.*
(b) *The union of a finite set and a (very) tame set is (very) tame.*
(c) *Every infinite discrete set in $\mathbb{C}^n$ is the union of two very tame sets.*

Proof Part (a) follows from part (b) in Theorem 4.6.2 with $k = 1$. To prove (b), it suffices to consider the union with the set N (4.40) and to apply Theorem 4.6.2. To prove (c), write $n = k + m$ with $k, m \in \mathbb{N}$ and put

$$E_1 = \{(z', z'') \in E : |z''| \leq |z'|\}, \qquad E_2 = \{(z', z'') \in E : |z''| > |z'|\}.$$

Theorem 4.6.2 directly applies to E_1, and it also applies to E_2 with the roles of π' and π'' reversed. Thus, both E_1 and E_2 are very tame in $\mathbb{C}^n$. If one of these two sets is finite, it suffices to apply part (b). □

Corollary 4.6.4 *If A is a proper algebraic subvariety of $\mathbb{C}^n$ for some $n > 1$, then every discrete set contained in A is very tame.*

Proof After a linear change of coordinates on $\mathbb{C}^n$ we achieve that the coordinate projection $\pi' : \mathbb{C}^n \to \mathbb{C}^{n-1}$, $\pi(z', z_n) = z'$ is proper when restricted to A. Hence, every discrete set $E \subset A$ satisfies condition (b) in Theorem 4.6.2. □

Corollary 4.6.5 *For every infinite discrete set in $\mathbb{C}^n$ ($n > 1$) there is an injective holomorphic map $H : \mathbb{C}^n \to \mathbb{C}^n$ such that $\mathrm{J}H \equiv 1$ and $H(E) = N$.*

Proof In the proof of Theorem 4.5.1 (p. 126) we found an injective holomorphic map $F : \mathbb{C}^n \to \mathbb{C}^n$ with $\mathrm{J}F \equiv 1$ such that the restriction of the first coordinate projection $\pi_1 : \mathbb{C}^n \to \mathbb{C}$ to the set $F(E)$ is one-to-one and $\pi_1(F(E))$ is discrete in $\mathbb{C}$. Theorem 4.6.2(b) (with $k = 1$) shows that $F(E)$ is very tame in $\mathbb{C}^n$. Let

$G \in \mathrm{Aut}_1(\mathbb{C}^n)$ be such that $G(F(E)) = N$. The composition $H = G \circ F$ then satisfies the conclusion of Corollary 4.6.5. □

The following result shows that a discrete set which goes to infinity sufficiently fast is tame. The first part is due to Winkelmann [578], and the second part to Kolarič [348, Theorem 2.3] who obtained several further results on tameness of discrete sets in the presence of an algebraic subvariety.

Theorem 4.6.6 *A discrete set $\{a_k\}_{k\in\mathbb{N}} \subset \mathbb{C}^n$ $(n > 1)$ satisfying the growth condition $\sum_{k=1}^{\infty} 1/|a_k|^{2n-2} < +\infty$ is tame. Furthermore, given an algebraic subvariety $A \subset \mathbb{C}^n$ of codimension ≥ 2 and discrete sets $\{a_j\}$, $\{b_j\}$ in $\mathbb{C}^n$ contained in $\mathbb{C}^n \setminus A$ and satisfying the above growth condition, there exists an $F \in \mathrm{Aut}(\mathbb{C}^n)$ such that $F(z) = z$ for all $z \in A$ and $F(a_j) = b_j$ for $j = 1, 2, \ldots$.*

4.7 Unavoidable and Rigid Discrete Sets

In this section we consider certain classes of nontame discrete sets that were introduced by Rosay and Rudin [477].

Definition 4.7.1 A subset E of $\mathbb{C}^n$ is *unavoidable* for a certain class $\mathcal{F}$ of holomorphic maps $\mathbb{C}^n \to \mathbb{C}^n$ if $F(\mathbb{C}^n) \cap E \neq \varnothing$ for every $F \in \mathcal{F}$.

By Picard's theorem, every set in $\mathbb{C}$ containing at least two points is unavoidable by nonconstant entire functions. On the other hand, complex hyperplanes in $\mathbb{C}^n$ for $n > 1$ are avoidable by injective holomorphic maps $\mathbb{C}^n \to \mathbb{C}^n$ (see Example 4.3.7 on p. 120), and hence every tame discrete set is avoidable by injective holomorphic maps $\mathbb{C}^n \to \mathbb{C}^n$. However, we now show that this fails for some discrete sets in $\mathbb{C}^n$.

Theorem 4.7.2 ([477, Theorem 4.5]) *Let $n > 1$. There exists a discrete set in $\mathbb{C}^n$ which is unavoidable by nondegenerate holomorphic self-maps of $\mathbb{C}^n$.*

We need a couple of lemmas. Denote by $\mathbb{B}^n$ the open unit ball in $\mathbb{C}^n$.

Lemma 4.7.3 *Let $1 \leq k \leq n$ be integers. Given numbers $0 < a_1 < a_2$, $0 < r_1 < r_2$, $c > 0$, let Γ be the class of all holomorphic maps $F = (f_1, \ldots, f_k)\colon a_2\mathbb{B}^n \to r_2\mathbb{B}^k$ such that*

(i) $|F(0)| \leq \frac{r_1}{2}$, *and*
(ii) $\left|\frac{\partial(f_1,\ldots,f_k)}{\partial(z_1,\ldots,z_k)}\right| \geq c$ *at some point of the closed ball $a_1\overline{\mathbb{B}^n}$.*

Then there is a finite set $E = E(a_1, a_2, r_1, r_2, c) \subset b(r_1\mathbb{B}^k)$ such that if $F \in \Gamma$ and $F(a_1\mathbb{B}^n)$ intersects $b(r_1\mathbb{B}^k)$, then $F(a_2\mathbb{B}^n)$ intersects E.

Proof Let $E_1 \subset E_2 \subset \cdots$ be finite subsets of the sphere $b(r_1\mathbb{B}^k)$ whose union $\bigcup_{j=1}^{\infty} E_j$ is dense in $b(r_1\mathbb{B}^k)$. To reach a contradiction, assume that no E_j satisfies the conclusion of the lemma. This means that there exist $F_j \in \Gamma$ and $z_j \in a_1\mathbb{B}^n$ $(j = 1, 2, \ldots)$ such that

$$F_j(z_j) \in b\big(r_1\mathbb{B}^k\big), \qquad F_j\big(a_2\mathbb{B}^n\big) \cap E_j = \varnothing, \quad j = 1, 2, \ldots.$$

Note that Γ is a normal family. Passing to a subsequence we may therefore assume that, as $j \to \infty$, $z_j \to w \in a_1\overline{\mathbb{B}^n}$ and $F_j \to F \in \Gamma$, uniformly on compacts in $a_2\mathbb{B}^n$. Hence, $F(w) = \lim_{j\to\infty} F_j(z_j) \in b(r_1\mathbb{B}^k)$. Condition (ii) in the lemma shows that the open set

$$\Omega = \big\{z \in a_2\mathbb{B}^n : \operatorname{rank} F'(z) = k\big\}$$

is not empty. The complement of Ω is clearly a subvariety of $a_2\mathbb{B}^n$, and hence $F(\Omega)$ is a connected open dense set in $F(a_2\mathbb{B}^n)$.

Since $w \in a_2\mathbb{B}^n$ and $|F(w)| = r_1$, the maximum principle shows that $F(a_2\mathbb{B}^n)$ contains points outside $r_1\overline{\mathbb{B}^k}$, and hence so does $F(\Omega)$. Further, condition (i) in the lemma shows that $F(\Omega)$ intersects $r_1\mathbb{B}^k$. Being connected, $F(\Omega)$ must therefore intersect $b(r_1\mathbb{B}^k)$. Choose a point $p \in \Omega$ such that $F(p) = q \in b(r_1\mathbb{B}^k)$. Since $\operatorname{rank} F'(p) = k$, there is a compact set K in $a_2\mathbb{B}^n$, contained in a k-dimensional affine complex subspace, such that F maps K bijectively onto a closed ball $\overline{\mathbb{B}^k(q,\delta)}$ for some $\delta > 0$. Since $F_j \to F$ uniformly on K, Rouché's theorem implies that $\mathbb{B}^k(q, \frac{\delta}{2}) \subset F_j(K)$ for all large $j \in \mathbb{N}$. But the ball $\mathbb{B}^k(q, \frac{\delta}{2})$ intersects E_j for large j (since $\bigcup_{j=1}^{\infty} E_j$ is dense in $b(r_1\mathbb{B}^k)$), a contradiction to our assumption. □

We now use the previous lemma with $k = n$. Set $\mathbb{B} = \mathbb{B}^n$ and $S = b\mathbb{B}$.

Lemma 4.7.4 *For each $m \in \mathbb{N}$ there is a discrete set $E_m \subset \mathbb{C}^n \setminus m\mathbb{B}$ such that for every holomorphic map $F : m\mathbb{B} \to \mathbb{C}^n$ satisfying*

(i) $|F(0)| \le \frac{m}{2}$,
(ii) $|JF(z)| \ge \frac{1}{m}$ *at some point* $z \in \frac{m}{2}\overline{\mathbb{B}}$, *and*
(iii) $F(m\mathbb{B}) \cap E_m = \varnothing$

we have $F(\frac{m}{2}\mathbb{B}) \subset m\mathbb{B}$.

Proof Choose real sequences a_j and r_j such that

$$\frac{m}{2} = a_1 < a_2 < a_3 < \cdots < \frac{3m}{4}, \qquad m = r_1 < r_2 < \cdots, \qquad \lim_{j\to\infty} r_j = +\infty.$$

Using the notation of Lemma 4.7.3, we define

$$E_m = \bigcup_{j=1}^{\infty} E(a_j, a_{j+1}, r_j, r_{j+1}, 1/m). \tag{4.41}$$

Since E_m is a union of finite sets lying in the spheres $r_j S$ and $r_j \to \infty$, we conclude that E_m is discrete. If F satisfies the conditions of the lemma, then F is bounded on $\frac{3m}{4}\mathbb{B}$, and hence $F(a_{j+1}\mathbb{B}) \subset r_{j+1}\mathbb{B}$ for some $j \in \mathbb{N}$. Condition (iii) implies that $F(a_{j+1}\mathbb{B})$ does not intersect $E(a_j, a_{j+1}, r_j, r_{j+1}, 1/m)$, and hence Lemma 4.7.3 shows that $F(a_j\mathbb{B}) \cap r_j S = \varnothing$. Thus $F(a_j\mathbb{B}) \subset r_j\mathbb{B}$. By a downward induction on j we conclude that $F(a_1\mathbb{B}) \subset r_1\mathbb{B}$. □

Proof of Theorem 4.7.2 Let E_m be given by (4.41) and set $E = \bigcup_{m=1}^{\infty} E_m$. Since $E_m \subset \mathbb{C}^n \setminus m\mathbb{B}$, $E \cap m\mathbb{B}$ is finite for each m, and hence E is discrete. We claim that the range of every nondegenerate entire map $F\colon \mathbb{C}^n \to \mathbb{C}^n$ intersects E. Assume to the contrary. Then F satisfies the hypotheses of Lemma 4.7.4 for all large $m \in \mathbb{N}$, and hence $F(\frac{m}{2}\mathbb{B}) \subset m\mathbb{B}$ for such m. Liouville's theorem implies that F is affine linear. Since F is nondegenerate, we have $F(\mathbb{C}^n) = \mathbb{C}^n$, a contradiction to the assumption. □

By a more elaborate version of the above construction, Rosay and Rudin also proved the following result.

Theorem 4.7.5 ([477, Theorem 5.1]) *For every $n \in \mathbb{N}$ there is a discrete set $D \subset \mathbb{C}^n$ such that the only nondegenerate holomorphic map $F\colon \mathbb{C}^n \to \mathbb{C}^n$ that maps $\mathbb{C}^n \setminus D$ to itself is the identity map. In particular, no automorphism of $\mathbb{C}^n$ other than identity can map D onto itself.*

A set D as in Theorem 4.7.5 is called *rigid*. The proof in [477, Sect. 5] actually produces a rigid set D with the following additional property.

Proposition 4.7.6 *For every $n > 1$ there exists a rigid set D in $\mathbb{C}^n$ such that, if $p' \neq p''$ are points in $\mathbb{C}^n \setminus D$, then there is no automorphism of $\mathbb{C}^n$ mapping $D' = D \cup \{p'\}$ onto $D'' = D \cup \{p''\}$.*

Since there exist continuum many choices of p'' for a given p', we get

Corollary 4.7.7 ([477, Corollary 5.3]) *There exist continuum many discrete sets in $\mathbb{C}^n$ no two of which are* $\mathrm{Aut}(\mathbb{C}^n)$*-equivalent.*

By replacing a discrete set of points in Theorem 4.7.2 by a discrete set of closed balls in affine subspaces of $\mathbb{C}^n$, we have the following theorem which summarizes results from [76], [188, Sect. 6], and [62].

Theorem 4.7.8 *Let $1 < k < n$ and $m \ge n-k$ be integers. There exists a closed set $B = \bigcup_{j=1}^{\infty} B_j$ in $\mathbb{C}^n$, where B_j are pairwise disjoint balls contained in k-dimensional affine subspaces Λ_j of $\mathbb{C}^n$, such that for every entire map $f\colon \mathbb{C}^m \to \mathbb{C}^n$ with* $\mathrm{rank}_p f \ge n-k$ *for some $p \in \mathbb{C}^m$, the set $\{z \in \mathbb{C}^m : f(z) \in B\}$ is infinite. It is possible to choose B such that $\mathbb{C}^n \setminus B$ is Eisenman $(n-k)$-hyperbolic.*

Furthermore, the balls B_j in Theorem 4.7.8 can be chosen such that any sufficiently small smooth perturbation B'_j of B_j (depending on j) still gives a set $D' = \bigcup_{j=1}^{\infty} B'_j \subset \mathbb{C}^n$ whose complement $\mathbb{C}^n \setminus B'$ is Eisenman $(n-k)$-hyperbolic. This is used in the proof of Theorem 4.18.7 on p. 187.

4.8 Algorithms for Computing Flows

We recall a result to the effect that the flow of a vector field can be approximated by iterates of a fixed map.

Definition 4.8.1 ([4, p. 254]) Let V be a continuous vector field on a manifold X, and let $A_t(x)$ be a continuous map from an open set in $\mathbb{R}_+ \times X$ containing $\{0\} \times X$ to X such that its t-derivative exists and is continuous. We say that A is an *algorithm for* V if for all $x \in X$ we have that

$$A_0(x) = x, \qquad \left.\frac{\partial}{\partial t}\right|_{t=0} A(t,x) = V_x. \tag{4.42}$$

In the following result we focus on positive time $t \geq 0$; the same result for $t \leq 0$ is obtained by replacing V by $-V$.

Theorem 4.8.2 ([4, Theorem 4.1.26, p. 254]) *Let V be a Lipschitz continuous vector field with flow ϕ_t on a manifold X. Let Ω be the fundamental domain of V* (1.42) *and $\Omega_+ = \Omega \cap (\mathbb{R}_+ \times X)$. If A is an algorithm for V, then for all $(t,x) \in \Omega_+$ the n-th iterate $A^n_{t/n}(x)$ of the map $A_{t/n}$ is defined for sufficiently large $n \in \mathbb{N}$ (depending on x and t), and we have that*

$$\lim_{n\to\infty} A^n_{t/n}(x) = \phi_t(x). \tag{4.43}$$

The convergence is uniform on compacts in Ω_+. Conversely, if $t_0 > 0$ is such that $A^n_{t/n}(x)$ is defined for all $t \in [0,t_0]$ and all sufficiently large $n \in \mathbb{N}$, and if $A^n_{t/n}(x)$ converges as $n \to \infty$, then $(t_0,x) \in \Omega_+$ and (4.43) *holds.*

Proof We give details for $X = \mathbb{R}^m$ as we only use this case; a similar argument applies in general. Fix a point $p \in \mathbb{R}^m$ and suppose that the flow $\phi_t(p)$ exists for $t \in [0,t_0]$. Let $C = \{\phi_t(p) \colon t \in [0,t_0]\}$. Choose compact sets $L_1 \subset L_2 \subset \mathbb{R}^m$ such that $C \subset \mathring{L}_1$ and $L_1 \subset \mathring{L}_2$. There is a compact neighborhood $K \subset \mathring{L}_1$ of p such that for every $x \in K$ the flow $\phi_t(x)$ exists for $t \in [0,t_0]$ and remains in L_1. From (4.42) we see that $|\phi_t(x) - A_t(x)| = o(t)$ as $t \to 0$, uniformly on L_2. Fix $n \in \mathbb{N}$. Given a point $x \in K$, assume for the moment that the orbit

$$y_0 = x, \quad y_1 = A_{t/n}(y_0), \quad y_2 = A_{t/n}(y_1), \quad \ldots, \quad y_n = A_{t/n}(y_{n-1}) \tag{4.44}$$

exists and lies in L_2. If β is a Lipschitz constant for V on L_2, then Lemma 1.9.3 (p. 34) gives the estimate $|\phi_t(x) - \phi_t(y)| \le e^{\beta t}|x-y|$. Since $\phi_t(x) = \phi^n_{t/n}(x)$, we have that

$$\phi_t(x) - A^n_{t/n}(x) = \sum_{j=1}^{n}\Big(\phi^{n-j}_{t/n}\big(\phi_{t/n}(y_{j-1})\big) - \phi^{n-j}_{t/n}\big(A_{t/n}(y_{j-1})\big)\Big).$$

Applying the inequality (1.46) (p. 34) to each term, we get

$$\big|\phi_t(x) - A^n_{t/n}(x)\big| \le \sum_{j=1}^{n} e^{\beta t(n-j)/n}\big|\phi_{t/n}(y_{j-1}) - A_{t/n}(y_{j-1})\big| \le n e^{\beta t} o(t/n).$$

This converges to 0 as $n \to \infty$, uniformly for $x \in K$. A similar estimate gives

$$\big|\phi_{kt/n}(x) - A^k_{t/n}(x)\big| \le k e^{\beta t} o(t/n), \quad k = 1, \ldots, n.$$

Using this estimate inductively, we see that for every sufficiently big $n \in \mathbb{N}$ the orbit (4.44) exists and lies in L_2 for each $x \in K$ and $t \in [0, t_0]$. This proves the main implication. The converse is left as an exercise. □

Proposition 4.8.3 *Let V and W be vector fields with flows ϕ_t, ψ_t. Then*

(i) *$\phi_t \circ \psi_t$ is an algorithm for $V + W$.*
(ii) *$\psi_{-\sqrt{t}} \circ \phi_{-\sqrt{t}} \circ \psi_{\sqrt{t}} \circ \phi_{\sqrt{t}}$ is an algorithm for $[V, W]$.*

Proof Part (i) follows from the Taylor expansion of flows (see (1.49) on p. 36), and (ii) follows from Lemma 1.9.7 on p. 37. □

By a repeated application of Proposition 4.8.3, using sums and Lie brackets, Theorem 4.8.2 gives the following result which is the key ingredient for the results in the next two sections.

Corollary 4.8.4 *Let $V_1, \ldots, V_m$ be $\mathbb{R}$-complete holomorphic vector fields on a complex manifold X. Denote by $\mathfrak{g}$ the Lie subalgebra of $\aleph(X)$ generated by the V_j's and let $V \in \mathfrak{g}$. Assume that K is a compact set in X and $t_0 > 0$ is such that the flow $\phi_t(x)$ of V exists for every $x \in K$ and for all $t \in [0, t_0]$. Then ϕ_{t_0} is a uniform limit on K of a sequence of compositions of time-forward maps of the vector fields $V_1, \ldots, V_m$. In particular, ϕ_{t_0} can be approximated uniformly on K by holomorphic automorphisms of X.*

Example 4.8.5 If $V = V_1 + \cdots + V_m$, then its flow ϕ_t is given by

$$\phi_t(x) = \lim_{n\to\infty}\big(\phi^1_{t/n} \circ \cdots \circ \phi^m_{t/n}\big)^n(x)$$

for all (t, x) in the fundamental domain of the vector field V. □

4.9 The Andersén-Lempert Theory

A major breakthrough in understanding the automorphism groups $\mathrm{Aut}(\mathbb{C}^n)$ was made by Andersén [23] (for automorphisms with Jacobian one) and Andersén and Lempert [25] in the early 1990's. We have already seen in Sect. 4.1 that the Euclidean space $\mathbb{C}^n$ of dimension $n > 1$ admits many complete holomorphic vector fields; in particular, the shear fields (4.9), (4.10). Their main observation is that every polynomial holomorphic vector field on $\mathbb{C}^n$ is the sum of finitely many polynomial shear vector fields (see Proposition 4.9.9). This leads to the proof of the following main result in their papers.

Theorem 4.9.1 *Let $n > 1$.*

(i) (Andersén and Lempert [25].) *Every holomorphic automorphism of $\mathbb{C}^n$ can be approximated uniformly on compacts by compositions of shears of the form* (4.7) *and* (4.8).
(ii) (Andersén [23].) *Every holomorphic automorphism of $\mathbb{C}^n$ with Jacobian one can be approximated uniformly on compacts by compositions of polynomial shears* (4.7).

In [25], Andersén and Lempert used Nevanlinna theory to prove that for every $n > 1$ there exists a holomorphic automorphism of $\mathbb{C}^n$ that is not a composition of finitely many shears. For example, the automorphism

$$(z_1, z_2) \mapsto \left(z_1 e^{z_1 z_2}, z_2 e^{-z_1 z_2}\right)$$

of $\mathbb{C}^2$ is not a composition of shears.

The proof of Theorem 4.9.1 given by Andersén and Lempert involved arguments on infinite dimensional complex manifolds; specifically, they considered the tangent space to the identity of the infinite dimensional group $\mathrm{Aut}(\mathbb{C}^n)$. Shortly thereafter, Rosay and the author [221] gave a somewhat different proof by using concatenations of flows of vector fields and Theorem 4.8.2. This made the proof conceptually simpler and amenable to generalizations. In this way, the original theorem of Andersén and Lempert got reformulated in the following stronger form which is now commonly used.

Theorem 4.9.2 (Forstnerič and Rosay [221, Theorem 1.1]) *Let Ω be a Runge domain in $\mathbb{C}^n$ for some $n > 1$, and let*

$$\Phi_t : \Omega \to \Omega_t = \Phi_t(\Omega) \subset \mathbb{C}^n, \quad t \in [0, 1]$$

be a $\mathscr{C}^1$ isotopy consisting of injective holomorphic maps, with Φ_0 the identity map on $\Omega = \Omega_0$. If $\Phi_t(\Omega)$ is Runge in $\mathbb{C}^n$ for each $t \in [0, 1]$, then Φ_1 can be approximated uniformly on compacts in Ω by holomorphic automorphisms of $\mathbb{C}^n$ (in fact, by compositions of shears (4.1), (4.2) *and their $SU(n)$-conjugates).*

If in addition Ω is a domain of holomorphy satisfying $H^{n-1}(\Omega;\mathbb{C})=0$ and $\mathrm{J}\Phi_t=1$ for every $t\in[0,1]$, then Φ_1 can be approximated by compositions of additive shears (4.1) *and their $SU(n)$-conjugates.*

In both cases, the automorphisms of $\mathbb{C}^n$ approximating the map Φ_1 can be chosen such that their finite order jets agree with the corresponding jets of Φ_1 at any given finite set of points in Ω.

Example 4.9.3 Theorem 4.9.2 fails in general for volume preserving maps if $H^{n-1}(\Omega;\mathbb{C})\neq 0$; here is an example. The family $\Phi_t(z,w)=(z,w+tz^{-1})$ is a one-parameter group of automorphism of $\mathbb{C}^*\times\mathbb{C}$ preserving the volume form $\mathrm{d}z\wedge\mathrm{d}w$. The circle $T=\{(z,\bar z)\in\mathbb{C}^2:|z|=1\}$ is polynomially convex and hence admits a pseudoconvex tubular neighborhood $\Omega\subset\mathbb{C}^*\times\mathbb{C}$ that is Runge in $\mathbb{C}^2$. Clearly, $H^1(\Omega;\mathbb{C})=H^1(T;\mathbb{C})=\mathbb{C}$. Thus, all conditions in Theorem 4.9.2 except the cohomological one are satisfied. However, Φ_t for $t\neq 0$ is not approximable by volume preserving automorphisms of $\mathbb{C}^2$ in any neighborhood of T. Indeed, $\int_{\Phi_t(T)} w\,\mathrm{d}z=2\pi\mathrm{i}(1+t)$ depends on t, while Stokes' theorem implies (in view of $\mathrm{d}(w\,\mathrm{d}z)=-\mathrm{d}z\wedge\mathrm{d}w$) that this integral is independent of t for any volume preserving automorphism Φ_t. □

Before proving Theorem 4.9.2, we indicate some of its consequences. Note that Theorem 4.9.1 is a special case in view of the following lemma.

Lemma 4.9.4 *For every $\Phi\in\mathrm{Aut}(\mathbb{C}^n)$ there is a smooth isotopy of automorphisms $\Phi_t\in\mathrm{Aut}(\mathbb{C}^n)$ ($t\in[0,1]$) such that $\Phi_0=\mathrm{Id}$ and $\Phi_1=\Phi$. When $\Phi_1\in\mathrm{Aut}_1(\mathbb{C}^n)$, the isotopy can be chosen in $\mathrm{Aut}_1(\mathbb{C}^n)$.*

Proof By a family of translations of $\mathbb{C}^n$ we connect Φ to $\Psi=\Phi-\Phi(0)$ satisfying $\Psi(0)=0$. The isotopy $\Psi_t(z)=\frac{1}{t}\Psi(tz)$ ($t\in(0,1]$) connects $\Psi=\Psi_1$ to $\Psi_0=\Phi'(0)\in GL_n(\mathbb{C})$. Finally, there is a path in $GL_n(\mathbb{C})$ from Ψ_0 to Id. Combining these isotopies and reparametrizing the interval $[0,1]$, we obtain a smooth isotopy of automorphisms of $\mathbb{C}^n$ connecting Φ to the identity. The analogous arguments apply when $\Phi\in\mathrm{Aut}_1(\mathbb{C}^n)$. □

The same proof gives the following corollary [25, Theorem 2.1]. Recall that every starshaped domain in $\mathbb{C}^n$ is Runge [142].

Corollary 4.9.5 *Let $n>1$. Every biholomorphic map $\Phi\colon D\to\Omega$ from a starshaped domain $D\subset\mathbb{C}^n$ onto a Runge domain $\Omega\subset\mathbb{C}^n$ can be approximated uniformly on compacts by holomorphic automorphisms of $\mathbb{C}^n$. This holds in particular for every biholomorphic map $\Phi\colon\mathbb{C}^n\to\Omega$ onto a Runge domain $\Omega\subset\mathbb{C}^n$.*

The converse to the last statement is also true: if a biholomorphic map $\Phi\colon\mathbb{C}^n\to\Omega\subset\mathbb{C}^n$ is a limit of a sequence of holomorphic automorphisms of $\mathbb{C}^n$, then its image $\Omega=\Phi(\mathbb{C}^n)$ is Runge in $\mathbb{C}^n$. Indeed, one can easily see that for any compact polynomially convex set K in $\mathbb{C}^n$, the image $\Phi(K)$ is also polynomially convex. Hence, we get the following corollary.

Corollary 4.9.6 *A Fatou-Bieberbach map $\Phi : \mathbb{C}^n \to \Omega \subset \mathbb{C}^n$ is a locally uniform limit of holomorphic automorphisms of $\mathbb{C}^n$ if and only if the image $\Phi(\mathbb{C}^n) = \Omega$ is Runge in $\mathbb{C}^n$.*

In Sect. 4.20 we give examples of non-Runge Fatou-Bieberbach domains.

Proof of Theorem 4.9.2 Let $\widetilde{\Omega}$ denote the trace of the isotopy $\{\Phi_t\}_{t\in[0,1]}$:

$$\widetilde{\Omega} = \{(t, z) : t \in [0, 1],\ z \in \Omega_t\} \subset \mathbb{R} \times \mathbb{C}^n.$$

We consider Φ_t as the flow of the time-dependent vector field

$$V(t, z) = \dot{\Phi}_t\big(\Phi_t^{-1}(z)\big), \quad (t, z) \in \widetilde{\Omega}.$$

(The dot denotes the derivative with respect to t.) Note that V is continuous on $\widetilde{\Omega}$ and is holomorphic on Ω_t for every fixed $t \in [0, 1]$. Choose an integer $N \in \mathbb{N}$ and subdivide $[0, 1]$ into subintervals $I_k = [t_k, t_{k+1}]$ of length $1/N$, where $t_k = k/N$ for $k = 0, 1, \ldots, N$. Consider the locally constant holomorphic vector field $\widetilde{V}(t, z)$ that equals $V(t_k, z)$ for $t \in I_k$ and $z \in \Omega_{t_k}$; let ϕ_t denote its flow. Since the flows of $V_{t_k} = V(t_k, \cdot)$ and of V are tangential at $t = t_k$, they differ on the interval $t \in I_k$ by a term of order $o(N^{-1})$. We see as in the proof of Theorem 4.8.2 (p. 133) that the total error is of size

$$\big|\phi_t(z) - \Phi_t(z)\big| = No\big(N^{-1}\big) = o(1), \quad t \in [0, 1],\ z \in \Omega.$$

As $N \to +\infty$, ϕ_t converges to Φ_t uniformly on compacts in Ω for all $t \in [0, 1]$. Further, given a compact set $K \subset \Omega$, for every N large enough we have

$$\big\{\phi_t(z) : z \in K,\ t \in I_k\big\} \Subset \Omega_{t_k}, \quad k = 0, \ldots, N - 1.$$

Hence, it suffices to show that the flow ϕ_t^k of the autonomous holomorphic vector field V_{t_k} (now chosen such that $\phi_0^k(z) = z$ for $z \in \Omega_{t_k}$) can be approximated, uniformly on any compact subset $L \subset \Omega_{t_k}$ whose trace $\widetilde{L} = \{\phi_t^k(z) : z \in L,\ t \in [0, 1/N]\}$ remains in Ω_{t_k}, by holomorphic automorphisms of $\mathbb{C}^n$. This is done in two steps. First, we approximate it by the flow of a polynomial vector field (the easy part); next, we approximate the latter by compositions of flows of shear vector fields. The second step depends on the main observation of the Andersén-Lempert theory.

Step 1: Approximation by the flow of a polynomial vector field. Since the domain Ω_{t_k} is Runge in $\mathbb{C}^n$, we can approximate the holomorphic vector field V_{t_k}, uniformly on a compact set in Ω_{t_k} that contains $\widetilde{L}$ in its interior, by a polynomial holomorphic vector field. Lemma 1.9.4 on p. 35 shows that ϕ_t can be approximated uniformly on L by flows of polynomial vector fields.

A more careful argument is needed when $J\Phi_t = 1$ for all $t \in [0, 1]$. Let $V_t = \sum_{i=1}^n V_{t,i} \frac{\partial}{\partial z_i}$ and $\omega = dz_1 \wedge \cdots \wedge dz_n$. We see from (4.13) (p. 110) that

$$J\Phi_t = 1\ \big(\forall t \in [0, 1]\big) \iff \mathrm{div}_\omega V_t = 0\ \big(\forall t \in [0, 1]\big);$$

that is, divergence zero vector fields correspond to volume preserving flows. From the Cartan formula $\mathrm{d}(V_t \rfloor \omega) = \mathrm{div}_\omega V_t \cdot \omega$ (see (4.12) on p. 110) we infer that $\mathrm{div}_\omega V_t = 0$ if and only if the holomorphic $(n-1)$-form $\alpha_t = V_t \rfloor \omega$ on Ω_t is closed; assume this is so. If Ω is a domain of holomorphy and $H^{n-1}(\Omega; \mathbb{C}) = 0$, then the same holds for every $\Omega_t = \Phi_t(\Omega)$. Since the de Rham cohomology classes of a Stein manifold are represented by holomorphic forms [496, Theorem 1], we get $\alpha_t = \mathrm{d}\beta_t$ for some holomorphic $(n-2)$-form β_t on Ω_t. Approximating the coefficients of β_t by holomorphic polynomials we get a $(n-2)$-form $\widetilde{\beta}_t$ on $\mathbb{C}^n$. The polynomial vector field W_t, defined by $W_t \rfloor \omega = \mathrm{d}\widetilde{\beta}_t$, has divergence zero and it approximates the vector field V_t.

Step 2: Approximating the flow of a polynomial vector field by holomorphic automorphisms. The following key lemma is due to Andersén [23] (for divergence zero vector fields) and Andersén and Lempert [25].

Proposition 4.9.7 *Let $n > 1$. Every polynomial holomorphic vector field V on $\mathbb{C}^n$ is a finite sum of complete polynomial shear fields* (4.9) *and* (4.10) *(p.* 109). *If* $\mathrm{div}_\omega V = 0$, *then V is a finite sum of shear fields* (4.9).

In view of Corollary 4.8.4 and Example 4.8.5, this implies the following.

Corollary 4.9.8 *Let $V = \sum_{i=1}^n V_i \frac{\partial}{\partial z_i}$ be a polynomial (or entire) vector field on $\mathbb{C}^n$ for some $n > 1$. Assume that $t_0 > 0$ and the flow $\phi_t(x)$ of V exists for all $t \in [0, t_0]$ and all points z in a compact set $K \subset \mathbb{C}^n$. Then $\phi_{t_0}|_K$ is a uniform limit of compositions of shears (of additive shears if* $\mathrm{div}_\omega V = 0$).

Note that Corollary 4.9.8 will complete the proof of the approximation statement in Theorem 4.9.2.

Proof of Proposition 4.9.7 Write $V = V_0 + V_1 + \cdots + V_m$ where each V_k is a vector field on $\mathbb{C}^n$ whose coefficients are homogeneous polynomials of degree k. Then, $\mathrm{div}_\omega V = 0$ if and only if $\mathrm{div}_\omega V_k = 0$ for all k, and hence it suffices to prove the result for homogeneous vector fields. This is accomplished by the following lemma in which we identify a vector $v = (v_1, \ldots, v_n) \in \mathbb{C}^n$ with the constant vector field $\sum_{j=1}^n v_j \frac{\partial}{\partial z_j}$.

Lemma 4.9.9 *For each integer $k \in \mathbb{N}$ there exist finitely many linear forms $\lambda_1, \ldots, \lambda_r$ on $\mathbb{C}^n$ and vectors $v_1, \ldots, v_r \in \mathbb{C}^n$, with $\lambda_i(v_i) = 0$ and $|v_i| = 1$ for all $i \in \{1, \ldots, r\}$, such that every holomorphic polynomial map $V\colon \mathbb{C}^n \to \mathbb{C}^n$ that is homogeneous of degree k is of the form*

$$V(z) = \sum_{i=1}^r c_i \, \lambda_i(z)^k \, v_i + d_i \, \lambda_i(z)^{k-1} \langle z, v_i \rangle \, v_i \tag{4.45}$$

for some numbers $c_i, d_i \in \mathbb{C}$. *(The* λ_i*'s and* v_i*'s may appear with repetition.) If* $\mathrm{div}_\omega V = 0$, *we may take* $d_i = 0$ *for all* i. *The* λ_i*'s can be chosen from any nonempty open subset of* $(\mathbb{C}^n)^*$ *(the dual of* $\mathbb{C}^n$*).*

Note that the summands in (4.45) are homogeneous shear vector fields of the form (4.9), (4.10).

Proof For each $k \in \mathbb{N}$ there exist linear forms $\lambda_i(z) = \sum_{j=1}^n a_{i,j} z_j$ on $\mathbb{C}^n$ ($i = 1, \ldots, m = m(k,n)$) such that every homogeneous polynomial $P \in \mathbb{C}[z_1, \ldots, z_n]$ of degree k can be written in the form

$$P(z) = \sum_{i=1}^m c_i \, \lambda_i(z)^k, \quad c_i \in \mathbb{C}. \tag{4.46}$$

The forms λ_i may be chosen from any nonempty open set $U \subset (\mathbb{C}^n)^*$. (See e.g. [23, Lemma 5.6].) For convenience, we replace U by a cone with vertex at the origin, so that any nonzero multiple of an element in U remains in U (this requires a rescaling of the coefficients in (4.45) and (4.46)). By a linear change of coordinates we may assume that $e_1^*(z) = z_1$ belongs to U. Choose $\lambda_1, \ldots, \lambda_m \in U$ satisfying (4.46) for homogeneous polynomials of degrees k and $k-1$ and such that $a_{i,n} = \lambda_i(e_n) = 1$ for $i = 1, \ldots, m$. (It suffices to ensure that $\lambda_i(e_n) \neq 0$ and then rescale them.) Applying (4.46) to the homogeneous polynomial $\mathrm{div}_\omega V$ of degree $k-1$, we get $(\mathrm{div}_\omega V)(z) = \sum_{i=1}^m d_i \, \lambda_i(z)^{k-1}$ for some $d_i \in \mathbb{C}$. Choose vectors $v_i \in \ker \lambda_i$ with $|v_i| = 1$ and set

$$V_i(z) = d_i \, \lambda_i(z)^{k-1} \langle z, v_i \rangle \, v_i, \quad i = 1, \ldots, m.$$

These vector fields are taken as the second part of the expression (4.45). Since $\mathrm{div}_\omega V_i = d_i \, \lambda_i(z)^{k-1}$, the vector field $X = V - \sum_{i=1}^m V_i$ (that is homogeneous of degree k) has divergence zero.

It remains to show that every divergence zero homogeneous polynomial vector field X on $\mathbb{C}^n$ of degree k is of the form

$$X(z) = \sum_{i=1}^r c_i \, \lambda_i(z)^k \, v_i, \quad c_i \in \mathbb{C}, \tag{4.47}$$

where $\lambda_i \in U$ and $v_i \in \ker \lambda_i$. Applying (4.46) to the component X_j of X for $j \in \{1, \ldots, n-1\}$, we get $X_j(z) = \sum_{i=1}^m c_{i,j} \lambda_i(z)^k$. Each vector field

$$V_{i,j}(z) = c_{i,j} \, \lambda_i(z)^k (e_j - a_{i,j} \, e_n)$$

is a shear (4.9), with zero divergence, that may be used as one of the summands in (4.47). Set $W = \sum V_{i,j} = (W_1, \ldots, W_n)$ (summation over $i = 1, \ldots, m$ and $j = 1, \ldots, n-1$). Then, $W_l = X_l$ for $l = 1, \ldots, n-1$, and hence $X = W + (X_n - W_n) \, e_n$. Since $\mathrm{div}_\omega X = 0 = \mathrm{div}_\omega W$, we get

$$0 = \mathrm{div}_\omega (X - W) = \partial (X_n - W_n) / \partial z_n.$$

Thus, $X_n - W_n$ is independent of z_n, and hence $(X_n - W_n)\,e_n$ is a shear field (4.9). By choosing additional linear forms $\lambda'_j \in U$ that only depend on the variables $z' = (z_1, \ldots, z_{n-1})$ (such exist due to our choice of coordinates) we write $X_n - W_n = \sum c'_j \lambda'_j(z')$. Adding the corresponding vector fields $c'_j \lambda'_j(z')\,e_n$ to W we obtain the desired expression (4.47) for X.

This proves Lemma 4.9.9 and Proposition 4.9.7. □

To complete the proof of Theorem 4.9.2, it remains to justify the claim concerning the jet interpolation at finitely many points $p_1, \ldots, p_m \in \Omega$. Choose a compact set $K \subset \Omega$ containing these points in the interior and an automorphism $F \in \mathrm{Aut}(\mathbb{C}^n)$ which is close to Φ_1 on K. The jets of order k of F and Φ_1 at are then also close at each of the points p_j. Corollary 4.15.5 (p. 174) furnishes an automorphism $G \in \mathrm{Aut}(\mathbb{C}^n)$ whose jets at the points $q_j = F(p_j)$ $(j = 1, \ldots, m)$ are chosen such that the composition $\widetilde{F} = G \circ F \in \mathrm{Aut}(\mathbb{C}^n)$ agrees to order k with Φ_1 at the points $p_1, \ldots, p_m$. (The cited result is proved by explicit calculations, so there is no problem with this forward reference.) Furthermore, G can be chosen close to the identity provided that the jets of F and Φ_1 at the points p_j were close, so $\widetilde{F}$ is still close to Φ_1.

This completes the proof of Theorems 4.9.1 and 4.9.2. □

The following parametric version of Theorem 4.9.2 was proved by Kutzschebauch [356, Theorem 2.3].

Theorem 4.9.10 *Let Ω be an open set in $\mathbb{C}^n = \mathbb{C}^k \times \mathbb{C}^m$ with $m > 1$. For every $t \in [0, 1]$ let $\Phi_t : \Omega \to \mathbb{C}^n$ be an injective holomorphic map of the form*

$$\Phi_t(z, w) = \big(z, \varphi_t(z, w)\big), \quad z \in \mathbb{C}^k, \ w \in \mathbb{C}^m \tag{4.48}$$

and of class $\mathscr{C}^1$ in $(t, z, w) \in [0, 1] \times \Omega$, with $\Phi_0 = \mathrm{Id}$. Assume that the domain $\Phi_t(\Omega)$ is Runge in $\mathbb{C}^n$ for every $t \in [0, 1]$. Then Φ_1 can be approximated uniformly on compacts in Ω by holomorphic automorphisms of the form (4.48).

Proof Consider a holomorphic vector field on $\mathbb{C}^n = \mathbb{C}^k \times \mathbb{C}^m$ of the form

$$V(z, w) = \sum_{j=1}^{m} a_j(z, w) \frac{\partial}{\partial w_j}, \quad z \in \mathbb{C}^k, \ w \in \mathbb{C}^m. \tag{4.49}$$

Its flow remains in the level sets $z = \mathrm{const}$, and V is complete on $\mathbb{C}^n$ if and only if $V(z, \cdot)$ is complete on $\mathbb{C}^m$ for each fixed $z \in \mathbb{C}^k$. We conclude the proof as in Theorem 4.9.2, replacing Proposition 4.9.7 with the following lemma.

Lemma 4.9.11 *If $m \geq 2$, then every polynomial vector field V* (4.49) *on $\mathbb{C}^n = \mathbb{C}^k \times \mathbb{C}^m$ is a finite sum of complete polynomial fields of the same type.*

Proof We have $V(z,w) = \sum_\alpha z^\alpha V_\alpha(w)$ (a finite sum), where $z^\alpha = z_1^{\alpha_1} \cdots z_k^{\alpha_k}$ and $V_\alpha = \sum_{j=1}^m a_{\alpha,j}(w) \frac{\partial}{\partial w_j}$ is a polynomial vector field on $\mathbb{C}^m$. By Corollary 4.9.7, each V_α is a finite sum of complete polynomial fields on $\mathbb{C}^m$. The products of such fields with z^α are complete on $\mathbb{C}^n$. □

A much more general result is due to Kutzschebauch and Ramos-Peon [366] who proved the infinite parametric transitivity property on Stein manifolds with the density property (see Definition 4.10.1).

4.10 The Density Property

The Andersén-Lempert theorem, presented in the previous section, had a major influence on the study of automorphism groups of other complex and algebraic manifolds with big automorphism groups. There are various notions of *big* in this context. The bottom line is that the group $\mathrm{Aut}(X)$ should be infinite dimensional. It is also reasonable to ask that it act transitively on the manifold X, and perhaps even k-transitively for every $k \in \mathbb{N}$. Finally, it is very useful to have the analogue of Theorem 4.9.2 hold.

A major step in extending the Andersén-Lempert theory was made by Varolin in his PhD dissertation and in the papers [558–560]. His vantage point is summarized by Corollary 4.8.4 (p. 134):

The flow of any holomorphic vector field which is a Lie combination of complete holomorphic vector fields can be approximated by compositions of holomorphic automorphisms of the given manifold.

The relevant *density property* of a complex manifold (see Definition 4.10.1), introduced by Varolin, conceptualizes the proof of the Andersén-Lempert theorem. The analogous notion in the class of affine algebraic manifolds was introduced by Kaliman and Kutzschebauch. It has since become one of the standard notions of elliptic geometry with close ties to Oka theory. In this and the next section we present some of the main developments on this subject. We provide insight where possible, but many specific results and examples are stated without proof. This topic is intimately connected to the Oka theory discussed in Chaps. 5–7, which is one of the reasons for including it here.

Recall that $\mathcal{T}_X$ denotes the tangent sheaf of a complex manifold X, $\aleph_{\mathcal{O}}(X) = \aleph(X) = \mathcal{T}_X(X)$ is the Lie algebra of all holomorphic vector fields on X, and $\aleph_{\mathcal{A}}(X)$ stands for the Lie algebra of all algebraic vector field on an algebraic manifold X. Let $\mathfrak{g}$ be a complex Lie subalgebra of $\aleph(X)$. Given a subset $\mathfrak{g}_0$ of $\mathfrak{g}$, we use the following notation:

- $\mathrm{Span}(\mathfrak{g}_0)$ is the linear span of $\mathfrak{g}_0$ (a linear subspace of $\mathfrak{g}$),
- $\overline{\mathrm{Span}}(\mathfrak{g}_0) = \overline{\mathrm{Span}(\mathfrak{g}_0)}$ is the closed linear span of $\mathfrak{g}_0$,
- $\mathrm{Lie}(\mathfrak{g}_0)$ is the Lie subalgebra of $\mathfrak{g}$ generated by $\mathfrak{g}_0$,

- $\overline{\mathrm{Lie}}(\mathfrak{g}_0)$ is the closure of $\mathrm{Lie}(\mathfrak{g}_0)$; this is the smallest closed Lie subalgebra of $\aleph(X)$ containing $\mathfrak{g}_0$.

Obviously, $\mathrm{Span}(\mathfrak{g}_0) \subset \mathrm{Lie}(\mathfrak{g}_0) \subset \overline{\mathrm{Lie}}(\mathfrak{g}_0)$. We also let

$$\mathfrak{g}_{\mathrm{int}} = \{V \in \mathfrak{g} : V \text{ is } \mathbb{C}\text{-complete}\}. \tag{4.50}$$

See Definition 1.9.9 on p. 39 and recall that every $\mathbb{R}$-complete holomorphic vector field on a Stein manifold without bounded plurisubharmonic functions (such as an affine algebraic manifold) is also $\mathbb{C}$-complete (see [187, Corollary 2.2]).

Definition 4.10.1 Let X be a complex manifold.

(a) A Lie algebra $\mathfrak{g} \subset \aleph(X)$ has the *density property* if $\bar{\mathfrak{g}} = \overline{\mathrm{Lie}}(\mathfrak{g}_{\mathrm{int}})$ (i.e., the Lie algebra $\mathrm{Lie}(\mathfrak{g}_{\mathrm{int}})$ generated by all complete vector fields in $\mathfrak{g}$ is dense in $\mathfrak{g}$). The manifold X has the density property if this holds for $\mathfrak{g} = \aleph(X)$.
(b) Let X be an algebraic manifold. A Lie algebra $\mathfrak{g} \subset \aleph_{\mathcal{A}}(X)$ has the *algebraic density property* if $\mathfrak{g} = \mathrm{Lie}(\mathfrak{g}_{\mathrm{int}})$. If this holds for $\mathfrak{g} = \aleph_{\mathcal{A}}(X)$, then X has the *algebraic density property*.
(c) A complex manifold X with a holomorphic volume form ω has the *volume density property* if the Lie algebra

$$\aleph^{\omega}(X) = \left\{V \in \aleph(X) : \mathrm{div}_{\omega} V = 0\right\}$$

of divergence zero vector fields (see (4.12)) has the density property.
(d) An algebraic manifold X with an algebraic volume form ω has the *algebraic volume density property* if the Lie algebra

$$\aleph^{\omega}_{\mathcal{A}}(X) = \left\{V \in \aleph_{\mathcal{A}}(X) : \mathrm{div}_{\omega} V = 0\right\}$$

has the algebraic density property.

Every compact complex manifold enjoys the density property since each vector field is complete; however, the Lie algebra $\aleph(X)$ is finite dimensional. On the other hand, the density property is a very restrictive condition on noncompact manifolds. In particular, a Stein manifold with the density property is elliptic (Definition 5.6.13, p. 229), hence an Oka manifold (Definition 5.4.1, p. 215).

In Definition 4.10.1(i), we can replace $\mathrm{Lie}(\mathfrak{g}_{\mathrm{int}})$ by $\mathrm{Span}(\mathfrak{g}_{\mathrm{int}})$ due to the following elementary observation.

Lemma 4.10.2 *For every Lie algebra* $\mathfrak{g} \subset \aleph(X)$ *we have*

$$\overline{\mathrm{Span}}(\mathfrak{g}_{\mathrm{int}}) = \overline{\mathrm{Lie}}(\mathfrak{g}_{\mathrm{int}}).$$

Proof Since $\overline{\mathrm{Lie}}(\mathfrak{g}_{\mathrm{int}})$ is the smallest closed Lie subalgebra of $\aleph(X)$ containing $\mathrm{Span}(\mathfrak{g}_{\mathrm{int}})$, it suffices to see that $\overline{\mathrm{Span}}(\mathfrak{g}_{\mathrm{int}})$ is a Lie algebra. This holds if the commutator of any two complete vector fields $V, W \in \mathfrak{g}_{\mathrm{int}}$ belongs to $\overline{\mathrm{Span}}(\mathfrak{g}_{\mathrm{int}})$. Let

$\{\phi_t\}_{t\in\mathbb{R}}$ denote the flow of V. The Lie bracket

$$[V, W] = L_V W = \lim_{t\to 0} \frac{1}{t}(\phi_t^* W - W)$$

can be approximated by $t^{-1}(\phi_t^* W - W)$ for small $t \neq 0$, uniformly on compacts in X. If V and W are complete, the desired conclusion follows from the fact that the vector fields $t^{-1}\phi_t^* W$ and $t^{-1} W$ are complete for each $t \neq 0$. □

Lemma 4.10.3 *If X is an affine algebraic manifold, then the algebraic density property of X implies the holomorphic density property. Furthermore, if ω is an algebraic volume form on X, then the algebraic volume density property of (X, ω) implies the holomorphic volume density property.*

Proof The first part is obvious. For the second part, it suffices to show that every holomorphic vector field V on X satisfying $\mathrm{div}_\omega V = 0$ can be approximated by algebraic vector fields W satisfying $\mathrm{div}_\omega W = 0$. The proof (see [324]) is similar to the argument in the proof of Theorem 4.9.2: From $\mathrm{d}(V \rfloor \omega) = \mathrm{div}_\omega V \cdot \omega = 0$ we see that $\alpha = V \rfloor \omega$ is a closed holomorphic $(n-1)$-form. By Grothendieck [270], there exists a closed algebraic $(n-1)$-form $\widetilde{\alpha}$ such that $\alpha = \widetilde{\alpha} + \mathrm{d}\beta$ for some holomorphic $(n-2)$-form β. Approximating β by an algebraic form $\widetilde{\beta}$ gives a closed algebraic $(n-1)$-form $\theta = \widetilde{\alpha} + \mathrm{d}\widetilde{\beta}$ approximating α. The algebraic vector field W, determined by the equation $W \rfloor \omega = \theta$, approximates V and satisfies $\mathrm{div}_\omega W = 0$. □

The significance of the density property is shown by the following result that follows from Corollary 4.8.4 and the proof of Theorem 4.9.2.

Proposition 4.10.4 *Let $\mathfrak{g} \subset \aleph(X)$ be a Lie algebra with the density property, and let $V \in \mathfrak{g}$. Assume that for some $t_0 > 0$ the flow $\phi_t(x)$ of V exists for all $t \in [0, t_0]$ and all x in an open set $\Omega \subset X$. Then, $\phi_{t_0} \colon \Omega \to X$ can be approximated uniformly on compacts in Ω by compositions of flows of complete vector fields contained in $\mathfrak{g}$. The same conclusion holds for the flow of a continuous time-dependent vector field $V_t \in \mathfrak{g}$ $(t \in [0, t_0])$.*

By using Proposition 4.10.4 with $\mathfrak{g} = \aleph(X)$, we obtain the following analogue of Theorem 4.9.2.

Theorem 4.10.5 *Let X be a Stein manifold with the density property. If*

$$\Phi_t \colon \Omega_0 \to \Omega_t = \Phi_t(\Omega_0) \subset X \quad \big(t \in [0, 1]\big)$$

is a $\mathscr{C}^1$ isotopy of biholomorphic maps between pseudoconvex Runge domains in X with $\Phi_0 = \mathrm{Id}_{\Omega_0}$, then Φ_1 can be approximated uniformly on compacts in Ω_0 by holomorphic automorphisms of X.

Proof Every holomorphic vector field on a pseudoconvex Runge domain $\Omega_t \subset X$ can be approximated by holomorphic vector fields on X in view of the Oka-Weil theorem (see Theorem 2.6.8 on p. 57). The rest of the proof of Theorem 4.9.2 goes through as before, taking into account Proposition 4.10.4. □

The analogous result holds in the volume preserving case by assuming a suitable cohomological condition on X (see Theorem 4.9.2). One can prove a similar result for other Lie algebras of holomorphic vector fields; however, the approximation problem for vector fields on Runge domains in X by vector fields on X may become rather delicate for proper Lie subalgebras of $\aleph(X)$ (see Example 4.9.3). The volume preserving and the symplectic case are considered in more detail in the papers [185, 186].

Proposition 4.9.7 (p. 138) (the main point of the Andersén-Lempert theory) says that $\mathbb{C}^n$ for $n > 1$ has the algebraic density property, and also the algebraic volume density property with respect to $\omega = \mathrm{d}z_1 \wedge \cdots \wedge \mathrm{d}z_n$. We give a short proof, due to Varolin [560] and Kaliman and Kutzschebauch [322, Corollary 2.2], that effectively uses commutators.

Proposition 4.10.6 *If $n > 1$ then $\mathbb{C}^n$ has the algebraic density property.*

Proof Let $\mathfrak{g} = \aleph_{\mathcal{A}}(\mathbb{C}^n)$, the Lie algebra of polynomial vector fields on $\mathbb{C}^n$. Let $z = (z_1, \ldots, z_n)$ be complex coordinates on $\mathbb{C}^n$. The constant vector field $V_i = \frac{\partial}{\partial z_i}$ is a derivation on the polynomial ring $\mathbb{C}[z_1, \ldots, z_n]$ whose kernel $\ker V_i$ consists of polynomials independent of z_i. For any $f_i \in \ker V_i$ the vector fields $f_i V_i$ and $z_j f_i V_i$ (any j) are complete shear fields of the form (4.11). Hence, the vector field

$$f_1 f_2 V_2 = [f_1 V_1, z_1 f_2 V_2] - [z_1 f_1 V_1, f_2 V_2]$$

belongs to the Lie subalgebra $\mathrm{Lie}(\mathfrak{g}_{\mathrm{int}})$ of $\mathfrak{g}$ generated by the complete algebraic vector fields. Since $\mathbb{C}[z]$ is spanned by elements $f_1 f_2 \in \ker V_1 \cdot \ker V_2$, we conclude that $\mathrm{Lie}(\mathfrak{g}_{\mathrm{int}})$ contains all polynomial vector fields proportional to V_2. Now repeat the same argument for $V_1, V_3, \ldots, V_n$. □

After the original Andersén-Lempert theorem, the first new example was the Lie algebra of Hamiltonian holomorphic vector fields on even dimensional Euclidean spaces [187]; this work predates Varolin's definition of the density property. Let ω denote the standard *symplectic holomorphic form* on $\mathbb{C}^{2n}$:

$$\omega = \mathrm{d}z_1 \wedge \mathrm{d}z_2 + \cdots + \mathrm{d}z_{2n-1} \wedge \mathrm{d}z_{2n}.$$

A holomorphic vector field V is *Hamiltonian* if $\mathrm{d}(V \rfloor \omega) = 0$. Every holomorphic function H determines a Hamiltonian vector field V_H by the equation $V_H \rfloor \omega = \mathrm{d}H$. We have the following analogue of Proposition 4.9.7.

Proposition 4.10.7 ([187, Proposition 5.2]) *The Lie algebra of all polynomial Hamiltonian vector fields on $(\mathbb{C}^{2n}, \omega)$ has the algebraic density property. Furthermore, every Hamiltonian vector field on $\mathbb{C}^{2n}$ can be approximated uniformly on* *compacts by Lie combinations of complete Hamiltonian vector fields of the form* (4.20).

The proof is similar to that of Proposition 4.9.7 and is omitted. Proposition 4.10.7 implies the following result.

Theorem 4.10.8 ([187, Theorem 5.1]) *For each $n \in \mathbb{N}$ the group of all finite compositions of symplectic shears* (4.19) *is dense in the group* $\mathrm{Aut}_{\mathrm{sp}}(\mathbb{C}^{2n})$ *of all symplectic holomorphic automorphisms of* $\mathbb{C}^{2n}$.

For further results on symplectic automorphisms, see [186, 400].

We now mention some examples of manifolds and Lie algebras with the density property. The following theorem summarizes the main results of Varolin's paper [560]. The proof involves elementary calculations using sums and Lie brackets, similar to those in the proof of Proposition 4.10.6.

Theorem 4.10.9 (Varolin [560])

(a) *If X and Y are Stein manifolds with the density property, then so are the manifolds $X \times Y$, $X \times \mathbb{C}$, and $X \times \mathbb{C}^*$.*
(b) *For every complex Lie group G, $G \times \mathbb{C}$ has the density property. In particular, $\mathbb{C}^k \times (\mathbb{C}^*)^l$ has the density property when $k \geq 1$ and $k + l \geq 2$.*
(c) *If a complex Lie group G has the volume density property, then so does $G \times \mathbb{C}^*$. In particular, $(\mathbb{C}^*)^n$ has the volume density property.*
(d) *If $0 < k < n$, then the Lie algebra $\mathfrak{g}_0^{n,k}$ of holomorphic vector fields on $\mathbb{C}^n = \mathbb{C}^k \times \mathbb{C}^{n-k}$ that vanish on $\mathbb{C}^k \times \{0\}^{n-k}$ has the density property.*
(e) *If $n > k \geq 2$, then the Lie algebra $\mathfrak{g}_T^{n,k}$ of holomorphic vector fields on $\mathbb{C}^n$ that are tangent to $\mathbb{C}^k \times \{0\}^{n-k}$ has the density property.*

Note that the volume density property of a complex Lie group G is meant with respect to the *Haar form*, the unique left invariant holomorphic volume form ω that equals the standard volume form on the Lie algebra $\mathfrak{g} = \mathrm{T}_1 G \cong \mathbb{C}^n$. The Haar form on $(\mathbb{C}^*)^n$ is

$$(z_1 \cdots z_n)^{-1} \mathrm{d}z_1 \wedge \cdots \wedge \mathrm{d}z_n. \tag{4.51}$$

The algebraic density property fails on $(\mathbb{C}^*)^n$ since every globally integrable algebraic (rational) vector field has zero divergence with respect to the form (4.51) (see the paper by Andersén [24] where a complete classification of such vector fields is given). The following is a long-standing open problem.

Problem 4.10.10 Does $(\mathbb{C}^*)^n$ enjoy the density property when $n > 1$? Does every $F \in \mathrm{Aut}(\mathbb{C}^*)^n$ preserve the volume form (4.51) up to a sign?

The next result is due to Tóth and Varolin; see also Theorem 4.10.22.

Theorem 4.10.11 ([553, 554]) *Every complex semisimple Lie group G has the density property. If the center of G is trivial and K is a reductive subgroup of G, then the homogeneous space G/K has the density property.*

Since every complex semisimple Lie group is Stein, Theorem 4.10.11, together with the main results of Varolin [559], gives the following conclusions.

Corollary 4.10.12 ([553, Corollaries 1–3]) *A complex semisimple Lie group G of dimension n admits an open subset biholomorphic to $\mathbb{C}^n$, and a proper open subset biholomorphic to G.*

We now present more advanced methods for establishing the density property that were introduced by Kaliman and Kutzschebauch in a series of papers [322–324, 326]. They have mainly treated the algebraic (volume) density property in the class of affine algebraic manifolds. Here, we focus on the holomorphic case on Stein manifolds, following [325] and [392]. The main result along these lines is Theorem 4.10.19.

The following definition (see [392, Definition 2.1]) was first introduced by Kaliman and Kutzschebauch [323] in the algebraic case.

Definition 4.10.13 Let X be a complex manifold and $p \in X$. A set $T \subset \mathrm{T}_p X$ is a *generating set* (for $\mathrm{T}_p X$) if the orbit of T under the induced action of the stabilizer $\mathrm{Aut}(X)_p$ contains a basis of $\mathrm{T}_p X$. If X has a volume form ω, then a set $T \subset \mathrm{T}_p X \wedge \mathrm{T}_p X$ is *ω-generating* if the orbit of T under the induced action of the stabilizer $\mathrm{Aut}_\omega(X)_p$ contains a basis of $\mathrm{T}_p X \wedge \mathrm{T}_p X$.

The manifold X is called *tangentially semihomogeneous* if it is $\mathrm{Aut}(X)$-homogeneous and it admits a generating set consisting of one vector.

We shall use the following notion; see p. 141 for the notation.

Definition 4.10.14 A complete holomorphic vector field $V \in \aleph_{\mathrm{int}}(X)$ is *stably complete* if for every $f \in \mathcal{O}(X)$ we have $fV \in \overline{\mathrm{Lie}\,\aleph_{\mathrm{int}}(X)}$.

Note that every constant vector field on $\mathbb{C}^n$ is stably complete in view of Proposition 4.9.7. Let us record the following observation.

Lemma 4.10.15

(a) *The $\mathcal{O}(X)$-submodule of $\aleph(X)$ generated by any collection of stably complete vector fields is contained in $\overline{\mathrm{Lie}\,\aleph_{\mathrm{int}}(X)}$.*
(b) *If V is stably complete and $\Phi \in \mathrm{Aut}(X)$, then $\Phi_* V$ is also stably complete.*

Proof Part (a) is obvious. For part (b), note that the vector field $\Phi_* V$ is complete and $\Phi_*(fV) = (f \circ \Phi^{-1})\Phi_* V$ holds for any $f \in \mathcal{O}(X)$. □

The next result provides a useful sufficient condition implying the density property. It is an adaptation of [323, Theorem 2]. For the volume preserving case, see [392, Proposition 2.7] which is an adaption of [326, Theorem 1].

Proposition 4.10.16 ([392, Proposition 2.5]) *Let X be a Stein manifold on which the automorphism group* $\mathrm{Aut}(X)$ *acts transitively. Assume that there are stably complete holomorphic vector fields $V_1, \ldots, V_m$ on X and a point $p \in X$ such that the vectors $V_1(p), \ldots, V_m(p) \in \mathrm{T}_p X$ are a generating set for* $\mathrm{T}_p X$ (*see Definition* 4.10.13). *Then X has the density property.*

Proof We must prove that every vector field $V \in \aleph(X)$ can be approximated, uniformly on any compact set $K \subset X$, by elements of $\mathrm{Lie}\,\aleph_{\mathrm{int}}(X)$. We may assume that K is $\mathcal{O}(X)$-convex. Let $\Omega \Subset X$ be a relatively compact Stein Runge neighborhood of K. The definition of a generating set implies that we can add to the existing collection $V_1, \ldots, V_m$ finitely many images of these vector fields by elements of $\mathrm{Aut}(X)$ in order to ensure that the vectors $V_i(x)$ span $\mathrm{T}_x X$ for every point $x \in \Omega$. By Lemma 4.10.15(b), every vector field in the enlarged collection $V_1, \ldots, V_N$ is stably complete. By the Nakayama lemma, these vector fields generate the tangent sheaf $\mathcal{T}_X$ over Ω. By Theorem 2.6.8 (p. 57), every holomorphic vector field on X can be approximated uniformly on K by elements of the form $\sum_{i=1}^N f_i V_i$ for some $f_i \in \mathcal{O}(X)$. Any such vector field belongs to $\overline{\mathrm{Lie}}\,\aleph_{\mathrm{int}}(X)$ by Lemma 4.10.15(a). □

We now describe a way to find vector fields as in Proposition 4.10.16. These results are due to Kaliman and Kutzschebauch [323, 326].

Definition 4.10.17

(i) A pair (V, W) of complete vector fields on a complex manifold X is a *semicompatible pair* if $\overline{\mathrm{Span}}(\ker V \cdot \ker W)$ contains a nontrivial ideal $I \subset \mathcal{O}(X)$.
(ii) A semicompatible pair (V, W) is a *compatible pair* if there exists $h \in \ker W \subset \mathcal{O}(X)$ such that $V(h) \in \ker V \setminus \{0\}$.

Recall that $\ker V$ is the kernel of the linear map $f \mapsto V(f)$. Condition (ii) implies that hV and hW are complete vector fields.

Lemma 4.10.18 *If (V, W) is a compatible pair with ideal I and function h, then the $\mathcal{O}(X)$-submodule $I \cdot V(h)W$ of $\aleph(X)$ is contained in* $\overline{\mathrm{Lie}}\,\aleph_{\mathrm{int}}(X)$.

Proof If $f \in \ker V$ and $g \in \ker W$, then $fV, fhV, gW, ghW \in \aleph_{\mathrm{int}}(X)$. Hence,

$$fgV(h)W = [fV, ghW] - [fhV, gW] \in \mathrm{Lie}\,\aleph_{\mathrm{int}}(X)$$

and thus every vector field $(\sum f_i g_i)V(h)W$ with $f_i \in \ker V$ and $g_i \in \ker W$ is contained in $\mathrm{Lie}\,\aleph_{\mathrm{int}}(X)$. Since the closure of the set of sums $\sum_i f_i g_i$ of this type contains the ideal I by Definition 4.10.17, the result follows. □

Theorem 4.10.19 *Let X be a Stein manifold of dimension n.*

(a) [392, Theorem 1.2] *Assume that* $\mathrm{Aut}(X)$ *acts transitively on X. If there are compatible pairs (V_i, W_i) and a point $p \in X$ where the vectors $W_i(p)$ form a generating set of* T_pX, *then X has the density property.*
(b) [325] *Assume that ω is a holomorphic volume form on X such that* $\mathrm{Aut}_\omega(X)$ *acts transitively on X and $H^{n-1}(X;\mathbb{C}) = 0$. If there are semicompatible pairs (V_i, W_i) of divergence zero vector fields and a point $p \in X$ where the vectors $V_i(p) \wedge W_i(p)$ form a generating set of* $\mathrm{T}_pX \wedge \mathrm{T}_pX$, *then X has the volume density property.*

Proof We first prove (a). Let I_i be the ideals and h_i the functions of the compatible pairs (V_i, W_i) (see Definition 4.10.17). Pick a function $f_i \in I_i \cdot V_i(h_i) \setminus \{0\}$ for every i. Since the set of points $p \in X$ where the vector fields $W_i(p)$ are a generating set is open and nonempty, there is a point $q \in X$ where the vectors $f_i(q)W_i(q)$ are a generating set for T_qX. By Lemma 4.10.18 each vector field f_iW_i is stably complete, and hence the $\mathscr{O}(X)$-module generated by the vector fields f_iW_i is contained in $\overline{\mathrm{Lie}\,\aleph_{\mathrm{int}}(X)}$. Proposition 4.10.16 then implies that X has the density property. The proof of (b) is similar, but it uses a different formula given in [326, Proposition 3.1] (see also [392, Proposition 2.7]) instead of Proposition 4.10.16. □

As an application of these results, Leuenberger [392] investigated the (volume) density property of a class of algebraic hypersurfaces of the form

$$X = \{(x, y, z) \in \mathbb{C}^{n+3} : x^2y = a(z) + xb(z)\}, \tag{4.52}$$

where $x, y \in \mathbb{C}$ and $a, b \in \mathbb{C}[z]$ are polynomials in $z = (z_0, \dots, z_n) \in \mathbb{C}^{n+1}$. Note that X has the algebraic volume form $\omega = x^{-2}\mathrm{d}x \wedge \mathrm{d}z_0 \wedge \cdots \wedge \mathrm{d}z_n$.

Theorem 4.10.20 (Leuenberger [392, Main theorem]) *Let $n \geq 0$, and let $a, b \in \mathbb{C}[z_0, \dots, z_n]$ be holomorphic polynomials such that $\deg_{z_0}(a) \leq 2$, $\deg_{z_0}(b) \leq 1$, and not both $\deg_{z_0}(a)$ and $\deg_{z_0}(b)$ are equal to zero. Then the hypersurface X* (4.52) *has the density property provided that the group* $\mathrm{Aut}(X)$ *acts transitively on X. Moreover, if $H^{n+1}(X;\mathbb{C}) = 0$ and the group* $\mathrm{Aut}_\omega(X)$ *acts transitively on X, then X has the volume density property.*

Leuenberger also provided sufficient conditions in terms of the polynomials a and b which ensure that the group of all (volume preserving) holomorphic automorphisms acts transitively on X.

The family of hypersurfaces covered by Theorem 4.10.20 includes the *Koras-Russell cubic threefold*

$$C = \{(x, y, z_0, z_1) \in \mathbb{C}^4 : x^2y + x + z_0^2 + z_1^3 = 0\} \tag{4.53}$$

and shows that C has the density and volume density property. This threefold is a famous example of an affine variety which is diffeomorphic to $\mathbb{R}^6$, but is not algebraically isomorphic to $\mathbb{C}^3$; in particular, the algebraic automorphism group does not act transitively on C [138, 406]. Theorem 4.10.20 shows that the situation in the holomorphic context is completely different. However, it remains an open question whether C is biholomorphic to $\mathbb{C}^3$.

The following result of Kaliman and Kutzschebauch from 2008 is the original algebraic version of Theorem 4.10.19. The proof is similar to that of Theorem 4.10.19 and is not included here.

Theorem 4.10.21 ([322, Theorem 2]) *Let X be an affine algebraic manifold on which the group $\mathrm{Aut}_{\mathrm{alg}}(X)$ of algebraic automorphisms acts transitively. If there exist finitely many compatible pairs $\{(V_{k,1}, V_{k,2})\}_{k=1}^m$ of algebraic vector fields such that for some $p \in X$ the set of vectors $\{V_{k,2}(p)\}_{k=1}^m \subset \mathrm{T}_p X$ is a generating set* (*Definition* 4.10.13), *then X has the algebraic density property.*

One of the main applications of Theorem 4.10.21, along with more elaborated techniques, is the following result of Kaliman and Kutzschebauch.

Theorem 4.10.22 ([321, Theorem 1.3]) *Let X be a connected affine homogeneous space of a linear algebraic group G.*

(1) *If X is not isomorphic to a line or a torus $(\mathbb{C}^*)^k$, then X has the algebraic density property.*
(2) *If X has a left-invariant (with respect to the G-action) volume form ω, then X has the algebraic volume density property with respect to ω.*

This result subsumes Theorem 4.10.11 of Tóth and Varolin, and also [322, Theorem 3] of Kaliman and Kutzschebauch. The details of proof are considerable and are not included here. Instead, we present a simple proof from [322] of the following special case.

Example 4.10.23 $SL_n(\mathbb{C})$ for $n > 1$ has the algebraic density and volume density property.
Every simple complex Lie group G is tangentially semihomogeneous (see Definition 4.10.13) since the adjoint action of G generates an irreducible representation on the Lie algebra $\mathfrak{g} = \mathrm{T}_1 G$ (i.e., any nonzero vector in $\mathfrak{g}$ is a generating set). In particular, $SL_n(\mathbb{C})$ for $n \ge 2$ is tangentially semihomogeneous. Every element of $SL_n(\mathbb{C})$ is a matrix $x = (c_{k,j})$ with determinant 1. We use the coefficients $c_{k,j}$ as coordinates. Consider the vector fields

$$V_1 = \sum_{j=1}^n c_{2,j} \frac{\partial}{\partial c_{1,j}}, \qquad V_2 = \sum_{j=1}^n c_{1,j} \frac{\partial}{\partial c_{2,j}}.$$

Note that constants and functions depending on $c_{k,j}$ for $k \neq 1$ are in $\ker V_1$, while constants and functions depending on $c_{k,j}$ for $k \neq 2$ are in $\ker V_2$; therefore, condition (i) of Definition 4.10.17 holds. Taking c_{11} as the function h in condition (ii) we see that V_1 and V_2 are compatible; hence Theorem 4.10.21 applies and shows that $SL_n(\mathbb{C})$ has the algebraic density property.

For the volume density property, we remark that for each semi-simple Lie group G, in particular for $SL_n(\mathbb{C})$, the condition $H^{N-1}(G;\mathbb{C})=0$ for $N=\dim G$ holds. Take the compatible pairs V_1 and V_2 as above for $SL_2(\mathbb{C})$. For any subgroup of $SL_n(\mathbb{C})$ isomorphic to $SL_2(\mathbb{C})$ the induced vector fields on $SL_n(\mathbb{C})$ are semicompatible (in fact, compatible). Consider these pairs over all SL_2-subgroups of $SL_n(\mathbb{C})$, we find a spanning set as in Theorem 4.10.19(b). This proves the algebraic volume density property of $SL_n(\mathbb{C})$.

We describe another type of affine hypersurfaces for which the density property was established by Kaliman and Kutzschebauch [323].

Theorem 4.10.24 *If f is a holomorphic function on $\mathbb{C}^n$ such that $\mathrm{d}f \neq 0$ on the zero locus $\{f=0\}$, then the smooth hypersurface*

$$X=\big\{(x,y,z_1,\ldots,z_n)\in\mathbb{C}^{n+2}: xy=f(z_1,\ldots,z_n)\big\} \tag{4.54}$$

has the holomorphic density property. If $f\in\mathbb{C}[z_1,\ldots,z_n]$ is a polynomial and $\mathrm{d}f\neq 0$ on $\{f=0\}$, then X enjoys the algebraic density property.

Proof (sketch) For any complete holomorphic (resp. polynomial) vector field V on $\mathbb{C}^n_z$ and for any $h\in\mathcal{O}(\mathbb{C}^n_z)$ (resp. $h\in\mathbb{C}[z_1,\ldots,z_n]$) we get the following complete vector fields on $\mathbb{C}^{n+2}$ that are clearly tangent to X:

$$xV+V(f)\frac{\partial}{\partial y},\qquad yV+V(f)\frac{\partial}{\partial x},\qquad h(z)\left(x\frac{\partial}{\partial x}-y\frac{\partial}{\partial y}\right).$$

It can be seen that vector fields of these types densely generate the Lie algebra of holomorphic (resp. algebraic) vector fields on X. □

If the zero fibre $\{f=0\}$ is contractible, then the hypersurface (4.54) is diffeomorphic to $\mathbb{C}^{n+1}$, but it is not known for which f it is biholomorphic (or algebraically isomorphic if f is a polynomial) to $\mathbb{C}^{n+1}$. When $n=1$, surfaces (4.54) are called *Danielewski surfaces*; their automorphism groups and complete vector fields were studied in the papers [33, 361, 363].

In [362], Kutzschebauch et al. investigated the algebraic density property for not necessarily smooth affine *toric varieties*. Every smooth toric variety is an Oka manifold (see Theorem 5.6.12 on p. 228).

Gizatullin surfaces are normal complex affine surfaces X that are not algebraically isomorphic to $\mathbb{C}^*\times\mathbb{C}^*$ or $\mathbb{C}\times\mathbb{C}^*$ and satisfy the following equivalent conditions (see [161, Theorem 4.3]; a third equivalent property is that the Makar-Limanov invariant of X is trivial):

- The group of algebraic automorphisms of X acts with an open orbit whose complement is finite.
- X admits a completion by a linear chain of smooth rational curves.

In [31], Andrist established the density property of all Gizatullin surfaces admitting a $\mathbb{C}_+$-action whose quotient is a $\mathbb{C}^1$-fibration with a reduced degenerate fibre; this contains all previous results on the density property for affine surfaces as special cases. Before that, Andrist et al. [34] showed the algebraic density property for smooth Gizatullin surfaces completed by four rational curves; these are algebraic surfaces in $\mathbb{C}^4_{x,y,u,v}$ that can be described by the equations

$$yu = xP(x), \qquad xv = uQ(u), \qquad yv = P(x)Q(u)$$

where P and Q are nonconstant polynomials. They also proved the holomorphic density property for smooth analytic surfaces in $\mathbb{C}^4$ given by these equations when P and Q are holomorphic functions.

In [32], Andrist and Kutzschebauch introduced the notion of the fibred density property. They proved that the natural fibration of the spectral ball over the symmetrized polydisc enjoys the fibred density property, and they described the automorphism group of the spectral ball. The definition of the spectral ball can be found in Sect. 6.14 where we describe Andrist's solution of the spectral Nevanlinna-Pick lifting problem.

4.11 Automorphisms Fixing a Subvariety

Let $n > 1$. We are interested in the question for which closed complex subvarieties A of $\mathbb{C}^n$ are the automorphism group $\mathrm{Aut}(\mathbb{C}^n \setminus A)$ and its subgroup

$$\mathrm{Aut}\big(\mathbb{C}^n, A\big) = \big\{F \in \mathrm{Aut}\big(\mathbb{C}^n\big) : F(z) = z,\ \forall z \in A\big\} \tag{4.55}$$

big in the sense of the Andersén-Lempert theory. To this end, we study the Lie algebra $\aleph(\mathbb{C}^n, A)$ of all holomorphic vector fields on $\mathbb{C}^n$ that vanish on A. If A is algebraic, we consider the corresponding algebraic automorphism group $\mathrm{Aut}_{\mathrm{alg}}(\mathbb{C}^n, A)$ and the Lie algebra $\aleph_{\mathcal{A}}(\mathbb{C}^n, A)$ of all algebraic vector fields on $\mathbb{C}^n$ that vanish on A. The main questions are the following:

Problem 4.11.1 When does the Lie algebra $\aleph(\mathbb{C}^n, A)$ enjoy the density property (see Definition 4.10.1 on p. 142)? Assuming that A is algebraic, when does $\aleph_{\mathcal{A}}(\mathbb{C}^n, A)$ enjoy the algebraic density property?

It is easily seen that $\aleph_{\mathcal{A}}(\mathbb{C}^n, A)$ has the algebraic density property if A is a complex affine subspace of $\mathbb{C}^n$ ([560]; see Theorem 4.10.9(iv)). On the other hand, there exist algebraic hypersurfaces A in $\mathbb{C}^n$ whose complement $\mathbb{C}^n \setminus A$ is hyperbolic, and consequently $\mathrm{Aut}(\mathbb{C}^n \setminus A)$ is small. For example, the complement of a generic set of $2n$ hyperplanes in $\mathbb{C}^n$ is hyperbolic by Green's theorem [262]. By removing additional hypersurfaces, one can find examples when the group $\mathrm{Aut}(\mathbb{C}^n \setminus A)$ is trivial. Looking at lower dimensional non-algebraic subvarieties, we have seen in Sect. 4.6 that there also exist discrete sets $A \subset \mathbb{C}^n$ for which $\mathrm{Aut}(\mathbb{C}^n \setminus A)$ is trivial.

Here is the main observation. Assume that $\pi : \mathbb{C}^n \to \mathbb{C}^{n-1}$ is a linear projection such that the restriction $\pi|_A : A \to \mathbb{C}^{n-1}$ is proper; then $\pi(A)$ is a closed analytic subvariety of $\mathbb{C}^{n-1}$. If $\dim A < n-1$ then $\pi(A)$ is a proper subvariety of $\mathbb{C}^{n-1}$, and hence there exist nonconstant holomorphic functions f on $\mathbb{C}^{n-1}$ (polynomials if A is algebraic) that vanish on A. Choose a vector $0 \neq v \in \ker \pi$ and consider it as a constant holomorphic vector field on $\mathbb{C}^n$. The shear vector field $(f \circ \pi)v$ on $\mathbb{C}^n$ then vanishes on A and generates the $\mathbb{C}_+$ action

$$F_t(z) = z + tf\big(\pi(z)\big)\, v, \quad t \in \mathbb{C}. \tag{4.56}$$

This action fixes the subvariety $\pi^{-1}(\pi(A)) = A + \mathbb{C}v$ of $\mathbb{C}^n$ and hence it belongs to the group $\mathrm{Aut}(\mathbb{C}^n, A)$ (resp. to $\mathrm{Aut}_{\mathrm{alg}}(\mathbb{C}^n, A)$ if A is algebraic).

This suggests that we consider subvarieties with $\dim A \le n-2$ such that, in suitable coordinates on $\mathbb{C}^n$, there exist sufficiently many linear projections $\pi : \mathbb{C}^n \to \mathbb{C}^{n-1}$ that are proper when restricted to A. We analyze this condition more carefully. Given a point $v \in \mathbb{C}^n \setminus \{0\}$, we denote by $[v]$ the complex line $\mathbb{C}v$ considered as a point in $\mathbb{P}^{n-1}$ (the hyperplane at infinity in $\mathbb{P}^n$). For a closed analytic subvariety $A \subset \mathbb{C}^n$ we consider the limit set of A at infinity:

$$\omega(A) = \left\{ \lim_{j\to\infty} \left[a^j\right] : a^j \in A,\ \left|a^j\right| \to \infty \right\} \subset \mathbb{P}^{n-1}. \tag{4.57}$$

Clearly, $\omega(A)$ is a closed subset of $\mathbb{P}^{n-1}$ and the closure of A in $\mathbb{P}^n$ equals $\overline{A} = A \cup \omega(A)$. A subvariety $A \subset \mathbb{C}^n$ of pure dimension p is algebraic if and only if $\omega(A)$ is an algebraic subset of $\mathbb{P}^{n-1}$ of dimension $p-1$; a complex hypersurface $A \subset \mathbb{C}^n$ is algebraic if and only if $\omega(A) \neq \mathbb{P}^{n-1}$ (see [98, Sect. I.7]).

Lemma 4.11.2 *If $0 \neq v \in \mathbb{C}^n$ is such that $[v] \notin \omega(A)$, then for any surjective linear map $\pi : \mathbb{C}^n \to \mathbb{C}^{n-1}$ with $\pi(v) = 0$ the map $\pi|_A : A \to \mathbb{C}^{n-1}$ is proper.*

Proof By a $\mathbb{C}$-linear change of coordinates we may assume that $v = e_n = (0, \ldots, 0, 1)$ and $\pi(z) = (z_1, \ldots, z_{n-1})$. If $\pi|_A$ is not proper, there exists a sequence $\{a^j\} \subset A$ such that $|a^j| \to \infty$ but $\pi(a^j)$ is bounded. Thus $|a^j_n| \to \infty$ and therefore $[a^j] \to [e_n]$ as $j \to \infty$. Hence $[v] = [e_n] \in \omega(A)$. □

We now introduce a class of varieties for which the above groups are big; this class contains all algebraic subvarieties of codimension at least two.

Definition 4.11.3 A closed analytic subset $A \subset \mathbb{C}^n$ is *tame* if there exists a holomorphic automorphism $\Phi \in \mathrm{Aut}(\mathbb{C}^n)$ such that $\omega(\Phi(A)) \neq \mathbb{P}^{n-1}$.

Every closed complex subvariety A of $\mathbb{C}^n$ contained in a proper algebraic subvariety is tame. (Such A is said to be *algebraically degenerate*.) A hypersurface A is tame if and only if there exists a $\Phi \in \mathrm{Aut}(\mathbb{C}^n)$ such that $\Phi(A)$ is algebraic. For discrete sets, this notion of tameness coincides with the one of Rosay-Rudin (see Definition 4.6.1 on p. 128) in view of Theorem 4.6.2(i) (p. 128) and the following elementary lemma that is left as an exercise.

Lemma 4.11.4 *An analytic subvariety A of $\mathbb{C}^n$ is tame if and only if there exist holomorphic coordinates $z = (z', z_n)$ on $\mathbb{C}^n$ and a $C > 0$ such that*

$$A \subset \{(z', z_n) \in \mathbb{C}^n : |z_n| \le C(1 + |z'|)\}. \tag{4.58}$$

Lemma 4.11.5 *Assume that $A \subset \mathbb{C}^n$ is a closed complex subvariety with* $\dim A \le n - 2$ *and $\omega(A) \neq \mathbb{P}^{n-1}$. For every point $p \in \mathbb{C}^n \setminus A$ and vector $0 \neq v \in \mathbb{C}^n$ such that $[v] \neq \omega(A)$ there exists a vector $w \in \mathbb{C}^n$ arbitrarily close to v such that the affine complex line $\Lambda = p + \mathbb{C}w$ does not intersect A, and we can move p to any point $q \in \Lambda$ by a single shear of the form* (4.56).

Proof We may assume that $p = 0$. Let $\mathbb{C}^n_* = \mathbb{C}^n \setminus \{0\}$, and let $\rho \colon \mathbb{C}^n_* \to \mathbb{P}^{n-1}$ denote the projection $\rho(z) = [z] \in \mathbb{P}^{n-1}$. For every neighborhood $U \subset \mathbb{C}^n_*$ of v we let $\Gamma_U = \{tw : w \in U,\ t \in \mathbb{C}^*\} \subset \mathbb{C}^n_*$, the open cone spanned by U. Since $[z] \notin \omega(A)$, we can choose U small enough such that the set $A \cap \Gamma_U$ is bounded, and hence the restricted projection $\rho \colon A \cap \Gamma_U \to \rho(U)$ is proper. Since $\dim A \le n - 2$, it follows that $\rho(A \cap \Gamma_U)$ is a proper subvariety of the open set $\rho(U) \subset \mathbb{P}^{n-1}$. For every point $w \in U$ such that $[w] \neq \rho(A \cap \Gamma_U)$ the complex line $\Lambda = \mathbb{C}w$ misses A. This proves the first statement; the second one is an immediate consequence. □

We shall need the following lemma which follows from the Taylor expansion of the flow (see (1.49) on p. 36).

Lemma 4.11.6 *Let V be a holomorphic vector field on a complex manifold X. Suppose that f is a holomorphic function such that $V(f) = 0$ and $f(p) = 0$ for some $p \in X$. Then, the flow of the vector field fV generates a linear action on the tangent space T_pX given by the formula*

$$\mathrm{T}_pX \ni v \mapsto v + \mathrm{d}f_p(v)V_p.$$

The following result is due to Gromov [266] and Winkelmann [576, Proposition 1] who considered algebraic subvarieties. The generalization to (not necessarily algebraic) tame subvarieties was given in [187, Proposition 9.1].

Proposition 4.11.7 *If $A \subset \mathbb{C}^n$ is a tame analytic subvariety of codimension at least two then* $\mathrm{Aut}(\mathbb{C}^n, A)$ *acts transitively on $\mathbb{C}^n \setminus A$, and for any $p \in \mathbb{C}^n \setminus A$ the image of any vector $v \in \mathrm{T}_p\mathbb{C}^n$ under the isotropy group* $\mathrm{Aut}(\mathbb{C}^n, A)_p$ *generates* $\mathrm{T}_p\mathbb{C}^n$ *(compare with Definition* 4.10.13). *If A is algebraic then the same conclusions hold for the group* $\mathrm{Aut}_{\mathrm{alg}}(\mathbb{C}^n, A)$.

Proof Replacing A by $\Phi(A)$ for a suitable $\Phi \in \mathrm{Aut}(\mathbb{C}^n)$, we can assume that $\mathbb{P}^{n-1} \setminus \omega(A) \neq \varnothing$. Since this set is open, we can choose a basis $v_1, \ldots, v_n$ of $\mathbb{C}^n$ such that $[v_j] \notin \omega(A)$ for $j = 1, \ldots, n$. Let $\pi_j \colon \mathbb{C}^n \to \mathbb{C}^{n-1}$ be a surjective linear map such that $\pi_j(v_j) = 0$. Lemma 4.11.5 shows that for every point $p \in \mathbb{C}^n \setminus A$ the vectors v_j can be chosen such that for all points q sufficiently near p, none of the affine lines $q + \mathbb{C}v_j$ intersect A. Choose $f_j \in \mathcal{O}(\mathbb{C}^{n-1})$ that vanishes on $\pi_j(A)$ but is nonzero at $\pi_j(p)$. The shears (4.56) generated by the vector fields $V_j = (f_j \circ \pi_j)\, v_j$ ($j = 1, \ldots, n$) then act transitively in a neighborhood of p, i.e., we can map p to any nearby point by their compositions.

Consider now the problem of moving p to q for an arbitrary pair of points $p, q \in \mathbb{C}^n \setminus A$. Choose $v_1, \ldots, v_n$ as above, write $q - p = \sum_{j=1}^n \lambda_j v_j$ ($\lambda_j \in \mathbb{C}$) and define the points $p_0, \ldots, p_n$ by

$$p_0 = p, \quad p_1 = p_0 + \lambda_1 v_1, \quad p_2 = p_1 + \lambda_2 v_2, \quad \ldots, \quad p_n = p_{n-1} + \lambda_n v_n = q.$$

We wish to move p_{j-1} to p_j by a shear in the direction v_j. The affine line through these two points may intersect A. To avoid this problem, choose small pairwise disjoint balls $U_j \ni p_j$ for $j = 0, 1, \ldots, n$ such that for every pair of points $a \in U_{j-1}$, $b \in U_j$ we have $[a - b] \neq \omega(A)$ (this is possible since $\omega(A)$ is a closed set and $[p_{j-1} - p_j] = [v_j] \neq \omega(A)$.) By Lemma 4.11.5 we can move $p = p_0$ to some point $p_1' \in U_1$ by a shear (4.56) that is fixed on A. By the same argument, we can move p_1' to some point $p_2' \in U_2$ by a shear that is fixed on A. Continuing in this way, we move p to some point $p_n' \in U_n$ by a composition of n shears. Assuming that U_n is a sufficiently small neighborhood of $p_n = q$, we can move p_n' to q by the first part of the proof. This proves that $\mathbb{C}^n \setminus A$ is $\mathrm{Aut}(\mathbb{C}^n, A)$-homogeneous. If A is algebraic, then all shears in the proof can be chosen polynomial.

Fix a point $p \in \mathbb{C}^n \setminus A$. Pick a vector $v \neq 0$ with $[v] \notin \omega(A)$ and choose a projection $\pi : \mathbb{C}^n \to \mathbb{C}^{n-1}$ with $\ker \pi = \mathbb{C}v$ such that $\pi(p) \notin \pi(A)$. Let $f \in \mathcal{O}(\mathbb{C}^{n-1})$ vanish on $\pi(A)$ and have a simple zero at $\pi(p)$. Then, p is a fixed point of the $\mathbb{C}_+$ action $(t, z) \mapsto z + tf(\pi(z))v$. By Lemma 4.11.6 (p. 153), the induced $\mathbb{C}_+$ action on $T_p\mathbb{C}^n$ is given by $w \mapsto w + df_p(w)\, v$. This shows that the isotropy group $\mathrm{Aut}(\mathbb{C}^n, A)_p$ induces an irreducible representation on $T_p\mathbb{C}^n$ which gives the second statement. □

The following result will be used in Chaps. 5 and 6. It says that, under the stated conditions, $\mathbb{C}^n \setminus A$ is an (algebraically) elliptic manifold.

Proposition 4.11.8 *If $A \subset \mathbb{C}^n$ is a tame subvariety of codimension at least two, then there exist finitely many shear vector fields* (4.9) *on $\mathbb{C}^n$ that vanish on A and span the tangent space $T_z\mathbb{C}^n$ at every point $z \in \mathbb{C}^n \setminus A$. If A is algebraic, then this holds with algebraic shear fields having algebraic flows.*

Proof Choose a basis $v_1, \ldots, v_n$ of $\mathbb{C}^n$, linear projections π_j, and functions f_j as in the proof of Proposition 4.11.7. The corresponding (complete) shear vector fields $V_j = (f_j \circ \pi_j)v_j$ then span $T_z\mathbb{C}^n$ at every point outside the proper complex subvariety $A' = \bigcup_{j=1}^n \pi_j^{-1}(\pi_j(A)) \supset A$ of $\mathbb{C}^n$. Choosing a point $p \in A' \setminus A$ and adding to the previous collection finitely many shears of the same form that span $T_p\mathbb{C}^n$, we reduce A' to a smaller subvariety containing A. In the algebraic case we can complete the proof in finitely many steps by reducing the dimension of $A' \setminus A$ at every step.

In the analytic case, the variety $A' \setminus A$ may have countably many connected components, and a more precise argument is needed to show that the dimension can be reduced at each step. The main point is that, due to its special form, the subvariety A' still has sufficiently many proper projections π to $\mathbb{C}^{n-1}$, and any discrete set of points in A' projects to a discrete set in $\mathbb{C}^{n-1}$ under such a projection. It is then possible to choose the coefficient function $f \in \mathcal{O}(\mathbb{C}^{n-1})$ in the corresponding shear $(f \circ \pi)v$ such that f vanishes on $\pi(A)$, but does not vanish at those projected points that belong to $\mathbb{C}^{n-1} \setminus \pi(A)$. □

Further evidence that a tame subvariety of codimension ≥ 2 in $\mathbb{C}^n$ has a holomorphically big complement is provided by the following result. (See also Corollaries 4.12.2 and 4.12.7.)

Proposition 4.11.9 *Let A be a tame subvariety of $\mathbb{C}^n$ of codimension at least two. For every point $p \in \mathbb{C}^n \setminus A$ there exists an injective holomorphic map $F\colon \mathbb{C}^n \to \mathbb{C}^n \setminus A$ (a Fatou-Bieberbach map) such that $F(0) = p$.*

Proof We may assume that $p = 0$. By Lemma 4.11.5 there is a basis $v_1, \ldots, v_n$ of $\mathbb{C}^n$ such that $[v_j] \notin \omega(A)$ and the complex line $\mathbb{C}v_j$ does not intersect A for $j = 1, \ldots, n$. After a linear change of coordinates, we may assume that these conditions hold for the standard basis $v_j = e_j$. Let $\pi_j\colon \mathbb{C}^n \to \mathbb{C}^{n-1}_j$ denote the projection onto the hyperplane $\mathbb{C}^{n-1}_j = \{z_j = 0\}$, so that $\pi_j(e_j) = 0$. By Lemma 4.11.2, the restricted projection $\pi_j|_A\colon A \to \mathbb{C}^{n-1}_j$ is proper, and hence $A_j = \pi_j(A)$ is a proper closed analytic subset of $\mathbb{C}^{n-1}_j$. By the construction we have $0 \notin A_j$ for $j = 1, \ldots, n$. Choose a holomorphic function g_j on $\mathbb{C}^{n-1}_j$ such that $g_j(0) = -\log 2$ and $g_j = 0$ on A_j, and set

$$\Phi_j(z) = \big(z_1, \ldots, z_{j-1}, e^{g_j(\pi_j(z))} z_j, z_{j+1}, \ldots, z_n\big).$$

Then, $\Phi = \Phi_1 \circ \Phi_2 \circ \cdots \circ \Phi_n \in \mathrm{Aut}(\mathbb{C}^n)$, $\Phi|_A$ is the identity, $\Phi(0) = 0$, and $\Phi'(0) = \frac{1}{2} I$. Thus, $0 \in \mathbb{C}^n$ is an attracting fixed point of Φ whose basin $\Omega \subset \mathbb{C}^n \setminus A$ is a Fatou-Bieberbach domain (see Theorem 4.3.2, p. 116). The corresponding Fatou-Bieberbach map $F\colon \mathbb{C}^n \to \Omega$ satisfies the conclusion. □

Remark 4.11.10 In the situation of Proposition 4.11.9, with A algebraic, Buzzard and Hubbard [79] constructed a Fatou-Bieberbach domain $\Omega \subset \mathbb{C}^n$ whose closure $\overline{\Omega}$ does not intersect A. □

The above results come close to, but still do not answer Problem 4.11.1 on p. 151. We now describe a partial answer in the algebraic case, due to Kaliman and Kutzschebauch [323], which enables one to approximate isotopies of biholomorphic maps in the complement of an algebraic subvariety A by holomorphic automorphisms of $\mathbb{C}^n$ that are fixed on A.

Given an ideal $I \subset \mathbb{C}^{[n]} = \mathbb{C}[z_1, \ldots, z_n]$, we denote by $\aleph_I(\mathbb{C}^n)$ the Lie algebra of vector fields $V = \sum_{j=1}^n a_j \frac{\partial}{\partial z_j}$ whose coefficient functions a_j belong to I. If I is the ideal of an algebraic subvariety $A \subset \mathbb{C}^n$, then $\aleph_I(\mathbb{C}^n) = \aleph_{\mathcal{A}}(\mathbb{C}^n, A)$. Let $\mathrm{Lie}_{\mathrm{alg}}(\mathbb{C}^n, A)$ denote the Lie subalgebra of $\aleph_{\mathcal{A}}(\mathbb{C}^n, A)$ generated by the complete algebraic vector fields on $\mathbb{C}^n$ that vanish on A.

Theorem 4.11.11 ([323, Theorem 4]) *Let A be a closed algebraic subvariety of codimension at least two in $\mathbb{C}^n$ and $I \subset \mathbb{C}^{[n]}$ be the ideal of functions vanishing on A. There is an ideal $L \subset \mathbb{C}^{[n]}$ with radical $\sqrt{L} = I$ such that $\aleph_L(\mathbb{C}^n) \subset \mathrm{Lie}_{\mathrm{alg}}(\mathbb{C}^n, A)$. That is, every vector field with coefficients in L is generated by complete algebraic vector fields that vanish on A.*

Proof Let $z = (z_1, \ldots, z_n)$ be coordinates on $\mathbb{C}^n$ and $\pi_i : \mathbb{C}^n \to \mathbb{C}^{n-1}$ the projection onto the hyperplane $\mathbb{C}_i^{n-1} = \{z_i = 0\}$. As in the proof of Proposition 4.11.9, we may assume that $\pi_i|_A$ is proper for every $i = 1, \ldots, n$.

Let $V_i = \frac{\partial}{\partial z_i}$. Given functions $f_i, h_i \in \ker V_i$ (i.e., independent of z_i) and such that h_i vanishes on $\pi_i(A)$, the algebraic vector field $f_i h_i V_i$ (a shear) is $\mathbb{C}$-complete and it vanishes on A; hence it generates a $\mathbb{C}_+$ action on $\mathbb{C}^n \setminus A$ by elements of the group $\mathrm{Aut}(\mathbb{C}^n, A)$. The vector field

$$[f_1 h_1 V_1, z_1 f_2 h_2 V_2] - [z_1 f_1 h_1 V_1, f_2 h_2 V_2] = f_1 f_2 h_1 h_2 V_2$$

then belongs to $\mathrm{Lie}_{\mathrm{alg}}(\mathbb{C}^n, A)$. Since $\ker V_1 \cdot \ker V_2$ spans the polynomial ring $\mathbb{C}[z]$, we see that $\mathrm{Lie}_{\mathrm{alg}}(\mathbb{C}^n, A)$ contains all algebraic fields proportional to $V_2 = \frac{\partial}{\partial z_2}$ whose coefficients belong to the principal ideal $(h_1 h_2)$ generated by $h_1 h_2$. (Note that A is contained in the zero locus of $(h_1 h_2)$.) Since one can perturb z_2 (as a linear function), we infer that $\mathrm{Lie}_{\mathrm{alg}}(\mathbb{C}^n, A)$ contains all algebraic vector fields whose coordinates are in some nonzero ideal L of $\mathbb{C}[z]$. Since $\mathbb{C}^n \setminus A$ is $\mathrm{Aut}(\mathbb{C}^n, A)$-homogenous, we can arrange (by moving the point and the projection direction) that the zero locus of L equals A. By the Nullstellensatz, this means that $\sqrt{L} = I$. □

Since the polynomial ring $\mathbb{C}^{[n]}$ is Noetherian, a simple argument shows that for any ideal $L \subset \mathbb{C}^{[n]}$ with radical $\sqrt{L} = I$ there exists $k \in \mathbb{N}$ such that $I^k \subset L$. Hence, Theorem 4.11.11 implies the following corollary.

Corollary 4.11.12 *If A is an algebraic subvariety of $\mathbb{C}^n$ with $\dim A \leq n - 2$ then every polynomial vector field V vanishing on A to a sufficiently high order is a Lie combination of complete polynomial vector fields vanishing on A.*

Kaliman and Kutzschebauch also proved the following more precise result.

Theorem 4.11.13 ([323, Theorem 6]) *If A is an algebraic subvariety of $\mathbb{C}^n$ of codimension at least two whose Zariski tangent space $\mathrm{T}_z A$ has dimension at most $n - 1$ for any point $z \in A$, then the Lie algebra $\aleph_{\mathcal{A}}(\mathbb{C}^n, A)$ of polynomial vector fields on $\mathbb{C}^n$ that vanish on A has the algebraic density property.*

4.12 Moving Polynomially Convex Sets

In this section we consider the problem of moving compact polynomially convex sets in $\mathbb{C}^n$ by holomorphic automorphisms.

We begin with the following analogue of Theorem 4.9.2 for compact sets. The first part is [221, Theorem 2.1]; the second part also uses results of Kaliman and Kutzschebauch discussed in the previous section.

Theorem 4.12.1 *Let Ω be an open subset of $\mathbb{C}^n$ $(n > 1)$ and $\Phi_t\colon \Omega \to \mathbb{C}^n$ $(t \in [0,1])$ be a $\mathcal{C}^1$-isotopy of injective holomorphic maps such that $\Phi_0 = \mathrm{Id}$. If K is a compact set in Ω such that $K_t = \Phi_t(K)$ is polynomially convex for every $t \in [0,1]$, then Φ_1 can be approximated uniformly on K by holomorphic automorphisms of $\mathbb{C}^n$.*

Furthermore, if A is an algebraic subvariety of $\mathbb{C}^n$ with $\dim A \le n-2$ and $\Phi_t(K) \cap A = \varnothing$ for all $t \in [0,1]$, then Φ_1 can be approximated uniformly on K by holomorphic automorphisms of $\mathbb{C}^n$ fixing A pointwise.

The approximating automorphisms can be chosen to agree with Φ_1 to a given finite order at any finite set of points in K.

Proof Since a compact polynomially convex set admits a basis of Stein neighborhoods that are Runge in $\mathbb{C}^n$, the proof of Theorem 4.9.2 (p. 135) easily adapts to give the first statement.

To get the second part, note that a holomorphic function on a neighborhood of a compact polynomially convex set K in the complement of an affine algebraic subvariety $A \subset \mathbb{C}^n$ can be approximated, uniformly on K, by polynomials that vanish to any given order on A. (Choose polynomials $h_1, \ldots, h_m$ whose common zero set equals A and let $\ell \in \mathbb{N}$. Since $A \cap K = \varnothing$ and K is polynomially convex, we can write the given function near K as $f = \sum_{j=1}^m f_j h_j^\ell$ by applying Cartan's division theorem, where f_j are holomorphic functions near K. It remains to approximate f_j by holomorphic polynomials.) The same holds for vector fields by applying this to each coefficient. By Corollary 4.11.12 (p. 156) every polynomial vector field on $\mathbb{C}^n$ that vanishes to a sufficiently high order on A is a Lie combination of complete algebraic vector fields vanishing on A. Hence, the proof of Theorem 4.9.2 (p. 135) gives approximation by automorphisms that restrict to the identity on A. □

By using the second part of Theorem 4.12.1 (instead of Theorem 4.9.2), one can strengthen many results in this and the subsequent sections. Here is an example

Corollary 4.12.2 *Let $A \subset \mathbb{C}^n$ be an algebraic subvariety with $\dim A \le n-2$. Given a compact convex set $K \subset \mathbb{C}^n \setminus A$ and $\epsilon > 0$, there is an injective holomorphic map $F\colon \mathbb{C}^n \to \mathbb{C}^n$ with $F(\mathbb{C}^n) \cap A = \varnothing$ and $\sup_{z\in K} |F(z) - z| < \epsilon$.*

Proof We may assume that $0 \in K$. Consider the isotopy $\Phi_t(z) = tz$ for z near K and $t \in (0, 1]$. By Theorem 4.12.1 there exists an automorphism $\Phi \in \mathrm{Aut}(\mathbb{C}^n, A)$ that approximates the dilation $z \mapsto \frac{1}{2}z$ uniformly on K. We can also ensure that $\Phi(0) = 0$. Then, 0 is an attracting fixed point of Φ whose basin Ω satisfies $K \subset \Omega \subset \mathbb{C}^n \setminus A$. Let $G\colon \mathbb{C}^n \to \Omega$ be a Fatou-Bieberbach map furnished by Theorem 4.3.2 (p. 116). Since G is a limit of automorphisms of $\mathbb{C}^n$, the set $L = G^{-1}(K)$ is polynomially convex. Since G^{-1} maps the convex set K biholomorphically onto a polynomially convex set L, we can approximate G^{-1} uniformly on a neighborhood of K by an automorphism $H \in \mathrm{Aut}(\mathbb{C}^n)$. Then, $F = G \circ H$ maps $\mathbb{C}^n$ biholomorphically onto Ω and it approximates the identity map on K. □

Applying Theorem 4.9.10 (p. 140) instead of Theorem 4.9.2 gives the following parametric version of Theorem 4.12.1.

Theorem 4.12.3 *Let Ω be an open set in $\mathbb{C}^n = \mathbb{C}^k \times \mathbb{C}^m$ $(m > 1)$, and let $\Phi_t\colon \Omega \to \mathbb{C}^n$ be a $\mathscr{C}^1$-isotopy of biholomorphic maps as in Theorem* 4.9.10 *such that Φ_0 is the identity map. Let $K \subset \Omega$ be a compact polynomially convex subset of $\mathbb{C}^n$. If $\Phi_t(K)$ is polynomially convex for every $t \in [0, 1]$, then for every $t \in [0, 1]$ the map Φ_t can be approximated uniformly on K by holomorphic automorphisms of $\mathbb{C}^n$ of the form* (4.48) *(p.* 140).

The following application of Theorem 4.12.1 shows that, under suitable conditions, we can simultaneously approximate a collection of automorphisms on disjoint compact sets by a single automorphism.

Corollary 4.12.4 *Let $K_1, K_2, \ldots, K_m$ be pairwise disjoint compact sets in $\mathbb{C}^n$ such that all but one are starshaped. Let $\Phi^j \in \mathrm{Aut}(\mathbb{C}^n)$ $(j = 1, \ldots, m)$ be such that the images $K'_j = \Phi^j(K_j)$ are pairwise disjoint. If the sets $K = \bigcup_{j=1}^m K_j$ and $K' = \bigcup_{j=1}^m K'_j$ are polynomially convex, then for every $\epsilon > 0$ there exists $\Phi \in \mathrm{Aut}(\mathbb{C}^n)$ such that $|\Phi(z) - \Phi^j(z)| < \epsilon$ for $z \in K_j$, $j = 1, \ldots, m$.*

Proof Assume that each of the sets $K_2, \ldots, K_m$ is starshaped with respect to a point $a_j \in K_j$. Since $K = \bigcup_{j=1}^m K_j$ is polynomially convex, each of the sets K_j is also polynomially convex. In view of Theorem 4.12.1, it suffices to find for each $j = 1, 2, \ldots, m$ an isotopy of biholomorphic maps Φ_t^j $(t \in [0, 1])$, defined on a neighborhood of K_j in $\mathbb{C}^n$, such that Φ_0^j is the identity map, $\Phi_1^j = \Phi^j$, and for each $t \in [0, 1]$ the sets $K_{j,t} = \Phi_t^j(K_j)$ $(j = 1, \ldots, m)$ are pairwise disjoint and their union $\bigcup_{j=1}^m K_{j,t}$ is polynomially convex.

Lemma 4.9.4 (p. 136) gives for each j a smooth isotopy of automorphisms $\Psi_t^j \in \mathrm{Aut}(\mathbb{C}^n)$ $(t \in [0, 1])$ such that $\Psi_0^j = \mathrm{Id}$ and $\Psi_1^j = \Phi^j$. We take $\Phi_t^1 = \Psi_t^1$, but the other isotopies need to be modified.

We first replace Ψ_t^j by a smooth isotopy $(z, t) \mapsto \widetilde{\Psi}_t^j(z) = \Psi_t^j(z) + \gamma_j(t)$, where the maps $\gamma_j\colon [0, 1] \to \mathbb{C}^n$ are chosen such that $\gamma_j(0) = 0 = \gamma_j(1)$ and the points

$b_{j,t} = \widetilde{\Psi}_t^j(a_j)$ for $j = 2, \ldots, m$ are pairwise distinct and contained in $\mathbb{C}^n \setminus \Phi_t^1(K_1)$. In the sequel, we drop the tildes.

Note that the union of a compact polynomially convex set and a finite set of points is still polynomially convex. The same holds if we replace points by small closed balls around them. Hence, we can find a small enough $\delta > 0$ such that the closed balls $\overline{\mathbb{B}}(b_{j,t}, \delta)$ $(j = 2, \ldots, m)$ and the set $\Psi_t^1(K_1)$ are pairwise disjoint and their union is polynomially convex for every $t \in [0, 1]$.

Choose a number $\eta > 0$ such that Ψ_t^j maps the ball $\overline{\mathbb{B}}(a_j, \eta)$ into the ball $\mathbb{B}(b_{j,t}, \delta)$ for every $t \in [0, 1]$ and $j = 2, \ldots, m$. Let $R > 0$ be sufficiently large such that for every $j = 2, \ldots, m$ the set K_j is contained in the ball $\mathbb{B}(a_j, R)$. Set $\Theta_t^j(z) = a_j + t(z - a_j)$. An isotopy Φ_t^j with the required properties is now obtained by first squeezing K_j into $\mathbb{B}(a_j, \eta)$ by using the contractions Θ_t^j, then following the isotopy Ψ_t^j restricted to the ball $\mathbb{B}(a_j, \eta)$, and finally expanding the image of K_j onto K'_j. To give explicit formulas, we determine $c > 0$ by $1 - \frac{c}{3} = \frac{\eta}{R}$ (hence $\Theta^j(K_j) \subset \mathbb{B}(a_j, \eta)$), and we define the isotopy Φ_t^j on a small open neighborhood of K_j as follows:

$$\Phi_t^j = \begin{cases} \Theta_{1-ct}^j & \text{if } 0 \le t \le \frac{1}{3}; \\ \Psi_{3t-1}^j \circ \Theta_{\eta/R}^j & \text{if } \frac{1}{3} < t \le \frac{2}{3}; \\ \Phi^j \circ \Theta_{1+c(t-1)}^j & \text{if } \frac{2}{3} < t \le 1. \end{cases}$$

For each $t \in [0, 1]$ the set $\bigcup_{j=1}^m \Phi_t^j(K_j)$ is polynomially convex: For $0 \le t \le \frac{1}{3}$ this holds since $\Phi_t^j(K_j) \subset K_j$ $(2 \le j \le m)$, and similarly for $\frac{2}{3} \le t \le 1$ we have $\Phi_t^j(K_j) \subset K'_j$. (Here, we use that the sets $K_2, \ldots, K_m$ are starshaped.) On $\frac{1}{3} \le t \le \frac{2}{3}$ we have $\Phi_t^j(K_j) \subset \overline{\mathbb{B}}(b_{j,t}, \delta)$, and the union of these balls together with $\Phi_t^1(K_1)$ is polynomially convex. By a reparametrization of the interval $[0, 1]$ we obtain a smooth isotopy with the same properties. □

Corollary 4.12.4 and the push-out method (Corollary 4.4.2 on p. 125) gives several useful results on separation of sets by Fatou-Bieberbach domains. We begin with the following result.

Proposition 4.12.5 *Let $n > 1$. Given disjoint compact convex sets $K, L \subset$ in $\mathbb{C}^n$, there is a Fatou-Bieberbach domain Ω such that $K \subset \Omega \subset \mathbb{C}^n \setminus L$; furthermore, a Fatou-Bieberbach map $F \colon \mathbb{C}^n \to \Omega$ can be chosen arbitrarily close to the identity map on K.*

Proof The union $K \cup L$ of two disjoint compact convex sets in $\mathbb{C}^n$ is polynomially convex and hence Corollary 4.12.4 applies. Recall that $\mathbb{B}$ denotes the unit ball in $\mathbb{C}^n$. Pick $N_1 \in \mathbb{N}$ such that $K \cup L \subset B_1 := N_1\mathbb{B}$. Choose an affine linear automorphism ψ_1 of $\mathbb{C}^n$ such that $\psi_1(L) \subset \mathbb{C}^n \setminus \bar{B}_1$ and the set $\bar{B}_1 \cup \psi_1(L)$ is polynomially convex. (The latter property holds when every $\psi_1(L)$ is contained in a sufficiently small ball around a point in $\mathbb{C}^n \setminus \bar{B}_1$; this can be arranged by a suitable choice of ψ_1.) Corollary 4.12.4 furnishes $\phi_1 \in \mathrm{Aut}(\mathbb{C}^n)$ which is close to the identity on K and is close to the map ψ_1 on L. Set $L_1 = \phi_1(L)$. If the approximations are close enough, then $\phi_1(K) \subset B_1$, $L_1 \cap \bar{B}_1 = \varnothing$, and $\bar{B}_1 \cup L_1$ is polynomially convex. Pick $N_2 \geq N_1 + 1$ such that $L_1 \subset B_2 := N_2\mathbb{B}$. By repeating the above argument (with K replaced by $\bar{B}_1$ and L replaced by L_1), we find $\phi_2 \in \mathrm{Aut}(\mathbb{C}^n)$ which approximates the identity on $\bar{B}_1$, it sends L_1 to $L_2 := \phi_2(L_1) \subset \mathbb{C}^n \setminus \bar{B}_2$, and $\bar{B}_2 \cup L_2$ is polynomially convex. Continuing this process, we inductively find automorphisms $\phi_k \in \mathrm{Aut}(\mathbb{C}^n)$ for $k = 1, 2, \ldots$ such that the sequence $\Phi_k = \phi_k \circ \phi_{k-1} \circ \cdots \circ \phi_1$ converges on a domain $\Omega \subset \mathbb{C}^n$ to a Fatou-Bieberbach map $\Phi \colon \Omega \to \mathbb{C}^n$ onto $\mathbb{C}^n$. The domain of convergence Ω consists of the points $z \in \mathbb{C}^n$ with bounded orbits $\{\Phi_k(z) : k \in \mathbb{N}\}$. (See Corollary 4.4.2.) By the construction, we have $K \subset \Omega$, $\Omega \cap L = \varnothing$, and Φ approximates the identity map on K. The inverse map $F = \Phi^{-1} \colon \mathbb{C}^n \to \Omega$ then satisfies the claim in the last sentence. □

The union of three pairwise disjoint compact convex sets in general fails to be polynomially convex [329, 474], unless we are dealing with three disjoint balls [329, 532] or three disjoint polydiscs in $\mathbb{C}^2$ with sides parallel to the coordinate axes [474]. This give results analogous to Proposition 4.12.5.

Corollary 4.12.6 *Assume that B_1, B_2, B_3 are pairwise disjoint closed balls in $\mathbb{C}^n$ $(n > 1)$, or pairwise disjoint closed polydiscs in $\mathbb{C}^2$ with sides parallel to the coordinate axes. Then, there exists a Fatou-Bieberbach domain $\Omega \subset \mathbb{C}^n$ containing two of these sets and disjoint from the third one.*

The first statement in the next result is a special case of Corollary 4.12.4. By pushing all the points $p_1, \ldots, p_m$ far enough from K, we can separate these two sets by a pair of disjoint balls; hence Proposition 4.12.5 implies the second statement. (Compare with Corollary 4.12.2 on p. 157.)

Corollary 4.12.7 *Let K be a compact polynomially convex set in $\mathbb{C}^n$, and let $p_1, \ldots, p_m$ and $q_1, \ldots, q_m$ be m-tuples of pairwise distinct points in $\mathbb{C}^n \setminus K$. Given $\epsilon > 0$, there exists $\Phi \in \mathrm{Aut}(\mathbb{C}^n)$ such that*

$$\left|\Phi(z) - z\right| < \epsilon \ (\forall z \in K), \qquad \Phi(p_j) = q_j \ (j = 1, \ldots, m).$$

Furthermore, there is a biholomorphic map F of $\mathbb{C}^n$ onto a Fatou-Bieberbach domain $\Omega \subset \mathbb{C}^n \setminus \{p_1, \ldots, p_m\}$ such that $\sup_{z \in K} |F(z) - z| < \epsilon$.

The next result is [220, Proposition 9]. One proves it by first applying Corollary 4.12.4 to separate the sets K and L by a pair of disjoint balls, and then applying Proposition 4.12.5.

Corollary 4.12.8 *Let $n > 1$. Given disjoint compact sets $K, L \subset \mathbb{C}^n$ such that L is starshaped and $K \cup L$ is polynomially convex, there exist Fatou-Bieberbach do-*

mains $\Omega, \Omega' \subset \mathbb{C}^n$ *satisfying*

$$K \subset \Omega \subset \mathbb{C}^n \setminus L \quad \text{and} \quad L \subset \Omega' \subset \mathbb{C}^n \setminus K.$$

The analogous result holds if L is a union of finitely many paiwise disjoint compact starshaped sets.

4.13 Moving Totally Real Submanifolds

The Andersén-Lempert theory, combined with the $\overline{\partial}$-methods, allows us to approximate isotopies of polynomially convex totally real submanifolds of $\mathbb{C}^n$ by holomorphic automorphisms of $\mathbb{C}^n$ (see [214, 215, 221]). The most precise result without any loss of derivatives is the following one.

Theorem 4.13.1 ([215, Theorem 1.3]) *Let* $n, r \geq 2$ *be integers, and let M be a compact smooth manifold. Given a* $\mathscr{C}^r$*-isotopy of embeddings* $f_t \colon M \hookrightarrow \mathbb{C}^n$ $(t \in [0,1])$ *such that the submanifold* $f_t(M) \subset \mathbb{C}^n$ *is totally real and polynomially convex for each* $t \in [0,1]$, *there exists a sequence* $\Phi_j \in \mathrm{Aut}(\mathbb{C}^n)$ *such that*

$$\lim_{j\to\infty} \|\Phi_j \circ f_0 - f_1\|_{\mathscr{C}^r(M)} = 0 \quad \text{and} \quad \lim_{j\to\infty} \left\|\Phi_j^{-1} \circ f_1 - f_0\right\|_{\mathscr{C}^r(M)} = 0.$$

The proof is similar to that of Theorem 4.13.2, the main difference being that one solves the $\overline{\partial}$-problem with the Henkin-type integral kernel developed in [215]. The following result from [214], which relies on the simpler L^2 method at the cost of losing finitely many derivatives, suffices for our applications in this book.

Theorem 4.13.2 ([214, Theorem 2.1]) *Let* $n \geq 2$ *and* $r \geq 3$. *Let* $K \subset \mathbb{C}^n$ *be a compact polynomially convex set, and let* $U \subset \mathbb{C}^n$ *be an open set containing* K. *For each* $t \in [0,1]$ *let* $M_t \subset \mathbb{C}^n$ *be a compact totally real submanifold of class* $\mathscr{C}^r$, *and let* $f_t \colon M_0 \to M_t$ *for* $t \in [0,1]$ *be a* $\mathscr{C}^r$ *diffeomorphism satisfying the following conditions*:

(i) $K \cup M_t$ *is polynomially convex for all* $0 \leq t \leq 1$,
(ii) $f_0(z) = z$ *for all* $z \in M_0$,
(iii) $f_t(z) = z$ *for all* $z \in M_0 \cap U$ *and* $t \in [0,1]$, *and*
(iv) f_t *and* $\partial f_t/\partial t$ *are of class* $\mathscr{C}^r$ *in* $(t,z) \in [0,1] \times M_0$.

Let $m = \dim M$. *Choose an integer k with* $1 \leq k \leq r - \frac{m+3}{2}$. *Given* $\epsilon > 0$, *there is an automorphism* $\Phi \in \mathrm{Aut}(\mathbb{C}^n)$ *satisfying*

(a) $\|\Phi|_{M_0} - f_1\|_{\mathscr{C}^k(M_0)} < \epsilon$,
(b) $\|\Phi^{-1}|_{M_1} - f_1^{-1}\|_{\mathscr{C}^k(M_1)} < \epsilon$,
(c) $\|\Phi - \mathrm{Id}\|_{\mathscr{C}^k(K)} < \epsilon$.

If E is a finite subset of $K \cup M_0$, then we can choose Φ such that, in addition to the above, it agrees with the identity to order k at each point of $E \cap K$, and $\Phi|_{M_0}$ agrees with f_1 to order k at each point of $E \cap M_0$.

Proof We think of the isotopy $\{f_t\}$ as the flow of a time-dependent vector field. We extend t to a complex variable. Thus, let $\mathbb{C}^{n+1} = \mathbb{C} \times \mathbb{C}^n$ be the extended phase space with complex coordinates $w = (t, z)$, with $t \in \mathbb{C}$ and $z \in \mathbb{C}^n$. Consider the following sets in $\mathbb{C}^{n+1}$:

$$L_0 = [0, 1] \times K, \qquad S = \bigcup_{0 \le t \le 1} \{t\} \times M_t, \qquad L = L_0 \cup S.$$

Since $K \cup M_t$ is polynomially convex for each $t \in [0, 1]$, L is also polynomially convex. Clearly, $S = L \setminus L_0$ is a totally real submanifold with boundary of class $\mathcal{C}^r$. We extend each f_t to $M_0 \cup U$ as the identity map on U. For each $t \in [0, 1]$, let $V_t \colon M_t \cup U \to \mathbb{C}^n$ be the velocity vector field of f_t defined by

$$\dot{f}_t(z) = V_t\big(f_t(z)\big), \quad z \in M_0 \cup U,\ t \in [0, 1].$$

The map $V(t, z) = V_t(z)$ with values in $\mathbb{C}^n$ is defined on $S \cup ([0, 1] \times U) \subset \mathbb{C}^{n+1}$. Property (iii) of $\{f_t\}$ implies that V vanishes on $[0, 1] \times U$, and the restriction $V|_S$ is of class $\mathcal{C}^r(S)$. After shrinking U slightly around K we can extend V to a $\mathcal{C}^r$ map $V \colon \mathbb{C}^{n+1} \to \mathbb{C}^n$ with compact support that vanishes on $\mathbb{C} \times U \subset \mathbb{C}^{n+1}$ and satisfies the estimates

$$\big|\partial^\alpha \overline{\partial} V(w)\big| = o\big(\mathrm{dist}(w, S)^{r-1-|\alpha|}\big), \quad 0 \le |\alpha| \le r - 1.$$

The derivative ∂^α (with respect to the real coordinates) is applied to each component of the $(0, 1)$-form $\overline{\partial} V$, and $|\partial^\alpha \overline{\partial} V|$ denotes the Euclidean norm.

The following lemma follows from Hörmander's L^2-method (see Theorem 2.7.1 on p. 58) and the interior regularity estimates (see Lemma 2.7.2 on p. 58). Let $L(\epsilon)$ denote the ϵ-neighborhood of L.

Lemma 4.13.3 ([214, Proposition 2.3]) *Let $r \in \mathbb{Z}_+$, $r \ge \frac{m+5}{2}$, and $k = r - \frac{m+3}{2} \ge 1$. For each sufficiently small $\epsilon > 0$ there exists an entire map $V^\epsilon \colon \mathbb{C}^{n+1} \to \mathbb{C}^n$ such that for all $\alpha \in \mathbb{Z}_+^{2n+2}$ with $|\alpha| \le k$ we have*

$$\big\|\partial^\alpha\big(V - V^\epsilon\big)\big\|_{L^\infty(L(\epsilon))} = o\big(\epsilon^{k-|\alpha|}\big), \quad \epsilon \to 0.$$

Let ψ_t^ϵ be the flow of the time-dependent holomorphic vector field V^ϵ:

$$\dot{\psi}_t^\epsilon(z) = V^\epsilon\big(t, \psi_t^\epsilon(z)\big), \quad \psi_0^\epsilon(z) = z.$$

From Lemma 4.13.3 and standard results on perturbations of flows (see Lemma 1.9.4 on p. 35) we obtain the following.

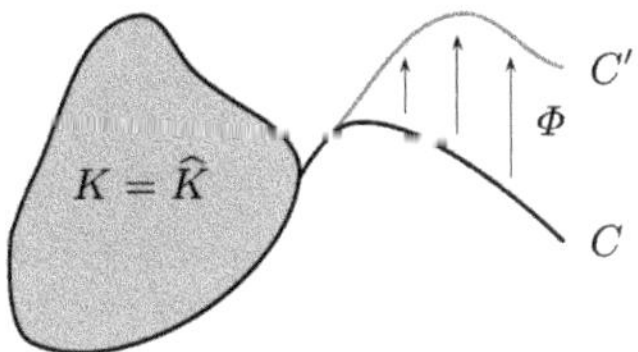

Fig. 4.1 Combing hair

Proposition 4.13.4 *There are a neighborhood U_0 of $K \cup M_0$ and an $\epsilon_0 > 0$ such that $\psi_t^\epsilon(z)$ exists for all $z \in U_0$, $t \in [0, 1]$ and $0 < \epsilon < \epsilon_0$, and it satisfies*

$$\lim_{\epsilon\to 0}\| f_t - \psi_t^\epsilon \|_{C^k(K\cup M_0)} = 0, \qquad \lim_{\epsilon\to 0}\| (f_t)^{-1} - (\psi_t^\epsilon)^{-1} \|_{C^k(K\cup M_t)} = 0.$$

Theorem 4.13.2 follows by combining Proposition 4.13.4 (for $t = 1$) and Proposition 4.9.8 (p. 138); the latter one says that for each fixed $\epsilon > 0$, ψ_1^ϵ is a uniform limit of automorphisms of $\mathbb{C}^n$ in a neighborhood of $K \cup M_0$. The requirement on matching f at finitely many points is a trivial addition. □

Corollary 4.13.5 (Combing hair by automorphisms) *Assume that K is a compact polynomially convex set in $\mathbb{C}^n$ $(n > 1)$, $U \subset \mathbb{C}^n$ is an open neighborhood of K, and $C \subset \mathbb{C}^n$ is a smooth embedded arc attached to K in a single point of K, or disjoint from K. Given $r \in \mathbb{Z}_+$, $\epsilon > 0$ and a smooth diffeomorphism $f : C \cup U \to C' \cup U \subset \mathbb{C}^n$ which equals the identity on U (see Fig. 4.1), there exist a neighborhood $W \subset U$ of K and an automorphism $\Phi \in \mathrm{Aut}(\mathbb{C}^n)$ satisfying*

$$\|\Phi - \mathrm{Id}\|_{C^r(W)} < \epsilon, \qquad \|\Phi - f\|_{C^r(C)} < \epsilon.$$

In particular, if $C \cap K = \varnothing$ and B is a ball in $\mathbb{C}^n \setminus K$, we can choose $\Phi \in \mathrm{Aut}(\mathbb{C}^n)$ arbitrarily close to the identity on K and such that $\Phi(C) \subset B$.

Proof The union $K \cup C$ of a compact polynomially convex set K and a smooth embedded arc C, attached to K at a single point or disjoint from K, is polynomially convex [528, 530]. Hence, the diffeomorphism $f : C \to C'$ can be connected to the identity map on C by a diffeotopy $f_t : C \to C_t$ $(t \in [0, 1])$ satisfying the hypotheses of Theorem 4.13.2. For the last part, note that the arc C can be moved within $\mathbb{C}^n \setminus K$ into the ball B. □

The following result from [183, Corollary 4.2] (see also [221, Theorem 5.2] for the special case of surfaces in $\mathbb{C}^3$) shows that the hypotheses of Theorem 4.13.2 are satisfied for generic submanifolds of $\mathbb{C}^n$ of dimension $\le 2n/3$.

Theorem 4.13.6 ([183, Corollary 4.2]) *Let $p \ge 2$ be an integer.*

(i) *For every compact submanifold $M \subset \mathbb{C}^n$ of class C^p with $\dim M \le 2n/3$ there exist arbitrarily small C^p deformations $\widetilde{M} \subset \mathbb{C}^n$ which are totally real and polynomially convex.*

(ii) *If* $f: M \to f(M) \subset \mathbb{C}^n$ *is a diffeomorphism of class* $\mathscr{C}^p$ *such that both* M *and* $f(M)$ *are totally real polynomially convex submanifolds of dimension* $\le 2n/3$, *then there exists an isotopy* $f_t: M \to M_t \subset \mathbb{C}^n$ $(t \in [0,1])$ *of* $\mathscr{C}^p$ *diffeomorphisms such that* $f_0 = \mathrm{Id}_M$, $f_1 = f$, *and the submanifold* $f_t(M)$ *of* $\mathbb{C}^n$ *is totally real and polynomialy convex for every* $t \in [0,1]$.

Total reality is easily achieved by using transversality (see Theorem 9.1.2), but polynomial convexity is a nontrivial addition. Note that the union of these two properties is stable under small smooth deformations.

The next corollary follows immediately from Theorems 4.13.2 and 4.13.6.

Corollary 4.13.7 *Let* $f: M \to M'$ *be a smooth diffeomorphism between a pair of smooth, compact, totally real and polynomially convex submanifolds* $M, M' \subset \mathbb{C}^n$ *of dimension at most* $2n/3$. *Given* $\epsilon > 0$ *and* $k \in \mathbb{N}$, *there exists a holomorphic automorphism* $\Phi \in \mathrm{Aut}(\mathbb{C}^n)$ *such that*

$$\|f - \Phi|_M\|_{\mathscr{C}^k(M)} < \epsilon \quad \text{and} \quad \left\| f^{-1} - \Phi^{-1}\big|_{M'}\right\|_{\mathscr{C}^k(M')} < \epsilon.$$

Further results on approximation of diffeomorphisms on smooth submanifolds by volume preserving and symplectic holomorphic automorphisms of $\mathbb{C}^n$ can be found in the papers [185, 186].

In conclusion, we mention a result of Løw and Wold [401] on generic polynomial convexity of totally real submanifolds of dimension $< n$ in $\mathbb{C}^n$. We generalize the notion of the polynomial hull to noncompact closed sets as follows. If $K \subset \mathbb{C}^n$ is closed, write $K = \bigcup_{j=1}^\infty K_j$, where K_j are compact sets satisfying $K_j \subset K_{j+1}$ for each $j \in \mathbb{N}$, and define $\widehat{K} = \bigcup_{j=1}^\infty \widehat{K}_j$. It is easily seen that $\widehat{K}$ is independent of the particular exhaustion of K. Let $h(K)$ denote the set $h(K) = \overline{\widehat{K} \setminus K}$. We say that K has *bounded exhaustion hulls* in $\mathbb{C}^n$ if the set $h(k\overline{\mathbb{B}^n} \cup K)$ is bounded for every $k \in \mathbb{N}$.

Theorem 4.13.8 (Løw and Wold [401]) *A generic closed totally real submanifold* M *of* $\mathbb{C}^n$ *with* $\dim_{\mathbb{R}} M < n$ *is polynomially convex and has bounded exhaustion hulls in* $\mathbb{C}^n$.

A recent generalization of this result, concerning totally real embeddings with prescribed (nontrivial) polynomial hulls, is due to Arosio and Wold [37].

4.14 Carleman Approximation by Automorphisms

In this section we present the Andersén-Lempert theorem with Carleman-type approximation on certain *noncompact* totally real submanifolds of $\mathbb{C}^n$ which are sufficiently tame near infinity.

The first set of results concerns Carleman approximation on unbounded arcs attached to a compact polynomially convex set. The condition on the curve near infinity is the *nice projection property*, see Definition 4.14.2. For simplicity, we work with $\mathscr{C}^\infty$-smooth arcs (this suffices for our applications), although much less regularity is needed. The main idea first appeared in the proof of the following result of Buzzard and the author.

Theorem 4.14.1 ([77, Theorem 1.1]) *Let $n > 1$ and $r \geq 0$ be integers. Given a proper smooth embedding $\lambda \colon \mathbb{R} \hookrightarrow \mathbb{C}^n$ and a continuous function $\eta \colon \mathbb{R} \to (0, \infty)$, there is a proper holomorphic embedding $f \colon \mathbb{C} \hookrightarrow \mathbb{C}^n$ such that*

$$\left| f^{(s)}(t) - \lambda^{(s)}(t) \right| < \eta(t), \quad t \in \mathbb{R},\ s = 0, 1, \ldots, r.$$

The proof uses the push-out method from Sect. 4.4, but with an addition that ensures the control of unbounded arcs at infinity. This method was developed and exploited by Wold [581, 582], and it was also used in [226]. The following notion, introduced in a similar form in [582], is the key geometric assumption in subsequent constructions.

Definition 4.14.2 A family $\ell = \bigcup_{i=1}^m \ell_i \subset \mathbb{C}^n$ of pairwise disjoint embedded smooth curves $\ell_i = \{\gamma_i(t) : t \in [0, \infty) \text{ or } t \in (-\infty, \infty)\}$ has the *nice projection property* if there is a linear projection $\pi : \mathbb{C}^n \to \mathbb{C}$ such that

(i) $\lim_{|t| \to \infty} |\pi(\gamma_i(t))| = \infty$ for $i = 1, 2, \ldots, m$, and
(ii) there is a number $r_0 > 0$ such π is injective on $\ell \setminus \pi^{-1}(\overline{\mathbb{D}}_{r_0})$.

This corresponds to the *immediate nice projection property* in [365, 404, 405]; the nice projection property for them means that there exists $\alpha \in \mathrm{Aut}(\mathbb{C}^n)$ such that $\alpha(\ell)$ has the immediate nice projection property. In most sources it is additionally required that for all sufficiently big $r > 0$ the set $\mathbb{C} \setminus (\pi(\ell) \cup \overline{\mathbb{D}}_r)$ has no bounded connected component. We replace the latter condition by the following lemma.

Lemma 4.14.3 *Assume that $\ell = \bigcup_{i=1}^m \ell_i \subset \mathbb{C}^n$ has the nice projection property with respect to a projection $\pi : \mathbb{C}^n \to \mathbb{C}$. For every $r > 0$ there exists $r' \geq r$ such that the polynomial hull of $\overline{\mathbb{D}}_r \cup \pi(\ell)$ is contained in $\overline{\mathbb{D}}_{r'} \cup \pi(\ell)$.*

Proof We may assume that $r \geq r_0$, where r_0 satisfies condition (ii) in Definition 4.14.2. Condition (i) implies that each of the projected curves $\pi(\ell_i) \subset \mathbb{C}$ can exit and reenter the disc $\overline{\mathbb{D}}_r$ at most finitely many times, and by (ii) these curves do not intersect outside $\overline{\mathbb{D}}_r$. It follows that the polynomial hull of $\overline{\mathbb{D}}_r \cup \pi(\ell)$ is the union of this set with at most finitely many (bounded) connected components of $\mathbb{C} \setminus (\overline{\mathbb{D}}_r \cup \pi(\ell))$. □

The following lemma is due to Majcen [405, Theorem 1.3]; similar results were proved beforehand in [77, Sect. 3] and [582, Lemma 1].

Lemma 4.14.4 *Let K and L be compact polynomially convex sets in $\mathbb{C}^n$ for some $n \geq 2$. Assume that $\ell = \{\gamma(t) : t \in [0, \infty)\}$ is a smooth embedded curve in $\mathbb{C}^n$ such that $K \cap \ell = \{\gamma(0)\}$ and $\alpha(\ell)$ has the nice projection property for some $\alpha \in \mathrm{Aut}(\mathbb{C}^n)$. Set $C = \gamma([0, 1])$, and let $F : C \to C'$ be a smooth diffeomorphism onto an arc $C' \subset \mathbb{C}^n$ such that $F|_{C \cap U} = \mathrm{Id}$ for some neighborhood U of K and $F(\gamma(1)) \notin L$. Given $\epsilon > 0$ and $s \in \mathbb{Z}_+$ there exists $\Phi \in \mathrm{Aut}(\mathbb{C}^n)$ satisfying the following conditions:*

$$\|\Phi - \mathrm{Id}\|_{\mathcal{C}^0(K)} < \epsilon, \quad \|\Phi - F\|_{\mathcal{C}^s(C)} < \epsilon, \quad \Phi(\ell \setminus C) \cap L = \varnothing. \tag{4.59}$$

The analogous result holds if $\ell = \bigcup_{i=1}^m \ell_i$ is a union of finitely many curves.

Proof We consider the case when ℓ is a single curve; the same proof applies to finitely many curves. It suffices to prove the result for $\alpha = \mathrm{Id}$; in the general case we obtain Φ satisfying (4.59) by taking $\Phi = \alpha^{-1} \circ \Phi' \circ \alpha$, where $\Phi' \in \mathrm{Aut}(\mathbb{C}^n)$ is chosen such that

- Φ' approximates the identity map on $\alpha(K)$,
- Φ' approximates the diffeomorphism $F' = \alpha \circ F \circ \alpha^{-1}$ on $\alpha(C)$, and
- $\Phi'(\alpha(\ell \setminus C)) \cap \alpha(L) = \varnothing$.

Thus, we assume that ℓ has the nice projection property with respect to the projection $\pi(z_1, \ldots, z_n) = z_1$. Let $r_0 > 0$ be as in Definition 4.14.2(ii). Choose $r \geq r_0$ such that $\pi_1(K \cup C) \subset \mathbb{D}_r$, and let $r' \geq r$ be such that Lemma 4.14.3 holds. Write $\gamma(t) = (\gamma_1(t), \gamma_2(t), \ldots)$. Pick a number $t_1 \geq 1$ such that $|\gamma_1(t)| > r'$ for all $t \geq t_1$ and $\gamma_1'(t_1) \neq 0$; this holds for almost every big t_1 since $\lim_{t\to\infty} |\gamma_1(t)| = \infty$. Set $C_1 = \gamma([0, t_1])$. We extend the given diffeomorphism $F : C \to C'$ to a smooth diffeomorphism $F : C_1 \to C_1'$, where the arc C_1' is an extension of C' across the endpoint $F(\gamma(1)) \in \mathbb{C}^n \setminus L$ such that $C_1' \setminus C_1 \subset \mathbb{C}^n \setminus L$. We seek $\Phi \in \mathrm{Aut}(\mathbb{C}^n)$ satisfying (4.59) as a composition $\Phi = \phi \circ \psi$ of two automorphisms. Corollary 4.13.5 furnishes $\phi \in \mathrm{Aut}(\mathbb{C}^n)$ such that

- ϕ is uniformly close to the identity on a neighborhood of K, and
- $\phi|_{C_1}$ approximates the diffeomorphism $F : C_1 \to C_1'$ in the $\mathcal{C}^s$ topology.

If the approximation is close enough, then $\phi(C_1 \setminus C) \cap L = \varnothing$. Note that $\phi(\ell \setminus C_1) \cap L$ may be nonempty, so the last condition in (4.59) need not hold. We remove this intersection by precomposing ϕ by a suitably chosen shear

$$\psi(z) = z + h(z_1)e_2 = \big(z_1, z_2 + h(z_1), z_3, \ldots\big) \tag{4.60}$$

with $h \in \mathcal{O}(\mathbb{C})$. Let $A = \phi^{-1}(L)$. It suffices to choose h such that

$$\psi \approx \mathrm{Id} \text{ on } \overline{\mathbb{D}}_r \cup \gamma_1\big([0, t_1]\big), \quad \psi\big(\gamma(t)\big) \notin A \text{ for all } t \geq t_1. \tag{4.61}$$

The first condition holds if h is small on $\overline{\mathbb{D}}_r \cup \gamma_1([0, t_1])$. Note that the polynomial hull of this set is contained in $\overline{\mathbb{D}}_{r'} \cup \gamma_1([0, t_1])$. To achieve the second condition, choose $M > r'$ such that A is contained in the polydisc $\overline{\mathbb{D}}^n_M$ of radius M centered at the origin, and let $t' > t_1$ be such that $|\gamma_1(t)| > M$ for $t \ge t'$. Let $c_2 = \sup\{|\gamma_2(t)| : t \in [0,t']\}$. Pick $b \in \mathbb{C}$ such that $M + c_2 < |b|$. Let $q = \gamma(t_1) = (q_1, q_2, \ldots, q_n)$ and $q' = (q_1, b, q_3, \ldots)$; then $q, q' \notin A$. Consider the affine line $\Lambda = q + \mathbb{C}e_2$. The intersection $A \cap \Lambda$ is polynomially convex in Λ, and hence $\Lambda \setminus A$ is connected. Therefore, the points q and q' can be connected in $\Lambda \setminus A$ by an arc $\{q + \lambda(s)e_2 : s \in [0,1]\}$, with $\lambda(0) = 0$ and $\lambda(1) = b - q_2$. Pick a number $t_2 > t_1$ close enough to t_1 such that the arc $\gamma_1([t_1,t_2]) \subset \mathbb{C}$ is embedded (this is possible since $\gamma_1'(t_1) \neq 0$) and

$$\big\{\gamma(t) + \lambda(s)e_2 : t \in [t_1,t_2],\ s \in [0,1]\big\} \subset \mathbb{C}^n \setminus A.$$

Choose a smooth increasing function $\chi : \mathbb{R}_+ \to [0,1]$ such that $\chi(t) = 0$ for $t \le t_1$ and $\chi(t) = 1$ for $t \ge t_2$. Each point $\zeta \in \gamma_1([t_1,t_2])$ equals $\zeta = \gamma_1(t)$ for a unique $t \in [t_1,t_2]$. Define a function $\tilde h \colon \overline{\mathbb{D}}_{r'} \cup \pi_1(\ell) \to \mathbb{C}$ by

$$\tilde h(\zeta) = \begin{cases} 0, & \text{if } \zeta \in \overline{\mathbb{D}}_{r'} \cup \gamma_1([0,t_1]); \\ \lambda(\chi(t)), & \text{if } \zeta = \gamma_1(t),\ t \in [t_1,t_2]; \\ \lambda(1), & \text{if } \zeta \in \gamma_1([t_2,\infty)). \end{cases}$$

Then, $\gamma(t) + \tilde h(\gamma_1(t))e_2 \notin A$ for all $t \in [0,\infty)$. By Mergelyan's theorem we can approximate $\tilde h$, uniformly on the polynomial hull of the set $\overline{\mathbb{D}}_r \cup \gamma([0,t'])$ (which is contained in $\overline{\mathbb{D}}_{r'} \cup \gamma([0,t'])$), by entire functions $h \in \mathcal{O}(\mathbb{C})$. If the approximation is close enough, then the shear (4.60) satisfies conditions (4.61). Note that the values of h on the unbounded arc $\gamma_1([t',\infty)) \subset \mathbb{C} \setminus \overline{\mathbb{D}}_M$ are not important since the set A projects to the disc $\mathbb{D}_M$. □

Corollary 4.14.5 *Let K and L be compact polynomially convex sets in $\mathbb{C}^n$ for some $n \ge 2$, and let $\ell = \{\gamma(t) : t \in [0,\infty)$ or $t \in (-\infty,\infty)\}$ be a smooth embedded curve in $\mathbb{C}^n$ such that $K \cap \ell = \varnothing$. Assume that there exists $\alpha \in \mathrm{Aut}(\mathbb{C}^n)$ such that $\alpha(\ell)$ has the nice projection property (see Definition 4.14.2). Given $\epsilon > 0$, there exists $\Phi \in \mathrm{Aut}(\mathbb{C}^n)$ satisfying*

$$\|\Phi - \mathrm{Id}\|_{\mathcal{C}^0(K)} < \epsilon \quad \text{and} \quad \Phi(\ell) \cap L = \varnothing.$$

The same holds if ℓ is a finite union of smooth pairwise disjoint curves.

Proof Assume that $\ell = \{\gamma(t) : t \in \mathbb{R}\}$. Choose numbers $t_1 < t_2$ such that $\gamma([t_1,t_2]) \cap L = \varnothing$. The set $K' = K \cup \gamma([t_1,t_2])$ is polynomially convex by Stolzenberg [530]. It now suffices to apply Lemma 4.14.4 with K' in place of K, and with the two arcs $\ell_1 = \gamma((-\infty,t_1])$ and $\ell_2 = \gamma([t_2,+\infty))$ attached to K'. We deal similarly with the other cases. □

Using Corollary 4.14.5 and the push-out method from Sect. 4.4, we now prove the following result of Wold [581]. This is used in Sect. 9.10 to construct proper holomorphic embeddings of certain bordered Riemann surfaces into $\mathbb{C}^2$.

Theorem 4.14.6 (Wold [581]) *Let $n > 1$. Assume that X is a complex curve in $\mathbb{C}^n$ (possibly with singularities) with smooth boundary $bX = \bigcup_{j=1}^m \ell_j$ consisting of finitely many unbounded curves $\ell_j \cong \mathbb{R}$. If $\alpha(bX)$ has the nice projection property for some $\alpha \in \mathrm{Aut}(\mathbb{C}^n)$, then there exists a Fatou-Bieberbach domain $\Omega \subset \mathbb{C}^n$ such that $X \subset \Omega$ and $bX \subset b\Omega$; furthermore, a Fatou-Bieberbach map $\Phi : \Omega \to \mathbb{C}^n$ can be chosen to be as close as desired to the identity on a given compact set in X. In particular, the restriction $\Phi|_X : X \hookrightarrow \mathbb{C}^n$ is a proper holomorphic embedding of X into $\mathbb{C}^n$.*

Proof We claim that X enjoys the following conditions (1)–(3) which are invariant under composition by holomorphic automorphisms.

(1) *X admits an exhaustion $K_1 \subset K_2 \subset \cdots \subset \bigcup_{j=1}^\infty K_j = X$ by compact polynomially convex sets K_j.*

To see this, it suffices to show that any smoothly bounded compact set $K \subset X$ that is holomorphically convex in X is also polynomially convex in $\mathbb{C}^n$. Since $\widehat{K} = \widehat{bK}$ and bK is a union of smooth curves, the set $A = \widehat{K} \setminus bK$ is an analytic subvariety of $\mathbb{C}^n \setminus bK$ containing $K \setminus bK$ (see Stolzenberg [530]). If $A \neq K \setminus bK$, then A contains a local extension of K in X near a boundary component of K; hence, $\widehat{K}$ contains at least one connected component of $X \setminus K$, a contradiction since each of these components is unbounded in $\mathbb{C}^n$.

A similar argument shows the following.

(2) *For any compact polynomially convex set $K \subset \mathbb{C}^n \setminus bX$, the set $K \cup K_j$ is polynomially convex for all big $j \in \mathbb{N}$.* (In fact, this holds when $K \cap X \subset K_j$.)

(3) *For every compact polynomially convex set L contained in $\mathbb{C}^n \setminus bX$ and for every pair of numbers $\epsilon > 0$ (small) and $R > 0$ (big) there exists $\phi \in \mathrm{Aut}(\mathbb{C}^n)$ such that*

$$\sup_{x \in L} \left|\phi(x) - x\right| < \epsilon \quad \text{and} \quad \phi(bX) \subset \mathbb{C}^2 \setminus R\mathbb{B}.$$

(Here, $\mathbb{B}$ is the unit ball in $\mathbb{C}^n$.) Such ϕ is furnished by Corollary 4.14.5.

Using these conditions, we find a sequence of automorphisms

$$\Phi_j = \phi_j \circ \phi_{j-1} \circ \cdots \circ \phi_1 \in \mathrm{Aut}\big(\mathbb{C}^n\big), \quad j = 1, 2, \ldots \tag{4.62}$$

whose domain of convergence is a Fatou-Bieberbach domain Ω with the stated properties. The induction step is the following. Fix $j \in \mathbb{N}$ and assume that $\Phi_j(bX) \cap j\overline{\mathbb{B}} = \varnothing$. (This trivially holds for $j = 0$ with $\Phi_0 = \mathrm{Id}$.) By condition (2) there exists $m_j \in \mathbb{N}$ such that $L_j = j\overline{\mathbb{B}} \cup \Phi_j(K_{m_j})$ is polynomially convex. Given $\epsilon_j > 0$, condition (3) furnishes $\phi_{j+1} \in \mathrm{Aut}(\mathbb{C}^n)$ such that

- $|\phi_{j+1}(x) - x| < \epsilon_j$ for all $x \in L_j$, and
- $|\phi_{j+1}(x)| > j + 1$ for all $x \in \Phi_j(bX)$.

Setting $\Phi_{j+1} = \phi_{j+1} \circ \Phi_j$ completes the induction step. Suitable choices of sequences $\epsilon_j \to 0$ and $m_j \to +\infty$ ensure that the sequence $\Phi_j \in \mathrm{Aut}(\mathbb{C}^n)$ converges locally uniformly on the domain $\Omega = \bigcup_{j=1}^{\infty} \Phi_j^{-1}(j\overline{\mathbb{B}}) \subset \mathbb{C}^n$ to a biholomorphic map $\Phi\colon \Omega \to \mathbb{C}^n$ onto $\mathbb{C}^n$ (see Corollary 4.4.2 on p. 125). By the construction, we have that $X \subset \Omega$ and $bX \subset b\Omega$. □

We now present without proof a result of Kutzschebauch and Wold [367] on Carleman approximation by holomorphic automorphisms of $\mathbb{C}^n$ on noncompact totally real submanifolds of dimension $< n$ which are flat at infinity.

Denote the coordinates on $\mathbb{C}^n$ by $z = x + \mathrm{i}y$ with $z_j = x_j + \mathrm{i}y_j$. Given $s \in \{1, \ldots, n\}$, we identify $\mathbb{R}^s$ with the real subspace of $\mathbb{C}^n$ given by

$$\mathbb{R}^s = \left\{x + \mathrm{i}y \in \mathbb{C}^n : x_j = 0 \text{ for } j = s+1, \ldots, n;\ \ y = 0\right\}.$$

Theorem 4.14.7 ([367, Theorem 1.1]) *Let $1 \le s < n$. Assume that $K \subset \mathbb{C}^n$ is a compact polynomially convex set, $\Omega \subset \mathbb{C}^n$ is an open set containing K, and $\phi_t : \Omega \cup \mathbb{R}^s \to \mathbb{C}^n$ $(t \in [0,1])$ is an isotopy of smooth embeddings, with $\phi_0 = \mathrm{Id}$, satisfying the following conditions for all $t \in [0,1]$:*

(1) *the map $\phi_t|_\Omega$ is holomorphic,*
(2) *the set $\phi_t(K \cup \mathbb{R}^s)$ is polynomially convex, and*
(3) *there is a compact set $C \subset \mathbb{R}^s$ (independent of t) such that $\phi_t|_{\mathbb{R}^s \setminus C} = \mathrm{Id}$.*

Then, for any $k \in \mathbb{N}$ the map ϕ_1 can be approximated in the fine $\mathscr{C}^k$-topology on $K \cup \mathbb{R}^s$ by holomorphic automorphisms of $\mathbb{C}^n$.

The authors show in addition that if the set $\phi_t(K)$ is polynomially convex for all $t \in [0,1]$, then there are arbitrarily small smooth perturbations of ϕ_t achieving condition (2) in the theorem (see [367, Proposition 4.1]). Furthermore, condition (3) can be relaxed to an approximate condition.

Corollary 4.14.8 ([367, Theorem 1.2]) *Let $\phi\colon \mathbb{R}^s \to \mathbb{R}^s$ be a smooth diffeomorphism, and assume that $s < n$. Then, ϕ can be approximated in the fine Whitney topology by holomorphic automorphisms of $\mathbb{C}^n$.*

4.15 Automorphisms with Given Jets

In this section we prove a jet interpolation theorem for automorphisms of $\mathbb{C}^n$ at finitely many points; see Proposition 4.15.3. This is a preparation for the Mittag-Leffler type interpolation theorems obtained in the following section. For interpolation theorems on more general affine algebraic varieties, we refer to the papers [38, 559].

Definition 4.15.1 Let $m, n \in \mathbb{N}$.

(i) An *A-jet of order m* at $0 \in \mathbb{C}^n$ is a holomorphic polynomial $P\colon \mathbb{C}^n \to \mathbb{C}^n$ of degree $\le m$ such that $P(0) = 0$ and $JP(0) = \det P'(0) \ne 0$.
(ii) P as in (i) is an *A_1-jet of order m* if $JP(z) = 1 + O(|z|^m)$ as $z \to 0$.

Given an open set $U \subset \mathbb{C}^n$, a holomorphic map $F\colon U \to \mathbb{C}^n$, a point $a \in U$, and an integer $m \in \mathbb{N}$, we denote by $F_{m,a}$ the Taylor polynomial of F of order m at a without the constant term:

$$F(z) = F(a) + F_{m,a}(z-a) + O\big(|z-a|^{m+1}\big). \tag{4.63}$$

If F is nondegenerate at a (i.e., $JF(a) \neq 0$), then $F_{m,a}$ is an A-jet, called *the A-jet of F of order m at a*. If $JF \equiv 1$, then $F_{m,a}$ is an A_1-jet for each $a \in U$. The following lemma is evident by composing the power series.

Lemma 4.15.2 *If F and G are holomorphic maps defined on open subsets of $\mathbb{C}^n$, with values in $\mathbb{C}^n$, then for each integer $m \in \mathbb{N}$ we have*

$$(G \circ F)_{m,a}(z) = G_{m,F(a)} \circ F_{m,a}(z) + O\big(|z|^{m+1}\big), \quad z \to 0$$

at each point a in the domain of $G \circ F$. If F is locally invertible at a, then

$$\big(F^{-1}\big)_{m,F(a)} \circ F_{m,a}(z) = z + O\big(|z|^{m+1}\big), \quad z \to 0.$$

The following is the main result of this section (see [25, Proposition 6.2] and [188, Proposition 2.1]). In [400] the authors obtained the analogous jet-interpolation theorem for symplectic holomorphic automorphisms of $\mathbb{C}^{2n}$.

Proposition 4.15.3 *Let $m, n, N \in \mathbb{N}$, $n > 1$. Assume that*

(a) *K is a compact polynomially convex set in $\mathbb{C}^n$,*
(b) *$\{a_j\}_{j=1}^s$ is a finite set of points in K,*
(c) *p and q are points in $\mathbb{C}^n \setminus K$, and*
(d) *$P\colon \mathbb{C}^n \to \mathbb{C}^n$ is an A-jet of order m.*

Given $\epsilon > 0$, there exists $\Phi \in \mathrm{Aut}(\mathbb{C}^n)$ satisfying the following conditions:

(i) *$\Phi(p) = q$ and $\Phi(z) = q + P(z-p) + O(|z-p|^{m+1})$ as $z \to p$,*
(ii) *$\Phi(z) = z + O(|z-a_j|^N)$ as $z \to a_j$ for each $j = 1, 2, \ldots, s$, and*
(iii) *$|\Phi(z) - z| + |\Phi^{-1}(z) - z| < \epsilon$ for each $z \in K$.*

If P is an A_1-jet of order m, then there exists a polynomial automorphism Φ with $J\Phi \equiv 1$ satisfying conditions (i)–(iii).

Proof Set $K_0 = K$. Choose compact polynomially convex sets $K_1 \subset K_2 \subset K_3 \subset \mathbb{C}^n \setminus \{p, q\}$ and a number $\epsilon_0 > 0$ such that

$$K_0 \subset K_1, \quad \mathrm{dist}\big(K_j, \mathbb{C}^n \setminus K_{j+1}\big) > \epsilon_0, \quad j = 0, 1, 2.$$

Choose a number $\epsilon \in (0, \epsilon_0)$. Condition (iii) holds if

$$\big|\Phi(z) - z\big| < \frac{\epsilon}{2}, \quad z \in K_1. \tag{4.64}$$

Indeed, if (4.64) holds then Rouché's theorem [98, p. 110] implies that $\Phi(K_1) \supset K$, and hence $\Phi^{-1}(z) \in K_1$ for each $z \in K_0$. Setting $w = \Phi^{-1}(z) \in K_1$, we have $|\Phi^{-1}(z) - z| = |w - \Phi(w)| < \frac{\epsilon}{2}$ by (4.64), and hence (iii) holds.

We seek $\Phi \in \mathrm{Aut}(\mathbb{C}^n)$ of the form

$$\Phi = H^{-1} \circ S \circ G \circ H. \tag{4.65}$$

Each of the automorphisms G, H, H^{-1}, and S will move points of K_2 by less than $\frac{\epsilon}{8}$; clearly this will imply (4.64).

Choose a ball B containing the set K_3, and then choose points $p', q' \in \mathbb{C}^n$ belonging to an affine complex hyperplane Σ that does not intersect $\bar{B}$. (If $p = q$, we choose $p' = q'$.) By Corollary 4.12.7 there exists a holomorphic automorphism $H \in \mathrm{Aut}(\mathbb{C}^n)$ satisfying $H(p) = p'$, $H(q) = q'$, and

$$\left|H(z) - z\right| + \left|H^{-1}(z) - z\right| < \frac{\epsilon}{8}, \quad z \in K_2.$$

Set $b_j = H(a_j)$ for $j = 1, \ldots, s$. Next, we construct a polynomial additive shear G satisfying $|G(z) - z| < \frac{\epsilon}{8}$ for all $z \in K_2$, $G(p') = q'$, and

$$G(z) = z + O\big(|z - b_j|^{N+1}\big) \quad \text{as } z \to b_j, \quad j = 1, \ldots, s.$$

To find G, choose a linear form $\lambda \colon \mathbb{C}^n \to \mathbb{C}$ that is constant on Σ; hence $\lambda(p') = \lambda(q') \notin \lambda(\bar{B})$. Choose a holomorphic polynomial $f \colon \mathbb{C} \to \mathbb{C}$ such that $f(\lambda(p')) = 1$, f vanishes to order N at each of the points $\lambda(b_j)$ $(j = 1, \ldots, s)$, and $|f(\zeta)(q' - p')| < \frac{\epsilon}{8}$ for each $\zeta \in \lambda(K_2)$. The shear

$$G(z) = z + f\big(\lambda(z)\big)\big(q' - p'\big)$$

then satisfies the required conditions.

It remains to find $S \in \mathrm{Aut}(\mathbb{C}^n)$ satisfying the following conditions:

(i) $S(q') = q'$,
(ii) $S(z) = z + O(|z - b_j|^{N+1})$ as $z \to b_j$ for $j = 1, \ldots, s$,
(iii) $|S(z) - z| < \frac{\epsilon}{8}$ for all $z \in K_2$, and
(iv) the A-jet $S_{m,q'}$ of S at the point q' is determined by

$$S_{m,q'}(z) = H_{m,q} \circ P \circ \big(H^{-1}\big)_{m,p'} \circ \big(G^{-1}\big)_{m,q'}(z) + O\big(|z|^{m+1}\big).$$

The automorphism Φ (4.65) then clearly satisfies (ii) and (iii) in Proposition 4.15.3; condition (i) follows by Lemma 4.15.2 and the choice of $S_{m,q'}$.

We may assume that $q' = 0$ and that Σ is a complex hyperplane through the origin. Choose $\eta > 0$ (to be fixed later). We seek S as a composition $S = S_m \circ S_{m-1} \circ \cdots \circ S_1$ of automorphisms $S_k \in \mathrm{Aut}(\mathbb{C}^n)$ satisfying the following conditions for $k = 1, \ldots, m$:

$$\begin{aligned} &S_k(0) = 0, \\ &S_k(z) = z + O\big(|z - b_j|^{N+1}\big), \quad z \to b_j,\ j = 1, \ldots, s, \\ &\big|S_k(z) - z\big| < \eta, \quad z \in K_3. \end{aligned} \tag{4.66}$$

Let Q be the A-jet $S_{m,q'}$ in condition (iv) above. Write $Q(z) = Q_1(z) + O(|z|^2)$ as $z \to 0$. For $k = 1$, we find an automorphism $S_1 \in \mathrm{Aut}(\mathbb{C}^n)$ such that $S_1(z) = Q_1(z) + O(|z|^2)$ as $z \to 0$. Then,

$$Q \circ S_1^{-1}(z) = z + Q_2(z) + O\big(|z|^3\big), \quad z \to 0,$$

where $Q_2 \colon \mathbb{C}^n \to \mathbb{C}^n$ is homogeneous of degree 2. Next, we find $S_2 \in \mathrm{Aut}(\mathbb{C}^n)$ satisfying (4.66) and $S_2(z) = z + Q_2(z) + O(|z|^3)$ as $z \to 0$. Then,

$$Q \circ S_1^{-1} \circ S_2^{-1}(z) = z + Q_3(z) + O\big(|z|^4\big), \quad z \to 0,$$

where Q_3 is homogeneous of degree 3. Continuing inductively, we obtain in m steps an automorphism $S = S_m \circ S_{m-1} \circ \cdots \circ S_1$ satisfying conditions (i)–(iv), provided that $\eta > 0$ was chosen small enough.

Construction of S_1*:* Choose $\mathbb{C}$-linearly independent forms $\lambda_1, \ldots, \lambda_n \in (\mathbb{C}^n)^*$ satisfying $0 \notin \lambda_j(\bar{B})$ for $j = 1, 2, \ldots, n$ (this holds if $\ker \lambda_j$ is close to Σ for every j), and then choose vectors $e_1, \ldots, e_n$ of $\mathbb{C}^n$ such that $\lambda_j(e_k) = \delta_{j,k}$. The group $SL_n(\mathbb{C})$ is generated by linear shears of the form

$$z \mapsto z + \alpha \lambda_j(z)\, e_k, \qquad 1 \le j \ne k \le n, \quad \alpha \in \mathbb{C}.$$

In order to satisfy (4.66), we interpolate each of these shears at the origin by a polynomial shear of the form $z \mapsto z + f(\lambda_j(z))\, e_k$, where f is a polynomial on $\mathbb{C}$ such that $f(\zeta) = \alpha\zeta + O(|\zeta|^2)$ as $\zeta \to 0$, f vanishes to order N at all points $\lambda_j(b_l)$ for $l = 1, 2, \ldots, s$, and $|f|$ is small on the set $\lambda_j(\bar{B}) \subset \mathbb{C} \setminus \{0\}$. A suitable composition of such shears will give a desired S_1 in the special case when $S_1'(0) = Q_1 \in SL_n(\mathbb{C})$.

In the general case when $Q_1 \in GL_n(\mathbb{C})$ we let S_0 be a multiplicative shear

$$S_0(z) = z + \big(\mathrm{e}^{f(\lambda_1(z))} - 1\big)\langle z, e_2\rangle\, e_2,$$

where $f \colon \mathbb{C} \to \mathbb{C}$ is a polynomial that vanishes to order N at all points $\lambda_1(b_l)$ $(l = 1, 2, \ldots, s)$, $|f|$ is small on $\lambda_1(\bar{B})$, and $\mathrm{e}^{f(0)} = \mathrm{J}Q(0)$. Since $\mathrm{J}S_0(0) = \mathrm{e}^{f(0)}$, we have $\mathrm{J}S_0(0) = \mathrm{J}Q(0)$. The map $Q \circ S_0^{-1}$ has Jacobian one at the origin and we are back in the previous case. Choosing $\widetilde{S}_1$ as in the special case above such that its derivative at $0 \in \mathbb{C}^n$ matches that of $Q \circ S_0^{-1}$, the composition $S_1 = \widetilde{S}_1 \circ S_0$ satisfies all the required conditions.

The induction step. Suppose that $2 \le k \le m$ and that we have already found automorphisms $S_1, \ldots, S_{k-1}$ satisfying (4.66) and such that

$$Q \circ S_1^{-1} \circ \cdots \circ S_{k-1}^{-1}(z) = z + V_z + O\big(|z|^{k+1}\big), \quad z \to 0 \tag{4.67}$$

for a homogeneous polynomial map $V\colon \mathbb{C}^n \to \mathbb{C}^n$ of degree k. We now construct the next automorphism S_k of $\mathbb{C}^n$ satisfying (4.66) and such that

$$S_k(z) = z + V_z + O\big(|z|^{k+1}\big), \quad z \to 0. \tag{4.68}$$

From (4.67) and (4.68) it follows that

$$Q \circ S_1^{-1} \circ \cdots \circ S_k^{-1}(z) = z + O\big(|z|^{k+1}\big), \quad z \to 0.$$

This will conclude the induction step in the construction of S.

To find S_k, we write V in the form (4.45):

$$V_z = \sum_{j=1}^{r} c_j \lambda_j(z)^k\, v_j + d_j \lambda_j(z)^{k-1} \langle z, v_j\rangle\, v_j$$

for some constants $c_j, d_j \in \mathbb{C}$, where the linear forms $\lambda_j \in (\mathbb{C}^n)^*$ are chosen such that $0 \notin \lambda_j(\bar{B})$. If $\operatorname{div} V = 0$, then the above holds with $d_j = 0$ for all j. Choose polynomials $f_j, g_j\colon \mathbb{C} \to \mathbb{C}$ $(j = 1, \ldots, r)$ that are close to 0 on $\lambda_j(\bar{B})$, that vanish to order N at points $\lambda_j(b_l)$ $(l = 1, \ldots, s)$, and that satisfy

$$f_j(\zeta) = c_j \zeta^k + O\big(|\zeta|^{k+1}\big), \qquad g_j(\zeta) = d_j \zeta^{k-1} + O\big(|\zeta|^k\big), \quad \zeta \to 0.$$

Define the automorphisms $\Phi_j, \Psi_j \in \operatorname{Aut}(\mathbb{C}^n)$ by

$$\begin{aligned} \Phi_j(z) &= z + f_j\big(\lambda_j(z)\big)\, v_j = z + c_j \lambda_j(z)^k\, v_j + O\big(|z|^{k+1}\big), \\ \Psi_j(z) &= z + \big(e^{g_j(\lambda_j(z))} - 1\big)\langle z, v_j\rangle\, v_j \\ &= z + d_j\big(\lambda_j(z)\big)^{k-1} \langle z, v_j\rangle\, v_j + O\big(|z|^{k+1}\big). \end{aligned}$$

These are the time-one maps of $f_j(\lambda_j z) v_j$, resp. of $g_j(\lambda_j(z))\langle z, v_j\rangle\, v_j$. Take S_k to be the composition of all maps Φ_j and Ψ_j (in any order). The Taylor expansion of each of these maps begins with z+homogeneous terms of order k. When composing such maps, their homogeneous parts of degree k add up and we get (4.68). This completes the induction step.

Since each of the m automorphism S_k moves points of $K_3 \subset \bar{B}$ by less than η, we can achieve (by choosing $\eta > 0$ sufficiently small) that their composition S moves points of K_2 by less than $\frac{\epsilon}{8}$ as required by (iii). This completes the proof of Proposition 4.15.3 when P is an A-jet.

Suppose now that P is an A_1-jet, $JP(z) = 1 + O(|z|^m)$. Recall that H and G are finite compositions of polynomial additive shears. The A-jet $Q = S_{m,q'}$, being conjugate to the A_1-jet P by a volume preserving automorphism, is itself an A_1-jet of order m. Thus $\mathrm{J}Q(0)=1$, and hence S_1 constructed above is volume preserving. Assuming inductively that map S_l is volume preserving for $l=1,\dots,k-1$ for some $k\in\{2,\dots,m\}$, it follows from chain rule that the Jacobian of the map in (4.67) agrees with 1 to order m at the origin. This implies

$$1+O\big(|z|^m\big)=\det\big(I+V'_z+O\big(|z|^k\big)\big)=1+(\operatorname{div}V)(z)+O\big(|z|^k\big).$$

Since $\operatorname{div}V$ is a homogeneous polynomial of degree $k-1$, we get $\operatorname{div}V=0$. Consequently, V is a sum of vector fields whose time-one maps are additive shears (4.1). This shows that each S_k for $k=1,2,\dots,m$ is a finite composition of additive shears, and hence the same is true for the maps S and Φ. □

Remark 4.15.4 It is easy to obtain the following improvement of Proposition 4.15.3. Let $\Psi\in\mathrm{Aut}(\mathbb{C}^n)$ be such that the set $K\cup\Psi(K)$ does not contain the points p and q. Then, there exists $\Phi\in\mathrm{Aut}(\mathbb{C}^n)$ satisfying

(i′) $\Phi(z)=q+P(z-p)+O(|z-p|^{m+1})$ as $z\to p$,
(ii′) $\Phi(z)=\Psi(z)+O(|z-a_j|^N)$ as $z\to a_j$ for each $j=1,2,\dots,s$, and
(iii′) $|\Phi(z)-\Psi(z)|<\epsilon$ for each $z\in K$, and $|\Phi^{-1}(z)-\Psi^{-1}(z)|<\epsilon$ for each $z\in\Psi(K)$.

We find such Φ as follows. Choose a compact set K_1 containing K in its interior and such that $p,q\notin K_1\cup\Psi(K_1)$. Let $p'=\Psi(p)$, $q'=\psi(q)$. Set $\Phi=G\circ H\circ\Psi$, where $G,H\in\mathrm{Aut}(\mathbb{C}^n)$ are chosen such that

- H is close to the identity on $\Psi(K_1)$, it interpolates the identity map to order N at each of the points $a'_j=\Psi(a_j)$, and $H(p')=p$, $H(q')=q$;
- G is close to the identity on $\Psi(K_1)$, it interpolates the identity map to order N at each of the points a'_j, $G(p)=q$, and its A-jet $G_{m,p}$ satisfies $P=G_{m,p}\circ H_{m,p'}\circ\Psi_{m,p}$.

Automorphisms G and H with these properties exist by Proposition 4.15.3. Similarly, given r-tuples $\{p_1,\dots,p_r\}$ and $\{q_1,\dots,q_r\}$ of distinct points in the complement of $K\cup\Psi(K)$, we can find $\Phi\in\mathrm{Aut}(\mathbb{C}^n)$ such that $\Phi(p_j)=q_j$ and the interpolation condition (i) above holds for each $j=1,\dots,r$ (with the polynomial $P=P_j$ depending on j). □

The following result is a corollary to Proposition 4.15.3; a special case was proved by Buzzard [73]. In Sect. 4.16 we prove analogous results for certain countable discrete sets $\{a_j\}$ and $\{b_j\}$ in $\mathbb{C}^n$.

Corollary 4.15.5 *Let $n>1$. Given finite subsets $\{a_j\}_{j=1}^r$ and $\{b_j\}_{j=1}^r$ of $\mathbb{C}^n$ and for each $j=1,\dots,r$ a holomorphic polynomial map $P_j\colon\mathbb{C}^n\to\mathbb{C}^n$ of degree $m_j\in\mathbb{N}$ satisfying $P_j(0)=0$ and $\mathrm{J}P_j(0)\neq 0$, there exists $\Phi\in\mathrm{Aut}(\mathbb{C}^n)$ such that for each $j=1,2,\dots,r$ we have*

$$\Phi(z)=b_j+P_j(z-a_j)+O\big(|z-a_j|^{m_j+1}\big),\quad z\to a_j.$$

If $JP_j(z) = 1 + O(|z|^{m_j})$ *as* $z \to 0$ *for each* j*, then we may choose* Φ *to be a polynomial automorphism with Jacobian one. In addition,* Φ *can be chosen as close as desired to the identity map on a given compact polynomially convex set* $K \subset \mathbb{C}^n$ *that does not contain any of the points* a_j *and* b_j*.*

Furthermore, if for every $j = 1, \ldots, r$ *the difference* $|a_j - b_j|$ *is small and* P_j *is close to the zero polynomial, then* Φ *can be chosen close to the identity map on a fixed compact set in* $\mathbb{C}^n$*.*

Proof Proposition 4.15.3 gives for each $j = 1, 2, \ldots, r$ an automorphism Φ_j of $\mathbb{C}^n$ satisfying the following conditions:

$$\Phi_j(z) = b_j + P_j(z - a_j) + O\big(|z - a_j|^{m_j+1}\big), \quad z \to a_j,$$

$$\Phi_j(z) = z + O\big(|z - b_k|^{m_k+1}\big), \quad z \to b_k, \ 1 \le k \le j-1,$$

$$\Phi_j(z) = z + O\big(|z - a_k|^{m_k+1}\big), \quad z \to a_k, \ j+1 \le k \le r.$$

The composition $\Phi = \Phi_r \circ \Phi_{r-1} \circ \cdots \circ \Phi_1$ then satisfies Corollary 4.15.5. The additional claims follow from the proof. □

4.16 Mittag-Leffler Theorem for Automorphisms of $\mathbb{C}^n$

In this section we construct automorphisms mapping tame discrete sequences of $\mathbb{C}^n$ one onto another with a prescribed finite jet at each point, thereby solving a general *Mittag-Leffler interpolation problem* for automorphisms.

Theorem 4.16.1 ([78, Theorem 1.1]) *Assume that* $n > 1$*, that* a_j *and* b_j $(j \in \mathbb{N})$ *are tame sequences in* $\mathbb{C}^n$ *without repetitions, and that* $P_j : \mathbb{C}^n \to \mathbb{C}^n$ *is an A-jet of order* $m_j \in \mathbb{N}$ *(see Definition* 4.15.1*). Then there exists* $F \in \mathrm{Aut}(\mathbb{C}^n)$ *such that for every* $j = 1, 2, \ldots$ *we have* $F(a_j) = b_j$ *and*

$$F(z) = b_j + P_j(z - a_j) + O\big(|z - a_j|^{m_j+1}\big), \quad z \to a_j. \tag{4.69}$$

If in addition every P_j *is an* A_1*-jet of order* m_j *and the sequences* a_j *and* b_j *are very tame, then there exists an* $F \in \mathrm{Aut}_1(\mathbb{C}^n)$ *with these properties.*

If $a_j = b_j = je_1$ for all $j \in \mathbb{N}$ and the degrees of the polynomials P_j are uniformly bounded, one obtains a desired automorphism F as a finite composition of shears (see [188, Theorem 3.1]). The restriction to tame sequences is justified in view of the results in Sect. 4.5. Under certain circumstances, the above interpolation result holds while at the same time approximating a given automorphism on a polynomially convex set as in the next theorem.

Theorem 4.16.2 ([77, Theorem 1.2]) *In addition to the hypotheses of Theorem* 4.16.1, *assume that* $\Phi \in \mathrm{Aut}(\mathbb{C}^n)$ *and* $K \subset \mathbb{C}^n \setminus \{a_j\}_{j=1}^\infty$ *is a compact polynomially convex set such that* $\Phi(K) \subset \mathbb{C}^n \setminus \{b_j\}_{j=1}^\infty$. *Given* $\epsilon > 0$, *there exists* $F \in \mathrm{Aut}(\mathbb{C}^n)$ *satisfying Theorem* 4.16.1 *and also*

$$\left|F(z) - \Phi(z)\right| < \epsilon \ (\forall z \in K), \quad \left|F^{-1}(w) - \Phi^{-1}(w)\right| < \epsilon \ \big(\forall w \in \Phi(K)\big). \tag{4.70}$$

If $\Phi \in \mathrm{Aut}_1(\mathbb{C}^n)$ *and the volume preserving assumptions in Theorem* 4.16.1 *hold, we may choose* $F \in \mathrm{Aut}_1(\mathbb{C}^n)$.

Corollary 4.16.3 *Let* $n > 1$. *For every tame discrete set* $\{a_j\} \subset \mathbb{C}^n$ *there exists* $\Phi \in \mathrm{Aut}(\mathbb{C}^n)$ *with an attracting fixed point at each* a_j.

A parametric version of Theorem 4.16.1, with the parameter space a Stein manifold, was proved by Ugolini [556]; we mention this generalization at the end of the section (see Theorem 4.16.8).

The following lemma is the key ingredient in Theorems 4.16.1 and 4.16.2.

Lemma 4.16.4 *Given* P_j *and* m_j *as in Theorem* 4.16.1, *there exist a discrete sequence* $\{c_j\}$ *contained in the* z_1*-axis and an automorphism* $F \in \mathrm{Aut}(\mathbb{C}^n)$ *such that for all* $j \in \mathbb{N}$ *we have*

$$F(z) = c_j + P_j(z - c_j) + O\big(|z - c_j|^{m_j+1}\big), \quad z \to c_j. \tag{4.71}$$

If each P_j *is an* A_1*-jet, there exists an* $F \in \mathrm{Aut}_1(\mathbb{C}^n)$ *with this property.*

Proof Let $\epsilon_j = 2^{-j}$ for $j \in \mathbb{N}$. We construct the sequence $\{c_j\}$ inductively and find F as the limit of a sequence of compositions of automorphisms.

Let $K_0 = \varnothing$, $K_1 = 2\overline{\mathbb{B}}$, and $c_1 = 3e_1$ where $e_1 = (1, 0, \ldots, 0)$. By Proposition 4.15.3, applied to $K = K_1$, there exists $\Psi_1 \in \mathrm{Aut}(\mathbb{C}^n)$ such that

$$\Psi_1(z) = c_1 + P_1(z - c_1) + O\big(|z - c_1|^{m_1+1}\big), \quad z \to c_1,$$
$$\left|\Psi_1(z) - z\right| + \left|\Psi_1^{-1}(z) - z\right| < \epsilon_1, \quad z \in K_1.$$

If P_1 is an A_1-jet, there exists a $\Psi_1 \in \mathrm{Aut}_1(\mathbb{C}^n)$ with these properties.

Let $F_0(z) = z$ and $F_1 = \Psi_1$. We shall inductively choose automorphisms $F_k = \Psi_k \circ \cdots \circ \Psi_1$. For the induction, suppose that we have the following:

1. Compact convex sets $K_0 \subset K_1 \subset \cdots \subset K_k$ with $j\mathbb{B} \cup F_{j-1}(j\mathbb{B}) \subset K_j$ and

$$\mathrm{dist}\big(K_{j-1} \cup F_{j-1}(j\mathbb{B}), \mathbb{C}^n \setminus K_j\big) > \epsilon_j, \quad j = 1, \ldots, k.$$

2. Points $c_j \in K_{j+1} \setminus K_j$ for $j = 1, \ldots, k-1$, and $c_k \in \mathbb{C}^n \setminus K_k$, such that each c_j is contained in the z_1-axis.

3. Automorphisms $\Psi_j \in \mathrm{Aut}(\mathbb{C}^n)$ for $j = 1, \ldots, k$ satisfying

$$\left|\Psi_j(z) - z\right| + \left|\Psi_j^{-1}(z) - z\right| < \epsilon_j, \quad z \in K_j,$$

with $\Psi_j \in \mathrm{Aut}_1(\mathbb{C}^n)$ if P_j is an A_1-jet.
4. Automorphisms $F_k = \Psi_k \circ \Psi_{k-1} \circ \cdots \circ \Psi_1$ satisfying (4.71) for $j = 1, \ldots, k$.

Let K_{k+1} be a compact convex set in $\mathbb{C}^n$ such that

$$(k+1)\mathbb{B} \cup K_k \cup F_k\big((k+1)\mathbb{B}\big) \cup \{c_k\} \subset K_{k+1},$$
$$\mathrm{dist}\big(K_k \cup F_k\big((k+1)\mathbb{B}\big), \mathbb{C}^n \setminus K_{k+1}\big) > \epsilon_{k+1}.$$

Choose a point c_{k+1} in the z_1-axis such that $c_{k+1}, F_k(c_{k+1}) \in \mathbb{C}^n \setminus K_{k+1}$. Let $N = \max\{m_1, \ldots, m_k\} + 1$, $d_{k+1} = F_k(c_{k+1})$, and $Q_{k+1} = P_{k+1} \circ (F_k^{-1})_{m_{k+1}, d_k}$. Proposition 4.15.3 furnishes $\Psi_{k+1} \in \mathrm{Aut}(\mathbb{C}^n)$ (resp. $\Psi_{k+1} \in \mathrm{Aut}_1(\mathbb{C}^n)$ if P_{k+1} is an A_1-jet and $F_k \in \mathrm{Aut}_1(\mathbb{C}^n)$) such that the following conditions hold:

(i) $\Psi_{k+1}(z) = c_{k+1} + Q_{k+1}(z - d_{k+1}) + O(|z - d_{k+1}|^{m_{k+1}+1})$ as $z \to d_{k+1}$,
(ii) $\Psi_{k+1}(z) = z + O(|z - c_j|^N)$ as $z \to c_j$ for $j = 1, \ldots, k$, and
(iii) $|\Psi_{k+1}(z) - z| + |\Psi_{k+1}^{-1}(z) - z| < \epsilon_{k+1}$ for each $z \in K_{k+1}$.

Taking $F_{k+1} = \Psi_{k+1} \circ F_k$, we obtain the induction hypotheses at stage $k+1$, and hence the induction proceeds.

By Proposition 4.4.1 (p. 124) the sequence F_j converges uniformly on compacts in $\Omega = \bigcup_{j=1}^{\infty} F_j^{-1}(K_j)$ to a biholomorphic map F from Ω onto the domain $D = \bigcup_{j=1}^{\infty} K_j$. Since $j\mathbb{B} \subset K_j$, we see that $D = \mathbb{C}^n$. Moreover, we have $F_j^{-1}(K_j) = F_{j-1}^{-1}\Psi_j^{-1}(K_j)$. By Rouché's theorem [98, p. 110] and the induction hypotheses (1) and (3) we see that $F_{j-1}(j\mathbb{B}) \subset \Psi_j^{-1}(K_j)$. Hence, $j\mathbb{B} \subset F_j^{-1}(K_j)$, so $\Omega = \mathbb{C}^n$. Thus, $F \in \mathrm{Aut}(\mathbb{C}^n)$, and F satisfies (4.71) for all $j \in \mathbb{N}$ since each F_k satisfies this condition for $j = 1, \ldots, k$. Finally, if all P_j are A_1-jets, then $\Psi_j \in \mathrm{Aut}_1(\mathbb{C}^n)$ for each j, and hence $F \in \mathrm{Aut}_1(\mathbb{C}^n)$. □

Remark 4.16.5 Given numbers $R > 0$ and $\epsilon \in (0, 1)$, we can replace the set K_1 by $(R+1)\overline{\mathbb{B}}$, the point c_1 by $(R+2)e_1$, and ϵ_j by $\epsilon\epsilon_j/2$, to construct an F satisfying Lemma 4.16.4 and also $|F(z) - z| + |F^{-1}(z) - z| < \epsilon$ on $R\mathbb{B}$. □

We now use a classical one-variable interpolation result to find an automorphism fixing each point je_1 and having a prescribed jet at this point.

Lemma 4.16.6 *Let P_j and m_j be as in Theorem* 4.16.1. *For each $R > 0$ and $\epsilon > 0$ there exists an automorphism $F \in \mathrm{Aut}(\mathbb{C}^n)$ such that*

$$F(z) = je_1 + P_j(z - je_1) + O\big(|z - je_1|^{m_j+1}\big), \quad z \to je_1, \ j > R,$$
$$\left|F(z) - z\right| + \left|F^{-1}(z) - z\right| < \epsilon, \quad |z| \le R.$$

If all P_j's are A_1-jets then we can choose $F \in \mathrm{Aut}_1(\mathbb{C}^n)$.

Proof We shall first prove the lemma without the approximation condition. By Lemma 4.16.4 there exist an automorphism $G \in \mathrm{Aut}(\mathbb{C}^n)$ and a sequence of points $\{c_j\} \subset \mathbb{C}^n$ in the z_1-axis such that

$$G(z) = c_j + P_j(z - c_j) + O\big(|z - c_j|^{m_j+1}\big), \quad z \to c_j.$$

(We can take $G \in \mathrm{Aut}_1(\mathbb{C}^n)$ if each P_j is an A_1-jet.) We must find an automorphism $\Psi \in \mathrm{Aut}_1(\mathbb{C}^n)$ mapping the point je_1 to the point c_j for each $j \in \mathbb{N}$, with $\Psi_{m_j, je_1}(z) = z - je_1$; Lemma 4.15.2 (p. 170) then shows that the automorphism $\Psi^{-1} \circ G \circ \Psi$ has the desired properties. To find such Ψ, let $\xi_j \in \mathbb{C}$ be such that $c_j = \xi_j e_1$. By a standard one-variable interpolation theorem (see [300, Corollary 1.5.4]) there exists an entire function f_1 on $\mathbb{C}$ satisfying

$$f_1(\zeta) = j + O\big(|\zeta - j|^{m_j+1}\big) \quad \text{as } \zeta \to j,\ \forall j \in \mathbb{N}.$$

Let $\Psi_1(z) = z + f_1(z_1)e_2$, where $e_2 = (0, 1, 0, \ldots, 0)$. Then $\Psi_1 \in \mathrm{Aut}_1(\mathbb{C}^n)$ and

$$\Psi_1(z) = j(e_1 + e_2) + (z - je_1) + O\big(|z - je_1|^{m_j+1}\big), \quad z \to je_1.$$

Thus Ψ_1 maps the point je_1 to the point $j(e_1 + e_2)$, and it agrees with a translation to order $m_j + 1$ at je_1.

Likewise, choosing f_2 entire with $f_2(\zeta) = \xi_j - j + O(|\zeta - j|^{m_j+1})$ as $\zeta \to j$, and taking $\Psi_2(z) = z + f_2(z_2)e_1$, we see that Ψ_2 maps the point $j(e_1 + e_2)$ to the point $\xi_j e_1 + je_2$, and it agrees with a translation to order $m_j + 1$ at $j(e_1 + e_2)$. Similarly, we can find $\Psi_3(z) = z_1 + f_3(z_1)e_2$ that maps $\xi_j e_1 + je_2$ to $\xi_j e_1 = c_j$ and that agrees with a translation to order $m_j + 1$ at $\xi_j e_1 + je_2$.

Let $\Psi = \Psi_3 \circ \Psi_2 \circ \Psi_1$. Then

$$\Psi(z) = c_j + (z - je_1) + O\big(|z - je_1|^{m_j+1}\big), \quad z \to je_1,$$

and by Lemma 4.15.2 we see that

$$\Psi^{-1}(z) = je_1 + (z - c_j) + O\big(|z - c_j|^{m_j+1}\big), \quad z \to c_j.$$

Also, $\Psi \in \mathrm{Aut}_1(\mathbb{C}^n)$ since each Ψ_l has Jacobian one. By Lemma 4.15.2, the automorphism $F = \Psi^{-1} \circ G \circ \Psi$ satisfies Lemma 4.16.6.

The proof also shows that the sequence $\{je_1\}$ could be replaced by any discrete sequence $\{d_j\}$ without repetition contained in the z_1-axis. Moreover, to get the approximation condition on F, suppose that $\{d_j\}$ and $\{c_j\}$ lie outside the ball $R\overline{\mathbb{B}}$ (in the z_1-axis). A simple argument (using Runge's and Weierstrass's theorem) shows that we can choose each f_l as above and such that $|f_l|$ is small on a neighborhood of the closed disc of radius R in $\mathbb{C}$. Hence, by the remark after the proof of Lemma 4.16.4, we see that we may choose F satisfying the conclusions of Lemma 4.16.6 with d_j in place of je_1 and such that $|F(z) - z| + |F^{-1}(z) - z| < \epsilon$ for $z \in R\mathbb{B}$. □

Proof of Theorem 4.16.1 Since $\{a_j\}$ and $\{b_j\}$ are tame sequences, there exist $H_1, H_2 \in \mathrm{Aut}(\mathbb{C}^n)$ such that $H_1(a_j) = H_2(b_j) = je_1$; if the sequences are very tame, this holds with $H_1, H_2 \in \mathrm{Aut}_1(\mathbb{C}^n)$. Hence, it suffices to find $G \in \mathrm{Aut}(\mathbb{C}^n)$ (resp. $G \in \mathrm{Aut}_1(\mathbb{C}^n)$) such that $G(je_1) = je_1$ and $(H_2^{-1}GH_1)_{m_j,a_j} = P_j$. By Lemma 4.15.2, the latter condition can be satisfied by making

$$G_{m_j,je_1}(z) = (H_2)_{m_j,b_j} \circ P_j \circ \left(H_1^{-1}\right)_{m_j,je_1}(z) + O\left(|z|^{m_j+1}\right), \quad z \to 0$$

for each $j \in \mathbb{N}$. By Lemma 4.16.6 we can find $G \in \mathrm{Aut}(\mathbb{C}^n)$ satisfying this condition and also $G(je_1) = je_1$. Moreover, if each P_j is an A_1-jet and if $H_1, H_2 \in \mathrm{Aut}_1(\mathbb{C}^n)$ then we can choose $G \in \mathrm{Aut}_1(\mathbb{C}^n)$. Finally, taking $F = H_2^{-1} \circ G \circ H_1$ we obtain the desired automorphism. $\square$

In the proof of Theorem 4.16.2 we need the following lemma.

Lemma 4.16.7 *Let $H \in \mathrm{Aut}(\mathbb{C}^n)$, let K be a compact polynomially convex set in $\mathbb{C}^n$, and let $\{a_j\}_{j=1}^{\infty}$ be a discrete sequence disjoint from K and contained in the z_1-axis. Given numbers $R > 0$ with $H(K) \subset R\mathbb{B}$ and $\epsilon > 0$, there exists $\Psi \in \mathrm{Aut}(\mathbb{C}^n)$ such that $|\Psi(z) - H(z)| < \epsilon$ on K, $|\Psi^{-1}(z) - H^{-1}(z)| < \epsilon$ on $H(K)$, and $\Psi(a_j) = (R + j)e_1$ for all $j \in \mathbb{N}$. If $H \in \mathrm{Aut}_1(\mathbb{C}^n)$, then we can choose $\Psi \in \mathrm{Aut}_1(\mathbb{C}^n)$.*

Proof Choose $r > 0$ such that $K \subset r\mathbb{B}$. Let $K_0 \subset r\mathbb{B}$ be a compact polynomially convex set such that $K \subset \mathring{K}_0$, $K_0 \cap \{a_j\} = \varnothing$, and $H(K_0) \subset R\mathbb{B}$. Let $a_{j_1}, \ldots, a_{j_m}$ be the points in $\{a_j\} \cap r\overline{\mathbb{B}}$, and let $\delta > 0$. Since the union of K_0 with finitely many points is again polynomially convex, we can apply Proposition 4.15.3 m times to find $\Psi_1 \in \mathrm{Aut}_1(\mathbb{C}^n)$ such that $|\Psi_1(z) - z| + |\Psi_1^{-1}(z) - z| < \delta$ for $z \in K_0$ and $\Psi_1(a_{j_k}) = H^{-1}((R + j_k)e_1)$ for $k = 1, \ldots, m$.

Let $\pi_2(z) = z_2$. For a fixed vector $v \in \mathbb{C}^n \setminus \{0\}$ and $j \in \mathbb{N}$ we consider the 1-variable function $g_j(\zeta) = \pi_2 H\Psi_1(a_j + \zeta v)$. Since the kernel of the derivative $d(\pi_2 H\Psi_1)_{a_j}$ is an $(n-1)$-dimensional subspace for each j, we may choose v arbitrarily near e_2 such that g_j is nonconstant for each $j \in \mathbb{N}$ (and hence the image of g_j omits at most one point in $\mathbb{C}$), and such that there exists a $\mathbb{C}$-linear form λ satisfying $\lambda(v) = 0$, $\lambda(e_1) = 1$, and $\lambda(a_j) \notin \lambda(r\overline{\mathbb{B}})$ if $j \notin \{j_1, \ldots, j_m\}$.

Choose f entire on $\mathbb{C}$ such that $|f| < \frac{\delta}{2}$ on $\lambda(r\mathbb{B})$, $f(\lambda(a_{j_k})) = 0$ for each $k = 1, \ldots, m$, and $|\pi_2 H\Psi_1(a_j + f(\lambda(a_j))v)| = R + j$ for each $j \notin \{j_1, \ldots, j_m\}$. Let $\Psi_2(z) = z + f(\lambda(z))v$. Then $\Psi_2 \in \mathrm{Aut}_1(\mathbb{C}^n)$, $|\Psi_2(z) - z| + |\Psi_2^{-1}(z) - z| < \delta$ on $r\mathbb{B}$, $H\Psi_1\Psi_2(a_{j_k}) = (R + j_k)e_1$ for $k = 1, \ldots, m$, and $|\pi_2 H\Psi_1\Psi_2(a_j)| = R + j$ for $j \notin \{j_1, \ldots, j_m\}$.

Using a composition of two shears as in the proof of Lemma 4.16.6, we can find $\Psi_3 \in \mathrm{Aut}_1(\mathbb{C}^n)$ such that $|\Psi_3(z) - z| + |\Psi_3^{-1}(z) - z| < \delta$ on $R\mathbb{B}$, $\Psi_3((R + j_k)e_1) = (R + j_k)e_1$ for $k = 1, \ldots, m$, and $\Psi_3 H\Psi_1\Psi_2(a_j) = (R + j)e_1$ for each $j \notin \{j_1, \ldots, j_m\}$.

Let $\Psi = \Psi_3 \circ H \circ \Psi_1 \circ \Psi_2$. Then $\Psi(a_j) = (R + j)e_1$ for all $j \in \mathbb{N}$, and for δ sufficiently small, we have $|\Psi(z) - H(z)| < \epsilon$ on K and $|\Psi^{-1}(z) - H^{-1}(z)| < \epsilon$ on

$H(K)$, so the lemma follows. Since each $\Psi_l \in \mathrm{Aut}_1(\mathbb{C}^n)$, we see that $\Psi \in \mathrm{Aut}_1(\mathbb{C}^n)$ if $H \in \mathrm{Aut}_1(\mathbb{C}^n)$. □

Proof of Theorem 4.16.2 Since the sequences $\{a_j\}$ and $\{b_j\}$ are tame, there exist automorphisms H_1, H_2 of $\mathbb{C}^n$ such that $H_1(a_j) = je_1$ and $H_2(b_j) = je_1$ for each $j \in \mathbb{N}$. Replacing Φ by $H_2 \circ \Phi \circ H_1^{-1}$, K by $H_1(K)$, and adjusting the jets P_j as in the proof of Theorem 4.16.1 we reduce the problem to the case when $a_j = b_j = je_1$ for all $j \in \mathbb{N}$.

Choose a polynomially convex set $L \subset \mathbb{C}^n \setminus \{je_1 \colon j \in \mathbb{N}\}$ such that $K \subset \mathring{L}$. Fix $\eta > 0$, and choose an integer $r > 0$ such that $L \cup \Phi(L) \subset r\mathbb{B}$. By Lemma 4.16.7 there exist automorphisms $\Psi, \Theta \in \mathrm{Aut}(\mathbb{C}^n)$ such that

$$\begin{aligned} &\Psi(je_1) = \Theta(je_1) = (r+j)e_1, \quad j = 1, 2, \ldots, \\ &\left|\Psi(z) - \Phi(z)\right| < \eta, \quad z \in L, \\ &\left|\Theta(z) - z\right| + \left|\Theta^{-1}(z) - z\right| < \eta, \quad z \in \Phi(L). \end{aligned}$$

By Lemma 4.16.6 there exists $G \in \mathrm{Aut}(\mathbb{C}^n)$ such that $G((r+j)e_1) = (r+j)e_1$ for $j \in \mathbb{N}$, $|G(z) - z| < \eta$ for $|z| \le r$, and for each $j \in \mathbb{N}$ the jet $Q_j = G_{m_j,(r+j)e_1}$ satisfies

$$P_j(z) = \left(\Theta^{-1}\right)_{m_j,(r+j)e_1} \circ Q_j \circ \Psi_{m_j, je_1}(z) + O\left(|z|^{m_j+1}\right), \quad z \to 0.$$

Let $F = \Theta^{-1} \circ G \circ \Psi$. Then $F(je_1) = je_1$ for each $j \in \mathbb{N}$, F satisfies (4.70) (p. 176) provided that $\eta > 0$ is chosen sufficiently small (depending on ϵ and $\mathrm{dist}(K, \mathbb{C}^n \setminus L)$), and Lemma 4.15.2 shows that F satisfies (4.69) (p. 175) with $a_j = b_j = je_1$. Finally, if the sequences a_j and b_j are very tame, the polynomials P_j are A_1-jets, and $\Phi \in \mathrm{Aut}_1(\mathbb{C}^n)$, then the automorphisms Ψ, Θ, and G can be chosen to have Jacobian one. □

Ugolini [556] proved the following parametric version of Theorem 4.16.1, where the parameter space is a Stein manifold.

Theorem 4.16.8 ([556]) *Let X be a Stein manifold, $\{a_j\}_{j\in\mathbb{N}}, \{b_j\}_{j\in\mathbb{N}} \subset \mathbb{C}^n$ $(n > 1)$ be tame sequences without repetition and $m_j \in \mathbb{N}$ for $j = 1, 2, \ldots$. For every $j \in \mathbb{N}$ let $P^j \colon X \to \mathrm{J}^{m_j}(\mathbb{C}^n)$ be a holomorphic family of A-jets of order m_j (i.e., polynomials of degree $\le m_j$ without the constant term) such that the linear part map $Q^j \colon X \to GL_n(\mathbb{C})$ of P^j is nullhomotopic. Then there exists a holomorphic map $F \colon X \to \mathrm{Aut}(\mathbb{C}^n)$ such that*

$$F_x(z) = b_j + P_x^j(z - a_j) + O\left(|z - a_j|^{m_j+1}\right) \quad \text{for } z \to a_j, \ j \in \mathbb{N}, \ x \in X.$$

Furthermore, if P_x^j is an A_1-jet of order m_j for every $x \in X$ and $j \in \mathbb{N}$, and if the sequences $\{a_j\}$, $\{b_j\}$ are very tame, then we can take $F_x \in \mathrm{Aut}_1(\mathbb{C}^n)$ to be volume preserving for all $x \in X$.

The nontrivial addition in the proof of Theorem 4.16.8 is to interpolate the linear parts $Q^j \cdot X \to GL_m(\mathbb{C})$ of the jets P^j. To this end, Ugolini uses the solution of the holomorphic Vaserstein problem; see Theorem 8.7.1 on p. 376. The rest of the proof closely follows the nonparametric case.

4.17 Interpolation by Fatou-Bieberbach Maps

In this section we prove the following interpolation theorem that summarizes several results in the literature; see [207] and [188, Theorem 6.1] for the first part, and [208, Theorem 1.3] for the second part.

Theorem 4.17.1 *Let X be a closed complex subvariety of $\mathbb{C}^n$ $(n > 1)$.*

(i) *For every discrete sequence $\{b_j\}_{j\in\mathbb{N}} \subset \mathbb{C}^n$ without repetition there exist a domain $\Omega \subset \mathbb{C}^n$ containing X and a biholomorphic map $\Phi\colon \Omega \to \mathbb{C}^n$ onto $\mathbb{C}^n$ such that $b_j \in \Phi(X)$ for $j = 1, 2, \ldots$.*
(ii) *If $\{a_j\}_{j\in\mathbb{N}} \subset X$ is a tame discrete sequence without repetition, then the map Φ in* (i) *can be chosen such that $\Phi(a_j) = b_j$ for $j = 1, 2, \ldots$.*

In case (i) *we can ensure that $b_j \in \Phi(X_{\mathrm{reg}})$ for all j. In both cases one can prescribe finite order jets of $\Phi(X)$ at all points $b_j \in \Phi(X_{\mathrm{reg}})$.*

The domain Ω in Theorem 4.17.1 is either all of $\mathbb{C}^n$, or a Fatou-Bieberbach domain. The image $\Phi(X)$ is a closed complex subvariety of $\mathbb{C}^n$ that is biholomorphic to X and contains the sequence $\{b_j\}_{j\in\mathbb{N}}$.

Example 4.17.2 Part (ii) of Theorem 4.17.1 fails if $\mathbb{C}^n \setminus \{a_j\}_{j\in\mathbb{N}}$ is volume hyperbolic while the sequence $\{b_j\}_{j\in\mathbb{N}}$ is tame in $\mathbb{C}^n$. Indeed, assume that there is a biholomorphic map $\Phi\colon \Omega \to \mathbb{C}^n$ from a domain $\Omega \subset \mathbb{C}^n$ containing $\{a_j\}$ onto $\mathbb{C}^n$ satisfying $\Phi(a_j) = b_j$ for all $j = 1, 2, \ldots$. The set $\Omega \setminus \{a_j\}_{j\in\mathbb{N}} \subset \mathbb{C}^n \setminus \{a_j\}_{j\in\mathbb{N}}$ is volume hyperbolic, and hence its Φ-image $\mathbb{C}^n \setminus \{b_j\}_{j\in\mathbb{N}}$ is volume hyperbolic as well, a contradiction since $\{b_j\}$ is tame. □

Proof of Theorem 4.17.1, part (i) Choose compact $\mathcal{O}(X)$-convex subsets $K_1 \subset K_2 \subset \cdots \subset \bigcup_{k=1}^\infty K_k = X$. Let $B_k = k\overline{\mathbb{B}}$ denote the closed ball of radius k in $\mathbb{C}^n$. Choose $\epsilon \in (0, 1)$ and let $\epsilon_k = 2^{-k}\epsilon$ for $k \in \mathbb{N}$. A Fatou-Bieberbach map $\Phi\colon \Omega \to \mathbb{C}^n$ satisfying Theorem 4.17.1 will be found as a limit $\Phi = \lim_{k\to\infty} \Phi_k$ of a sequence $\Phi_k = \Psi_k \circ \Psi_{k-1} \circ \cdots \circ \Psi_1 \in \mathrm{Aut}(\mathbb{C}^n)$.

We begin by choosing $\Psi_1 \in \mathrm{Aut}(\mathbb{C}^n)$ such that the regular locus of the subvariety $X_1 = \Psi_1(X)$ contains all points of the b_j sequence that belong to the first ball B_1, and X_1 contains none of the points b_j in the shell $B_2 \setminus B_1$; in addition, we can prescribe finite order jets of X_1 at all points $b_j \in B_1$. Such Ψ_1 exists by Proposition 4.15.3 (p. 170): choose a suitable number of points $a_j \in X_{\mathrm{reg}}$ and bring them to the points $b_j \in B_1$ by an automorphism with a prescribed jet at every point; if the image variety contains any of the points $b_j \in B_2 \setminus B_1$, it is pushed away by another automorphism close to the identity on B_1 and fixed to a high order at the points $b_j \in B_1$.

Choose $k_1 \geq 1$ such that $B_1 \cap X_1 \subset \Psi_1(K_{k_1})$; the set $B_1 \cup \Psi_1(K_{k_1})$ is then polynomially convex. In the second step, we apply Proposition 4.15.3 (p. 170) to get an automorphism $\Psi_2 \in \mathrm{Aut}(\mathbb{C}^n)$ such that

(i) $|\Psi_2(z) - z| < \epsilon_2$ for all $z \in B_1 \cup \Psi_1(K_{k_1})$,
(ii) Ψ_2 is tangent to the identity at every point $b_j \in B_1$,
(iii) the subvariety $X_2 = \Psi_2(X_1)$ contains all points $b_j \in B_2 \setminus B_1$ in its regular locus, with prescribed jets at these points, and
(iv) X_2 does not contain any of the points $b_j \in B_3 \setminus B_2$.

Set $\Phi_2 = \Psi_2 \circ \Psi_1$. The subvariety $X_2 = \Phi_2(X) = \Psi_2(X_1)$ then contains all points $b_j \in B_2$ (with the correct jets at these points) and none of the points $b_j \in B_3 \setminus B_2$. We now choose $k_2 > k_1$ such that $B_2 \cap X_2 \subset \Phi_2(K_{k_2})$, and then choose $\Psi_3 \in \mathrm{Aut}(\mathbb{C}^n)$ satisfying the above conditions with all indexes increased by one. Proceed inductively. Proposition 4.4.1 (p. 124) shows that the sequence $\Phi_k = \Psi_k \circ \Psi_{k-1} \circ \cdots \circ \Psi_1 \in \mathrm{Aut}(\mathbb{C}^n)$ converges to a limit map with the stated properties. This proves part (i) of Theorem 4.17.1. □

The following lemma provides the key step in the proof of part (ii).

Lemma 4.17.3 *Let $\{a_j\} \subset X \subset \mathbb{C}^n$ and $\{b_j\} \subset \mathbb{C}^n$ satisfy the hypotheses of Theorem 4.17.1(ii). Let $B \subset B'$ be closed balls in $\mathbb{C}^n$ and set $L = X \cap B'$. Assume that all points of the $\{b_j\}$ sequence that belong to $B \cup L$ coincide with the corresponding points of the $\{a_j\}$ sequence, and all remaining points of the $\{a_j\}$ sequence are contained in $X \setminus L$. Given $\epsilon > 0$ and a compact set $K \subset X$, there exist a ball $B'' \subset \mathbb{C}^n$ containing B' (B'' may be chosen as big as desired), a compact polynomially convex set $M \subset X$ with $K \cup L \subset M$, and a holomorphic automorphism θ of $\mathbb{C}^n$ satisfying the following conditions:*

(i) *$|\theta(z) - z| < \epsilon$ for all $z \in B \cup L$,*
(ii) *if $a_j \in M$ for some index j, then $\theta(a_j) = b_j \in B''$,*
(iii) *if $b_j \in B' \setminus (B \cup L)$ for some j, then $a_j \in M$ and $\theta(a_j) = b_j$,*
(iv) *$\theta(M) \subset \mathring{B}''$, and*
(v) *if $a_j \in X \setminus M$ for some j, then $\theta(a_j) \in \mathbb{C}^n \setminus B''$.*

Proof An automorphism θ with these properties is constructed in two steps, $\theta = \psi \circ \phi$ with $\phi, \psi \in \mathrm{Aut}(\mathbb{C}^n)$. The proof is illustrated in Fig. 4.2.

Since $X \cap B \subset L$ and the sets B and L are polynomially convex, $B \cup L$ is also polynomially convex (see e.g. [188, Lemma 6.5, p. 111]). By applying a preliminary automorphism of $\mathbb{C}^n$ that is close to the identity on $B \cup L$, we may assume that X does not contain any points of the $\{b_j\}$ sequence, except those that coincide with the corresponding points $a_j \in X$. The same procedure is repeated whenever necessary during later stages of the construction without mentioning it again.

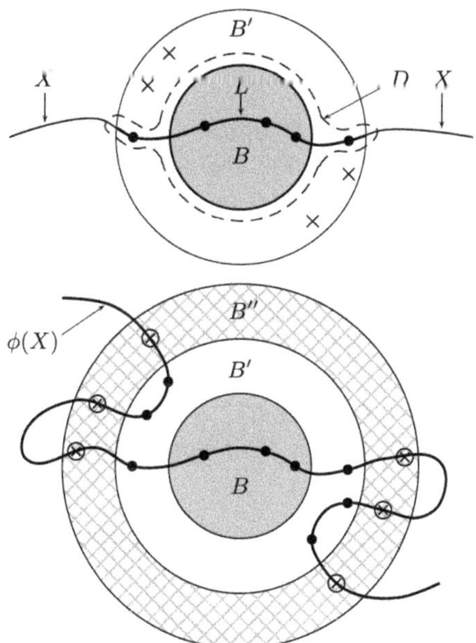

Fig. 4.2 The proof of Lemma 4.17.3. ([208, p. 550, Fig. 1])

Choose a pair of compact polynomially convex neighborhoods $D_0 \subset D \subset \mathbb{C}^n$ of $B \cup L$, with $D_0 \subset \mathring{D}$, such that D does not contain any additional points of the $\{a_j\}$ or the $\{b_j\}$ sequence. Choose $\epsilon_0 > 0$ so small that

$$\mathrm{dist}\big(B \cup L, \mathbb{C}^n \setminus D_0\big) > \epsilon_0 \quad \text{and} \quad \mathrm{dist}\big(D_0, \mathbb{C}^n \setminus D\big) > \epsilon_0.$$

We may assume that $0 < \epsilon < \epsilon_0$. Choose a compact polynomially convex set $M \subset X$ containing $K \cup (X \cap D)$ (and hence L), and also containing all points of the $\{a_j\}$ sequence for which the corresponding point b_j lies in the ball B'. (Of course M may also contain some other points of the $\{a_j\}$ sequence for which $b_j \in \mathbb{C}^n \setminus B'$.) Theorem 4.16.2, applied with $\Phi = \mathrm{Id}$, furnishes an automorphism ϕ of $\mathbb{C}^n$ satisfying the following conditions:

(a) $\sup_{z\in D} |\phi(z) - z| < \frac{\epsilon}{2}$ and $\sup_{z\in D} |\phi^{-1}(z) - z| < \frac{\epsilon}{2}$,
(b) $\phi(a_j) = b_j$ for all $a_j \in M$, and
(c) $\phi(a_j) = a_j$ for all $a_j \in X \setminus M$.

Condition (a) and the choice of ϵ imply $\phi(D_0) \subset D$ and $\phi(\mathbb{C}^n \setminus D) \cap D_0 = \varnothing$; the latter condition also implies that $\phi(X) \cap D_0 \subset \phi(M)$. Since the sets $\phi(M) \subset \phi(X)$ and D_0 are polynomially convex, their union $\phi(M) \cup D_0$ is also polynomially convex. Choose a ball $B'' \subset \mathbb{C}^n$ containing $\phi(M) \cup B'$. Theorem 4.16.2 furnishes an automorphism ψ of $\mathbb{C}^n$ satisfying the following conditions:

(a′) $|\psi(z) - z| < \frac{\epsilon}{2}$ when $z \in \phi(M) \cup D_0$,
(b′) $\psi(\phi(a_j)) = \phi(a_j) = b_j$ for all $a_j \in M$, and
(c′) $\psi(a_j) \in \mathbb{C}^n \setminus B''$ for all $a_j \in X \setminus M$.

We can also ensure that ψ fixes all points $\phi(a_j) \in \phi(X) \setminus B''$. It is immediate that $\theta = \psi \circ \phi$ satisfies Lemma 4.17.3. □

The proof of Lemma 4.17.3 is illustrated in Fig. 4.2. The first drawing shows the initial situation; the dots on X indicate the points $b_j \in B \cup L$ that agree with the corresponding points a_j, while the crosses indicate the remaining points $b_j \in B'$ that are matched with the images of a_j by applying the automorphism ϕ. The second drawing shows the situation after the application of ϕ; the black dots in $\phi(X) \cap B'$ indicate the points $b_j = \phi(a_j) \in B'$, while the crossed dots on the subvariety $\phi(X)$ inside the set $B'' \setminus B'$ will be expelled from the ball B'' by the next automorphism ψ.

Proof of Theorem 4.17.1(ii) Choose an exhaustion $K_1 \subset K_2 \subset \cdots \subset \bigcup_{j=1}^{\infty} K_j = X$ by compact sets. Fix a number ϵ with $0 < \epsilon < 1$. We inductively construct the following:

(a) a sequence of holomorphic automorphisms Φ_k of $\mathbb{C}^n$ $(k \in \mathbb{N})$,
(b) an exhaustion $L_1 \subset L_2 \subset \cdots \subset \bigcup_{k=1}^{\infty} L_k = X$ by compact polynomially convex sets, and
(c) a sequence of balls $B_1 \subset B_2 \subset \cdots \subset \bigcup_{k=1}^{\infty} B_k = \mathbb{C}^n$ centered at $0 \in \mathbb{C}^n$ whose radii satisfy $r_{k+1} > r_k + 1$ for $k = 1, 2, \ldots$

such that the following hold for all $k = 1, 2, \ldots$ (conditions (iv) and (v) are void for $k = 1$):

(i) $\Phi_k(L_k) = \Phi_k(X) \cap B_{k+1}$,
(ii) if $a_j \in L_k$ for some j then $\Phi_k(a_j) = b_j$,
(iii) if $b_j \in \Phi_k(L_k) \cup B_k$ for some j then $a_j \in L_k$ and $\Phi_k(a_j) = b_j$,
(iv) $L_{k-1} \cup K_{k-1} \subset \mathring{L}_k$,
(v) $|\Phi_k(z) - \Phi_{k-1}(z)| < 2^{-k}\epsilon$ for all $z \in B_{k-1} \cup L_{k-1}$.

To begin, set $B_0 = \varnothing$ and choose a pair of balls $B_1 \subset B_2 \subset \mathbb{C}^n$ whose radii satisfy $r_2 \geq r_1 + 1$. Theorem 4.16.2 (p. 176) furnishes an automorphism Φ_1 of $\mathbb{C}^n$ such that $\Phi_1(a_j) = b_j$ for all those (finitely many) indexes j for which $b_j \in B_2$, and $\Phi_1(a_j) \in \mathbb{C}^n \setminus B_2$ for the remaining indexes j. Setting $L_1 = \{z \in X : \Phi_1(z) \in B_2\}$, conditions (i), (ii) and (iii) hold for $k = 1$, and conditions (iv), (v) are void.

Assume inductively that we have found sets $L_1, \ldots, L_k \subset X$, balls $B_1, \ldots, B_{k+1} \subset \mathbb{C}^n$, and automorphisms $\Phi_1, \ldots, \Phi_k$ of $\mathbb{C}^n$ such that conditions (i)–(v) hold up to index k. We then apply Lemma 4.17.3 with $B = B_k$, $B' = B_{k+1}$, X replaced by $X_k = \Phi_k(X)$, and $L = \Phi_k(L_k) \subset X_k$. This gives a compact polynomially convex set $M = M_k \subset X_k$ containing $\Phi_k(K_k \cup L_k)$, an automorphism $\theta = \theta_k$ of $\mathbb{C}^n$, and a ball $B'' = B_{k+2} \subset \mathbb{C}^n$ of radius $r_{k+2} \geq r_{k+1} + 1$ such that the conclusion of Lemma 4.17.3 holds. In particular, $\theta_k(M_k) \subset B_{k+2}$, the interpolation condition holds for all points $b_j \in \theta_k(M_k) \cup B_{k+1}$, and the remaining points in the sequence $\{\Phi_k(a_j)\}_{j \in \mathbb{N}}$ are sent by θ_k out of the ball B_{k+2}. Setting

$$\Phi_{k+1} = \theta_k \circ \Phi_k, \qquad L_{k+1} = \left\{z \in X : \Phi_{k+1}(z) \in B_{k+2}\right\}$$

one easily checks that properties (i)–(v) hold for the index $k+1$ as well. The induction may now continue.

Let $\Omega = \bigcup_{k=1}^{\infty} \Phi_k^{-1}(B_k)$ consist of all points $z \in \mathbb{C}^n$ for which the sequence $\{\Phi_k(z)\}_{k\in\mathbb{N}}$ is bounded. Proposition 4.4.1 (p. 124) implies that the limit $\lim_{k\to\infty} \Phi_k = \Phi$ exists on Ω, and $\Phi\colon \Omega \to \mathbb{C}^n$ is a biholomorphic map of Ω onto $\mathbb{C}^n$. From (v) we see that $X \subset \Omega$, and properties (ii), (iii) imply that $\Phi(a_j) = b_j$ for $j = 1, 2, \ldots$. This completes the proof of Theorem 4.17.1. □

4.18 Twisted Holomorphic Embeddings into $\mathbb{C}^n$

The first part of this section is a survey of results on the existence of nonstraightenable holomorphic embeddings of Euclidean spaces to higher dimensional Euclidean spaces. The main tool used in these constructions is Theorem 4.17.1 (p. 181). We also consider the situation in the algebraic category, for embeddings of affine varieties. In the second part we discuss the existence of complete complex submanifolds of the unit ball in $\mathbb{C}^n$.

Definition 4.18.1 Let $0 < k < n$ be integers.
(A) A proper holomorphic embedding $f\colon \mathbb{C}^k \hookrightarrow \mathbb{C}^n$ is *straightenable* if there is a holomorphic automorphism $\Phi \in \mathrm{Aut}(\mathbb{C}^n)$ such that

$$\Phi\big(f(\mathbb{C}^k)\big) = \mathbb{C}^k \times \{0\}^{n-k}.$$

A proper algebraic embedding $f\colon \mathbb{C}^k \hookrightarrow \mathbb{C}^n$ is *algebraically straightenable* if the above holds for some polynomial automorphism $\Phi \in \mathrm{Aut}_{\mathrm{alg}}\mathbb{C}^n$.
(B) Holomorphic embeddings $f, g\colon \mathbb{C}^k \hookrightarrow \mathbb{C}^n$ are *$\mathrm{Aut}(\mathbb{C}^n)$-equivalent* if there exists $\Phi \in \mathrm{Aut}(\mathbb{C}^n)$ such that $\Phi \circ f = g$. They are $(\mathrm{Aut}(\mathbb{C}^k), \mathrm{Aut}(\mathbb{C}^n))$-equivalent if there exist $\Phi \in \mathrm{Aut}(\mathbb{C}^n)$ and $\Psi \in \mathrm{Aut}(\mathbb{C}^k)$ such that $\Phi \circ f = g \circ \Psi$.

Thus, a proper holomorphic embedding $\mathbb{C}^k \hookrightarrow \mathbb{C}^n$ is straightenable if and only if it is $(\mathrm{Aut}(\mathbb{C}^k), \mathrm{Aut}(\mathbb{C}^n))$-equivalent to the inclusion $z \mapsto (z, 0)$. The analogous notions make sense in the algebraic case.

Let us first survey the situation in the algebraic category.

Problem 4.18.2 Let $0 < k < n$. Is every algebraic (polynomial) embedding $\mathbb{C}^k \hookrightarrow \mathbb{C}^n$ algebraically straightenable?

Theorem 4.18.3 (Abhyankar and Moh [2], Suzuki [540]) *Every polynomial embedding $\mathbb{C} \hookrightarrow \mathbb{C}^2$ is algebraically straightenable.*

There are several proofs of this result in the literature, including a relatively simple one by Rudolph [484] that uses elementary knot theory.

A positive answer to Problem 4.18.2 is known if $n > 2k+1$ (Kaliman [319], Jelonek [310]). In particular, every polynomial embedding $\mathbb{C} \hookrightarrow \mathbb{C}^n$ for $n \geq 4$ is algebraically straightenable. Every polynomial embedding $\mathbb{C} \hookrightarrow \mathbb{C}^3$ is holomorphically straightenable (see Kaliman [319]), but it is an open problem whether it is algebraically straightenable. Kaliman also proved the following results for general affine algebraic varieties.

Theorem 4.18.4 (Kaliman [318, 319]) *Assume that X is an affine algebraic variety and $f, g \colon X \hookrightarrow \mathbb{C}^n$ are proper algebraic embeddings.*

(a) *If $n \geq \max\{1 + 2\dim X, \dim TX\}$, then f and g are isotopic through proper algebraic embeddings.* (TX *is the Zariski tangent bundle of* X.)
(b) *If $n > \max\{1 + 2\dim X, \dim TX\}$, then there exists an algebraic automorphism $\alpha \in \mathrm{Aut}_{\mathrm{alg}}(\mathbb{C}^n)$ such that $f = \alpha \circ g$.*

In codimension one, Problem 4.18.2 is also known as the

Conjecture 4.18.5 (Abhyankar-Sathaye Conjecture, [1, 485]) *Every polynomial embedding $\mathbb{C}^{n-1} \hookrightarrow \mathbb{C}^n$ is algebraically straightenable.*

A more general version of Problem 4.18.2 considered in the literature is to understand which algebraic subvarieties A of $\mathbb{C}^n$ have the *Abhyankar-Moh property* (AMP): For every algebraic embedding $f \colon A \hookrightarrow \mathbb{C}^n$ there exists a polynomial automorphism $\Phi \in \mathrm{Aut}_{\mathrm{alg}}\mathbb{C}^n$ such that $f = \Phi|_A$. In this formulation, Conjecture 4.18.5 says that an algebraic hypersurface in $\mathbb{C}^n$ has the AMP. Examples of such hypersurfaces were given by Jelonek [311].

We now show that the holomorphic analogue of Problem 4.18.2 has a negative answer for every pair of integers $0 < k < n$. Let us begin with embeddings $\mathbb{C} \hookrightarrow \mathbb{C}^n$. Parts (a) and (b) of the following theorem were proved by Rosay and Rudin [477] for $n > 2$ (see also [319, Theorem 7]), and by Globevnik, Rosay and the author [207] for $n = 2$. Part (c) is due to Buzzard and Fornæss [76]. The proof for $n = 2$ uses the version of the Andersén-Lempert theorem due to Rosay and the author (see Theorem 4.9.2 on p. 135).

Theorem 4.18.6 *Let $n > 1$.*

(a) *There exists a nonstraightenable proper holomorphic embedding $\mathbb{C} \hookrightarrow \mathbb{C}^n$.*
(b) *The set of $\mathrm{Aut}(\mathbb{C}^n)$-equivalence classes of proper holomorphic embeddings $\mathbb{C} \hookrightarrow \mathbb{C}^n$ is uncountable.*
(c) *There exists a proper holomorphic embedding $f \colon \mathbb{C} \hookrightarrow \mathbb{C}^2$ whose complement $\mathbb{C}^2 \setminus f(\mathbb{C})$ is Kobayashi hyperbolic.*

Proof (a) By Theorem 4.7.2 (p. 130) there exists a rigid discrete set $D = \{a_1, a_2, \ldots\}$ in $\mathbb{C}^n$. Theorem 4.17.1 (p. 181) furnishes a proper holomorphic embedding $f \colon \mathbb{C} \hookrightarrow \mathbb{C}^n$ such that $f(j) = a_j$ for $j = 1, 2, \ldots$. If $\Phi \in \mathrm{Aut}(\mathbb{C}^n)$ satisfies $\Phi(f(\mathbb{C})) = \mathbb{C} \times \{0\}^{n-1}$, then Φ maps D to the affine line $\mathbb{C} \times \{0\}^{n-1}$. This implies that D is tame, thereby contradicting our assumption.

(b) Let $D = \{a_1, a_2, \ldots\}$ be a rigid discrete set in $\mathbb{C}^n$ furnished by Proposition 4.7.6 (p. 132). Choose a pair of points $p \neq q$ in $\mathbb{C}^n \setminus D$. By Theorem 4.17.1 there exist proper holomorphic embeddings $f, g: \mathbb{C} \hookrightarrow \mathbb{C}^n$ such that

(i) $f(0) = p$ and $f(j) = a_j$ for $j = 1, 2, \ldots$;
(ii) $g(0) = q$ and $g(j) = a_j$ for $j = 1, 2, \ldots$.

If f and g are $\mathrm{Aut}(\mathbb{C}^n)$-equivalent, then there is an automorphism of $\mathbb{C}^n$ which fixes D and maps p to q, contradicting the choice of D. This gives a continuum of pairwise nonequivalent embeddings $\mathbb{C} \hookrightarrow \mathbb{C}^n$, one for each choice of $q \in \mathbb{C}^n \setminus D \cup \{p\}$.

(c) We explain the main idea and refer to [76] for the details. One begins with a suitably chosen discrete sequence of pairwise disjoint closed complex discs $\Delta_1, \Delta_2, \ldots$, contained in affine lines in $\mathbb{C}^2$ parallel to the coordinate axes, such that for each sufficiently small holomorphic perturbation Δ'_j of Δ_j in $\mathbb{C}^2$ the set $\mathbb{C}^2 \setminus \bigcup_{j=1}^{\infty} \Delta'_j$ is Kobayashi hyperbolic. By the proof of Theorem 4.17.1, there is a proper holomorphic embedding $f: \mathbb{C} \hookrightarrow \mathbb{C}^2$ such that $f(\mathbb{C})$ contains a sequence of disjoint complex discs Δ'_j (images of suitably chosen discs in $\mathbb{C}$) that are as close as desired to the linear discs Δ_j for all $j = 1, 2, \ldots$. If the approximations are close enough, then $\mathbb{C}^2 \setminus f(\mathbb{C})$ is Kobayashi hyperbolic since it is contained in the hyperbolic domain $\mathbb{C}^2 \setminus \bigcup_{j=1}^{\infty} \Delta'_j$. □

The idea from part (a) above also gives the following result.

Theorem 4.18.7 ([62], [188, Corollary 5.3]) *For every pair of integers $1 \leq k < n$ there exists a proper holomorphic embedding $f: \mathbb{C}^k \hookrightarrow \mathbb{C}^n$ such that $\mathbb{C}^n \setminus f(\mathbb{C}^k)$ is $(n-k)$-hyperbolic in the sense of Eisenman* [141].

If f is as in Theorem 4.18.7, then every entire map $g: \mathbb{C}^{n-k} \to \mathbb{C}^n$ such that $g(\mathbb{C}^{n-k}) \cap f(\mathbb{C}^k) = \varnothing$ is degenerate, in the sense that its rank is less than $n-k$ at every point of $\mathbb{C}^{n-k}$.

These results imply that the set of $\mathrm{Aut}(\mathbb{C}^n)$-equivalence classes of proper holomorphic embeddings $\mathbb{C}^k \hookrightarrow \mathbb{C}^n$ is uncountable. The following result, applied with $X = \mathbb{C}$, shows that the smaller set of $(\mathrm{Aut}(\mathbb{C}^k), \mathrm{Aut}(\mathbb{C}^n))$-equivalence classes of such embeddings is also uncountable.

Theorem 4.18.8 (Borell and Kutzschebauch [62, Theorem 1.2]) *Assume that X is a positive dimensional Stein space whose group of holomorphic automorphisms is a Lie group, possibly with countably many components. If there is a proper holomorphic embedding of X into $\mathbb{C}^n$ for some $n > \dim X$, then for any $k \geq 0$ there exist uncountably many proper holomorphic embeddings $X \times \mathbb{C}^k \hookrightarrow \mathbb{C}^n \times \mathbb{C}^k$ such that any two are nonequivalent with respect to* $(\mathrm{Aut}(X \times \mathbb{C}^k), \mathrm{Aut}(\mathbb{C}^n \times \mathbb{C}^k))$.

Kutzschebauch and Lodin [364] constructed holomorphic families of proper holomorphic embeddings $\mathbb{C}^k \hookrightarrow \mathbb{C}^n$ $(0 < k < n-1)$ such that no two embeddings in the family are $(\mathrm{Aut}(\mathbb{C}^k), \mathrm{Aut}(\mathbb{C}^n))$-equivalent. As an application, they derived the existence of families of holomorphic $\mathbb{C}^*$ actions on $\mathbb{C}^n$ for $n \geq 5$ such that different actions in the family are not conjugate.

The idea in part (c) of Theorem 4.18.6 applies to any complex subvariety of $\mathbb{C}^n$. The following result of Borell and Kutzschebauch uses the construction developed in the proof of [188, Theorem 5.1].

Theorem 4.18.9 ([62, Proposition 2.5]) *Let X be a closed k-dimensional complex subvariety of $\mathbb{C}^n$ with $0 < k = \dim X < n$. Given a discrete set $A = \{a_1, a_2, \ldots\}$ in $\mathbb{C}^n$, there are an open set $\Omega \subset \mathbb{C}^n$ containing X and a biholomorphic map $\Phi \colon \Omega \to \mathbb{C}^n$ onto $\mathbb{C}^n$ such that $A \subset \Phi(X)$ and $\mathbb{C}^n \setminus \Phi(X)$ is $(n-k)$-hyperbolic in the sense of Eisenman.*

These results do not answer the question whether there exist topologically twisted proper holomorphic embeddings of $\mathbb{C}$ in $\mathbb{C}^n$.

Problem 4.18.10 Is there a proper holomorphic embedding $f \colon \mathbb{C} \hookrightarrow \mathbb{C}^n$ such that $\mathbb{C}^n \setminus f(\mathbb{C})$ is not homeomorphic to $\mathbb{C}^n \setminus (\mathbb{C} \times \{0\}^{n-1})$?

A positive answer is known for embedded discs in $\mathbb{C}^2$.

Theorem 4.18.11 (Baader et al. [40]) *There exists a topologically knotted proper holomorphic embedding of the unit disc $\mathbb{D}$ into $\mathbb{C}^2$.*

We mentioned in Sect. 4.4 that Globevnik [238] constructed Fatou-Bieberbach domains $\Omega \subset \mathbb{C}^2$ whose intersections with $\mathbb{C} \times \mathbb{D}$ are small $\mathcal{C}^1$ perturbations of the bidisc $\mathbb{D}^2 = \mathbb{D} \times \mathbb{D}$. In [40] the authors use the theory of complex algebraic curves in $\mathbb{C}^2$ to find a knotted holomorphic embedding $\phi \colon \mathbb{D} \hookrightarrow \Omega$ such that $\phi(b\mathbb{D}) \subset b\Omega$. Composing ϕ with a biholomorphism $\Omega \to \mathbb{C}^2$ furnishes a knotted proper embedding $\mathbb{D} \hookrightarrow \mathbb{C}^2$.

Problem 4.18.12 Does there exist an *unknotted* proper holomorphic embedding $\mathbb{D} \hookrightarrow \mathbb{C}^2$?

Holomorphic automorphisms have also been used in the construction of highly twisted bounded complex submanifolds of $\mathbb{C}^n$. A locally closed submanifold $M \subset \mathbb{C}^n$ is said to be *complete* if every divergent path in M has infinite Euclidean length. (A path $\lambda \colon [0,1) \to M$ is divergent if the point $\lambda(t)$ leaves every compact subset of M when $t \to 1$.) Let $\mathbb{B}^n$ denote the ball in $\mathbb{C}^n$.

Theorem 4.18.13 (Globevnik [240]) *For every pair of integers $1 \leq k < n$ there exists a complete, closed, k-dimensional complex submanifold of $\mathbb{B}^n$.*

This gives an affirmative answer to a problem posed by P. Yang in 1977 (see the references in [240]). In [241], Globevnik proved the same result for any pseudoconvex domain in $\mathbb{C}^n$ in place of the ball.

Globevnik obtained Theorem 4.18.13 as a consequence of his main result in [240] to the effect that for every $n \geq 2$ there is a holomorphic function f on the ball $\mathbb{B}^n$ whose real part $\Re f$ is unbounded on every path of finite length that ends on $b\mathbb{B}^n$. It follows that every level set $M_c = \{f = c\}$ is a closed complete complex hypersurface in $\mathbb{C}^n$, and M_c is smooth for most values of c in view of Sard's lemma. To find a function f with this property, he first constructs a labyrinth $\Lambda \subset \mathbb{B}^n$, consisting of pairwise disjoint closed polygonal domains in real affine hyperplanes, such that every curve in $\mathbb{B}^n \setminus \Lambda$ which terminates on $b\mathbb{B}^n$ has infinite length. Furthermore, Λ is exhausted by an increasing sequence of compact polynomially convex sets that are finite unions of connected components of Λ. The construction of the labyrinth is very delicate. He then uses Runge's approximation theorem to find a function $f \in \mathcal{O}(\mathbb{B}^n)$ such that the restriction of $\Re f$ to Λ grows to infinity as we approach $b\mathbb{B}^n$. Hence, every level set $\{\Re f = c\}$ intersects at most finitely many connected components of Λ. It follows that every divergent path in $\{\Re f = c\}$ has infinite length. By this method one cannot control the topology or the complex structure of the embedded complete submanifolds.

A different proof of Theorem 4.18.13, closer to the methods developed in this chapter, was given by Alarcón, Globevnik and López [18]. They provide a simpler construction of a suitable labyrinth $\Lambda \subset \mathbb{B}^n$, consisting of a sequence of balls in well chosen affine real hyperplanes, with the same property regarding paths in $\mathbb{B}^n \setminus \Lambda$. Next, they use a sequence of shear automorphisms of $\mathbb{C}^n$ to inductively push a given closed complex subvariety $X \subset \mathbb{C}^n$ off Λ. The sequence of automorphisms converges on a certain pseudoconvex Runge domain $D \subset X \cap \mathbb{B}^N$, and the limit map embeds D onto a closed complex submanifold $M \subset \mathbb{B}^n \setminus \Lambda$ which is proper in $\mathbb{B}^n$ and is necessarily complete by the choice of Λ. In particular, taking $X = \mathbb{C}^k \hookrightarrow \mathbb{C}^n$ with $1 \leq k < n$, they find a complete closed complex submanifold $M \subset \mathbb{B}^n$ which is biholomorphic to a pseudoconvex Runge domain in $\mathbb{C}^k$.

In dimensions $k = 1$ and $n = 2$, the most precise result in this direction was given by Alarcón and Globevnik [17] who constructed smooth complete complex curves $M \subset \mathbb{B}^2$ that are homeomorphic to any given open oriented surface. Besides automorphisms, they also used the technique of Wold and the author [226] for embedding Riemann surfaces in $\mathbb{C}^2$ (see Sect. 9.10).

By using a different method, the following result was shown earlier by Alarcón and the author [14]. The difference with respect to the aforementioned results is that their technique allows one to control the conformal structure (and not just the topology) of M, but it does not give embeddings into $\mathbb{C}^2$.

Theorem 4.18.14 *Every bordered Riemann surface M admits a complete proper holomorphic immersion $M \to \mathbb{B}^2$ and a complete proper holomorphic embedding $M \hookrightarrow \mathbb{B}^3$.*

4.19 Nonlinearizable Periodic Automorphisms of $\mathbb{C}^n$

A holomorphic automorphism $\Phi \in \mathrm{Aut}(\mathbb{C}^n)$ is said to be *linearizable* if there exist $A \in GL(n, \mathbb{C})$ and $\Psi \in \mathrm{Aut}(\mathbb{C}^n)$ such that $\Phi = \Psi^{-1} \circ A \circ \Psi$. Similarly one defines linearizability of a group of holomorphic automorphisms of $\mathbb{C}^n$—the entire group should be conjugate to a subgroup of $GL_n(\mathbb{C})$ by some $\Psi \in \mathrm{Aut}(\mathbb{C}^n)$. The following result settled a long-standing problem.

Theorem 4.19.1 (Derksen and Kutzschebauch [119]) *For every $n \geq 2$ there exists a nonlinearizable holomorphic automorphism of period n on $\mathbb{C}^{2+n}$. In particular, there is a nonlinearizable holomorphic involution on $\mathbb{C}^4$.*

Derksen and Kutzschebauch actually proved that every nontrivial complex reductive Lie group G admits an effective nonlinearizable holomorphic action of G on $\mathbb{C}^N$ for all sufficiently big N (depending on G). An important ingredient in their proof are proper holomorphic embeddings $\phi: \mathbb{C} \hookrightarrow \mathbb{C}^2$ such that every holomorphic map $\mathbb{C}^2 \to \mathbb{C}^2 \setminus \phi(\mathbb{C})$ is degenerate (see Theorem 4.18.6 on p. 186). Choose such ϕ. For every $k \in \mathbb{Z}_+$ define the embedding

$$\phi_k : \mathbb{C} \times \mathbb{C}^k \hookrightarrow \mathbb{C}^2 \times \mathbb{C}^k, \quad \phi_k(z, y) = \big(\phi(z), y\big).$$

Corollary 4.19.2 *The embedding ϕ_k is nonstraightenable for every $k \in \mathbb{Z}_+$.*

Proof If ϕ_k is straightenable (see Definition 4.18.1), then

$$\mathbb{C}^{2+k} \setminus \phi_k\big(\mathbb{C}^{1+k}\big) \cong \big(\mathbb{C}^2 \setminus \phi(\mathbb{C})\big) \times \mathbb{C}^k$$

is biholomorphic to $\mathbb{C}^* \times \mathbb{C}^{k+1}$. Hence, there exists a nondegenerate holomorphic map $f: \mathbb{C}^{2+k} \to \mathbb{C}^{2+k} \setminus \phi_k(\mathbb{C}^{1+k})$. By composing f with the coordinate projection $(\mathbb{C}^2 \setminus \phi(\mathbb{C})) \times \mathbb{C}^k \to \mathbb{C}^2 \setminus \phi(\mathbb{C})$ we get an entire map $\mathbb{C}^{k+2} \to \mathbb{C}^2 \setminus \phi(\mathbb{C})$ of rank two, contradicting the choice of ϕ. □

We now describe certain pseudo-affine modifications that are used in the proof of Theorem 4.19.1. Let $X \subset Y$ and Z be Stein spaces, and let f be a holomorphic function on Z. Suppose that the ideal sheaf of X in Y is generated by finitely many functions $h_1, \ldots, h_r \in \mathcal{O}(Y)$ (this always holds if X and Y are nonsingular, see [170, Satz 5.5] or Theorem 8.5.4). Consider the map

$$\psi : Y \times \big(Z \setminus \{f = 0\}\big) \to Y \times Z \times \mathbb{C}^r, \quad \psi(y, z) = \left(y, z, \frac{h_1(y)}{f(z)}, \ldots, \frac{h_r(y)}{f(z)}\right)$$

and let

$$\mathcal{R}(X, Y, Z, f) = \overline{\mathrm{Im}(\psi)} \subset Y \times Z \times \mathbb{C}^r$$

denote the closure of the image of ψ. (Note that the topological closure equals the holomorphic Zariski closure. In [119, p. 44] these are called *Rees spaces*.) It is easily seen that, up to a biholomorphism, $\mathcal{R}(X, Y, Z, f)$ does not depend on the choice of the generators h_j for the ideal $\mathcal{J}_X(Y)$.

Lemma 4.19.3 *The spaces $\mathcal{R}(X, Y, Z, f)$ have the following properties:*

1. $\mathcal{R}(X, Y, Z, f) \cong \mathcal{R}(\psi(X), Y, Z, f)$ *for every* $\phi \in \mathrm{Aut}(Y)$.
2. $\mathcal{R}(X \times W, Y \times W, Z, f) \cong \mathcal{R}(X, Y, Z, f)$ *for any Stein space* W.
3. $\mathcal{R}(X \times \{0\}, Y \times \mathbb{C}^m, Z, f) \cong \mathcal{R}(X, Y, Z, f) \times \mathbb{C}^m$.

Proof The first two properties are immediate. To show the third property, suppose that the ideal $\mathcal{J}_X(Y)$ is generated by $h = (h_1, \ldots, h_r) \in \mathcal{O}(Y)^r$. Then, $\mathcal{J}_{X \times \{0\}}(Y \times \mathbb{C}^m)$ is generated by (h, u), where $u = (u_1, \ldots, u_m)$ are the coordinates on $\mathbb{C}^m$. Consider the map

$$\psi : Y \times \mathbb{C}^m \times \big(Z \setminus \{f = 0\}\big) \to Y \times \mathbb{C}^m \times Z \times \mathbb{C}^r \times \mathbb{C}^m,$$

$$\psi(y, u, z) = \big(y, u, z, f(z)^{-1} h, f(z)^{-1} u\big).$$

Define $\phi \in \mathrm{Aut}(Y \times \mathbb{C}^m \times Z \times \mathbb{C}^r \times \mathbb{C}^m)$ by

$$\phi(y, u, z, v, w) = \big(y, u - f(z) w, z, v, w\big).$$

Then, we have $\phi \circ \psi(y, u, z) = (y, 0, z, f(z)^{-1} h, f(z)^{-1})$. It follows that

$$\mathcal{R}\big(X \times \{0\}, Y \times \mathbb{C}^m, Z, f\big) = \overline{\mathrm{Im}(\psi)} \cong \overline{\mathrm{Im}(\phi \circ \psi)} \cong \mathcal{R}(X, Y, Z, f) \times \mathbb{C}^m.$$

This proves Lemma 4.19.3. □

The following example will be used in the proof of Theorem 4.19.1. Suppose that the hypersurface $X \subset Y$ is given by $X = \{h = 0\}$ for some $h \in \mathcal{O}(Y)$ which generates the ideal $\mathcal{J}_X(Y)$. Let $\psi : Y \times Z \setminus \{f = 0\} \to Y \times Z \times \mathbb{C}$ be given by $\psi(y, z) = (y, z, h(y)/f(z))$. Then

$$\mathcal{R}(X, Y, Z, f) = \overline{\mathrm{Im}(\psi)} = \big\{(y, z, w) \in Y \times Z \times \mathbb{C} : h(y) = f(z) w\big\}.$$

Lemma 4.19.4 *If $\phi_1 : X \hookrightarrow \mathbb{C}^n$ and $\phi_2 : X \hookrightarrow \mathbb{C}^m$ are proper holomorphic embeddings, then*

$$\mathcal{R}\big(\phi_1(X), \mathbb{C}^n, Z, f\big) \times \mathbb{C}^m \cong \mathcal{R}\big(\phi_2(X), \mathbb{C}^m, Z, f\big) \times \mathbb{C}^n.$$

Proof It is easily see that there is an automorphism $\alpha \in \mathrm{Aut}(\mathbb{C}^n \times \mathbb{C}^m)$ such that $\alpha \circ (\phi_1 \times \{0\}^m) = \{0\}^n \times \phi_2$. From Lemma 4.19.3 it follows that

$$\begin{aligned}\mathcal{R}\big(\phi_1(X), \mathbb{C}^n, Z, f\big) &\cong \mathcal{R}\big(\phi_1(X) \times \{0\}, \mathbb{C}^n \times \mathbb{C}^m, Z, f\big) \\ &\cong \mathcal{R}\big(\{0\} \times \phi_2(X), \mathbb{C}^n \times \mathbb{C}^m, Z, f\big) \cong \mathcal{R}\big(\phi_2(X), \mathbb{C}^m, Z, f\big).\end{aligned}$$

This proves Lemma 4.19.4. □

Applying Lemma 4.19.4 with $X = \mathbb{C}^k$ for some $1 \le k < n$ and letting ϕ_2 be the identity map on $\mathbb{C}^k$, we obtain

Corollary 4.19.5 *For an embedding* $\phi\colon \mathbb{C}^k \hookrightarrow \mathbb{C}^n$ *and* $f \in \mathcal{O}(\mathbb{C}^l)$ *we have*

$$\mathcal{R}\big(\phi(X), \mathbb{C}^n, \mathbb{C}^l, f\big) \times \mathbb{C}^k \cong \mathcal{R}\big(\mathbb{C}^k, \mathbb{C}^k, \mathbb{C}^l, f\big) \times \mathbb{C}^n \cong \mathbb{C}^{k+l+n}.$$

Proof of Theorem 4.19.1 Let $G = \langle g \rangle$ be a cyclic group of order $n > 1$. Choose a nonstraightenable proper holomorphic embedding $\phi\colon \mathbb{C} \hookrightarrow \mathbb{C}^2$ as in Corollary 4.19.2. Let $h \in \mathcal{O}(\mathbb{C}^2)$ generate the ideal sheaf of the smooth curve $\phi(\mathbb{C}) \subset \mathbb{C}^2$. For every $n \in \mathbb{N}$ the set

$$Y_n = \big\{(x, y, z) \in \mathbb{C}^3 : h(x, y) = z^n\big\}$$

is a smooth complex hypersurface in $\mathbb{C}^3$. Let ζ be a primitive n-th root of 1. Define the action $\tau : G \times Y_n \to Y_n$ by

$$\tau\big(g^j, (x, y, z)\big) = \big(x, y, \zeta^j z\big), \quad j = 0, 1, \ldots, n-1.$$

For each $k = 1, 2, \ldots$ we extend τ to an action $\tau_k\colon G \times (Y_n \times \mathbb{C}^k) \to Y_n \times \mathbb{C}^k$ acting trivially on $\mathbb{C}^k$. Note that the fixed point set

$$\operatorname{Fix} \tau_k = (\operatorname{Fix} \tau) \times \mathbb{C}^k = \phi(\mathbb{C}) \times \{0\} \times \mathbb{C}^k$$

is a smooth complex hypersurface in $Y_n \times \mathbb{C}^k$.

Lemma 4.19.6 *If* $Y_n \times \mathbb{C}^k \cong \mathbb{C}^{2+k}$, *then the action* τ_k *is nonlinearizable.*

Proof To reach a contradiction, assume that τ_k is linearizable. The linearization map then takes the fixed point set of τ_k onto a linear hyperplane in $\mathbb{C}^{2+k}$, and hence the set

$$\widetilde{Y} = \big(Y_n \times \mathbb{C}^k\big) \setminus \operatorname{Fix} \tau_k = (Y_n \setminus \operatorname{Fix} \tau) \times \mathbb{C}^k$$

is biholomorphic to $\mathbb{C}^* \times \mathbb{C}^{1+k}$. The projection $\pi : \mathbb{C}^{3+k} \to \mathbb{C}^{2+k}$, $\pi(x, y, z, w) = (x, y, w)$ gives a proper ramified n-to-one covering $\pi\colon Y_n \times \mathbb{C}^k \to \mathbb{C}^{2+k}$ whose branch locus equals $\phi(\mathbb{C}) \times \{0\} \times \mathbb{C}^k = \operatorname{Fix} \tau_k$. The restriction

$$\pi : \widetilde{Y} = \big(Y_n \times \mathbb{C}^k\big) \setminus \operatorname{Fix} \tau_k \to \big(\mathbb{C}^2 \setminus \phi(\mathbb{C})\big) \times \mathbb{C}^k$$

is an unramified covering, and hence both sets have the same universal covering space X. Let $\eta\colon X \to (\mathbb{C}^2 \setminus \phi(\mathbb{C})) \times \mathbb{C}^k$ be the universal covering projection. Since $\widetilde{Y} \cong \mathbb{C}^* \times \mathbb{C}^{1+k}$, we have $X \cong \mathbb{C}^{2+k}$. Hence, the restriction of η to a suitable 2-plane in X, followed by the projection $\mathbb{C}^2 \times \mathbb{C}^k \to \mathbb{C}^2$, is an entire map $\mathbb{C}^2 \to \mathbb{C}^2 \setminus \phi(\mathbb{C})$ of rank 2. This contradicts the choice of ϕ. □

Lemma 4.19.7 *For each pair of integers* $n \geq 2$ *and* $k \geq 1$ *there exists an action* $\sigma_k\colon G \times \mathbb{C}^{1+n+k} \to \mathbb{C}^{1+n+k}$ *by holomorphic automorphisms with the fixed point set* $\operatorname{Fix} \sigma_k \cong Y_n \times \mathbb{C}^k$.

Proof Let ϕ and h be as above. The set

$$X_n = \{(x, y, w_1, w_2, \ldots, w_n) \in \mathbb{C}^{2+n} : h(x, y) = w_1 w_2 \cdots w_n\} \tag{4.72}$$

is a smooth hypersurface in $\mathbb{C}^{2+n}$. Note that $X_n = \mathcal{R}(\phi(\mathbb{C}), \mathbb{C}^2, \mathbb{C}^{n-1}, f)$, where $f \in \mathcal{O}(\mathbb{C}^{n-1})$ equals $f(w_1, \ldots, w_{n-1}) = w_1 w_2 \cdots w_{n-1}$.

Let $\sigma : G \times \mathbb{C}^{2+n} \to \mathbb{C}^{2+n}$ be the action defined by

$$g \cdot (x, y, w_1, w_2, \ldots, w_n) = (x, y, w_2, \ldots, w_{n-1}, w_1).$$

Since X_n is clearly invariant under σ, we have an induced action on X_n which we still denote by σ. For each integer $k \geq 1$ we extend σ to an action

$$\sigma_k : G \times (X_n \times \mathbb{C}^k) \to X_n \times \mathbb{C}^k$$

by acting trivially on $\mathbb{C}^k$. We have that

$$\begin{aligned} \mathrm{Fix}\,\sigma &= \{(x, y, w) \in \mathbb{C}^{2+n} : w_1 = w_2 = \cdots = w_n,\ h(x, y) = (w_1)^n\} \cong Y_n, \\ \mathrm{Fix}\,\sigma_k &= \mathrm{Fix}\,\sigma \times \mathbb{C}^k \cong Y_n \times \mathbb{C}^k. \end{aligned}$$

To complete the proof, it suffices to observe that $X_n \times \mathbb{C}^k$ is biholomorphic to $\mathbb{C}^{1+n+k}$ for each $k \geq 1$. Indeed, by Corollary 4.19.5 (p. 192) we have

$$X_n \times \mathbb{C}^k \cong \mathcal{R}(\phi(\mathbb{C}), \mathbb{C}^2, \mathbb{C}^{n-1}, f) \times \mathbb{C}^k \cong \mathbb{C}^{1+n+k}.$$

This proves Lemma 4.19.7. □

Comparing Lemmas 4.19.6 and 4.19.7 we draw the following conclusion.

Corollary 4.19.8 *For each $n \geq 2$ and $k \geq 1$, at least one of the actions σ_k and τ_k is a nonlinearizable action of the cyclic group $\mathbb{Z}/n\mathbb{Z}$ by holomorphic automorphisms on $\mathbb{C}^{1+n+k}$ and $\mathbb{C}^{2+k}$, respectively.*

Theorem 4.19.1 is an immediate consequence of Corollary 4.19.8. □

Problem 4.19.9 Is the hypersurface X_n (4.72) biholomorphic to $\mathbb{C}^{n+1}$?

This is a special case of the *holomorphic Zariski cancellation problem*:

Problem 4.19.10 Let X be a complex manifold such that $X \times \mathbb{C}$ is biholomorphic to $\mathbb{C}^{n+1}$ $(n \geq 2)$; is X necessarily biholomorphic to $\mathbb{C}^n$?

In conclusion, we describe a construction of nonlinearizable $\mathbb{C}^*$ actions on $\mathbb{C}^l$ for every $l \geq 4$ (see [119, Sect. 4]). Let $\phi : \mathbb{C} \hookrightarrow \mathbb{C}^2$ be a nonstraightenable proper holomorphic embedding furnished by Theorem 4.18.6 (p. 186). Choose an entire function $h \in \mathcal{O}(\mathbb{C}^2)$ with $\{h=0\} = \phi(\mathbb{C})$ and $\mathrm{d}h \neq 0$ on $\{h=0\}$. Let X be the hypersurface in $\mathbb{C}^4$ defined by

$$X = \{(x,y,z,w) \in \mathbb{C}^4 : h(x,y) = zw\}. \tag{4.73}$$

Note that $X = \mathcal{R}(\phi(\mathbb{C}), \mathbb{C}^2, \mathbb{C}, z)$. By Corollary 4.19.5 we have $X \times \mathbb{C} \cong \mathbb{C}^4$. Now, $X \times \mathbb{C}^k \subset \mathbb{C}^4 \times \mathbb{C}^k$ is stable under the linear $\mathbb{C}^*$ action on $\mathbb{C}^4 \times \mathbb{C}^k$,

$$\lambda \cdot (x,y,z,w,t_1,\dots,t_k) = (x,y,\lambda z,\lambda^{-1}w,t_1,\dots,t_k).$$

The restriction of this action to $X \times \mathbb{C}^k$ induces via the biholomorphism $\alpha_k \colon X \times \mathbb{C}^k \xrightarrow{\cong} \mathbb{C}^{3+k}$ a holomorphic $\mathbb{C}^*$ action $\sigma_k \colon \mathbb{C}^* \times \mathbb{C}^{k+3} \to \mathbb{C}^{k+3}$.

Proposition 4.19.11 *The action σ_k is not linearizable for every $k \geq 1$. Hence, for every $l \geq 4$ there exists a non-linearizable action of $\mathbb{C}^*$ on $\mathbb{C}^l$.*

Proof Suppose that $\alpha_k \colon X \times \mathbb{C}^k \xrightarrow{\cong} \mathbb{C}^{3+k}$ is a biholomorphic $\mathbb{C}^*$-equivariant map, where $\mathbb{C}^*$ acts linearly on $\mathbb{C}^{3+k}$. This representation of $\mathbb{C}^*$ on $\mathbb{C}^{k+3}$ must be isomorphic to the representation of $\mathbb{C}^*$ on the tangent space of some fixed point of $X \times \mathbb{C}^k$. With respect to some coordinates, this action is given by

$$\lambda \cdot (z,w,u_1,\dots,u_{k+1}) = (\lambda z,\lambda^{-1}w,u_1,\dots,u_{k+1}).$$

The categorical quotient of $X \times \mathbb{C}^k$ is given by $\pi_1 \colon X \times \mathbb{C}^k \to \mathbb{C}^{2+k}$,

$$\pi_1(x,y,z,w,t_1,\dots,t_k) = (x,y,t_1,\dots,t_k),$$

and the categorical quotient of $\mathbb{C}^{3+k}$ is given by $\pi_2 \colon \mathbb{C}^{3+k} \to \mathbb{C}^{2+k}$ where

$$\pi_2(z,w,u_1,\dots,u_{k+1}) = (zw,u_1,\dots,u_{k+1}).$$

The fixed point set $(X \times \mathbb{C}^k)^{\mathbb{C}^*}$ of the action λ equals

$$\{(x,y,z,w,t_1,\dots,t_k) \in X \times \mathbb{C}^k : f(x,y) = z = w = 0\}.$$

Its image under π_1 is $\phi(\mathbb{C}) \times \mathbb{C}^k \subset \mathbb{C}^2 \times \mathbb{C}^k$. On the other hand, $(\mathbb{C}^{3+k})^{\mathbb{C}^*}$ is

$$\{(z,w,u_1,\dots,u_{k+1}) \in \mathbb{C}^{3+k} : z = w = 0\},$$

and its image under π_2 is $\{0\} \times \mathbb{C}^{1+k} \subset \mathbb{C}^{2+k}$. Now, $\alpha_k \colon X \times \mathbb{C}^k \xrightarrow{\cong} \mathbb{C}^{3+k}$ induces a biholomorphism $\gamma \colon \mathbb{C}^{2+k} \xrightarrow{\cong} \mathbb{C}^{2+k}$ of the categorical quotients such that $\gamma(\phi(\mathbb{C}) \times \mathbb{C}^k) = \{0\} \times \mathbb{C}^{k+1}$. In view of Corollary 4.19.2 this contradicts our choice of ϕ. □

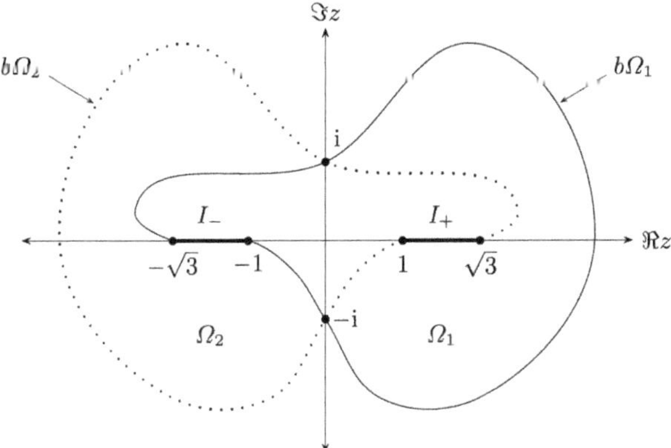

Fig. 4.3 Domains Ω_1 and Ω_2. (Modified from [584, p. 777, Fig. 1])

4.20 A Non-Runge Fatou-Bieberbach Domain

By Corollary 4.9.5 on p. 136, a Fatou-Bieberbach domain $\Omega \subset \mathbb{C}^n$ is Runge in $\mathbb{C}^n$ if and only if any Fatou-Bieberbach map $F: \mathbb{C}^n \to \Omega$ onto Ω is a limit of a sequence of holomorphic automorphisms of $\mathbb{C}^n$. It was a long-standing problem whether every Fatou-Bieberbach domain in $\mathbb{C}^n$ is Runge. We describe a counterexample given by Wold [584] in 2008. The following is the key lemma.

Lemma 4.20.1 ([584]) *There exists a compact set Y in $\mathbb{C}^* \times \mathbb{C}$ such that*

(i) *Y is holomorphically convex in $\mathbb{C}^* \times \mathbb{C}$,*
(ii) *the polynomial hull $\widehat{Y}$ contains the origin $(0,0) \in \mathbb{C}^2$, and*
(iii) *for every nonempty open set $U \subset \mathbb{C}^* \times \mathbb{C}$ there exists an automorphism $G \in \mathrm{Aut}(\mathbb{C}^* \times \mathbb{C})$ such that $Y \subset G(U)$.*

Proof A set Y with properties (i) and (ii) is found as a union of two disjoint totally real discs, following [529] or [532, pp. 392–396]. Choose smoothly bounded simply connected domains $\Omega_1, \Omega_2 \subset \mathbb{C}$ as in Fig. 4.3. (The set Ω_2 is obtained by reflecting Ω_1 by the map $x + \mathrm{i}y \mapsto -x + \mathrm{i}y$.) Their boundaries intersect only at the points $\pm\mathrm{i}$, the interval $I_+ = [1, \sqrt{3}]$ is contained in $\Omega_1 \cap b\Omega_2$, and the interval $I_- = [-\sqrt{3}, -1]$ is contained in $\Omega_2 \cap b\Omega_1$.

Consider the following sets in $\mathbb{C}^* \times \mathbb{C}$:

$$X_1 = \big\{(z,w) : z \in b\Omega_1,\ w = z^2 - c,\ c \in [0,1]\big\},$$
$$X_2 = \big\{(z,w) : z \in b\Omega_2,\ w \in [1,2]\big\}.$$

Note that X_1 and X_2 are disjoint totally real annuli, X_1 is foliated by the boundaries of analytic discs $\Gamma_c = \{(z, z^2 - c) : z \in \overline{\Omega}_1\}$ $(c \in [0,1])$, and X_2 is foliated by the boundaries of analytic discs $\Lambda_c = \{(z,c) : z \in \overline{\Omega_2}\}$ $(c \in [1,2])$. Hence, the curves

Γ_c and Λ_c lie in the polynomial hulls $\widehat{X}_1$, $\widehat{X}_2$, respectively. In particular, the origin $(0,0) \in \Gamma_0$ belongs to $\widehat{X}_1$. Note that all these curves intersect the complex line $\{0\} \times \mathbb{C}$, and it can be seen that the union $X_1 \cup X_2$ is $\mathcal{O}(\mathbb{C}^* \times \mathbb{C})$-convex. We now remove from X_1 and X_2 suitably chosen discs such that the remaining sets become discs whose union still contains $(0,0)$ in its polynomial hull. Take

$$S_1 = \{(z,w) : z \in I_-,\ w \in [1,2]\},$$

$$S_2 = \{(z,w) : z \in I_+,\ w = z^2 - c,\ c \in [0,1]\}$$

and set

$$Y_1 = \overline{X_1 \setminus S_1}, \quad Y_2 = \overline{X_2 \setminus S_2}, \quad Y = Y_1 \cup Y_2.$$

It is easily verified that Y_1 and Y_2 are discs and that $S_1 \subset \widehat{X}_2$, $S_2 \subset \widehat{X}_1$. Hence, neither S_1 nor S_2 can contain any peak points for the polynomial algebra $\mathcal{P}(X_1 \cup X_2)$. It follows that $\widehat{X_1 \cup X_2} = \widehat{Y_1 \cup Y_2} = \widehat{Y}$; in particular, $(0,0) \in \widehat{Y}$. (Note however that each of the two discs Y_1 and Y_2 is polynomially convex.) This establishes properties (i) and (ii) in the lemma.

Let us now show that condition (iii) holds as well.

The manifold $\mathbb{C}^* \times \mathbb{C}$ enjoys the density property by Theorem 4.10.9(b). Let U be any nonempty open set in $\mathbb{C}^* \times \mathbb{C}$. Since Y is the disjoint union of two totally real discs, there are a $\mathcal{O}(\mathbb{C}^* \times \mathbb{C})$-convex neighborhood $V \subset \mathbb{C}^* \times \mathbb{C}$ of Y and an isotopy $\theta_t \colon V \to V_t \subset \mathbb{C}^* \times \mathbb{C}$ $(t \in [0,1])$ of injective holomorphic maps such that $\theta_0 = \mathrm{Id}_V$, $\theta_1(Y) \subset U$, and V_t is $\mathcal{O}(\mathbb{C}^* \times \mathbb{C})$-convex for every $t \in [0,1]$. (Essentially, we contract each of the two discs Y_1, Y_2 to a small neighborhood of a point and drag them into U.) By Theorem 4.9.2 (p. 135) we can approximate θ_1 by an automorphism $g \in \mathrm{Aut}(\mathbb{C}^* \times \mathbb{C})$. If the approximation is close enough on Y, then $g(Y) \subset U$ and the inverse map $G = g^{-1}$ satisfies condition (iii). □

Part (a) of the following corollary follows from Lemma 4.20.1. Part (b) is obtained by choosing a Fatou-Bieberbach map $\theta \colon \mathbb{C}^n \hookrightarrow \mathbb{C}^* \times \mathbb{C}^{n-1}$, applying part (a) to the compact set $L = \theta(K)$, and setting $\phi = \psi \circ \theta$.

Corollary 4.20.2 *Let $n > 1$.*

(a) *Given a compact holomorphically convex set $L \subset \mathbb{C}^* \times \mathbb{C}^{n-1}$ with nonempty interior, there exists $\psi \in \mathrm{Aut}(\mathbb{C}^* \times \mathbb{C}^{n-1})$ such that $0 \in \widehat{\psi(L)}$.*
(b) *For every compact set $K \subset \mathbb{C}^n$ with nonempty interior there exists an injective holomorphic map $\phi \colon \mathbb{C}^n \hookrightarrow \mathbb{C}^n$ such that $\widehat{\phi(K)} \not\subset \phi(\mathbb{C}^n)$.*

Clearly, the image of any map $\phi \colon \mathbb{C}^n \hookrightarrow \mathbb{C}^n$ satisfying Corollary 4.20.2(b) fails to be Runge in $\mathbb{C}^n$, so we have the following corollary due to Wold.

Corollary 4.20.3 ([584]) *For every $n > 1$ there exists a non-Runge Fatou-Bieberbach domain in $\mathbb{C}^n$.*

The method used in the proof of Lemma 4.20.1 and Corollary 4.20.2 was introduced by Wold in [584]; it will henceforth be called the *Wold process*.

4.21 A Long $\mathbb{C}^2$ Without Holomorphic Functions

A complex manifold X of dimension n is said to be a *long* $\mathbb{C}^n$ if it is the union of an increasing sequence of domains

$$X_1 \subset X_2 \subset \cdots \subset \bigcup_{j=0}^{\infty} X_j = X \tag{4.74}$$

such that each X_j is biholomorphic to $\mathbb{C}^n$. Denote by $\iota_j \colon X_j \hookrightarrow X_{j+1}$ the inclusion map. Choosing biholomorphic maps $\psi_j \colon \mathbb{C}^n \xrightarrow{\cong} X_j$, we have for each $j \in \mathbb{N}$ a commutative diagram

$$\begin{array}{ccc} \mathbb{C}^n & \xrightarrow{\phi_j} & \mathbb{C}^n \\ \psi_j \downarrow & & \downarrow \psi_{j+1} \\ X_j & \xrightarrow{\iota_j} & X_{j+1} \end{array} \tag{4.75}$$

in which $\phi_j \colon \mathbb{C}^n \hookrightarrow \mathbb{C}^n$ is a Fatou-Bieberbach map. Conversely, every sequence of Fatou-Bieberbach maps $\phi_j \colon \mathbb{C}^n \hookrightarrow \mathbb{C}^n$ determines a long $\mathbb{C}^n$, X, whose elements are represented by infinite strings $x = (x_i, x_{i+1}, \ldots)$, where $i \in \mathbb{N}$ and for every $k = i, i+1, \ldots$ we have $x_k \in \mathbb{C}^n$ and $x_{k+1} = \phi_k(x_k)$. Another string $y = (y_j, y_{j+1}, \ldots)$ determines the same element of X if and only if one of the following possibilities holds:

- $i = j$ and $x_i = y_i$ (and hence $x_k = y_k$ for all $k > i$);
- $i < j$ and $y_j = \phi_{j-1} \circ \cdots \circ \phi_i(x_i)$;
- $j < i$ and $x_i = \phi_{i-1} \circ \cdots \circ \phi_j(y_j)$.

For each $j \in \mathbb{N}$, let $\psi_j \colon \mathbb{C}^n \hookrightarrow X$ be the injective map sending $z \in \mathbb{C}^n$ to the equivalence class of the string $(z, \phi_j(z), \phi_{j+1}(\phi_j(z)), \ldots)$. Set $X_j = \psi_j(\mathbb{C}^n)$ and let $\iota_j \colon X_j \hookrightarrow X_{j+1}$ be the inclusion map induced by the left shift $(x_j, x_{j+1}, x_{j+2}, \ldots) \mapsto (x_{j+1}, x_{j+2}, \ldots)$. This gives the diagram (4.75).

It is immediate that any long $\mathbb{C}$ is biholomorphic to $\mathbb{C}$. However, for $n > 1$ this class of complex manifolds is rather poorly understood. The long-standing open problem, whether there exists a long $\mathbb{C}^n$ which is not biholomorphic to $\mathbb{C}^n$, was answered affirmatively for every $n > 1$ by Wold [585] in 2010. His construction is based on examples of non-Runge Fatou-Bieberbach domains in [584]; see Corollary 4.20.3. In 2016, Boc Thaler and the author [60] improved Wold's construction to obtain the following result.

Theorem 4.21.1 ([60]) *For every integer $n > 1$ there exists a long $\mathbb{C}^n$ without any nonconstant holomorphic or plurisubharmonic function.*

The crucial ingredient is the following lemma whose proof is based on Wold's construction of a non-Stein long $\mathbb{C}^2$ (see Sect. 4.20), but with an additional twist that eliminates nonconstant plurisubharmonic functions.

Lemma 4.21.2 *Let K be a compact set with nonempty interior in $\mathbb{C}^n$ for some $n > 1$. Given a point $a \in \mathbb{C}^n$ there exists an injective holomorphic map $\phi \colon \mathbb{C}^n \hookrightarrow \mathbb{C}^n$ such that the polynomial hull of $\phi(K)$ contains the point $\phi(a)$.*

Proof We give the proof for $n = 2$; the same idea works for any $n > 1$.

Let $Y \subset \mathbb{C}^* \times \mathbb{C}$ be the compact set in Lemma 4.20.1; in particular, Y is $\mathcal{O}(\mathbb{C}^* \times \mathbb{C})$-convex, while its polynomial hull $\widehat{Y}$ contain the origin $0 \in \mathbb{C}^2$. Replacing K by $\widehat{K}$, we may assume that K is polynomially convex. Choose a Fatou-Bieberbach map $\theta \colon \mathbb{C}^2 \hookrightarrow \mathbb{C}^* \times \mathbb{C}$ whose image is Runge in $\mathbb{C}^2$. Then, $\theta(K)$ is polynomially convex, and hence $\mathcal{O}(\mathbb{C}^* \times \mathbb{C})$-convex. By Lemma 4.20.1 there exists $\psi \in \mathrm{Aut}(\mathbb{C}^* \times \mathbb{C})$ such that $Y \subset \psi(\theta(\mathring{K}))$. Note that the set $K' = \psi(\theta(K))$ is $\mathcal{O}(\mathbb{C}^* \times \mathbb{C})$-convex, while its polynomial hull $\widehat{K}'$ contains a neighborhood of $\widehat{Y}$, and hence a neighborhood $V \subset \mathbb{C}^2$ of $0 \in \mathbb{C}^2$.

Let $a \in \mathbb{C}^2$ and set $b = \psi(\theta(a))$. If $b \in \widehat{K}'$, take $\phi = \psi \circ \theta$. Otherwise, choose a point $a' \in V \cap (\mathbb{C}^* \times \mathbb{C}) \setminus K'$ and apply Theorem 4.10.5 (p. 143) to find $\tau \in \mathrm{Aut}(\mathbb{C}^* \times \mathbb{C})$ which is close to the identity on K' and satisfies $\tau(b) = a'$. (Such τ exists since $K' \cup \{b\}$ is $\mathcal{O}(\mathbb{C}^* \times \mathbb{C})$-convex, so it suffices to apply the cited result to an isotopy of injective holomorphic maps which equals the identity on a neighborhood of K' and moves the point b to a' within $\mathbb{C}^* \times \mathbb{C} \setminus K'$.) Assuming that τ is sufficiently close to the identity on K', we have $Y \subset \tau(K')$ and hence $a' \in \widehat{Y} \subset \widehat{\tau(K')}$. The injective holomorphic map $\phi = \tau \circ \psi \circ \theta \colon \mathbb{C}^2 \hookrightarrow \mathbb{C}^* \times \mathbb{C}$ then satisfies $\phi(a) \in \widehat{\phi(K)}$. □

Proof of Theorem 4.21.1 Pick a countable dense set $\{a_j\}_{j \in \mathbb{N}}$ in $\mathbb{C}^n$. Lemma 4.21.2 furnishes an injective holomorphic map $\phi_1 \colon \mathbb{C}^n \hookrightarrow \mathbb{C}^n$ such that

$$\phi_1(a_1) \in \widehat{\phi_1(K)} =: K_2.$$

Applying Lemma 4.21.2 to the set K_2 and the point $\phi_1(a_2)$ gives an injective holomorphic map $\phi_2 \colon \mathbb{C}^n \hookrightarrow \mathbb{C}^n$ such that

$$\phi_2\big(\phi_1(a_2)\big) \in \widehat{\phi_2(K_2)} =: K_3.$$

From the first step we also have $\phi_1(a_1) \in K_2$ and hence $\phi_2(\phi_1(a_1)) \in K_3$.

Continuing inductively, we obtain a sequence $\phi_j \colon \mathbb{C}^n \hookrightarrow \mathbb{C}^n$ of injective holomorphic maps and a sequence of compacts sets $K = K_1 \subset K_2 \subset \cdots$ in $\mathbb{C}^n$ such that, setting $\Phi_k = \phi_k \circ \cdots \circ \phi_1$, we have for every $k \in \mathbb{N}$ that

$$\Phi_k(a_j) \in K_{k+1} = \widehat{\phi_k(K_k)}, \quad j = 1, \ldots, k. \tag{4.76}$$

Note that for every holomorphic map $\phi\colon \mathbb{C}^n \to \mathbb{C}^n$ and compact set $K \subset \mathbb{C}^n$ we have $\widehat{\phi(K)} = \widehat{\phi(\widehat{K})}$. By using this inductively, we see that

$$K_{k+1} = \widehat{\Phi_k(K)}, \quad k = 1, 2, \ldots. \tag{4.77}$$

In the limit manifold $X = \bigcup_{k=1}^{\infty} X_k$ (a long $\mathbb{C}^n$) determined by the sequence $(\phi_k)_{k=1}^{\infty}$, the set $K \subset \mathbb{C}^n = X_1 \subset X$ coincides with $\Phi_k(K) \subset X_{k+1}$ for each $k = 1, 2, \ldots$. (We have identified the k-th copy of $\mathbb{C}^n$ in the sequence with its image $X_k \subset X$.) It follows from (4.76) and (4.77) that the hull $\widehat{K}_{\mathcal{O}(X)}$ of K in X contains the set $\{a_j\}_{j\in\mathbb{N}} \subset \mathbb{C}^n = X_1$. Since the latter set is dense in $\mathbb{C}^n$, every holomorphic function on X is bounded on $X_1 = \mathbb{C}^n$, whence constant. By the identity principle the function is constant on X.

The same argument shows that the plurisubharmonic hull $\widehat{K}_{\mathrm{Psh}(X)}$ of K contains the set $A_1 := \{a_j\}_{j\in\mathbb{N}} \subset \mathbb{C}^n \cong X_1$, and hence every $u \in \mathrm{Psh}(X)$ is bounded from above on A_1. Since A_1 is dense in X_1, it follows that u is bounded from above on X_1, and hence is constant on X_1 by Liouville's theorem. In order to ensure that u is constant on each copy $X_k \cong \mathbb{C}^n$ in the given exhaustion of X, we modify the construction as follows. After choosing the first map $\phi_1\colon \mathbb{C}^n \hookrightarrow \mathbb{C}^n$ such that $\phi_1(a_1) \in \widehat{\phi_1(K)}$, we choose a countable dense set $A'_2 = \{a'_{2,1}, a'_{2,2}, \ldots\}$ in $\mathbb{C}^n \setminus \phi_1(\mathbb{C}^n)$ and set $A_2 = \phi_1(A_1) \cup A'_2$ to get a countable dense set in $X_2 \cong \mathbb{C}^n$. Next, we find a Fatou-Bieberbach map $\phi_2\colon \mathbb{C}^n \hookrightarrow \mathbb{C}^n$ such that the first two points $\phi_1(a_1), \phi_1(a_2)$ of the set $\phi_1(A_1)$, and also the first point $a'_{2,1}$ of A'_2 are mapped by ϕ_2 into the polynomial hull K_3 of $\phi_2(\phi_1(K))$. Continue inductively. At the k-th stage we have chosen a Fatou-Bieberbach map $\phi_k\colon \mathbb{C}^n \hookrightarrow \mathbb{C}^n$, and we take $A_{k+1} = \phi_k(A_k) \cup A'_{k+1}$ where A'_{k+1} is a countable dense set in $\mathbb{C}^n \setminus \phi_k(A_k)$. In the limit manifold X we thus get an increasing sequence of countable sets $A_1 \subset A_2 \subset \cdots$ whose union $A := \bigcup_{k=1}^{\infty} A_k$ is dense in X and every point of A lies in the hull $\widehat{K}_{\mathcal{O}(X_k)} = \widehat{K}_{\mathrm{Psh}(X_k)}$ for all sufficiently big $k \in \mathbb{N}$. Hence, $\widehat{K}_{\mathrm{Psh}(X)}$ contains the set A. We conclude as before that any plurisubharmonic function on X is bounded on every $X_k \cong \mathbb{C}^n$, and hence is constant. □

In the proof of Theorem 4.21.1, every pair (X_j, X_{j+1}) in the exhaustion (4.74) of X is non-Runge. On the other hand, assuming a Runge pair hypothesis, we get the following observation due to Wold [580, Proposition 3].

Proposition 4.21.3 *If $X = \bigcup_{j=1}^{\infty} X_j$ is a long $\mathbb{C}^n$ such that (X_j, X_{j+1}) is a Runge pair for every $j \in \mathbb{N}$, then X is biholomorphic to $\mathbb{C}^n$.*

Proof Choose biholomorphisms $\psi_j\colon \mathbb{C}^n \to X_j$ ($j \in \mathbb{N}$). Since X_j is Runge in X_{j+1}, the map $\phi_j := \psi_{j+1}^{-1} \circ \psi_j\colon \mathbb{C}^n \hookrightarrow \mathbb{C}^n$ has Runge image. Let $\mathbb{B}$ denote the unit ball in $\mathbb{C}^n$. By Theorem 4.9.1, we can approximate ϕ_j uniformly on $j\mathbb{B}$ by automorphisms $\theta_j \in \mathrm{Aut}(\mathbb{C}^n)$; the biholomorphism $\psi_{j+1} \circ \theta_j\colon \mathbb{C}^n \to X_{j+1}$ then approximates ψ_j on $j\mathbb{B}$. Replacing ψ_{j+1} by $\psi_{j+1} \circ \theta_j$, one can inductively build a sequence ψ_j which converges uniformly on compacts in $\mathbb{C}^n$ to a biholomorphism $\psi = \lim_{j\to\infty} \psi_j\colon \mathbb{C}^n \to X$. □

On the other hand, the Runge pair condition in Proposition 4.21.3 may be violated even if $X = \mathbb{C}^n$. In fact, for any $n > 1$ there exists an increasing exhausting sequence $X_1 \subset X_2 \subset \cdots \subset \bigcup_{k=1}^{\infty} X_k = \mathbb{C}^n$ of non-Runge Fatou-Bieberbach domains in $\mathbb{C}^n$ (see [60, Proposition 5.1]).

The construction in the proof of Theorem 4.21.1 is quite flexible and allows one to find many nonequivalent long $\mathbb{C}^n$'s. In order to distinguish them from one another, we use biholomorphic invariants of a complex manifold introduced in [60], the *stable core* and the *strongly stable core*. The definition of these invariants is based on the following notion.

Definition 4.21.4 A compact set K in a complex manifold X has the *stable hull property*, SHP, if X admits an exhaustion $K_1 \subset K_2 \subset \cdots \subset \bigcup_{j=1}^{\infty} K_j = X$ by compact sets such that $K \subset K_1$, $K_j \subset \mathring{K}_{j+1}$ for every $j \in \mathbb{N}$, and the increasing sequence of hulls $\widehat{K}_{\mathcal{O}(K_j)}$ stabilizes, i.e., there is a $j_0 \in \mathbb{N}$ such that

$$\widehat{K}_{\mathcal{O}(K_j)} = \widehat{K}_{\mathcal{O}(K_{j_0})}, \quad j \geq j_0. \tag{4.78}$$

Obviously, SHP is a biholomorphically invariant property: if $K \subset X$ satisfies condition (4.78) and $F\colon X \to Y$ is biholomorphic, then $F(K) \subset Y$ satisfies (4.78) with respect to the exhaustion $L_j = F(K_j)$ of Y. It is easily seen that SHP is independent of the choice of the exhaustion; see [60, Lemma 4.1].

Definition 4.21.5 Let X be a complex manifold.

(i) The *stable core* of X, $SC(X)$, is the set of all points $x \in X$ which admit a compact neighborhood $K \subset X$ with the stable hull property.
(ii) A regular compact set $B \subset X$ ($B = \overline{\mathring{B}}$) is called the *strongly stable core* of X, denoted $SSC(X)$, if B has the stable hull property, but no compact set $K \subset X$ with $\mathring{K} \setminus B \neq \varnothing$ has the stable hull property.

Clearly, the stable core always exists and is a biholomorphic invariant, in the sense that a biholomorphic map $X \to Y$ maps $SC(X)$ onto $SC(Y)$. In particular, every holomorphic automorphism of X maps $SC(X)$ onto itself. The strongly stable core $SSC(X)$ need not exist in general; if it does, then its interior equals the stable core $SC(X)$ and $SSC(X) = \overline{SC(X)}$.

Theorem 4.21.6 ([60, Theorem 1.6]) *Let $n > 1$.*

(a) *For every regular compact polynomially convex set B in $\mathbb{C}^n$ there exists a long $\mathbb{C}^n$, $X(B)$, which admits no nonconstant plurisubharmonic functions and whose strongly stable core equals B: $SSC(X(B)) = B$.*
(b) *For every open set $U \subset \mathbb{C}^n$ there exists a long $\mathbb{C}^n$, $X(U)$, which admits no nonconstant holomorphic functions and satisfies $SC(X(U)) \subset U$ and $\overline{U} = \overline{SC(X(U))}$.*

By taking the set B in part (a) to be the closure of a strongly pseudoconvex domain, we see that the moduli space of long $\mathbb{C}^n$'s contains the moduli space of germs of smooth strongly pseudoconvex real hypersurfaces in $\mathbb{C}^n$. It was shown by Poincaré [456] that most pairs of such hypersurfaces are not biholomorphic (and not even formally equivalent) to each other when $n > 1$. A complete set of local biholomorphic invariants of real-analytic strongly pseudoconvex hypersurfaces is provided by the Chern-Moser normal form [96]. Hence, Theorem 4.21.6 implies the following corollary.

Corollary 4.21.7 *For every $n > 1$ there is a continuum of pairwise nonequivalent long $\mathbb{C}^n$'s with no nonconstant holomorphic or plurisubharmonic functions and no nontrivial holomorphic automorphisms.*

Let us outline the main idea behind the proof of Theorem 4.21.6; further details can be found in [60].

We construct manifolds with these properties by an inductive application of the Wold process (see Lemmas 4.20.1, 4.21.2 and Corollary 4.20.2). At every step of the induction, we perform the Wold process simultaneously on finitely many pairwise disjoint compact sets $K_1, \ldots, K_m$ in the complement of the given regular polynomially convex set $B \subset \mathbb{C}^n$, chosen such that $\bigcup_{j=1}^m K_j \cup B$ is polynomially convex, thereby ensuring that polynomial hulls of their images $\phi(K_j)$ escape from the range $\phi(\mathbb{C}^n)$ of the injective holomorphic map $\phi\colon \mathbb{C}^n \hookrightarrow \mathbb{C}^n$ constructed in the induction step. At the same time, we ensure that ϕ is close to the identity on B, and hence $\phi(B)$ is polynomially convex. In practice, the sets K_j are small pairwise disjoint closed balls in the complement of B whose number increases during the process. We devise the induction so that every point in a countable dense set $A = \{a_j\}_{j=1}^\infty \subset X \setminus B$ is the center of a decreasing sequence of balls whose $\mathcal{O}(X_k)$-hulls escape from each compact set in X; hence none of these balls has the stable hull property. This implies that B is the strongly stable core of X.

To prove part (b) of Theorem 4.21.6, we modify the process by introducing a new small ball $B' \subset U \setminus B$ at every stage of the induction. The sequence of added balls is chosen such that their union is dense in the given open subset $U \subset \mathbb{C}^n$, while the sequence of sets K_j on which the Wold process is performed densely fills the complement $X \setminus \overline{U}$. It follows that the stable core of the limit manifold $X = \bigcup_{k=1}^\infty X_k$ is contained in U and is dense in U.

This completes the outline of proof of Theorem 4.21.6.

Problem 4.21.8

(a) Does there exist a long $\mathbb{C}^2$ which admits a nonconstant holomorphic function, but is not Stein?
(b) To what extent is it possible to prescribe the algebra $\mathcal{O}(X)$ of a long $\mathbb{C}^n$?
(c) Does there exist a long $\mathbb{C}^n$ which is a Stein manifold different from $\mathbb{C}^n$?
(d) Does there exist a long $\mathbb{C}^n$ without nonconstant meromorphic functions?

(e) What can be said about the existence of complex analytic subvarieties of positive dimension in non-Stein long $\mathbb{C}^n$'s? What about the (non-)existence of holomorphic line bundles?

(f) Is there a holomorphic family X_s of long $\mathbb{C}^2$'s, parametrized by the disc $\mathbb{D} = \{s \in \mathbb{C} : |s| < 1\}$ or the plane $\mathbb{C}$, such that X_s is not biholomorphic to $X_{s'}$ whenever $s \neq s'$?

For partial results concerning (f), see [203] and [60, Theorem 1.7]. The latter result shows that there are holomorphic families of long $\mathbb{C}^2$'s such that near every parameter value s_0 there are both values s for which X_s is a $\mathbb{C}^2$, and also values for which X_s has no nonconstant holomorphic functions.

'But the novel, the novel,' she shouted to the Master, 'take the novel with you wherever you are flying.'
'I don't have to,' replied the Master, 'I remember it by heart.'

M.A. Bulgakov, *The Master and Margarita*

Part II
Oka Theory

Chapter 5
Oka Manifolds

We begin this chapter by outlining the main ideas and results of the classical Oka-Grauert theory and then explain how these lead to the theory of Oka manifolds. The main result, Theorem 5.4.4 on p. 216, subsumes the classical Oka-Grauert principle. Gromov's Oka theory for elliptic holomorphic submersions and its generalizations is treated in the following chapter.

5.1 A Historical Introduction to the Oka Principle

The Oka principle has its origin in the seminal works of Kiyoshi Oka, one of the pioneers of modern complex analysis in several variables. In a series of papers during 1936–1953, Oka invented new methods of constructing global analytic objects from local ones. In the sequel we mention several of his key papers; they are also available in Springer's 1984 edition [444]. For more information on Oka's contributions we refer the reader to Noguchi's monograph [435].

The notion of cohomology with coefficients in a sheaf originates in the paper of Cousin [107] after whom the *Cousin problems* are named. Following the motivating works of Henri Cartan [84, 85], Oka solved the first Cousin problem on a reasonably general class of domains [444, Chap. I] in 1936. The sheaf theoretic formulation of this result is a special case of the Oka-Cartan Theorem B: On a domain of holomorphy and, more generally, on a Stein space X, we have that $H^1(X;\mathcal{O}_X)=0$.

The beginning of Oka theory is Oka's theorem from 1939 [443] (see also [444, Chap. III]) that a second (multiplicative) Cousin problem on a domain of holomorphy is solvable with holomorphic functions if it is solvable with continuous functions. Equivalently, a holomorphic line bundle over such a domain is holomorphically trivial if it is topologically trivial. In modern language, the inclusion $\mathcal{O}_X^* \hookrightarrow \mathcal{C}_X^*$ of the sheaf of nonvanishing holomorphic functions into the sheaf of nonvanishing continuous functions induces an isomorphism $H^1(X;\mathcal{O}_X^*) \cong H^1(X;\mathcal{C}_X^*)$ on any complex space X satisfying $H^1(X;\mathcal{O}_X) = H^2(X;\mathcal{O}_X)=0$; in particular, on any Stein space (see Sect. 5.2).

F. Forstnerič, *Stein Manifolds and Holomorphic Mappings*, Ergebnisse der Mathematik und ihrer Grenzgebiete. 3. Folge / A Series of Modern Surveys in Mathematics 56,
DOI 10.1007/978-3-319-61058-0_5

In the 1930's, major new ideas were brought to analysis and topology by the introduction of Morse theory and fibre bundles. In 1940, Cartan started studying local bases of ideals of holomorphic functions and matrices transforming one set of generators to another set of generators [86]. Based on these developments and on the cohomological formulation of Oka's result [443] by Cartan and Serre, certain non-abelian generalizations of Oka's theorem were obtained by Frenkel [232].

In his seminal papers from 1957–1958, Grauert extended Oka's and Frenkel's theorem to principal holomorphic fibre bundles over Stein spaces with an arbitrary Lie group as fibre, showing that the holomorphic classification of such bundles coincides with the topological classification [252–254]. Cartan called this the *Oka-Grauert principle*, and Thom extolled it as *the most beautiful principle in analysis.* An exposition of Grauert's work is found in Cartan's paper [88] which is also available in [90]. In [260, p. 145], Grauert and Remmert give the following heuristic formulation of the Oka-Grauert principle:

Analytic problems on Stein manifolds which can be cohomologically formulated have only topological obstructions.

During the 1960's, Forster and Ramspott and others found many extensions and applications of the Oka-Grauert principle; see Sects. 8.4–8.5.

Another major event of the mid-1960's was the development of L^2-methods for solving the nonhomogeneous $\overline{\partial}$-equation by Hörmander [299], Kohn [345, 346], and Andreotti and Vesentini [28]. (See [441] for this subject.) This is a powerful tool for solving analytic problems by correcting a smooth solution to a holomorphic one via an inhomogeneous Cauchy-Riemann equation.

In spite of the progress in the 1960's, many natural problems in Stein geometry remained elusive since they could not be formulated as cohomological problems on fibre bundles with homogeneous fibres. Forster's 1970 paper [171] on embedding Stein manifolds into Euclidean spaces contains the correct conjecture on the minimal embedding dimension which was confirmed only much later by Eliashberg and Gromov [150] and Schürmann [492].

In 1986 (published in 1998) Henkin and Leiterer [287] gave an important contribution from the conceptual point of view by providing a new proof of Grauert's theorem using the bumping method. However, their work remains in the realm of principal fibre bundles and related structures.

The subject was revitalized in 1989 by Mikhael Gromov whose paper [266] marks the beginning of modern Oka theory. The emphasis shifted from the cohomological to homotopy theoretic aspect, focusing on analytic properties of a complex manifold Y which ensure that every continuous map from a Stein space X to Y is homotopic to a holomorphic map, with natural additions concerning approximation and interpolation of maps that are motivated by the classical function theory on Stein spaces. Specifically, for which complex manifolds Y is the natural inclusion $\mathcal{O}(X,Y) \hookrightarrow \mathcal{C}(X,Y)$ of the space of holomorphic maps into the space of continuous maps a weak homotopy equivalence for all Stein spaces X? The analogous questions are considered for sections of holomorphic submersions $Z \to X$ over Stein spaces.

Gromov's main result in [266] is that the existence of a holomorphic dominating fibre spray on $Z|_U$ over small open subsets U of the Stein base manifold X implies all forms of the Oka principle for sections $X \to Z$ (see Theorem 6.2.2 on p. 265). This condition is called *ellipticity* of the submersion $Z \to X$. An ostensibly weaker condition, *subellipticity*, which means the existence of local dominating families of fibre sprays, was implicitly present in Gromov's paper and was made explicit by the author in [190] where the Oka principle was proved for *subelliptic submersions*. Dominating (families of) holomorphic sprays are used as a replacement for the exponential map in the approximation and linearization problems which arise in the proof.

Although ellipticity and subellipticity are useful geometric sufficient conditions implying the Oka principle, it is not known whether they are necessary in general. Gromov asked in [266, 3.4.(D), p. 881] whether the Oka principle for maps from Stein spaces to a complex manifold Y could be characterized by a Runge approximation property for entire maps $\mathbb{C}^n \to Y$ on a certain family of simple compact subsets of $\mathbb{C}^n$. This was confirmed by the author in a series of papers [195, 197, 199, 202]. These developments are summarized by the main result of this chapter, Theorem 5.4.4, which says that a complex manifold Y satisfies all natural Oka properties if (and only if) any holomorphic map $K \to Y$ from a compact convex set K in $\mathbb{C}^n$ is a uniform limit of entire maps $\mathbb{C}^n \to Y$. Each of the Oka properties in Theorem 5.4.4 is natural in the sense that it is modeled on a classical result of complex analysis concerning holomorphic functions on Stein manifolds, considered as holomorphic mappings to the complex number field $\mathbb{C}$. In Sect. 5.15 we identify fifteen such properties and show that they are all equivalent to each other; a complex manifold enjoying any of these equivalent properties is said to be an *Oka manifold*. Since every complex homogeneous manifold is an Oka manifold, Theorem 5.4.4 includes the classical Oka-Grauert theory. A model category approach to Oka manifolds, due to Finnur Lárusson, is presented in Sect. 7.5.

Since the first edition of this book was published in 2011, the notion of an Oka manifold became well accepted in the theory, and a rich array of connections with other holomorphic flexibility properties have been discovered; see Chap. 7 for some of these connections. This notion is particularly fruitful among affine algebraic manifolds where there is a strong interplay with the algebraic density property and algebraic flexibility, the latter meaning the existence of many globally integrable algebraic vector fields with algebraic flows. I wish to thank numerous colleagues whose contributions made Oka manifold a standard term in complex geometry.

5.2 Cousin Problems and Oka's Theorem

In order to motivate Oka's theorem [443] (see also [444, Chap. III]), we begin by considering the *first Cousin problem* (see [107]) whose origin is the following *Mittag-Leffler problem*: Given a collection $\{(U_j, m_j)\}$, where $\mathcal{U} = \{U_j\}$ is an open cover of a complex space X and $m_j \in \mathcal{M}(U_j)$ is a meromorphic function on U_j such that every difference $f_{ij} = m_i|_{U_{ij}} - m_j|_{U_{ij}}$ is holomorphic on $U_{ij} = U_i \cap U_j$, find a global meromorphic function $m \in \mathcal{M}(X)$ on X such that $m|_{U_j} - m_j$ is holomorphic on U_j for every j.

The standard solution of Mittag-Leffler's problem proceeds via the *Cousin I problem*. The collection $f_{ij} \in \mathcal{O}(U_{ij})$ is a 1-cocycle on $\mathcal{U}$ with values in the sheaf $\mathcal{O}$, meaning that for each triple of indexes i, j, k we have that

$$f_{ij} + f_{jk} + f_{ki} = 0 \quad \text{on } U_{ijk} = U_i \cap U_j \cap U_k.$$

Given such a 1-cocycle $\{f_{ij}\}$, the Cousin I problem asks for a collection of holomorphic functions $f_j \in \mathcal{O}(U_j)$ (a 0-cochain) satisfying

$$f_i - f_j = f_{ij} \quad \text{on } U_{ij}.$$

If a solution exists, then $\{f_{ij}\}$ is said to be a *1-coboundary*, $\{f_{ij}\} = \delta(\{f_j\})$. From $f_i - f_j = f_{ij} = m_i - m_j$ on U_{ij} we see that the collection of meromorphic functions $m'_j = m_j - f_j \in \mathcal{M}(U_j)$ satisfies $m'_i = m'_j$ on U_{ij}, and hence it defines a global meromorphic function on X solving the Mittag-Leffler problem.

Let us denote by $Z^1(\mathcal{U}; \mathcal{O})$ the Abelian group of all 1-cocycles on $\mathcal{U}$ with coefficients in $\mathcal{O}$. By $C^0(\mathcal{U}; \mathcal{O})$ we denote the Abelian group of all collections $\{f_j\}$ of holomorphic functions $f_j \in \mathcal{O}(U_j)$ (these are 0-cochains on $\mathcal{U}$ with coefficients in $\mathcal{O}$). The coboundary homomorphism $\delta\colon C^0(\mathcal{U}; \mathcal{O}) \to Z^1(\mathcal{U}; \mathcal{O})$ is $\delta(\{f_j\}) = \{f_{ij}\}$, where $f_{ij} = f_i - f_j$. The quotient group

$$H^1(\mathcal{U}; \mathcal{O}) = Z^1(\mathcal{U}; \mathcal{O})/\delta\big(C^0(\mathcal{U}; \mathcal{O})\big)$$

is the *first Čech cohomology group* of the cover $\mathcal{U}$ with coefficients in $\mathcal{O}$. Given a refinement $\mathcal{V} = \{V_k\}$ of the cover $\mathcal{U}$ (every set $V_k \in \mathcal{V}$ is contained in some $U_{i(k)} \in \mathcal{U}$), there is a natural homomorphism $H^1(\mathcal{U}; \mathcal{O}) \to H^1(\mathcal{V}; \mathcal{O})$ induced by restricting a cochain (or a cocycle) over $\mathcal{U}$ to one over $\mathcal{V}$. The *first cohomology group of X with coefficients in $\mathcal{O}$* is the direct limit

$$H^1(X; \mathcal{O}) = \lim_{\longrightarrow \mathcal{U}} H^1(\mathcal{U}; \mathcal{O}).$$

Thus $H^1(X; \mathcal{O}) = 0$ means that every 1-cocycle becomes a 1-coboundary on a refinement, so every Mittag-Leffler problem is solvable. If X is a complex manifold, then the group $H^1(X; \mathcal{O})$ is isomorphic to the Dolbeault group $H^{0,1}_{\overline{\partial}}(X)$, and these groups vanish on any Stein manifold (see Theorem 2.6.7). If $\mathcal{U}$ is a *Leray cover* (meaning that the cohomology groups of all sets $U_j \in \mathcal{U}$ and of their finite intersections vanish; this holds if every U_j is Stein) then the natural homomorphism $H^1(\mathcal{U}; \mathcal{O}) \to H^1(X; \mathcal{O})$ is an isomorphism.

Let us now look at the *second (multiplicative) Cousin problem* which arises from the problem of finding holomorphic functions with prescribed zeros, or meromorphic functions with prescribed zeros and poles.

Consider an open cover $\mathcal{U} = \{U_i\}_{i \in I}$ of a complex space X and a collection of holomorphic functions $g_i \in \mathcal{O}(U_i)$ that do not vanish identically on any connected component of U_i. The collection $\{g_i\}_{i \in I}$ defines a *divisor D* on X if for any pair of indexes $i, j \in I$ there exists a nowhere vanishing holomorphic function $f_{ij} \in \mathcal{O}^*(U_{ij})$ such that $g_i = f_{ij} g_j$ holds on U_{ij}. (See Example 1.5.7 on p. 19.) The problem is to find a holomorphic function $f \in \mathcal{O}(X)$ that defines the divisor D; that is, such that f/g_i is a nonvanishing holomorphic function on U_i for every $i \in I$. A solution proceeds via the second Cousin problem: Given a collection $\{f_{ij}\}_{i,j\in I}$ of nonvanishing holomorphic functions $f_{ij}: U_{ij} \to \mathbb{C}^*$ satisfying the 1-cocycle condition

$$f_{ii} = 1, \quad f_{ij} f_{ji} = 1, \quad f_{ij} f_{jk} f_{ki} = 1$$

on the respective sets where all functions are defined, find nonvanishing holomorphic functions $f_j: U_j \to \mathbb{C}^*$ such that

$$f_i = f_{ij} f_j \quad \text{holds on } U_{ij}.$$

If such f_i exist, then $g_i/f_i = g_j/f_j$ on U_{ij} for every $i, j \in I$ which gives a solution $f \in \mathcal{O}(X)$. As before, we define the cohomology groups

$$H^1(\mathcal{U}; \mathcal{O}^*) = Z^1(\mathcal{U}; \mathcal{O}^*)/\delta(C^0(\mathcal{U}; \mathcal{O}^*)), \qquad H^1(X; \mathcal{O}^*) = \varinjlim H^1(\mathcal{U}; \mathcal{O}^*).$$

We can now state (an extension of) Oka's theorem from [443].

Theorem 5.2.1 *Assume that X is a complex space satisfying $H^1(X; \mathcal{O}) = 0$. Let $\mathcal{U} = \{U_i\}_{i\in I}$ be a cover of X by simply connected Stein open sets, and let $\{f_{ij}\} \in Z^1(\mathcal{U}; \mathcal{O}^*)$ be a multiplicative 1-cocycle. If there exist nonvanishing continuous functions $c_j: U_j \to \mathbb{C}^*$ satisfying $c_i = f_{ij} c_j$ on U_{ij} for every $i, j \in I$, then there also exist nonvanishing holomorphic functions $f_j: U_j \to \mathbb{C}^*$ such that $f_i = f_{ij} f_j$ on U_{ij}.*

Proof Since the sets U_j are simply connected, there exist continuous functions $g_j: U_j \to \mathbb{C}$ such that $c_j = e^{g_j}$. Setting $g_{ij} = g_i - g_j$ on U_{ij}, we have $e^{g_{ij}} = e^{g_i - g_j} = c_i/c_j = f_{ij}$, and hence g_{ij} is holomorphic. The collection g_{ij} clearly satisfies the additive 1-cocycle condition, so $\{g_{ij}\} \in Z^1(\mathcal{U}; \mathcal{O})$. From $H^1(X; \mathcal{O}) = 0$ and the assumption that the U_j are Stein we infer that $H^1(\mathcal{U}; \mathcal{O}) = 0$. Hence, $g_{ij} = h_i - h_j$ for some holomorphic functions $h_j: U_j \to \mathbb{C}$. The collection $f_j = e^{h_j}: U_j \to \mathbb{C}^*$ of nonvanishing holomorphic functions then satisfies $f_i/f_j = e^{h_i - h_j} = e^{g_{ij}} = f_{ij}$. □

We now give a cohomological formulation and proof of Oka's theorem. Let $\mathcal{C} = \mathcal{C}_X$ denote the sheaf of continuous functions and $\mathcal{C}^* = \mathcal{C}_X^*$ the sheaf of nonvanishing continuous functions on a complex space X.

Theorem 5.2.2 *If $(X, \mathcal{O})$ is a complex space satisfying $H^1(X; \mathcal{O}) = 0$, then the homomorphism $H^1(X; \mathcal{O}^*) \to H^1(X; \mathcal{C}^*)$, induced by the sheaf inclusion $\mathcal{O}^* \hookrightarrow \mathcal{C}^*$, is injective. In particular, if a Cousin II problem on X is solvable by continuous functions, then it is solvable by holomorphic functions. If in addition $H^2(X; \mathcal{O}) = 0$, then $H^1(X; \mathcal{O}^*) \to H^1(X; \mathcal{C}^*)$ is an isomorphism.*

Proof Let $\sigma(f) = \mathrm{e}^{2\pi \mathrm{i} f}$. Consider the *exponential sheaf sequence*:

$$\begin{array}{ccccccccc} 0 & \longrightarrow & \mathbb{Z} & \longrightarrow & \mathcal{O} & \stackrel{\sigma}{\longrightarrow} & \mathcal{O}^* & \longrightarrow & 1 \\ & & \downarrow & & \downarrow & & \downarrow & & \\ 0 & \longrightarrow & \mathbb{Z} & \longrightarrow & \mathcal{C} & \stackrel{\sigma}{\longrightarrow} & \mathcal{C}^* & \longrightarrow & 1 \end{array} \tag{5.1}$$

The vertical arrows are natural inclusions. Since $H^1(X;\mathcal{C}) = 0 = H^2(X;\mathcal{C})$, the relevant portion of the long exact sequence on cohomology reads:

$$\begin{array}{ccccccccc} H^1(X;\mathbb{Z}) & \longrightarrow & H^1(X;\mathcal{O}) & \longrightarrow & H^1(X;\mathcal{O}^*) & \stackrel{c_1}{\longrightarrow} & H^2(X;\mathbb{Z}) & \longrightarrow & H^2(X;\mathcal{O}) \\ & & \downarrow & & \downarrow & & \| & & \downarrow \\ & & 0 & \longrightarrow & H^1(X;\mathcal{C}^*) & \stackrel{c_1}{\longrightarrow} & H^2(X;\mathbb{Z}) & \longrightarrow & 0 \end{array} \tag{5.2}$$

Hence the map in the bottom row is an isomorphism $H^1(X;\mathcal{C}^*) \cong H^2(X;\mathbb{Z})$. If $H^1(X;\mathcal{O}) = 0$ then the *first Chern class map* c_1 in the first row is injective:

$$0 \longrightarrow H^1\big(X;\mathcal{O}^*\big) \stackrel{c_1}{\longrightarrow} H^2(X;\mathbb{Z}) \cong H^1\big(X;\mathcal{C}^*\big).$$

If also $H^2(X;\mathcal{O})$ then this map is an isomorphism. (If $H^q(X;\mathcal{O}) = 0$ for all $q > 0$ then we have $H^p(X;\mathcal{O}^*) \cong H^p(X;\mathcal{C}^*) \cong H^{p+1}(X;\mathbb{Z})$ for all $p > 0$.) □

Recall from Sect. 1.5 that

$$\mathrm{Vect}^1_{\mathrm{holo}}(X) = H^1\big(X;\mathcal{O}^*\big) = \mathrm{Pic}(X)$$

is the group of all equivalence classes of holomorphic line bundles over X, also called the *Picard group* of X. Likewise, $\mathrm{Vect}^1_{\mathrm{top}}(X) = H^1(X;\mathcal{C}^*)$ is the group of all equivalence classes of topological complex line bundles over X. Hence. Theorem 5.2.2 has the following interpretation. The last statement follows from $H^2(X;\mathbb{Z}) = 0$ when X is an open Riemann surface.

Corollary 5.2.3 *If X is a complex space with $H^1(X;\mathcal{O}) = H^2(X;\mathcal{O}) = 0$, then the natural map*

$$\mathrm{Vect}^1_{\mathrm{holo}}(X) \longrightarrow \mathrm{Vect}^1_{\mathrm{top}}(X)$$

is an isomorphism. In particular, a holomorphic line bundle over X is topologically trivial if and only if it is holomorphically trivial.

5.3 The Oka-Grauert Principle

Oka's theorem (see Theorem 5.2.2) was extended by Grauert [253, 254] to vector bundles of arbitrary rank and, more generally, to principal holomorphic fibre bundles over Stein spaces. (See also Cartan's exposition of Grauert's theorem in [88].) These generalizations are highly nontrivial since the structure group is nonabelian, and hence the cohomological methods do not apply directly. Since the classical Oka-Grauert principle is a special case of the main result of this chapter, Theorem 5.4.4, we shall not prove it separately; the main statements and applications of the classical theory can be found in Sects. 8.1–8.5. However, for historical reasons and by way of motivation we state here the following important special case concerning vector bundles.

Theorem 5.3.1 (Grauert [253, 254])

(i) *Every topological complex vector bundle over a Stein space admits a compatible holomorphic vector bundle structure.*
(ii) *Two holomorphic vector bundles over a Stein space that are equivalent as topological complex vector bundles are also equivalent as holomorphic vector bundles.*
(iii) *Every holomorphic vector bundle over an open Riemann surface and, more generally, over a* 1*-dimensional Stein space is holomorphically trivial.*

Theorem 5.3.1 can be expressed by saying that for every Stein space X, the natural map

$$\mathrm{Vect}^r_{\mathrm{holo}}(X) \to \mathrm{Vect}^r_{\mathrm{top}}(X)$$

of equivalence classes of rank r complex vector bundles over X is a bijection for every $r \in \mathbb{N}$. For part (iii), see Corollary 8.3.3 on p. 361.

For vector bundles of rank $r > 1$ the cohomological proof presented in Sect. 5.2 fails due to noncommutativity. We describe an alternative approach. Assume that $\pi: E \to X$ is a topological complex vector bundle with fibre $\mathbb{C}^r$. By using a partition of unity one can embed E as a topological complex vector subbundle in a trivial bundle $X \times \mathbb{C}^N$. Let $G_{r,N}$ denote the complex Grassmann manifold whose elements are r-dimensional complex linear subspaces of $\mathbb{C}^N$ (see Example 1.2.8 on p. 9). Consider the universal bundle $U_{r,N} \to G_{r,N}$ whose fibre over $\lambda \in G_{r,N}$ consists of all vectors $v \in \mathbb{C}^N$ belonging to the subspace $\lambda \subset \mathbb{C}^N$ (see Example 1.5.3, p. 17). Let $f: X \to G_{r,N}$ denote the continuous map sending the point $x \in X$ to the subspace $E_x = \pi^{-1}(x) \subset \mathbb{C}^N$, considered as an element of $G_{r,N}$. Then, E is isomorphic to the pullback $f^*U_{r,N}$ of the universal bundle $U_{r,N} \to G_{r,N}$. If $f_t: X \to G_{r,N}$ $(t \in [0, 1])$ is a homotopy of continuous maps, then the bundles $f_t^*U_{r,N}$ are topologically isomorphic to each other (see [305, 522]). Furthermore, the pullback by a holomorphic map is a holomorphic vector bundle over X. Therefore, Theorem 5.3.1(i) holds if we can prove that

Every continuous map $X \to G_{r,N}$ is homotopic to a holomorphic map.

Since the Grassmann manifold $G_{r,N}$ is $GL_N(\mathbb{C})$-homogeneous, the above statement is a special case of Grauert's results; see Theorem 5.3.2.

Consider now the equivalence problem for holomorphic vector bundles E, E' of rank r over X. There is an open cover $\mathcal{U} = \{U_j\}$ of X and holomorphic vector bundle isomorphisms $\theta_j: E|_{U_j} \to U_j \times \mathbb{C}^r$, $\theta'_j: E'|_{U_j} \to U_j \times \mathbb{C}^r$. Let

$$g_{ij} : U_{ij} \to GL_r(\mathbb{C}), \qquad g'_{ij} : U_{ij} \to GL_r(\mathbb{C})$$

denote the holomorphic transition maps for the two bundles:

$$\theta_i \circ \theta_j^{-1}(x, v) = \big(x, g_{ij}(x)v\big), \quad x \in U_{ij}, \ v \in \mathbb{C}^r,$$

and likewise for E'. A complex vector bundle isomorphism $\Phi: E \to E'$ is given by a collection of isomorphisms $\Phi_j: U_j \times \mathbb{C}^r \to U_j \times \mathbb{C}^r$ of the form

$$\Phi_j(x, v) = \big(x, \phi_j(x)v\big), \quad x \in U_j, \ v \in \mathbb{C}^r$$

with $\phi_j(x) \in GL_r(\mathbb{C})$, satisfying the compatibility conditions

$$\phi_i = g'_{ij}\phi_j g_{ij}^{-1} = g'_{ij}\phi_j g_{ji} \quad \text{on } U_{ij}. \tag{5.3}$$

Let $h: Z \to X$ denote the holomorphic fibre bundle with fibre $G = GL_r(\mathbb{C})$ and transition maps (5.3). Precisely, $Z|_{U_j} \cong U_j \times G$ for each j, and an element $(x, v) \in U_j \times G$ with $x \in U_{ij}$ is identified with the element $(x, v') \in U_i \times G$, where

$$v' = g'_{ij}(x)\, v\, g_{ji}(x).$$

A collection of maps $\phi_j: U_j \to G$ satisfying (5.3) is then a section $X \to Z$; hence, complex vector bundle isomorphisms $E \to E'$ correspond to sections $X \to Z$. Part (ii) of Theorem 5.3.1 now follows from the following result of Grauert [253, 254] (for Lie group fibres) and Grauert and Kerner [257] and Ramspott [466] (for complex homogeneous fibres).

Theorem 5.3.2 *If X is a Stein space and $h: Z \to X$ is a holomorphic fibre bundle with a complex homogeneous fibre whose structure group is a complex Lie group acting transitively on the fibre, then the inclusion $\Gamma_{\mathcal{O}}(X; Z) \hookrightarrow \Gamma(X; Z)$ of the space of holomorphic sections into the space of continuous sections is a weak homotopy equivalence.*

Theorem 5.3.2 is a special case of Theorem 5.4.4 in the following section.

Further results can be found in Sects. 8.2–8.6. The most direct development of the Oka-Grauert theory was given by Leiterer [386], mainly following Cartan [88]. Heunemann [291], Leiterer [383, 384], Pankov [450] and Sebbar [494] obtained the Oka-Grauert principle for holomorphic vector bundles on strongly pseudoconvex Stein domains which are continuous or smooth up to the boundary. We refer to Sect. 8.12 for results of modern Oka theory in this setting. A different proof and an extension of the classical theory to 1-convex manifolds is due to Henkin and Leiterer [287]; see also the example by Colţoiu [104]. For complex Banach-valued vector bundles over Stein spaces, see Bungart [72].

Equivariant versions of the Oka-Grauert principle were obtained by Heinzner and Kutzschebauch [286], Kutzschebauch, Lárusson and Schwarz [357–360], and Schwarz [493]. Their results involve a considerable amount of geometric invariant theory and will not be treated here.

5.4 What is an Oka Manifold?

The class of Oka manifolds was introduced in the author's paper [199] in 2009. The simplest characterization of this class is given by the following Runge approximation property for holomorphic maps from compact convex subsets of complex Euclidean spaces; we adopt it as the definition.

Definition 5.4.1 A complex manifold Y is an *Oka manifold* if every holomorphic map $K \to Y$ from (a neighborhood of) a compact convex set $K \subset \mathbb{C}^n$ for any $n \in \mathbb{N}$ can be approximated uniformly on K by entire holomorphic maps $\mathbb{C}^n \to Y$. The class of Oka manifolds is denoted Oka.

The main result of this chapter, Theorem 5.4.4, says that maps $X \to Y$ from any reduced Stein space X to an Oka manifold Y satisfy all forms of the Oka principle considered in the literature; the same holds for sections of stratified holomorphic fibre bundles with Oka fibres over reduced Stein spaces. Furthermore, all Oka properties which include either approximation or interpolation condition are pairwise equivalent; see Proposition 5.15.1. Hence, the term *Oka manifold* can be used either in the sense of Definition 5.4.1, or as a manifold Y satisfying the conclusion of Theorem 5.4.4 and Corollary 5.4.5.

It is sometimes convenient to reduce the testing family of compact convex sets in Definition 5.4.1 to the following subclass. Let $z=(z_1,\dots,z_n)$ be complex coordinates on $\mathbb{C}^n$, with $z_j = x_j + \mathrm{i}y_j$.

Definition 5.4.2 A compact convex set K in $\mathbb{C}^n$ is *special* if

$$K = \big\{z \in Q : y_n \le h(z_1,\dots,z_{n-1},x_n)\big\}, \tag{5.4}$$

where Q is a cube

$$Q = \big\{z \in \mathbb{C}^n : |x_j| \le a_j,\ |y_j| \le b_j,\ j=1,\dots,n\big\} \tag{5.5}$$

and h is a continuous concave function with values in $(-b_n, b_n)$. Such a pair (K,Q) is called a *special convex pair* in $\mathbb{C}^n$.

The following condition was introduced in [195, 197].

Definition 5.4.3 ([197]) Let $n \in \mathbb{N}$. A complex manifold Y enjoys the *n-dimensional Convex Approximation Property* (CAP_n) if for every special convex pair (K,Q) in $\mathbb{C}^n$, any holomorphic map $K \to Y$ can be approximated uniformly on K by holomorphic maps $Q \to Y$. A manifold Y enjoys CAP if it enjoys CAP_n for all $n \in \mathbb{N}$.

By definition, an Oka manifold satisfies CAP. The converse implication follows from Theorem 5.4.4 applied with $X=\mathbb{C}^n$, K a compact convex set in $\mathbb{C}^n$ and $P=\{p\}$ a singleton. One may further reduce the testing family for CAP to special polyhedral pairs (see Definition 5.15.3 and Lemma 5.15.4).

The following is the main result of Oka theory for sections of stratified fibre bundles with Oka fibres over reduced Stein spaces.

Theorem 5.4.4 *Let X be a reduced Stein space, Z be a complex space, and $\pi: Z \to X$ be a stratified holomorphic fibre bundle whose fibres enjoy* CAP. *Then, sections $X \to Z$ satisfy the following* Parametric Oka Property with Approximation and (Jet) Interpolation (POPAI, resp. POPAJI):

Given a compact $\mathcal{O}(X)$-convex subset K of X, a closed complex subvariety X' of X, compact sets $P_0 \subset P$ in a Euclidean space $\mathbb{R}^m$, and a continuous map $f: P \times X \to Z$ such that

(a) *for every $p \in P$, $f(p,\cdot): X \to Z$ is a section of $Z \to X$ that is holomorphic on a neighborhood of K (independent of p) and such that $f(p,\cdot)|_{X'}$ is holomorphic on X', and*
(b) *$f(p,\cdot)$ is holomorphic on X for every $p \in P_0$,*

there is a homotopy $f_t: P \times X \to Z$ $(t \in [0,1])$, with $f_0 = f$, such that f_t enjoys properties (a) *and* (b) *for all $t \in [0,1]$ and also the following hold:*

(i) *$f_1(p,\cdot)$ is holomorphic on X for all $p \in P$,*
(ii) *f_t is uniformly close to f on $P \times K$ for all $t \in [0,1]$, and*
(iii) *$f_t = f$ on $(P_0 \times X) \cup (P \times X')$ for all $t \in [0,1]$.*

If, in addition to (a) *and* (b), *every section $f(p,\cdot): X \to Z$ is holomorphic on a neighborhood of X' (independent of $p \in P$), then for every coherent analytic sheaf of ideals $\mathcal{S} \subset \mathcal{O}_X$ supported on X' the homotopy f_t can be chosen such that $\delta(f_t(p,\cdot), f(p,\cdot)) \in \mathcal{S}$ for all $t \in [0,1]$ and $p \in P$ (see Definition* 1.3.2). *In particular, f_t may be chosen tangent to f to a given finite order along X'.*

If the Stein space X has finite dimension n, then the conclusions hold if each fibre Y of π enjoys CAP_N *with* $N = \dim X + \dim Y$.

Theorem 5.4.4 has a complex genesis spanning seven decades, from 1939 to 2010. Its proof is outlined at the end of this section; the details are provided in Sects. 5.8–5.13, proceeding from the simplest to the most general case. The proof given here is based on my papers [195, 197, 199, 202] which in turn build upon the works of Grauert [253, 254], Henkin and Leiterer [287], Gromov [266], and my joint papers with Prezelj [217–219]. An up-to-the-boundary version of this result (on strongly pseudoconvex domains) is given in Sect. 8.12.

Theorem 5.4.4 provides an affirmative answer to Gromov's question [266, p. 881, 3.4.(D)] whether Runge approximation on a certain class of compact sets in Euclidean spaces suffices to infer the Oka property.

An important special case of Theorem 5.4.4 concerns maps $X \to Y$ from a reduced Stein space X to an Oka manifold Y, considered as sections of the trivial bundle $X \times Y \to X$. Since this is how the result is most often used, we state separately the following special case of Theorem 5.4.4 which is called the *basic Oka property with approximation and interpolation*, abbreviated BOPAI, for maps from Stein spaces to Oka manifolds.

Corollary 5.4.5 (BOPA(J)I for maps to Oka manifolds) *Assume that X is a reduced Stein space, K is a compact $\mathcal{O}(X)$-convex subset of X, and X' is a closed complex subvariety of X. Let Y be an Oka manifold endowed with a Riemannian distance function* dist. *Given a continuous map $f: X \to Y$ which is holomorphic on* *a neighborhood of K and on the subvariety X', there exists for every $\epsilon > 0$ a homotopy of continuous maps $f_t : X \to Y$ $(t \in [0,1])$ satisfying the following conditions:*

(a) *$f_0 = f$ and the map $f_1 : X \to Y$ is holomorphic,*
(b) *there is a neighborhood U of K such that for every $t \in [0,1]$ the restriction $f_t|_U$ is holomorphic and satisfies $\sup_{p \in K} \mathrm{dist}(f_t(p), f(p)) < \epsilon$, and*
(c) *$f_t|_{X'} = f|_{X'}$ holds for all $t \in [0,1]$.*

If in addition f is holomorphic on a neighborhood of the subvariety X', then we also have the following jet interpolation condition:

(c′) *given $r \in \mathbb{N}$, there is a neighborhood V of $K \cup X'$ such that f_t is holomorphic on V and agrees with $f = f_0$ to order r on X' for every $t \in [0,1]$.*

If $\dim X < \infty$, then the above conclusions hold provided that the manifold Y enjoys $\mathrm{CAP}_{\dim X + \dim Y}$ (see Definition 5.4.3).

Corollary 5.4.5 can be seen as a nonlinear version of the Cartan-Oka-Weil theorem; see Theorem 2.8.4 on p. 62. Intuitively speaking, it says that an Oka manifold has big (flexible) complex structure, in the sense that it can receive nontrivial holomorphic maps from any source Stein manifold, with approximation and interpolation conditions. This is just the opposite of Kobayashi hyperbolic manifolds whose structure is small (rigid) in a precise sense.

By combining Corollary 5.4.5 with Theorem 3.8.1, we get the following Mergelyan type theorem for maps from Stein spaces to Oka manifolds.

Corollary 5.4.6 *In addition to the assumptions in Corollary 5.4.5, assume that E is a smooth compact totally real submanifold of $X_{\mathrm{reg}} \setminus X'$ such that the compact set $K \cup E$ is $\mathcal{O}(X)$-convex and the restriction of the map $f : X \to Y$ to E is of class $\mathcal{C}^r(E)$ for some $r \in \mathbb{Z}_+$. Then there is a homotopy $f_t : X \to Y$ $(t \in [0,1])$ satisfying the conclusion of Corollary 5.4.5 such that $f_t|_E \in \mathcal{C}^r(E)$ approximates $f|_E$ in the $\mathcal{C}^r(E)$ topology for every $t \in [0,1]$.*

Let us also state the special case of this result for open Riemann surfaces.

Corollary 5.4.7 *Let X be an open Riemann surface, K be a compact set in X, E be a finite union of smooth compact curves in X such that $X \setminus (K \cup E)$ has no relatively compact connected components, and $p_1, \ldots, p_m \in K$ be points. Assume that Y is an Oka manifold and $f : X \to Y$ is a continuous map which is holomorphic in a neighborhood of K and satisfies $f|_E \in \mathcal{C}^r(E)$. Then there is a holomorphic map $F : X \to Y$ which approximates f uniformly on K such that $F|_E$ approximates $f|_E$ in $\mathcal{C}^r(E)$ and F agrees with f to a given finite order at each of the points $p_1, \ldots, p_m$.*

Intuitively speaking, these results say the following:

In the absence of topological obstructions, holomorphic maps from reduced Stein spaces to Oka manifolds enjoy the same properties as holomorphic functions on Stein spaces.

Hence, Stein spaces are natural sources of holomorphic maps, while Oka manifolds are natural targets. Indeed, the class of Oka manifolds is dual to the class of Stein manifolds in a sense that has been made precise by means of abstract homotopy theory; see Lárusson [370–373]. This homotopy-theoretic point of view on Oka theory is presented in Sect. 7.5.

The individual parts of Theorem 5.4.4 (such as approximation, interpolation, jet interpolation, the combination of both, with or without parameters) define the corresponding *Oka properties* of the manifold Y. We show in Sect. 5.15 that these Oka properties are pairwise equivalent, and hence any one of them characterizes the class of Oka manifolds (see Proposition 5.15.1 on p. 259). When $Y = \mathbb{C}$ is the complex number field, the Oka property with approximation is just the Oka-Weil approximation theorem, while the Oka property with interpolation is the Oka-Cartan extension theorem. When X and Y have nontrivial topology, we must of course assume that the given map from a subset of X to Y extends to a countinuous map $X \to Y$.

The full conclusion of Theorem 5.4.4 is expressed by saying that sections $X \to Z$ satisfy the *parametric Oka principle with approximation and (jet) interpolation*. By omitting approximation and interpolation, it means that a continuous map $f: P \to \Gamma_{\mathcal{C}}(X, Z)$ with $f(P_0) \subset \Gamma_{\mathcal{O}}(X, Z)$ can be deformed to a continuous map $f_1: P \to \Gamma_{\mathcal{O}}(X, Z)$ by a homotopy that is fixed on P_0:

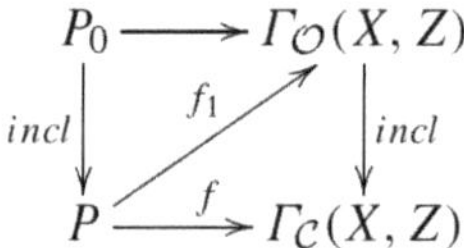

In the *nonparametric case* (with P a singleton and $P_0 = \varnothing$) the conclusion is called the *basic Oka principle with approximation and (jet) interpolation*.

Since every complex homogeneous manifold is an Oka manifold in view of Grauert's approximation theorem (see Proposition 5.6.1), Theorem 5.4.4 includes the classical Oka-Grauert theory (see Theorem 5.3.2 and the results in Sects. 8.1–8.5). A comparison between the Oka condition and other holomorphic flexibility properties of complex manifolds is discussed in Chap. 7 where the reader can also find a discussion of *Oka maps* (see Sect. 7.4).

In the following section we collect a few basic properties of Oka manifolds and some elementary consequences of Theorem 5.4.4.

Outline of proof of Theorem 5.4.4. Assume first that X is a Stein manifold and $\pi: Z \to X$ is a holomorphic fibre bundle whose fibre Y satisfies CAP.

We exhaust X by an increasing sequence of compact strongly pseudoconvex domains A_k such that $A_{k+1} = A_k \cup B_k$ is obtained by attaching to A_k either a small convex bump B_k, or a special handle whose core is a totally real disc attached to bA_k along a complex tangential sphere, so that the fibre bundle is trivial over a neighborhood of B_k. A holomorphic section $f: X \to Z$ in a given homotopy class is found as a locally uniform limit $f = \lim_{k\to\infty} f_k$ of a sequence of continuous sections $f_k: X \to Z$ such that f_k is holomorphic over a neighborhood of A_k. In the induction step we find a section f_{k+1} that approximates f_k uniformly on a neighborhood of A_k and is homotopic to it. We treat separately the extension across a convex bump (the *noncritical case*) and the crossing of a critical level of an exhaustion function (the *critical case*).

The *noncritical case* is considered in Sect. 5.10. (This is the only part of the proof of Theorem 5.4.4 where we use the convex approximation property of the fibres of $Z \to X$.) We begin by approximating f_k on a neighborhood of the attaching set $C_k = A_k \cap B_k$ by a holomorphic section g_k defined on a neighborhood of B_k. Since the bundle is trivial there, this is a Runge approximation problem for maps to the fibre Y, and it is here that CAP of the fibre is invoked. We then glue f_k and g_k into a section f_{k+1} which is holomorphic over a neighborhood of $A_{k+1} = A_k \cup B_k$. In the classical case when f_k and g_k are maps to a complex Lie group G, one has $f_k = g_k \cdot \gamma$ on C_k, where $\gamma = g_k^{-1} \cdot f_k \colon C_k \to G$ is a holomorphic map close to the identity. By Cartan's lemma [260, p. 88] we split γ as a product $\gamma = \beta \cdot \alpha^{-1}$ of holomorphic maps $\alpha \colon A_k \to G$, $\beta \colon B_k \to G$, and we take $f_k \cdot \alpha = g_k \cdot \beta$ as the next map f_{k+1}. In the general case considered here, Cartan's lemma no longer applies. We work instead with *thick sections*, also called *local holomorphic sprays of sections*. These are families of holomorphic sections, depending holomorphically on a parameter in a neighborhood of the origin in a Euclidean space, which are submersive with respect to the parameter. Given a thick section f_k over A_k, we approximate f_k over C_k by a thick section g_k over B_k as before. We then find a fibre preserving biholomorphic transition map γ close to the identity map and satisfying $f_k = g_k \circ \gamma$. Next, we split $\gamma = \beta \circ \alpha^{-1}$ as a composition of two holomorphic maps, α over A_k, resp. β over B_k (see Proposition 5.8.1, p. 235). Then, $f_k \circ \alpha = g_k \circ \beta$ holds over C_k, and hence these two thick sections amalgamate into a thick section f_{k+1} that is holomorphic over a neighborhood of A_{k+1}. The induction may proceed.

The *critical case* (see Sect. 5.11) is obtained by using Theorem 3.8.1 (p. 88) to approximately extend the holomorphic section f_k across the stable manifold at the critical point, thereby reducing the problem to the noncritical case.

In the case of a stratified fibre bundle, the proof proceeds by induction on strata; the main step is furnished by Proposition 5.12.1 on p. 247. Every step of the proof can also be done in the parametric case, and this shows that a parametric version of CAP (called PCAP) of the fibres of $Z \to X$ implies the parametric Oka principle for sections $X \to Z$. The proof is then completed by showing that CAP implies PCAP. We refer to Sect. 5.13 for the details.

5.5 Basic Properties of Oka manifolds

In this section we collect some basic properties of Oka manifolds and observe a few consequences of Theorem 5.4.4. We begin with the following summary.

Theorem 5.5.1 *Let X be a reduced Stein space of positive dimension and Y be a connected Oka manifold. Then, the following hold.*

(a) *The set of holomorphic maps $X \to Y$ with dense image is dense in $\mathcal{O}(X, Y)$. (Recall that $\mathcal{O}(X, Y)$ is a separable Baire space, see Sect.* 1.1.)

(b) *If* X *is smooth and* $\dim Y \geq 2\dim X$, *then the set of immersions* $X \to Y$ *is residual in* $\mathcal{O}(X, Y)$.
(c) *If* X *is smooth and* $\dim Y \geq 2\dim X + 1$, *then the set of injective immersions* $X \to Y$ *is residual in* $\mathcal{O}(X, Y)$.
(d) *If* $\dim X \geq \dim Y$, *then every continuous map* $X \to Y$ *is homotopic to a surjective strongly dominating holomorphic map* $F: X \to Y$, *i.e., such that for every* $y \in Y$ *there exists* $x \in X_{\mathrm{reg}}$ *such that* $\mathrm{d}F_x: \mathrm{T}_x X \to \mathrm{T}_y Y$ *is surjective.*
(e) Y *admits a strongly dominating holomorphic map* $\mathbb{C}^{\dim Y} \to Y$.

Proof Let dist be a Riemannian distance function on Y.

Proof of (a): Fix a holomorphic map $f_0: X \to Y$, a compact set $K \subset X$ and a number $\epsilon > 0$. Replacing K by $\widehat{K}_{\mathcal{O}(X)}$, we may assume that K is $\mathcal{O}(X)$-convex. Choose a countable discrete set $\{a_j\}$ in X contained in $X \setminus K$ and a dense set $\{b_j\}_{j \in \mathbb{N}} \subset Y$. Since Y is connected, it is elementary to find a homotopy of continuous maps $f_t: X \to Y$ ($t \in [0, 1]$) such that f_0 is the given map, f_t agrees with f_0 on a neighborhood U of K for all $t \in [0, 1]$, and $f_1(a_j) = b_j$ holds for all $j \in \mathbb{N}$. By Corollary 5.4.5, we can deform f_1 to a holomorphic map $F: X \to Y$ such that $\sup_{x \in K} \mathrm{dist}(F(x), f_0(x)) < \epsilon$ and $F(a_j) = f_1(a_j) = b_j$ for all $j \in \mathbb{N}$. This shows that $F(X)$ is dense in Y. Since F was chosen arbitrarily close to a given map f_0 on any chosen compact K, we see that the set of such maps is dense in $\mathcal{O}(X, Y)$.

Proof of (b) and (c): This is Corollary 8.9.3 (p. 388) which is a consequence of Theorem 5.4.4 and the general position results in Theorem 8.9.1.

Proof of (d): If Y is compact, the proof is an easy consequence of Corollary 5.4.5; we give a sketch and refer to [179, Theorem 1.1] for the details.

Assume for simplicity that X is smooth and of dimension $n = \dim Y$. We cover Y with finitely many open sets $V_1, \ldots, V_m$ biholomorphic to the ball $\mathbb{B}^n \subset \mathbb{C}^n$. Choose slightly smaller compact balls $L_j \Subset V_j$ such that $\bigcup_{j=1}^m L_j = Y$. Pick pairwise disjoint closed coordinate balls $K_1, \ldots, K_m \subset X$ whose union $K = \bigcup_{j=1}^m K_j$ is $\mathcal{O}(X)$-convex. (This holds if the sets K_j are chosen small enough and $\mathcal{O}(X)$-convex.) Let $f: X \to Y$ be a continuous map. It is easy to deform f to a continuous map $f_0: X \to Y$ which is holomorphic on a neighborhood of K and maps a neighborhood of each K_j biholomorphically onto a neighborhood of L_j. By Corollary 5.4.5, we can approximate f_0 uniformly on a neighborhood of K by a holomorphic map $F: X \to Y$ homotopic to f_0. It is immediate that F is strongly dominating provided that the approximation is close enough. A similar proof applies when $\dim X > \dim Y$. If Y is noncompact, we apply the same argument inductively on a countable family of coordinate balls in X and Y; see [179, Theorem 1.1] for the details.

Proof of (e): Apply part (d) to a constant map $\mathbb{C}^{\dim Y} \to y_0 \in Y$. □

Recall that a holomorphic map $f: \mathbb{C}^n \to Y$ is said to be *dominating* at the point $y_0 = f(0) \in Y$ if the differential $\mathrm{d}f_0: \mathrm{T}_0\mathbb{C}^n \to \mathrm{T}_{y_0} Y$ is surjective; if such a map exists, the manifold Y is said to be *dominable at* y_0. A complex manifold which is dominable at every point is called *strongly dominable* (see Definition 7.1.7). Theorem 5.5.1(e) shows that every Oka manifold is strongly dominable; furthermore, there is a holomorphic map $f: \mathbb{C}^n \to Y$ with $n = \dim Y$ such that the family of maps $\{f \circ \phi_a\}_{a \in \mathbb{C}^n}$, where $\phi_a : \mathbb{C}^n \to \mathbb{C}^n$ is the translation $z \mapsto z + a$, dominates at every point of Y. The converse is not known. We discuss this topic in more depth in Sect. 7.1.

Theorem 5.5.1(e) has the following important consequence.

Corollary 5.5.2 *The Kobayashi pseudometric and the Kobayashi-Eisenman volume forms of an Oka manifold vanish identically. In particular, a compact Oka manifold is not of Kodaira general type.*

The last statement follows from the result of Kobayashi and Ochiai [342] that a compact connected complex manifold Y which is dominable by $\mathbb{C}^{\dim Y}$ is not of Kodaira general type.

Remark 5.5.3 In connection with part Theorem 5.5.1(e), we do not know whether every Oka manifold Y is the image of a *locally biholomorphic* map $\mathbb{C}^n \to Y$ with $n = \dim Y$. A related problem is whether locally biholomorphic self-maps of $\mathbb{C}^n$ for $n > 1$ satisfy the Runge approximation theorem; see Problem 9.12.3 and Theorem 9.13.4. □

We now discuss some other consequences of Theorem 5.4.4.

The following result, whose basic version is due to Ramspott [466], follows from Theorem 5.4.4 and Proposition 5.6.1 to the effect that every complex homogeneous manifold is Oka. In Ramspott's theorem the structure group of the bundle is assumed to be a complex Lie group. This condition is superfluous, it is only the fibre that matters.

Corollary 5.5.4 *If X is a Stein space and $\pi : Z \to X$ is a stratified holomorphic fibre bundle with complex homogeneous fibres, then its sections $X \to Z$ satisfy the parametric Oka principle with approximation and jet interpolation.*

Corollary 5.5.5 *Assume that X is a Stein space, Y is an Oka manifold, K is a compact $\mathcal{O}(X)$-convex subset of X, U is an open neighborhood of K in X, and $f : U \to Y$ is a null-homotopic holomorphic map. Then f can be approximated uniformly on K by holomorphic maps $X \to Y$.*

Proof Since $f : U \to Y$ is null-homotopic, there exists a continuous map $F : X \to Y$ that equals f in some neighborhood of K. The conclusion then follows from Corollary 5.4.5. □

Corollary 5.5.6 (The weak homotopy equivalence principle) *If $\pi : Z \to X$ is a stratified holomorphic fibre bundle over a reduced Stein space X whose fibres are Oka manifolds, then the inclusion $\iota : \Gamma_{\mathcal{O}}(X, Z) \hookrightarrow \Gamma_{\mathcal{C}}(X, Z)$ of the space of holomorphic sections of π into the space of continuous sections is a* weak homotopy equivalence, *i.e., the induced map of homotopy groups*

$$\pi_k(\iota) : \pi_k\big(\Gamma_{\mathcal{O}}(X, Z)\big) \longrightarrow \pi_k\big(\Gamma_{\mathcal{C}}(X, Z)\big)$$

is an isomorphism for every $k = 0, 1, 2, \ldots$.

In particular, if X is a reduced Stein space and Y is an Oka manifold then the inclusion $\mathcal{O}(X, Y) \hookrightarrow \mathcal{C}(X, Y)$ is a weak homotopy equivalence.

Proof Denote by D^{k+1} the closed unit ball in $\mathbb{R}^{k+1}$ and by $S^k = bD^{k+1}$ the k-sphere. Applying Theorem 5.4.4 with $P = S^k$, $P_0 = \varnothing$, $K = \varnothing$ and $X' = \varnothing$ we see that every continuous map $S^k \to \Gamma_{\mathcal{C}}(X, Z)$ can be deformed to a continuous map $S^k \to \Gamma_{\mathcal{O}}(X, Z)$, so $\pi_k(\iota)$ is surjective. Applying Theorem 5.4.4 with $P = D^{k+1}$, $P_0 = S^k$, $K = \varnothing$ and $X' = \varnothing$ we conclude that every map $f: D^{k+1} \to \Gamma_{\mathcal{C}}(X, Z)$ such that $f(S^k) \subset \Gamma_{\mathcal{O}}(X, Z)$ can be deformed to a map $D^{k+1} \to \Gamma_{\mathcal{O}}(X, Z)$ by a homotopy that is fixed on S^k. This means that $\pi_k(\iota)$ is injective. □

The analogous result holds for the space of sections that agree with a given holomorphic section $f: X' \to Z|_{X'}$ on a closed complex subvariety X' of a Stein space X. Let us denote by

$$\iota: \Gamma_{\mathcal{O}}\big(X, Z; X', f\big) \hookrightarrow \Gamma_{\mathcal{C}}\big(X, Z; X', f\big) \tag{5.6}$$

the inclusion of the space of holomorphic sections $F: X \to Z$ with $F|_{X'} = f$ into the space of continuous sections satisfying the same condition. The following result is proved in the same way as Corollary 5.5.6.

Corollary 5.5.7 *If $\pi: Z \to X$ satisfies the hypotheses of Theorem* 5.4.4 *and $f: X' \to Z|_{X'}$ is a holomorphic section over a closed complex subvariety X' of X, then the inclusion* (5.6) *is a weak homotopy equivalence.*

Problem 5.5.8 Are the inclusions in Corollary 5.5.6 homotopy equivalences?

While this is not known in general, Lárusson [376] obtained the following affirmative result for Stein manifolds of finite type.

Theorem 5.5.9 *Assume that X is a Stein manifold which admits a strongly plurisubharmonic Morse exhaustion function with finitely many critical points. Then, for any Oka manifold Y the space of holomorphic mappings $\mathcal{O}(X, Y)$ is a deformation retract of the space of continuous mappings $\mathcal{C}(X, Y)$. This holds in particular if X is an affine algebraic manifold.*

Applying Theorem 5.4.4 with $K = \varnothing$, $P = \{p\}$ and $P_0 = \varnothing$ gives

Corollary 5.5.10 (The Oka principle for extensions) *Suppose that X is a reduced Stein space, X' is a closed complex subvariety of X, and Y is an Oka manifold. Given a continuous map $F: X \to Y$ such that $F|_{X'} = f: X' \to Y$ is holomorphic, there is a homotopy $F_t: X \to Y$ ($t \in [0, 1]$) such that $F_0 = F$, F_1 is holomorphic, and the homotopy is fixed on X'.*

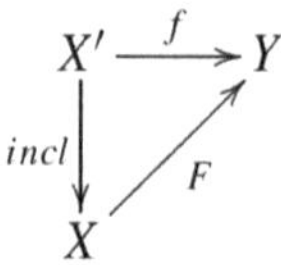

Given a map $\pi: E \to B$, we say that a map $F: X \to E$ is a *lifting* of a map $f: X \to B$ if $\pi \circ F = f$. Similarly one defines homotopies of liftings.

Corollary 5.5.11 (The Oka principle for liftings) *Assume that $\pi: E \to B$ is a stratified holomorphic fibre bundle all of whose fibres are Oka manifolds. If X is a reduced Stein space and $f: X \to B$ is a holomorphic map, then any continuous lifting $F_0: X \to E$ of f admits a homotopy of liftings $F_t: X \to E$ ($t \in [0, 1]$) such that F_1 is holomorphic. Furthermore, if F_0 is holomorphic on a subset $K \cup X' \subset X$ as in Theorem* 5.4.4, *then the homotopy $\{F_t\}$ can be chosen to satisfy properties* (i) *and* (ii) *in Theorem* 5.4.4.

Proof Assume first that $\pi: E \to B$ is a holomorphic fibre bundle with Oka fibre Y. Let $h: f^*E \to X$ denote the pullback bundle whose fibre over a point $x \in X$ is $E_{f(x)} \cong Y$.

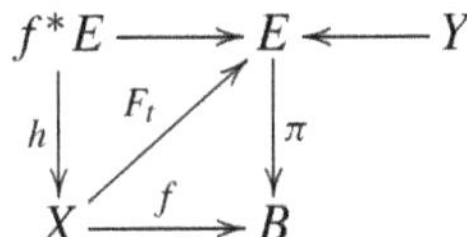

Sections $X \to f^*E$ are in one-to-one correspondence with liftings $F: X \to E$ of the map $f: X \to B$. Since the fibre Y of $h: f^*E \to X$ is Oka, the conclusion follows from Theorem 5.4.4. In the general case we stratify X so that the strata are mapped by f to the strata of B; then f^*E is also a stratified fibre bundle over X and we conclude the proof as before. □

5.6 Examples of Oka Manifolds

This section contains a survey of examples of Oka manifolds; additional ones can be found in Sect. 6.4 where we consider elliptic and subelliptic manifolds, and also in Sect. 7.6. We adopt a historical and pedagogical approach, providing simple direct proofs where possible in spite of subsequent generalizations.

Complex Homogeneous Manifolds

Proposition 5.6.1 (Grauert [252]) *Every complex Lie group and, more generally, every complex homogeneous manifold is an Oka manifold.*

Proof Let G be a complex Lie group. Denote by $1 \in G$ the identity element and by $\exp: \mathfrak{g} = \mathrm{T}_1 G \cong \mathbb{C}^n \to G$ the exponential map. Assume that K is a compact convex set in $\mathbb{C}^n$, $U \subset \mathbb{C}^n$ is an open convex neighborhood of K, and $f: U \to G$ is a holomorphic map.

If $f(K)$ lies sufficiently close to $1 \in G$ then there is a unique holomorphic map $h: U' \to \mathfrak{g}$ on a smaller neighborhood of K such that $f = \exp h$. Approximating h

uniformly on K by an entire map $\widetilde{h}: \mathbb{C}^n \to \mathfrak{g}$ and taking $\widetilde{f} = \exp \widetilde{h}: \mathbb{C}^n \to G$ gives a desired entire approximation of f.

In general we proceed as follows. We may assume that the origin $0 \in \mathbb{C}^n$ is contained in $\mathring{K}$. Set $f_t(z) = f(tz)$ for $t \in [0,1]$ and $z \in U$; then $f_1 = f$ and f_0 is the constant map $\mathbb{C}^n \ni z \to f(0) \in G$. Choose $N \in \mathbb{N}$. Using the group structure on G we can write

$$f = f_1 = f_1 (f_{\frac{N-1}{N}})^{-1} \cdot f_{\frac{N-1}{N}} (f_{\frac{N-2}{N}})^{-1} \cdots f_{\frac{1}{N}} (f_0)^{-1} \cdot f_0.$$

If N is sufficiently big then each of the quotients $f_{\frac{k}{N}} (f_{\frac{k-1}{N}})^{-1}$ $(k = 1, \ldots, N)$ is so close to 1 so that it admits a holomorphic logarithm $h_k: K \to \mathfrak{g}$. Approximating h_k by an entire map $\widetilde{h}_k: \mathbb{C}^n \to \mathfrak{g}$ and taking $\widetilde{g}_k = \exp \widetilde{h}_k$ and $\widetilde{f} = \widetilde{g}_N \widetilde{g}_{N-1} \cdots \widetilde{g}_1 f_0: \mathbb{C}^n \to G$ gives a desired approximation of f.

The proof for a homogeneous manifold Y is quite similar. Let G be a complex Lie group acting holomorphically and transitively on Y. Note that the holomorphic map $s: Y \times \mathfrak{g} \to Y$, $s(y, v) = \exp v \cdot y$, is submersive in a neighborhood of the zero section in $Y \times \mathfrak{g}$. Let $K \subset \mathbb{C}^n$ be a compact convex set and $f: K \to Y$ be a holomorphic map. Let f_t be a homotopy defined as above. If $N \in \mathbb{N}$ is big enough then for every $k = 1, 2, \ldots, N$ the implicit function theorem gives a holomorphic map $h_k: K \to \mathfrak{g}$ such that $f_{\frac{k}{N}} = \exp h_k \cdot f_{\frac{k-1}{N}}$. Approximating each h_k by a holomorphic map $\widetilde{h}_k: \mathbb{C}^n \to \mathfrak{g}$ completes the proof as before. □

Example 5.6.2 The following hypersurface in $\mathbb{C}^n$ is called the *null quadric*:

$$\mathcal{A} = \{(z_1, \ldots, z_n) \in \mathbb{C}^n : z_1^2 + z_2^2 + \cdots + z_n^2 = 0\}. \tag{5.7}$$

This variety plays a major role in the theory of conformal minimal immersions of open Riemann surfaces into $\mathbb{R}^n$ for any $n \geq 3$; see e.g. Osserman [446] for a classical survey of this subject. It is easily seen that the punctured null quadric $\mathcal{A}_* = \mathcal{A} \setminus \{0\}$ is a homogeneous space of the complex Lie group $SO(n, \mathbb{C}) \times \mathbb{C}^*$, where $SO(n, \mathbb{C})$ is the connected component of the identity in the Lie group $O(n, \mathbb{C}) = \{A \in GL(n, \mathbb{C}) : A \cdot A^t = \mathrm{Id}\}$. Hence, $\mathcal{A}_*$ is an Oka manifold. This enables effective applications of Oka theory in the classical theory of minimal surfaces. For recent results, see [13, 15, 16] and the references therein. □

Ascent and Descent of Oka Property in Fibre Bundles

We prove that the Oka property passes both up and down in a holomorphic fibre bundle with Oka fibre (see Theorem 5.6.5). The proof is especially simple for holomorphic covering projections.

Proposition 5.6.3 *If $\pi: E \to B$ is a holomorphic covering map of complex manifolds, then B is an Oka manifold if and only if E is an Oka manifold.*

Proof Assume that E is an Oka manifold. Let K be compact convex set in $\mathbb{C}^n$ and $f: K \to B$ a holomorphic map. By the lifting property for coverings there exists a holomorphic map $h: K \to E$ such that $\pi \circ h = f$. Since E is an Oka manifold, we can approximate h, uniformly on K, by a holomorphic map $\widetilde{h}: \mathbb{C}^n \to E$. The map $\widetilde{f} = \pi \circ \widetilde{h}: \mathbb{C}^n \to B$ then approximates f on K, so B is an Oka manifold. A similar argument proves the converse implication. □

Corollary 5.6.4 *The following are equivalent for a Riemann surface Y:*

(a) *Y is an Oka manifold.*
(b) *Y is not Kobayashi hyperbolic.*
(c) *Y is one of the surfaces $\mathbb{P}^1$, $\mathbb{C}$, $\mathbb{C}^*$, or a torus.*

Proof The universal covering of any Riemann surface is one of the Riemann surfaces $\mathbb{P}^1$, $\mathbb{C}$, or the disc $\mathbb{D} = \{z \in \mathbb{C} : |z| < 1\}$. $\mathbb{P}^1$ is homogeneous and hence Oka; it has no nontrivial complex quotients. The complex plane $\mathbb{C}$ covers $\mathbb{C}^*$ and the tori, so these are Oka. The disc and its quotients are hyperbolic. □

Proposition 5.6.3 is a special case of the following result.

Theorem 5.6.5 *If E and X are complex manifolds and $\pi: E \to X$ is a holomorphic fibre bundle whose fibre is an Oka manifold, then the base X is an Oka manifold if and only if the total space E is an Oka manifold. In particular, the Cartesian product of Oka manifolds is Oka.*

Proof Assume first that E is an Oka manifold. Let $K \subset Q$ be a pair of compact convex sets in $\mathbb{C}^n$ and $f: U \to X$ be a holomorphic map from an open convex neighborhood $U \subset \mathbb{C}^n$ of K. By the homotopy lifting theorem there exists a continuous lifting $h: U \to E$ of f. By Corollary 5.5.11 we can replace h by a holomorphic lifting. Since E is Oka, we can approximate h uniformly on K by a holomorphic map $\widetilde{h}: Q \to E$. The map $\widetilde{f} = \pi \circ \widetilde{h}: Q \to X$ then approximates f on K. This shows that X is an Oka manifold.

Conversely, assume that X is Oka. Choose $K \subset Q \subset \mathbb{C}^n$ and a holomorphic map $h: U \to E$ as above. Let $f = \pi \circ h: U \to X$. Since X is Oka, we can approximate f uniformly on K by a holomorphic map $f_1: V \to X$ in an open set $V \supset Q$. If the approximation is close enough, we can also approximate h, uniformly on K, by a holomorphic map $h_1: K \to E$ that is a lifting of f_1. To do this, recall that Lemma 3.3.5 (p. 75) provides for every $x \in V$ a holomorphic retraction ρ_x of an open neighborhood $\Omega_x \subset E$ of the fibre $R_x = \pi^{-1}(f_1(x)) \subset E$ onto R_x, with ρ_x depending holomorphically on $x \in V$. If f_1 is sufficiently uniformly close to f on K, then for every point x in an open neighborhood $U' \subset V$ of K we have $h(x) \in \Omega_x$, and for such x we can define $h_1(x) = \rho_x(h(x)) \in R_x$. The map $h_1: U' \to E$ is holomorphic, it is uniformly close to h on K, and $\pi \circ h_1(x) = f_1(x)$ for $x \in U'$ as required.

Since $\pi: E \to X$ is a fibre bundle and the set $K \subset \mathbb{C}^n$ is convex, the map h_1 extends from a smaller neighborhood of K in $\mathbb{C}^n$ to a continuous map $h_1: Q \to E$ satisfying $\pi \circ h_1(x) = f_1(x)$ for all $x \in Q$. Since the fibre of π is an Oka manifold, Corollary 5.5.11 (p. 223) shows that we can deform h_1 to a holomorphic lifting

$\widetilde{h}: Q \to E$ of $f_1: Q \to X$ by a homotopy of liftings which remains uniformly close to h_1 on K. In particular, $\widetilde{h}$ approximates h uniformly on K. This shows that E is an Oka manifold. □

Corollary 5.6.6 *If $\pi: E \to X$ is a holomorphic fibre bundle whose base X and fibre Y are one of the Riemann surfaces $\mathbb{P}^1$, $\mathbb{C}$, $\mathbb{C}^*$, or a torus $\mathbb{T} = \mathbb{C}/\Gamma$, then the total space E is an Oka manifold. In particular, all Hirzebruch surfaces H_l ($l = 0, 1, 2, \ldots$) are Oka manifolds.*

Proof The list $\{\mathbb{P}^1, \mathbb{C}, \mathbb{C}^*, \mathbb{T} = \mathbb{C}/\Gamma\}$ contains all Riemann surfaces which are Oka; the others are uniformized by the disc $\mathbb{D}$ and therefore hyperbolic. Hence the first statement follows from Proposition 5.6.3. For the second statement, note that Hirzebruch surfaces H_l are $\mathbb{P}^1$-bundles over $\mathbb{P}^1$ (see [45, p. 191]). Each H_l is birationally equivalent to $\mathbb{P}^2$. □

Increasing Unions and Holomorphic Retracts

Proposition 5.6.7 *If a complex manifold Y is exhausted by open domains $Y_1 \subset Y_2 \subset \cdots \subset \bigcup_{j=1}^{\infty} Y_j = Y$ such that every Y_j is an Oka manifold, then Y is an Oka manifold. In particular, every long $\mathbb{C}^n$ is an Oka manifold.*

Recall that a *long* $\mathbb{C}^n$ is a complex manifold exhausted by an increasing sequence of domains biholomorphic to $\mathbb{C}^n$. For each $n > 1$, Theorems 4.21.1 and 4.21.6 furnish a continuum of pairwise nonequivalent long $\mathbb{C}^n$'s which do not admit any nonconstant holomorphic or plurisubharmonic functions. Hence, long $\mathbb{C}^n$'s provide a rich source of exotic Oka manifolds.

Proof Let $f: K \to Y$ be a holomorphic map from a compact convex set $K \subset \mathbb{C}^n$. Then $f(K)$ is compact and hence is contained in Y_j for some j. As Y_j is Oka, f can be approximated uniformly on K by entire maps $\mathbb{C}^n \to Y_j \subset Y$. Thus, Y is an Oka manifold. □

The following observation is immediate, but we lack interesting examples.

Proposition 5.6.8 *A holomorphic retract of an Oka manifold is Oka.*

Complements of Thin Algebraic Subvarieties

Question: If $\widehat{Y}$ is an algebraic Oka manifold and A is a closed algebraic subvariety of $\widehat{Y}$, when is the complement $Y = \widehat{Y} \setminus A$ an Oka manifold?

The complement of an algebraic hypersurface in an algebraic Oka manifold is not Oka in general. In fact, the general belief is that for a generic projective hypersurface

A of sufficiently high degree in $\mathbb{P}^n$, both A and $\mathbb{P}^n \setminus A$ are hyperbolic. (See the papers [114–116, 122, 505, 506, 508], among others, and the recent monograph by Diverio and Rousseau [123].) Similarly, the complement of an ample divisor A in an abelian variety is hyperbolic [507, 508]. We discuss this topic in Sect. 7.6.

We now consider subvarieties of codimension > 1.

Definition 5.6.9 A closed complex subvariety A in a complex manifold X is *thin* if at every point $p \in A$ we have $\dim_p A \le \dim_p X - 2$.

Proposition 5.6.10 *Let $\widehat{Y}$ denote $\mathbb{C}^k$, $\mathbb{P}^k$, or a complex Grassmannian. If A is a thin algebraic subvariety of $\widehat{Y}$, then $Y = \widehat{Y} \setminus A$ is an Oka manifold.*

Each of the manifolds in the proposition is of Class $\mathcal{A}$ (see Definition 6.4.4), so the result follows from Proposition 6.4.5. We give a simple direct proof.

Proof Each of the manifolds $\mathbb{C}^k$, $\mathbb{P}^k$ or a complex Grassmannian is complex homogeneous and hence Oka (see Proposition 5.6.1 on p. 223).

Consider first the case $\widehat{Y} = \mathbb{C}^k$. Given a special convex pair (K, Q) in $\mathbb{C}^n$ (see Definition 5.4.2 on p. 215) and a holomorphic map $f_0 : U \to Y$ from an open neighborhood U of K to $Y = \mathbb{C}^k \setminus A$, we need to find a holomorphic map $\widetilde{f} : Q \to Y$ that approximates f_0 as closely as desired on K; this will verify that Y enjoys CAP, and hence is an Oka manifold. Choose a compact convex set $L \subset U$ with $K \subset \mathring{L}$. By the Oka-Weil theorem (see Theorem 2.3.1) we can approximate f_0 uniformly on L by a holomorphic polynomial map $f : \mathbb{C}^n \to \mathbb{C}^k$. By the jet transversality theorem (see Corollary 8.8.7, p. 380), f can be chosen such that the algebraic subvariety $\Sigma = f^{-1}(A) \subset \mathbb{C}^n$ is thin (of codimension at least two) in $\mathbb{C}^n$. Assuming that the approximations were sufficiently close, we have $L \cap \Sigma = \varnothing$. By Corollary 4.12.2 (p. 157) there exists for any $\delta > 0$ a holomorphic automorphism ψ of $\mathbb{C}^n$ satisfying $\sup_{z \in K} |\psi(z) - z| < \delta$ and $\psi(Q) \cap \Sigma = \varnothing$. If $\delta > 0$ is small enough, then $\widetilde{f} = f \circ \psi : \mathbb{C}^n \to \mathbb{C}^k$ maps Q into Y and it approximates the initial map f_0 uniformly on K.

Next, we consider the case $\widehat{Y} = \mathbb{P}^k$ for $k \ge 2$. The quotient projection $\pi : \mathbb{C}^{k+1}_* = \mathbb{C}^{k+1} \setminus \{0\} \to \mathbb{P}^k$ is a holomorphic fibre bundle with fibre $\mathbb{C}^* = \mathbb{C} \setminus \{0\}$, and by adding the zero section we obtain the universal line bundle $L \to \mathbb{P}^k$. Assume that (K, Q) is a special convex pair in $\mathbb{C}^n$, $U \subset \mathbb{C}^n$ is an open convex neighborhood of K, and $f : U \to Y = \mathbb{P}^k \setminus A$ is a holomorphic map. Since U is contractible, the bundle $f^* L \to U$ is topologically trivial, and hence it admits a nowhere vanishing section. Pushing this section forward gives a continuous lifting $h : U \to \mathbb{C}^{k+1}_*$ of f, i.e., $\pi \circ h = f$. Since the fibre $\mathbb{C}^*$ of π is a Lie group and hence enjoys CAP, Corollary 5.5.11 (p. 223) allows us to replace h by a holomorphic lifting of f (still denoted h). The set $A' = \pi^{-1}(A) \cup \{0\}$ is a thin algebraic subvariety of $\mathbb{C}^{k+1}$ satisfying $h(U) \cap A' = \varnothing$. By the first part of the proposition, we can approximate h arbitrarily well on K by a holomorphic map $\widetilde{h} : Q \to \mathbb{C}^{k+1} \setminus A'$. The map $\widetilde{f} = \pi \circ \widetilde{h} : Q \to Y$ is then holomorphic and close to f on K, so Y enjoys CAP.

The case when $\widehat{Y} = G_{k,m}$ for some $1 \le k < m$ is proved in essentially the same way. Consider the fibre bundle $\pi: V_{k,m} \to G_{k,m}$ where $V_{k,m}$ is the Stiefel manifold of k-frames in $\mathbb{C}^m$ (see Example 1.2.7). We have $V_{k,m} = \mathbb{C}^{km} \setminus \Sigma$ where $\Sigma \subset \mathbb{C}^{km}$ is an algebraic subvariety of codimension $n - k + 1 \ge 2$, hence is thin in $\mathbb{C}^{km}$. Therefore, the algebraic subset $A' = \pi^{-1}(A) \cup \Sigma \subset \mathbb{C}^{km}$ is also thin and we can complete the proof as in the projective case. □

Corollary 5.6.11 *Every Hopf manifold is an Oka manifold.*

Proof A Hopf manifold is a holomorphic quotient of $\mathbb{C}^n_*$ for some $n > 1$. Since $\mathbb{C}^n_*$ is an Oka manifold by Proposition 5.6.10, the conclusion follows from Proposition 5.6.3. □

Proposition 5.6.10 fails in general for non-algebraic subvarieties of $\mathbb{C}^n$ irrespectively of their dimension (however, see Proposition 5.6.17). Indeed, Theorem 4.7.2 (p. 130) furnishes a discrete infinite set in $\mathbb{C}^n$ whose complement is volume hyperbolic, so it fails to be Oka. See also Example 7.1.1 on p. 320.

Smooth Toric Varieties

The following result is due to Lárusson (see [204, Theorem 2.17]).

Theorem 5.6.12 *Every smooth complex toric variety is an Oka manifold.*

Proof Let X be a smooth toric variety over $\mathbb{C}$. If X has a torus factor, say X is isomorphic to $Y \times (\mathbb{C}^*)^k$, where $k \ge 1$ and Y is another smooth toric variety, then, since $(\mathbb{C}^*)^k$ is Oka, Theorem 5.6.5 (p. 225) shows that X is Oka if and only if Y is Oka. Hence we may assume that X has no torus factor, so the construction in [108, Sect. 5.1] applies; see in particular Theorem 5.1.11 in the cited source. We can write X as a geometric quotient $X = (\mathbb{C}^m \setminus Z)/G$, where Z is a union of coordinate subspaces of $\mathbb{C}^m$ and G is a complex subgroup of $(\mathbb{C}^*)^m$ acting on $\mathbb{C}^m \setminus Z$ by diagonal matrices. In fact, G is isomorphic to the product of a torus and a finite abelian group [108, Lemma 5.1.1], so G is reductive. Furthermore, $\operatorname{codim} Z \ge 2$ [108, p. 217], so $\mathbb{C}^m \setminus Z$ is Oka by Proposition 5.6.10. Since X is smooth, G acts freely on $\mathbb{C}^m \setminus Z$ [108, Exercise 5.1.10].

We claim that the projection $\mathbb{C}^m \setminus Z \to X$ is a holomorphic fibre bundle; since the fibre G is Oka by Proposition 5.6.1 (p. 223), this will imply that X is Oka in view of Theorem 5.6.5.

Note that Z, being a union of coordinate subspaces, is the intersection of unions of coordinate hyperplanes. Thus, $\mathbb{C}^m \setminus Z$ is the union of Zariski-open sets of the form $U = \mathbb{C}^m \setminus (H_1 \cup \cdots \cup H_k)$, where $H_1, \ldots, H_k$ are coordinate hyperplanes. Each U is affine algebraic (hence Stein), as well as G-invariant. By slice theory for actions of reductive groups, the quotient map $U \to U/G$ is a holomorphic fibre bundle [519, Corollary 5.5], or, from the algebraic point of view, a locally trivial fibration in the étale sense [402, Corollaire 5]. It follows that $\mathbb{C}^m \setminus Z \to X$ is a holomorphic fibre bundle. □

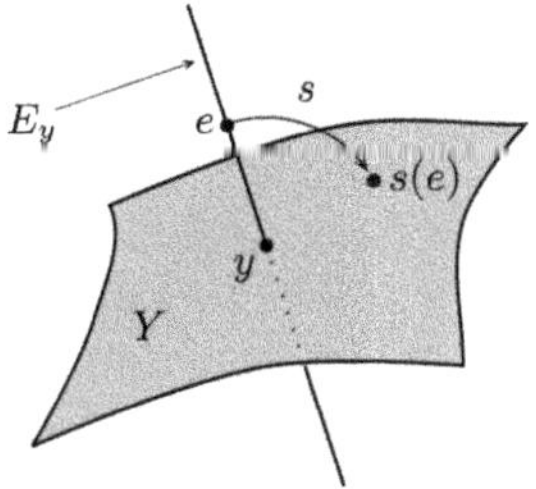

Fig. 5.1 A spray on Y. (Modified from [211, p. 13, Fig. 2])

Elliptic and Subelliptic Manifolds

Dominating sprays were introduced to Oka theory by Gromov [266]. We motivate their definition by the following consideration.

Choose an integer $n \ge \dim Y$ and let $\mathbb{B} = \mathbb{B}^n$ denote the open unit ball in $\mathbb{C}^n$. For any point $y \in Y$ there is a holomorphic submersion $g_y : \mathbb{B} \to Y$ with $g_y(0) = y$ (simply map $\mathbb{B}$ to a coordinate neighborhood of y). If Y is an Oka manifold then g_y is approximable, uniformly on any smaller ball $r\mathbb{B}$ $(0 < r < 1)$, by entire maps $s_y : \mathbb{C}^n \to Y$. If the approximation is sufficiently close then s_y is also a submersion on $r\mathbb{B}$, and we can arrange that $s_y(0) = y$. In summary, $s_y : \mathbb{C}^n \to Y$ is an entire map satisfying

(i) $s_y(0) = y$, and
(ii) the differential $\mathrm{d}s_y(0) : \mathrm{T}_0\mathbb{C}^n \cong \mathbb{C}^n \to \mathrm{T}_y Y$ is surjective.

More generally, given an open Stein subset U of Y, there are an integer $n \in \mathbb{N}$, an open Stein set $\Omega \subset U \times \mathbb{C}^n$ containing $U \times \{0\}$, and a holomorphic map $g : \Omega \to Y$ such that $g_y = g(y, \cdot)$ satisfies properties (i) and (ii) for each $y \in U$. We extend g to a continuous map $U \times \mathbb{C}^n \to Y$ without changing its values near $U \times \{0\}$. If Y is an Oka manifold, then there is a holomorphic map $s : U \times \mathbb{C}^n \to Y$ which agrees with g to second order at $U \times \{0\}$; hence the entire map $s_y = s(y, \cdot) : \mathbb{C}^n \to Y$ satisfies properties (i) and (ii) above. Such an s is called a *dominating holomorphic spray* on Y over the subset $U \subset Y$.

Definition 5.6.13 Let Y be a complex manifold.

(a) A *(holomorphic) spray* on Y is a triple (E, π, s) consisting of a holomorphic vector bundle $\pi : E \to Y$ (*a spray bundle*) and a holomorphic map $s : E \to Y$ (*a spray map*) such that for each $y \in Y$ we have $s(0_y) = y$. (See Fig. 5.1.) A spray (E, π, s) is algebraic if $\pi : E \to Y$ is an algebraic vector bundle over an algebraic manifold Y and $s : E \to Y$ is an algebraic map.
(b) A spray (E, π, s) on Y is *dominating* on a subset $U \subset Y$ if the differential $\mathrm{d}s_{0_y} : \mathrm{T}_{0_y} E \to \mathrm{T}_y Y$ maps the vertical subspace E_y of $\mathrm{T}_{0_y} E$ surjectively onto $\mathrm{T}_y Y$ for every $y \in U$; s is *dominating* if this holds for all $y \in Y$.
(c) A family of sprays (E_j, π_j, s_j) on Y $(j = 1, \ldots, m)$ is *dominating* on a subset $U \subset Y$ if for all $y \in U$ we have

$$(\mathrm{d}s_1)_{0_y}(E_{1,y}) + (\mathrm{d}s_2)_{0_y}(E_{2,y}) + \cdots + (\mathrm{d}s_m)_{0_y}(E_{m,y}) = \mathrm{T}_y Y. \tag{5.8}$$

The family is *dominating* if this holds for every $y \in Y$.

(d) A complex manifold Y is *elliptic* if it admits a dominating holomorphic spray, and is *subelliptic* if it admits a finite dominating family of holomorphic sprays.
(e) An algebraic manifold Y is *algebraically elliptic (A-elliptic)* if it admits a dominating algebraic spray, and is *algebraically subelliptic (A-subelliptic)* if it admits a finite dominating family of algebraic sprays.
(f) A complex manifold Y is *weakly elliptic* (resp. *weakly subelliptic*) if for every compact set $K \subset Y$ there exists a holomorphic spray on Y (resp. a family of holomorphic sprays) that is dominating at every point of K.

The class of elliptic manifolds was introduced by Gromov [266, Sect. 0.5], while subelliptic manifolds were introduced by the author in [190]. Clearly, we have the following implications, where the first row pertains to algebraic manifolds:

$$\begin{array}{ccccc} A\text{-elliptic} & \Longrightarrow & A\text{-subelliptic} & & \\ \Downarrow & & \Downarrow & & \\ \text{elliptic} & \Longrightarrow & \text{subelliptic} & \Longrightarrow & \text{weakly subelliptic} \end{array}$$

The relevance of these notions is shown by the following result which is essentially a corollary to Theorem 6.6.1 (p. 283).

Corollary 5.6.14 *Every weakly subelliptic manifold is an Oka manifold. In particular, every elliptic manifold is an Oka manifold.*

Proof Let K be a compact convex set in $\mathbb{C}^n$ and let $f: U \to Y$ be a holomorphic map from an open convex neighborhood $U \subset \mathbb{C}^n$ of K. We may assume that $0 \in K$. Let $f_t(z) = f(tz)$ for $z \in U$ and $t \in [0, 1]$; this is a homotopy from the constant map $f_0(z) = f(0) \in Y$ ($z \in \mathbb{C}^n$) to the map $f = f_1$. If Y is subelliptic then Theorem 6.6.1 (p. 283), applied to the projection $Z = \mathbb{C}^n \times Y \to \mathbb{C}^n$, shows that f can be approximated uniformly on K by entire maps $\mathbb{C}^n \to Y$. By Remark 6.6.4 (p. 286) the same holds if Y is weakly subelliptic. In each case we conclude that Y is an Oka manifold. □

It is not known whether the implications elliptic $\Rightarrow$ subelliptic $\Rightarrow$ Oka can be reversed. (See Sect. 7.1 and especially p. 325 for a further discussion.) However, these conditions are equivalent on a Stein manifold.

Proposition 5.6.15 ([266, 3.2.A]) *Every Stein Oka manifold is elliptic.*

Proof Let Y be a Stein manifold. Its tangent bundle $E = \mathrm{T}Y$ is also a Stein manifold. We identify Y with the zero section of E. By Proposition 3.3.2 (p. 74) there are a neighborhood $\Omega \subset E$ of Y and a holomorphic spray $s: \Omega \to Y$; that is, for every $y \in Y$ we have $s(y) = y$ and $ds_y: \mathrm{T}_y E \to \mathrm{T}_y Y$ maps the vertical subspace E_y of $\mathrm{T}_y E$ surjectively onto $\mathrm{T}_y Y$. We may assume that s extends to a continuous map $E \to Y$. Since E is Stein and Y is Oka, Corollary 5.4.5 gives a holomorphic map $\widetilde{s}: E \to Y$ that agrees with s to the second order along Y. Clearly, $\widetilde{s}$ is then a dominating spray on Y. □

We give some elementary examples of elliptic and subelliptic manifolds; for further examples see Sect. 6.4.

Example 5.6.16 (A) *Every complex homogeneous manifold is elliptic.* Indeed, assume that a complex Lie group G acts on a complex manifold Y transitively by holomorphic automorphisms of Y. Let $\mathfrak{g} \cong \mathbb{C}^p$ denote the Lie algebra of G and $\exp\colon \mathfrak{g} \to G$ the exponential map. The holomorphic map $s\colon Y \times \mathfrak{g} \cong Y \times \mathbb{C}^p \to Y$ given by

$$s(y, v) = \exp v \cdot y \in Y, \quad y \in Y,\ v \in \mathfrak{g} \tag{5.9}$$

is a dominating holomorphic spray on Y.

(B) (Gromov [266, Sect. 4.6.B]) *If the tangent bundle of a complex manifold Y is spanned by finitely many $\mathbb{C}$-complete holomorphic vector fields, then Y is elliptic.* (Such a manifold is flexible, see Definition 5.6.20.)

Indeed, let $V_1, \ldots, V_m$ be complete holomorphic vector fields on Y. Denote by $\mathbb{C} \ni t \mapsto \phi_j^t(y)$ the flow of V_j. The map $s\colon Y \times \mathbb{C}^m \to Y$ given by

$$s(y, t) = s(y, t_1, \ldots, t_m) = \phi_1^{t_1} \circ \phi_2^{t_2} \circ \cdots \circ \phi_m^{t_m}(y) \tag{5.10}$$

satisfies $s(y, 0) = y$ (hence it is a spray on Y), and

$$\frac{\partial}{\partial t_j} s(y, 0) = V_j(y), \quad y \in Y,\ j = 1, \ldots, m.$$

Thus, s is dominating at the point y precisely when the vectors $V_1(y), \ldots, V_m(y)$ span the tangent space $T_y Y$. In particular, if these vector fields span the tangent space at every point then s is a dominating spray on Y.

(C) The following example is related to Proposition 5.6.10.

Proposition 5.6.17 *If $A \subset \mathbb{C}^n$ is a tame (see Definition 4.11.3, p. 152) analytic subvariety with $\dim A \le n - 2$, then $\mathbb{C}^n \setminus A$ is elliptic. If in addition A is algebraic, then $\mathbb{C}^n \setminus A$ is algebraically elliptic.*

Proof By the proof of Proposition 4.11.8 (p. 154), there exist finitely many shear vector fields $V_1, \ldots, V_m$ on $\mathbb{C}^n$ that vanish on A and span the tangent space $T_z\mathbb{C}^n$ at every point $z \in \mathbb{C}^n \setminus A$. Let ϕ_j^t denote the flow of V_j. The associated spray $s\colon \mathbb{C}^n \times \mathbb{C}^m \to \mathbb{C}^n$, defined by (5.10), satisfies $s(z, t) \equiv z$ for $z \in A$, and $s(z, t) \in \mathbb{C}^n \setminus A$ for $z \in \mathbb{C}^n \setminus A$ and $t \in \mathbb{C}^m$. Since the vectors $V_j(z)$ span $T_z\mathbb{C}^n$ for every $z \in \mathbb{C}^n \setminus A$, the restricted map $s\colon (\mathbb{C}^n \setminus A) \times \mathbb{C}^m \to \mathbb{C}^n \setminus A$ is a dominating spray over $\mathbb{C}^n \setminus A$. If A is algebraic, then this construction can be done with polynomial shear vector fields having algebraic flows. □

Corollary 5.6.18 *Let $X = \mathbb{C}^n/\Gamma$ where Γ is a lattice in $\mathbb{C}^n$ ($n \ge 2$). Then, the complement $X \setminus \{x_1, \ldots, x_m\}$ of any finite set in X is an Oka manifold.*

Proof Let $\pi\colon\mathbb{C}^n\to X$ denote the quotient projection. Choose points $q_j\in\mathbb{C}^n$ with $\pi(q_j)=x_j$ $(j=1,\dots,m)$. The discrete set $\Gamma_0=\bigcup_{j=1}^m(\Gamma+q_j)\subset\mathbb{C}^n$ is tame (see [80, Proposition 4.1] and [74]). Therefore, $\widetilde{Y}=\mathbb{C}^p\setminus\Gamma_0$ is elliptic and hence Oka. Since $\pi\colon\widetilde{Y}\to Y=X\setminus\{x_1,\dots,x_m\}$ is a covering projection, Y is an Oka manifold by Proposition 5.6.3 (p. 224). □

(D) *If Y is a complex Grassmann manifold and $A\subset Y$ is a closed algebraic subvariety of codimension at least two, then $Y\setminus A$ is algebraically subelliptic* (Proposition 6.4.1). It is not known whether all such manifolds are elliptic.

(E) Further examples of elliptic manifolds are basins of uniformly attracting random sequences of automorphisms of $\mathbb{C}^n$; see Fornæss and Wold [168]. Let $f=(f_j)_{j\in\mathbb{N}}\subset\mathrm{Aut}(\mathbb{C}^n)$ be such that $f_j(0)=0$ for all $j\in\mathbb{N}$. The *non-autonomous basin* of f at 0 is $\Omega_f=\{z\in\mathbb{C}^n:\lim_{j\to\infty}f_j\circ\cdots\circ f_1(z)=0\}$. The sequence f is said to be *uniformly attracting* if there exist numbers $0<a<b<1$ such that $a|z|\le|f_j(z)|\le b|z|$ holds for all $z\in\mathbb{B}^n$ and $j\in\mathbb{N}$; in such case we say that Ω_f is a non-autonomous basin with uniform bounds. The following is [168, Theorem 1.1].

Theorem 5.6.19 *A non-autonomous basin with uniform bounds is elliptic, hence an Oka manifold.*

This is false in general for non-autonomous basins without uniform bounds (see Fornæss [162]). It is an open question (the *Bedford conjecture*) whether every non-autonomous basin with uniform bounds is biholomorphic to $\mathbb{C}^n$. For autonomous basins, this holds in view of Theorem 4.3.2. □

Flexible Manifolds

Definition 5.6.20 (Arzhantsev et al. [38]) A complex manifold Y is *flexible* if the $\mathbb{C}$-complete holomorphic vector fields on Y span the tangent space T_yY at every point $y\in Y$. An algebraic manifold is *algebraically flexible* if this holds for algebraic vector fields having algebraic flows.

If Y is algebraically flexible and A is a thin algebraic subvariety of Y, then $Y\setminus A$ is also algebraically flexible (see Flenner et al. [160]).

Remark 5.6.21 (on terminology) In this book, the term *flexibility property* is used for a vast array of anti-hyperbolicity and anti-general type properties of complex manifolds, in keeping with the terminology from the papers [196, 266]. To avoid possible confusion, we shall say *flexible in the sense of Arzhantsev et al.* when deemed necessary. □

Proposition 5.6.22

(a) *A flexible manifold is weakly elliptic, and hence an Oka manifold.*

(b) *A flexible Stein manifold is elliptic.*
(c) *An algebraically flexible algebraic manifold is algebraically elliptic.*

Proof Assume that Y is flexible. Given a compact set $K \subset Y$, there exist finitely many complete holomorphic vector fields on Y which span the tangent space $\mathrm{T}_y Y$ for every $y \in K$. Then, the associated flow-spray (5.10) is dominating on K. Hence, Y is weakly elliptic (see Definition 5.6.13(f)), and therefore Oka by Corollary 5.6.14. This proves (a).

The proof of (c) is obtained by adding new complete algebraic vector fields to an existing finite collection in order to reduce the dimension of the exceptional variety $A \subset Y$ where the given collection fails to span the tangent space of Y. Since A is algebraic, it has finitely many irreducible components and hence the process ends in finitely many steps.

In case (b), we proceed as in (c); however, the analytic set A may have countably many irreducible components A_i, $i \in \mathbb{N}$. The following proof is due to Kaliman and Kutzschebauch [323, Theorem 4]. Assume that Y is Stein, flexible and connected. Choose complete holomorphic vector fields $V_1, \ldots, V_k$ which span $\mathrm{T}_y Y$ for all points y outside of a proper complex subvariety $Y' \subset Y$. It suffices to find $\Phi \in \mathrm{Aut}(Y)$ such that the complete vector fields $\Phi_*(V_j)$ for $j = 1, \ldots, k$ span $\mathrm{T}_y Y$ at a generic point $y \in A_i$ for all $i \in \mathbb{N}$. Clearly, this holds if $\Phi^{-1}(A_i) \not\subset Y'$ for all $i \in \mathbb{N}$. The set $B_i \subset \mathrm{Aut}(Y)$ of all $\Phi \in \mathrm{Aut}(Y)$ satisfying this condition for a fixed $i \in \mathbb{N}$ is clearly open, and by flexibility of Y it is also dense (use compositions of any given $\Phi \in \mathrm{Aut}(\mathbb{C}^n)$ with short time flows of complete vector fields not tangential to A_i). Since Y is Stein, $\mathrm{Aut}(Y)$ is a Baire space (see Sect. 4.1), and hence the intersection $\bigcap_i B_i$ is nonempty. □

Manifolds with the Density Property

Recall (see Definition 4.10.1 on p. 142) that a complex manifold Y enjoys the *density property* if the Lie algebra $\mathrm{Lie}(Y)$, generated by all $\mathbb{C}$-complete holomorphic vector fields on Y, is dense in the Lie algebra $\aleph(Y)$ of all holomorphic vector fields in the compact-open topology.

Proposition 5.6.23 *A complex manifold with the density property is flexible in the sense of Arzhantsev et al. (see Definition 5.6.20), and hence an Oka manifold. In particular, a Stein manifold with the density property is elliptic.*

Proof Lemma 4.10.2 on p. 142 says that for any Lie algebra $\mathfrak{g}$ of vector fields on a complex manifold Y we have $\overline{\mathrm{Span}(\mathfrak{g}_{\mathrm{int}})} = \overline{\mathrm{Lie}(\mathfrak{g}_{\mathrm{int}})}$, where $\mathfrak{g}_{\mathrm{int}}$ is the set of all complete vector fields in $\mathfrak{g}$. Assuming that Y has the density property, we see by taking $\mathfrak{g} = \aleph(Y)$ that every holomorphic vector field on Y can be approximated by $\mathbb{C}$-linear combinations of complete holomorphic vector fields. It follows that the tangent space $\mathrm{T}_y Y$ at any point $y \in Y$ is spanned by complete holomorphic vector fields on Y. Hence Y is flexible and the conclusion follows from Proposition 5.6.22. □

5.7 Cartan Pairs

In this section we introduce the notion of a Cartan pair. Such pairs will be used in the process of gluing pairs of holomorphic sprays.

Definition 5.7.1 (I) A pair (A, B) of compact subsets in a complex space X is a *Cartan pair* if it satisfies the following two conditions:

(i) $A, B, D = A \cup B$ and $C = A \cap B$ are Stein compacts, and
(ii) A, B are *separated* in the sense that $\overline{A \setminus B} \cap \overline{B \setminus A} = \varnothing$.

(II) A pair (D_0, D_1) of open relatively compact subsets in a complex manifold X is a *strongly pseudoconvex Cartan pair of class* $\mathcal{C}^\ell$ $(\ell \geq 2)$ if $(\bar{D}_0, \bar{D}_1)$ is a Cartan pair and $D_0, D_1, D = D_0 \cup D_1, D_{0,1} = D_0 \cap D_1$ are strongly pseudoconvex Stein domains with $\mathcal{C}^\ell$ boundaries.

The following lemma is an immediate consequence of Siu's theorem; see Theorem 3.1.1 on p. 66.

Lemma 5.7.2 *If X is a complex subvariety of a complex space $\widetilde{X}$, then a pair of compact subsets $A, B \subset X$ is a Cartan pair in X if and only if it is a Cartan pair in $\widetilde{X}$.*

We shall need the following result on approximating Cartan pairs in complex manifolds by smooth strongly pseudoconvex Cartan pairs.

Proposition 5.7.3 *Given a Cartan pair (A, B) in a complex manifold X and open sets $U \supset A$, $V \supset B$ in X, there exists a smooth strongly pseudoconvex Cartan pair (D_0, D_1) satisfying $A \subset D_0 \Subset U$ and $B \subset D_1 \Subset V$.*

Proof Fix a distance function dist on X induced by a smooth Riemannian metric. Given a subset $A \subset X$ and a number $r > 0$, we set

$$A(r) = \big\{x \in X : \mathrm{dist}(x, y) < r \text{ for some } y \in A\big\}.$$

Lemma 5.7.4 *Given $A, B \subset X$ and $r > 0$ we have*

$$(A \cup B)(r) = A(r) \cup B(r), \qquad (A \cap B)(r) \subset A(r) \cap B(r).$$

If A and B are (relatively) compact and separated in X, then for all sufficiently small $r > 0$ we also have that $(A \cap B)(r) = A(r) \cap B(r)$, and the sets $\overline{A(r)}$, $\overline{B(r)}$ are separated.

Proof The first two properties are immediate. Now, write $A = (A \setminus B) \cup (A \cap B)$ and $B = (B \setminus A) \cup (A \cap B)$. Applying the first property, we get

$$A(r) = (A \setminus B)(r) \cup (A \cap B)(r), \qquad B(r) = (B \setminus A)(r) \cup (A \cap B)(r).$$

If $\overline{A \setminus B} \cap \overline{B \setminus A} = \varnothing$, then for every sufficiently small $r > 0$ we have

$$(A \setminus B)(r) \cap (B \setminus A)(r) = \varnothing$$

(In fact, even the closures of the sets $(A \setminus B)(r)$ and $(B \setminus A)(r)$ are disjoint.) Hence, the previous display gives $A(r) \cap B(r) = (A \cap B)(r)$ and the separation property for the pair $\overline{A(r)}, \overline{B(r)}$. □

Since A and B are Stein compacts, there exist smooth strongly pseudoconvex domains D'_0, D'_1 in X such that $A \subset D'_0 \Subset U$ and $B \subset D'_1 \Subset V$. Choose $r > 0$ such that $A(r) \Subset D'_1$, $B(r) \Subset D'_1$ and the conclusions of Lemma 5.7.4 hold. Since $A \cup B$ is a Stein compact, there is a closed strongly pseudoconvex Stein domain $\Omega \subset X$ such that $A \cup B \subset \Omega \Subset A(r) \cup B(r)$. Set

$$A' = D'_0 \cap \Omega, \qquad B' = D'_1 \cap \Omega.$$

Then $A' \cup B' = \Omega$ and the sets A' and B' are separated. By a small deformation of Ω we may assume that the intersections of $b\Omega$ with bD'_0 and bD'_1 are transverse, and hence A' and B' are piecewise smooth strongly pseudoconvex domains. By smoothing the corners of A' and B' we get a pair of smooth strongly pseudoconvex domains D_0, D_1 satisfying the stated conditions. □

5.8 A Splitting Lemma

One of the main analytic ingredients in the proof of Theorem 5.4.4 is a method of splitting and gluing holomorphic sprays of maps or sections. Such sprays can be thought of as *thick holomorphic sections* and should not be confused with global sprays in the sense of Definition 5.6.13. This technique has found a number of applications; see Sects. 8.10–8.13, 9.8 and the papers [135–137, 198]. For similar gluing methods in almost complex and symplectic geometry, see Donaldson [131] and McDuff and Salamon [412].

The key result is Proposition 5.8.1 which gives a compositional splitting of any fibre-preserving holomorphic map that is close enough to the identity, with control up to the boundary. It generalizes Cartan's lemma on splitting invertible holomorphic matrices [260, p. 88].

Given a compact subset K in a complex space X and an open set $W \subset \mathbb{C}^N$, we consider maps $\gamma\colon K \times W \to K \times \mathbb{C}^N$ of the form

$$\gamma(x, w) = \big(x, \psi(x, w)\big), \quad x \in K,\ w \in W. \tag{5.11}$$

We say that $\gamma \in \mathcal{A}(K \times W)$ if γ is continuous on $K \times W$ and holomorphic in $\mathring{K} \times W$. Let $\mathrm{Id}(x, w) = (x, w)$ be the identity map on $X \times \mathbb{C}^N$. Set

$$\mathrm{dist}_{K\times W}(\gamma, \mathrm{Id}) = \sup\big\{\big|\psi(x, w) - w\big| : x \in K,\ w \in W\big\}.$$

Proposition 5.8.1 ([135, 198]) *Let (D_0, D_1) be a strongly pseudoconvex Cartan pair of class $\mathcal{C}^2$ in a complex manifold X (see Definition 5.7.1 on p. 234). Set*

$D_{0,1} = D_0 \cap D_1$ and $D = D_0 \cup D_1$. Given a bounded open convex set $0 \in W \subset \mathbb{C}^N$ and a number $r \in (0,1)$, there is a $\delta > 0$ satisfying the following. For every map $\gamma \colon \bar{D}_{0,1} \times W \to \bar{D}_{0,1} \times \mathbb{C}^N$ of the form (5.11) and of class $\mathcal{A}(D_{0,1} \times W)$, satisfying $\mathrm{dist}_{D_{0,1}\times W}(\gamma, \mathrm{Id}) < \delta$, there exist maps

$$\alpha_\gamma : \bar{D}_0 \times rW \to \bar{D}_0 \times \mathbb{C}^N, \qquad \beta_\gamma : \bar{D}_1 \times rW \to \bar{D}_1 \times \mathbb{C}^N$$

of the form (5.11) and of class $\mathcal{A}(D_0 \times rW)$ and $\mathcal{A}(D_1 \times rW)$, respectively, depending smoothly on γ, such that $\alpha_{\mathrm{Id}} = \mathrm{Id}$, $\beta_{\mathrm{Id}} = \mathrm{Id}$, and

$$\gamma \circ \alpha_\gamma = \beta_\gamma \quad \text{holds on } \bar{D}_{0,1} \times rW.$$

If γ agrees with Id to order $m \in \mathbb{N}$ along $w = 0$ then so do α_γ and β_γ. Furthermore, if X' is a closed complex subvariety of X such that $X' \cap \bar{D}_{0,1} = \varnothing$, then we can choose α_γ and β_γ to be tangent to Id to any given finite order along $(X' \cap \bar{D}_0) \times rW$ and $(X' \cap \bar{D}_1) \times rW$, respectively.

Proof Denote by C_r and Γ_r the Banach spaces consisting of all continuous maps $\phi \colon \bar{D}_{0,1} \times rW \to \mathbb{C}^N$ which are holomorphic in $D_{0,1} \times rW$ and satisfy

$$\|\phi\|_{C_r} = \sup\big\{ |\phi(x,w)| : x \in \bar{D}_{0,1},\ w \in rW \big\} < +\infty,$$
$$\|\phi\|_{\Gamma_r} = \sup\big\{ |\phi(x,w)| + |\partial_w \phi(x,w)| : x \in \bar{D}_{0,1},\ w \in rW \big\} < +\infty.$$

Here, ∂_w denotes the partial differential with respect to the variable $w \in \mathbb{C}^N$, and $|\partial_w \phi(x,w)|$ is the Euclidean operator norm. Similarly we denote by A_r, B_r the Banach spaces of all continuous maps $\bar{D}_0 \times rW \to \mathbb{C}^N$, $\bar{D}_1 \times rW \to \mathbb{C}^N$, respectively, that are holomorphic in the interior, endowed with the sup-norm.

By the hypotheses, we are given a map $\gamma(x,w) = (x, \psi(x,w))$ with $\psi \in C_1$. Set $\psi_0(x,w) = w$. Choose a number $r_1 \in (r,1)$. By the Cauchy estimates, the restriction map $C_1 \to \Gamma_{r_1}$ is continuous, and hence $\psi|_{\bar{D}_{0,1}\times r_1 W} \in \Gamma_{r_1}$ and $\|\psi - \psi_0\|_{\Gamma_{r_1}} \le \mathrm{const}\|\psi - \psi_0\|_{C_1}$.

The following lemma provides a solution of the Cousin-I problem with bounds on a strongly pseudoconvex Cartan pair.

Lemma 5.8.2 *There exist bounded linear operators $\mathcal{A} \colon C_r \to A_r$, $\mathcal{B} \colon C_r \to B_r$ satisfying*

$$c = \mathcal{A}c - \mathcal{B}c \quad \text{for every } c \in C_r. \tag{5.12}$$

If c vanishes to order $m \in \mathbb{N}$ at $w = 0$, then so do $\mathcal{A}c$ and $\mathcal{B}c$. If X' is a closed complex subvariety of X such that $X' \cap \bar{D}_{0,1} = \varnothing$, then we can also ensure that $\mathcal{A}c$ and $\mathcal{B}c$ vanish to a given finite order along $(X' \cap \bar{D}_0) \times rW$ and $(X' \cap \bar{D}_1) \times rW$, respectively.

Proof By the separation condition (ii) in Definition 5.7.1, there is a smooth function $\chi \colon X \to [0,1]$ such that $\chi = 0$ in a neighborhood of $\overline{D_0 \setminus D_1}$ and $\chi = 1$ in a neighborhood of $\overline{D_1 \setminus D_0}$. Thus, for any $c \in C_r$ the product $\chi(x)c(x,w)$ extends to a continuous function on $\bar{D}_0 \times W$ that vanishes on $\overline{D_0 \setminus D_1} \times W$, and $(\chi(x) - 1)c(x,w)$

extends to a continuous function on $\bar{D}_1 \times W$ that vanishes on $\overline{D_1 \setminus D_0} \times W$. Furthermore,

$$\overline{\partial}(\chi c) = \overline{\partial}\big((\chi - 1)c\big) = c\,\overline{\partial}\chi$$

is a (0, 1)-form on $\bar{D} \times W$ with continuous coefficients and with support in $\bar{D}_{0,1} \times W$, depending holomorphically on $w \in W$.

Choose functions $f_1, \ldots, f_m \in \mathcal{O}(X)$ that vanish to order k along X' and have no common zeros on $\bar{D}_{0,1}$. Since $D_{0,1}$ is strongly pseudoconvex, the Oka-Cartan division theorem (see Corollary 2.6.4, p. 56) gives holomorphic functions $g_1, \ldots, g_m$ in a neighborhood of $\bar{D}_{0,1}$ such that $\sum_{j=1}^m f_j g_j = 1$.

Since the domain $D = D_0 \cup D_1$ is strongly pseudoconvex, there exists a sup-norm bounded linear solution operator $T : \mathcal{C}^0_{0,1}(D) \to \mathcal{C}^0(D)$ to the $\overline{\partial}$-equation on D at the level of (0, 1)-forms (see Theorem 2.7.3, p. 59). For any $c \in C_r$ and $w \in rW$ we set

$$(\mathcal{A}c)(x, w) = \chi(x)\, c(x, w) - \sum_{j=1}^m f_j(x)\, T\big(g_j\, c(\cdot, w)\overline{\partial}\chi\big)(x), \quad x \in \bar{D}_0;$$

$$(\mathcal{B}c)(x, w) = \big(\chi(x) - 1\big)c(x, w) - \sum_{j=1}^m f_j(x)\, T\big(g_j\, c(\cdot, w)\overline{\partial}\chi\big)(x), \quad x \in \bar{D}_1.$$

Then, $\mathcal{A}c - \mathcal{B}c = c$ on $\bar{D}_{0,1} \times rW$ and $\overline{\partial}_x(\mathcal{A}c) = 0$, $\overline{\partial}_x(\mathcal{B}c) = 0$ in the interior of their respective domains. Since $\overline{\partial}_w(c(x, w)\overline{\partial}\chi(x)) = 0$ and the operator T commutes with $\overline{\partial}_w$, we also have that $\overline{\partial}_w(\mathcal{A}c) = 0$ and $\overline{\partial}_w(\mathcal{B}c) = 0$. The estimates follow from boundedness of T. □

Given $\psi \in \Gamma_{r_1}$ sufficiently near ψ_0 and $c \in C_r$ sufficiently near 0, we define for $x \in \bar{D}_{0,1}$ and $w \in rW$:

$$\Phi(\psi, c)(x, w) = \psi\big(x, w + (\mathcal{A}c)(x, w)\big) - \big(w + (\mathcal{B}c)(x, w)\big).$$

Then, $(\psi, c) \mapsto \Phi(\psi, c)$ is a smooth map from an open neighborhood of $(\psi_0, 0)$ in the Banach space $\Gamma_{r_1} \times C_r$ to the Banach space C_r. Indeed, Φ is linear in ψ, and its partial differential with respect to the second variable equals

$$\partial_c \Phi(\psi, c_0)c(x, w) = \partial_w\, \psi\big(x, w + (\mathcal{A}c_0)(x, w)\big)\cdot (\mathcal{A}c)(x, w) - (\mathcal{B}c)(x, w).$$

This is again linear in ψ and continuous in all variables. A similar argument applies to higher order differentials of Φ. By (5.12), we have

$$\Phi(\psi_0, c) = \mathcal{A}(c) - \mathcal{B}(c) = c, \quad c \in C_r,$$

and hence $\partial_c \Phi(\psi_0, 0)$ is the identity on C_r. The implicit function theorem gives a smooth map $\psi \to \mathcal{C}(\psi) \in C_r$ in a neighborhood of ψ_0 in Γ_{r_1} such that

$$\Phi\big(\psi, \mathcal{C}(\psi)\big) = 0 \quad \text{and} \quad \mathcal{C}(\psi_0) = 0.$$

The maps a_ψ and b_ψ defined by

$$a_\psi(x, w) = w + \mathcal{A} \circ \mathcal{C}(\psi)(x, w), \qquad b_\psi(x, w) = w + \mathcal{B} \circ \mathcal{C}(\psi)(x, w)$$

then satisfy $a_\psi \in A_r$, $a_{\psi_0} = \psi_0$, $b_\psi \in B_r$, $b_{\psi_0} = \psi_0$ and

$$\psi\big(x, a_\psi(x,w)\big) = b_\psi(x,w), \quad (x,w) \in \bar{D}_{0,1} \times rW.$$

The associated maps

$$\alpha_\gamma(x,w) = \big(x, a_\psi(x,w)\big), \qquad \beta_\gamma(x,w) = \big(x, b_\psi(x,w)\big)$$

depend smoothly on γ and satisfy the conclusion of Proposition 5.8.1. □

Remark 5.8.3 (A) The solutions $\alpha_\gamma = \alpha(\gamma)$ and $\beta_\gamma = \beta(\gamma)$ in Proposition 5.8.1 are given by smooth nonlinear operators defined in a neighborhood of the identity map. If γ depends continuously or smoothly on some parameters, we get the same type of dependence for the solutions.

(B) The analogous splitting lemma holds in any Banach spaces of holomorphic functions for which there exists a bounded linear solution operator to the $\overline{\partial}$-equation on the level of (0, 1)-forms on strongly pseudoconvex domains; an example is H^∞ or Sobolev spaces. Further, if (D_0, D_1) is a strongly pseudoconvex Cartan pair of class $\mathcal{C}^\ell$ for some $\ell \geq 2$, then the analogous result holds for maps α, β, γ of class $\mathcal{A}^l$ ($l \in \{0, 1, \ldots, \ell\}$) on their respective domains, with $\mathcal{C}^l$ estimates [135, Theorem 3.2]. One uses appropriate function spaces and (in the proof of Lemma 5.8.2) a bounded linear solution operator $T: \mathcal{C}^l_{0,1}(D) \to \mathcal{C}^l(D)$ to the $\overline{\partial}$-equation, furnished by Theorem 2.7.3 on p. 59.

(C) Proposition 5.8.1 easily extends to the case when $\bar{D} \times \mathbb{C}^N$ is replaced by the total space of a complex vector bundle $\pi: E \to \bar{D}$ which is holomorphic over D and continuous (or smooth) over $\bar{D}$. Such E can be embedded as a complex vector subbundle of class $\mathcal{A}^l(D) = \mathcal{C}^l(\bar{D}) \cap \mathcal{O}(D)$ in a trivial bundle $\bar{D} \times \mathbb{C}^N$, and we have $\bar{D} \times \mathbb{C}^N = E \oplus E'$ for another subbundle E' of the same class. (These facts follows from Cartan's Theorem B for $\mathcal{A}^l(D)$-bundles; see the papers [291, 293, 383–385].) □

We now prove a version of Proposition 5.8.1 in which (A, B) is a Cartan pair in an arbitrary complex space.

Proposition 5.8.4 *Assume that (A, B) is a Cartan pair in a reduced complex space X. Set $C = A \cap B$. Given an open set $U \supset C$ in X, a bounded open convex set $0 \in W \subset \mathbb{C}^N$ and a number $r \in (0,1)$, there exist arbitrarily small open neighborhoods $U_A \supset A$, $U_B \supset B$ with $U_{A,B} = U_A \cap U_B \subset U$, and a number $\delta > 0$ satisfying the following. Given a holomorphic map $\gamma: U \times W \to U \times \mathbb{C}^N$* (5.11) *with* $\mathrm{dist}_{U \times W}(\gamma, \mathrm{Id}) < \delta$, *there exist holomorphic maps*

$$\alpha_\gamma : U_A \times rW \to U_A \times \mathbb{C}^N, \qquad \beta_\gamma : U_B \times rW \to U_B \times \mathbb{C}^N$$

of the form (5.11), *depending continuously on γ, with* $\alpha_{\mathrm{Id}} = \mathrm{Id}$, $\beta_{\mathrm{Id}} = \mathrm{Id}$ *and*

$$\gamma \circ \alpha_\gamma = \beta_\gamma \quad \text{on } U_{A,B} \times rW.$$

If γ agrees with Id *to order $m \in \mathbb{N}$ along $w = 0$ then so do α_γ and β_γ. Further, if X' is a closed complex subvariety of X such that $X' \cap C = \varnothing$ then we can choose α_γ to be tangent to* Id *to any given finite order along $(X' \cap A) \times rW$.*

Proof Since $D = A \cup B$ is a Stein compact in X, an open Stein neighborhood $U_0 \subset X$ of D embeds holomorphically as a closed complex analytic subvariety in an open set $\widetilde{D} \subset \mathbb{C}^n$. We replace X by U_0 and identify it with the image subvariety in the domain $\widetilde{D} \subset \mathbb{C}^n$; then $A \cup B$ is a Stein compact in $\mathbb{C}^n$. We may assume that the open set $U \supset C$ in the proposition is of the form $U = X \cap \widetilde{U}$ for some pseudoconvex domain $\widetilde{U} \subset \mathbb{C}^n$.

Choose a number $r_0 \in (r, 1)$ and a pseudoconvex domain $\widetilde{C} \Subset \widetilde{U}$ such that $C \subset \widetilde{C} \cap X \Subset U$. Lemma 2.8.3 furnishes a bounded extension operator $S\colon H^\infty(U \times W) \to H^\infty(\widetilde{C} \times r_0 W)$. Given a map $\gamma(x, w) = (x, w + c(x, w))$ $(x \in U,\ w \in W)$ as in Proposition 5.8.4, we set

$$\widetilde{\gamma}(x, w) = \big(x, w + S(c)(x, w)\big), \quad x \in \widetilde{C},\ w \in r_0 W.$$

Note that $\widetilde{\gamma}$ is a holomorphic map of the same type (5.11), it agrees with γ on the set $\widetilde{C} \cap X$, and we have an estimate

$$\mathrm{dist}_{\widetilde{C} \times (r_0 W)}(\widetilde{\gamma}, \mathrm{Id}) \le M \mathrm{dist}_{(\widetilde{C} \cap X) \times W}(\gamma, \mathrm{Id}) \le M\delta,$$

where the constant $M < +\infty$ comes from the bounded extension operator.

Choose open sets $\widetilde{A}, \widetilde{B} \subset \mathbb{C}^n$ with $A \subset \widetilde{A} \subset \widetilde{D}$ and $B \subset \widetilde{B} \subset \widetilde{D}$. By Lemma 5.7.3 (p. 234) there exists a smooth strongly pseudoconvex Cartan pair (D_0, D_1) in $\mathbb{C}^n$ such that $A \subset D_0 \subset \widetilde{A}$ and $B \subset D_1 \subset \widetilde{B}$. If $\delta > 0$ is small enough then Proposition 5.8.1, applied to $\widetilde{\gamma}$ on the Cartan pair (D_0, D_1), furnishes holomorphic maps

$$\widetilde{\alpha}\colon D_0 \times rW \to D_0 \times \mathbb{C}^N, \qquad \widetilde{\beta}\colon D_1 \times rW \to D_1 \times \mathbb{C}^N$$

satisfying the condition

$$\widetilde{\gamma} \circ \widetilde{\alpha} = \widetilde{\beta} \quad \text{on } (D_0 \cap D_1) \times rW.$$

Take $U_A = D_0 \cap X$, $U_B = D_1 \cap X$ and denote by α (resp. β) the restriction of $\widetilde{\alpha}$ (resp. $\widetilde{\beta}$) to $U_A \times rW$ (resp. to $U_B \times rW$). This completes the proof. □

5.9 Gluing Holomorphic Sprays

In this section we show how Proposition 5.8.1 (the splitting lemma) can be used to glue holomorphic sprays over a Cartan pair in the base space.

Assume that X and Z are reduced complex spaces and $\pi\colon Z \to X$ is a holomorphic submersion. For each point $z \in Z$ we denote by $\mathrm{VT}_z Z = \ker d\pi_z$ the vertical tangent space of Z (the tangent space to the fibre $\pi^{-1}(\pi(z))$) at z.

Definition 5.9.1 Let D be a domain in X. A *holomorphic spray of sections* of $\pi\colon Z \to X$ over D, or a *fibre spray*, is a holomorphic map $f\colon D \times P \to Z$, where P is an open set in some $\mathbb{C}^N$ containing the origin, satisfying

$$\pi\big(f(x, w)\big) = x, \quad x \in D,\ w \in P. \tag{5.13}$$

The spray f is *dominating* on a subset $K \subset D$ if the partial differential

$$\partial_w|_{w=0} f(x, w) : \mathrm{T}_0\mathbb{C}^N \cong \mathbb{C}^N \to \mathrm{VT}_{f(x,0)} Z$$

is surjective for all $x \in K$. The section $f_0 = f(\cdot, 0)$ is the *core* of f. If D is a relatively compact domain with boundary of class $\mathcal{C}^1$ contained in X_{reg}, then a *spray of sections of class* $\mathcal{A}^l(D)$ (with the parameter set P) is a $\mathcal{C}^l$ map $f: \bar{D} \times P \to Z$ that is holomorphic on $D \times P$ and satisfies (5.13).

The following result is our main *Heftungslemma* for holomorphic sprays; it can be viewed as a solution of a nonlinear Cousin-I problem.

Proposition 5.9.2 ([136, Proposition 2.4]) *Assume that Z is a complex manifold, X is a Stein manifold, $\pi: Z \to X$ is a holomorphic submersion, and (D_0, D_1) is a strongly pseudoconvex Cartan pair of class $\mathcal{C}^\ell$ ($\ell \ge 2$) in X (see Definition 5.7.1 on p. 234). Set $D = D_0 \cup D_1$ and $D_{0,1} = D_0 \cap D_1$. Let $l \in \{0, 1, \dots, \ell\}$. Given a holomorphic spray of sections $f: \bar{D}_0 \times P_0 \to Z$ ($P_0 \subset \mathbb{C}^N$) of class $\mathcal{A}^l(D_0)$ which is dominating on $\bar{D}_{0,1}$, there is an open set $P \subset \mathbb{C}^N$ with $0 \in P \subset P_0$ satisfying the following conditions:*

1. *For every holomorphic spray of sections $g: \bar{D}_1 \times P_0 \to Z$ of class $\mathcal{A}^l(D_1)$ which is sufficiently close to f in $\mathcal{C}^l(\bar{D}_{0,1} \times P_0)$ there exists a holomorphic spray of sections $f': \bar{D} \times P \to Z$ of class $\mathcal{A}^l(D)$, close to f in $\mathcal{C}^l(\bar{D}_0 \times P)$ (depending on the $\mathcal{C}^l$-distance between f and g on $\bar{D}_{0,1} \times P_0$), whose core f'_0 is homotopic to f_0 on $\bar{D}_0$ and is homotopic to g_0 on $\bar{D}_1$.*
2. *If f and g agree to order $m \in \mathbb{Z}_+$ along $\bar{D}_{0,1} \times \{0\}$, then f' can be chosen to agree to order m with f along $\bar{D}_0 \times \{0\}$, and with g along $\bar{D}_1 \times \{0\}$.*
3. *If σ is the zero set of finitely many $\mathcal{A}^l(D_0)$ functions and $\sigma \cap \bar{D}_{0,1} = \varnothing$, then f' can be chosen such that f'_0 agrees with f_0 to a finite order on σ.*

We need the following lemma which provides a transition map between a pair of holomorphic sprays. The hypotheses are as in Proposition 5.9.2.

Lemma 5.9.3 *Let $\epsilon > 0$. There exists a domain P_1 in $\mathbb{C}^N$, with $0 \in P_1 \subset P_0$, such that the following holds. If g approximates f sufficiently closely in the $\mathcal{C}^l$ topology on $\bar{D}_{0,1} \times P_0$, then there is a map $\psi: \bar{D}_{0,1} \times P_1 \to \mathbb{C}^N$ of class $\mathcal{A}^l(D \times P_1)$ satisfying the following conditions:*

$$\psi(x, w) = w + c(x, w), \qquad \|c\|_{\mathcal{C}^l(D_{0,1} \times P_1)} < \epsilon, \tag{5.14}$$

$$f(x, w) = g\big(x, \psi(x, w)\big), \quad (x, w) \in \bar{D}_{0,1} \times P_1. \tag{5.15}$$

If f and g agree to order m along $\bar{D} \times \{0\}$, then ψ can be chosen of the form $\psi(x, w) = w + \sum_{|J|=m} \widetilde{c}_J(x, w) w^J$ with $\widetilde{c}_J \in \mathcal{A}^l(D_{0,1} \times P_1)^N$.

Proof Let E denote the $\mathcal{A}^l(D_{0,1})$-subbundle of $\bar{D}_{0,1} \times \mathbb{C}^N$ with fibres

$$E_x = \ker\big(\partial_w f(x, w)|_{w=0} : \mathbb{C}^N \to \mathrm{VT}_{f(x,0)} Z\big), \quad x \in \bar{D}_{0,1}. \tag{5.16}$$

By Theorem B for $\mathcal{A}^l$-bundles [293, 385], E is complemented in the sense that there is an $\mathcal{A}^l(D_{0,1})$-subbundle E' of $\bar{D}_{0,1} \times \mathbb{C}^N$ such that

$$\bar{D}_{0,1} \times \mathbb{C}^N = E \oplus E'. \tag{5.17}$$

For each $x \in \bar{D}_{0,1}$ we write $\mathbb{C}^N \ni w = t_x \oplus t'_x \in E_x \oplus E'_x$. The map

$$\partial_w|_{w=0} f(\cdot, w) : E'|_{\bar{D}_{0,1}} \to \mathrm{VT}_{f_0(\bar{D}_{0,1})} Z$$

is an isomorphism. The implicit function theorem gives an open neighborhood $P_1 \Subset P_0$ of $0 \in \mathbb{C}^N$ such that for each holomorphic spray $g: \bar{D}_{0,1} \times P_0 \to Z$ which is sufficiently $\mathcal{C}^l$-close to f on $\bar{D}_{0,1} \times P_0$ there is a unique map

$$\widetilde{\psi}(x, w) = \widetilde{\psi}(x, t_x \oplus t'_x) = t_x \oplus (t'_x + \widetilde{c}(x, w)) \in E_x \oplus E'_x = \mathbb{C}^N$$

of class $\mathcal{A}^l(D_{0,1} \times P_1)$ satisfying the condition

$$f(x, \widetilde{\psi}(x, w)) = g(x, w)$$

and such that $\|\widetilde{c}\|_{\mathcal{A}^l(D_{0,1} \times P_1)}$ is controlled by the $\mathcal{C}^l$ distance between f and g on $\bar{D}_{0,1} \times P_0$. After shrinking P_1, the map $\widetilde{\psi}$ admits a fibrewise inverse $\psi(x, w) = t_x \oplus (t'_x + c'(x, w))$ which satisfies Lemma 5.9.3. □

Proof of Proposition 5.9.2 By Lemma 5.9.3 there exist a domain $P_1 \Subset P_0$ containing the origin and a transition map $\gamma: \bar{D}_{0,1} \times P_1 \to \bar{D}_{0,1} \times \mathbb{C}^N$ of the form $\gamma(x, w) = (x, \psi(x, w))$ and of class $\mathcal{A}^l(D_{0,1} \times P_1)$, close to the identity map $\mathrm{Id}(x, w) = (x, w)$ in the $\mathcal{C}^l$ topology (the closeness depending on the $\mathcal{C}^l$ distance between f and g on $\bar{D}_{0,1} \times P_0$), satisfying

$$f = g \circ \gamma \quad \text{on } \bar{D}_{0,1} \times P_1.$$

Let $P \Subset P_1$ be a domain containing the origin $0 \in \mathbb{C}^N$. If γ is sufficiently $\mathcal{C}^l$-close to the identity on $\bar{D}_{0,1} \times P_1$, then by Proposition 5.8.1 and Remark 5.8.3(B) there exist maps

$$\alpha: \bar{D}_0 \times P \to \bar{D}_0 \times \mathbb{C}^N, \qquad \beta: \bar{D}_1 \times P \to \bar{D}_1 \times \mathbb{C}^N,$$

of class $\mathcal{A}^l$ on their respective domains and satisfying

$$\gamma \circ \alpha = \beta \quad \text{on } \bar{D}_{0,1} \times P.$$

From this and $f = g \circ \gamma$ it follows that

$$f \circ \alpha = g \circ \beta \quad \text{on } \bar{D}_{0,1} \times P.$$

Hence, $f \circ \alpha$ and $g \circ \beta$ amalgamate into a holomorphic spray $f': \bar{D} \times P \to Z$ with the stated properties. □

Remark 5.9.4 (A) If the core maps of the two sprays in Proposition 5.9.2 agree on $D_{0,1}$ (as in part 2), then it suffices to assume that g is defined on $(\bar{D}_{0,1} \times P_0) \cup (\bar{D}_1 \times P')$ for some domain $P' \subset \mathbb{C}^N$ containing the origin, but possibly much smaller than P_0.

(B) Proposition 5.9.2 holds with the same additions as Proposition 5.8.1; see Remark 5.8.3 on p. 238. In particular, one can glue sprays defined on not necessarily trivial holomorphic vector bundles $E \to \bar{D}$.

(C) The same proof gives the analogue of Proposition 5.9.2 for gluing holomorphic sprays of sections over open neighborhoods of an arbitrary Cartan pair (A, B) in a Stein space; these neighborhoods are allowed to shrink during the gluing process. To this end, one uses Proposition 5.8.4 instead of Proposition 5.8.1. The proof of Lemma 5.9.3 applies in this setting without any changes and is even simpler. □

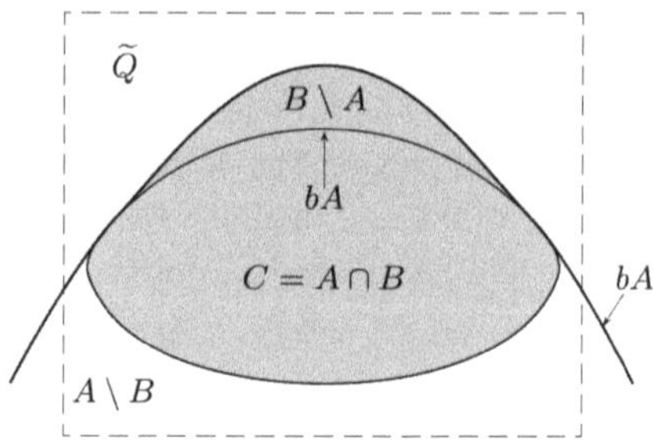

Fig. 5.2 A special Cartan pair. (Modified from [202, p. 860, Fig. 2])

5.10 Noncritical Strongly Pseudoconvex Extensions

We now apply the Heftungslemma from the previous section to obtain the following major ingredient in the proof of Theorem 5.4.4.

Proposition 5.10.1 *Assume that $A \subset A'$ are compact strongly pseudoconvex domains in a Stein manifold X such that there exists a strongly plurisubharmonic function ρ in an open set $V \supset \overline{A' \setminus A}$ in X, with $d\rho \neq 0$ on V, and numbers $c < c'$ such that*

$$A \cap V = \{x \in V : \rho(x) \le c\}, \qquad A' \cap V = \{x \in V : \rho(x) \le c'\}.$$

(*Such A' is said to be a* noncritical strongly pseudoconvex extension *of A.*) *Assume that $\pi\colon Z \to X$ is a holomorphic submersion such that $Z|_V \to V$ is a holomorphic fibre bundle whose fibre enjoys* CAP. *Then, every holomorphic section $f\colon U \to Z|_U$ over an open set $U \supset A$ can be approximated, uniformly on A, by sections f' which are holomorphic on an open neighborhood of A'.*

The main step is to extend a holomorphic section to a *special convex bump*. Proposition 5.10.1 follows immediately from Lemmas 5.10.3 and 5.10.5.

Definition 5.10.2 (Special Cartan pairs) A Cartan pair (A, B) in a complex space X (see Definition 5.7.1(I), p. 234) is a *special Cartan pair*, and B is a *special convex bump on A*, if there exist a compact set $\widetilde{Q} \subset X_{\mathrm{reg}}$ (which is called a *window for B*) that contains B in its interior and a holomorphic coordinate map ϕ from an open neighborhood of $\widetilde{Q}$ to $\mathbb{C}^n$ such that the following conditions hold (see Fig. 5.2):

(i) $Q = \phi(\widetilde{Q}) \subset \mathbb{C}^n$ is a cube (5.5) (p. 215), and
(ii) the sets $K = \phi(A \cap \widetilde{Q})$ and $K' = \phi((A \cup B) \cap \widetilde{Q})$ are special convex sets in Q (see Definition 5.4.2, p. 215).

Lemma 5.10.3 *Assume that $A \subset A' \subset X$ are as in Proposition 5.10.1, and let $\mathcal{U} = \{U_j\}$ be an open cover of $\overline{A' \setminus A}$. There are compact strongly pseudoconvex domains $A = A_0 \subset A_1 \subset \cdots \subset A_m = A'$ such that for every $k = 0, 1, \ldots, m-1$ we have $A_{k+1} = A_k \cup B_k$, where B_k is a special convex bump on A_k (see Definition 5.10.2) with a window $\widetilde{Q}_k$ contained in some U_j.*

Proof Given a pair of numbers a, b with $c \le a \le b \le c'$ we set

$$X_{a,b} = \{x \in V : a < \rho(x) < b\}.$$

If $b - a > 0$ is sufficiently small, Narasimhan's lemma (on local convexification of a strongly pseudoconvex hypersurface) shows that we can cover $X_{a,b}$ by the interiors of finitely many compact sets $\widetilde{Q}_1, \dots, \widetilde{Q}_N \subset V$ such that for every $j = 1, \dots, N$ there is a holomorphic coordinate map $\phi_j : U_j \to \phi_j(U_j) \subset \mathbb{C}^n$ on a neighborhood U_j of $\widetilde{Q}_j$ satisfying the following conditions:

(a) $\phi_j(\widetilde{Q}_j) = Q \subset \mathbb{C}^n$ is a cube of the form (5.5),
(b) the function $\widetilde{\rho}_j = \rho \circ \phi_j^{-1} : Q \to \mathbb{R}$ is strongly convex, and
(c) if $Q = Q_1 \times Q_2$ with $Q_1 \subset \mathbb{C}^{n-1} \times \mathbb{R}$ and $Q_2 \subset \mathrm{i}\mathbb{R}$, then

$$\phi_j(X_{a,b} \cap \widetilde{Q}_j) \cap (Q_1 \times bQ_2) = \varnothing.$$

Condition (c) shows that for every $t \in [a, b]$ the set

$$\phi_j\big(\{\rho \le t\} \cap \widetilde{Q}_j\big) = \big\{z \in Q : \widetilde{\rho}_j(z) \le t\big\}$$

is a special convex set (5.4) in Q.

By compactness we can partition $[c, c']$ into finitely many subintervals $c = c_0 < c_1 < \cdots < c_m = c'$ such that the above conditions hold on each subinterval $[c_{l-1}, c_l]$. It suffices to explain the construction of the bumps on each of the subintervals, and hence we may assume that the initial interval $[a, b] = [c, c']$ satisfies the above properties. Choose a smooth partition of unity $\{\chi_j\}$ in a neighborhood of $X_{a,b} = \overline{A' \setminus A}$ with $\operatorname{supp} \chi_j$ contained in the interior of $\widetilde{Q}_j$ for $j = 1, \dots, N$. For a sufficiently small $\epsilon > 0$ each of the functions $\rho_k = \rho - \sum_{j=1}^{k} \epsilon \chi_j$ is strongly plurisubharmonic, and the functions $\widetilde{\rho}_{j,k} = \rho_k \circ \phi_j^{-1} : \widetilde{Q}_j \to \mathbb{R}$ are strongly convex. Furthermore, for $\epsilon > 0$ small, the sets $\{z \in P : \widetilde{\rho}_{j,k} \le c\}$ for $c \in [a, b]$ are special convex sets of the form (5.4) (p. 215). Decreasing $\epsilon > 0$ if necessary we ensure that $b - a = \epsilon l$ for some $l \in \mathbb{N}$.

Let $A_k = A \cup \{x \in V : \rho_k(x) \le a\}$ for $k = 0, 1, \dots, N$. Since $\rho_k = \rho_{k-1} - \epsilon \chi_k$ and $\operatorname{supp} \chi_k$ contained in the interior of $\widetilde{Q}_k$, we see that

$$A_k = A \cup \big\{x \in V : \rho_{k-1}(x) \le a + \epsilon \chi_k(x)\big\} = A_{k-1} \cup B_k$$

where B_k is a special convex bump in the window $\widetilde{Q}_k$. Since $\sum_{j=1}^{N} \chi_j = 1$ in $X_{a,b}$, we have $A_N = A \cup \{\rho \le a + \epsilon\}$. We now repeat the same procedure of attaching bumps to A_N. Explicitly, we take

$$A_{N+1} = A \cup \big\{x \in V : \rho(x) \le a + \epsilon + \epsilon \chi_1(x)\big\}$$

etc. By attaching N convex bumps we obtain the set $A_{2N} = A \cup \{\rho \le a + 2\epsilon\}$. In l steps of this kind we reach the set $A' = A \cup \{\rho \le b\}$. □

Lemma 5.10.4 (Thickening a section) *Assume that $\pi : Z \to X$ is a holomorphic submersion onto a reduced complex space X, X' is a closed complex subvariety of X, and $A_0 \subset A$ are Stein compacts in X such that $A_0 \subset X \setminus X'$. Given open Stein sets $V \Subset V_0 \subset X$ containing A, a holomorphic section $f : V_0 \to Z$ and an* *integer $r \in \mathbb{N}$, there exist an integer $N \in \mathbb{N}$, an open set $W \subset \mathbb{C}^N$ containing 0, and a holomorphic spray of sections $F: V \times W \to Z$ such that*

(i) *$F_w = F(\cdot, w)$ is a section of $Z|_V$ for every fixed $w \in W$, with $F_0 = f$,*
(ii) *F_w agrees with f to order r along $V \cap X'$ for every $w \in W$, and*
(iii) *$\partial_w F(x, w)|_{w=0} : \mathbb{C}^N \to \mathrm{VT}_{f(x)} Z$ is surjective for every $x \in A_0$.*

If A_0 is $\mathcal{O}(A)$-convex and it admits a contractible Stein neighborhood in $V_0 \setminus X'$, then the above conclusion holds with $N = \dim \pi^{-1}(x)$ $(x \in A_0)$.

Proof The image $f(V_0)$ is a closed Stein subvariety of $Z|_{V_0}$, and hence it admits an open Stein neighborhood $\Omega \subset Z$. By Cartan's Theorem A there exist finitely many holomorphic vector fields $V_1, \ldots, V_N$ on Ω that are tangent to the fibres of π, that span VTZ at every point of $f(A_0)$, and that vanish to order r on the subvariety $\pi^{-1}(X') \cap \Omega$. Let ϕ_t^j denote the flow of V_j. The map

$$F(x, w_1, \ldots, w_N) = \phi_{w_1}^1 \circ \cdots \circ \phi_{w_N}^N \big(f(x)\big)$$

is defined and holomorphic for all $x \in V$ and for all $w = (w_1, \ldots, w_N)$ in an open set $W \subset \mathbb{C}^N$ containing the origin. Since $\partial_{w_j} F(x, w)|_{w=0} = V_j(f(x))$ and the vector fields V_j span VTZ on $f(A_0)$, we see that the differential $\partial_w F(x, w)|_{w=0}: \mathbb{C}^N \to \mathrm{VT}_{f(x)} Z$ is surjective for every $x \in A_0$.

This proves Lemma 5.10.4, except for the last claim that one can take $N = \dim \pi^{-1}(x)$ with $x \in A_0$. To prove this, we need the following special case of Theorem 5.3.1 on p. 213:

Every holomorphic vector bundle on a contractible Stein space is holomorphically trivial.

This is a well known consequence of the classical Oka-Grauert theory, but we wish to point out that it also follows from our proof since the particular value of N is not needed in the remainder of the proof of Proposition 5.10.5 (and hence of the proof of Theorem 5.4.4). Since a complex homogeneous space satisfies CAP (see Proposition 5.6.1 on p. 223), we thus obtain Theorem 5.3.1 and in particular the above statement.

We now complete the proof of Lemma 5.10.4. Since the set A_0 admits a contractible Stein neighborhood, $f^*\mathrm{VT}Z$ is a trivial holomorphic vector bundle of rank $N = \dim \pi^{-1}(x)$ $(x \in A_0)$ in a neighborhood of A_0. Therefore, VTZ is generated over a neighborhood of $f(A_0)$ in Z by N holomorphic sections. Since the set A_0 is $\mathcal{O}(A)$-convex, we can approximate them by sections (vertical holomorphic vector fields) in a neighborhood of $f(A)$ that vanish to order r on $\pi^{-1}(X')$. If the approximation is sufficiently close, then these new sections still generate VTZ on $f(A_0)$. Hence, Lemma 5.10.4 holds with $N = \dim \pi^{-1}(x)$. (This will prove Theorem 5.4.4 when each fibre Y satisfies CAP_{n+N} with $n = \dim X$ and $N = \dim Y$.) □

Lemma 5.10.5 (Extension across a special convex bump) *Assume that X is a reduced complex space, $X' \subset X$ is closed complex subvariety containing X_{sing}, and*

(A, B) is a special Cartan pair in X with a window $B \subset \widetilde{Q} \subset X \setminus X'$ (Definition 5.10.2 on p. 242). Let $\pi: Z \to X$ be a holomorphic submersion whose restriction $Z|_U$ to an open set $U \supset \widetilde{Q}$ is equivalent to a trivial bundle $U \times Y$ whose fibre Y enjoys CAP. *Then, every holomorphic section $f: V_0 \to Z$ over an open set $V_0 \supset A$ can be approximated uniformly on A by sections $\tilde{f}$ that are holomorphic in an open neighborhood of $A \cup B$ and agree with f to a given order $m \in \mathbb{N}$ along the subvariety X'. If in addition the set $A \cap B$ is $\mathcal{O}(A)$-convex, then the same conclusion holds if Y satisfies* CAP_{n+N}, *where $N = \dim Y$ and n is the dimension of the component of X_{reg} containing B.*

Proof Step 1: Thickening. Lemma 5.10.4 furnishes an open set V in X with $A \subset V \subset V_0$, an open set $0 \in W \subset \mathbb{C}^N$, and a holomorphic spray of sections $F: V \times W \to Z$ with the core $f = F(\cdot, 0)$. By using the trivialization $Z|_U \cong U \times Y$, we can write

$$F(x, w) = \big(x, F'(x, w)\big) \in X \times Y$$

for $x \in U \cap V$ and $w \in W$, where F' is a holomorphic map $(U \cap V) \times W \to Y$.

Step 2: Approximation. Set $K = \phi(A \cap \widetilde{Q})$ and $Q = \phi(\widetilde{Q})$; then, (K, Q) is a special convex pair in $\mathbb{C}^n$. Choose a compact cube $Q' \subset W$ containing $0 \in \mathbb{C}^N$ in its interior. Then, $(K \times Q', Q \times Q')$ is a special convex pair in $\mathbb{C}^n \times \mathbb{C}^N$. Since Y enjoys CAP, we can approximate F' uniformly on a neighborhood of $(A \cap \widetilde{Q}) \times Q'$ by a holomorphic map $G': U' \times Q' \to Y$, where U' is an open set satisfying $\widetilde{Q} \subset U' \subset U$. (For this approximation to hold, we must assume that Y enjoys CAP_{n+N}. This is the unique place in the proof of Theorem 5.4.4 where CAP of the fibre is invoked.)

Step 3: Gluing. Set $G(x, w) = (x, G'(x, w))$. If the approximation in Step 2 is sufficiently close, then by Proposition 5.9.2 and Remark 5.9.4(C) we can glue F and G into a holomorphic spray of sections $\widetilde{F}: V' \times rQ' \to Z$ over an open neighborhood V' of $A \cup B$ in X for some $0 < r < 1$. The section $\tilde{f} = \widetilde{F}(\cdot, 0): V' \to Z$ then satisfies the conclusion of Lemma 5.10.5.

To get the last statement, we use Lemma 5.10.4 with $N = \dim Y$. □

5.11 Proof of Theorem 5.4.4: The Basic Case

In this section we prove Theorem 5.4.4 in the special case when $\pi: Z \to X$ is a holomorphic fibre bundle over a Stein manifold X whose fibre Y enjoys CAP, $X' = \varnothing$, P is a singleton, and $P_0 = \varnothing$.

Assume that $K \subset X$ is a compact $\mathcal{O}(X)$-convex subset and $f_0: X \to Z$ is a continuous section that is holomorphic in an open set $U \subset X$ containing K.

Choose a smooth strongly plurisubharmonic Morse exhaustion function $\rho: X \to \mathbb{R}$ with nice critical points (see Definition 3.10.2, p. 95) such that $\rho < 0$ on K and $\rho > 0$ on $X \setminus U$. Let $p_1, p_2, p_3, \ldots$ be the critical points of ρ in $\{\rho > 0\}$, ordered so that $0 < \rho(p_1) < \rho(p_2) < \rho(p_3) < \cdots$. Choose a sequence

$$0 = c_0 < c_1 < c_2 < \cdots, \qquad \lim_{j \to \infty} c_j = +\infty$$

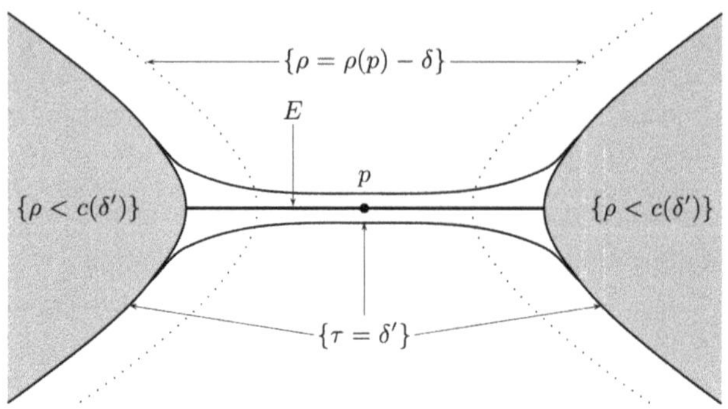

Fig. 5.3 The level sets of τ. (A modification of Fig. 3.5 on p. 100)

such that $c_{2j-1} < \rho(p_j) < c_{2j}$ for every $j = 1, 2, \ldots$. If there are only finitely many p_j's, we choose the remainder of the sequence c_j arbitrarily such that $\lim_{j\to\infty} c_j = +\infty$. We subdivide the parameter interval $[0, 1]$ of the homotopy into subintervals $I_j = [t_j, t_{j+1}]$ with $t_j = 1 - 2^{-j}$ $(j = 0, 1, 2, \ldots)$.

Choose a complete Riemannian distance function dist on Z. Fix an $\epsilon > 0$. We shall construct a homotopy of sections $f_t : X \to Z$ $(0 \le t < 1)$ such that for every $j \in \mathbb{Z}_+$ and $t \in [t_j, t_{j+1}]$ the section f_t is holomorphic in a neighborhood of the set $K_j = \{x \in X : \rho(x) \le c_j\}$ and satisfies

$$\sup\bigl\{\mathrm{dist}\bigl(f_t(x), f_{t_j}(x)\bigr) : x \in K_j,\ t \in [t_j, t_{j+1}]\bigr\} < 2^{-j-1}\epsilon.$$

The limit section $f_1 = \lim_{t\to 1} f_t : X \to Z$ is then holomorphic on X and satisfies $\sup\{\mathrm{dist}(f_1(x), f_0(x)) : x \in K_0\} < \epsilon$.

Assuming inductively that a homotopy $\{f_t\}$ with the stated properties has been constructed for $t \in [0, t_j]$, we explain how to find it for $t \in [t_j, t_{j+1}]$.

The noncritical case: If j is even, then ρ has no critical points in $K_{j+1} \setminus K_j$. By Proposition 5.10.1 (p. 242) we can approximate f_{t_j} as closely as desired, uniformly on a neighborhood of K_j, by a section $f_{t_{j+1}}$ that is holomorphic in a neighborhood of K_{j+1}. We also get a homotopy of global continuous sections f_t $(t \in [t_j, t_{j+1}])$ which are holomorphic in a neighborhood of K_{t_j}.

The critical case: If j is odd, ρ has a unique critical point $p \in K_{j+1} \setminus K_j$. Denote by E the stable manifold of p for the gradient flow of ρ (see Fig. 5.3). For a small $\delta > 0$, Proposition 3.11.4 on p. 102 (for $q = 1$) gives a smooth strongly plurisubharmonic function τ on $\{x \in X : \rho(x) < \rho(p) + 3\delta\}$ satisfying

(i) $\{\rho \le c_j\} \cup E \subset \{\tau \le 0\} \subset \{\rho \le \rho(p) - \delta\} \cup E$,
(ii) $\{\rho \le \rho(p) + \delta\} \subset \{\tau \le 2\delta\} \subset \{\rho < \rho(p) + 3\delta\}$, and
(iii) τ has no critical values in $(0, 3\delta)$.

A typical level set $\{\tau = \delta'\}$ for small $\delta' > 0$ is shown on Fig. 5.3. Outside of a neighborhood of the critical point, the level set $\{\tau = \delta'\}$ coincides with a certain level set $\{\rho = c(\delta')\}$ of ρ.

We proceed in four steps, hence dividing $[t_j, t_{j+1}]$ into four subintervals.

Step 1: By the noncritical case we deform f_{t_j} to a section that is holomorphic on a neighborhood of $\{\rho \le \rho(p) - \delta\}$. The deformation consists of sections which are holomorphic and close to f_{t_j} near K_j.

Step 2: Theorem 3.8.1 on p. 88 allows us to deform the final section from Step 1 to a section that is holomorphic in a neighborhood of $\{\rho \le \rho(p) - \delta\} \cup E$. The deformation consists of sections which are holomorphic and close to the initial section on a neighborhood of $\{x \in X : \rho(x) \le \rho(p) - \delta\}$.

Step 3: By (i), the set $\{\tau \le \delta'\}$ for a small $\delta' > 0$ is contained in the region where the final section from Step 2 is holomorphic. Applying the noncritical case (see Proposition 5.10.1 on p. 242) with the function τ we deform it to a holomorphic section on a neighborhood of $\{\tau \le 2\delta\}$. (We disregard a part of the domain of the section from Step 2 on which $\tau > \delta'$.)

Step 4: By (ii), the section from Step 3 is holomorphic in a neighborhood of $\{\rho \le \rho(p) + \delta\}$. Applying the noncritical case with ρ, we deform it to a holomorphic section on a neighborhood of $K_{j+1} = \{\rho \le c_{j+1}\}$.

These four steps together give a homotopy $\{f_t : t \in [t_j, t_{j+1}]\}$ consisting of sections which are holomorphic and close to f_{t_j} near K_j and continuous elsewhere on X. The induction may proceed.

This completes the proof of the special case of Theorem 5.4.4.

5.12 Proof of Theorem 5.4.4: Stratified Fibre Bundles

In this section we prove Theorem 5.4.4 for stratified fibre bundles over Stein spaces in the nonparametric case $(P, P_0) = (\{p\}, \varnothing)$. The following result [202, Proposition 5.1] provides the main step.

Proposition 5.12.1 *Assume that X is a reduced Stein space, $M_1 \subset M_0$ are closed complex subvarieties of X such that $S = M_0 \setminus M_1$ is a complex manifold with $\overline{S} = M_0$, and $\pi : Z \to X$ is a holomorphic submersion such that $Z|_S \to S$ is a holomorphic fibre bundle whose fibre enjoys* CAP. *Given compact $\mathcal{O}(X)$-convex subsets $K \subset L$ of X and a continuous section $f : X \to Z$ that is holomorphic in an open neighborhood of $K_1 = K \cup (L \cap M_1)$, there is for every $\epsilon > 0$ and $\ell \in \mathbb{N}$ a homotopy of continuous sections $f_t : X \to Z$ ($t \in [0, 1]$) that are holomorphic in a neighborhood of K_1 and satisfy the following conditions:*

(i) *f_t agrees with $f_0 = f$ to order ℓ along $M_1 \cap L$ for each $t \in [0, 1]$,*
(ii) $\sup_{x \in K,\ t \in [0,1]} \mathrm{dist}(f_t(x), f_0(x)) < \epsilon$, *and*
(iii) *f_1 is holomorphic in a neighborhood of $K_0 = K \cup (L \cap M_0)$ in X.*

Assume Proposition 5.12.1 for the time being.

Proof of Theorem 5.4.4, the nonparametric case The proof proceeds by induction over the strata in a suitable stratification of X. Let dist be a complete distance function on Z. Choose a sequence of compact $\mathcal{O}(X)$-convex sets

$$K = K_0 \subset K_1 \subset K_2 \subset \cdots \subset \bigcup_{k=0}^{\infty} K_k = X$$

Set $t_k = 1 - 2^{-k}$ and $I_k = [t_k, t_{k+1}]$ for $k = 0, 1, \ldots$; thus $\bigcup_{k=0}^{\infty} I_k = [0, 1)$. Let $f = f_0 \colon X \to Z$ be a continuous section that is holomorphic on a complex subvariety X' of X and in an open neighborhood of $K = K_0$. Fix a number $\epsilon > 0$. By induction on $k \in \mathbb{Z}_+$ we shall construct homotopies of sections $f_t \colon X \to Z$, $t \in I_k$, that agree at the common endpoint t_{k+1} of the adjacent intervals I_k, I_{k+1} and satisfy the following conditions:

- for every $k = 0, 1, \ldots$ and $t \in I_k$ the section f_t is holomorphic in an open neighborhood of the set K_k and it satisfies the condition

$$\sup_{x \in K_k} \mathrm{dist}\big(f_t(x), f_{t_k}(x)\big) < 2^{-k-1}\epsilon,$$

- the homotopy $\{f_t\}_{t \in [0,1)}$ is fixed on the subvariety X'.

These conditions clearly imply that the limit section $f_1 = \lim_{t \to 1} f_t \colon X \to Z$ exists and is holomorphic on X, and it satisfies

$$\sup_{x \in K_0} \mathrm{dist}\big(f_1(x), f_0(x)\big) < \epsilon, \quad f_1|_{X'} = f_0|_{X'}.$$

Thus, the homotopy $\{f_t\}_{t \in [0,1]}$ satisfies the conclusion of Theorem 5.4.4.

Since all inductive steps are of the same kind, we shall explain how to get the first homotopy for $t \in I_0 = [0, \frac{1}{2}]$. Set $K = K_0$ and $L = K_1$. By the assumption, there exists an open set $U \subset X$ containing L such that $Z|_U$ is a stratified holomorphic fibre bundle whose strata satisfy CAP. Since L is $\mathcal{O}(X)$-convex, there is a relatively compact Stein domain Ω in X with $L \subset \Omega \subset U$. Choose a stratification

$$\Omega = X_0 \supset X_1 \supset \cdots \supset X_m = \varnothing$$

such that the restriction of $\pi \colon Z|_\Omega \to \Omega$ to each stratum $S \subset X_k \setminus X_{k+1}$ is a fibre bundle whose fibre enjoys CAP. Taking $X'_k = X_k \cup (X' \cap \Omega)$ we obtain a stratification

$$\Omega = X'_0 \supset X'_1 \supset \cdots \supset X'_m = \Omega \cap X'$$

with regular strata $X'_k \setminus X'_{k+1} = X_k \setminus (X_{k+1} \cup X')$, ending with $X'_m = \Omega \cap X'$.

By Theorem 3.4.1 (p. 75) we can assume that the initial section f_0 is holomorphic in an open neighborhood of $K \cup (L \cap X')$. Let $\{f_t\}_{t \in [0,1/2m]}$ be a homotopy furnished by Proposition 5.12.1 for the pair of subvarieties $M_1 = X'_m$ and $M_0 = X'_{m-1}$ of Ω, with ϵ replaced by $\frac{\epsilon}{2m}$. Then, $f_{\frac{1}{2m}}$ is holomorphic in a neighborhood of $K \cup (L \cap X'_{m-1})$, the homotopy is fixed on X', and $\sup_{x \in K_0} \mathrm{dist}(f_t(x), f_0(x)) < \frac{\epsilon}{2m}$ for all $t \in [0, \frac{1}{2m}]$.

Next, we apply Proposition 5.12.1 with the 'initial' section $f = f_{\frac{1}{2m}}$ and the pair of subvarieties $M_1 = X'_{m-1}$, $M_0 = X'_{m-2}$ in order to find a homotopy $\{f_t\}_{t\in[\frac{1}{2m},\frac{2}{2m}]}$ that is fixed on X' such that the section $f_{\frac{2}{2m}} = f_{\frac{1}{m}}$ is holomorphic in a neighborhood of $K \cup (L \cap X'_{m-2})$ and we have that $\sup_{x\in K_0} \mathrm{dist}(f_t(x), f_{\frac{1}{2m}}(x)) < \frac{\epsilon}{2m}$ for all $t \in [\frac{1}{2m}, \frac{2}{2m}]$.

Continuing in this way, we obtain after m steps a homotopy $\{f_t\}_{t\in[0,\frac{1}{2}]}$ with the required properties. In particular, the section $f_{\frac{1}{2}}$ is holomorphic in a neighborhood of $L = K_1$ and it agrees with f_0 on the subvariety X'. By using a cut-off function in the parameter, we can extend this homotopy to all of X without changing it near $L = K_1$. When the initial section $f = f_0$ is already holomorphic in a neighborhood of $K \cup X'$, we can use Theorem 3.4.5 (p. 78) and Remark 3.4.4 to keep the sections f_t holomorphic in a neighborhood of $K \cup X'$ and tangent to f_0 to a given order ℓ along X'.

This proves Theorem 5.4.4 in the nonparametric case. □

Proof of Proposition 5.12.1 We first consider the case when X is a Stein manifold and $M_0 = X$; the general case will be explained at the end.

We may assume that $L = \{x \in X : \rho(x) \le 0\}$, where $\rho\colon X \to \mathbb{R}$ is a smooth strongly plurisubharmonic exhaustion function on X such that $\rho|_K < 0$ and $d\rho \neq 0$ on $bL = \{\rho = 0\}$. By the assumption, f is holomorphic in an open set $U \supset K \cup M_1$. Since the compact set $K' = K \cup (M_1 \cap \{\rho \le 1\}) \subset U$ is $\mathcal{O}(X)$-convex, there is a smooth strongly plurisubharmonic exhaustion function $\tau\colon X \to \mathbb{R}$ such that $\tau < 0$ on K' and $\tau > 0$ on $X \setminus U$. By general position we may assume that 0 is a regular value of τ and the hypersurfaces $\{\rho = 0\} = bL$ and $\{\tau = 0\}$ intersect transversely along $\Sigma = \{\rho = 0\} \cap \{\tau = 0\}$. Hence, $D_0 = \{\tau \le 0\} \subset U$ is a strongly pseudoconvex domain with smooth boundary. For each $s \in [0, 1]$ let

$$\rho_s = \tau + s(\rho - \tau) = (1-s)\tau + s\rho, \qquad D_s = \{\rho_s \le 0\} = \big\{\tau \le s(\tau - \rho)\big\}. \tag{5.18}$$

We have that $D_0 = \{\tau \le 0\}$ and $D_1 = \{\rho \le 0\} = L$. Let

$$\Omega = \{\rho < 0,\ \tau > 0\} \subset D_1 \setminus D_0, \qquad \Omega' = \{\rho > 0,\ \tau < 0\} \subset D_0 \setminus D_1.$$

As s increases from 0 to 1, $D_s \cap L$ increases to $D_1 = L$ while $D_s \setminus L \subset D_0$ decreases to $\varnothing$. All hypersurfaces $bD_s = \{\rho_s = 0\}$ intersect along Σ. Since $d\rho_s = (1-s)d\tau + s d\rho$ and the differentials $d\tau$, $d\rho$ are linearly independent along Σ, bD_s is smooth near Σ. Finally, bD_s is strongly pseudoconvex at every smooth point, in particular, at every point where $d\rho_s \neq 0$. (See Fig. 5.4.)

We investigate the singular points of $bD_s = \{\rho_s = 0\}$ inside Ω. The defining equation of $D_s \cap \Omega$ can be written as $\tau \le s(\tau - \rho)$ and, after dividing by $\tau - \rho > 0$, as

$$D_s \cap \Omega = \left\{ x \in \Omega : h(x) = \frac{\tau(x)}{\tau(x) - \rho(x)} \le s \right\}.$$

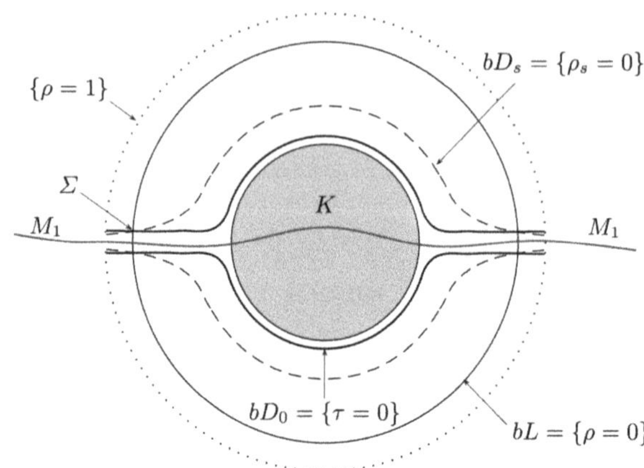

Fig. 5.4 The sets D_s. (Modified from [135, p. 245, Fig. 4] and [202, p. 859, Fig. 1])

The critical point equation $\mathrm{d}h = 0$ is equivalent to

$$(\tau - \rho)\mathrm{d}\tau - \tau(\mathrm{d}\tau - \mathrm{d}\rho) = \tau\mathrm{d}\rho - \rho\,\mathrm{d}\tau = 0.$$

A generic choice of ρ and τ ensures that there are at most finitely many solutions $p_1, \ldots, p_m \in \Omega$ and no solution on $b\Omega$. A calculation shows that at each critical point the complex Hessians satisfy

$$(\tau - \rho)^2 H_h = \tau H_\rho - \rho H_\tau.$$

Since $\tau > 0$ and $-\rho > 0$ on Ω, we conclude that $H_h > 0$ at such points. By a small modification of h we can assume that all its critical points are nice (see Definition 3.10.2, p. 95) and belong to different levels of h.

Let s_0, s_1 with $0 \le s_0 < s_1 \le 1$ be regular values of h on Ω such that h has at most one critical point in the domain

$$\Omega_{s_0,s_1} = \big\{x \in \Omega : s_0 < h(x) < s_1\big\}.$$

Assume inductively that we have found a homotopy $f_t\colon X \to Z$, $t \in [0, s_0]$, satisfying the conditions in Proposition 5.12.1 and such that f_{s_0} is holomorphic in a neighborhood of D_{s_0}. We wish to deform f_{s_0} to a section f_{s_1} that is holomorphic in a neighborhood of D_{s_1} by a homotopy which is fixed on M_1 and consists of sections that are holomorphic near $D_{s_0} \cap D_{s_1}$; the proof is then completed by a finite induction. We consider two cases as in Sect. 5.11.

The noncritical case: h has no critical values in Ω_{s_0,s_1}. By subdividing $[s_0, s_1]$ into finitely many subintervals and replacing $[s_0, s_1]$ by any such subinterval, we can assume that D_{s_1} is obtained from $D_{s_0} \cap D_{s_1}$ by attaching to the latter set finitely many special convex bumps contained in $X \setminus M_1$ (see Definition 5.10.2 on p. 242 and Fig. 5.5). On each bump, we apply Proposition 5.10.5 exactly as in Sect. 5.11. In finitely many steps we accomplish our task.

The critical case: h has a unique critical point $p \in \Omega_{c_0,c_1}$. We apply the critical case construction in Sect. 5.11, using the function h instead of ρ.

This proves the special case of Proposition 5.12.1.

Fig. 5.5 A special bump B on $D_{s_0} \cap D_{s_1}$. (Modified from [202, p. 861, Fig. 3])

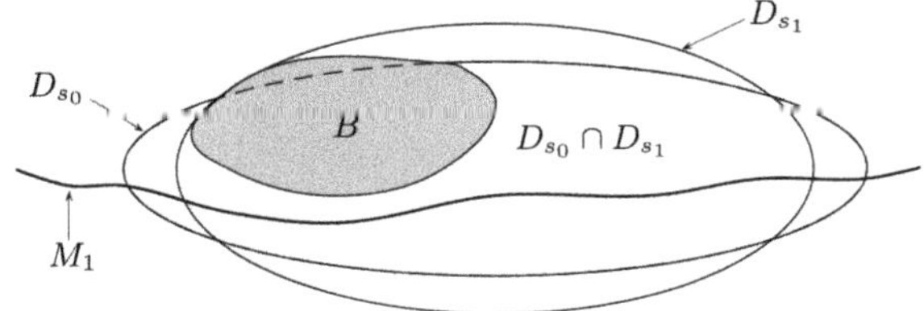

The general case: X is a Stein space and $M_1 \subset M_0$ are closed complex subvarieties of X whose difference $S = M_0 \setminus M_1$ is a complex manifold such that $\overline{S} = M_0$. By the assumption, f is holomorphic in an open set $U \subset X$ containing $K_1 = K \cup (M_1 \cap L)$. We may assume that $L = \{x \in X : \rho(x) \le 0\}$, where $\rho: X \to \mathbb{R}$ is a smooth strongly plurisubharmonic exhaustion function whose restriction to S has no critical points on $bL \cap S = \{\rho = 0\} \cap S$. Note that only finitely many connected components of S intersect the compact set L, and hence we can refine the stratification and assume that S is connected. We embed a relatively compact neighborhood of L in X holomorphically into $\mathbb{C}^N$. Since the set K_1 is $\mathcal{O}(X)$-convex, there is a smooth strongly plurisubharmonic function $\tau: X \to \mathbb{R}$ such that $\tau < 0$ on K_1 and $\tau > 0$ on $X \setminus U$. Set

$$D_0 = \big\{x \in M_0 : \tau(x) \le 0\big\} \subset U.$$

By general position, we may assume that 0 is a regular value of $\tau|_S$ and that the hypersurfaces $\{\rho = 0\} \cap S = bL \cap S$ and $\{\tau = 0\} \cap S$ in S intersect transversely along the submanifold $\Sigma = \{\rho = 0\} \cap \{\tau = 0\} \cap S$ of S. We define ρ_s as in (5.18) and set

$$D_s = \big\{x \in M_0 : \rho_s(x) \le 0\big\} = \big\{x \in M_0 : \tau(x) \le s\big(\tau(x) - \rho(x)\big)\big\}.$$

As s increases from 0 to 1, $D_s \cap L$ increases from $D_0 \cap L \subset M_0$ to $D_1 = L \cap M_0$. As in the special case, we successively attach to the set $A_0 = K \cup (D_0 \cap L)$ convex bumps and handles contained in the submanifold S, thereby reaching the set $K_0 = K \cup (L \cap M_0)$ in finitely many steps. Note that A_0 is $\mathcal{O}(X)$-convex and contains a collar around the set $K_1 = K \cup (L \cap M_1)$ in $L \cap M_0$.

Consider a typical step in the *noncritical case.* (For details, see Sect. 5.11.) Assume that (A, B) is a Cartan pair in X such that A is obtained by attaching to A_0 finitely many bumps and handles contained in $S \cap L$ and $B \subset S \cap L$ is a convex bump attached to $A \cap S$. Then, (A, B) is also a Cartan pair in $\mathbb{C}^N$. Assume inductively that a section $f: X \to Z$ is holomorphic in a neighborhood of $K_1 = K \cup (L \cap M_1)$ and in a relative neighborhood of $A \cap S$ in S. By using Lemma 5.10.4 (p. 243), we embed f (on its domain of holomorphicity) into a family of holomorphic sections $F(x, w)$ of Z, depending holomorphically on a parameter w in an open neighborhood W of the origin in some $\mathbb{C}^p$, such that $f = F(\cdot, 0)$, $F(x, w) = f(x)$ for $x \in M_1$ and $w \in W$, and F is submersive in the w-variable for all x in a neighborhood of $A \cap B$ in S.

Since the fibre of the bundle $Z|_S \to S$ satisfies CAP, we can approximate F, uniformly in a neighborhood of $A \cap B$ in S, by a family G of holomorphic sections defined in a relative neighborhood of B in S and depending holomorphically on $w \in W$. If the approximation is sufficiently close, then Lemma 5.9.3 (p. 240) furnishes a holomorphic transition map of the form

$$\gamma(x, w) = \big(x, \psi(x, w)\big),$$

close to the identity map and satisfying $F = G \circ \gamma$ on the intersection of their domains in $S \times \mathbb{C}^p$. Applying Proposition 5.8.4 (p. 238) we decompose

$$\gamma = \beta \circ \alpha^{-1}$$

over a pair of neighborhoods $(\widetilde{A}, \widetilde{B})$ of the Cartan pair (A, B) in X, where α and β are biholomorphic maps of the same type as γ and close to the identity on $\widetilde{A}$ and $\widetilde{B}$, respectively. (The parameter set W again shrinks a little.) In addition, the map α can be chosen to agree with the identity map to order ℓ along the intersection of its domain with the subvariety M_1. From

$$F = G \circ \gamma = G \circ \beta \circ \alpha^{-1}$$

we see that

$$F \circ \alpha = G \circ \beta \quad \text{holds near } A \cap B.$$

Hence, the two sides define a spray of holomorphic sections $\widetilde{F}$ of $Z \to X$ over a neighborhood of $A \cup B$. By the construction, the section $\widetilde{F}(\cdot, 0)$ approximates f uniformly on K, it agrees with f to order ℓ along the subvariety M_1, and is homotopic to f by a homotopy satisfying the required properties. The induction may now proceed.

We deal with the critical points of the function $h = \frac{\tau}{\tau - \rho}$ in $S \cap L$ exactly as before by reducing to the noncritical case (see the critical case above).

In finitely many such steps we obtain a homotopy $\{f_t\}_{t \in [0,1]}$ with the required properties such that f_1 is holomorphic in a neighborhood of K_1 in X, and also in a relative neighborhood of $L \cap M_0$ in the subvariety M_0.

By Theorem 3.4.1 (p. 75) there is a holomorphic section $\widetilde{f_1}$ in a neighborhood of $K_0 = K \cup (L \cap M_0)$ in X such that $\widetilde{f_1}$ is as close as desired to f_1 on K, $\widetilde{f_1} = f_1$ on $L \cap M_0$, and $\widetilde{f_1}$ agrees with f_1 to order ℓ along $L \cap M_1$. Replace f_1 by $\widetilde{f_1}$ and adjust the homotopy $\{f_t\}$ accordingly. By using a cut-off function in the parameter of the homotopy, we can extend $\{f_t\}$ continuously to all of X without changing it near K_0 and on M_1.

This complete the proof of Proposition 5.12.1. □

5.13 Proof of Theorem 5.4.4: The Parametric Case

The following proposition shows that one can replace the parameter subset P_0 of P in Theorem 5.4.4 by a neighborhood of P_0. This will enable us to use cut-off functions on P in order to patch together families of sections.

Proposition 5.13.1 *Let $\pi: Z \to X$ and (K, X', P_0, P, f) be as in Theorem 5.4.4 (p. 216). Given a domain $D \Subset X$, there exist a neighborhood P_0' of P_0 in P and a* *homotopy* $f_t: P \times X \to Z$ $(t \in [0, 1])$ *satisfying the conclusion of Theorem* 5.4.4, *except that condition* (*i*) *is replaced by the following one*:

(i′) $f_1(p, \cdot)$ *is holomorphic on* D *for all* $p \in P'_0$.

Proof The special case when $\pi: Z = X \times \mathbb{C} \to X$ is the trivial line bundle is covered by Theorem 2.8.4. The general case reduces to this special case by using that for every $p_0 \in P_0$ the Stein subvariety $\{f(p_0, x) : x \in X\}$ of Z admits an open Stein neighborhood in Z (see Theorem 3.1.1, p. 66). Embedding these neighborhoods in Euclidean spaces and using holomorphic retractions onto fibres of π, furnished by Theorem 3.3.5 (p. 75), the special case gives neighborhoods $U_{p_0} \subset U'_{p_0}$ of p_0 in P and a continuous family of sections $f_1: P \times \bar{D} \to Z$, homotopic to f, satisfying the following conditions:

(i) $f_1(p, \cdot)$ is holomorphic on $\bar{D}$ for $p \in U_{p_0}$,
(ii) $f_1(p, \cdot) = f(p, \cdot)$ for $p \in P_0 \cup (P \setminus U'_{p_0})$,
(iii) $f_1(p, x) = f(p, x)$ for all $x \in X' \cap D$ and $p \in P$, and
(iv) f_1 approximates f uniformly on $K \times P$.

The special case is first used for the values p in a neighborhood U'_{p_0} of p_0; the resulting family of holomorphic maps $\bar{D} \times U'_{p_0} \to \mathbb{C}^N$ is then patched with f by using a cut-off function $\chi(p)$ with support in U'_{p_0} that equals one on a smaller neighborhood U_{p_0} of p_0, and applying holomorphic retractions onto the fibres of π. In finitely many steps of this kind we complete the proof. □

We introduce a parametric version of the convex approximation property.

Definition 5.13.2 A complex manifold Y enjoys the *Parametric Convex Approximation Property* (PCAP) for a certain pair of compact Hausdorff spaces $P_0 \subset P$ if the following holds. Let (K, Q) be a special convex pair in $\mathbb{C}^n$ (see Definition 5.4.2, p. 215). Given a continuous map $f: P \times Q \to Y$ such that $f(p, \cdot): Q \to Y$ is holomorphic for every $p \in P_0$ and is holomorphic on K for every $p \in P$, f can be approximated uniformly on $P \times K$ by continuous maps $\widetilde{f}: P \times Q \to Y$ such that $\widetilde{f}(p, \cdot)$ is holomorphic on Q for all $p \in P$ and $\widetilde{f} = f$ on $P_0 \times Q$.

Theorem 5.13.3 *If* $\pi: Z \to X$ *is a stratified holomorphic fibre bundle over a reduced Stein space* X *such that all fibres* $Z_x = \pi^{-1}(x)$ $(x \in X)$ *enjoy* PCAP *for a certain pair of compact Hausdorff spaces* $P_0 \subset P$, *then sections* $X \to Z$ *satisfy the parametric Oka property with approximation and jet interpolation for this pair* $P_0 \subset P$.

Proof We follow the proof of Theorem 5.4.4, using parametric versions of all main ingredients. Each of the basic steps is of the following two types:

- extension to a special bump (see Proposition 5.10.5 on p. 244), or
- extension across a totally real disc (see Theorem 3.8.1 on p. 88).

The first step consists of the following four substeps:

(i) embedding a holomorphic section as the core of a dominating fibre spray (see Lemma 5.10.4 on p. 243),
(ii) approximation of the fibre spray on a special Cartan pair (by invoking CAP of the fibre),
(iii) finding a transition map between the two fibre sprays (see Lemma 5.9.3 on p. 240), and
(iv) splitting the transition map (see Proposition 5.8.1 on p. 235) and gluing the pair of fibre sprays.

Each of these substeps can be performed simultaneously for all sections in the family with continuous dependence on the parameter:

- In substep (i), we apply a continuous partition of unity on P to find a family of holomorphic sprays depending continuously on $p \in P$.
- Substep (ii) is a tautology by the definition of PCAP. The sections belonging to the values of the parameter $p \in P_0$ are left unchanged.
- For substep (iii), note that the transition map furnished by Lemma 5.9.3 is uniquely determined by a choice of a complementary bundle E' (5.17) to the bundle E (5.16). In the parametric case, the subbundle E_p depends continuously on the parameter $p \in P$, and we need a continuous family of complementary subbundles E'_p. Clearly, the same subbundle E' can be used for an open set of points in the parameter space P, and we find the family E'_p by a partition of unity on P. (See also Lemma 6.5.2 on p. 281.)
- Substep (iv) is immediate since Proposition 5.8.1 includes parameters.

Similar arguments apply in the second step. □

The parametric case of Theorem 5.4.4 now follows from Theorem 5.13.3 and the following result from [199].

Theorem 5.13.4 *If a complex manifold Y enjoys* CAP, *then it also enjoys* PCAP *for each pair $P_0 \subset P$ of compact subsets of a Euclidean space $\mathbb{R}^m$.*

Proof Assume that Y enjoys CAP; we proved that the nonparametric case of Theorem 5.4.4 then holds for maps from Stein manifolds to Y.

Assume that $P_0 \subset P$ are compacts in $\mathbb{R}^m \subset \mathbb{C}^m$, (K, Q) is a special convex pair in $\mathbb{C}^n$, $U \supset K$ and $V \supset Q$ are open convex neighborhoods of K resp. of Q in $\mathbb{C}^n$, and $f : P \times V \to Y$ is a continuous map such that

(i) the map $f(p, \cdot) : V \to Y$ is holomorphic for every $p \in P_0$, and
(ii) $f(p, \cdot)$ is holomorphic on U for every $p \in P$.

We shall obtain the conclusion of Definition 5.13.2 for this f.

After shrinking the set $V \supset Q$, we can assume by Proposition 5.13.1 (p. 252) that $f(p, \cdot)$ is holomorphic on V for all p in a neighborhood $P'_0 \subset \mathbb{C}^m$ of P_0. We may assume that $0 \in \mathbb{C}^n$ belongs to $\mathring{K}$. Choose a continuous function $\tau : P \to [0, 1]$ such that $\tau = 0$ on P_0 and $\tau = 1$ on $P \setminus P_0'$. Set

$$f_t(p, z) = f\big(p, \big(1 - (1-t)\tau(p)\big)z\big) \in Y, \quad p \in P,\ z \in V,\ t \in [0, 1].$$

Then, f_t has the same properties as $f = f_1$, the homotopy f_t is fixed for $p \in P_0$, and the map $f_0(p, \cdot)$ is holomorphic on V for all $p \in P$.

Set $Z = \mathbb{C}^m \times \mathbb{C}^n \times Y$. Let (p, z) be the coordinates on $\mathbb{C}^m \times \mathbb{C}^n$ and let $\pi_Y : Z \to Y$ denote the projection. For every $t \in [0, 1]$ let

$$F_t(p, z) = \big(p, z, f_t(p, z)\big), \quad \Sigma_t = F_t(P \times K) \subset Z, \quad S_0 = F_0(P \times Q) \subset Z.$$

By Corollary 3.6.6 on p. 85, the sets S_0 and Σ_t ($t \in [0, 1]$) are Stein compacts in Z. Hence, there are numbers $0 = t_0 < t_1 < \cdots < t_N = 1$ and Stein domains $\Omega_0, \ldots, \Omega_{N-1} \subset Z$ such that

$$\Sigma_t \subset \Omega_j \quad \text{when } t_j \le t \le t_{j+1} \text{ and } j = 0, 1, \ldots, N-1. \tag{5.19}$$

Let $E \to Z$ denote the vertical tangent bundle of the projection map $Z \to \mathbb{C}^m \times \mathbb{C}^n$. By Proposition 3.3.2 (p. 74), for every Stein domain $\Omega \subset Z$ there exist a Stein neighborhood $W \subset E|_\Omega$ of the zero section $\Omega \subset E|_\Omega$ and a holomorphic map $s : W \to Z$ taking the fibre $W_{(p,z,y)}$ over $(p, z, y) \in Z$ biholomorphically onto a neighborhood of this point in $\{(p, z)\} \times Y$ and preserving the zero section Ω of $E|_\Omega$. (Such a map s is a local dominating fibre spray in the sense of Definition 5.9.1, p. 239.) We may assume that W is Runge in $E|_\Omega$ and that its fibres are convex domains in the fibres of E.

Since $E|_\Omega$ is Stein and Y enjoys the Basic Oka Property, s can be approximated uniformly on compacts in W by a holomorphic dominating fibre spray $\widetilde{s} : E|_\Omega \to Z$ that agrees with s to the second order along the zero section Ω. Thus, $\widetilde{s}$ maps each fibre $E_{(p,z,y)}$ into the corresponding fibre $\{(p, z)\} \times Y$, and its differential in the fibre direction is an isomorphism along the zero section.

This shows that, after refining our subdivision $\{t_j\}$ of $[0, 1]$ and shrinking the set $U \supset K$, there are Stein domains $\Omega_0, \ldots, \Omega_{N-1} \subset Z$ as in (5.19), sprays $s_j : E|_{\Omega_j} \to Z$, and homotopies of z-holomorphic sections ξ_t ($t \in [t_j, t_{j+1}]$) of the restricted bundle $E|_{F_{t_j}(P \times U)}$ such that ξ_{t_j} is the zero section, $\xi_t(p, \cdot)$ is independent of t when $p \in P_0$ (hence it equals the zero section), and

$$s_j \circ \xi_t \circ F_{t_j} = F_t \quad \text{on } P \times U,\ t \in [t_j, t_{j+1}]. \tag{5.20}$$

Furthermore, the existence of such liftings ξ_t is stable under sufficiently small perturbations of the homotopy F_t. (See Fig. 5.6.)

Consider the homotopy of sections ξ_t of $E|_{F_0(P \times U)}$ for $t \in [0, t_1]$. By the parametric Oka-Weil theorem (see Theorem 2.8.4 on p. 62) we can approximate ξ_t uniformly on $F_0(P \times K)$ by z-holomorphic sections $\widetilde{\xi}_t$ of $E|_{F_0(P \times V')}$, where $V' \subset \mathbb{C}^n_z$ is an open convex set with $Q \subset V' \subset V$, such that $\widetilde{\xi}_0 = \xi_0$ and $\widetilde{\xi}_t = \xi_t$ for all $t \in [0, 1]$ on the set $F_0(P_0 \times V')$.

By Corollary 3.6.6 (p. 85) there is a Stein neighborhood $\Omega \subset Z$ of $S_0 = F_0(P \times Q)$ such that $\Sigma_0 = F_0(P \times K)$ is $\mathcal{O}(\Omega)$-convex. Hence, $E|_{\Sigma_0}$ is exhausted by $\mathcal{O}(E|_\Omega)$-convex compact sets. Since $E|_\Omega$ is a Stein manifold and Y enjoys the

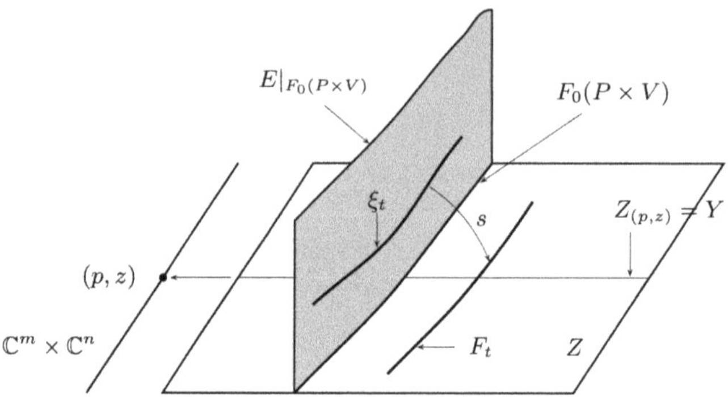

Fig. 5.6 Lifting sections F_t. (Modified from [201, p. 148, Fig. 1])

Basic Oka Property, the fibre spray $s_0\colon E|_{\Omega_0} \to Z$ can be approximated on the range of the homotopy $\{\xi_t : t \in [0, t_1]\}$ by a fibre spray $\widetilde{s}_0\colon E|_{\Omega} \to Z$ that agrees with s_0 to the second order along the zero section. The maps

$$\widetilde{f}_t = \pi_Y \circ \widetilde{s}_0 \circ \widetilde{\xi}_t \circ F_0 : P \times V' \to Y, \quad t \in [0, t_1]$$

are then z-holomorphic on the set $V' \supset Q$ and they approximate f_t uniformly on $P \times K$. If the approximation is sufficiently close, we obtain a new homotopy $\{f_t : t \in [0, 1]\}$ that agrees with $\widetilde{f}_t$ for $t \in [0, t_1]$ (hence, it is holomorphic on Q for $t \in [0, t_1]$), and that agrees with the initial homotopy for $t \in [t_1', 1]$ for some $t_1' > t_1$ close to t_1.

We repeat the same argument with the interval $[t_1, t_2]$, using f_{t_1} as the new reference map. Let F_{t_1} denote the section corresponding to f_{t_1}. We must ensure that the new homotopy f_t for $t \in [t_1, t_2]$ can be lifted by the spray s_1 to a family of sections ξ_t $(t \in [t_1, t_2])$ of $E|_{F_{t_1}(P\times U)}$ such that (5.20) holds. But this is so if the approximations in the first step were sufficiently close since the existence of liftings is stable under small perturbations of the homotopy F_t. This gives a new homotopy that is z-holomorphic on Q for $t \in [0, t_2]$. After finitely many steps of this kind we obtain a desired homotopy whose final map $\widetilde{f}$ at $t = 1$ satisfies Definition 5.13.2. This proves Theorem 5.13.4, and hence completes the proof of the parametric case of Theorem 5.4.4. □

5.14 Existence Theorems for Holomorphic Sections

Theorem 5.4.4 gives a holomorphic section, homotopic to a given continuous section, under suitable assumptions on the fibres. We now add a connectivity condition on the fibres to ensure the existence of sections.

Theorem 5.14.1 *Assume that X is a finite dimensional reduced Stein space, Z is a complex space, $\pi\colon Z \to X$ is a holomorphic submersion, and $X = X_0 \supset X_1 \supset \cdots \supset X_m = \varnothing$ is a stratification such that for each connected component S of $X_k \setminus X_{k+1}$ the restriction $Z|_S \to S$ is a holomorphic fibre bundle whose fibre Y_S is an Oka* *manifold satisfying* $\pi_q(Y_S) = 0$ *for all* $q < \dim S$. *Then, there exists a holomorphic section* $X \to Z$.

Furthermore, given a closed complex subvariety $X' \subset X$, *a compact* $\mathcal{O}(X)$*-convex subset* $K \subset X$, *an open set* $U \supset K$, *and a holomorphic section* $f: U \cup X' \to Z$, *there exists a holomorphic section* $\tilde{f}: X \to Z$ *such that* $\tilde{f}|_{X'} = f|_{X'}$ *and* $\tilde{f}$ *approximates* f *as closely as desired uniformly on* K. *If* f *is holomorphic in a neighborhood of* $K \cup X'$, *then* $\tilde{f}$ *can in addition be chosen to agree with* f *to any given finite order along the subvariety* X'.

Proof The only place in the proof of Theorem 5.4.4 where a topological condition on the fibre is needed is to cross a critical point p of index $k \geq 1$ of a strongly plurisubharmonic Morse function ρ on a stratum S. (See the *critical case* in the proof of Proposition 5.12.1 on p. 247.) At such a point $p \in S$ we must be able to extend a given holomorphic section, defined on a sublevel set $\{\rho \leq c\}$ for some $c < \rho(p)$ close to the critical level $\rho(p)$, to a continuous section over a k-dimensional totally real disc $E \subset S$, attached with its boundary $(k-1)$-sphere bE to $\{\rho = c\}$, such that $\{\rho \leq c\} \cup E$ is a strong deformation retraction of a sublevel set $\{\rho \leq c'\}$ for some $c' > \rho(p)$.

Since $Z|_S \to S$ is a fibre bundle with fibre Y_S, sections coincide with maps to the fibre. We see that a required extension exists if and only if the map $f: bE \to Y_S$ is null-homotopic in Y_S, and this is certainly the case if the homotopy group $\pi_{k-1}(Y_S)$ vanishes. Since all Morse indexes of a strongly plurisubharmonic function on S are $\leq \dim S$, the condition that $\pi_q(Y_S) = 0$ for $q < \dim S$ therefore ensures the existence of a continuous extension of a section at each critical point on every stratum. □

Corollary 5.14.2 *Assume that* X *is a reduced Stein space of dimension* n *and* Y *is an Oka manifold satisfying* $\pi_q(Y) = 0$ *for* $q < \dim X$. *Given a compact* $\mathcal{O}(X)$*-convex subset* K *of* X, *a closed complex subvariety* X' *of* X, *an open set* U *in* X *containing* K *and a holomorphic map* $f: U \cup X' \to Y$, *we can approximate* f *uniformly on* K *by holomorphic maps* $\tilde{f}: X \to Y$ *such that* $\tilde{f}|_{X'} = f|_{X'}$. *This holds in particular if* $Y = \mathbb{C}^N \setminus A$ *where* A *is a tame complex subvariety of codimension* $q \geq \max\{2, \frac{n+1}{2}\}$.

Proof The first conclusion follows from Theorem 5.14.1. For the second part, note that $\mathbb{C}^N \setminus A$ is elliptic and hence Oka (see Proposition 6.4.1, p. 271), and $\pi_k(\mathbb{C}^N \setminus A) = 0$ for $k \leq 2q - 2$ by the transversality theorem. (For algebraic subvarieties A we can appeal to Proposition 5.6.10 on p. 227.) □

Essentially the same argument gives the following result.

Corollary 5.14.3 *Assume that* $X = X_0 \supset X_1 \supset \cdots \supset X_m = X'$ *is a stratification of a reduced Stein space* X *and* $\pi: Z \to X$ *is a stratified holomorphic fibre bundle whose fibre* Y_S *over every stratum* $S \subset X_k \setminus X_{k+1}$ *is of the form* $Y_S = \mathbb{C}^N \setminus A_S$, *where* $A_S \subset \mathbb{C}^N$ *is a tame subvariety whose codimension* $q_S = \mathrm{codim}_{\mathbb{C}} A_S$ *satisfies* $q_S \geq 2$ *and* $2q_S > \dim S$. *Then, every holomorphic section* $f': X' \to Z|_{X'}$ *extends to a holomorphic section* $f: X \to Z$.

More refined results on the existence and homotopy classification of continuous sections of fibre bundles are obtained by methods of *obstruction theory*; see e.g. Spanier [520], Steenrod [522], or Whitehead [569].

5.15 Equivalences Between Oka Properties

Recall that a complex manifold Y is said to enjoy a certain Oka property if the corresponding form of the Oka principle holds for maps from every reduced Stein space X to Y (see Theorem 5.4.4 and Corollary 5.4.5). Every Oka property comes in two forms, the *basic* and the *parametric*. When speaking of the basic properties, we let the parameter set P be a singleton and $P_0 = \varnothing$. The parametric properties refer to pairs $P_0 \subset P$ of Euclidean compacts.

We describe fourteen ostensibly different Oka properties of a complex manifold Y and then show that they are all equivalent (see Proposition 5.15.1). The fifteenth property, discovered by Kusakabe [354], is given by Theorem 5.15.2 below.

CAP–*Convex Approximation Property*. (See Definition 5.4.3 on p. 215.)

BOPA—*Basic Oka Property with Approximation*: Every continuous map $f_0 : X \to Y$ from a Stein space X that is holomorphic on (a neighborhood of) a compact $\mathcal{O}(X)$-convex subset $K \subset X$ can be deformed to a holomorphic map $f_1 : X \to Y$ by a homotopy of maps that are holomorphic near K and arbitrary close to f_0 on K.

Note that CAP is obtained by applying BOPA in the model case when K is a special compact convex set in $X = \mathbb{C}^n$.

BOPI—*Basic Oka Property with Interpolation*: Given a diagram

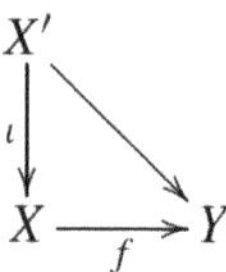

where $\iota : X' \hookrightarrow X$ is the inclusion of a closed complex subvariety into a Stein space X and $f : X \to Y$ is a continuous map such that $f|_{X'} : X' \to Y$ is holomorphic, there is a homotopy $f_t : X \to Y$ ($t \in [0,1]$) that is fixed on X' from $f = f_0$ to a holomorphic map $f_1 : X \to Y$.

BOPJI—*Basic Oka Property with Jet Interpolation*: This is the same as BOPI, except that the given continuous map $f : X \to Y$ is assumed to be holomorphic in a neighborhood of a subvariety X', and we ask that there exist a homotopy $f_t : X \to Y$ ($t \in [0,1]$) as above that is fixed to a given finite order along X'.

We introduce a restricted version of BOPI as follows. Let D be a convex domain in $\mathbb{C}^m$ and $\phi : D \hookrightarrow \mathbb{C}^k$ be a proper holomorphic embedding. Let

$$X' = \{(x, \phi(x)) : x \in D\} \subset \mathbb{C}^n = \mathbb{C}^m \times \mathbb{C}^k.$$

CIP—*Convex Interpolation Property*: Given a subvariety $X' \subset \mathbb{C}^n$ as above, every holomorphic map $f: X' \to Y$ extends to a holomorphic map $\mathbb{C}^n \to Y$. (The existence of a continuous extension is trivial in this case.) This property was introduced by Lárusson [374]. Note that CIP plays the same role versus BOPI as CAP does versus BOPA.

BOPAI—*Basic Oka Property with Approximation and Interpolation*: Combine BOPA and BOPI, i.e., we demand that both the approximation and the interpolation conditions hold.

BOPAJI—*Basic Oka Property with Approximation and Jet Interpolation*: Combine BOPA and BOPJI. This means the full conclusion of Theorem 5.4.4 for the trivial parameter space $(P, P_0) = (\{p\}, \varnothing)$.

The corresponding parametric Oka properties are introduced in an obvious way (see Theorem 5.4.4):

PCAP—*Parametric Convex Approximation Property* (see Definition 5.13.2, p. 253).

POPA—*Parametric Oka Property with Approximation*. Note that PCAP is a restricted version of POPA.

POPI—*Parametric Oka Property with Interpolation*.

POPJI—*Parametric Oka Property with Jet Interpolation*.

PCIP—*Parametric Convex Interpolation Property*. This is the restricted version of POPI, analogous to CIP.

POPAI—*Parametric Oka Property with Approximation and Interpolation*.

POPAJI—*Parametric Oka Property with Approximation and Jet Interpolation*. This means the full conclusion of Theorem 5.4.4, and is equivalent to Gromov's Ell_∞-*property* [266, Sect. 3.1] when the parameter pairs (P, P_0) are finite polyhedra and their subpolyhedra.

We thus have fourteen ostensibly different Oka properties.

Proposition 5.15.1 ([195, 199, 372]) *The following properties of a complex manifold are equivalent, and hence any of them characterizes the class of Oka manifolds*: CAP, CIP, BOPA, BOPI, BOPJI, BOPAI, BOPAJI, PCAP, PCIP, POPA, POPI, POPAI, POPJI, POPAJI.

Proof By Theorem 5.4.4, CAP implies all the other properties, including the parametric ones. Conversely, each of the properties containing the letter A in its name trivially implies CAP; likewise, each of the properties containing the letter I trivially implies CIP. It remains to show that CIP implies CAP.

Let K be a compact convex set in $\mathbb{C}^n$ and $f: U \to Y$ be a holomorphic map from an open convex neighborhood $U \subset \mathbb{C}^n$ of K. Choose a proper holomorphic embedding $\phi: U \hookrightarrow \mathbb{C}^r$. Its graph

$$\Sigma = \{(z, \phi(z)) : z \in U\} \subset \mathbb{C}^{n+r}$$

is a closed complex subvariety of $\mathbb{C}^{n+r}$ as in the definition of CIP. Consider the holomorphic map $g\colon \Sigma \to Y$ defined by

$$g\big(z, \phi(z)\big) = f(z), \quad z \in \Sigma.$$

Assuming that Y enjoys CIP, there exists a holomorphic map $G\colon \mathbb{C}^{n+r} \to Y$ such that $G|_\Sigma = g$. Thus,

$$G\big(z, \phi(z)\big) = f(z), \quad z \in U.$$

By the Oka-Weil theorem there is a sequence $\phi_k\colon \mathbb{C}^n \to \mathbb{C}^r$ $(k = 1, 2, \ldots)$ of holomorphic maps converging to ϕ uniformly on K. Then,

$$f_k(z) = G\big(z, \phi_k(z)\big), \quad z \in \mathbb{C}^n,\ k = 1, 2, \ldots$$

is a sequence of holomorphic maps $\mathbb{C}^n \to Y$ such that $\lim_{k\to\infty} f_k = f$ on K. This shows that Y enjoys CAP. □

Comparison with other holomorphic flexibility properties is considered in Chap. 7; see in particular Sect. 7.1. Oka properties can also be introduced for holomorphic maps; see Sect. 7.4 for the notion of an *Oka map*. In fact, a complex manifold Y is an Oka manifold if and only if the constant map $Y \to point$ is an Oka map.

We now describe another characterization of the class of Oka manifolds which was discovered by Y. Kusakabe [354]. The following statement is equivalent to the nontrivial part of [354, Theorem 3.2].

Theorem 5.15.2 (Kusakabe [354, Theorem 3.2]) *Let Y be a complex manifold with a Riemannian distance function* dist_Y. *Assume that for every compact convex set $K \subset \mathbb{C}^n$ $(n \in \mathbb{N})$ and holomorphic maps $f_0, f_1 \in \mathcal{O}(K, Y)$ there exist a neighborhood $U \subset \mathbb{C}^n$ of K and a holomorphic map $\Phi \in \mathcal{O}(\mathbb{C} \times U, Y)$ such that* $\sup_{z\in K} \mathrm{dist}_Y(\Phi(t, z), f_t(z)) < \epsilon$ *holds for $t = 0, 1$. Then Y is an Oka manifold.*

We may think of the map $\mathbb{C} \ni t \mapsto \Phi(t, \cdot) \in \mathcal{O}(U, Y)$ as a complex line in the mapping space $\mathcal{O}(U, Y)$ which approximates f_0 at $t = 0$ and f_1 at $t = 1$. Kusakabe's condition implies that Y enjoys CAP. The proof (see below) is nontrivial and uses some of the main tools developed in the proof of Theorem 5.4.4 to the effect that CAP implies Oka. The converse to Theorem 5.15.2 (in a stronger form) is an immediate consequence of the interpolation statement (BOPI) in Theorem 5.4.4: if X is a Stein manifold, Y is an Oka manifold and $f_0, f_1 \in \mathcal{O}(X, Y)$ are homotopic as maps in $\mathcal{C}(X, Y)$, then there exists a holomorphic map $\Phi \in \mathcal{O}(\mathbb{C} \times X, Y)$ such that $\Phi(t, \cdot) = f_t$ for $t = 0, 1$.

Proof We will show that the hypothesis of Theorem 5.15.2 implies that Y enjoys CAP (see Definition 5.4.3), and hence it is Oka by Theorem 5.4.4.

A *convex polyhedron* in a finite dimensional vector space is a compact set which is the intersection of finitely many closed affine half-spaces. We first note that it suffices to test CAP on special pairs of convex polyhedra.

Definition 5.15.3 A pair $K \subset L$ of compact convex sets in $\mathbb{R}^N$ is a *special polyhedral pair* if L is a polyhedron and

$$K = \{z \in L : \lambda(z) \le 0\} \tag{5.21}$$

for some affine linear function $\lambda: \mathbb{R}^N \to \mathbb{R}$.

Lemma 5.15.4 *Suppose that Y is a complex manifold such that for each special polyhedral pair $K \subset L$ in $\mathbb{C}^n$ (for any $n \in \mathbb{N}$), every holomorphic map $K \to Y$ can be approximated uniformly on K by holomorphic maps $L \to Y$. Then Y enjoys* CAP, *and hence is an Oka manifold.*

Proof Every compact convex set $K \subset \mathbb{R}^N$ can be approximated from the outside by convex polyhedra. Furthermore, given a pair $K \subset Q \subset \mathbb{R}^N$ of compact convex sets, with Q a polyhedron, and an open set $U \subset \mathbb{R}^N$ containing K, there exists a finite sequence of polyhedra $K_1 \subset K_2 \subset \cdots \subset K_m = Q$ such that $K \subset K_1 \subset U$ and (K_i, K_{i+1}) is a special polyhedral pair for every $i = 1, \ldots, m-1$. Assume now that $K \subset Q \subset \mathbb{C}^n$ is special convex pair (see Definition 5.4.2) and $f: U \to Y$ is a holomorphic map on an open set containing K. Choosing a sequence of polyhedra as above and applying the hypothesis of the lemma to every pair (K_i, K_{i+1}) we infer that Y enjoys CAP. □

By Lemma 5.15.4 it suffices to consider the approximation problem on a special polyhedral pair $K \subset L \subset \mathbb{C}^n$ (5.21). Let $U \subset \mathbb{C}^n$ be an open convex set containing K, and let $f: U \to Y$ be a holomorphic map. We must find a holomorphic map $F: L \to Y$ approximating f as closely as desired uniformly on K. Choose a small $\epsilon > 0$ and set

$$A = \{z \in L : \lambda(z) \le 2\epsilon\} \subset U, \qquad B = \{z \in L : \lambda(z) \ge \epsilon\}.$$

The compact convex polyhedra A and B form a Cartan pair with $A \cup B = L$ and $C = A \cap B = \{z \in L : \epsilon \le \lambda(z) \le 2\epsilon\}$. Note that K and C are disjoint compact convex sets and hence $K \cup C$ is polynomially convex.

Let $d = \dim Y$, and let $P \subset \mathbb{C}^d$ be a closed cube centered at the origin. We extend $f: U \to Y$ to a map $f: U \times \mathbb{C}^d \to Y$ which is independent of the $\mathbb{C}^d$-coordinate. Let $h: W \to Y$ be a holomorphic embedding of a small open neighborhood $W \subset \mathbb{C}^d$ of P into a coordinate chart in Y. Denote by $\pi: \mathbb{C}^n \times \mathbb{C}^d \to \mathbb{C}^d$ the projection and let $g = h \circ \pi: \mathbb{C}^n \times W \to Y$. Thus, g is a local dominating spray over the constant map $\mathbb{C}^n \to h(0) \in Y$.

By the hypothesis of the theorem applied with the compact convex polyhedron $A \times P \subset \mathbb{C}^{n+d}$, and after shrinking the neighborhoods $U \supset A$ and $W \supset P$, there exists a holomorphic map $\Phi: \mathbb{C} \times U \times W \to Y$ such that $\Phi(0, \cdot): U \times W \to Y$ approximates f and $\Phi(1, \cdot): U \times W \to Y$ approximates g as closely as desired uniformly on $A \times P$.

Since $K \cup C$ is polynomially convex, there is a function $\phi \in \mathcal{O}(\mathbb{C}^n)$ which is uniformly close to 0 on K and to 1 on C. Let (z, w) denote the coordinates on $\mathbb{C}^n \times \mathbb{C}^d$. Consider the map $\tilde{f}: U \times W \to Y$ defined by

$$\tilde{f}(z, w) = \Phi\big(\phi(z), z, w\big), \quad z \in U, \ w \in W.$$

If $(z, w) \in K \times P$ then $\phi(z)$ is close to 0, so $\tilde{f}(z, w)$ is close to $\Phi(0, z, w)$ which is close to $f(z)$. If on the other hand $(z, w) \in C \times P$, then $\tilde{f}(z, w)$ is close to $\Phi(1, z, w)$ which is close to $g(w)$. Since g is a dominating spray and (A, B) is a Cartan pair with $A \cap B = C$ and $A \cup B = L$, we can apply Proposition 5.9.2 to glue $\tilde{f}$ and g into a spray of maps $F: L \times P' \to Y$, where $P' \subset P$ is a neighborhood of the origin in $\mathbb{C}^d$. By the construction, the map $F(\cdot, 0): L \to Y$ then approximates f on K. □

Theorems 5.4.4 and 5.15.2 imply that a connected complex manifold Y is Oka if and only if for any bounded convex domain $U \subset \mathbb{C}^n$ there exists a dense entire curve $\mathbb{C} \to \mathcal{O}(U, Y)$. Kusakabe showed that *dense* can be replaced by *somewhere dense* in the above statement (see [354, Corollary 4.3]). He also characterized the Oka property of a complex manifold Y by the existence of universal maps $X \to Y$ from Stein source manifolds (see [354, Theorem 1.4]).

Chapter 6
Elliptic Complex Geometry and Oka Theory

In this chapter we consider the problem of finding global holomorphic sections of holomorphic maps that are more general than fibre bundles. The main results are Gromov's Oka principle for sections of elliptic submersions over Stein manifolds, and its extension to stratified subelliptic submersions over reduced Stein spaces; see Theorem 6.2.2 on p. 265.

A holomorphic submersion $Z \to X$ is elliptic in the sense of Gromov [266] if there exists a dominating holomorphic fibre spray on the restricted submersion $Z|_U \to U$ over small open sets $U \subset X$. Subellipticity, an ostensibly weaker condition introduced in [190], means the existence of a finite dominating family of fibre sprays on $Z|_U$ over small open sets $U \subset X$ (see Definition 6.1.2 on p. 264). Each of these conditions implies the *Homotopy Approximation Property* (HAP) on the approximation of homotopies of holomorphic sections of the restricted submersion $Z|_U$ (see Definition 6.6.5 on p. 287). The crux of the matter is the following result; see Theorem 6.6.6 on p. 287:

Let $Z \to X$ be a holomorphic submersion onto a reduced Stein space. If every point $x \in X$ admits an open neighborhood $U \subset X$ such that the restricted submersion $Z|_U \to U$ satisfies HAP, *then sections $X \to Z$ satisfy all forms of the Oka principle. The same holds for a stratified holomorphic submersion which satisfies* HAP *over small open sets in every stratum.*

The proof uses similar analytic techniques as in the locally trivial case considered in Chap. 5 (where HAP coincides with CAP), but the induction scheme is substantially more involved. Some of the applications in Chaps. 8 and 9, most notably the solution of the holomorphic Vaserstein problem in Sect. 8.7, depend on this more advanced version of the Oka principle.

In Sect. 6.15 we prove the basic version of the homotopy Runge approximation theorem (HAP) also for algebraic maps from affine algebraic varieties to algebraically subelliptic manifolds. In Sect. 6.13 we present a generalization of these results to 1-convex spaces, due to Prezelj [461, 462]. In Sect. 6.14 we consider the Oka principle for multi-valued sections of branched holomorphic maps.

F. Forstnerič, *Stein Manifolds and Holomorphic Mappings*, Ergebnisse der Mathematik und ihrer Grenzgebiete. 3. Folge / A Series of Modern Surveys in Mathematics 56, DOI 10.1007/978-3-319-61058-0_6

6.1 Fibre Sprays and Elliptic Submersions

The notion of a dominating spray on a complex manifold was introduced in Sect. 5.6 (see p. 229). We now define the notion of a dominating fibre spray.

Definition 6.1.1 (Gromov [266, Sect. 1.1.B]) Let X and Z be complex spaces and $h\colon Z \to X$ be a holomorphic submersion.

(a) A *fibre spray* on $h\colon Z \to X$ is a triple (E, π, s), where $\pi : E \to Z$ is a holomorphic vector bundle and $s\colon E \to Z$ is a holomorphic map such that for each $z \in Z$ we have

$$s(0_z) = z \quad \text{and} \quad s(E_z) \subset Z_{h(z)} = h^{-1}\big(h(z)\big).$$

(b) A fibre spray (E, π, s) is *dominating* at a point $z \in Z$ if the differential $ds_{0_z} : T_{0_z}E \to T_z Z$ maps the subspace $E_z \subset T_{0_z}E$ surjectively onto the vertical tangent space $VT_z Z = \ker dh_z$. The fibre spray is *dominating* if it is dominating at every point $z \in Z$.

(c) A family of fibre sprays (E_j, π_j, s_j) $(j = 1, \dots, m)$ on Z is dominating at the point $z \in Z$ if

$$(ds_1)_{0_z}(E_{1,z}) + (ds_2)_{0_z}(E_{2,z}) + \cdots + (ds_m)_{0_z}(E_{m,z}) = VT_z Z. \tag{6.1}$$

The family is *dominating* if (6.1) holds at every point $z \in Z$.

We denote the restriction of the differential $ds_{0_z} : T_{0_z}E \to T_z Z$ to the vertical subspace $E_z \subset T_{0_z}E$ by

$$Vds_z = ds_{0_z}|_{E_z} : E_z \longrightarrow VT_z Z \tag{6.2}$$

and call it the *vertical derivative* of s at the point $z \in Z$.

Definition 6.1.2 A surjective holomorphic submersion $h\colon Z \to X$ is *elliptic* (resp. *subelliptic*) if every point $x_0 \in X$ has an open neighborhood $U \subset X$ such that the restricted submersion $h\colon Z|_U \to U$ admits a dominating fibre spray (resp. a finite dominating family of fibre sprays).

Comparing with Definition 5.6.13 (p. 229), we see that a spray on a manifold Y is the same thing as a fibre spray on the trivial submersion $Y \to point$. A holomorphic fibre bundle map is (sub-)elliptic if and only if its fibre is such.

Example 6.1.3 If U is an open set in X and $V_1, \dots, V_n$ are $\mathbb{C}$-complete holomorphic vector fields on $Z|_U$ that are tangential to the fibres Z_x and span $VT_z Z$ at every point of $z \in Z|_U$, then the composition of their flows

$$s(z, t_1, \dots, t_n) = \phi_1^{t_1} \circ \phi_2^{t_2} \circ \cdots \circ \phi_n^{t_n}(z) \tag{6.3}$$

is a dominating fibre spray on $Z|_U$. (Compare with Example 5.6.16(B).) □

Further examples of elliptic and subelliptic submersions are given in Sect. 6.4.

6.2 The Oka Principle for Sections of Stratified Subelliptic Submersions

We now describe the main results of this chapter.

Definition 6.2.1 Let X and Z be complex spaces. A holomorphic submersion $h\colon Z \to X$ is *stratified elliptic (resp. stratified subelliptic)* if there exists a stratification of X by closed complex subvarieties

$$X = X_0 \supset X_1 \supset \cdots \supset X_m = \varnothing \tag{6.4}$$

such that every difference $S_k = X_k \setminus X_{k+1}$ is nonsingular and the restricted submersion $h\colon Z|_{S_k} \to S_k$ is elliptic (resp. subelliptic) (see Definition 6.1.2).

Theorem 6.2.2 ([266, Sect. 4.5], [190, 202]) *If $h\colon Z \to X$ is a stratified subelliptic submersion onto a reduced Stein space X, then sections $X \to Z$ satisfy the parametric Oka property with approximation and interpolation (i.e., the conclusion of Theorem 5.4.4 holds). In particular, the inclusion $\Gamma_{\mathcal{O}}(X, Z) \hookrightarrow \Gamma(X, Z)$ of the space of holomorphic sections into the space of continuous sections is a weak homotopy equivalence.*

The parametric Oka property now refers to an arbitrary pair of compact Hausdorff parameter space $P_0 \subset P$. However, some of the arguments can be simplified by restricting to Euclidean compacts and using Stein neighborhoods constructed in Sect. 3.6 as was done in the proof of Theorem 5.13.3.

Theorem 6.2.2 has a complex genesis. For elliptic submersions over Stein manifolds, and without the interpolation condition, it is due to Gromov [266, 4.5]; a detailed exposition was given by Prezelj and the author [219]. Interpolation on subvarieties of Stein manifolds was added in [218], and the proof for stratified elliptic submersions was outlined without details in [218, Sect. 7]. The notion of a subelliptic submersion was introduced in [190], although dominating families of sprays were already considered by Gromov in [266]. The general case of Theorem 6.2.2 was proved in [202].

The following main induction step in the proof of Theorem 6.2.2 is quite useful and is sometimes applied independently.

Theorem 6.2.3 *Let $h\colon Z \to X$ be a holomorphic submersion of a complex space Z onto a reduced Stein space X. Assume that $X_1 \subset X_0$ are closed complex subvarieties of X and K is a compact $\mathcal{O}(X)$-convex subset of X. Let $f\colon X \to Z$ be a continuous section that is holomorphic in an open neighborhood of K and on X_1. If the submersion $Z \to X$ is subelliptic over $X_0 \setminus (X_1 \cup K)$, then there is a homotopy of section $f_t\colon X \to Z$ ($t \in [0, 1]$) which is fixed on X_1 such that $f_0 = f$, every section f_t is holomorphic and close to f in a neighborhood of K, and the section f_1 is holomorphic on X_0. The analogous conclusion holds for a continuous family of sections with these properties.*

The proof of these results occupies Sects. 6.9–6.12. We use an induction scheme outlined in [266] and elaborated in [219]. Our exposition mainly follows [218, 219], with improvements in many places. In Sects. 6.3–6.5 we collect the preparatory material on holomorphic sprays and give examples of (sub-)elliptic submersions. In Sect. 6.6 we prove the main analytic ingredient, an Oka-Weil type approximation theorem for homotopies of holomorphic sections of subelliptic submersions over Stein spaces; this is the only place in the proof where subellipticity is used. In Sect. 6.7 we explain how to glue homotopic families of holomorphic sections. In Sect. 6.8 we introduce the notion of a holomorphic complex, the main bookkeeping tool that is used in the proof.

Outline of proof of Theorem 6.2.2 For simplicity of exposition, we focus on the nonparametric case and without the interpolation condition. The main problem is then the following:

Given a compact $\mathcal{O}(X)$-convex set K in X and a continuous section $a\colon X \to Z$ that is holomorphic in an open set $U_0 \supset K$, find a homotopy of sections $H^s\colon X \to Z$ ($s \in [0,1]$) such that $H^0 = a$, the section $f = H^1$ is holomorphic on X, and every section H^s is holomorphic in a neighborhood of K and it approximates a uniformly on K (also uniformly in the parameter).

We begin by constructing a locally finite cover $\mathcal{A} = \{A_0, A_1, A_2, \ldots\}$ of X by compact sets A_j, and a family of holomorphic sections $a_{(j)}\colon U_j \to Z$ over open sets $U_j \supset A_j$, satisfying the following conditions (see Proposition 6.10.1):

- $K \subset A_0 \subset U_0$;
- for each $n \ge 1$ the sequence $(A_0, A_1, \ldots, A_n)$ is a $\mathcal{C}$*-string* (see Definition 6.9.1);
- for every $j = 1, 2, \ldots$ the restricted submersion $h\colon Z|_{U_j} \to U_j$ admits a dominating family of fibre sprays;
- for each $j \in \mathbb{Z}_+$ the section $a_{(j)}\colon U_j \to Z$ is homotopic to $a|_{U_j}$;
- for each pair of indexes $i \ne j$ such that $A_i \cap A_j \ne \varnothing$ there is a holomorphic homotopy between $a_{(i)}$ and $a_{(j)}$ in $U_{(i,j)} = U_i \cap U_j$;
- more generally, for each multi-index $J = (j_0, j_1, \ldots, j_n)$ such that $A_J = A_{j_0} \cap \cdots \cap A_{j_n} \ne \varnothing$ there is an n-dimensional homotopy $a_J(t)$ of holomorphic sections on $U_J = U_{j_0} \cap \cdots \cap U_{j_n}$, with the parameter t belonging to the standard n-simplex $\Delta^n \subset \mathbb{R}^n$, such that for every t in a boundary face of Δ^n determined by a sub-multi-index $I \subset J$, we have $a_J(t) = a_I(t)|_{U_J}$.

The parameter space of our collection of holomorphic sections and homotopies between them is the geometric realization of a simplicial complex called the *nerve* of the cover $\mathcal{A}$ (see Sect. 6.8). The sets U_j will shrink around A_j, but the A_j's will remain fixed during the entire construction.

The family $\{a_{(j)}\colon j \in \mathbb{Z}_+\}$, together with all the connecting homotopies, is a puzzle whose pieces are systematically reassembled into a holomorphic section $f\colon X \to Z$ that is homotopic to the initial continuous section a. All modifications of the complex will consist of holomorphic homotopies.

Assume inductively that the sections $a_{(0)}, \ldots, a_{(n-1)}$ have already been joined into a holomorphic section f^{n-1} over a neighborhood of $A^{n-1} = A_0 \cup A_1 \cup \cdots \cup$

A_{n-1}. The inductive step amounts to gluing f^{n-1} with the next section $a_{(n)}$ in the initial collection. To accomplish this, we need a holomorphic homotopy between the two sections in a neighborhood of $A^{n-1} \cap A_n$. Our inductive construction is such that for each $j = 0, 1, \ldots, n-1$ we have a holomorphic homotopy between f^{n-1} and $a_{(n)}$ in a neighborhood of $A_j \cap A_n$, inherited from the initial homotopy between $a_{(j)}$ and $a_{(n)}$. We now patch these n partial homotopies into a homotopy defined in a neighborhood of $A^{n-1} \cap A_n = \bigcup_{j=0}^{n-1}(A_j \cap A_n)$. This can be done by induction on n, provided that the sequence $(A_0 \cap A_n, A_1 \cap A_n, \ldots, A_{n-1} \cap A_n)$ is also a $\mathcal{C}$-string. Once this is achieved, we use that (A^{n-1}, A_n) is a $\mathcal{C}$-pair to glue f^{n-1} and $a_{(n)}$ into a section f^n over A^n by appealing to Proposition 6.7.2, thereby completing the induction step. The main induction step is furnished by Proposition 6.11.1. The resulting sequence of sections f^n converges uniformly on compacts in X to a holomorphic section $f: X \to Z$ which solves the problem.

In the stratified case, we apply the above argument within a stratum in X and proceed from one stratum to the next one just as was done in the case of a stratified fibre bundle in Sect. 5.12.

6.3 Composed and Iterated Sprays

In this section we explain some basic operations with sprays, in particular the notions of *composed* and *iterated* sprays, which will be used in the proof of the Oka-Weil approximation theorem for homotopies of holomorphic sections (see Sect. 6.6). These constructions are due to Gromov [266].

Lemma 6.3.1 *Let Y be a complex manifold. If $s_j : E_j \to Y$ $(j = 1, 2, \ldots, k)$ is a dominating family of sprays on Y defined on trivial bundles $E_j \cong Y \times \mathbb{C}^{m_j}$, then Y admits a dominating spray. The analogous result holds for fibre sprays.*

Proof Assume that s_j is defined on $E_j = Y \times \mathbb{C}^{m_j}$ for each j. We inductively define sprays $s^{(j)} : E_1 \oplus \cdots \oplus E_j = Y \times \mathbb{C}^{m_1+\cdots+m_j} \to Y$ by $s^{(1)} = s_1$ and

$$s^{(j)}(y, e_1, \ldots, e_j) = s_j\big(s^{(j-1)}(y, e_1, \ldots, e_{j-1}), e_j\big), \quad j = 2, \ldots, k.$$

Clearly, we have

$$\mathrm{d}s^{(k)}_{0_y}(E_{1,y} \oplus \cdots \oplus E_{k,y}) = (\mathrm{d}s_1)_{0_y}(E_{1,y}) + \cdots + (\mathrm{d}s_k)_{0_y}(E_{k,y}).$$

Hence, $(s_1, \ldots, s_k)$ is a dominating family of sprays on Y (see Definition 5.6.13) if and only if $s^{(k)}$ is a dominating spray. $\square$

Lemma 6.3.2 *Assume that (E_j, π_j, s_j) $(j = 1, 2, \ldots, k)$ is a dominating family of sprays on a complex manifold Y. If every bundle $E_j \to Y$ is generated by finitely many holomorphic sections over Y, then there exists a dominating spray on Y. The analogous result holds for fibre sprays.*

Proof Assume that E_j is generated by sections $f_{j,l}$ $(l = 1, \ldots, m_j)$. Let $\sigma_j : E'_j = Y \times \mathbb{C}^{m_j} \to E_j$ be the holomorphic vector bundle epimorphism which takes the l-th standard basis section of $Y \times \mathbb{C}^{m_j}$ to $f_{j,l}$. Then, the map $s'_j = s_j \circ \sigma_j : E'_j \to Y$ is a holomorphic spray, and Lemma 6.3.1 applies to the family (E'_j, π'_j, s'_j) $(j = 1, 2, \ldots, k)$. □

Lemma 6.3.3 *A subelliptic Stein manifold is elliptic. More generally, if Z is a Stein space then every subelliptic submersion $h : Z \to X$ is elliptic.*

Proof Since every holomorphic vector bundle over a Stein manifold is generated by finitely many holomorphic sections by Cartan's Theorem A, the first claim follows from Lemma 6.3.2. Assume now that Z is a Stein space and $h : Z \to X$ is a subelliptic submersion. Fix a point $x_0 \in X$ and choose a small Stein neighborhood $U \subset X$ of x_0 such that $Z|_U \to U$ admits a finite dominating family of fibre sprays (E_j, p_j, s_j). Since Z is Stein, it follows that $Z|_U = h^{-1}(U)$ is also Stein. Hence, Lemma 6.3.2 applied to the family (E_j, p_j, s_j) gives a dominating fibre spray on $Z|_U \to U$. □

Lemma 6.3.4 *The Cartesian product of any finite family of elliptic (resp. subelliptic) manifolds is elliptic (resp. subelliptic).*

Proof It suffices to prove the result for the Cartesian product of two manifolds. Let $Y = Y_1 \times Y_2$ and let $\pi_j : Y \to Y_j$ $(j = 1, 2)$ denote the projection $\pi_j(y_1, y_2) = y_j$. If (E_j, π_j, s_j) is a spray on Y_j for $j = 1, 2$, we get a spray map $s_1 \times s_2$ defined on the bundle $E = \pi_1^* E_1 \oplus \pi_2^* E_2 \to Y$ by

$$(s_1 \times s_2)(y_1, y_2, e_1, e_2) = \big(s_1(y_1, e_1), s_2(y_2, e_2)\big).$$

If s_1 is dominating on Y_1 and s_2 is dominating on Y_2, then $s_1 \times s_2$ is dominating on $Y_1 \times Y_2$. Similarly, if a family of sprays $\{s_j : j = 1, \ldots, j_0\}$ is dominating on Y_1 and a family of sprays $\{\sigma_k : k = 1, \ldots, k_0\}$ is dominating on Y_2, then the family $\{s_j \times \sigma_k : j = 1, \ldots, j_0,\ k = 1, \ldots, k_0\}$ is dominating on $Y_1 \times Y_2$. □

Definition 6.3.5 Let (E, π, s), (E_1, π_1, s_1) and (E_2, π_2, s_2) be fibre sprays for a holomorphic submersion $h : Z \to X$.

(a) The *composed fibre spray* $(E_1 * E_2, \pi_1 * \pi_2, s_1 * s_2)$ is defined by

$$E_1 * E_2 = \big\{(e_1, e_2) \in E_1 \times E_2 : s_1(e_1) = \pi_2(e_2)\big\},$$

$$\pi_1 * \pi_2(e_1, e_2) = \pi_1(e_1), \quad s_1 * s_2(e_1, e_2) = s_2(e_2).$$

(See Fig. 6.1.)

(b) For each integer $k = 1, 2, \ldots$ the k-th *iterated spray* $(E^{(k)}, \pi^{(k)}, s^{(k)})$ is

$$E^{(k)} = \big\{e = (e_1, e_2, \ldots, e_k) : e_j \in E \quad \text{for } j = 1, 2, \ldots, k,$$
$$s(e_j) = \pi(e_{j+1}) \quad \text{for } j = 1, 2, \ldots, k-1\big\},$$

$$\pi^{(k)}(e) = \pi(e_1), \quad s^{(k)}(e) = s(e_k).$$

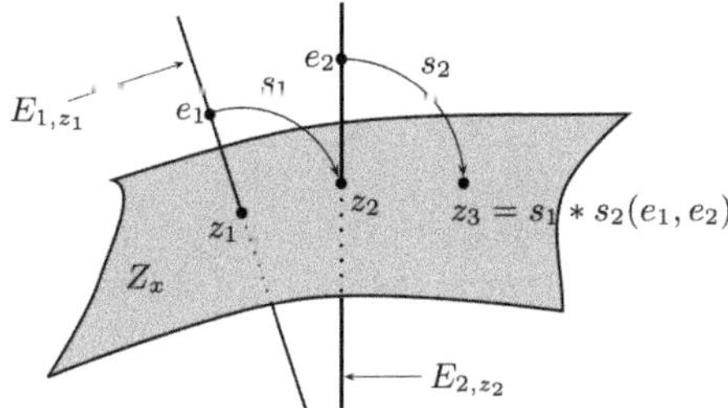

Fig. 6.1 A composed spray $s_1 * s_2$ on a fibre Z_x

Note that $E_1 * E_2 = s_1^*(E_2)$ is the pullback of the vector bundle $\pi_2 \colon E_2 \to Z$ by the spray map $s_1 \colon E_1 \to Z$.

The following diagram shows all relevant maps in a composed spray; ι is the natural map over s_1, and $s_1 * s_2 = s_2 \circ \iota \colon E_1 * E_2 \to Z$ is the composed spray map.

$$
\begin{array}{ccccc}
s_1^* E_2 = E_1 * E_2 & \xrightarrow{\iota} & E_2 & \xrightarrow{s_2} & Z \\
\downarrow & & \downarrow \pi_2 & & \\
E_1 & \xrightarrow{s_1} & Z & & \\
\pi_1 \downarrow & & & & \\
Z & & & &
\end{array}
$$

Each of the two projections in $E_1 * E_2 \longrightarrow E_1 \xrightarrow{\pi_1} Z$ in the first column is a holomorphic vector bundle, but their composition $\pi_1 * \pi_2 \colon E_1 * E_2 \to Z$ does not have a natural holomorphic vector bundle structure. Nevertheless, it has a well defined zero section which we identify with the base space Z.

The composition of sprays extends to any finite number of factors and is associative. In particular, the k-th iterate $(E^{(k)}, \pi^{(k)}, s^{(k)})$ is obtained by composing k copies of the spray (E, π, s).

Let $h \colon Z \to X$ be a holomorphic submersion and (E, π, s) be a composed fibre spray on Z. We denote by $\mathrm{VT}(E) = \ker d\pi$ the vertical tangent bundle of E and by $\mathrm{VT}(E)|_Z$ its restriction to the zero section $Z \subset E$. If (E, π, s) is obtained by composing the fibre sprays (E_j, π_j, s_j) $(j = 1, \ldots, m)$, then

$$\mathrm{VT}E|_Z \cong E_1 \oplus E_2 \oplus \cdots \oplus E_m.$$

Since $s \colon E \to Z$ preserves the zero section and maps the fibre $E_z = \pi^{-1}(z)$ to the fibre $Z_{h(z)}$ for every $z \in Z$, its differential ds at the zero section maps $\mathrm{VT}E|_Z$ to $\mathrm{VT}Z$. In analogy to (6.2) we denote this map by

$$\mathrm{V}ds : \mathrm{VT}E|_Z \to \mathrm{VT}Z \tag{6.5}$$

and call it the *vertical derivative* of s. A composed fibre spray (E, π, s) is said to be *dominating* if $\mathrm{V}ds$ is surjective; in such case, $\ker \mathrm{V}ds \subset \mathrm{VT}E|_Z$ is a holomorphic vector subbundle of the vertical tangent bundle $\mathrm{VT}E|_Z$.

Lemma 6.3.6 *A family of fibre sprays* (E_j, π_j, s_j) $(j = 1, \ldots, m)$ *is dominating at* $z \in Z$ *if and only if the composed fibre spray* $s = s_1 * \cdots * s_m : E_1 * \cdots * E_m \to Z$ *is dominating at* z.

Proof Note that $s(z, 0, \ldots, e_j, \ldots, 0) = s_j(z, e_j)$, where $e_j \in E_{j,z}$ appears at the j-th spot. Hence, the range of the vertical derivative of the composed spray $s = s_1 * \cdots * s_m$ at the point $z \in E_1 * \cdots * E_m$ equals the vector sum of the vertical derivatives $\mathrm{V}ds_j$ of the individual sprays s_j at the points $0_z \in E_{j,z}$. The lemma now follows from (6.1). □

Lemma 6.3.7 *Assume that* $h\colon Z \to X$ *is a holomorphic submersion. Given a family* (E_j, π_j, s_j) $(j = 1, \ldots, m)$ *of fibre sprays on* Z, *let* (E, π, s) *be the associated composed spray. For any Stein subset* $\Omega \subset Z$ *(either an open Stein subset, or a Stein subvariety) there is a fibre-preserving biholomorphic map*

$$\Theta\colon E|_\Omega = E_1 * \cdots * E_m|_\Omega \to E_1 \oplus E_2 \oplus \cdots \oplus E_m|_\Omega$$

which preserves the zero section and whose differential at the zero section equals the identity map.

Proof It suffices to prove the result for $m = 2$ and apply induction. Recall that $E_1 * E_2 = s_1^*(E_2)$ is the pullback of $\pi_2\colon E_2 \to Z$ by $s_1\colon E_1 \to Z$. Let $E_1' = E_1|_\Omega$, and let $\pi_1'\colon E_1' \to \Omega$ (resp. $s_1'\colon E_1' \to Z$) denote the restriction of π_1 (resp. of s_1) to E_1'. Consider the homotopy of holomorphic maps

$$g_t\colon E_1' \to Z, \qquad g_t(z, e) = s_1(z, te), \quad t \in \mathbb{C}.$$

Note that $g_0 = \pi_1'$ and $g_1 = s_1'$. Since Ω is Stein, the total space E_1' is also a Stein space. By pulling the bundle $\pi_2\colon E_2 \to Z$ back to E_1' by $g_t\colon E_1' \to Z$, we obtain a family of holomorphically isomorphic vector bundles according to Corollary 8.3.10 (which follows from the classical Oka-Grauert theory proved in Chap. 5). In particular, we have an isomorphism between $g_1^* E_2 = (s_1')^* E_2 = E_1 * E_2|_\Omega$ and $g_0^* E_2 = (\pi_1')^*(E_2|_\Omega)$ (as holomorphic vector bundles over E_1'). The total space of $g_0^* E_2$ is clearly isomorphic to $E_1 \oplus E_2|_\Omega$ which is a holomorphic vector bundle over Ω. This furnishes the desired isomorphism $\Theta\colon E_1 * E_2|_\Omega \to E_1 \oplus E_2|_\Omega$. □

Pullback of sprays. Given a holomorphic submersion $h\colon Z \to X$ and a holomorphic map $f\colon Y \to X$, consider the submersion $\widetilde{h}\colon \widetilde{Z} \to Y$ defined by

$$\widetilde{Z} = \big\{(y, z) \in Y \times Z : f(y) = \pi(z)\big\}, \quad \widetilde{h}(y, z) = y. \tag{6.6}$$

Let $\iota\colon \widetilde{Z} \to Z$ denote the map $\iota(y, z) = z$ for $(y, z) \in \widetilde{Z}$, covering the base map $f\colon Y \to X$. Given a fibre spray (E, π, s) on $h\colon Z \to X$, let $\widetilde{\pi}\colon \widetilde{E} \to \widetilde{Z}$ denote the pullback of the vector bundle $\pi\colon E \to Z$ by the map ι, and let $\widetilde{\iota}\colon \widetilde{E} \to E$ be the natural map over ι. There is a fibre spray $\tilde{s}\colon \widetilde{E} \to \widetilde{Z}$ associated to the submersion $\tilde{h}\colon \widetilde{Z} \to Y$ such that the following diagram commutes:

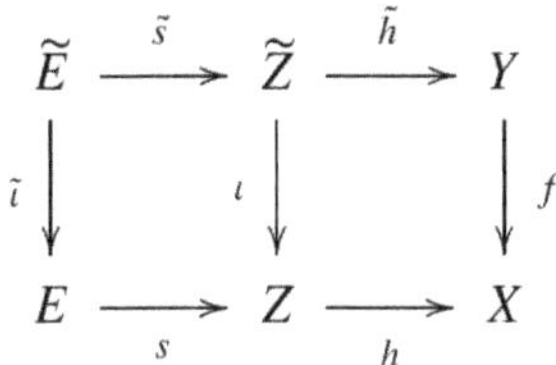

The triple $(\widetilde{E}, \tilde{\pi}, \tilde{s})$ is the pullback of the spray (E, p, s) by the map $f\colon Y \to X$. It is easily verified that a dominating fibre spray (resp. a dominating family of fibre sprays) on $Z \to X$ pulls back to a dominating fibre spray (resp. to a dominating family of fibre sprays) on $\widetilde{Z} \to Y$. Similarly, a fibre spray on $Z|_U$, where U is an open subset of X, pulls back to a fibre spray on $\widetilde{Z}|_{\widetilde{U}}$, where $\widetilde{U} = f^{-1}(U) \subset Y$. In conclusion, we have the following result.

Proposition 6.3.8 *If $h\colon Z \to X$ is an elliptic (resp. a subelliptic) submersion and $f\colon Y \to X$ is a holomorphic map, then the pullback submersion $\widetilde{Z} = f^*Z \to Y$* (6.6) *is also elliptic (resp. subelliptic).*

6.4 Examples of Subelliptic Manifolds and Submersions

Complements of Tame Subvarieties

Proposition 6.4.1

(a) *If A is an algebraic subvariety of $\mathbb{C}^n$ of dimension* $\dim A \le n-2$, *then $\mathbb{C}^n \setminus A$ is algebraically elliptic.*
(b) *If A is a tame analytic subvariety of $\mathbb{C}^n$ (see Definition* 4.11.3) *with* $\dim A \le n-2$, *then $\mathbb{C}^n \setminus A$ is elliptic.*
(c) *If Y is a complex Grassmann manifold and $A \subset Y$ is a closed complex (=algebraic) subvariety of codimension at least two, then $Y \setminus A$ is algebraically subelliptic. This holds in particular when $Y = \mathbb{P}^n$.*

Proof Parts (a) and (b) are Proposition 5.6.17. (See also Example 5.6.16(C).)

Assume now that $Y = \mathbb{P}^n$. Choose a hyperplane $\Lambda \subset \mathbb{P}^n$ and homogeneous coordinates $z = [z_0 \colon z_1 \colon \cdots \colon z_n]$ on $\mathbb{P}^n$ such that $\Lambda = \{z_0 = 0\}$. For every $j = 0, 1 \ldots, n$ let $U_j = \{z \in \mathbb{P}^n : z_j \neq 0\} \cong \mathbb{C}^n$; hence $\mathbb{P}^n = U_0 \cup \Lambda$. Let $L = [\Lambda]^{-1} \to \mathbb{P}^n$ be the inverse of the holomorphic line bundle $[\Lambda]$ determined by the divisor of Λ (see Example 1.5.7 on p. 19). The bundle L admits holomorphic vector bundle trivializations $\phi_j \colon L|_{U_j} \to U_j \times \mathbb{C}$ with the transition maps

$$\phi_{ik}(z,t) = \phi_i \circ \phi_k^{-1}(z,t) = (z, t z_i/z_k), \quad z \in U_{ik} = U_i \cap U_k,\ t \in \mathbb{C}.$$

On $U_0 \cong \mathbb{C}^n$ we use the affine coordinates $\zeta = (\zeta_1, \ldots, \zeta_n)$ with $\zeta_j = z_j/z_0$. Choose $v = (v_1, \ldots, v_n) \in \mathbb{C}^n$ such that the projection $\pi\colon \mathbb{C}^n \to \mathbb{C}^{n-1}$ with kernel $\mathbb{C}v$ is proper when restricted to $A \cap U_0$. (Since $\dim A \le n-2$, this holds for almost every $v \in \mathbb{C}^n$.) Then, $A' = \pi(A \cap U_0) \subset \mathbb{C}^{n-1}$ is a proper algebraic subvariety of $\mathbb{C}^{n-1}$. If p is a holomorphic polynomial on $\mathbb{C}^{n-1}$ that vanishes on A', then the map $U_0 \times \mathbb{C} \to U_0$, given by

$$s(\zeta, t) = \zeta + tp(\pi\zeta)v = \zeta + tf(\zeta), \quad \zeta \in \mathbb{C}^n,\ t \in \mathbb{C},$$

is an algebraic spray satisfying $\frac{\partial}{\partial t} s(\zeta, 0) = p(\pi\zeta)v = f(\zeta)$. Although s does not extend to a spray $\mathbb{P}^n \times \mathbb{C} \to \mathbb{P}^n$ on the trivial line bundle due to singularities on $\Lambda = \{z_0 = 0\}$, we now show that it induces an algebraic spray $\tilde{s}\colon L^{\otimes m} \to \mathbb{P}^n$, where $m \in \mathbb{N}$ is the degree of the polynomial p.

The line bundle $E = L^{\otimes m}$ admits vector bundle charts $\theta_i\colon E|_{U_i} \to U_i \times \mathbb{C}$ ($i = 0, 1, \ldots, n$) with the transition maps

$$\theta_{ik}(z, t) = \big(z, t(z_i/z_k)^m\big), \quad z \in U_{ik},\ t \in \mathbb{C}.$$

Set $\tilde{s} = s \circ \theta_0\colon E|_{U_0} \to \mathbb{P}^n$. We claim that $\tilde{s}$ extends to an algebraic spray $E \to \mathbb{P}^n$. Writing $z = [z_0 : z']$, we see that the spray s has the following expression in the homogeneous coordinates $z \in U_0 \subset \mathbb{P}^n$:

$$s(z, t) = \big[1 : s\big(z' z_0^{-1}, t\big)\big] = \big[1 : z' z_0^{-1} + tf\big(z' z_0^{-1}\big)\big] = \big[z_0 : z' + tz_0 f\big(z' z_0^{-1}\big)\big].$$

Hence we get for $k = 1, \ldots, n$ and $z \in U_0 \cap U_k$ that

$$\tilde{s} \circ \theta_k^{-1}(z, t) = s \circ \theta_{0k}(z, t) = s\big(z, t\big(z_0 z_k^{-1}\big)^m\big) = \big[z_0 : z' + t\, z_0^{m+1} z_k^{-m} f\big(z' z_0^{-1}\big)\big].$$

By the choice of m, the function $z_0^{m+1} z_k^{-m} f(z' z_0^{-1})$ vanishes on $\{z_0 = 0\} \cap U_k$, and hence $\tilde{s} \circ \theta_k^{-1}$ is holomorphic on U_k. Thus, $\tilde{s}\colon E \to \mathbb{P}^n$ is a spray satisfying

- $\tilde{s}(E|_{U_0 \setminus A}) \subset U_0 \setminus A$, and
- $\tilde{s}(z, t) = z$ for all $z \in A \cup \Lambda$ and $t \in E_z$.

For each point $p \in U_0 \setminus A$ we can find finitely many sprays of this kind which together dominate at p, and hence they dominate over a Zariski open set containing the point p. Repeating this construction at other points and for different choices of the hyperplane $\Lambda \subset \mathbb{P}^n$, we obtain a finite dominating family of algebraic sprays on $\mathbb{P}^n \setminus A$.

The same proof applies to a Grassmannian $Y = G_{k,n}$ since it is covered by finitely many Zariski open sets $U_j \cong \mathbb{C}^{k(n-k)}$ (see Example 1.2.8). $\square$

Localization of Algebraic Subellipticity

Algebraic subellipticity can be localized as follows (see [266, 3.5.B, 3.5.C] and [190, Proposition 1.3]). No comparable result is known for ellipticity.

Proposition 6.4.2 *If Y is an algebraic manifold such that each point $y_0 \in Y$ has a Zariski open neighborhood $U \subset Y$ and algebraic sprays $s_j \colon E_j \to Y$ $(j = 1, 2, \ldots, k)$ defined on algebraic vector bundles $p_j \colon E_j \to U$ and satisfying*

$$(\mathrm{d}s_1)_{0_y}(E_{1,y}) + (\mathrm{d}s_2)_{0_y}(E_{2,y}) + \cdots + (\mathrm{d}s_k)_{0_y}(E_{k,y}) = \mathrm{T}_y Y \qquad (6.7)$$

for every point $y \in U$, then Y is algebraically subelliptic.

Proof We repeatedly use the fact that for every closed algebraic subvariety A in an algebraic manifold Y and for every point $y_0 \in Y \setminus A$ there exists an algebraic hypersurface $\Lambda \subset Y$ such that $A \subset \Lambda$ but $y_0 \notin \Lambda$ (see Borelli [63, Theorem 4.1]).

Fix a point $y_0 \in Y$ and let $U \subset Y$ be a Zariski neighborhood of y_0 with finitely many algebraic sprays $s_j \colon E_j \to Y$ $(j = 1, \ldots, k)$ such that condition (6.7) holds at the point $y = y_0$. Replacing U by a smaller Zariski neighborhood of y_0, we may assume that $\Lambda = Y \setminus U$ is an algebraic hypersurface in Y and that the bundle $E_j|_U \to U$ is algebraically trivial for each j. Composing an algebraic trivialization of $E_j|_U$ with the spray s_j, we may therefore assume that s_j is defined on the product bundle $U \times \mathbb{C}^{N_j}$ and has values in Y. To remove the singularities of s_j along the hypersurface Λ, we replace the product bundle by $N_j[\Lambda]^{-m_j}$ for a sufficiently big integer $m_j \in \mathbb{N}$ (see the proof of Proposition 6.4.1). This gives finitely many algebraic sprays on Y which together dominate at y_0, and hence over a Zariski neighborhood of y_0. Finitely many such families of sprays then dominate on all of Y. □

Problem 6.4.3 Let Y be an algebraically subelliptic manifold. Is the complement $Y \setminus A$ of every thin algebraic subvariety algebraically subelliptic?

Zariski Locally Affine Manifolds

We consider a class of algebraically subelliptic manifolds that contains all manifolds in Proposition 6.4.1.

Definition 6.4.4 Let Y be an algebraic manifold.

(a) Y is of *Class* $\mathcal{A}_0$ if it is covered by finitely many Zariski open sets that are biregularly isomorphic to the affine space $\mathbb{C}^p$ with $p = \dim Y$.
(b) Y is of *Class* $\mathcal{A}$ if $Y = \widehat{Y} \setminus A$, where $\widehat{Y}$ is of class $\mathcal{A}_0$ and A is a thin (of codimension at least two) algebraic subvariety of $\widehat{Y}$.

Class $\mathcal{A}_0$ includes all complex affine spaces, projective spaces, and Grassmannians. Another example is the total space W of a holomorphic fibre bundle $\pi : W \to Y$ where the base Y is a manifold of Class $\mathcal{A}_0$, the fibre is $F = \pi^{-1}(y)$ is $\mathbb{C}^m$ or $\mathbb{P}^m$, and the structure group is $GL_m(\mathbb{C})$ or $PGL_m(\mathbb{C})$, respectively. Every such bundle is algebraic by the GAGA principle [498], and its restriction to any affine Zariski open set $\mathbb{C}^p \cong U \subset Y$ is algebraically trivial, $\pi^{-1}(U) \cong U \times F$. Hence, W is covered by Zariski open sets biregularly isomorphic to $\mathbb{C}^{p+m}$, i.e., W is of Class $\mathcal{A}_0$. An example are the Hirzebruch surfaces H_l for $l = 0, 1, 2, \ldots$ (see [45, p. 191]); these are $\mathbb{P}^1$-bundles over $\mathbb{P}^1$, and each of them is birationally equivalent to $\mathbb{P}^2$.

Proposition 6.4.5 *Every manifold of Class $\mathcal{A}$ is algebraically subelliptic, and hence an Oka manifold.*

Proof By Proposition 6.4.1(a), or by Example 5.6.16(C), the manifold $\mathbb{C}^p \setminus A$ is algebraically elliptic for each closed affine algebraic subvariety $A \subset \mathbb{C}^p$ of codimension at least two. The conclusion now follows from the localization principle furnished by Proposition 6.4.2. □

Manifolds of Class $\mathcal{A}$ have been considered by Gromov who called them *Ell-regular*. This class is stable under blowing up points.

Proposition 6.4.6 ([266, Sect. 3.5.D"]) *If Y is a manifold of Class $\mathcal{A}$ (respectively of Class $\mathcal{A}_0$) then any manifold $\widetilde{Y}$, obtained from Y by blowing up finitely many points, is also of Class $\mathcal{A}$ (respectively of Class $\mathcal{A}_0$).*

Proof By localization, it suffices to show that the manifold L obtained by blowing up $\mathbb{C}^n$ at the origin is of Class $\mathcal{A}$. Note that L is the total space of a holomorphic line bundle $\pi : L \to \mathbb{P}^{n-1}$ (the universal bundle); L is trivial over the complement of each hyperplane $\mathbb{P}^{n-2} \subset \mathbb{P}^{n-1}$ (which equals $\mathbb{C}^{n-1}$), and hence every point in L has a Zariski neighborhood of the form $\pi^{-1}(\mathbb{P}^{n-1} \setminus \mathbb{P}^{n-2})$ that is biregularly isomorphic to $\mathbb{C}^n$. □

Corollary 6.4.7 *Every compact rational surface is of Class $\mathcal{A}_0$ and hence an Oka manifold.*

Proof Every compact rational surface X is obtained by blowing up a certain Hirzebruch surface Y at finitely many points. Now, Y is either a $\mathbb{P}^2$ or a holomorphic fibre bundle $Y \to \mathbb{P}^1$ with fibre $\mathbb{P}^1$ over $\mathbb{P}^1$. In either case, Y is easily seen to be of class $\mathcal{A}_0$. It remains to apply Proposition 6.4.6. □

The following result, generalizing Proposition 6.4.6, was proved by Lárusson and Truong [378, Main Theorem].

Theorem 6.4.8 *A manifold of Class $\mathcal{A}$ blown up along any algebraic submanifold (not necessarily connected) is algebraically subelliptic, and hence an Oka manifold.*

A more general result in this direction was proved by Kaliman, Kutzschebauch and Truong [327]. They showed that the blowup of a locally stably flexible algebraic manifold at a smooth algebraic submanifold (not necessarily equidimensional or connected) is subelliptic, and hence Oka. We refer to the cited paper for the definition of a locally stably flexible manifold; this class contains all manifolds of Class $\mathcal{A}$ and is stable under removal of thin algebraic subvarieties.

Problem 6.4.9 Is the blowup of any algebraic Oka manifold along an algebraic submanifold Oka? Does this hold for complex manifolds?

Covering Spaces

By Proposition 5.6.3, the Oka property passes up and down in unramified holomorphic coverings. We consider the same question for ellipticity.

Proposition 6.4.10 *Let $\pi : \widetilde{Y} \to Y$ be an unramified holomorphic covering map. If Y is elliptic (resp. subelliptic) then so is $\widetilde{Y}$.*

Proof Let (E, p, s) be a spray on Y. Denote by $\widetilde{E} = \pi^*(E) \to \widetilde{Y}$ the pullback of the bundle $p: E \to Y$ by the map $\pi : \widetilde{Y} \to Y$:

$$\widetilde{E} = \big\{(\tilde{y}, e) : \tilde{y} \in \tilde{Y},\ e \in E,\ \pi(\tilde{y}) = p(e)\big\}.$$

Let $\sigma : \widetilde{E} \to Y$ be defined by $\sigma(\tilde{y}, e) = s(y, e)$ where $y = \pi(\tilde{y}) \in Y$. Fix $\tilde{y} \in \widetilde{Y}$. Since the fibre $\widetilde{E}_{\tilde{y}}$ is simply connected and π is a holomorphic covering, the map $\sigma(\widetilde{y}, \cdot): \widetilde{E}_{\tilde{y}} \to Y$ has a unique holomorphic lifting $\tilde{s}(\tilde{y}, \cdot): \widetilde{E}_{\tilde{y}} \to \widetilde{Y}$ (i.e., $\pi(\tilde{s}(\tilde{y}, e)) = \sigma(\tilde{y}, e))$ with $\tilde{s}(\tilde{y}, 0) = \tilde{y}$. Clearly, $\tilde{s}: \widetilde{E} \to \widetilde{Y}$ is a spray on $\widetilde{Y}$ which is dominating when the spray s is dominating. Hence, $\widetilde{Y}$ is elliptic whenever Y is elliptic. A similar argument applies to families of sprays, showing that subellipticity of Y implies subellipticity of $\widetilde{Y}$. □

Problem 6.4.11 Let $\pi : \widetilde{Y} \to Y$ be an unramified holomorphic covering map. Does (sub-)ellipticity of $\widetilde{Y}$ imply (sub-)ellipticity Y ?

A good test case may be complex tori $X = \mathbb{C}^n/\Gamma$. Denote by $\pi : \mathbb{C}^n \to X$ the quotient map and let $p_0 = \pi(0) \in X$; then $\pi^{-1}(p_0) = \Gamma$. By the proof of Corollary 5.6.18 (p. 231), the complement $\mathbb{C}^n \setminus \Gamma$ is elliptic. Is the punctured torus $X \setminus \{p_0\}$ also elliptic? (A torus with finitely many punctures is an Oka manifold by Corollary 5.6.18.)

Blowing Up a Tame Sequence in $\mathbb{C}^n$

Proposition 6.4.12 *Any manifold X, obtained from $\mathbb{C}^n$ $(n \geq 2)$ by blowing up each point of a tame discrete set (see Definition 4.6.1 on p. 128), is weakly subelliptic and hence an Oka manifold.*

Proof For simplicity we consider the case $n = 2$. Let (z, w) be complex coordinates on $\mathbb{C}^2$. It suffices to prove the result for the tame sequence $e_k = (k, 0)$, $k = 0, 1, 2, \ldots$. Let $X_0 = \widehat{\mathbb{C}^2}$ be the surface obtained by blowing up $\mathbb{C}^2$ at $e_0 = (0, 0)$. Denote by $\pi_0: X_0 \to \mathbb{C}^2$ the natural projection, and let $S_0 = \pi_0^{-1}(e_0) \cong \mathbb{P}^1$ be the exceptional fibre with the homogeneous coordinates $[z : w]$. Let $\Lambda_0 \subset X_0$ be the proper transform of the line $\{w = 0\}$:

$$\Lambda_0 = \overline{\pi_0^{-1}\big(\{(z, 0) : z \neq 0\}\big)} \cong \mathbb{C}.$$

Note that $\Lambda_0 \cap S_0 = [1:0] \in S_0$ and $S_0 \setminus \Lambda_0 = S_0 \setminus \{[1:0]\} \cong \mathbb{C}$. The complement $X_0 \setminus \Lambda_0$ is a holomorphic line bundle over $S_0 \setminus \Lambda_0 \cong \mathbb{C}$, hence it is biholomorphic to $\mathbb{C}^2$. As in the proof of Proposition 6.4.1 we obtain an algebraic spray (E_0, p_0, s_0) on X_0 which is dominating over $X_0 \setminus \Lambda_0 \cong \mathbb{C}^2$ and has values in this set, and that vanishes over Λ_0.

Let X_1 be obtained by blowing up the point $e_1 = (1,0) \in X_0$ (we are identifying $e_1 \in \mathbb{C}^2$ with $\pi_0^{-1}(e_1) \in X_0$). Let $\pi_1 \colon X_1 \to X_0$ be the associated projection which is biholomorphic over $X_0 \setminus \{e_1\}$ and has the exceptional fibre $S_1 \cong \mathbb{P}^1$ over e_1. We may consider $X_0 \setminus \Lambda_0$ as an open subset of $X_1 \setminus \pi_1^{-1}(\Lambda_0)$. Since the spray (E_0, p_0, s_0) vanishes over Λ_0, it extends to a spray on X_1 that vanishes on the curve $\pi_1^{-1}(\Lambda_0)$.

Next, let $\pi_2 \colon X_2 \to X_1$ be obtained by blowing up X_1 at $e_2 = (2,0)$. The same argument as before shows that $X_0 \setminus \Lambda_0 \cong \mathbb{C}^2$ is an open subset of X_2 and (E_0, p_0, s_0) extends to a spray on X_2 that vanishes on the curve $(\pi_2 \circ \pi_1)^{-1}(\Lambda_0)$.

Continuing this process, we obtain a complex surface X ($\mathbb{C}^2$ blown up at each point e_k, $k = 0, 1, \ldots$) with a projection $\pi : X \to X_0$, and a complex curve $\Sigma_0 = \pi^{-1}(\Lambda_0)$ such that $U_0 = X \setminus \Sigma_0 \cong X_0 \setminus \Lambda_0 \cong \mathbb{C}^2$. Note that Σ_0 is a normal crossing divisor consisting of a copy of $\mathbb{C}$ that is intersected by a rational curve $\mathbb{P}^1$ at each of the points $e_1, e_2, \ldots$, and these rational curves are pairwise disjoint. Furthermore, we get a spray (E_0, p_0, s_0) on X that is dominating over the open set $U_0 \cong \mathbb{C}^2$ in X, with values in this set and vanishing over $\Sigma_0 = X \setminus U_0$. Considering the projection $\pi_0 \circ \pi : X \to \mathbb{C}^2$, we see that Σ_0 is the total transform of the line $\{w = 0\}$ at each point $e_1, e_2, \ldots$, and is the exact transform of this line at the point e_0.

In the same way, we get for every $k \in \mathbb{Z}_+$ a complex curve $\Sigma_k \subset X$ by taking the exact transform of the line $\{w = 0\}$ at e_k and its total transform at every other point e_j ($j \neq k \in \mathbb{Z}_+$). We also get a spray (E_k, p_k, s_k) on X that is dominating on $U_k = X \setminus \Sigma_k$ and vanishes over Σ_k.

Let $\Sigma \subset X$ denote the exact transform of $\{w = 0\}$ at every point e_k for $k \in \mathbb{Z}_+$. Then, $X \setminus \Sigma = \bigcup_{k=0}^{\infty} U_k$. The collections of sprays (E_k, p_k, s_k) dominates over $X \setminus \Sigma$. We now add another family of sprays to get domination over all of X. For this purpose, consider the shear $\phi \in \mathrm{Aut}(\mathbb{C}^2)$ defined by

$$\phi(z, w) = \big(z, w + \sin(\pi z)\big), \quad (z, w) \in \mathbb{C}^2.$$

Clearly, there is an induced automorphism $\Phi \in \mathrm{Aut}(X)$ such that the following diagram commutes:

$$\begin{array}{ccc} X & \xrightarrow{\Phi} & X \\ \downarrow & & \downarrow \\ \mathbb{C}^2 & \xrightarrow{\phi} & \mathbb{C}^2 \end{array}$$

Since $\sin(\pi z)$ vanishes at the integer values of z, Φ maps the exceptional fibre $S_k \cong \mathbb{P}^1$ of $X \to \mathbb{C}^2$ over e_k to itself for every $k \in \mathbb{Z}_+$, and $\Phi \colon S_k \to S_k$ is the nontrivial automorphism of $S_k \cong \mathbb{P}^1$ sending $[1:0]$ to $[1:(-1)^k \pi]$ and fixing the point

$[0:1]$. Hence, $\Phi(\Sigma) \cap \Sigma = \varnothing$ in X. By adding to the family of sprays (E_k, p_k, s_k) $(k \in \mathbb{Z}_+)$ all their pullbacks by Φ (see Sect. 6.3) we thus obtain a collection of sprays that dominates at each point of X, and a finite subcollection dominates over a given compact in X. $\square$

Let us call a *complex torus* any quotient $\mathbb{C}^n/\Gamma$ of $\mathbb{C}^n$ by a lattice, not necessarily of maximal rank.

Corollary 6.4.13 *A complex torus of dimension* > 1 *blown up at finitely many points is an Oka manifold.*

Proof Let Γ be a lattice in $\mathbb{C}^n$ and $\pi : \mathbb{C}^n \to \mathbb{T} = \mathbb{C}^n/\Gamma$ be the quotient projection. Fix points $p_1, \ldots, p_m \in \mathbb{T}$ and choose points $z_1, \ldots, z_m \in \mathbb{C}^n$ such that $\pi(z_k) = p_k$ for $k = 1, \ldots, m$. Then, $\pi^{-1}(p_k) = z_k + \Gamma$, and the union

$$\Lambda = \pi^{-1}\big(\{p_1, \ldots, p_m\}\big) = \bigcup_{k=1}^{m} (z_k + \Gamma) \subset \mathbb{C}^n$$

is a tame discrete set in $\mathbb{C}^n$ (see [80, Proposition 4.1] and [74]). Let Y be obtained from $\mathbb{T}$ by blowing up each of the points $p_1, \ldots, p_m$, and let X be obtained from $\mathbb{C}^n$ by blowing up each point of Λ. Every translation of $\mathbb{C}^n$ by an element $\gamma \in \Gamma$ induces a biholomorphism $\widetilde{\gamma} : X \to X$. In this way, Γ becomes a discrete subgroup of $\mathrm{Aut}(X)$, and we have a covering map $\widetilde{\pi} : X \to X/\Gamma \cong Y$ such that the following diagram commutes:

$$\begin{array}{ccc} X & \xrightarrow{\ \widetilde{\pi}\ } & Y \cong X/\Gamma \\ \big\downarrow & & \big\downarrow \\ \mathbb{C}^n & \xrightarrow{\ \pi\ } & \mathbb{T} \cong \mathbb{C}^n/\Gamma \end{array}$$

The manifold X is weakly subelliptic by Proposition 6.4.12, and hence an Oka manifold by Corollary 5.6.14. Proposition 5.6.3 implies that Y is Oka. $\square$

Complements of Subvarieties in Fibre Bundles

Assume that X and Z are reduced complex spaces and $\pi : Z \to X$ is a holomorphic fibre bundle with an elliptic fibre Y. Let A be a closed complex subvariety of Z whose fibre $A_x = A \cap Z_x$ has codimension at least two in $Z_x \cong Y$ for every $x \in X$. Under which condition on the subvariety A is the restricted submersion $\pi : Z \setminus A \to X$ elliptic or subelliptic?

Proposition 6.4.1 gives sufficient conditions for the trivial submersion $Z = Y \setminus A \to \mathit{point}$ when Y equals $\mathbb{C}^n$ or $\mathbb{P}^n$. We now extend it to fibre bundles.

Proposition 6.4.14 *Assume that $\pi: Z \to X$ is a holomorphic fibre bundle with fibre $\mathbb{C}^n$ ($n \geq 2$). Let $A \subset Z$ be a* locally uniformly tame subvariety *in the following sense: For every point $x_0 \in X$ there are a neighborhood $U \subset X$, a fibre bundle isomorphism $\Phi: Z|_U \to U \times \mathbb{C}^n$, and a number $C > 0$ such that for every point $x \in U$ we have*

$$\Phi_x(A_x) \subset \{(z', z_n) \in \mathbb{C}^n : |z_n| \leq C(1 + |z'|)\}. \tag{6.8}$$

If every fibre $A_x = A \cap \pi^{-1}(x)$ has codimension at least two in Z_x, then the restricted submersion $\pi: Z \setminus A \to X$ is elliptic. Therefore, sections $X \to Z$ whose images avoid A satisfy all forms of the Oka principle.

Proof Let $A' = \Phi(A \cap Z|_U) \subset U \times \mathbb{C}^n$. Condition (6.8) implies that

$$\omega(A'_x) \subset \Gamma = \{[v' : v_n] \in \mathbb{P}^{n-1} : |v_n| \leq C|v'|\}, \quad x \in U.$$

(Here, $\omega(A'_x)$ is the limit set at infinity, see (4.57) on p. 152.) Since Γ is a proper closed subset of $\mathbb{P}^{n-1}$, the proof of Proposition 5.6.17 gives complete holomorphic vector fields $V_1, \ldots, V_m$ on $U \times \mathbb{C}^n$ that are tangent to the fibres $\{x\} \times \mathbb{C}^n$, vanish on A', and span the tangent space to the fibre at every point in $(U \times \mathbb{C}^n) \setminus A'$. The composition of their flows (see (5.10) on p. 231) is a dominating fibre spray on $(U \times \mathbb{C}^n) \setminus A'$, so $Z \setminus A \to X$ is elliptic. □

To a holomorphic vector bundle $\pi: E \to X$ of rank n we associate a holomorphic fibre bundle $\widehat{\pi}: \widehat{E} \to X$ whose fibre $\widehat{E}_x \cong \mathbb{P}^n$ is obtained by compactifying $E_x \cong \mathbb{C}^n$ with the hyperplane at infinity $\Lambda_x \cong \mathbb{P}^{n-1}$.

Proposition 6.4.15 *Let $\pi: E \to X$ be a holomorphic vector bundle of rank n over a Stein space X, and let $\widehat{\pi}: \widehat{E} \to X$ be the associated fibre bundle with fibre $\mathbb{P}^n$. Assume that A is a closed analytic subset of E such that*

(a) *the fibre A_x has codimension ≥ 2 in $E_x \cong \mathbb{C}^n$ for every $x \in X$, and*
(b) *the closure of A in $\widehat{E}$ does not contain any hyperplane at infinity Λ_x.*

Then, the restricted submersion $\pi: E \setminus A \to X$ is elliptic.

Proof Fix a point $p \in X$ and choose a local trivialization for $\widehat{E}|_U \cong U \times \mathbb{P}^n$ in a neighborhood $U \subset X$ of p such that $E|_U = U \times \mathbb{C}^n$. By the hypothesis the set $\Lambda_p \setminus \bar{A}$ is nonempty. Since the set $\bar{A} \cap (\widehat{E} \setminus E)$ is closed, after shrinking U around p there is a closed set $\sigma \subset \mathbb{P}^{n-1}$ with nonempty interior such that $\bar{A} \cap (U \times \sigma) = \varnothing$. By a linear change coordinates, we may assume that the point $[0 : \cdots : 0 : 1] \in \mathbb{P}^{n-1}$ lies in the interior of σ. Let $\Gamma \subset \mathbb{C}^n$ be the open complex cone determined by the set $\mathbb{P}^{n-1} \setminus \sigma$. It is easily seen that $A_x \subset R\mathbb{B} \cup \Gamma$ for a big $R > 0$ and for all $x \in U$. This means that (6.8) holds for a suitably big constant $C > 0$ and hence Proposition 6.4.14 applies. □

In a similar way, one can generalize to fibre bundles the fact that the complement of a thin subvariety in a complex Grassmannian is subelliptic (Proposition 6.4.1). The proof of the following result is left to the reader.

Proposition 6.4.16 *Let $\pi : Z \to X$ be a holomorphic fibre bundle whose fibre is a projective space or a Grassmannian. If A is a closed complex subvariety of Z whose fibre $A_x = A \cap Z_x$ has complex codimension at least two in Z_x for every $x \in X$, then $\pi : Z \setminus A \to X$ is a stratified subelliptic submersion.*

The following result, which generalizes Corollary 5.14.3 on p. 257, follows by combining Theorem 6.2.2 and Proposition 6.4.15. The topological condition needed to extend the section is verified as in Corollary 5.14.3.

Corollary 6.4.17 *Let $\pi : E \to X$ be a holomorphic vector bundle of rank n over a reduced Stein space X, and let A be a closed complex subvariety of E. Assume that $X = X_0 \supset X_1 \supset \cdots \supset X_m = X'$ is a stratification such that for every stratum $S \subset X_k \setminus X_{k+1}$ the intersection $A \cap E|_S$ is locally uniformly tame in $E|_S$ (see Proposition 6.4.14), and for every point $x \in S$ we have*

$$n - \dim A_x = \operatorname{codim} A_x \geq 2 \quad \textit{and} \quad 2 \operatorname{codim} A_x > \dim S.$$

Given a compact $\mathcal{O}(X)$-convex subset K in X, a neighborhood U of $K \cup X'$ and a holomorphic section $f : U \to E|_U$ such that $f(K \cup X') \cap A = \varnothing$, there exists a holomorphic section $\widetilde{f} : X \to E$ that approximates f uniformly on K, agrees with f to a given finite order along X', and satisfies $\widetilde{f}(X) \cap A = \varnothing$.

Open Problems

Problem 6.4.18 Is there a subelliptic manifold that is not elliptic? In particular, is there an algebraic subvariety $A \subset \mathbb{P}^n$ of codimension ≥ 2 such that $\mathbb{P}^n \setminus A$ fails to be elliptic?

Recall that a map $X \to Y$ is a *Serre fibration* if it enjoys the homotopy lifting property for cells (see [569, p. 8]). If X and Y are manifolds then a Serre fibration $X \to Y$ is also a *Hurewicz fibration* [520, 569].

Problem 6.4.19 Let X and Y be complex manifolds and $h : X \to Y$ a surjective elliptic Serre fibration. Assuming that Y is elliptic, does it follows that X is elliptic? What is the answer if elliptic is replaced by subelliptic?

Problem 6.4.20 Let $h : X \to Y$ be a holomorphic covering map. If X is elliptic (resp. subelliptic), does it follow that Y is elliptic (resp. subelliptic)?

It was shown by Ritter [472] that the quotient of $\mathbb{C}^n$ by a group of affine transformations acting freely and properly discontinuously is elliptic.

Problem 6.4.21 Is every Oka manifold elliptic or subelliptic? This is true for a Stein Oka manifold, see Proposition 5.6.15 on p. 230.

6.5 Lifting Homotopies to Spray Bundles

The following result will be used in Sect. 6.6.

Proposition 6.5.1 *Assume that X is a reduced Stein space, $h\colon Z \to X$ is a holomorphic submersion, and (E, π, s) is a dominating composed fibre spray on Z (see Definition 6.1.1). Given a homotopy of holomorphic sections $f_t\colon X \to Z$ $(t \in [0, 1])$ and an open relatively compact subset $V \Subset X$, there are $k \in \mathbb{N}$ and a homotopy of holomorphic sections ξ_t $(t \in [0, 1])$ of the iterated spray bundle $E^{(k)}$ over $f_0(V) \subset Z$ such that for every $z \in f_0(V)$ and $t \in [0, 1]$ we have*

$$\xi_0(z) = z \quad \textit{and} \quad s^{(k)}\big(\xi_t(z)\big) = f_t\big(h(z)\big).$$

More generally, given a compact Hausdorff space P and a family of holomorphic sections $f_{p,t}\colon X \to Z$ depending continuously on $(p, t) \in P \times [0, 1]$, there are $k \in \mathbb{N}$ and a continuous family of holomorphic sections $\xi_{p,t}$ of $E^{(k)}|_{f_{p,0}(V)}$ such that for all $z \in f_{p,0}(V)$, $p \in P$, and $t \in [0, 1]$ we have

$$\xi_{p,0}(z) = z \quad \textit{and} \quad s^{(k)}\big(\xi_{p,t}(z)\big) = f_{p,t}\big(h(z)\big).$$

Here we assume the existence of a dominating composed fibre spray over all of Z. In a subelliptic submersion $Z \to X$, this result will be applied over small open sets $U \subset X$.

Proof Consider first the basic case. We shall find numbers $0 = t_0 < t_1 < \cdots < t_k = 1$, and for each $j = 0, 1, \ldots, k-1$ a homotopy $\{\xi_t^j : t \in [t_j, t_{j+1}]\}$ of holomorphic sections of the restricted bundle $E|_{f_{t_j}(V)}$ satisfying

$$s \circ \xi_t^j\big(f_{t_j}(x)\big) = f_t(x), \quad x \in V,\ t \in [t_j, t_{j+1}]. \tag{6.9}$$

In particular, $s \circ \xi_{t_{j+1}}^j(f_{t_j}(x)) = f_{t_{j+1}}(x)$ for $j = 0, 1, \ldots, k-1$. Comparing these compatibility conditions with those defining the iterated spray bundle $E^{(k)}$, we see that these k homotopies join into a homotopy of holomorphic sections $\xi_t\colon f_0(V) \to E^{(k)}|_{f_0(V)}$ $(t \in [0, 1])$ satisfying Proposition 6.5.1. Explicitly, using the notation in Definition 6.3.5(b), we define for $x \in V$ and $t \in [t_j, t_{j+1}]$

$$\xi_t\big(f_0(x)\big) = \big(\xi_{t_1}^0\big(f_0(x)\big), \xi_{t_2}^1\big(f_{t_1}(x)\big), \ldots, \xi_{t_j}^{j-1}\big(f_{t_{j-1}}(x)\big), \xi_t^j\big(f_{t_j}(x)\big), \overbrace{0, \ldots, 0}^{k-j-1}\big).$$

To construct the homotopies ξ_t^j, we begin by the following consideration. Assume that Ω is a Stein open subset of Z. By Lemma 6.3.7 the restricted bundle $E|_\Omega$ admits a holomorphic vector bundle structure. Since $\mathrm{V}ds\colon \mathrm{VT}E|_Z \cong E \to \mathrm{VT}Z$ is surjective, we have a holomorphic direct sum splitting $E|_\Omega = \ker \mathrm{V}ds|_\Omega \oplus E'$ such that $\mathrm{V}ds\colon E' \xrightarrow{\cong} \mathrm{VT}Z|_\Omega$ is an isomorphism. Given an open subset $U \subset X$ and a

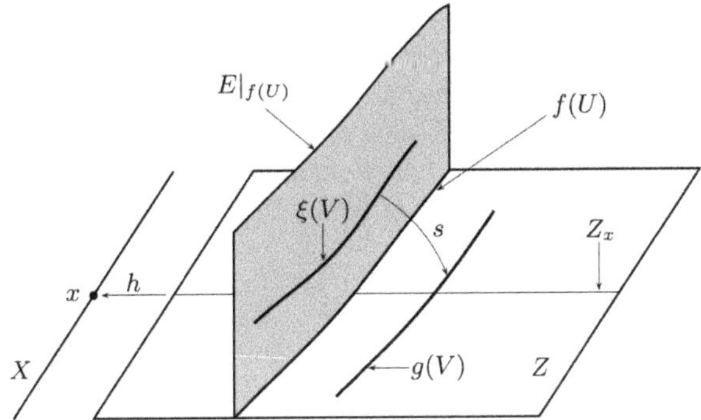

Fig. 6.2 Lifting a section to the spray bundle. (Modified from [201, p. 148, Fig. 1])

holomorphic section $f: U \to Z|_U$ such that $\overline{f(U)} \subset \Omega$, we get an isomorphism $\mathrm{Vd}s: E'|_{f(U)} \to \mathrm{VT}Z|_{f(U)}$. By the inverse function theorem, the spray s maps an open neighborhood of the zero section in $E'|_{f(U)}$ biholomorphically onto an open neighborhood of $f(U)$ in Z. Assuming that $V \Subset U$ and that $g: V \to Z$ is a holomorphic section which is sufficiently close to $f|_V$, there is a unique holomorphic lifting $\xi: f(V) \to E'|_{f(V)}$ satisfying $s \circ \xi(f(x)) = g(x)$ for $x \in V$ (see Fig. 6.2). In particular, given a homotopy of holomorphic sections $f_t: X \to Z$ and an open set $U \Subset X$ such that $f_{t_0}(\overline{U}) \subset \Omega$ for some $t_0 \in [0,1]$, we can lift $f_t|_U$ for all t sufficiently near t_0 to a holomorphic homotopy $\xi_t: U \to E|_{f_{t_0}(U)}$.

Assume now that $f_t: X \to Z$ ($t \in [0,1]$) is a homotopy of holomorphic sections and $V \Subset X$ is an open set. Choose a compact $\mathcal{O}(X)$-convex set $K \subset X$ containing V in its interior. Then, $f_t(K)$ is $\mathcal{O}(f_t(X))$-convex. Since $f_t(X)$ is a closed Stein subspace of Z, $f_t(K)$ admits an open Stein neighborhood in Z by Theorem 3.2.1. By compactness of the set $\{f_t(x) : x \in K,\ t \in [0,1]\}$ there are numbers $0 < c_1 < c_2 < \cdots < c_m = 1$ and Stein open sets $\Omega_1, \ldots, \Omega_m \subset Z$ such that $f_t(K) \subset \Omega_j$ for all $t \in [c_{j-1}, c_j]$ and $j = 1, \ldots, m$. The above lifting argument now shows that by subdividing each segment $[c_{j-1}, c_j]$ into smaller segments we obtain a subdivision $0 = t_0 < t_1 < \cdots < t_k = 1$ and homotopies ξ_t^j satisfying condition (6.9). This proves the basic case.

The parametric case is not a straightforward extension since several steps in the proof depend on the choice of a Stein domain $\Omega \subset Z$ containing a part of our homotopy. We need a couple of lemmas.

Lemma 6.5.2 *Let $h: Z \to X$ and (E, π, s) be as in Proposition 6.5.1. Assume that P is a compact Hausdorff space and $f_p: X \to Z$ is a family of holomorphic sections depending continuously on $p \in P$. For any open relatively compact subset $U \Subset X$ there exist a family $G_p: \mathrm{VT}Z|_{f_p(U)} \to \mathrm{VT}E|_{f_p(U)}$ of holomorphic splittings of $\mathrm{Vd}s: \mathrm{VT}E|_Z \to \mathrm{VT}Z$, and holomorphic direct sums*

$$\mathrm{VT}E|_{f_p(U)} = \ker \mathrm{Vd}s|_{f_p(U)} \oplus E'_p, \tag{6.10}$$

both depending continuously on the parameter $p \in P$.

Proof For each fixed $p \in P$ the set $f_p(X)$ is a closed Stein subspace of Z, and hence is contained in an open Stein domain $D_p \subset Z$ by Theorem 3.2.1. By compactness of P and $\overline{U}$ there exist an open cover $P_1, \ldots, P_{j_0}$ of P and a family of open Stein subsets $D_1, \ldots, D_{j_0}$ of Z such that $f_{p,0}(U) \subset D_j$ for all $p \in P_j$. Choose a continuous partition of unity $\{\chi_j\}$ on P subordinate to the cover $\{P_j\}$. Since D_j is Stein, there is a holomorphic splitting $H_j : \mathrm{VTZ}|_{D_j} \to \mathrm{VT}E|_{D_j}$ of Vds over D_j. The family

$$G_p = \sum_{j=1}^{j_0} \chi_j(p) H_j : \mathrm{VTZ}|_{f_p(U)} \to \mathrm{VT}E|_{f_p(U)}, \quad p \in P$$

then satisfies Lemma 6.5.2. Note that G_p is well defined since $\chi_j(p) = 0$ if $f_{p,0}(U) \not\subset D_j$, and it is a splitting of Vds since it is a convex linear combination of splittings. Taking $E'_p = G_p(\mathrm{VTZ}|_{f_p(U)})$ we get (6.10). □

In the following lemma we keep the assumptions from Lemma 6.5.2, identifying Z with the zero section of a composed spray bundle $\pi : E \to Z$.

Lemma 6.5.3 *Let $E'_p \subset \mathrm{VT}E|_{f_p(X)}$ be a family of holomorphic vector subbundles of $\mathrm{VT}E$ depending continuously on $p \in P$. For each compact set K in X there exist an open set $\Omega \subset E$ containing $\bigcup_{p \in P} f_p(K)$ and a family of not necessarily closed complex subspaces Σ_p of $\Omega \cap E|_{f_p(X)}$, depending continuously on $p \in P$ and with regular fibres, such that $\Omega \cap f_p(X) \subset \Sigma_p$ and the vertical tangent bundle to Σ_p along $\Omega \cap f_p(X)$ equals E'_p for each $p \in P$.*

Proof Since $f_p(X)$ admits an open Stein neighborhood in Z for each $p \in P$, we can find a cover $P_1, \ldots, P_m$ of P by compact sets and open Stein sets $D_1, \ldots, D_m \subset Z$ such that $f_p(K) \subset D_j$ for every $p \in P_j$ and $j = 1, \ldots, m$. By continuity there is an open set $U \subset X$ containing K and compact sets $P'_j \subset P$ such that $P_j \subset \mathring{P}'_j$ and $f_p(U) \subset D_j$ for every $p \in P'_j$. By Lemma 6.3.7 there is for each $j = 1, \ldots, m$ a biholomorphic map $\Theta_j : E|_{D_j} \to \mathrm{VT}E|_{D_j}$ that is tangent to the identity map along the zero section of the two bundles. We replace E'_p by its restriction to $f_p(U)$ and consider it as a complex vector subbundle of $\mathrm{VT}E|_{f_p(U)}$. Then,

$$\Sigma_{p,j} := \Theta_j^{-1}\big(E'_p\big) \subset E|_{f_p(U)}, \quad p \in P'_j$$

is a family of complex subspaces satisfying Lemma 6.5.3 for the values $p \in P'_j$. In particular, the vertical tangent bundle of $\Sigma_{p,j}$ along the zero section equals E'_p. Set $\Phi_{j,p} = \Theta_j^{-1}|_{E'_p} : E'_p \hookrightarrow E|_{f_p(U)}$.

The families of subspaces $\Sigma_{j,p}$ do not necessarily agree on the intersections $P_j \cap P_k$. We combine them by the *method of successive patching* which was first applied on p. 78; here are the details. Choose a compact set $\widetilde{P}_1 \subset P$ such that $P_1 \subset \mathrm{Int}\,\widetilde{P}_1$ and $\widetilde{P}_1 \subset \mathrm{Int}\,P'_1$. Also choose a continuous function $\chi : P \to [0,1]$ such that $\chi = 1$ on $\widetilde{P}_1$ and $\mathrm{supp}\,\chi \subset P'_1$. For $p \in \widetilde{P}_1 \cup P'_2$ we set

$$\Phi_p = \chi(p)\Phi_{1,p} + \big(1 - \chi(p)\big)\Phi_{2,p}, \qquad \Sigma_p = \Phi_p\big(E'_p\big) \subset E|_{f_p(U)}.$$

For $p \in \widetilde{P}_1$ we have $\chi(p) = 1$ and hence $\Phi_p = \Phi_{1,p}$. The convex linear combination therefore only occurs for $p \in P_2' \setminus \widetilde{P}_1 \subset P_2'$ where we use the holomorphic vector bundle structure on $E|_{D_2}$ induced by Θ_2. Since the two maps are tangent at the zero section, the resulting family of subspaces Σ_p satisfies Lemma 6.5.3 for $p \in \widetilde{P}_1 \cup P_2'$, and the latter set contains $P_1 \cup P_2$ in its interior.

Next, we patch $\{\Phi_{3,p} : p \in P_3'\}$ with the family $\{\Phi_p : p \in \widetilde{P}_1 \cup P_2'\}$ by using the complex vector bundle structure on $E|_{D_3}$. This gives a new family satisfying the required properties for p in an open neighborhood of $P_1 \cup P_1 \cup P_3$. After m steps of this kind we obtain a solution for all $p \in P$. □

We now conclude the proof of Proposition 6.5.1. Choose an open set $U \Subset X$ with $\overline{V} \subset U$. Let $\Sigma_{p,t}$ for $(p,t) \in P \times [0,1]$ be a family of complex submanifolds of $E|_{f_{p,t}(U)}$ given by Lemma 6.5.3. By the construction, the vertical derivative $\mathrm{V}ds \colon \mathrm{VT}E|_Z \to \mathrm{VT}Z$ maps the vertical tangent bundle to $\Sigma_{p,t}$ (namely $E'_{p,t}$) isomorphically onto $\mathrm{VT}Z|_{f_{p,t}(U)}$. By the inverse function theorem, the restriction of the spray s to $\Sigma_{p,t}$ maps a neighborhood of $f_{p,t}(U)$ in $\Sigma_{p,t}$ biholomorphically onto a neighborhood of $f_{p,t}(U)$ in Z. Denote its inverse by $u_{p,t}$. By the continuous dependence on $p \in P$ and the compactness of P, the domains of $u_{p,t}$ are uniform with respect to $p \in P$. Hence, there are numbers $0 = t_0 < t_1 < \cdots < t_k = 1$ such that for all $t \in [t_j, t_{j+1}]$ and $p \in P$, the set $f_{p,t}(V)$ belongs to the domain of u_{p,t_j}. The sections

$$\xi_{p,t} = u_{p,t_j} \circ f_{p,t} \circ h : f_{p,t_j}(V) \to E$$

for $t \in [t_j, t_{j+1}]$ and $j = 0, 1, \ldots, k-1$ then satisfy (6.9) for every $p \in P$. □

6.6 Runge Theorem for Sections of Subelliptic Submersions

In this section we prove that the existence of a finite dominating family of fibre sprays on a holomorphic submersion $Z \to X$ over a Stein space X implies the homotopy approximation property (HAP) for sections $X \to Z$; see Theorem 6.6.2. This generalizes the approximation results of Grauert et al. [252, 253, 257] for sections of fibre bundles with complex homogeneous fibres. This is the first step in the proof of Theorem 6.2.2, and it is the only step where a nontrivial assumption on the submersion is needed. In Sect. 6.15 we prove analogous approximation results for algebraic maps from affine algebraic varieties to algebraically subelliptic manifolds.

We begin with the following basic case without parameters.

Theorem 6.6.1 ([190, 217, 266]) *Assume that X is a reduced Stein space and $h \colon Z \to X$ is a holomorphic submersion which admits a finite dominating family of fibre sprays (see Definition* 6.1.1*). Let K be a compact $\mathcal{O}(X)$-convex set in X. Given an open set $U \Subset X$ containing K and a homotopy of holomorphic sections $f_t \colon U \to Z$ $(t \in [0,1])$ such that f_0 extends to a holomorphic section*

$f_0\colon X\to Z$, there exists for every $\epsilon>0$ a homotopy of holomorphic sections $\tilde f_t\colon X\to Z$ $(t\in[0,1])$ such that $\tilde f_0=f_0$ and

$$\sup\bigl\{\mathrm{dist}\bigl(\tilde f_t(x),f_t(x)\bigr):x\in K,\ t\in[0,1]\bigr\}<\epsilon.$$

Proof By the assumption, there are fibre sprays (E_j,π_j,s_j) $(j=1,2,\ldots,m)$ on Z satisfying the domination condition (6.1) at each point $z\in Z$. Let (E,π,s), with $\pi\colon E=E_1*E_2\cdots*E_m\to Z$ and $s=s_1*s_2*\cdots*s_m\colon E\to Z$, denote the composed fibre spray and $(E^{(k)},\pi^{(k)},s^{(k)})$ its k-th iterate (see Definition 6.3.5). Choose an open set V in X with $K\subset V\Subset U$. By Proposition 6.5.1, the homotopy f_t $(t\in[0,1])$ lifts to a homotopy ξ_t of sections of the bundle $E^{(k)}|_{f_0(V)}\to f_0(V)$ for some $k\in\mathbb{N}$ such that ξ_0 is the zero section.

By Lemma 6.3.7, the restricted bundle $E^{(k)}|_{f_0(X)}$ admits the structure of a holomorphic vector bundle. By the parametric Oka-Weil theorem (see Theorem 2.8.4 on p. 62) we can approximate $\{\xi_t\}_{t\in[0,1]}$ uniformly on $f_0(K)$ by a homotopy of global holomorphic sections $\tilde\xi_t\colon f_0(X)\to E^{(k)}|_{f_0(X)}$ such that $\tilde\xi_0$ is the zero section. Then,

$$\tilde f_t(x){:}=s^{(k)}\circ\tilde\xi_t\circ f_0(x)\in Z,\quad x\in X,\ t\in[0,1]$$

is a homotopy of holomorphic sections $X\to Z$ satisfying Theorem 6.6.1. □

The following is a parametric version of Theorem 6.6.1.

Theorem 6.6.2 ([190, 217, 266]) *Let $h\colon Z\to X$ and $K\subset U\subset X$ be as in Theorem 6.6.1, and let $P_0\subset P$ be compact Hausdorff spaces. Set*

$$Q=P\times[0,1]\quad and\quad Q_0=\bigl(P\times\{0\}\bigr)\cup\bigl(P_0\times[0,1]\bigr).$$

Assume that $f\colon X\times Q\to Z$ is a continuous map such that

(i) *for every $q=(p,t)\in Q$, $f_q=f(\cdot,q)\colon X\to Z$ is a continuous section of $Z\to X$ that is holomorphic on U, and*
(ii) *for every $q\in Q_0$ the section f_q is holomorphic on X.*

Given $\epsilon>0$, there is a continuous map $\tilde f\colon X\times Q\to Z$ such that for each $q\in Q$, $\tilde f_q=\tilde f(\cdot,q)\colon X\to Z$ is a holomorphic section, $\tilde f_q=f_q$ for $q\in Q_0$, and

$$\mathrm{dist}\bigl(\tilde f(x,q),f(x,q)\bigr)<\epsilon,\quad x\in K,\ q\in Q.$$

Furthermore, there is a homotopy $f^\tau\colon X\times Q\to Z$ $(\tau\in[0,1])$ that is fixed on Q_0 such that each f^τ satisfies the same conditions as $f^0=f$ and $f^1=\tilde f$.

It will be convenient to use the following terminology.

Definition 6.6.3 Let P be a nonempty compact Hausdorff space and let P_0 be a closed subset of P (possibly empty). Let $h\colon Z\to X$ be a holomorphic map.

(a) A *P-section* of $h\colon Z \to X$ is a continuous map $f\colon X \times P \to Z$ such that $f_p = f(\cdot, p)\colon X \to Z$ is a section of h for each $p \in P$. Such f is *holomorphic* if f_p is holomorphic on X for each fixed $p \in P$. If K is a compact set in X and if X' is a closed complex subvariety of X, then f is *holomorphic on* $K \cup X'$ if there is an open set $U \subset X$ containing K such that the restrictions $f_p|_U$ and $f_p|_{X'}$ are holomorphic for every $p \in P$.

(b) A *homotopy of P-sections* is a $P \times [0,1]$-section, i.e., a continuous map $H\colon X \times P \times [0,1] \to Z$ such that $H_t = H(\cdot,\cdot,t)\colon X \times P \to Z$ is a P-section for each $t \in [0,1]$. Such homotopy H is holomorphic if $H_{p,t} = H(\cdot, p, t)\colon X \to Z$ is holomorphic for each fixed $(p,t) \in P \times [0,1]$.

(c) A *(P, P_0)-section* of h is a P-section $f\colon X \times P \to Z$ such that $f_p = f(\cdot, p)\colon X \to Z$ is holomorphic on X for each $p \in P_0$. A (P, P_0)-section is holomorphic on a subset $U \subset X$ if $f_p|_U$ is holomorphic for every $p \in P$.

In the sense of Definition 6.6.3, a map f as in Theorem 6.6.2 is a (Q, Q_0)-section of $Z \to X$ which is holomorphic on U, $\tilde f$ is a Q-section which is holomorphic on X, and $\{f^\tau\}_{\tau\in[0,1]}$ is a homotopy of (Q, Q_0)-sections which are holomorphic on an open neighborhood of K.

Proof of Theorem 6.6.2 Choose an open Stein set $V \Subset X$ that is Runge in X such that $K \subset U \Subset V$. It suffices to find a map $\tilde f\colon V \times Q \to Z$ satisfying the conclusion of Theorem 6.6.2 on V and then apply induction over an increasing sequence of Runge Stein domains $V = V_0 \subset V_1 \subset \cdots \subset \bigcup_{j=0}^{\infty} V_j = X$.

By Proposition 5.13.1, we may assume that there is an open set P' in P, with $P_0 \subset P'$, such that $f_{p,t}$ is holomorphic on X for all (p,t) in the set

$$Q' = \big(P' \times [0,1]\big) \cup \big(P \times \{0\}\big).$$

By the hypothesis, there exists a dominating composed fibre spray (E, π, s) on Z. For each $k \in \mathbb{N}$ we denote by $(E^{(k)}, \pi^{(k)}, s^{(k)})$ the k-th iterated spray. After shrinking the open sets $P' \supset P_0$ (in P) and $U \supset K$ (in X), Proposition 6.5.1 gives an integer $k \in \mathbb{N}$ and a continuous family of sections $\xi_{p,t}$ of $E^{(k)}|_{f_{p,0}(V)}$ satisfying the following conditions:

- $\xi_{p,0}$ is the zero section of $E^{(k)}|_{f_{p,0}(V)}$ for all $p \in P$,
- $\xi_{p,t}$ is holomorphic on $f_{p,0}(U)$ for all $(p,t) \in Q$,
- $\xi_{p,t}$ is holomorphic on $f_{p,0}(V)$ for all $(p,t) \in Q'$, and
- $s^{(k)} \circ \xi_{p,t}(f_{p,0}(x)) = f_{p,t}(x)$ for all $(x,p,t) \in (V \times Q') \cup (U \times Q)$.

As in the proof of Lemma 6.5.3, we find a cover of P by compact sets $P_1, \ldots, P_m$ and open Stein sets $D_2, \ldots, D_m \subset Z$ such that

- $P_0 \subset \mathring{P}_1 \subset P'$,
- $P_0 \cap P_j = \varnothing$ for $j = 2, \ldots, m$, and
- $f_{p,0}(K) \subset D_j$ for all $p \in P_j$ and $j = 2, \ldots, m$.

Since $E^{(k)}|_{D_j}$ admits a holomorphic vector bundle structure for each $j = 2, \ldots, m$ (see Lemma 6.3.7), we can apply the parametric Oka-Weil theorem (see Theorem 2.8.4) to approximate $\xi_{p,t}$ for $(p,t) \in P_j \times [0,1]$, uniformly on $f_{p,0}(K)$, by a continuous family of holomorphic sections $\xi^j_{p,t} \colon f_{p,0}(V) \to E^{(k)}|_{f_{p,0}(V)}$ such that $\xi^j_{p,0}$ is the zero section for each $p \in P_j$. For $j = 1$ we take $\xi^1_{p,t} = \xi_{p,t}$ since these sections are already holomorphic over $f_{p,0}(V)$.

Finally, we glue the families $\{\xi^j_{p,t}\}$ $(j = 1, \ldots, m)$ into a single family $\tilde\xi_{p,t}$ by successive patching as on p. 282. The family of sections

$$\tilde f_{p,t} = s^{(k)} \circ \tilde\xi_{p,t} \circ f_{p,0} : V \to Z, \quad p \in P,\ t \in [0,1]$$

satisfies the conclusion of Theorem 6.6.2. The existence of a homotopy $\{f^\tau\}_{\tau\in[0,1]}$ from $f^0 = f$ to $f^1 = \tilde f$ follows from the proof since all steps were made by homotopies. □

Remark 6.6.4 Theorems 6.6.1 and 6.6.2 can also be proved by the following somewhat more intuitive argument.

Since the set $\bigcup_{t\in[0,1]} f_t(K) \subset Z$ is compact, we can find an open set V in X with $K \subset V \Subset U$, numbers $0 = t_0 < t_1 < \cdots < t_k = 1$, and Stein domains $\Omega_1, \ldots, \Omega_k \subset Z$ such that $f_t(V) \subset \Omega_j$ for every $t \in [t_{j-1}, t_j]$ and $j = 1, \ldots, k$. The restriction $E|_{\Omega_j}$ of the composed spray bundle E to each of the Stein domains Ω_j admits a holomorphic vector bundle structure according to Lemma 6.3.7. The domination property of the composed spray $s \colon E \to Z$ implies that, after refining our subdivision of the parameter interval $[0,1]$, the homotopy f_t $(t \in [t_j, t_{j+1}],\ j = 0, 1, \ldots, k-1)$ lifts to a homotopy of holomorphic sections ξ_t of $E|_{f_{t_j}(V)}$ satisfying

$$s \circ \xi_t \circ f_{t_j}(x) = f_t(x), \quad x \in V,\ t \in [t_j, t_{j+1}].$$

By the parametric Oka-Weil theorem (see Theorem 2.8.4) we can approximate $\{\xi_t\}_{t\in[0,t_1]}$ uniformly on a neighborhood of $f_0(K)$ in $f_0(X)$ by a homotopy of holomorphic sections $\tilde\xi_t \colon f_0(X) \to E|_{f_0(X)}$ $(t \in [0, t_1])$ such that $\tilde\xi_0$ is the zero section. The holomorphic sections $\tilde f_t = s \circ \tilde\xi_t \circ f_0$ $(t \in [0, t_1])$ are then close to f_t near K. In particular, we may assume that $\tilde f_{t_1}(K)$ is contained in a Stein tubular neighborhood of $f_{t_1}(V)$ in Z, and hence we can connect these two sections by a homotopy of nearby holomorphic sections over a neighborhood of K. We thus get a homotopy of sections $\tilde f_t \colon X \to Z$ $(t \in [0,1])$, close to the original one in a neighborhood of K, such that $\tilde f_0 = f_0$, $\tilde f_t$ is holomorphic on X for $t \in [0, t_1]$, and it is holomorphic in a neighborhood of K for $t \in (t_1, 1]$.

We now use $\tilde f_{t_1}$ as the new base section and repeat the same procedure for the next interval $t \in [t_1, t_2]$. Assuming that $\tilde f_{t_1}$ is sufficiently close to f_{t_1} near K, the choice of t_2 implies that we can lift all section $\tilde f_t$ for $t \in [t_1, t_2]$ to sections ξ_t of $E|_{\tilde f_{t_1}(X)}$. Repeating the above procedure, we obtain a new homotopy of sections which are holomorphic on X for $t \in [0, t_2]$, and are holomorphic on a neighborhood of K for $t \in (t_2, 1]$. After k steps the proof is complete.

The same proof works in the parametric case (Theorem 6.6.2), with P a compact set in a Euclidean space $\mathbb{R}^m$, by using Stein neighborhoods constructed in Sect. 3.6. (See the proof of Theorem 5.13.4.)

Since this proof only uses the domination property of sprays on the range of the given homotopy, which is a compact subset of Z, it shows that Theorems 6.6.1 and 6.6.2 hold under the following weaker hypothesis:

For every compact set L in Z there exists a composed fibre spray (E, π, s) on Z that is dominating over L.

This holds for the trivial submersion $Z = X \times Y \to X$ if the fibre Y is weakly subelliptic (see Definition 5.6.13 on p. 229). □

For future purposes, we axiomatise the conclusion of Theorem 6.6.2.

Definition 6.6.5 A holomorphic submersion $h\colon Z \to X$ onto a reduced Stein space satisfies the *Homotopy Approximation Property (HAP)* if the conclusion of Theorem 6.6.2 holds for all data $K \subset X$, $P_0 \subset P$, and $f\colon X \times Q \to Z$.

HAP is a natural replacement for (and a generalization of) the convex approximation property, CAP, that was used in the case of stratified holomorphic fibre bundles (see Definition 5.4.3 and Theorem 5.4.4). Our proof of Theorem 6.2.2 will give the following analogue of Theorem 5.4.4.

Theorem 6.6.6 *If X is a reduced Stein space and $h\colon Z \to X$ is a stratified holomorphic submersion whose restriction $Z|_S \to S$ to every stratum S satisfies* HAP *over a neighborhood of any point in S, then sections $X \to Z$ enjoy the parametric Oka property with approximation and jet interpolation.*

6.7 Gluing Holomorphic Sections on $\mathcal{C}$-Pairs

In this section we prove the main *Heftungslemma* that will be used in this chapter; see Proposition 6.7.2. This result plays the analogous role as Proposition 5.9.2 does in Chap. 5.

Definition 6.7.1 An ordered pair of compact sets (A, B) in a complex space X is a $\mathcal{C}$*-pair* if

(i) each of the sets $A, B, D = A \cup B$ is a Stein compact,
(ii) $\overline{A \setminus B} \cap \overline{B \setminus A} = \varnothing$ (the separation condition), and
(iii) the set $C = A \cap B$ is $\mathcal{O}(B)$-convex. (C may be empty.)

As a consequence of (i) the set $C = A \cap B$ is also a Stein compact. Comparing with Definition 5.7.1 (p. 234), we see that (A, B) is a $\mathcal{C}$-pair if and only if it is a Cartan pair and, in addition, A and B are Stein compacts and (iii) holds.

Proposition 6.7.2 *Assume that $h\colon Z \to X$ is a holomorphic submersion onto a reduced Stein space X, X' is a closed complex subvariety of X, and (A, B) is a $\mathcal{C}$-pair in X (see Definition 6.7.1) such that $C = A \cap B \subset X \setminus X'$. Suppose that $\widetilde{A} \supset A$ and $\widetilde{B} \supset B$ are open sets such that the restriction $Z|_{\widetilde{B}} \to \widetilde{B}$ admits a finite dominating family of fibre sprays (Definition 6.1.1(iii), p. 264). Let $P_0 \subset P$ be compact Hausdorff spaces. Assume that $a\colon \widetilde{A} \times P \to Z|_{\widetilde{A}}$ and $b\colon \widetilde{B} \times P \to Z|_{\widetilde{B}}$ are holomorphic P-sections (see Definition 6.6.3, p. 284) whose restrictions to $\widetilde{C} = \widetilde{A} \cap \widetilde{B}$ are homotopic by a homotopy of P-sections. Then, there exist open sets A', B' in X, with $A \subset A' \subset \widetilde{A}$ and $B \subset B' \subset \widetilde{B}$, and for every $\epsilon > 0$ there exist homotopies of holomorphic P-sections $a^t : A' \times P \to Z$, $b^t : B' \times P \to Z$ $(t \in [0, 1])$, satisfying the following conditions:*

(i) $a^0 = a|_{A' \times P}$, $b^0 = b|_{A' \times P}$,
(ii) $a^1 = b^1$ *on* $(A' \cap B') \times P$,
(iii) $\mathrm{dist}(a^t_p(x), a_p(x)) < \epsilon$ *for all* $x \in C' = A' \cap B'$, $p \in P$ *and* $t \in [0, 1]$,
(iv) *the homotopy a^t is fixed to a given finite order over $A' \cap X'$, and*
(v) *if the homotopy between a and b is fixed for all $p \in P_0$, then a^t_p and b^t_p can be chosen independent of $t \in [0, 1]$ for all $p \in P_0$.*

Proof The homotopies a^t, b^t are found in two steps. In the first step, we deform b to a new P-section $\tilde{b}$ which is close to a on a neighborhood of C; this is done by applying Theorem 6.6.2 (p. 284) to the given homotopy between a and b, with b serving as the initial section. In the second step, we glue a and $\tilde{b}$ into a holomorphic P-section over a neighborhood of $D = A \cup B$. This is done by 'thickening' both P-sections and applying the gluing techniques from Sects. 5.8–5.9. All moves are made by homotopies of P-sections.

We now turn to the details. We may assume that $\widetilde{A}$, $\widetilde{B}$ and $\widetilde{C} = \widetilde{A} \cap \widetilde{B}$ are relatively compact Stein domains and $\widetilde{C} \cap X' = \varnothing$.

We begin by considering the nonparametric case.

Thus, we are given holomorphic sections $a\colon \widetilde{A} \to Z$, $b\colon \widetilde{B} \to Z$, and a homotopy of holomorphic sections $b_t : \widetilde{C} \to Z$ $(t \in [0, 1])$ such that $b_0 = a|_{\widetilde{C}}$ and $b_1 = b|_{\widetilde{C}}$. We first thicken all sections b_t in the given homotopy by using local flows of vector fields. As in Lemma 5.10.4 (p. 243), we find an open Stein neighborhood $\Omega \subset Z$ of $a(\widetilde{A})$ and finitely many holomorphic vector fields $V_1, \ldots, V_N$ on Ω, tangent to the fibres of h, which span VTZ at every point of $a(\widetilde{C})$ and vanish to a chosen order $m \in \mathbb{N}$ on the subvariety $h^{-1}(X') \cap \Omega$. Let $\theta_{j,y}$ denote the flow of V_j for time $y \in \mathbb{C}$, with $\theta_{j,0}$ being the identity map on Z. We extend our submersion $h\colon Z \to X$ by adding the variables $y = (y_1, \ldots, y_N) \in \mathbb{C}^N$; that is, we take $\widetilde{X} = X \times \mathbb{C}^N$, $\widetilde{Z} = Z \times \mathbb{C}^N$, and let $\tilde{h}\colon \widetilde{Z} \to \widetilde{X}$ be defined by $\tilde{h}(z, y) = (h(z), y)$ for $(z, y) \in \widetilde{Z}$. To simplify the notation, we identify a point $z \in Z$ over $x = h(z) \in X$ with the point $(z, y) \in \widetilde{Z}$ over $(x, y) \in \widetilde{X}$ when there is no danger of confusion. In particular, a holomorphic section $a(x, y)$ of $\widetilde{Z} \to \widetilde{X}$ may be considered as a family of sections $a(\cdot, y)$ of $Z \to X$ depending holomorphically on the parameter y. A dominating family of fibre sprays on $Z|_U$ trivially extends to a family with the same property on $\widetilde{Z}|_{\widetilde{U}}$, where $\widetilde{U} = U \times \mathbb{C}^N$.

By shrinking the set $\widetilde{A} \supset A$ we may assume that the closure of $a(\widetilde{A})$ is contained in Ω. Choose a number $0 < t_1 \le 1$ such that $b_t(\widetilde{C}) \subset \Omega$ for $t \in [0, t_1]$. Then, choose a continuous function $\chi : [0, 1] \to [0, 1]$ which equals one near $t = 0$ and has support contained in $[0, t_1)$. Consider the following sections of $\widetilde{Z} \to \widetilde{X}$ for $y = (y_1, \ldots, y_N)$ in a small ball $r\mathbb{B} \subset \mathbb{C}^N$:

$$a'(x, y) = \theta_{1,y_1} \circ \cdots \circ \theta_{N,y_N} \circ a(x), \qquad x \in \widetilde{A}, \tag{6.11}$$

$$b'_t(x, y) = \theta_{1,\chi(t)y_1} \circ \cdots \circ \theta_{N,\chi(t)y_N} \circ b_t(x), \quad x \in \widetilde{C}. \tag{6.12}$$

Using the above mentioned identification, we have

$$a'(x, 0) = a(x), \quad b'_t(x, 0) = b_t(x), \quad a'(x, y) = b'_0(x, y) \quad \text{for } x \in \widetilde{C}.$$

If $t \in [t_1, 1]$ then $\chi(t) = 0$, hence $\theta_{j,\chi(t)y} = \theta_{j,0} = \mathrm{Id}$ and we get

$$b'_t(x, y) = b_t(x) \quad \text{for } x \in \widetilde{C},\ y \in \mathbb{C}^N,\ t \in [t_1, 1].$$

Now, apply the h-Runge approximation theorem (Theorem 6.6.1 on p. 283) to the homotopy b'_t $(t \in [0, 1])$, with $b'_1(x, y) = b(x)$ $(x \in \widetilde{B},\ y \in r\mathbb{B})$ serving as the initial section. This gives a homotopy $\tilde{b}'_t$ of holomorphic sections over $\widetilde{B} \times r\mathbb{B} \subset \widetilde{X}$ such that $\tilde{b}'_1 = b'_1 = b$, and $\tilde{b}'_t$ approximates b'_t as close as desired on $\widetilde{C} \times r\mathbb{B}$. (We must shrink the domains a little.) At $t = 0$, the section $b' := \tilde{b}'_0$ approximates a' as close as desired on $\widetilde{C} \times r\mathbb{B}$.

Choose an open Stein set C', with $C \subset C' \Subset \widetilde{C}$, and open sets $A' = U_A$ and $B' = U_B$ as in Proposition 5.8.4 (p. 238) such that

$$A \subset A' \Subset \widetilde{A}, \quad B \subset B' \Subset \widetilde{B}, \quad C' = A' \cap B' \Subset \widetilde{C}.$$

Note that the family of sections $a'(\cdot, y)$ is a local dominating spray around the core section $a = a'(\cdot, 0)$ as in Lemma 5.10.4 (p. 243); the analogous observation holds for b'. If b' is close enough to a' over $\widetilde{C} \times r\mathbb{B}$ then we can glue them into a family of holomorphic sections over the union $D' = A' \cup B'$ by following the proof of Proposition 5.9.2 (p. 240). (See also Remark 5.9.4(C).) We first use Lemma 5.9.3 to find a transition map between them, and then Proposition 5.8.4 to split the transition map. Since all steps are made by homotopies, we also get homotopies a^t, b^t as in Proposition 6.7.2. Interpolation over the subvariety X' is built into the construction in view of Proposition 5.8.4.

All tools that we have used are available with a continuous dependence on a parameter $p \in P$. We cover the image of the initial P-section a by a finite family of open Stein sets $D_j \subset Z$ and choose a finite cover $\{P_j\}$ of P such that $a_p(\widetilde{A}) \subset D_j$ for all $p \in P_j$. For each j we choose holomorphic vector fields $V_1^j, \ldots, V_{N_j}^j$ on D_j as above and multiply them by a continuous partition of unity on P subordinate to $\{P_j\}$. This allows us to define families of sections $a'(x, y, p)$ (resp. $b'_t(x, y, p)$) by (6.11) (resp. by (6.12)), depending continuously on the parameter $p \in P$ and such that a' is submersive in the y-variable for all $(x, p) \in \widetilde{C} \times P$. The subsequent steps (pushing them close together and gluing them) can also be made continuously with respect to the parameter $p \in P$ as is clear from the results used in the process. $\square$

6.8 Complexes of Holomorphic Sections

A *holomorphic complex* is a collection of holomorphic sections of a certain submersion $Z \to X$, defined on subdomains of X, and of multi-parameter homotopies between these sections on the intersections of their domains (see Definition 6.8.3 on p. 291). This is a convenient bookkeeping device which is used in the proof of Theorem 6.2.2. In the following sections we describe the operations on complexes which will enable us to assemble the collection of local sections into a global section, thereby proving Theorem 6.2.2.

We begin by describing the nerve complex and its geometric realization.

Definition 6.8.1 Let $\mathcal{A} = \{A_0, A_1, A_2, \ldots\}$ be a locally finite family of nonempty subsets of a set X. The *nerve* of $\mathcal{A}$ is the combinatorial simplicial complex, $\mathcal{K}(\mathcal{A})$, consisting of all multi-indexes $J = (j_0, j_1, \ldots, j_k) \in \mathbb{Z}_+^{k+1}$ $(k \in \mathbb{Z}_+)$, with $0 \le j_0 < j_1 < \cdots < j_k$, for which

$$A_J = A_{j_0} \cap A_{j_1} \cap \cdots \cap A_{j_k} \neq \varnothing.$$

Recall [304] that the *geometric realization*, $K(\mathcal{A})$, of a simplicial complex $\mathcal{K}(\mathcal{A})$ is a topological space obtained as follows. Each multi-index $J = (j_0, j_1, \ldots, j_k) \in \mathcal{K}(\mathcal{A})$ of length $k+1$ determines a closed k-dimensional face $|J| \subset K(\mathcal{A})$, homeomorphic to the standard k-simplex $\Delta^k \subset \mathbb{R}^k$ (the closed convex hull of the set $\{0, e_1, e_2, \ldots, e_k\} \subset \mathbb{R}^k$ containing the origin and the standard basis vectors), and any k-dimensional face of $K(\mathcal{A})$ is of this form. The face $|J|$ is called the *body* (or *carrier*) of J, and J is the *vertex scheme* of $|J|$. The vertexes of $K(\mathcal{A})$ correspond to the individual sets $A_j \in \mathcal{A}$, i.e., to singletons $\{j\} \in \mathcal{K}(\mathcal{A})$. Given $I, J \in \mathcal{K}(\mathcal{A})$ we have $|I| \cap |J| = |I \cap J|$. For any two (bodies of) simplexes in $K(\mathcal{A})$, either one is a subset of the other, or else their intersection is a simplex of lower dimension, possibly empty.

For each $n \in \mathbb{Z}_+$ we denote by

$$\mathcal{K}^n(\mathcal{A}) = \mathcal{K}(A_0, A_1, \ldots, A_n) \subset \mathcal{K}(\mathcal{A})$$

the nerve of the finite subfamily $\mathcal{A}_n = \{A_0, \ldots, A_n\}$, and by $K^n(\mathcal{A})$ its body. Clearly $\mathcal{K}^n(\mathcal{A}) \subset \mathcal{K}^{n+1}(\mathcal{A})$ for each n, and $\mathcal{K}(\mathcal{A}) = \bigcup_{n=0}^{\infty} \mathcal{K}^n$. More generally, for any multi-index $J = (j_0, j_1, \ldots, j_k) \in \mathbb{Z}_+^{k+1}$ (not necessarily belonging to $\mathcal{K}(\mathcal{A})$) we denote by

$$\mathcal{K}_J(\mathcal{A}) = \mathcal{K}(A_{j_0}, A_{j_1}, \ldots, A_{j_k})$$

the nerve of the indicated family of sets, and by $K_J(\mathcal{A})$ its body. Note that $\mathcal{K}_J(\mathcal{A})$ is a finite subcomplex of $\mathcal{K}(\mathcal{A})$ whose body is a k-dimensional simplex if and only if $J \in \mathcal{K}(\mathcal{A})$; otherwise it is a union of simplexes of lower dimension. Occasionally we shall write simply $\mathcal{K}$ instead of $\mathcal{K}(\mathcal{A})$ when it is clear from the context which family $\mathcal{A}$ is meant.

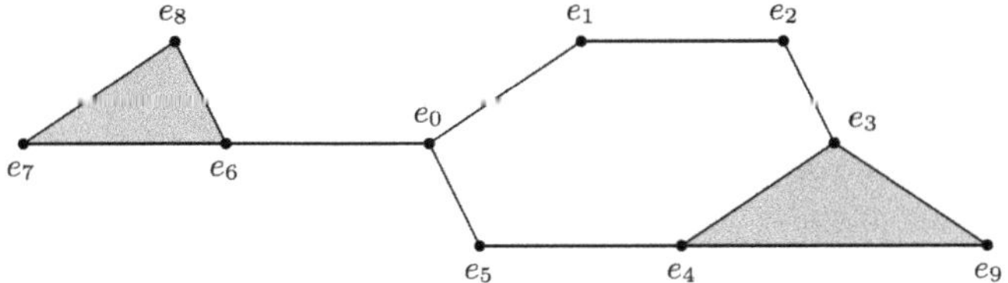

Fig. 6.3 The body $K(A_0, \dots, A_9)$

Example 6.8.2 Figure 6.3 shows the body $K(A_0, \dots, A_9)$ in which the nonempty intersections of two sets are shown by firm lines, and the nonempty triple intersections are $A_{3,4,9}$ and $A_{6,7,8}$.

Assume now that X is a locally compact Hausdorff space and $\mathcal{A} = \{A_0, A_1, A_2, \dots\}$ is a (finite or countable) locally finite family of compact subsets of X. An open neighborhood of $\mathcal{A}$ is a family $\mathcal{U} = \{U_0, U_1, U_2, \dots\}$ of open subsets of X such that $A_j \subset U_j$ for each index j. Such a neighborhood $\mathcal{U}$ is said to be *faithful* if $\mathcal{K}(\mathcal{U}) = \mathcal{K}(\mathcal{A})$, that is, the two families have the same nerve complex. It is easily seen that any locally finite family $\mathcal{A}$ of compact subsets admits an open faithful neighborhood. As before, for each multi-index $J = (j_0, j_1, \dots, j_k)$ we write

$$U^J = U_{j_0} \cup U_{j_1} \cup \cdots \cup U_{j_k}, \qquad U_J = U_{j_0} \cap U_{j_1} \cap \cdots \cap U_{j_k}.$$

We now explain how the body of a simplicial complex is used to parametrize a collection of sections. Assume that Z and X are complex spaces and $h\colon Z \to X$ is a holomorphic submersion.

Given an open set $U \subset X$, we denote by $\Gamma_{\mathcal{O}}(U, Z)$ (resp. $\Gamma(U, Z)$) the set of all holomorphic (resp. continuous) sections $f\colon U \to Z|_U$ of the submersion $h\colon Z \to X$ over the set U.

Definition 6.8.3 Assume that $h\colon Z \to X$ is a holomorphic submersion and $\mathcal{A} = \{A_0, A_1, A_2, \dots\}$ is a locally finite family of compacts in X.

1. A *holomorphic $\mathcal{K}(\mathcal{A})$-complex* with values in Z is a continuous family of holomorphic sections

$$f_* = \big\{ f_J : |J| \to \Gamma_{\mathcal{O}}(U_J, Z),\ J \in \mathcal{K}(\mathcal{A}) \big\},$$

where $\mathcal{U} = \{U_0, U_1, U_2, \dots\}$ is a faithful neighborhood of $\mathcal{A}$, which satisfy the following compatibility conditions:

$$I, J \in \mathcal{K}(\mathcal{A}),\ I \subset J \implies f_J(t) = f_I(t)|_{U_J}, \quad t \in |I|.$$

2. A *continuous $\mathcal{K}(\mathcal{A})$-complex* with values in Z is a continuous family

$$f_* = \big\{ f_J : |J| \to \Gamma(U_J, Z),\ J \in \mathcal{K}(\mathcal{A}) \big\}$$

satisfying the same compatibility conditions as in (i).

3. If f_* is a $\mathcal{K}(\mathcal{A})$-complex and $\mathcal{K}'$ is a subcomplex of $\mathcal{K}(\mathcal{A})$, we denote by $f_*|_{\mathcal{K}'}$ the restriction of f_* to $\mathcal{K}'$.
4. A $\mathcal{K}(\mathcal{A})$-complex f_*, defined on a faithful neighborhood $\mathcal{U} = \{U_j\}$ of $\mathcal{A}$, is *constant* if there is a section $g\colon \bigcup_j U_j \to Z$ such that $f_J(t)|_{U_J} = g|_{U_J}$ for each $J \in \mathcal{K}(\mathcal{A})$ and $t \in |J|$.

Thus, a complex of sections of $Z \to X$ is a family of sections, depending continuously on the parameter $t \in K(\mathcal{A})$, such that the domain of the section corresponding to a point $t \in K(\mathcal{A})$ is the union of all sets U_J for which $t \in |J|$; equivalently, it equals U_J where $J \in \mathcal{K}(\mathcal{A})$ is the shortest multi-index for which $t \in |J|$. For each $J = (j_0, \ldots, j_k) \in \mathcal{K}(\mathcal{A})$ we have a family of holomorphic sections $f_J(t)\colon U_J \to Z$, depending continuously on $t \in |J|$, which we consider as a homotopy of sections over the set $U_J = U_{j_0} \cap U_{j_1} \cap \cdots \cap U_{j_k}$, with the parameter t belonging to the simplex $|J| \subset K(\mathcal{A})$. For each face I of J and for every $t \in |I| \subset |J|$ the section $f_J(t)$ agrees with $f_I(t)$, restricted from its original domain U_I to the subdomain U_J.

We consider $\mathcal{K}(\mathcal{A})$-complexes in the sense of their germs: two $\mathcal{K}(\mathcal{A})$-complexes f_* and g_* are considered *equivalent* if there is an open faithful neighborhood $\mathcal{U} = \{U_j\}$ of $\mathcal{A} = \{A_j\}$ such that for each $J \in \mathcal{K}(\mathcal{A})$ and $t \in |J|$, the sections $f_J(t)$ and $g_J(t)$ are defined and equal in U_J.

We also need the notion of a multi-parameter homotopy of $\mathcal{K}(\mathcal{A})$-complexes. A suitable concept is the following.

Definition 6.8.4 Let $h\colon Z \to X$ and $\mathcal{A}$ be as in Definition 6.8.3 and let $k \in \mathbb{Z}_+$.

1. A *holomorphic* $(\mathcal{K}(\mathcal{A}), k)$*-prism*, or a k*-prism over* $\mathcal{K}(\mathcal{A})$, with values in Z is a family of continuous maps

$$f_* = \big\{ f_J\colon |J| \times [0,1]^k \to \Gamma_{\mathcal{O}}(U_J, Z),\ J \in \mathcal{K}(\mathcal{A}) \big\},$$

where $\mathcal{U}$ is a faithful open neighborhood of $\mathcal{A}$, such that for each fixed $y \in [0,1]^k$ the associated family

$$f_{*,y} = \big\{ f_{J,y} = f_J(\cdot, y) : |J| \to \Gamma_{\mathcal{O}}(U_J, Z),\ J \in \mathcal{K}(\mathcal{A}) \big\}$$

is a holomorphic $\mathcal{K}(\mathcal{A})$-complex.
2. A *continuous* $(\mathcal{K}(\mathcal{A}), k)$*-prism* with values in Z is a family

$$f_* = \big\{ f_J : |J| \times [0,1]^k \to \Gamma(U_J, Z),\ J \in \mathcal{K}(\mathcal{A}) \big\}$$

such that $f_{*,y}$ is a continuous $\mathcal{K}(\mathcal{A})$-complex for each $y \in [0,1]^k$.
3. A prism f_* is *sectionally constant* if there is an open set $U \subset X$ containing $\bigcup_{j\ge 0} A_j$ such that the complex $f_{*,y}$ is represented by a section $f_y : U \to Z|_U$ for each fixed $y \in [0,1]^k$. If this holds for all y in a subset $Y \subset [0,1]^k$, we say that f_* is sectionally constant on Y.

Thus, a $(\mathcal{K}(\mathcal{A}), 0)$-prism is a $\mathcal{K}(\mathcal{A})$-complex, a $(\mathcal{K}(\mathcal{A}), 1)$-prism is the same thing as a homotopy of $\mathcal{K}(\mathcal{A})$-complexes, a 2-prism is a homotopy of 1-prisms, etc. A sectionally constant $(\mathcal{K}(\mathcal{A}), k)$-prism is a k-parameter homotopy of sections over an open neighborhood of the union $\bigcup_j A_j$ of all the sets in $\mathcal{A}$.

6.9 $\mathcal{C}$-Strings

In order to be able to glue a collection of sections and homotopies between them in a holomorphic $\mathcal{K}(\mathcal{A})$-complex, the sets in $\mathcal{A} = (A_0, A_1, \ldots)$ must satisfy certain analytic conditions generalizing those of a $\mathcal{C}$-pair (Definition 6.7.1). The following notion was introduced by Gromov [266, 4.2.D'].

Definition 6.9.1 An ordered collection $(A_0, A_1, \ldots, A_n)$ of compact sets in a complex space X is a $\mathcal{C}$-*string of length* $n+1$ if

(i) $(A_0 \cup \cdots \cup A_{n-1}, A_n)$ is a $\mathcal{C}$-pair (see Definition 6.7.1 on p. 287), and
(ii) $(A_0, \ldots, A_{n-1})$ and $(A_0 \cap A_n, \ldots, A_{n-1} \cap A_n)$ are $\mathcal{C}$-strings of length n.

An ordered locally finite cover $\mathcal{A} = (A_0, A_1, \ldots)$ of X by compact sets is a $\mathcal{C}$-*cover* if $(A_0, A_1, \ldots, A_n)$ is a $\mathcal{C}$-string for each $n \in \mathbb{N}$.

The order of sets in a $\mathcal{C}$-string $(A_0, A_1, \ldots, A_n)$ is clearly important. Each of the sets A_k, and the union $A^k = \bigcup_{i=0}^k A_i$, is a Stein compact in X.

$\mathcal{C}$-strings enjoy the following hereditary property.

Proposition 6.9.2 *If $(A_0, A_1, \ldots, A_n)$ is a $\mathcal{C}$-string in X and if B is a Stein compact in X, then $(A_0 \cap B, \ldots, A_n \cap B)$ is also a $\mathcal{C}$-string in X.*

Proof We proceed by induction on the length $n \in \mathbb{N}$. Note that, if A and B are Stein compacts then so is $C = A \cap B$. For $n = 1$ we have a $\mathcal{C}$-pair (A_0, A_1), and we claim that $(A_0 \cap B, A_1 \cap B)$ is also a $\mathcal{C}$-pair. Clearly it satisfies properties (i) and (ii) in Definition 6.7.1. Property (iii) follows from the following lemma.

Lemma 6.9.3 *Let $D_0 \subset D_1$ and Ω be open Stein domains in a complex space X. If D_0 is Runge in D_1, then $D_0 \cap \Omega$ is Runge in $D_1 \cap \Omega$.*

Proof Choose a compact set $K \Subset D_0 \cap \Omega$. We denote by $\widehat{K}_D$ the holomorphically convex hull of K with respect to a domain $D \subset X$ containing K. Clearly, $\widehat{K}_{D_1 \cap \Omega} \subset \widehat{K}_{D_1} \cap \widehat{K}_\Omega$. Since D_0 is Runge in D_1 and both domains are Stein, we have $\widehat{K}_{D_0} = \widehat{K}_{D_1}$ (see [300, Theorem 2.7.3]). Therefore,

$$\widehat{K}_{D_1 \cap \Omega} \subset \widehat{K}_{D_1} \cap \widehat{K}_\Omega = \widehat{K}_{D_0} \cap \widehat{K}_\Omega \Subset D_0 \cap \Omega.$$

It follows that $D_0 \cap \Omega$ is Runge in $D_1 \cap \Omega$ (see [300, Theorem 2.7.3]). $\square$

This settles the case $n = 1$. Assume inductively that the result holds for some n. Let $(A_0, A_1, \ldots, A_{n+1})$ be a $\mathcal{C}$-string of length $n+2$. To see that $(A_0 \cap B, \ldots, A_{n+1} \cap B)$ is also a $\mathcal{C}$-string, we must verify the following:

- The pair of sets $(A_0 \cap B) \cup \cdots \cup (A_n \cap B) = (A_0 \cup \cdots \cup A_n) \cap B$ and $A_{n+1} \cap B$ is a $\mathcal{C}$-pair. Since $(A_0 \cup \cdots \cup A_n, A_{n+1})$ is a $\mathcal{C}$-pair, this follows from the case $n = 1$ proved above.
- $(A_0 \cap B, \ldots, A_n \cap B)$ and $(A_0 \cap A_{n+1} \cap B, \ldots, A_n \cap A_{n+1} \cap B)$ are $\mathcal{C}$-strings of length $n+1$. This follows immediately from Definition 6.9.1 and from the inductive hypothesis. □

Corollary 6.9.4 *If $\mathcal{A} = (A_0, A_1, A_2, \ldots)$ is a sequence of compact sets in a complex space X such that for each $n \in \mathbb{N}$ the pair $(A_0 \cup \cdots \cup A_{n-1}, A_n)$ is a $\mathcal{C}$-pair, then $(A_0, A_1, \ldots, A_n)$ is a $\mathcal{C}$-string for each $n \in \mathbb{N}$.*

Proof This follows from Proposition 6.9.2 by induction on n. □

Another hereditary property of $\mathcal{C}$-strings is the following; the straightforward proof is left to the reader.

Proposition 6.9.5 *If $(A_0, A_1, \ldots, A_n)$ is a $\mathcal{C}$-string in a complex space X then for each closed complex subspace X' of X, $(A_0 \cap X', \ldots, A_n \cap X')$ is a $\mathcal{C}$-string in X'.*

We now show that each Stein space admits arbitrarily fine $\mathcal{C}$-covers.

Proposition 6.9.6 *For each open cover $\mathcal{U} = \{U_j\}$ of a Stein space X there exists a $\mathcal{C}$-cover $\mathcal{A} = (A_0, A_1, \ldots)$ of X which is subordinate to $\mathcal{U}$, in the sense that each A_i is contained in some U_j. Moreover, if $K \subset X$ is a compact $\mathcal{O}(X)$-convex subset and $U_0 \subset X$ is an open set containing K, then we can choose $\mathcal{A}$ such that $K \subset A_0 \subset U_0$ and $A_i \cap K = \varnothing$ for all $i \ge 1$.*

Proof We consider first the case when X is a Stein manifold. The conditions imply that there is a strongly plurisubharmonic Morse exhaustion function $\rho \colon X \to \mathbb{R}$ with nice critical points (see Definition 3.10.2 on p. 95) such that $K \subset \{\rho < 0\} \Subset U_0$ and 0 is a regular value of ρ. Let $p_1, p_2, p_3, \ldots$ be the critical points of ρ in $\{\rho > 0\}$, ordered so that $0 < \rho(p_1) < \rho(p_2) < \cdots$. Choose a sequence of numbers $0 = c_0 < c_1 < c_2 < \cdots$, with $\lim_{j\to\infty} c_j = +\infty$, such that $c_{2j-1} < \rho(p_j) < c_{2j}$ for every $j = 1, 2, \ldots$. The numbers c_{2j-1} and c_{2j} can be chosen as close as desired to $\rho(p_j)$. If there are only finitely many critical points p_j, we choose the remainder of the sequence c_j arbitrarily.

Set $A_0 = \{\rho \le 0\}$. Since the set $\{\rho \le c_1\}$ is a noncritical strongly pseudoconvex extension of A_0, Lemma 5.10.3 provides a finite sequence of compact strongly pseudoconvex domains $A_0 \subset A'_1 \subset \cdots \subset A'_{m_1-1} = \{\rho \le c_1\}$ such that for every $k = 0, 1, \ldots, m_1 - 2$ we have $A'_{k+1} = A'_k \cup A_k$, where A_k is a special bump on A'_k contained in one of the sets $U_j \in \mathcal{U}$. Hence, (A'_k, A_k) is a $\mathcal{C}$-pair for each k, and $(A_0, \ldots, A_{m_1-1})$ is a $\mathcal{C}$-string by Corollary 6.9.4.

To pass the critical point p_1, we attach to the set $A'_{m_1-1} = \{\rho \le c_1\}$ a strongly pseudoconvex handle (a 'bone') A_{m_1-1}, described in [287, Sect. 2], such that (A'_{m_1-1}, A_{m_1-1}) is a $\mathcal{C}$-pair, and the sublevel set $\{\rho \le c_2\}$ is a noncritical strongly pseudoconvex extension of $A'_{m_1} = A'_{m_1-1} \cup A_{m_1-1}$ for a possibly different strongly plurisubharmonic function τ. (This means that the above sets are sublevel sets of τ, and τ has no critical points in between.) If c_1 and c_2 were chosen sufficiently close to $\rho(p_1)$ then A_{m_1-1} can be chosen small enough to be contained in a set $U_j \in \mathcal{U}$. Using the same procedure as above with τ (instead of ρ) we continue attaching small bumps to reach the next level $\{\rho \le c_2\}$; then we revert back to the original function ρ and continue attaching bumps to reach the level $\{\rho \le c_3\}$. Next we attach a bone to pass the critical level of ρ at p_2, etc. Corollary 6.9.4 shows that the resulting string $\mathcal{A} = (A_0, A_1, A_2, \ldots)$ obtained in this way is a $\mathcal{C}$-cover of X.

To do the same on a Stein space X, we choose an injective proper holomorphic map $\Phi\colon X \hookrightarrow \mathbb{C}^N$ which is a topological embedding. Using the above procedure, we find a $\mathcal{C}$-cover $(\widetilde{A}_0, \widetilde{A}_1, \ldots)$ of $\mathbb{C}^N$ such that the $\mathcal{C}$-string $(A_0, A_1, A_2, \ldots)$, where $A_k = X \cap \widetilde{A}_k$, is a $\mathcal{C}$-cover of X with the stated properties (see Corollary 6.9.5). $\square$

6.10 Construction of the Initial Holomorphic Complex

The following proposition provides a homotopy of complexes $a_{*,s}$ (a 1-prism) connecting the initial continuous section $a\colon X \to Z$ (which is thought of as a constant continuous complex $a_{*,0}$) to a certain holomorphic complex $a_{*,1}$.

Proposition 6.10.1 (The initial holomorphic complex) *Let X be a reduced Stein space, $K \subset X$ a compact $\mathcal{O}(X)$-convex subset, and $h\colon Z \to X$ a holomorphic submersion. Given a continuous section $a\colon X \to Z$ which is holomorphic in an open set $U_0 \supset K$, there exist a $\mathcal{C}$-cover $\mathcal{A} = (A_0, A_1, \ldots)$ of X and a continuous $(\mathcal{K}(\mathcal{A}), 1)$-prism $a_* = \{a_{*,s} : s \in [0, 1]\}$ with values in Z (see Definition* 6.8.4) *satisfying the following conditions*:

(i) $K \subset A_0 \subset U_0$, $K \cap A_i = \varnothing$ *for* $i \ge 1$, *and* $a_{(0),s} = a|_{U_0}$ *for* $s \in [0, 1]$,
(ii) $a_{*,0}$ *is the constant $\mathcal{K}(\mathcal{A})$-complex given by the section $a\colon X \to Z$, and*
(iii) $a_{*,1}$ *is a holomorphic $\mathcal{K}(\mathcal{A})$-complex.*

If the submersion $Z \to X$ is subelliptic over $X \setminus K$ (see Definition 6.1.2 *on p.* 264) *then $\mathcal{A}$ can be chosen such that, in addition to the above, there exists for each $j \ge 1$ an open set $U_j \supset A_j$ such that the restricted submersion $Z|_{U_j} \to U_j$ admits a finite dominating family of fibre sprays.*

Proof For simplicity of exposition we consider the case when X is a Stein manifold; the extension to a Stein space is immediate, either directly or by using local holomorphic embeddings of X into $\mathbb{C}^n$.

Let $N = \dim Z = n + m$, where $n = \dim X$ and m is the dimension of the fibres $h^{-1}(x)$, $x \in X$. Denote by P^N the unit open polydisc in $\mathbb{C}^N$ with the complex coordinates $\zeta = (\zeta', \zeta'')$, where $\zeta' \in \mathbb{C}^n$ and $\zeta'' \in \mathbb{C}^m$. Let $\pi: P^N \to P^n$ be the projection $\pi(\zeta', \zeta'') = \zeta'$.

Since $h: Z \to X$ is a submersion, there exist for each point $z_0 \in Z$ open neighborhoods $V \subset Z$ of z_0, $U \subset X$ of $x_0 = h(z_0)$, and biholomorphic maps $\Phi: V \to P^N$, $\phi: U \to P^n$, such that $\pi \circ \Phi = \phi \circ h$ on V and $\Phi(z_0) = 0$. Such Φ induces a linear structure on the fibres of $h|_V$ which lets us add sections of $h: V \to U$ and take their convex linear combinations.

If z_0 belongs to the graph of a, we can choose the above neighborhoods and maps such that $a(U) \subset V$. In this case, Φ maps $a(U)$ onto the graph of a section $\tilde{a}(\zeta') = (\zeta', a''(\zeta'))$ $(\zeta' \in P^n)$ of the projection $\pi: P^N \to P^n$. The family $a_s: U \to V$, given by

$$a_s(x) = \Phi^{-1}\big(\phi(x), (1-s)a''(\phi(x))\big), \quad x \in U, \ s \in [0,1],$$

is a homotopy of continuous sections of h over U such that $a_s(U) \subset V$ for each $s \in [0,1]$, $a_0 = a|_U$, and the section a_1 is holomorphic. By shrinking U around x_0 and replacing V by $V \cap h^{-1}(U)$ we may ensure in addition that the graph of the homotopy $\{a_s\}$ lies in a given open neighborhood of $\overline{a(U)}$.

Let $U_0 \subset X$ be an open set containing K such that a is holomorphic in U_0. Set $V_0 = h^{-1}(U_0) \subset Z$ and $a_{(0),s} = a|_{U_0}$ for $s \in [0,1]$. Using the above argument, we can cover the graph of a outside of V_0 by open neighborhoods $V_j \subset Z$ $(j = 1, 2, \ldots)$, biholomorphic to P^N and with $U_j = h(V_j) \subset X$ biholomorphic to P^n, such that for each $j \in \mathbb{N}$ we have a homotopy of continuous sections $a_{(j),s}: U_j \to V_j$ $(s \in [0,1])$ of h, satisfying the following conditions:

- $a_{(j),0} = a|_{U_j}$,
- the section $a_{(j),1}$ is holomorphic in U_j,
- for each $j \geq 1$, $U_j \cap K = \varnothing$ and $Z|_{U_j} \to U_j$ admits a finite dominating family of fibre sprays, and
- if $U_{(i,j)} = U_i \cap U_j \neq \varnothing$ then $a_{(i),s}(U_{(i,j)}) \subset V_j$ for each $s \in [0,1]$.

The last condition ensures that for each pair of indexes $i, j \in \mathbb{Z}_+$ such that $U_{(i,j)} \neq \varnothing$ there is a homotopy of sections $a_{(i,j),s}(t): U_{(i,j)} \to Z$, depending continuously on $t, s \in [0,1]$ and satisfying the following conditions:

- $a_{(i,j),s}(0) = a_{(i),s}|_{U_{(i,j)}}$,
- $a_{(i,j),s}(1) = a_{(j),s}|_{U_{(i,j)}}$,
- $a_{(i,j),0}(t) = a|_{U_{(i,j)}}$, and
- the section $a_{(i,j),1}(t)$ is holomorphic on $U_{(i,j)}$ for each $t \in [0,1]$.

We get $a_{(i,j),s}(t)$ by taking the convex linear combinations (with respect to $t \in [0,1]$) of sections $a_{(i),s}$ and $a_{(j),s}$, restricted to $U_{(i,j)}$. At least one of the indexes i, j is positive, say $i > 0$, and the convex combinations are taken by using the linear structure on the fibres of $h|_{V_i}$ induced by Φ_i.

Likewise, if $U_{(i,j,k)} \neq \varnothing$ for some $J = (i,j,k)$, we can use the linear structure on the fibres of one of the sets V_i, V_j, V_k to get a homotopy of sections $a_{J,s}(t): U_J \to Z$, with t belonging to the standard 2-simplex $\Delta^2 \subset \mathbb{R}^2$, whose restriction to the sides of the simplex equals the respective homotopy obtained in the previous step. Continuing in this way, we build a 1-prism a_* on the cover $\mathcal{U} = \{U_0, U_1, U_2, \ldots\}$ of X. By Proposition 6.9.6 there is a $\mathcal{C}$-cover $\mathcal{A} = (A_0, A_1, \ldots)$ of X subordinate to $\mathcal{U}$ such that $K \subset A_0 \subset U_0$. Then, a_* defines a $(\mathcal{K}(\mathcal{A}), 1)$-prism with the required properties. □

6.11 The Main Modification Lemma

The following result is the key step in the proof of Theorem 6.2.2.

Proposition 6.11.1 *Let $h\colon Z \to X$ be a holomorphic submersion of a reduced complex space Z onto a reduced complex space X, and let X' be a closed complex subvariety of X. Given a $\mathcal{C}$-string $(A_0, \ldots, A_n)$ in X with $A_1, \ldots, A_n \subset X \setminus X'$, we denote by $\mathcal{K}^n = \mathcal{K}(A_0, \ldots, A_n)$ its nerve. Assume that for each $i = 1, \ldots, n$ there is an open set $U_i \supset A_i$ in $X \setminus X'$ such that $h\colon Z|_{U_i} \to U_i$ admits a finite dominating family of fibre sprays. If f_* is a holomorphic $(\mathcal{K}^n, k)$-prism with values in Z which is sectionally constant (Definition 6.8.4) on a compact set $Y \subset [0, 1]^k$, then there is a homotopy f_*^u $(u \in [0, 1])$ of holomorphic $(\mathcal{K}^n, k)$-prisms such that the following conditions hold:*

(i) *$f_*^0 = f_*$ is the given prism,*
(ii) *the prism f_*^1 is sectionally constant,*
(iii) *the section $f_{(0),y}^u$ approximates $f_{(0),y}$ on A_0 as closely as desired uniformly in $y \in [0, 1]^k$ and $u \in [0, 1]$,*
(iv) *$f_{*,y}^u = f_{*,y}$ for all $y \in Y$ and $u \in [0, 1]$, and*
(v) *the homotopy $f_{*,y}^u$ is fixed to a given finite order along X'.*

If the restriction of f_* to the subcomplex $\mathcal{K}^{n-1} = \mathcal{K}(A_0, \ldots, A_{n-1})$ is sectionally constant, then f_*^u can be chosen such that, in addition to the above, the prism $f_*^u|_{\mathcal{K}^{n-1}}$ is sectionally constant for each $u \in [0, 1]$, and the corresponding sections $f_{*,y}^u|_{\mathcal{K}^{n-1}}$ $(y \in [0, 1]^k)$, which are holomorphic in a neighborhood of A^{n-1}, approximate $f_{*,y}|_{\mathcal{K}^{n-1}}$ uniformly on A^{n-1}.

Proof Replacing X by a Stein neighborhood of $A^n = \bigcup_{j=0}^n A_j$ we may assume that X is Stein. The proof goes by induction on $n \geq 0$, and for $n = 0$ there is nothing to prove.
The case $n = 1$. Our data consists of a $\mathcal{C}$-pair (A_0, A_1) in X, with $A_1 \subset X \setminus X'$, and a holomorphic $(\mathcal{K}(A_0, A_1), k)$-prism f_* which is sectionally constant on a compact set $Y \subset [0, 1]^k$. Such f_* is determined by the following data:

- a pair of open sets $U_0 \supset A_0$ and $U_1 \supset A_1$ with $U_1 \Subset X \setminus X'$,
- families of holomorphic sections

$$a_y = f_{(0),y}\colon U_0 \to Z|_{U_0}, \qquad b_y = f_{(1),y}\colon U_1 \to Z|_{U_1},$$

depending continuously on $y \in [0, 1]^k$, and

- a family of holomorphic sections

$$c_{y,t} = f_{(0,1),y}(t) : U_{(0,1)} = U_0 \cap U_1 \to Z|_{U_0 \cap U_1}$$

depending continuously on $y \in [0,1]^k$ and $t \in [0,1]$,

such that $a_y|_{U_{(0,1)}} = c_{y,0}$, $b_y|_{U_{(0,1)}} = c_{y,1}$, and for each $y \in Y$ the section $c_{y,t}$ is independent of $t \in [0,1]$. Hence, for $y \in Y$ the family $\{c_{y,t} : t \in [0,1]\}$ determines a holomorphic section $c_y : U_0 \cup U_1 \to Z$ such that $c_y|_{U_0} = a_y$ and $c_y|_{U_1} = b_y$. Write $f_* = (a_*, b_*, c_*)$, where $*$ indicates the missing parameters.

Our goal is to construct a homotopy $f_*^u = (a_*^u, b_*^u, c_*^u)$ $(u \in [0,1])$ of holomorphic $(\mathcal{K}(A_0, A_1), k)$-prisms over smaller open sets $U_0 \supset A_0$ and $U_1 \supset A_1$ such that $f_*^0 = f_*$ and f_*^1 is a constant prism, that is, a family of holomorphic sections $f_y^1 : U_0 \cup U_1 \to Z$. Furthermore, the homotopy must be fixed for $y \in Y$, it must approximate the sections a_y over A_0 for $y \in [0,1]^k$, and the homotopy a_*^u must be fixed to a finite order along the subvariety X'. We denote the data in the homotopy f_*^u by the same letters as above, adding the upper index u.

The homotopy f_*^u will be found in two steps. For convenience we use the parameter interval $u \in [0,2]$ and later rescale it to $[0,1]$. In the first step we apply the h-Runge approximation theorem (see Theorem 6.6.2 on p. 284) to obtain a homotopy $\{f_*^u\}_{u \in [0,1]}$ from $f_*^0 = f_*$ to another prism f_*^1 such that we do not move the section a_y (i.e., $a_y^u = a_y : U_0 \to Z$ for all u and y) and the section $b_y^1 : U_1 \to Z$ approximates a_y in a neighborhood of $A_0 \cap A_1$ for each $y \in [0,1]^k$. In the second step, we apply Proposition 6.7.2 on p. 288 to obtain homotopies of sections

$$a_y^u : U_0 \to Z, \quad b_y^u : U_1 \to Z, \quad c_{y,t}^u : U_{(0,1)} \to Z, \quad u \in [1,2]$$

such that at $u = 2$, $a_y^2 = b_y^2$ over $U_{(0,1)}$ for each $y \in [0,1]^k$; hence these two sections amalgamate to a holomorphic section $f_y^2 : U_0 \cup U_1 \to Z$.

Consider the first step. Since the set $A_{(0,1)} = A_0 \cap A_1$ is $\mathcal{O}(A_1)$-convex and the submersion $h : Z \to X$ admits a finite dominating family of fibre sprays over a neighborhood of A_1, the h-Runge approximation theorem (Theorem 6.6.2 on p. 284) gives (after shrinking the sets $U_0 \supset A_0$ and $U_1 \supset A_1$) a homotopy of holomorphic sections $g_{y,t}^s : U_{0,1} \to Z$ $(s \in [0,1])$, depending continuously on t, s, y and satisfying the following conditions:

- $g_{y,t}^0 = c_{y,t}$ for each y and t,
- $g_{y,1}^s = c_{y,1} = b_y|_{U_0 \cap U_1}$ for each s and y,
- $g_{y,t}^1$ extends to a holomorphic section over U_1 for each y and t,
- the homotopy is fixed on Y, i.e., for $y \in Y$ we have $g_{y,t}^s = c_y|_{U_0 \cap U_1}$ for each s and t, and
- $g_{y,t}^s$ approximates $c_{y,t}$ in a neighborhood of $A_{(0,1)}$ as closely as desired, uniformly with respect to all parameters.

We define $f_*^u = (a_*^u, b_*^u, c_*^u)$ for $u \in [0, 1]$ by

$$a_y^u = a_y, \quad b_y^u = g_{y,1-u}^1,$$

$$c_{y,t}^u = \begin{cases} c_{y,2t(1-u)} & \text{if } 0 \le t \le \frac{1}{2}; \\ g_{y,1-u}^{2t-1} & \text{if } \frac{1}{2} \le t \le 1. \end{cases}$$

This satisfies all required conditions. In particular, at $u = 1$, the section $b_y^1 = g_{y,0}^1$ approximates $c_{y,0} = a_y$ in a neighborhood of $A_{(0,1)}$.

We now apply Proposition 6.7.2 (p. 288) to get homotopies of sections $a_y^u \colon U_0 \to Z$ and $b_y^u \colon U_1 \to Z$ for $u \in [1, 2]$ such that a_y^u approximates $a_y^1 = a_y$ on A_0 for each $u \in [1, 2]$, the homotopy is fixed along X', and $a_y^2 = b_y^2$ in $U_{(0,1)}$. Moreover, over a neighborhood of the set A_0, the graphs of all sections a_y^u, b_y^u ($u \in [1, 2]$) and $c_{y,t}^1$ ($t \in [0, 1]$) lie in small tubular neighborhood of the image of the section a_y in Z. By using a vector bundle structure on such a neighborhood, we see that the triangle of homotopies formed by these families is contractible and can be filled by a 2-parameter homotopy $c_{y,t}^u$ ($t \in [0, 1]$, $u \in [1, 2]$) over a neighborhood of $A_{(0,1)}$. This proves Proposition 6.11.1 for $n = 1$.

The induction step $n \Rightarrow n + 1$. Suppose that Proposition 6.11.1 holds for all $\mathcal{C}$-strings of length $n + 1$ for some $n \ge 1$ and for all $k \ge 0$. Let $\mathcal{A} = (A_0, \dots, A_{n+1})$ be a $\mathcal{C}$-string of length $n + 2$ with the nerve $\mathcal{K}^{n+1} = \mathcal{K}(\mathcal{A})$, and let f_* be a holomorphic $(\mathcal{K}^{n+1}, k)$-prism with values in Z that is sectionally constant on a compact subset $Y \subset [0, 1]^k$. Let $\mathcal{K}^n = \mathcal{K}(A_0, \dots, A_n) \subset \mathcal{K}^{n+1}$. The proof consists of the following three steps, each of which is accomplished by constructing a suitable homotopy of prisms.

Step 1: Reduction to the case when $f_*|_{\mathcal{K}^n}$ is a sectionally constant prism.
Step 2: Reduction to the case when f_* represents a $(k + 1)$-prism over the $\mathcal{C}$-pair (A^n, A_{n+1}), where $A^n = A_0 \cup A_1 \cup \cdots \cup A_n$.
Step 3: Applying the case $n = 1$ to the prism in Step 2 to get a sectionally constant $(\mathcal{K}^{n+1}, k)$-prism.

We begin with some general considerations. We denote the coordinates on $\mathbb{R}^{n+1}$ by $t = (t', t_{n+1})$, where $t' = (t_1, \dots, t_n) \in \mathbb{R}^n$, and identify $\mathbb{R}^n$ with the coordinate hyperplane $\mathbb{R}^n \times \{0\} \subset \mathbb{R}^{n+1}$. The body K^{n+1} of the nerve $\mathcal{K}^{n+1}$ can be represented as the union of certain faces of the standard simplex $\Delta^{n+1} \subset \mathbb{R}^{n+1}$. (In fact, $K^{n+1} = \Delta^{n+1}$ if and only if $A_0 \cap A_1 \cap \cdots \cap A_{n+1} \neq \varnothing$.) The body $K^n \subset \mathbb{R}^n$ of the subcomplex $\mathcal{K}^n = \mathcal{K}(A_0, \dots, A_n) \subset \mathcal{K}^{n+1}$ is precisely the base $K^{n+1} \cap \{t_{n+1} = 0\}$ of K^{n+1}. We shall also need the complex

$$\mathcal{K}_1^n = \mathcal{K}(A_0 \cap A_{n+1}, \dots, A_n \cap A_{n+1}) \subset \mathcal{K}^n. \tag{6.13}$$

Note that $\mathcal{K}_1^n = \{J \in \mathcal{K}^n : (J, n + 1) \in \mathcal{K}^{n+1}\}$. Its body K_1^n is a subset of K^n which equals $\overline{(K^{n+1} \setminus K^n)} \cap K^n$. Moreover, for each $s \in [0, 1]$ the section $K^{n+1} \cap \{t_{n+1} = s\}$ is homeomorphic to K_1^n. The map

$$r \colon \mathbb{R}^{n+1} \to \mathbb{R}^{n+1}, \quad r(t, s) = \big(t(1 - s), s\big), \quad t \in \mathbb{R}^n, \ s \in \mathbb{R}$$

maps the prism $\Delta^n \times [0,1]$ onto the standard simplex Δ^{n+1} (it is homeomorphic outside the level $s=1$), and it maps $K_1^n \times \{s\}$ homeomorphically onto $K^{n+1} \cap \{t_{n+1}=s\}$ for each $s \in (0,1)$. (See Fig. 6.4 on p. 303.)

Proof of Step 1. Since $\tilde f_*^0 = f_*|_{\mathcal{K}^n} = \{f_J : J \in \mathcal{K}^n\}$ is a k-prism over a $\mathcal{C}$-string of length $n+1$, the induction hypothesis provides a holomorphic homotopy $\tilde f_* = \{\tilde f_*^u : u \in [-1,0]\}$ such that each $\tilde f_*^u$ is a $(\mathcal{K}^n, k)$-prism, the homotopy is fixed for all $y \in Y$, and the prism $\tilde f_*^{-1}$ is sectionally constant. The parameter space of the prism f_* is $K^{n+1} \times [0,1]^k$, while the parameter space of $\tilde f_*$ is $K' \times [0,1]^k$ where

$$K' = \big\{(t',u) \in \mathbb{R}^n \times \mathbb{R} : t' \in K^n,\ -1 \le u \le 0\big\}.$$

Note that f_* and $\tilde f_*$ agree on the intersection of their domains $K^n \times [0,1]^k$ and hence define a family of sections parametrized by the set $L \times [0,1]^k$, where $L = K^{n+1} \cup K' \subset \mathbb{R}^{n+1}$. We denote this family by $\{g_y(t) : t \in L,\ y \in [0,1]^k\}$.

For each $s \in [0,1]$ we denote by $L_s \subset \mathbb{R}^{n+1}$ the set

$$L_s = \big(K^{n+1} \setminus K^n\big) \cup \big\{(t', t_{n+1}) : t' \in K_1^n,\ -s \le t_{n+1} \le 0\big\} \cup \big\{(t',-s) : t' \in K^n\big\}.$$

Intuitively speaking, L_s is obtained by pushing the base K^n of K^{n+1} for s units in the negative t_{n+1} direction and adding the vertical sides $K_1^n \times [-s,0]$. Clearly, $L_0 = K^{n+1}$, and L_s is homeomorphic to K^{n+1} for each $s \in [0,1]$. In fact, there is a continuous family of homeomorphisms $\Theta_s : K^{n+1} \to L_s$ $(0 \le s \le 1)$ such that Θ_0 is the identity, each Θ_s preserves the top vertex $(0,\dots,0,1) \in K^{n+1}$ and the cellular structure of the two sets, and Θ_s maps K^n (the base of K^{n+1}) onto $K^n \times \{-s\}$ (the base of L_s) by a downward shift for s units. By 'respecting the cellular structure', we mean the following. Each face $J \in \mathcal{K}_1^n$ determines a face $\tilde J = (J, n+1) \in \mathcal{K}^{n+1}$, and Θ_s maps its body $|\tilde J| \subset K^{n+1}$ onto the set

$$|\tilde J| \cup \big\{(t', t_{n+1}) : t' \in |J|,\ -s \le t_{n+1} \le 0\big\} \subset L_s.$$

We define a homotopy H_*^u $(0 \le u \le 1)$ of $(\mathcal{K}^{n+1}, k)$-prisms

$$H_J^u(t) = g_y\big(\Theta_u(t)\big), \quad J \in \mathcal{K}^{n+1},\ t \in |J| \subset K^{n+1}.$$

Clearly, $H_*^0 = f_*$ and $H_*^1|_{\mathcal{K}^n} = \tilde f_*^{-1}|_{\mathcal{K}^n}$ is sectionally constant.

Proof of Step 2: By Step 1 we may assume that the prism f_* is such that $f_*|_{\mathcal{K}^n}$ is sectionally constant. The next step is to modify f_* by a homotopy of holomorphic prisms into another prism which is sectionally constant also in the direction of the last variable t_{n+1}. Let $\mathcal{K}_1^n$ be the complex (5.1). We associate to f_* a holomorphic $(\mathcal{K}_1^n, k+1)$-prism

$$F_* = \big\{F_{J,(y,s)} : |J| \to \mathcal{O}_h(U_{(J,n+1)}, Z),\ J \in \mathcal{K}_1^n,\ y \in [0,1]^k,\ s \in [0,1]\big\}$$

where

$$F_{J,(y,s)}(t) = f_{(J,n+1),y}\big(r(t,s)\big), \quad t \in |J|, \ y \in [0,1]^k, \ s \in [0,1].$$

Consider the following compact subset of $[0,1]^{k+1}$:

$$Y_1 = \big(Y \times [0,1]\big) \cup \big([0,1]^k \times \{0,1\}\big). \tag{6.14}$$

Since $f_*|_{\mathcal{K}^n}$ is sectionally constant, the prism F_* (which is associated to the complex $\mathcal{K}_1^n$ of length $n+1$) satisfies the induction hypothesis with respect to the set Y_1 (6.14). Hence, there is a homotopy F_*^u ($u \in [0,1]$) of holomorphic $(\mathcal{K}_1^n, k+1)$-prisms such that $F_*^0 = F_*$, the homotopy is fixed for $(y,s) \in Y_1$, and F_*^1 is sectionally constant. The latter condition means that for each fixed $(y,s) \in [0,1]^{k+1}$ the $\mathcal{K}_1^n$-complex $F_{*,(y,s)}^1$ is constant, so it represents a holomorphic section $F_y^1(s) \colon V \to Z|_V$ over an open set $V \supset A^n \cap A_{n+1}$. Since the homotopy is fixed for $s=0$ and $s=1$, $F_y^1(0)$ coincides with the section represented by the constant complex $f_{*,y}|_{\mathcal{K}^n}$, and $F_y^1(1)$ coincides with the section $f_{(0,\ldots,0,1),y}$ associated to A_{n+1}.

Proof of Step 3: We consider the family of sections obtained in Step 2,

$$F_*^1 = \big\{F_y^1(s) \colon V \to Z|_V : s \in [0,1], \ y \in [0,1]^k\big\},$$

as a holomorphic k-prism over the complex $\mathcal{K}' = \mathcal{K}(A^n, A_{n+1})$ determined by the $\mathcal{C}$-pair (A^n, A_{n+1}). The parameter s is the variable in the body $|\mathcal{K}'| = [0,1]$. For each $y \in [0,1]^k$ the section $F_y^1(0)$ extends holomorphically to a neighborhood of A^n, and $F_y^1(1)$ extends to a neighborhood of A_{n+1}.

The case $n=1$ of Proposition 6.11.1 gives a homotopy G_*^u ($u \in [1,2]$) of holomorphic $(\mathcal{K}', k)$-prisms such that $G_*^1 = F_*^1$, G_y^2 is a constant $\mathcal{K}'$-complex for each $y \in [0,1]^k$ (i.e., a holomorphic section over an open neighborhood of $A^{n+1} = A_0 \cup \cdots \cup A_{n+1}$), the homotopy is fixed for $y \in Y$ (where $G_y^1 = F_y^1$ is already a section over A^{n+1}), it is fixed to a finite order along X', and the section $G_y^u(0)$ (which is holomorphic over a neighborhood of A^n) approximates the section $F_y^1(0) = f_{*,y}|_{\mathcal{K}^n}$ on the set A^n for each $u \in [1,2]$.

The collection $\{f_*^u : u \in [0,1]\}$ defined by

$$f_*^u = \begin{cases} F_*^{2u} & \text{if } u \in [0, \frac{1}{2}]; \\ G_*^{2u} & \text{if } u \in [\frac{1}{2}, 1], \end{cases}$$

is a homotopy of holomorphic $(\mathcal{K}^{n+1}, k)$-prisms, beginning at $u=0$ with f_* and ending at $u=1$ with the sectionally constant prism G_*^2.

If we assume in addition that the restriction $f_*|_{\mathcal{K}^n}$ is sectionally constant on $[0,1]^k$ (so $f_{*,y}|_{\mathcal{K}^n}$ is a holomorphic section in a neighborhood of A^n for each $y \in [0,1]^k$), then we can skip the initial step in the proof of the inductive step. By the construction, the restriction $F_*^u|_{\mathcal{K}^n}$ is independent of $u \in [0,1]$ since the homotopy F_*^u is fixed on the set Y_1 (6.14), and the homotopy G_*^u is such that the complex $G^u_{*,y}|_{\mathcal{K}^n}$ is represented by a holomorphic section in a neighborhood of A^n which approximates the section $F^1_{*,y}|_{\mathcal{K}^n} = f_{*,y}|_{\mathcal{K}^n}$ on A^n, uniformly with respect to $u \in [0, 1]$ and $y \in [0, 1]^k$. Hence, the section $f^u_{*,y}|_{\mathcal{K}^n}$ approximates $f_{*,y}|_{\mathcal{K}^n}$ on A^n, the approximation being uniform with respect to $u \in [0, 1]$ and $y \in [0, 1]^k$. This concludes the induction step and thereby completes the proof of Proposition 6.11.1. □

The next proposition shows that a holomorphic 1-prism can be extended from a finite subcomplex to the entire complex such that the 0-level of the prism matches a given complex. This does not require any analytic tools, and hence the result applies to any locally finite family $\mathcal{A}$.

Proposition 6.11.2 *Assume that $h\colon Z \to X$ is a holomorphic submersion onto a complex space X. Let $\mathcal{A} = \{A_0, A_1, \ldots\}$ be a locally finite family of compact sets in X. Denote its nerve by $\mathcal{K}(\mathcal{A})$, and let $\mathcal{K}^n = \mathcal{K}(A_0, \ldots, A_n) \subset \mathcal{K}(\mathcal{A})$ for each $n \in \mathbb{N}$. If f_* is a holomorphic $\mathcal{K}(\mathcal{A})$-complex with values in Z, and if g_* is a holomorphic $(\mathcal{K}^n, 1)$-prism for some $n \in \mathbb{N}$ such that $g_{*,0} = f_*|_{\mathcal{K}^n}$, there exists a holomorphic $(\mathcal{K}(\mathcal{A}), 1)$-prism G_* such that $G_{*,0} = f_*$ and $G_*|_{\mathcal{K}^n} = g_*$. Similarly, if f_* is a holomorphic $(\mathcal{K}(\mathcal{A}), k)$-prism and g_* is a holomorphic $(\mathcal{K}^n, k+1)$-prism with the base $f_*|_{\mathcal{K}^n}$, then g_* extends to a holomorphic $(\mathcal{K}(\mathcal{A}), k+1)$-prism G_* such that $G_{*,0} = f_*$.*

Proof We choose representatives of f_* and g_* defined on a faithful open neighborhood $\mathcal{U}$ of $\mathcal{A}$ (see Definition 6.8.3). Write $A^n = A_0 \cup \cdots \cup A_n$ as before. Let $m \geq n$ be the smallest integer such that $A_k \cap A^n = \varnothing$ for all $k \geq m$; such m exists since the family $\{A_j\}$ is locally finite. We represent the body $K^m = K(A_0, \ldots, A_m)$ of the subcomplex $\mathcal{K}^m \subset \mathcal{K}(\mathcal{A})$ as a subset of $\mathbb{R}^m$. Denote the coordinates on $\mathbb{R}^{m+1}$ by (t, s), with $t \in \mathbb{R}^m$ and $s \in \mathbb{R}$, and identify $\mathbb{R}^m$ with $\mathbb{R}^m \times \{0\} = \{s = 0\} \subset \mathbb{R}^{m+1}$. Similarly, we identify a set $K \subset \mathbb{R}^m$ with $K \times \{0\} \subset \mathbb{R}^{m+1}$ and write $K \times [0, 1] = \{(t, s) : t \in K,\ s \in [0, 1]\}$. For each face $J \in \mathcal{K}^m$ we denote by $b|J| \subset K^m$ the boundary of its body $|J|$.

We need the following lemma.

Lemma 6.11.3 *There exists a retraction*

$$r\colon K^m \times [0, 1] \to K^m \cup \big(K^n \times [0, 1]\big) \subset \mathbb{R}^{m+1}$$

such that for each face $J \in \mathcal{K}^m \setminus \mathcal{K}^n$ we have

(i) *$r(|J| \times [0, 1]) \subset |J| \cup (b|J| \times [0, 1])$, and*
(ii) *if $|J| \cap K^n = \varnothing$ then $r(t, s) = t$ for each $t \in |J|$ and $s \in [0, 1]$.*

Proof The retraction r is shown on Fig. 6.4. We first define r over those faces $J \in \mathcal{K}^m$ for which either $|J| \subset K^n$ (in such case we let r be the identity on $|J| \times [0, 1]$), or $|J| \cap K^n = \varnothing$ (we let $r(t, s) = t$ for $t \in |J|$). We also define r to be the identity map on the bottom side $K^m = K^m \times \{0\}$. On the remaining

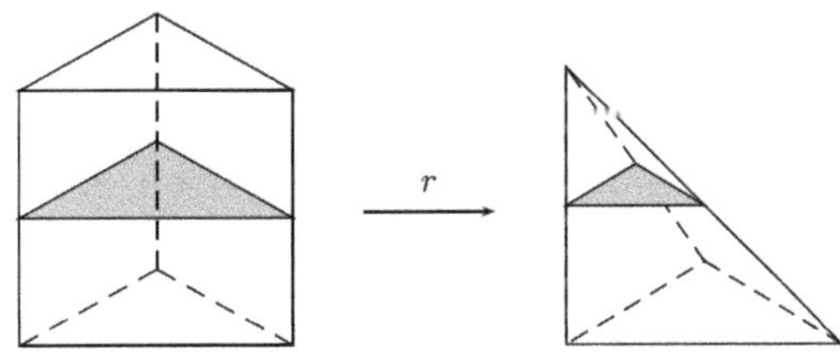

Fig. 6.4 Retracting a prism onto a simplex

faces $|J|$, $J \in \mathcal{K}^m$, we define r inductively with respect to the dimension of J. Suppose that r has already been defined on all faces of dimension $< k$ and let $J = (j_0, \ldots, j_k) \in \mathcal{K}^m$. Then r is already defined on $|J| \cup (b|J| \times [0,1])$ and it satisfies (i); it also satisfies (ii) on those sides of $b|J|$ which are disjoint from K^n. Moreover, r is the identity on $|J| = |J| \times \{0\}$. It is now clear that r extends from $|J| \cup (b|J| \times [0,1])$ to $|J| \times [0,1]$ so that (i) holds. □

Let r be as in Lemma 6.11.3. Write $r(t,s) = (r_0(t,s), u(t,s))$, where $r_0(t,s) \in K^m$ and $u(t,s) \in [0,1]$. We define a holomorphic $(\mathcal{K}^m, 1)$-prism G_* by setting for each $J \in \mathcal{K}^m$, $t \in |J|$ and $s \in [0,1]$:

$$G_{J,s}(t) = \begin{cases} f_J\big(r_0(t,s)\big) & \text{if } u(t,s) = 0; \\ g_{J,u(t,s)}\big(r_0(t,s)\big) & \text{if } u(t,s) > 0. \end{cases}$$

Condition (i) in Lemma 6.11.3 implies that the section $G_{J,s}(t)$ for $t \in |J|$ is defined and holomorphic in the set U_J. (It may even be holomorphic in a larger set if $r_0(t,s) \in b|J|$, but in such case we restrict it to U_J.) The family $G_* = \{G_{J,s} : J \in \mathcal{K}^m,\ s \in [0,1]\}$ is then a holomorphic $(\mathcal{K}^m, 1)$-prism which extends g_* and satisfies $G_{*,0} = f_*$. Property (ii) of the retraction r lets us extend G_* to a prism over the entire complex $\mathcal{K}(\mathcal{A})$ by observing that for those faces $J \in \mathcal{K}(\mathcal{A})$ which do not belong to $\mathcal{K}^m$ we have $|J| \cap K^n = \varnothing$ (by the definition of m), and therefore $r(t,s) = t$ for $t \in |J| \cap K^m$. Thus, we can simply take $G_{J,s}(t) = f_J(t)$ for $t \in |J|$ and $s \in [0,1]$. □

6.12 Proof of Theorems 6.2.2 and 6.6.6

We begin with the basic case when $Z \to X$ is a subelliptic submersion onto a Stein space X and $X' = \varnothing$ (i.e., without the interpolation condition). For simplicity of exposition we concentrate on a single section; the proof in the parametric case is an immediate extension since all tools have been prepared in the required generality. Given a compact $\mathcal{O}(X)$-convex subset K of X and a continuous section $a \colon X \to Z$ which is holomorphic in an open set $U_0 \supset K$, our goal is to find a homotopy $H_s \colon X \to Z$ ($s \in [0,1]$) of continuous sections such that $H_0 = a$, the section H_1 is holomorphic on X, and for each $s \in [0,1]$ the section H_s is holomorphic near K and it approximates a on K.

Let $\mathcal{A} = \{A_0, A_1, \ldots\}$ be a $\mathcal{C}$-cover of X such that $K \subset A_0 \subset U_0$ and $K \cap A_i = \varnothing$ for all $i = 1, 2, \ldots$ (see Proposition 6.9.6 on p. 294). Let $a_* = \{a_{*,s}\}_{s\in[0,1]}$ be a continuous $(\mathcal{K}(\mathcal{A}), 1)$-prism furnished by Proposition 6.10.1 (p. 295). The 0-level complex $a_{*,0}$ agrees with the initial section a, the final complex $a_{*,1}$ is a holomorphic, and $a_{(0),s} = a|_{U_0}$ for each $s \in [0, 1]$.

Let dist be a complete distance function on Z. Fix an $\epsilon > 0$. We construct a sequence of holomorphic $\mathcal{K}(\mathcal{A})$-complexes f_*^n and a sequence of holomorphic $(\mathcal{K}(\mathcal{A}), 1)$-prisms $G_*^n = \{G_{*,s}^n\}_{s\in[0,1]}$ $(n = 0, 1, \ldots)$ satisfying the following:

(a) $f_*^0 = a_{*,1}$ is the initial holomorphic complex,
(b) $G_{*,0}^n = f_*^n$ and $G_{*,1}^n = f_*^{n+1}$ for each $n \in \mathbb{Z}_+$ (hence G_*^n is a homotopy of complexes connecting f_*^n and f_*^{n+1}),
(c) for each $n, k \in \mathbb{Z}_+$, $n \ge k$, and each $s \in [0, 1]$ the complexes $f_*^n|_{\mathcal{K}^k}$ and $G_{*,s}^n|_{\mathcal{K}^k}$ are constant, i.e., they are given by holomorphic sections f^n, resp. G_s^n, in an open neighborhood of $A^k = A_0 \cup \cdots \cup A_k$, and
(d) for each $n \in \mathbb{Z}_+$ and $s \in [0, 1]$ we have

$$\operatorname{dist}\big(G_s^n(x), f^n(x)\big) < \epsilon 2^{-n-1}, \quad x \in A^n.$$

In particular, $\operatorname{dist}(f^{n+1}(x), f^n(x)) < \epsilon 2^{-n-1}$ for $x \in A^n$. (Here we are using the notation for sections established in property (c) above.)

Condition (d) implies that the sequence of holomorphic sections $f^n : A^n \to Z$ converges, uniformly on compacts in X, to a holomorphic section $f^\infty = \lim_{n\to\infty} f^n : X \to Z$ satisfying

$$\operatorname{dist}\big(f^\infty(x), a(x)\big) = \operatorname{dist}\big(f^\infty(x), f^0(x)\big) < \epsilon, \quad x \in A_0.$$

To find a homotopy $H_s : X \to Z$ $(s \in [0, 1])$ with $H_0 = a$ and $H_1 = f^\infty$, we first construct a continuous $(\mathcal{K}(\mathcal{A}), 1)$-prism h_* with $h_{*,0} = a$ and $h_{*,1} = f^\infty$. It suffices to collect all individual 1-prisms a_* and G_*^n $(n \in \mathbb{Z}_+)$ into a single 1-prism as follows. For each $n \in \mathbb{Z}_+$, set $I_n = [1 - 2^{-n}, 1 - 2^{-n-1}]$ and let $\lambda_n : I_n \to [0, 1]$ be the linear bijection $\lambda_n(s) = 2^{n+1}(s - 1 + 2^{-n})$. Then, $\bigcup_{n=0}^\infty I_n = [0, 1)$. For $s \in [0, 1]$ we define

$$h_{*,s} = \begin{cases} a_{*,2s} & \text{if } s \in I_0 = [0, \frac{1}{2}]; \\ G_{*,\lambda_n(s)}^{n-1} & \text{if } s \in I_n,\ n \ge 1; \\ f^\infty & \text{if } s = 1. \end{cases}$$

The two definitions of $h_{*,s}$ at the values $s = 1 - 2^{-n}$ are compatible by (b).

Conditions (c) and (d) imply that $\lim_{s\to 1} h_{*,s} = f^\infty$, uniformly on compacts in X. Indeed, each compact set $L \Subset X$ is contained in some A^m, and for $n \ge m$ the complex $G_{*,s}^n$ is constant on A^m, i.e., it is represented there by a holomorphic section. Hence, for $1 - 2^{-n-1} \le s < 1$ the complex $h_{*,s}|_{\mathcal{K}^n}$ is a holomorphic section in a neighborhood of A^n, and these sections converge uniformly on compacts to f^∞ when $s \to 1$. This proves that $h_* = \{h_{*,s}\}_{s\in[0,1]}$ is a continuous $(\mathcal{K}(\mathcal{A}), 1)$-prism. Note also that the restriction of h_* to the trivial subcomplex $\mathcal{K}(A_0)$ is a homotopy of holomorphic sections h_s ($s \in [0,1]$) in a neighborhood of A_0, connecting $h_0 = a$ to $h_1 = f^\infty$, such that

$$\mathrm{dist}\big(h_s(x), a(x)\big) < \epsilon, \quad x \in A_0,\ s \in [0,1].$$

To complete the proof, we apply Proposition 6.11.1 (p. 297) to modify the 1-prism h_* by a homotopy of 1-prisms (keeping the ends $s=0$ and $s=1$ fixed) into a 1-prism H_* which is *sectionally constant*, i.e., such that H_* represents a homotopy of continuous sections $H_s\colon X \to Z$ ($s \in [0,1]$). Moreover, we can achieve that H_s and h_s agree in a neighborhood of A_0.

This concludes the proof of Theorem 6.2.2 in the basic case.

The parametric case is proved by the same tools by introducing the parameter space P into the definition of holomorphic (and continuous) complexes and prisms and repeating the same arguments in this setting.

It remains to consider the stratified case; this will also yield Theorem 6.2.3. We follow the scheme of proof of Theorem 5.4.4. Given a stratification (6.4), we replace each X_k by $X_k \cup X'$ and get another stratification

$$X = X_0 \supset X_1 \supset \cdots \supset X_m = X'$$

with regular strata $S_j = X_j \setminus X_{j+1}$ such that $Z|_{S_j}$ is a subelliptic submersion for $j = 0, \ldots, m-1$.

In a typical induction step, we have compact $\mathcal{O}(X)$-convex sets $K \subset L$ in X and a section $f\colon X \to Z$ which is holomorphic in an open neighborhood of K and whose restriction to the subvariety X' is holomorphic. (For the jet interpolation case, f is holomorphic in a neighborhood of $K \cup X'$.) By a homotopic deformation which is fixed on X' we must find a new section that is holomorphic on a neighborhood of the bigger set L.

As in Sect. 5.12 we inductively extend the section across the strata $X_k \setminus X_{k+1}$. (We replace every stratum by its intersection with the compact subset L, so there are only finitely many connected components to consider.) By the inductive hypothesis, we have a section $f\colon X \to Z$ that is holomorphic in an open set $U \subset X$ containing $K \cup (L \cap X_{k+1})$. Choose a $\mathcal{C}$-string $\mathcal{A} = (A_0, A_1, \ldots, A_n)$ in X such that

- $K \cup (L \cap X_{k+1}) \subset A_0 \subset U$,
- the sets $A_1, \ldots, A_n$ are contained in the stratum S_k, and
- $K \cup (L \cap X_k) \subset K \cup (\bigcup_{j=1}^n A_j)$.

The set A_0 is chosen as the union of $K \cup (L \cap X_{k+1})$ and a small collar of this set inside the stratum $S_k = X_k \setminus X_{k+1}$. Each of the sets $A_1, \ldots, A_n$ is a convex bump or a handle contained in S_k and attached to the previous set so that they form a $\mathcal{C}$-pair (see the *general case* in the proof of Proposition 5.12.1). Then $\mathcal{A}$ is a $\mathcal{C}$-string by Corollary 6.9.4 on p. 294.

Proposition 6.10.1 (p. 295) furnishes a holomorphic $\mathcal{K}(\mathcal{A})$-complex in which f is the holomorphic section associated to the initial set A_0.

By following the proof of the basic case of Theorem 6.2.2 and using Proposition 6.11.1, we obtain a new section $\widetilde{f}\colon X \to Z$, homotopic to f relative to the subvariety X_{k+1}, which is holomorphic over a neighborhood of A_0 in X, and also over a relative neighborhood of $\bigcup_{j=1}^n A_j$ in the stratum S_k. This is accomplished in finitely many moves. Finally, by Theorem 3.4.1 on p. 75 we can deform $\widetilde{f}$ to a section that is holomorphic in an open neighborhood of the set $\bigcup_{j=0}^n A_j$ in X, thereby completing the induction step. The remaining details are exactly as in Sect. 5.12.

This complete the proof of Theorems 6.2.2, 6.2.3 and 6.6.6.

6.13 Relative Oka Principle on 1-Convex Manifolds

The notion of a 1-convex complex manifold is given by Definition 3.1.3 on p. 66; see also Remark 3.1.4. The following result is due to Prezelj [461, 462].

Theorem 6.13.1 ([461, Theorem 1.1]) *Let X be a 1-convex complex manifold with the exceptional variety S, Z be a complex manifold, and $h\colon Z \to X$ be a holomorphic submersion whose restriction $h\colon Z\setminus h^{-1}(S) \to X\setminus S$ is a stratified subelliptic submersion (see Definition 6.2.1), or a stratified fibre bundle with Oka fibres (see Definition 1.4.1). Let K be a compact $\mathcal{O}(X)$-convex subset of X containing S. Given a continuous section $f\colon X \to Z$ that is holomorphic in a neighborhood of K, there exists a homotopy of sections $f_t\colon X \to Z$ ($t \in [0,1]$) such that the following conditions hold:*

(i) *$f_t(x) = f_0(x)$ for all $x \in S$ and $t \in [0,1]$,*
(ii) *for every $t \in [0,1]$ the sections f_t is holomorphic near K and it approximates f_0 uniformly on K as closely as desired, and*
(iii) *the section f_1 is holomorphic on X.*

The analogous result holds for families of sections depending continuously on a parameter in a compact Hausdorff space (compare with Theorem 6.2.2).

When $h\colon Z \to X$ is a principal holomorphic fibre bundle with a Lie group fibre, this relative Oka principle is due to Henkin and Leiterer [287] and Colţoiu [104] (when X is nonsingular), and to Leiterer and Vâjâitu [387] for 1-convex spaces with singularities. This classical case implies the relative Oka-Grauert principle concerning the classification of holomorphic vector bundles over 1-convex spaces (see Corollary 8.3.11 on p. 363).

The proof of the general case of Theorem 6.13.1 is much more subtle. Compared to the proof of Theorem 6.2.2, the main new difficulty is to construct local dominating fibre sprays around a given section in the presence of a compact subvariety S of the base manifold X. Two key new results are developed for this purpose: (i) the construction of conical Stein neighborhoods in Z of the graph of a holomorphic section over the complement of a subvariety of X containing its exceptional subvariety (see [461] and Theorem 3.2.3 on p. 73), and (ii) obtaining a solution with estimates of the $\overline{\partial}$-equation for $(0,1)$-forms on conical Stein neighborhoods; see [462]. (Although the main theorem in [462] is stated and proved for the case when the exceptional variety S is smooth, the proof given there holds without this additional assumption.)

Proof (sketch) We follow the proof of Theorem 6.2.2.

Recall from Sect. 6.7 that, for the purpose of gluing a pair of holomorphic sections over a Cartain pair (A, B) in X, the initial holomorphic section $f: U \subset X \to Z$ defined on a neighborhood U of the set A is embedded as a core of a local dominating holomorphic fibre spray. If X is Stein, such a spray is furnished by Lemma 5.10.4 on p. 243; this depends on Siu's theorem on the existence of a Stein neighborhood of the section (see Theorem 3.1.1 on p. 66).

If X is only 1-convex, then the graph $f(U) \subset Z$ does not have any Stein neighborhood in Z if U contains a compact subvariety of X. To overcome this problem, choose a complex hypersurface $X' \subset X$ containing the exceptional subvariety S of X and a conical Stein neighborhood $\Omega \subset Z$ of $f(U \setminus X')$ constructed in [461, Theorem 3.2] (see Theorem 3.2.3 on p. 73). On such domain Ω, there are finitely many vertical (tangent to VTZ) holomorphic vector fields which span the vertical tangent space at every point and vanish to a high order along the edge $f(U \cap X')$ of the cone Ω. This highly nontrivial construction, which was explained in Prezelj's second paper on this subject [462], uses precise estimates for the solution of the $\overline{\partial}$-equation on conical Stein domains. Granted such vector fields, the vanishing property along the edge $f(U \cap X')$ implies that their flows exist for each initial point in a finer cone $\Omega' \subset \Omega$ and for the time variable in a fixed neighborhood of the origin in $\mathbb{C}$. The composition of their local flows then gives a local holomorphic spray of sections with the core f that is fixed along the subvariety $X' \cap U$ and is dominating over $U \setminus X'$. Finitely many such sprays (with respect to different conical neighborhoods) are dominating over $U \setminus S$. This enables the gluing of the given spray with another local spray, provided that the intersection of their domains in X does not meet the exceptional set S. The proof then proceeds exactly as in Theorem 6.2.2. □

6.14 The Oka Principle for Sections of Branched Maps

We explain a version of the Oka principle, obtained in [192], which concerns multi-valued sections of branched holomorphic maps over Stein spaces; see Theorem 6.14.4. We outline the main ideas without the details.

Let X and Z be reduced complex spaces and $h: Z \to X$ be a surjective holomorphic map. A point $z \in Z$ is a *branch point* if h fails to be a submersion in any neighborhood of z (see Definition 1.3.1). The set $\mathrm{br}_h \subset Z$ of all branch points is the *branch locus* of the map h. Note that a smooth point $z \in Z_{\mathrm{reg}}$ belongs to br_h if and only if the differential $dh_z: T_z Z \to T_{h(z)} X$ fails to be surjective, and this holds precisely when there is no local $\mathcal{C}^1$ section of h passing through z. (Indeed, if f is a local $\mathcal{C}^1$ section, we see from $h(f(x)) = x$ that dh_z maps the tangent space to the graph of f at z onto $T_{h(z)} X$, and hence h is a submersion at z.) The following simple example is illuminative.

Example 6.14.1 Let $p, q \in \mathbb{N}$ be a pair of coprime integers. The function $h: \mathbb{C}^2 \to \mathbb{C}$, $h(z, w) = z^p w^q$, is branched at $(0, 0)$. Pick integers $p', q' \in \mathbb{Z}$ with

$pp' + qq' = 1$. The map $f\colon \mathbb{C} \to \mathbb{C}^2$, $f(re^{i\theta}) = (r^{1/2p}e^{ip'\theta}, r^{1/2q}e^{iq'\theta})$ is a Hölder continuous section of h, but clearly h does not admit any local $\mathcal{C}^1$ section with $f(0) = (0,0)$. □

Natural objects associated to branched maps are *multi-valued sections*. Let $d \in \mathbb{N}$. A d-valued section of $Z \to X$ assigns to every point $x \in X$ a d-tuple of points in the fibre $Z_x = h^{-1}(x)$. To discuss them properly, we use the notion of a symmetric product; see [574, Appendix V] and [192, Sect. 4] for the details.

Let Z^d denote the d-fold Cartesian power of a set Z. The group Π_d of all permutations on d elements acts on Z^d by permuting the entries; denote this action by ρ. The quotient space $Z^d/\sigma = Z^d_{\mathrm{sym}}$ is the *d-fold symmetric power* of Z. For $z = (z_1, \ldots, z_d) \in Z^d$ we write

$$\pi(z_1, \ldots, z_d) = [z_1, \ldots, z_d] \in Z^d_{\mathrm{sym}}.$$

A *d-valued map* from X to Z is a map $F\colon X \to Z^d_{\mathrm{sym}}$, and $d = \deg F$ is the *degree* of F. Such F is said to be *unbranched* at $x \in X$ if its restriction to an open neighborhood $U \subset X$ of x is a union of d single valued maps. We denote by br_F the *branch locus* of F, i.e. the set of all points $x \in X$ such that F is not a union of single valued maps in any neighborhood of x.

Assume that X and Z are reduced complex spaces. The quotient projection $\pi\colon Z^d \to Z^d_{\mathrm{sym}}$ induces a natural complex structure on Z^d_{sym} such that holomorphic functions on Z^d_{sym} correspond to ρ-invariant holomorphic functions on Z^d. In particular, if $F = [f_1, \ldots, f_d]\colon X \to Z^d_{\mathrm{sym}}$ is a holomorphic map and P is a ρ-invariant holomorphic function on Z^d, then $P(f_1, \ldots, f_d)$ is a well defined holomorphic function on X.

Similarly, if $h\colon Z \to X$ is a holomorphic map of complex spaces, then a holomorphic (resp. continuous) *d-valued section* of h is a holomorphic (resp. a continuous) map $F\colon X \to Z^d_{\mathrm{sym}}$ such that $F(x) \subset h^{-1}(x)$ for each $x \in X$.

Let $\#F(x)$ denote the number of distinct points in $F(x)$. Set

$$\mu_F = \max\{\#F(x) : x \in X\} \le d, \qquad \delta_F = \{x \in X : \#F(x) < \mu_F\}.$$

The set δ_F is the *discriminant locus* of F. Clearly, $\mathrm{br}_F \subset \delta_F$ and both sets are closed when F is continuous. If F is holomorphic, then br_F and δ_F are closed complex subvarieties of X.

The following result is due to Whitney [574, Appendix V]. The notation $\mathcal{H}^k(A) = 0$ means that the set A has the k-dimensional Hausdorff measure zero; this notion makes sense for subsets of analytic spaces.

Proposition 6.14.2 ([574, Appendix V]) *Assume that $F\colon X \to Z^d_{\mathrm{sym}}$ is a continuous (resp. a holomorphic) map such that δ_F is nowhere dense in X and $X \setminus \delta_F$ is pathwise connected and locally pathwise connected. Then, F has a unique decomposition $F = \oplus m_j F_j$ (up to the ordering of terms) where the continuous (resp. holomorphic) maps $F_j\colon X \to Z^{d_j}_{\mathrm{sym}}$ are irreducible. Furthermore, $\mathrm{br}F_j \subset \mathrm{br}F$ and*

$\delta_{F_j} \subset \delta_F$ hold for each j. Such a decomposition exists in particular if X is an irreducible and locally irreducible n-dimensional complex space and $\mathcal{H}^{2n-1}(\delta_F) = 0$.

An *analytic chain* in Z is a locally finite combination $V = \sum m_j V_j$ of closed complex subvarieties $V_j \subset Z$ with integer coefficients $m_j \in \mathbb{Z}$. We have $[V] = \sum_j m_j [V_j]$ in the sense of currents. The chain V is called *effective* if $m_j \geq 0$ for all j. The following result shows that multi-valued holomorphic sections of $Z \to X$ correspond to effective analytic chains V in Z such that $h|_V \colon V \to X$ is a proper finite projection onto X, i.e., an *analytic cover*. (For general properties of analytic covers we refer to [574].)

Proposition 6.14.3 ([192, Proposition 4.2]) *Assume that X and Z are reduced complex spaces, the space X is irreducible and locally irreducible, and $h \colon Z \to X$ is a surjective holomorphic map. Then, the following hold:*

(a) *If $F \colon X \to Z^d_{\mathrm{sym}}$ is a holomorphic d-valued section of h, then its graph $V(F)$ is an effective holomorphic chain in Z and the projection $h \colon V(F) \to X$ is a d-sheeted analytic cover.*
(b) *Conversely, if V is an effective holomorphic chain in Z such that $h|_V \colon V \to X$ is a d-sheeted analytic cover, then V is the graph of a holomorphic d-valued section of h.*
(c) *The decomposition of $F \colon X \to Z^d_{\mathrm{sym}}$ into irreducible components, given by Proposition 6.14.2, corresponds to the decomposition of its graph $V(F)$ into irreducible complex subvarieties of Z.*

The following result is a relative Oka principle for multi-valued sections of branched holomorphic maps.

Theorem 6.14.4 ([192, Theorem 1.1]) *Let X be an irreducible and locally irreducible reduced Stein space, and let $h \colon Z \to X$ be a surjective holomorphic map. Assume that K is a compact $\mathcal{O}(X)$-convex set in X and X' is a closed complex subvariety of X with $h(\mathrm{br}_h) \subset X'$. Let $\Omega = X \setminus (K \cup X')$. Assume that the restriction $h : Z|_\Omega \to \Omega$ is a submersion satisfying* HAP *over a neighborhood of any point in Ω (see Definition* 6.6.5*). Given a continuous d-valued section F of h that is holomorphic in a neighborhood of $K \cup X'$, unbranched over $X \setminus X'$ and satisfies $\mathcal{H}^{2n-1}(\delta_F) = 0$, there is a homotopy $F_t \colon X \to Z^d_{\mathrm{sym}}$ $(t \in [0, 1])$ of continuous d-valued sections of h such that $F_0 = F$, F_1 is holomorphic on X, and for every $t \in [0, 1]$, F_t is holomorphic in a neighborhood of $K \cup X'$, unbranched over $X \setminus X'$, uniformly close to F on K, and satisfies $F_t(x) = F(x)$ for all $x \in X'$.*

Recall that the condition HAP in the theorem is satisfied in particular if $h : Z|_\Omega \to \Omega$ is a stratified subelliptic submersion (see Definition 6.1.2), or a stratified holomorphic fibre bundle with Oka fibres (see Definition 1.4.1). Theorem 6.14.4 is a generalization of Theorem 6.6.6 (and of Theorem 7.4.5) even for single valued sections since the latter result only applies when $h \colon Z \to X$ is non-branched. Theorem 6.14.4 can be seen as an existence theorem for complex subvarieties $\Lambda \subset Z$ such that $h \colon \Lambda \to X$ is a finite branched cover.

Remark 6.14.5 If X is locally irreducible, then the condition $\mathcal{H}^{2n-1}(\delta_F) = 0$ in Theorem 6.14.4 implies that the discriminat locus δ_F is nowhere dense in X and $X \setminus \delta_F$ is path connected and locally path connected. This guaranties a unique decomposition of F into irreducible components by Proposition 6.14.2. This might fail if $\mathcal{H}^{2n-1}(\delta_F) > 0$. An example is the 2-valued map $F: \mathbb{C} \to \mathbb{C}^2_{\mathrm{sym}}$ given by $F(x+iy) = [\,|x|, -|x|\,]$ (with $\delta_F = \{x = 0\}$ of Hausdorff dimension one). There are two splittings into single valued continuous maps:

$$\text{(a) } F_1(z) = |x|,\ F_2(z) = -|x|; \qquad \text{(b) } F_1(z) = x,\ F_2(z) = -x.$$

A jet transversality argument shows that for a generic smooth multi-valued section F, the set $\delta_F \cap X_{\mathrm{reg}}$ is a smooth real submanifold of real codimension at least two, and hence $\mathcal{H}^{2n-1}(\delta_F) = 0$. However, a generic perturbation of F may introduce additional branch points. □

Theorem 6.14.4 easily reduces to the following result concerning single valued sections (see [192, Theorem 2.1]). The slightly improved formulation used here was proposed by Andrist [30, Theorem 2.5].

Theorem 6.14.6 *Let $h: Z \to X$ be a holomorphic map of a reduced complex space Z onto a reduced Stein space X. Assume that Z' is a closed complex subvariety of Z containing* br_h *such that the restricted submersion $h: Z \setminus Z' \to X$ satisfies* HAP *over a neighborhood of any point in $h(Z \setminus Z')$ (see Definition* 6.6.5*). Assume that X' is a closed complex subvariety of X and $F: X \to Z$ is a continuous section which is holomorphic in a neighborhood of X' and satisfies $F(X \setminus X') \subset Z \setminus Z'$. Given $k \in \mathbb{N}$ there exist an open neighborhood $U \subset X$ of X' and a homotopy of continuous sections $F_t: X \to Z$ $(t \in [0,1])$ such that $F_0 = F$, F_t is holomorphic in U and tangent to F to order k along X' for each $t \in [0,1]$, and F_1 is holomorphic on X. If furthermore F is holomorphic in a neighborhood of $K \cup X'$ for some compact $\mathcal{O}(X)$-convex subset K of X, then F_1 can be chosen uniformly close to F on K.*

The proof of Theorem 6.14.6 follows the same scheme as the proof of Theorem 6.2.2. The only nontrivial additional ingredient is the existence of a local holomorphic spray around a holomorphic section F which is tangent to F to a given finite order along the subvariety X' and is dominating in the complement of X'; see [192, Proposition 2.2].

Theorem 6.14.6 also has a version for liftings of holomorphic maps; see [192, Theorem 1.3].

We give two applications of Theorem 6.14.6. The first one, due to Andrist [30], concerns the Nevanlinna-Pick interpolation problem for the spectral ball.

Example 6.14.7 Let $M_n(\mathbb{C}) \cong \mathbb{C}^{n^2}$ denote the set of all complex $n \times n$ matrices. The *spectral ball* is the set $\Omega_n \subset M_n(\mathbb{C})$ of matrices with spectral radius less than 1. Given distinct points $a_1, \ldots, a_m$ in the unit disc $\mathbb{D}$ and matrices $\{A_i\}_{i=1}^m \subset \Omega_n$, the *Nevanlinna-Pick interpolation problem* is to find a holomorphic map $F: \mathbb{D} \to \Omega_n$ such that $F(a_i) = A_i$ for $i = 1, \ldots, m$.

Let π associate to a matrix $A \in M_n(\mathbb{C})$ the n elementary symmetric functions of its eigenvalues. After symmetrization, π induces a holomorphic surjection $h\colon \Omega_n \to \mathbb{G}_n$ onto the symmetrized polydisc $\mathbb{G}_n = \mathbb{D}^n_{\mathrm{sym}}$. A generic fibre of h, i.e. a fibre above a base point with no multiple eigenvalues, consists of exactly one equivalence class of similar matrices; hence it is a $SL_n(\mathbb{C})$-homogeneous manifold where the group $SL_n(\mathbb{C})$ acts by conjugation. A singular fibre decomposes into several strata which are $SL_n(\mathbb{C})$-homogeneous manifolds as well, but not necessarily connected. The following is the *spectral Nevanlinna-Pick lifting problem*:

Given matrices $A_1, \ldots, A_m \in \Omega_n$ and a holomorphic map $f\colon \mathbb{D} \to \mathbb{G}_n$ with $f(a_i) = h(A_i)$ for $i = 1, \ldots, m$, decide whether there exists a holomorphic map $F\colon \mathbb{D} \to \Omega_n$ such that $F(a_i) = A_i$ for $i = 1, \ldots, m$ and $h \circ F = f$. That is, F is a holomorphic lifting of f with respect to $h\colon \Omega_n \to \mathbb{G}_n$.

In the paper [30], Andrist used Theorem 6.14.6 to show that a necessary and sufficient condition for the existence of such a lifting F is the existence of a local holomorphic lifting at every point a_i. Indeed, the lifting problem reduces in a standard way to the problem of finding a section of the pullback map $f^*h\colon \mathbb{D} \to \Omega_n$ (see the proof of Corollary 5.5.11). The local lifting problem was studied by several authors and we refer to [30] for the references. □

In the second example, we describe a special case of results of Guralnick [275] and Leiterer [382] on similarity of holomorphic matrix-valued functions.

Example 6.14.8 Let $M_n(\mathbb{C}) \cong \mathbb{C}^{n^2}$ denote the set of all complex $n \times n$ matrices. Consider the algebraic subvariety

$$\Sigma = \big\{(A, C) : A \in M_n(\mathbb{C}),\ C \in GL_n(\mathbb{C}),\ CA = AC\big\}. \tag{6.15}$$

Let $\pi : \Sigma \to M_n(\mathbb{C})$ denote the projection $\pi(A, C) = A$. It is easily seen that the fibre $\Sigma(A) = \pi^{-1}(A)$ over any $A \in M_n(\mathbb{C})$ is a connected complex subgroup of $GL_n(\mathbb{C})$ (see [382, Lemma 4.2]). Furthermore, we can stratify $M_n(\mathbb{C})$ by a finite descending sequence of closed algebraic subvarieties such that π is an elliptic submersion over every stratum; a dominating fibre-spray is provided by the exponential map on the fibres (see Example 5.6.16(A)).

Matrices $A, B \in M_n(\mathbb{C})$ are said to be *similar* if there exists an invertible matrix $C \in GL_n(\mathbb{C})$ such that $CA = BC$. If $D \in GL_n(\mathbb{C})$ is another matrix satisfying $DA = BD$, then $CAC^{-1} = B = DAD^{-1}$ and hence the matrix $\Phi = D^{-1}C \in GL_n(\mathbb{C})$ satisfies $\Phi A = A\Phi$, i.e., $\Phi \in \Sigma(A)$. Suppose now that X is a complex space and $A, B\colon X \to M_n(\mathbb{C})$ are holomorphic matrix-valued functions. We say that A and B are *locally holomorphically similar* if every point $p \in X$ admits an open neighborhood $U \subset X$ and a holomorphic map $H\colon U \to GL_n(\mathbb{C})$ such that $HA = BH$ holds on U. If this holds with $U = X$ then A and B are *globally holomorphically similar*.

The following result is due to Guralnick [275] in the special case when X is an open Riemann surface, and to Leiterer [382] in the general case.

Theorem 6.14.9 *Let X be a one-dimensional Stein space. If $A, B\colon X \to GL_n(\mathbb{C})$ are holomorphic matrix-valued functions which are locally holomorphically similar at every point, then they are globally holomorphically similar.*

Leiterer also proved results of this type when X is a Stein space of dimension > 1; however, the conditions under which local holomorphic similarity implies global holomorphic similarity are more complicated. He showed in particular that the existence of a $\mathcal{C}^\infty$-smooth map $C\colon X \to GL_n(\mathbb{C})$ conjugating A to B implies the existence of a holomorphic map doing the same, but maps of class $\mathcal{C}^\ell$ for any finite $\ell \in \mathbb{N}$ need not suffice.

Leiterer's proof of Theorem 6.14.9 uses the Forster-Ramspott Oka principle for Oka pairs of sheaves [176] and a reduction to Guralnick's theorem. Here we obtain it as an application of Theorem 6.14.6.

Proof Consider the complex subvariety $Z \subset X \times GL_n(\mathbb{C})$ defined by

$$Z = \big\{(x, C) : x \in X,\ C \in GL_n(\mathbb{C}),\ CA(x) = B(x)C\big\}.$$

Let $h\colon Z \to X$ denote the projection $h(x, C) = x$. By the hypothesis, every point $p \in X$ admits a neighborhood $U \subset X$ and a holomorphic map $H\colon U \to GL_n(\mathbb{C})$ satisfying $HA = BH$ on U. Thus, $x \mapsto (x, H(x))$ is a holomorphic section of $h\colon Z|_U \to U$. If $U \ni x \to (x, C(x))$ is another section of h over U, then $\Phi(x) = H(x)^{-1}C(x) \in \Sigma(A(x))$ for $x \in U$. This shows that $h\colon Z \to X$ is locally fibrewise isomorphic to the pullback $A^*\Sigma \to X$ of $\pi : \Sigma \to X$ (6.15) by $A\colon X \to M_n(\mathbb{C})$; in particular, all its fibres are isomorphic to connected complex Lie groups. Since X is 1-dimensional, there is a discrete set of points $X' = \{p_j\} \subset X$ such that $Z \to X$ is an elliptic submersion over $X \setminus X'$. The existence of a holomorphic section on a neighborhood of $X' = \{p_j\}$ is guaranteed by the assumption of the theorem. Since the fibres are connected and X has the homotopy type of a 1-dimensional CW complex (see Corollary 3.12.2), it is immediate that this holomorphic section extends from a neighborhood of X' to a global continuous section $X \to Z$. By Theorem 6.14.6 we can deform it to a global holomorphic section by a deformation which is fixed to any given finite order on X'. This completes the proof. □

6.15 Approximation by Algebraic Maps

The problem of approximating holomorphic maps by algebraic maps is of central importance in algebraic geometry. Algebraic approximations in general do not exist even for maps between very simple affine algebraic manifolds. For instance, there are no nontrivial algebraic morphisms $\mathbb{C} \to \mathbb{C} \setminus \{0\}$. However, we establish the following analogue of Theorem 6.6.1 (p. 283) for algebraic maps from affine algebraic varieties to algebraically subelliptic manifolds. All algebraic maps are assumed to be morphisms (without singularities).

Theorem 6.15.1 ([196, Theorem 3.1]) *Assume that X is an affine algebraic variety and Y is an algebraically subelliptic manifold (see Definition 5.6.13(e)). Given an algebraic map $f_0\colon X \to Y$, a compact $\mathcal{O}(X)$-convex subset K of X, an open set $U \subset X$ containing K, a homotopy $f_t\colon U \to Y$ of holomorphic maps ($t \in [0,1]$) and an $\epsilon > 0$, there exists an algebraic map $F\colon X \times \mathbb{C} \to Y$ such that $F(\cdot, 0) = f_0$ and*

$$\sup_{x\in K,\, t\in[0,1]} \operatorname{dist}\big(F(x,t), f_t(x)\big) < \epsilon.$$

In particular, a holomorphic map $X \to Y$ that is homotopic to an algebraic map through a family of holomorphic maps is a limit of algebraic maps.

Corollary 6.15.2 (Algebraic CAP) *If Y is an algebraically subelliptic manifold, then every holomorphic map $K \to Y$ from a compact convex subset $K \subset \mathbb{C}^n$ can be approximated, uniformly on K, by algebraic maps $\mathbb{C}^n \to Y$.*

Proof We may assume that $0 \in K$ and that f is defined on a convex open set $U \supset K$. Let $f_t(z) = f(tz)$ ($z \in U$, $t \in [0,1]$). The map $f_0(z) = f(0)$ is constant and therefore algebraic, so Theorem 6.15.1 applies. □

Theorem 6.15.1 is a special case of the following result.

Theorem 6.15.3 *Let $h\colon Z \to X$ be an algebraic submersion from an algebraic variety Z onto an affine algebraic variety X. Assume that h is* algebraically subelliptic, *in the sense that Z admits a family of algebraic fibre sprays (E_j, π_j, s_j) ($j = 1, \ldots, k$) satisfying the fibre-domination property (6.1) (p. 264) at every point $z \in Z$. Let $K \subset X$ be a compact $\mathcal{O}(X)$-convex set and $f_t\colon K \to Z$ ($t \in [0,1]$) be a homotopy of holomorphic sections such that f_0 extends to an algebraic section $X \to Z$. For every $\epsilon > 0$ there is an algebraic map $F\colon X \times \mathbb{C} \to Z$ such that*

- *$h(F(x,t)) = x$ for all $x \in X$ and $t \in \mathbb{C}$,*
- *$F(\cdot, 0) = f_0$, and*
- $\operatorname{dist}(F(x,t), f_t(x)) < \epsilon$ *for every $x \in K$ and $t \in [0,1]$.*

Proof Let (E, p, s) be the composed algebraic fibre spray on Z obtained from the fibre sprays (E_j, p_j, s_j) ($j = 1, \ldots, k$) (see Definition 6.3.5, p. 268). Choose open Stein neighborhoods $V \Subset U$ of K in X such that the homotopy f_t ($t \in [0,1]$) in Theorem 6.15.3 is defined in U. Set $V_t = f_t(V) \subset Z$ for $t \in [0,1]$.

Lemma 6.15.4 *There are numbers $l \in \mathbb{N}$ and $0 = t_0 < t_1 < \cdots < t_l = 1$ such that for every $j = 0, 1, \ldots, l-1$ there exists a homotopy of holomorphic sections ξ_t of the restricted bundle $E|_{V_{t_j}} \to V_{t_j}$ ($t \in I_j = [t_j, t_{j+1}]$) such that ξ_{t_j} is the zero section and $s(\xi_t(z)) = f_t(h(z))$ for all $t \in I_j$ and $z \in V_{t_j}$.*

Proof Assume first that there exists a Stein open set $\Omega \subset Z$ containing $\bigcup_{t\in[0,1]} \overline{V}_t$. By Lemma 6.3.7 (p. 270), the restriction $E|_\Omega \to \Omega$ admits the structure of a holomorphic vector bundle. Choose a holomorphic direct sum splitting $E|_\Omega = H \oplus H'$, where H' is the kernel of $\mathrm{d}s$ at the zero section of E. The domination condition

$$(\mathrm{d}s)_{0_z}(\mathrm{T}_{0_z}E_z) = \mathrm{VT}_z Z, \quad z \in Z$$

implies that for every $z \in \Omega$ the restriction $s\colon H_z \to Z_{h(z)} = h^{-1}(h(z))$ maps a neighborhood of 0_z in H_z biholomorphically onto a relative neighborhood of z in the fibre $Z_{h(z)}$. The size of this neighborhood, and of its image in the corresponding fibre of Z, can be chosen uniform for points in the compact set $\bigcup_{t\in[0,1]} \overline{V}_t \subset \Omega$. Hence, there is a $\delta > 0$ such that for every $t \in [0,1]$ the local inverse of $s\colon H|_{V_t} \to Z$ at the zero section gives a homotopy of sections ξ_τ of $H|_{V_t}$ ($\tau \in J_t = [t, t+\delta] \cap [0,1]$), with ξ_t being the zero section, such that

$$s\big(\xi_\tau(z)\big) = f_\tau\big(h(z)\big), \quad \tau \in J_t,\ z \in V_t.$$

This proves Lemma 6.15.4 in the special case. For the general case observe that $f_t(U)$, being a closed Stein submanifold of $Z|_U$, admits an open Stein neighborhood in Z by Theorem 3.1.1 (p. 66). By compactness there are Stein open sets $\Omega_j \subset Z$ ($j = 1, 2, \ldots, m$) and a partition $[0,1] = \bigcup_{j=1}^m I_j$ into adjacent closed subintervals I_j such that $\bigcup_{t\in I_j} \overline{V}_t \subset \Omega_j$. It remains to apply the above argument separately for each I_j. □

Let $(E^{(l)}, p^{(l)}, s^{(l)})$ be the l-th iterated bundle of (E, p, s) (Definition 6.3.5(b), p. 268). Let (E', p', s') denote the pullback of $(E^{(l)}, p^{(l)}, s^{(l)})$ to X by the algebraic map $f_0\colon X \to Z$; this is an algebraic composed spray bundle over X.

Lemma 6.15.5 *There is a homotopy $\eta_t\colon V \to E'|_V$ ($t \in [0,1]$), consisting of holomorphic sections of the restricted bundle $E'|_V \to V$, such that η_0 is the zero section and $s'(\eta_t(x)) = f_t(x)$ for every $x \in V$ and $t \in [0,1]$.*

Proof It suffices to assemble the individual homotopies $\{\xi_t : t \in [t_j, t_{j+1}]\}$ ($j = 0, \ldots, l-1$), furnished by Lemma 6.15.4, into a homotopy of sections $\widetilde{\xi}_t\colon V_0 \to E^{(l)}|_{V_0}$ ($t \in [0,1]$) of the iterated bundle $E^{(l)}$ over the open subset $V_0 = f_0(V)$ in the algebraic submanifold $f_0(X)$ of Z. Clearly, $\widetilde{\xi}_t$ corresponds to a holomorphic sections $\eta_t\colon V \to E' = f_0^*(E^{(l)})$ of the pullback bundle, with η_0 being the zero section. (For details see the proof of Proposition 6.5.1.) □

Lemma 6.15.6 *Let $\{\eta_t\}_{t\in[0,1]}$ be as in Lemma 6.15.5. For every $\delta > 0$ there is an algebraic map $\eta' : X \times \mathbb{C} \to E'$ satisfying*

(i) $\eta'(x,0) = 0_x \in E'_x$ $(x \in X)$, *and*
(ii) $\mathrm{dist}(\eta'(x,t), \eta_t(x)) < \delta$ *for all $x \in K$ and $t \in [0,1]$.*

Proof By construction of the composed bundle $E' \to X$ there is a sequence

$$E' = E^{m,0} \longrightarrow E^{m-1,0} \longrightarrow \ldots \longrightarrow E^{1,0} \longrightarrow X, \qquad (6.16)$$

with $m = kl$ and $E^{(1,0)} = f_0^* E_1 \to X$, in which every map $E^{j,0} \to E^{j-1,0}$ is an algebraic vector bundle projection. Here, k is the number of the initial sprays in Theorem 6.15.3 and l is the number in Lemma 6.15.4.

Since X is an affine algebraic manifold, the algebraic vector bundle $E^{1,0} \to X$ is generated by finitely many (say n_1) algebraic sections according to Serre's Theorem A. This gives a surjective algebraic map $\pi_1 : E^{1,1} = X \times \mathbb{C}^{n_1} \to E^{1,0}$ from the trivial rank n_1 bundle onto $E^{1,0}$. Pulling back the sequence (6.16) to the new total space $E^{1,1}$ we obtain a commutative diagram

$$\begin{array}{ccccccccc} E^{m,1} & \longrightarrow & E^{m-1,1} & \longrightarrow \cdots \longrightarrow & E^{2,1} & \longrightarrow & E^{1,1} & \longrightarrow & X \\ \downarrow \pi_m & & \downarrow \pi_{m-1} & & \downarrow \pi_2 & & \downarrow \pi_1 & & \| \\ E^{m,0} & \longrightarrow & E^{m-1,0} & \longrightarrow \cdots \longrightarrow & E^{2,0} & \longrightarrow & E^{1,0} & \longrightarrow & X \end{array}$$

in which all horizontal maps are algebraic vector bundle projections and the vertical maps π_j for $j \ge 2$ are the induced natural maps which are bijective on fibres. More precisely, let $E^{2,1} \to E^{1,1}$ be the pullback of the vector bundle $E^{2,0} \to E^{1,0}$ (in the bottom row) by the vertical morphism $\pi_1 : E^{1,1} \to E^{1,0}$, and denote by $\pi_2 : E^{2,1} \to E^{2,0}$ the associated natural map which makes the respective diagram commute. Moving one step to the left, $E^{3,1} \to E^{2,1}$ is the pullback of the bundle $E^{3,0} \to E^{2,0}$ in the bottom row by the vertical morphism $\pi_2 : E^{2,1} \to E^{2,0}$, and $\pi_3 : E^{3,1} \to E^{3,0}$ is the associated natural map; etc. There is an algebraic spray map $s^1 : E^{m,1} \to Z$ which is the composition of $\pi_m : E^{m,1} \to E^{m,0}$ with the initial spray $s : E^{m,0} = E \to Z$.

We claim that the homotopy η_t of holomorphic sections of $E^{m,0}|_V = E'|_V \to V$, furnished by Lemma 6.15.5, lifts to a homotopy η_t^1 of holomorphic sections of $E^{m,1}|_V \to V$ such that $s^1(\eta_t^1) = f_t$ for all $t \in [0,1]$, and η_0^1 is the zero section. It suffices to see that the $E^{1,0}$-component of η_t (i.e., the projection of η_t under the composed projection $E^{m,0} \to E^{1,0}$) lifts to $E^{1,1}$; the rest of the lifting is then obtained by applying the inverses of the fibrewise isomorphic vertical maps. But this follows from the fact that the surjective vector bundle map $\pi_1 : E^{1,1} \to E^{1,0}$ admits a holomorphic splitting $\sigma_1 : E^{1,0} \to E^{1,1}$ over X, with $\pi_1 \circ \sigma_1$ the identity on $E^{1,0}$ (see Corollary 2.6.6).

Repeating the same argument with the bundle $E^{2,1} \to E^{1,1} = X_1$ over the affine manifold $X_1 = X \times \mathbb{C}^{n_1}$ we obtain a surjective algebraic vector bundle map $E^{2,2} = X_1 \times \mathbb{C}^{n_2} = X \times \mathbb{C}^{n_1+n_2} \to E^{2,1}$. As before, we lift the top line in the above diagram to a new level

$$E^{m,2} \longrightarrow E^{m-1,2} \longrightarrow \ldots \longrightarrow E^{2,2} \longrightarrow E^{1,1} = X_1 = X \times \mathbb{C}^{n_1}.$$

Note that $E^{2,2} = X_1 \times \mathbb{C}^{n_2} = X \times \mathbb{C}^{n_1+n_2}$ (algebraic equivalence). The homotopy η_t^1 lifts to a homotopy $\eta_t^2 \colon V \to E^{m,2}|_V$, with η_0^2 the zero section, and we have a new spray map $s^2 \colon E^{m,2} \to Z$ satisfying $s^2(\eta_t^2) = f_t$ for all $t \in [0,1]$.

Continuing inductively we obtain after m steps a lifting of the homotopy η_t to a homotopy $\eta_t^m \colon V \to E^{m,m}|_V = V \times \mathbb{C}^N$ $(t \in [0,1])$ of holomorphic sections of $E^{m,m} = X \times \mathbb{C}^N$ $(N = n_1 + n_2 + \cdots + n_m)$ over the open subset $V \subset X$, with η_0^m being the zero section. By construction there is an algebraic spray $s^m \colon E^{m,m} \to Z$ such that $s^m(\eta_t^m) = f_t \colon V \to Z|_V$ for all $t \in [0,1]$.

Recall that X is a closed algebraic submanifold of an affine space $\mathbb{C}^n$. The $\mathcal{O}(X)$-convex set $K \subset X$ is then polynomially convex in $\mathbb{C}^n$, and $K \times [0,1]$ is polynomially convex in $\mathbb{C}^{n+1}$. (We have identified $[0,1] \subset \mathbb{R}$ with its image in $\mathbb{C}$.) By the Oka-Weil theorem (see Theorem 2.8.4) we can approximate the homotopy $\{\eta_t^m\}_{t\in[0,1]}$ (which is continuous in $(x,t) \in V \times [0,1]$ and holomorphic with respect to $x \in V$ for every fixed $t \in [0,1]$), uniformly on the set $K \times [0,1]$, by a holomorphic polynomial map $\mathbb{C}^n \times \mathbb{C} \to \mathbb{C}^n \times \mathbb{C}^N$ of the form $\widetilde{g}(x,t) = (x, g(x,t))$, with $g(x,0) = 0$ for $x \in \mathbb{C}^n$. By projecting the point $\widetilde{g}(x,t) \in E^{m,m} = X \times \mathbb{C}^N$ $(x \in X,\ t \in \mathbb{C})$ back to the bundle $E' = E^{m,0}$ we obtain an algebraic map $\eta'(x,t)$ satisfying Lemma 6.15.6. □

If s', η' and δ are as in Lemmas 6.15.5 and 6.15.6 with $\delta > 0$ chosen small enough, then the algebraic map

$$F(x,t) = s'\big(\eta'(x,t)\big) \in Z, \quad (x,t) \in X \times \mathbb{C}$$

satisfies the conclusion of Theorem 6.15.3. □

There are examples of holomorphic maps from an affine algebraic manifold to an algebraically subelliptic manifold which are not homotopic to an algebraic map, so the absolute form of the Oka principle for algebraic maps fails in this context. I wish to thank Burt Totaro for pointing out the following examples in a private communication.

Example 6.15.7 (Loday [399]) Let Σ^n denote the complex n-sphere, i.e., the affine variety

$$\Sigma^n = \big\{(z_0,\ldots,z_n) \in \mathbb{C}^{n+1} : z_0^2 + \cdots + z_n^2 = 1\big\}.$$

Then Σ^n retracts onto the real n-sphere S^n. Also, Σ^n is algebraically subelliptic for $n \geq 2$ because it is homogeneous for the complex Lie group $SO(n+1,\mathbb{C})$, and $\mathrm{Hom}(SO(n+1,\mathbb{C}),\mathbb{C}^*) = 1$ for $n \geq 2$. Loday [399] showed that every algebraic map $\Sigma^p \times \Sigma^q \to \Sigma^{p+q}$ is null-homotopic when p and q are odd, but there always exists a homotopically nontrivial continuous map $\Sigma^p \times \Sigma^q \to \Sigma^{p+q}$ (in fact, an isomorphism of the homology groups H_{p+q}). □

Example 6.15.8 Consider an affine algebraic surface $X = M - D$, where M is a smooth quartic surface in $\mathbb{P}^3$ (a K3 surface) and D is a smooth hyperplane section in M. Let $Y = \mathbb{P}^n$ where $n \geq 3$. Then, every continuous map $X \to Y$ is homotopic to a holomorphic map, but in general not to an algebraic map. Indeed, to ask whether every continuous map $X \to Y$ is homotopic to an algebraic map amounts to asking whether every topological complex line bundle on X is isomorphic to an algebraic line bundle. (See the discussion on p. 213.) Equivalently, we are asking whether all of $H^2(X;\mathbb{Z})$ is algebraic, that is, spanned by the Chern classes of algebraic curves in X. There is an obstruction coming from the Hodge decomposition

$$H^2(M;\mathbb{C}) = H^{2,0}(M) \oplus H^{1,1}(M) \oplus H^{0,2}(M).$$

Since any K3 surface M has $H^{0,2}(M) = H^2(M;\mathcal{O}) \cong \mathbb{C}$, not all of $H^2(M;\mathbb{Q})$ is algebraic. (The algebraic classes all map into $H^{1,1}(M)$ which equals $\mathbb{C}^{20}$ for any K3 surface.) We have $H^2(X;\mathbb{Z}) = H^2(M;\mathbb{Z})/(D)$, and the closure of any algebraic curve in $X = M \setminus D$ is an algebraic curve in M. Therefore, not all of $H^2(X;\mathbb{Q})$ is algebraic (otherwise all of $H^2(M;\mathbb{Q})$ would be algebraic). Hence, $H^2(X;\mathbb{Z})$ contains a non-torsion class which is not represented by an algebraic curve in X. □

Chapter 7
Flexibility Properties of Complex Manifolds and Holomorphic Maps

In Chap. 5 we have seen that all Oka properties considered so far are equivalent to each other; a complex manifold satisfying these Oka properties is called an Oka manifold. We have also described a number of flexibility properties of a complex manifold which imply the Oka property. It is now time to consider the place of the Oka property in the general framework of holomorphic flexibility properties of complex manifolds that have been studied in the literature, and in particular to determine the place of Oka manifolds in the classification of compact complex manifolds.

In Sect. 7.1 we introduce and compare a number of holomorphic flexibility properties which are implied by the Oka property: the basic Oka property (BOP), specialness in the sense of Campana, stratified Oka property, $\mathbb{C}$-connectedness dominability, strong dominability, and others. A summary of known relations between these properties is given by Corollary 7.1.12.

In Sect. 7.2 we prove that every stratified Oka manifold is strongly dominable; see Theorem 7.1.8 on p. 323. Among examples of stratified Oka manifolds we mention Kummer surfaces. We do not know whether every stratified Oka manifold is an Oka manifold; see Problem 7.6.4 on p. 343.

In Sect. 7.3 we summarize what we know about which compact complex surfaces are Oka. Compact Oka surfaces belong to one of the classes with the Kodaira dimension $-\infty$, 0, or 1; these are precisely the classes that have been studied most extensively in the literature.

In Sect. 7.4 we introduce and study the class of *Oka maps*; these are holomorphic maps which are topological fibrations and enjoy the Oka property for lifting holomorphic maps from Stein source manifolds.

In Sect. 7.5, which was contributed by Finnur Lárusson, it is explained how Oka manifolds and Oka maps naturally fit into an abstract homotopy-theoretic framework, not merely by way of analogy, but in precise, rigorous terms.

In Sect. 7.6 we collect some additional results, open problems and promising directions for future research.

F. Forstnerič, *Stein Manifolds and Holomorphic Mappings*, Ergebnisse der Mathematik und ihrer Grenzgebiete. 3. Folge / A Series of Modern Surveys in Mathematics 56,
DOI 10.1007/978-3-319-61058-0_7

7.1 Hierarchy of Holomorphic Flexibility Properties

In this section we compare the Oka property with several other holomorphic flexibility properties of complex manifolds.

Convex Approximation Properties

By Theorem 5.4.4, a complex manifold Y of dimension p enjoys all Oka properties for maps $X \to Y$ from Stein spaces of dimension $\dim X \le n$ if and only if Y enjoys CAP_{n+p} (see Definition 5.4.3). In particular, Y is an Oka manifold if and only if it enjoys CAP, which means CAP_k for all $k \in \mathbb{N}$. We have the obvious implications

$$\mathrm{CAP}_1 \Leftarrow \mathrm{CAP}_2 \Leftarrow \mathrm{CAP}_3 \Leftarrow \cdots \Leftarrow \mathrm{CAP} = \mathrm{Oka}.$$

The following example shows that the properties CAP_k are pairwise nonequivalent in the range $1 \le k \le \dim Y$. It is not clear what happens for $k > \dim Y$.

Example 7.1.1 For every pair of integers $1 \le k \le p$ there exists a p-dimensional complex manifold which enjoys CAP_{k-1} but not CAP_k.

For $k = p$ we can take $Y = \mathbb{C}^p \backslash A$, where A is a discrete subset of $\mathbb{C}^p$ which is *rigid* in the sense of Rosay and Rudin (see Theorem 4.7.2, p. 130), i.e., every holomorphic map $\mathbb{C}^p \to \mathbb{C}^p \setminus A$ is degenerate. Thus, CAP_p fails but CAP_{p-1} holds by a general position argument. For $1 \le k < p$ we take $Y = \mathbb{C}^p \backslash \phi(\mathbb{C}^{p-k})$, where $\phi\colon \mathbb{C}^{p-k} \hookrightarrow \mathbb{C}^p$ is a proper holomorphic embedding whose complement is Eisenman k-hyperbolic (see Theorem 4.18.7, p. 187); hence, CAP_k fails, but CAP_{k-1} holds by general position. □

The property CAP_p with $p = \dim Y$ is especially interesting since it clearly implies that Y is strongly dominable by $\mathbb{C}^p$ (see Definition 7.1.7). If such Y is compact, then it cannot be of Kodaira general type (see Kobayashi and Ochiai [342, Theorem 2]). Assuming also that Y is compact Kähler, Campana proved by an extension of the Kobayashi-Ochiai argument that Y is *special* [81, Corollary 8.11]. Special manifolds are important in Campana's structure theory of compact Kähler manifolds; see the next subsection for more information.

Recently, Diverio and Trapani [124] and Wu and Yau [589, 590] proved that a compact connected complex manifold Y, which admits a Kähler metric whose holomorphic sectional curvature is everywhere nonpositive and is strictly negative at least at one point, has positive canonical bundle K_Y. (See also Tosatti and Yang [552] and Nomura [436].) Hence, such Y is projective and of general type, so it does not enjoy CAP_p with $p = \dim Y$.

Problem 7.1.2

(a) Suppose that Y is a complex manifold of dimension p which enjoys CAP_p; does it follow that Y is an Oka manifold?

(b) Suppose that a complex manifold Y enjoys CAP_k for some $k \in \mathbb{N}$. Does Y enjoy the Oka property (i.e., the conclusion of Corollary 5.4.5) for maps $X \to Y$ from Stein spaces of dimension $\dim X \leq k$?

Basic Oka Property and Special Manifolds

The following property was introduced by Gromov [266].

Definition 7.1.3 A complex manifold Y enjoys the *basic Oka property (BOP)* if every continuous map $X \to Y$ from a Stein manifold X to Y is homotopic to a holomorphic map.

The difference between BOP and Oka is that the former does not include any approximation or interpolation condition. Thus, every Oka manifold satisfies BOP, but the converse fails. For example, every contractible manifold Y (such as the disc $\mathbb{D}$ or, more generally, a convex domain in $\mathbb{C}^n$) enjoys BOP since every map $X \to Y$ is homotopic to a constant map $X \to y_0 \in Y$. Similarly, the product of a contractible manifold and an Oka manifold enjoys BOP but is not necessarily Oka; an example is $\mathbb{D} \times \mathbb{C}$.

In [577], Winkelmann gave a complete answer to the question which pairs of Riemann surfaces (X, Y) (including the compact ones) satisfy BOP, in the sense that every continuous map $X \to Y$ is homotopic to a holomorphic map.

Campana and Winkelmann proved [83, Main Theorem] that a compact projective manifold Y which enjoys BOP is *special* in the sense of Campana [81, 82], and every holomorphic map from such a manifold Y to a Brody hyperbolic Kähler manifold is constant. We thus have the implications

$$\text{Oka} \Longrightarrow \text{BOP} \Longrightarrow \text{special},$$

where the second one holds for compact projective manifolds (and is expected to be true for all compact Kähler manifolds).

Remark 7.1.4 (on terminology) In the terminology of Campana and Winkelmann [83], BOP is called the *h-principle*, and they use the notation $hP(Y)$ to indicate that the manifold Y satisfies it. We prefer our terminology for two reasons. Firstly, it fits naturally into the scheme of Oka properties discussed in Chap. 5. Secondly, the term h-principle is used in the literature for a wide range of phenomena way beyond complex analysis. □

Being special is an anti-hyperbolicity or anti-general type notion defined for compact Kähler manifolds. The definition is fairly technical and we refer the reader to the cited sources (see e.g. [83, Sect. 2.1]). Roughly speaking, a compact Kähler manifold Y is special if its finite étale covers do not admit any non-degenerate meromorphic maps to orbifolds of general type. Campana proved that Y is special if it is either dominable, rationally connected, or has Kodaira dimension 0; on the other hand, no compact manifold of general type is special. Campana conjectured that Y is special if and only if it is $\mathbb{C}$-connected (see Definition 7.1.7(e)), if and only if the Kobayashi pseudometric of Y vanishes identically.

Concerning the relationship between BOP and Oka, one has the feeling that these two classes are essentially the same after eliminating the obvious counterexamples provided by contractible manifolds with a nontrivial Kobayashi or Eisenman pseudometric; these can be used as building blocks for BOP manifolds, but not for Oka manifolds. With this in mind, we proposed in [179] the following Oka property.

Definition 7.1.5 A connected complex manifold Y satisfies the *basic Oka property with surjectivity*, BOPS, if every continuous map $X \to Y$ from a Stein manifold X with $\dim X \geq \dim Y$ to Y is homotopic to a *surjective* holomorphic map $X \to Y$.

Theorem 5.5.1(d) says that Oka $\Rightarrow$ BOPS. A connected complex manifold Y enjoying BOPS admits a surjective holomorphic map $\mathbb{C}^{\dim Y} \to Y$ (apply the BOPS axiom to a constant map $\mathbb{C}^{\dim Y} \to y_0 \in Y$). In particular, the Kobayashi pseudometric of such manifold Y vanishes identically, and if Y is compact then it is not of general type.

Problem 7.1.6

(a) Is every manifold enjoying BOPS an Oka manifold?
(b) Do the properties BOP, BOPS and Oka coincide for compact (or compact Kähler, or compact projective) manifolds?

A Survey of Holomorphic Flexibility Properties

Several items in the following definition have been introduced before, but we take this opportunity to summarize them in one place.

Definition 7.1.7 Let Y be a connected complex manifold.

(a) Y is a *stratified Oka manifold* if it admits a stratification $Y = Y_0 \supset Y_1 \supset \cdots \supset Y_m = \varnothing$ by closed complex submanifolds Y_k such that each connected component of every difference $Y_{j-1} \setminus Y_j$ is an Oka manifold.
(b) Y is *dominable at* $y \in Y$ if there is a holomorphic map $f\colon \mathbb{C}^{\dim Y} \to Y$ such that $f(0) = y$ and f is a local biholomorphism at 0.
(c) Y is *dominable* if it is dominable at some point.
(d) Y is *strongly dominable* if it is dominable at every point.
(e) Y is *$\mathbb{C}$-connected* if any two points in Y can be joined by a finite chain of entire curves (holomorphic images of $\mathbb{C}$) in Y.
(f) Y enjoys the *jet transversality theorem* (JTT) if Theorem 8.8.5 (p. 380) holds for holomorphic maps $X \to Y$ from any Stein manifold X.
(g) Y is *Liouville* if it carries no nonconstant negative plurisubharmonic functions, and is *strongly Liouville* if the universal covering space of Y is Liouville.

In connection with (d), note that a family $\{f_y\}_{y\in Y}$ of dominating maps $f_y\colon \mathbb{C}^n \to Y$ with $f_y(0) = y$, depending holomorphically on the base point $y \in Y$, is a dominating spray on Y (see Definition 5.6.13 on p. 229).

All properties in Definition 7.1.7 except (g) are anti-hyperbolic in the sense that the only Kobayashi hyperbolic manifold satisfying any of them is the point. For the weakest property (g), this fails. Indeed, a simply connected compact Kobayashi hyperbolic manifold (for example, a smooth hyperbolic surface in $\mathbb{P}^3$) is obviously strongly Liouville. Removing a point, we obtain a noncompact manifold with the same properties.

An Oka manifold is obviously stratified Oka; the converse is open. The following result of Lárusson and the author is proved in the next section.

Theorem 7.1.8 ([212, Theorem 2]) *A stratified Oka manifold is strongly dominable.*

The following proposition summarizes the known relationships between the properties in Definition 7.1.7.

Proposition 7.1.9 *Let Y be a connected complex manifold.*

(a) *If Y is Oka, then Y is strongly dominable and enjoys JTT.*
(b) *If Y enjoys JTT, then Y is dominable.*
(c) *If Y is stratified Oka, then Y is strongly dominable.*
(d) *If Y is strongly dominable, then Y is $\mathbb{C}$-connected.*
(e) *If Y is $\mathbb{C}$-connected, then Y is strongly Liouville.*
(f) *If Y is compact and dominable, then it is not of Kodaira general type.*

Proof The first statement in (a) follows from Theorem 5.4.4 (p. 216), and the second one from Corollary 8.8.7 (p. 380).

To see (b), assume that Y satisfies the 1-jet transversality theorem for holomorphic maps $\mathbb{C}^n \to Y$ with $n = \dim Y$ (see Theorem 8.8.5 on p. 380). Let $B \subset \mathrm{J}^1(\mathbb{C}^n, Y)$ consist of all 1-jets of rank $< n$. For a generic holomorphic map $f\colon \mathbb{C}^n \to Y$, its 1-jet extension $\mathrm{j}^1 f\colon \mathbb{C}^n \to \mathrm{J}^1(\mathbb{C}^n, Y)$ is then transverse to B, and hence the set of points $z \in \mathbb{C}^n$ at which f has rank less than n is a proper subvariety of $\mathbb{C}^n$. Thus, f is a dominant map.

Part (c) is the content of Theorem 7.1.8 proved in the next section.

For (d), assume that Y is strongly dominable. Hence, for every point $y \in Y$ there is a holomorphic map $f\colon \mathbb{C}^n \to Y$ with $n = \dim Y$ such that $f(0) = y$ and f maps a small ball in $\mathbb{C}^n$ centered at 0 biholomorphically onto a neighborhood of y in Y. If follows that for any path $\lambda\colon [0,1] \to Y$ there is a subdivision $0 = t_0 < t_1 < t_2 < \cdots < t_k = 1$ such that for every $i = 1, \ldots, k$ the segment $\lambda([t_{i-1}, t_i])$ is contained in the image of a holomorphic map $f_i\colon \mathbb{C}^n \to Y$. Hence, there is an affine complex line $\Lambda_i \subset \mathbb{C}^n$ such that $f_i(\Lambda_i) \subset Y$ contains the points $\lambda(t_{i-1})$ and $\lambda(t_i)$. This shows that Y is $\mathbb{C}$-connected.

Part (e) follows from the fact that $\mathbb{C}$ is Liouville, so the restriction of a bounded plurisubharmonic function on Y to any complex line in Y is constant. Furthermore, if $\widetilde{Y} \to Y$ is a holomorphic covering then any holomorphic map $\mathbb{C} \to Y$ lifts to a holomorphic map $\mathbb{C} \to \widetilde{Y}$.

Part (f) is due to Kobayashi and Ochiai [342, Theorem 2]. □

The following example shows that a dominable manifold need not be strongly dominable or Liouville. (The implication dominable ⇒ Liouville was erroneously claimed in [212], item (4) on p. 3716.)

Example 7.1.10 Let $X \subset \mathbb{C}^n$ be a Fatou-Bieberbach domain with $\mathcal{C}^1$ boundary (such domains were constructed by Globevnik [238] and Stensønes [526]; see Theorem 4.4.3 on p. 125). Let $Y = X \cup B \subset \mathbb{C}^n$ where B is a small open ball centered at a boundary point of X. Clearly, Y is dominable since X is biholomorphic to $\mathbb{C}^n$, but it is not strongly dominable. Indeed, if B is chosen small enough, there clearly exists a nonconstant plurisubharmonic function $\rho \colon B \to [-1, 0]$ that equals -1 on $X \cap B$. Extending ρ by -1 to X gives a nonconstant negative plurisubharmonic function on Y. Hence, Y is not Liouville, and it is not dominable at any point where $\rho > -1$. □

Problem 7.1.11 Is every compact dominable manifold strongly dominable?

The following corollary summarizes the known relationships between the flexibility properties considered so far. Several of them are listed in Proposition 7.1.9 or are immediate consequences of definitions. For the implication homogeneous ⇒ elliptic, see Example 5.6.16 (p. 231); subelliptic ⇒ Oka is the content of Corollary 5.6.14 (p. 230); weakly subelliptic ⇒ Oka is Corollary 5.6.14 (p. 230), and the equivalence CAP ⇔ Oka is Theorem 5.4.4 (p. 216). Furthemore, flexibility in the sense of Arzhantsev et al. implies weak ellipticity by Proposition 5.6.22 (p. 232). For algebraic manifolds, CAP means the conclusion of Corollary 6.15.2 (p. 313).

Corollary 7.1.12 *The following implications hold for any complex manifold*:

$$\begin{array}{ccccccc}
\text{homogeneous} & \Longrightarrow & \text{elliptic} & \Longrightarrow & \text{subelliptic} & & \\
\Downarrow & & \Downarrow & & \Downarrow & & \\
\text{flexible} & \Longrightarrow & \text{weakly elliptic} & \Longrightarrow & \text{weakly subelliptic} & \Longrightarrow & \text{CAP} = \text{Oka}
\end{array}$$

$$\begin{array}{ccccccc}
\text{Oka} & \Longrightarrow & \text{stratified Oka} & \Longrightarrow & \text{strongly dominable} & \Longrightarrow & \mathbb{C}\text{-connected} \\
\Downarrow & & & & & & \Downarrow \\
\text{JTT} & \Longrightarrow & \text{dominable} & & & & \text{strongly Liouville}
\end{array}$$

In the algebraic category we have the implications

$$\text{elliptic} \Longrightarrow \text{subelliptic} \Longrightarrow \text{CAP} \Longrightarrow \text{strongly dominable.}$$

Complex homogeneous manifolds form a proper subclass of the class of elliptic or flexible manifolds. For example, the complement $\mathbb{C}^n \setminus A$ of a thin (of codimension ≥ 2) algebraic subvariety is elliptic and flexible (see Proposition 6.4.1 on p. 271), but very few such complements are homogeneous. The same proposition provides examples of subelliptic manifolds (i.e., complements of thin algebraic subvarieties in $\mathbb{P}^n$) which are not known to be elliptic in general. An example of a weakly subelliptic manifold is $\mathbb{C}^n$ blown up at all points of a tame discrete set (see Proposition 6.4.12 on p. 275).

It is not known whether the implications elliptic $\Rightarrow$ subelliptic $\Rightarrow$ Oka can be reversed. By Proposition 5.6.15 (p. 230) these properties are equivalent for Stein manifolds. In this connection, we mention the following result of Andrist, Shcherbina and Wold [35]. Recall that $\mathbb{B}^n$ is the ball in $\mathbb{C}^n$.

Proposition 7.1.13 *If $n \geq 3$, then $\mathbb{C}^n \setminus \overline{\mathbb{B}^n}$ is not weakly subelliptic.*

Proof (sketch) By Siu [503], any holomorphic vector bundle $E \to \mathbb{C}^n \setminus \overline{\mathbb{B}^n}$ extends to a coherent analytic sheaf on $\mathbb{C}^n$ which is locally free in the complement of finitely many points. In [35] one can find a precise proof of this result. (This fails for $n < 3$.) A holomorphic spray map $s\colon E \to \mathbb{C}^n \setminus \overline{\mathbb{B}^n}$ also extends across most points $p \in b\mathbb{B}^n$ by Hartogs's theorem. If $p \in b\mathbb{B}^n$ and the map $s|_{E_p}\colon E_p \to \mathbb{C}^n$ intersects the sphere $b\mathbb{B}^n$ transversely at $0_p \in E_p$, then for points $q \in \mathbb{C}^n \setminus \overline{\mathbb{B}^n}$ sufficiently close to p, the range of the map $s\colon E_q \to \mathbb{C}^n$ also intersects the ball $\mathbb{B}^n$, a contradiction. This shows that the vertical derivative of any such spray s at most points $p \in b\mathbb{B}^n$ is complex tangent to the sphere $b\mathbb{B}^n$. It follows from the identity principle that no finite family of such sprays is dominating anywhere, so $\mathbb{C}^n \setminus \overline{\mathbb{B}^n}$ is not weakly subelliptic.

The same proof shows that the complement $\mathbb{C}^n \setminus K$ of any compact set $K \subset \mathbb{C}^n$ ($n \geq 3$) with a nonempty interior and at least one $\mathcal{C}^1$-smooth boundary point fails to be weakly subelliptic. □

By Proposition 4.12.5, the set $\mathbb{C}^n \setminus \overline{\mathbb{B}^n}$ is a union of Fatou-Bieberbach domains, and hence is strongly dominable. Thus, Proposition 7.1.13 implies that at least one of the following implications cannot be reversed:

$$\text{subelliptic} \Longrightarrow \text{Oka} \Longrightarrow \text{strongly dominable.}$$

Which one? In my opinion, neither of them can be reversed.

7.2 Stratified Oka Manifolds and Kummer Surfaces

In this section we prove Theorem 7.1.8 to the effect that every stratified Oka manifold is strongly dominable. The proof, taken from the paper [212] by Lárusson and the author, is based on the following special case of Theorem 8.6.1 on p. 373 (the *Oka principle for elimination of intersections*).

Theorem 7.2.1 *Let X be a Stein manifold, Y be a complex manifold, Y' be a closed complex subvariety of Y, and $f\colon X \to Y$ be a continuous map that is holomorphic on an open neighborhood of the subvariety $X' = f^{-1}(Y') \subset X$. If $Y \setminus Y'$ is an Oka manifold, then for each $k \geq 1$ there exists a homotopy of continuous maps*

$f_t: X \to Y$, $t \in [0,1]$, *such that* $f = f_0$, f_1 *is holomorphic on* X, *and for each* $t \in [0,1]$ *the map* f_t *is holomorphic in a neighborhood of* X', *it agrees with* $f = f_0$ *to order k along* X', *and it maps* $X \setminus X'$ *into* $Y \setminus Y'$.

Proof of Theorem 7.1.8 Let $Y = Y_0 \supset Y_1 \supset \cdots \supset Y_l = \varnothing$ be a stratification by closed complex submanifolds Y_k such that $S_j = Y_j \setminus Y_{j+1}$ is an Oka manifold for $j = 0, \ldots, l-1$. Given a point $y_0 \in Y$, we wish to find a holomorphic map $F: \mathbb{C}^n \to Y$ ($n = \dim Y$) such that $F(0) = y_0$ and F is dominating at 0.

If $y_0 \in S_0 = Y_0 \setminus Y_1$, then the Oka property of S_0 implies that there is a holomorphic map $F: \mathbb{C}^n \to S_0$ with $F(0) = y_0$ and rank n at 0.

Now suppose $y_0 \in S_1 = Y_1 \setminus Y_2$. Let Σ be the connected component of S_1 containing y_0. Set $m = \dim \Sigma$ and let $d = n - m$. Since Σ is Oka by assumption, there is a holomorphic map $g: \mathbb{C}^m \to \Sigma$ such that $g(0) = y_0$ and g is a local biholomorphism at 0. We identify $\mathbb{C}^m$ with the subspace $\mathbb{C}^m \times \{0\}^d$ of $\mathbb{C}^n$. Write the coordinates on $\mathbb{C}^n$ as $z = (z', z'')$ with $z' = (z_1, \ldots, z_m)$ and $z'' = (z_{m+1}, \ldots, z_n)$. We shall construct a holomorphic map $F: \mathbb{C}^n \to Y$ such that $F(z', 0) = g(z')$ for all $z' \in \mathbb{C}^m$, F maps $\mathbb{C}^n \setminus (\mathbb{C}^m \times \{0\}^d)$ into S_0, and F is a local biholomorphism at 0.

Let $N \to \Sigma$ be the holomorphic normal bundle of Σ in Y. By Grauert's Oka principle (see Theorem 5.3.1 on p. 213), the pullback $g^*N \to \mathbb{C}^m$ is a trivial holomorphic vector bundle of rank d over $\mathbb{C}^m$. A trivialization of this bundle is given by d linearly independent holomorphic vector fields $V_1, \ldots, V_d$ on Y along g that are normal to Σ. The graph of g,

$$G = \{(z', g(z')) : z' \in \mathbb{C}^m\} \subset \mathbb{C}^m \times Y$$

has a Stein open neighborhood Ω in $\mathbb{C}^m \times Y$ by Siu's theorem (see Theorem 3.1.1 on p. 66). We identify each V_j with a vector field on $\mathbb{C}^m \times Y$, defined along G, that is tangent to the fibres of the projection $\pi_1: \mathbb{C}^m \times Y \to \mathbb{C}^m$. After shrinking Ω around G if necessary, we can assume that $V_1, \ldots, V_d$ extend to holomorphic vector fields on Ω that are tangent to the fibres of the projection π_1. Denote by ϕ_t^j the flow of V_j. Let $\pi_2: \mathbb{C}^m \times Y \to Y$ be the projection onto the second factor. The formula

$$f(z', z'') = \pi_2 \circ \phi^1_{z_{m+1}} \circ \cdots \circ \phi^d_{z_n}(z', g(z'))$$

defines a holomorphic map f from an open neighborhood of $\mathbb{C}^m \times \{0\}^d$ in $\mathbb{C}^n$ into Y such that $f(z', 0'') = g(z')$ for $z' \in \mathbb{C}^m$ and

$$\frac{\partial}{\partial z_j} f(z', z'')\big|_{z''=0} = V_{j-m}(z') \quad \text{for } z' \in \mathbb{C}^m \text{ and } j = m+1, \ldots, n.$$

Hence, the differential $\mathrm{d}f_{(z',0'')}: \mathrm{T}_{(z',0'')}\mathbb{C}^n \to \mathrm{T}_{g(z')}Y$ is an isomorphism for every $z' \in \mathbb{C}^m$ near the origin. In particular, f is dominating at $0 \in \mathbb{C}^n$ and $f(0) = g(0') = y_0$. Furthermore, as the vector fields $V_1, \ldots, V_d$ trivialize the normal bundle to Σ in Y, the above implies that there is a neighborhood U of $\mathbb{C}^m \times \{0\}^d$ in $\mathbb{C}^n$ such that

$$f(U \setminus \mathbb{C}^m \times \{0\}^d) \subset S_0 = Y_0 \setminus Y_1.$$

We may contract $\mathbb{C}^n$ into U by a smooth contraction that equals the identity on a smaller open neighborhood $V \subset U$ of $\mathbb{C}^m \times \{0\}^d$. Precomposing f with this contraction yields a continuous map $\mathbb{C}^n \to Y$ which agrees with f on V and maps $\mathbb{C}^n \setminus (\mathbb{C}^m \times \{0\}^d)$ into S_0. Theorem 7.2.1 now provides an entire map $F: \mathbb{C}^n \to Y$ which agrees with f to the second order along $\mathbb{C}^m \times \{0\}^d$ and maps $\mathbb{C}^n \setminus (\mathbb{C}^m \times \{0\}^d)$ into S_0. In particular, F is dominating at 0 and $F(0) = y_0$. This completes the proof when $y_0 \in S_1$.

In general, if $y_0 \in S_k = Y_k \setminus Y_{k+1}$ for some $k \in \{1, \ldots, l-1\}$, we choose strata $\Sigma_j \subset S_j$ for $j = 0, \ldots, k$ such that $y_0 \in \Sigma_k$ and $\Sigma_j \subset \overline{\Sigma}_{j-1}$ for $j = 1, \ldots, k$. Let $m_j = \dim \Sigma_j$, so $m_0 = n > m_1 > \cdots > m_k$. Set $d_j = m_j - m_{j+1}$ for $j = 0, \ldots, k-1$. Since Σ_k is Oka, there is an entire map $g_k: \mathbb{C}^{m_k} \to \Sigma_k$ which sends 0 to y_0 and is dominating at 0. By the above argument and downward induction over $j = 0, \ldots, k-1$, there are entire maps $g_j: \mathbb{C}^{m_j} \to \Sigma_j$ such that $g_j = g_{j+1}$ on $\mathbb{C}^{m_{j+1}} \times \{0\}^{d_j}$, and g_j is dominating at 0 (as a map into Σ_j). For $j = 0$ we thus get an entire map $F = g_0: \mathbb{C}^n \to Y$ which is dominating at 0 with $F(0) = y_0$. □

As an application of Theorem 7.1.8 we now show that all Kummer surfaces are stratified Oka manifolds, and hence strongly dominable in view of Theorem 7.1.8. We also prove a variant of the Oka property for maps of Stein surfaces to Kummer surfaces (see Theorem 7.2.5).

Let us recall the structure of Kummer surfaces (see [45]). Let $\mathbb{T}$ be a complex 2-torus, the quotient of $\mathbb{C}^2$ by a lattice $\mathbb{Z}^4 \cong \Gamma \subset \mathbb{C}^2$ of rank 4, acting on $\mathbb{C}^2$ by translations. Let $\pi: \mathbb{C}^2 \to \mathbb{T} = \mathbb{C}^2/\Gamma$ be the quotient map. The involution $\mathbb{C}^2 \to \mathbb{C}^2$, $(z_1, z_2) \mapsto (-z_1, -z_2)$ descends to an involution $\sigma: \mathbb{T} \to \mathbb{T}$ with 16 fixed points $p_1, \ldots, p_{16}$. In fact, if $\omega_1, \ldots, \omega_4 \in \mathbb{C}^2$ are generators for Γ, then $p_1, \ldots, p_{16}$ are the images under π of the 16 points $c_1\omega_1 + \cdots + c_4\omega_4$, where $c_1, \ldots, c_4 \in \{0, \frac{1}{2}\}$. The quotient space $\mathbb{T}/\{1, \sigma\}$ is a 2-dimensional complex space with 16 singular points $q_1, \ldots, q_{16}$. The singularities can be resolved by blowing them up, yielding a smooth compact surface Y containing 16 mutually disjoint smooth rational curves $C_1, \ldots, C_{16}$. This is the Kummer surface associated to the torus $\mathbb{T}$ or to the lattice Γ.

Here is an alternative description. Let X denote the surface obtained by blowing up $\mathbb{T}$ at each of the points $p_1, \ldots, p_{16}$. Let $E_j \cong \mathbb{P}_1$ denote the exceptional divisor over p_j. The involution σ of $\mathbb{T}$ lifts to an involution $\tau: X \to X$ with the fixed point set $E = E_1 \cup \cdots \cup E_{16}$. The eigenvalues of the differential $\mathrm{d}\tau$ at any point of E are ± 1. Hence the quotient $X/\{1, \tau\}$ is smooth and contains 16 rational (-2)-curves $C_j \cong \mathbb{P}_1$, the images of the rational (-1)-curves E_j in X. The quotient is the Kummer surface Y. Denoting by $\widehat{\mathbb{C}}^2$ the surface obtained by blowing up $\mathbb{C}^2$ at every point of the discrete set $\widetilde{\Gamma} = \pi^{-1}(\{p_1, \ldots, p_{16}\})$, we have the following diagram [45, p. 224]:

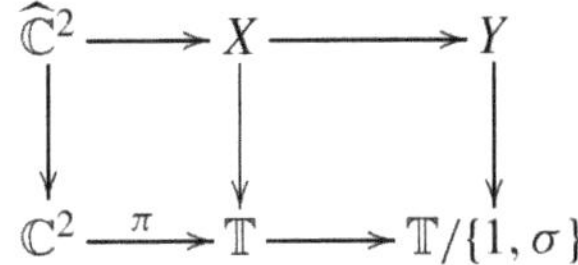

$$
\begin{array}{ccccc}
\widehat{\mathbb{C}}^2 & \longrightarrow & X & \longrightarrow & Y \\
\downarrow & & \downarrow & & \downarrow \\
\mathbb{C}^2 & \xrightarrow{\pi} & \mathbb{T} & \longrightarrow & \mathbb{T}/\{1, \sigma\}
\end{array}
$$

Lemma 7.2.2 *Let Y be a Kummer surface with the exceptional rational curves $C_1, \ldots, C_{16}$. Then, $Y \supset C = C_1 \cup \cdots \cup C_{16} \supset \varnothing$ is a stratification whose strata are Oka.*

Proof Since each curve $C_j \cong \mathbb{P}_1$ is Oka, we only need to prove that $Y \setminus C$ is Oka. Note that the involution $\tau: X \to X$ acts without fixed points on $X \setminus E$, so $X \setminus E$ is an unbranched double-sheeted covering space of $Y \setminus C$. Now $X \setminus E$ is universally covered by $\mathbb{C}^2 \setminus \widetilde{\Gamma}$. Buzzard and Lu [80, Proposition 4.1] showed that the discrete set $\widetilde{\Gamma}$ is tame in $\mathbb{C}^2$ in the sense of Rosay and Rudin (see Sect. 4.6). Hence, $\mathbb{C}^2 \setminus \widetilde{\Gamma}$ is Oka by Proposition 5.6.17 (p. 231). Since the Oka property passes down along unbranched holomorphic covering maps (see Proposition 5.6.3 on p. 224), $X \setminus E$ and $Y \setminus C$ are also Oka. □

Corollary 7.2.3 *Every Kummer surface is strongly dominable.*

Problem 7.2.4 Does every Kummer surface Y enjoy the Oka property for maps $X \to Y$ from all Stein surfaces? More generally, is Y an Oka manifold?

By inspecting the proof of Theorem 7.1.8, we obtain the following positive result for subcritical Stein surfaces, i.e., those with CW decomposition without cells of Morse index 2.

Theorem 7.2.5 *Let Y be a Kummer surface with the rational curves $C = C_1 \cup \cdots \cup C_{16}$. Let X be a Stein surface and Σ be a smooth (possibly disconnected) complex curve in X. If X is obtained from Σ by adding only cells of index 0 or 1, then every holomorphic map $f: \Sigma \to C$ extends to a holomorphic map $F: X \to Y$ such that $F(X \setminus \Sigma) \subset Y \setminus C$. If in addition f is dominating (of rank one) at some point $x \in \Sigma$, then the extension F can be chosen dominating (of rank two) at x.*

Proof By the proof of Theorem 7.1.8, f extends to a holomorphic map $f: U \to Y$ from an open tubular neighborhood U of Σ in S such that $f(U \setminus \Sigma) \subset Y \setminus C$, and the rank of f at any point $x \in \Sigma$ equals the rank of $f|_\Sigma$ plus 1. (The normal bundle of Σ in X is trivial by the Oka-Grauert principle, see Theorem 5.3.1 on p. 213.) Since $Y \setminus C$ is connected, the topological assumption on the pair (X, Σ) implies that f extends to a continuous map $X \to Y$ taking $X \setminus \Sigma$ to $Y \setminus C$. Since $Y \setminus C$ is Oka by Lemma 7.2.2, Theorem 7.2.1 enables us to deform f to a holomorphic map $F: X \to Y$ which agrees with f to second order along Σ and maps $X \setminus \Sigma$ to $Y \setminus C$. However, since $Y \setminus C$ is not simply connected, it may be impossible to extend f across cells of index 2 in a relative CW-complex representing the pair (X, Σ). □

7.3 Oka Properties of Compact Complex Surfaces

In this section we summarize what we know about which minimal compact complex surfaces are Oka, following [212]. Here, $\kappa \in \{-\infty, 0, 1, 2\}$ denotes the Kodaira dimension of the surface.

Surfaces of general type ($\kappa = 2$) are not dominable (see Proposition 7.1.9), and hence not Oka. A complete list of compact complex surfaces, classified according to the value of their Kodaira dimension $\kappa < 2$, can be found in [45] (see Table 10 on p. 244).

$\kappa = -\infty$: Rational surfaces are Oka. A ruled surface is Oka if and only if its base is Oka. Theorem 7.3.2 below covers surfaces of class VII if the global spherical shell conjecture is true.

$\kappa = 0$: Bielliptic surfaces, Kodaira surfaces, and tori are Oka. It is unknown whether any or all K3 surfaces or Enriques surfaces are Oka.

$\kappa = 1$: Any such surface is an elliptic surface; these are often called properly elliptic surfaces. Buzzard and Lu determined which of them are dominable [80]. Nothing further is known about the Oka property for these surfaces.

Let us go through this list and justify the claims.

$\kappa = -\infty$: Every non-singular rational surface can be obtained by repeatedly blowing up a minimal rational surface. The minimal rational surfaces are the projective plane $\mathbb{P}^2$ (which is complex homogeneous and therefore Oka) and the Hirzebruch surfaces Σ_r for $r \in \mathbb{Z}_+$ (here $\Sigma_0 = \mathbb{P}^1 \times \mathbb{P}^1$). The latter are holomorphic $\mathbb{P}^1$-bundles over $\mathbb{P}^1$, so they are Oka by Theorem 5.6.5 (p. 225). Repeated blowups preserve the Oka property for surfaces in this class (Proposition 6.4.6 on p. 274), so non-minimal rational surfaces are also Oka.

A *ruled surface* is a compact surface X which admits a ruling, that is, X is the total space of a holomorphic fibre bundle with fibre $\mathbb{P}^1$ over a compact curve C (see [45, p. 189]). By Theorem 5.6.5 such X is Oka if and only if the base C is Oka, which is so if and only if C is either $\mathbb{P}^1$ or a torus (Corollary 5.6.4, p. 225). Minimal ruled surfaces over $\mathbb{P}^1$ are just the Hirzebruch surfaces.

For surfaces of class VII, see Theorem 7.3.2 below.

$\kappa = 0$: Tori are complex homogeneous and hence Oka.

According to [45, p. 245], every bielliptic surface, and also every primary Kodaira surface, is the total space of a holomorphic fibre bundle with torus fibre over a torus, so it is Oka by Theorem 5.6.5 (p. 225). These surfaces have trivial canonical bundle; we have $b_1(X) = 2$ for bielliptic surfaces and $b_3(X) = 3$ for primary Kodaira surfaces. Primary Kodaira surfaces can also be seen as quotients $\mathbb{C}^2/G$, where G is a properly discontinuous non-abelian group of affine transformations without fixed points on $\mathbb{C}^2$ preserving the volume form $dz \wedge dw$; such a quotient is Oka by Proposition 5.6.3.

Secondary Kodaira surfaces are proper unramified holomorphic quotients of primary Kodaira surfaces, so they are Oka by Proposition 5.6.3. They are elliptic fibrations over $\mathbb{P}^1$ with $b_1(X) = 1$ and with nontrivial canonical bundle.

In this connection it is worth mentioning [45, Theorem 15.4, p. 127]: If X is a compact complex surface and $X \to S$ is a fibration without singular fibres onto a compact complex curve S which is either rational or elliptic, then the fibration is locally trivial, i.e., a holomorphic fibre bundle.

A K3 surface is a surface X with trivial canonical bundle and $b_1(X) = 0$. Examples include Kummer surfaces considered in Sect. 7.2 and most elliptic surfaces, i.e., surfaces admitting a fibration onto a torus with a torus as generic fibre. All elliptic fibrations in the K3 class are ramified, and at this time we have no way of deciding whether any or all of them are Oka.

$\kappa_X = 1$: These are properly elliptic surfaces. Nothing is known about the Oka property in this class; for dominability see Buzzard and Lu [80]. The following example of a non-dominable properly elliptic surface was communicated to me by J. Winkelmann.

Example 7.3.1 We construct an elliptic fibration $\pi: X \to \mathbb{P}^1$ such that $\dim X = 2$, $\kappa_X = 1$, and X is not dominable by $\mathbb{C}^2$.

Let Z be a hyperelliptic Riemann surface of genus $g \geq 2$ with involution $\sigma \in \mathrm{Aut}(Z)$. Then, Z is hyperbolic and $Z/\langle\sigma\rangle \cong \mathbb{P}^1$. Let E be an elliptic curve considered with its group structure, and let $\tau \in E$ be a non-neutral element of order 2, so $2\tau = 0$ in E. Then, $x \to x + \tau$ is a fixed point free involution on E. Take $X = (Z \times E)/\Gamma$, where Γ is the cyclic group of automorphisms generated by $\gamma(z, x) = (\sigma(z), x + \tau)$ which acts without fixed points on $Z \times E$. Let $\pi: X \to \mathbb{P}^1$ be the ramified elliptic fibration induced by the projection $p: Z \times E \to Z$. We have a commutative diagram

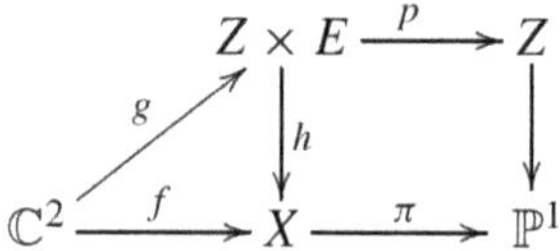

where h is a holomorphic covering. A holomorphic map $f: \mathbb{C}^2 \to X$ lifts to a map $g: \mathbb{C}^2 \to Z \times E$ such that $f = h \circ g$. Since Z is hyperbolic, the image of g is contained in a fibre of p, and hence the map $\pi \circ f: \mathbb{C}^n \to \mathbb{P}^1$ is constant. Thus, X is not dominable by $\mathbb{C}^2$. Note that $\kappa_X = 1$. □

Class VII in the Enriques-Kodaira classification comprises the non-algebraic compact complex surfaces of Kodaira dimension $\kappa = -\infty$. Minimal surfaces of class VII fall into several mutually disjoint classes. For second Betti number $b_2 = 0$, we have *Hopf surfaces* and *Inoue surfaces*. For $b_2 \geq 1$, there are *Enoki surfaces*, *Inoue-Hirzebruch surfaces*, and *intermediate surfaces*; together they form the class of *Kato surfaces*. By the *global spherical shell conjecture*, currently proved only for $b_2 = 1$ by Teleman [545], every minimal surface of class VII with $b_2 \geq 1$ is a Kato surface. Assuming that the conjecture holds, it is possible to determine which minimal surfaces of class VII are Oka.

Theorem 7.3.2 ([212, Theorem 4]) *Minimal Hopf surfaces and minimal Enoki surfaces are Oka. Inoue surfaces, Inoue-Hirzebruch surfaces, and intermediate surfaces, minimal or blown up, are not strongly Liouville, and hence not Oka.*

Proof If X is Hopf, its universal covering space is $\mathbb{C}^2 \setminus \{0\}$ which is Oka (see Proposition 5.6.10, p. 227), so X is Oka by Proposition 5.6.3 (p. 224).

If X is Inoue, then the universal covering space of X is $\mathbb{D} \times \mathbb{C}$, where $\mathbb{D}$ is the disc. Clearly, $\mathbb{D} \times \mathbb{C}$ is not Oka, and hence an Inoue surface is not Oka.

If X is intermediate, then X is not strongly Liouville by [127, Corollary 2.13], and hence is not Oka.

Let X be Inoue-Hirzebruch and D be the union of the finitely many rational curves in X. Let $\widetilde{D}$ be the preimage of D in the universal covering space $\widetilde{X}$ of X. The complement $\widetilde{X} \setminus \widetilde{D}$ is described in [127], proof of Theorem 2.16, and in [591, pp. 400–401]: it is isomorphic to the image by the map

$$\mathbb{C} \times \mathbb{C} \to \mathbb{C}^* \times \mathbb{C}^*, \qquad (\zeta_1, \zeta_2) \mapsto \left(e^{\zeta_1}, e^{\zeta_2}\right),$$

of the half-space in $\mathbb{C} \times \mathbb{C}$ defined by the inequality $-d\Re\zeta_1 + c\Re\zeta_2 < 0$, where $d < 0 < c$. Thus, $-d \log|z_1| + c \log|z_2|$ defines a nonconstant negative plurisubharmonic function on $\widetilde{X} \setminus \widetilde{D}$, which extends across $\widetilde{D}$ to a plurisubharmonic function on $\widetilde{X}$. It follows that X is not strongly Liouville and hence not Oka.

It is easily seen that not being strongly Liouville is preserved by blowing up. It follows that blown-up Inoue, Inoue-Hirzebruch, and intermediate surfaces are not strongly Liouville, and hence not Oka.

By Enoki's construction of the surfaces that now bear his name ([153], [154, Sect. 3]), the universal covering space Y of an Enoki surface X is obtained as follows. Let $W_0 = \mathbb{P}_1 \times \mathbb{C}$ and $\Gamma = \{\infty\} \times \mathbb{C} \subset W_0$. For each $k \geq 0$, W_{k+1} is W_k blown up at two distinct points p_k and p_{-k-1}, such that, when $k \geq 1$, p_k lies in the total transform of p_{k-1}, and p_{-k-1} lies in the total transform of p_{-k}. We take $p_0 = (\infty, 0)$ and $p_{-1} = (a, 0)$ with $a \in \mathbb{C}$. Also, p_k lies in the proper transform Γ_k of Γ, but p_{-k-1} lies outside the total transform of Γ. (We interpret the proper transform and the total transform of Γ in W_0 as Γ itself.) Then $p_{-1}, p_{-2}, \ldots$ lie in the total transforms of the line $\{a\} \times \mathbb{C}$. Let $Y_k = W_k \setminus (\Gamma_k \cup \{p_{-k-1}\})$. Then Y_k may be viewed as an open subset of Y_{k+1}, and Y is the colimit of the sequence $Y_0 \subset Y_1 \subset Y_2 \subset \cdots$. (There is a misprint on [154, p. 459]: the total transform of p_{-k-1} is C_{-k-1}, not C_{-k-2}.)

It suffices to show that Y_k is Oka for all $k \geq 0$; their union Y is then Oka (see Proposition 5.6.7, p. 226), so X is Oka (see Proposition 5.6.7, p. 226). We claim that $W_k \setminus \Gamma_k$ is Zariski-locally affine (affine meaning algebraically isomorphic to $\mathbb{C}^2$); then Y_k is Oka by Proposition 6.4.5 on p. 274.

Being Zariski-locally affine is preserved by blowing up points (Proposition 6.4.6 on p. 274). Since $W_0 \setminus \Gamma = \mathbb{C} \times \mathbb{C}$ is affine, the complement in W_k of the total transform of Γ is Zariski-locally affine. Thus we need to show that every point in $W_{k+1} \setminus \Gamma_{k+1}$, $k \geq 0$, that lies in the total transform of Γ has an affine Zariski-open neighborhood in $W_{k+1} \setminus \Gamma_{k+1}$.

We claim that every point in W_k, $k \geq 0$, that lies in the total transform of Γ has an affine Zariski neighborhood U in W_k containing p_k but not p_{-k-1}, in which Γ_k appears as a straight line. Namely, for $k = 0$, let $U = (\mathbb{P}_1 \setminus \{a\}) \times \mathbb{C}$. Suppose the claim is true for k and let $w \in W_{k+1}$ lie in the total transform of Γ. Let V be an affine Zariski-open neighborhood of the image of w in W_k containing p_k but not p_{-k-1}, in which Γ_k appears as a straight line. Blowing up V at p_k yields a Zariski-open neighborhood V' of w in W_{k+1}. Take a line L in V through p_k different from Γ_k, whose proper transform L' contains neither w nor p_{k+1}, and set $U = V' \setminus L'$. Then U is a Zariski neighborhood of w in W_{k+1} containing p_{k+1}

but not p_{-k-2}. Moreover, U is algebraically isomorphic to the total space of an algebraic line bundle over $\mathbb{C}$, so U is affine, and Γ_{k+1} appears as a straight line in U.

Finally, let $w \in W_{k+1} \setminus \Gamma_{k+1}$, $k \geq 0$, lie in the total transform of Γ. Let U be an affine Zariski-open neighborhood of the image of w in W_k containing p_k but not p_{-k-1}, in which Γ_k appears as a straight line. Blowing up U at p_k and removing the proper transform of Γ_k yields an affine Zariski-open neighborhood of w in $W_{k+1} \setminus \Gamma_{k+1}$.

This shows that minimal Enoki surfaces are Oka. □

Enoki surfaces are generic among Kato surfaces. Inoue-Hirzebruch surfaces and intermediate surfaces can be obtained as degenerations of Enoki surfaces (see Dloussky [126]). Thus, Theorem 7.3.2 yields the following corollary.

Corollary 7.3.3 ([212, Corollary 5]) *Compact complex surfaces that are Oka can degenerate to a surface that is not strongly Liouville, and hence is not Oka. Consequently, the following properties are in general not closed in holomorphic families of compact complex manifolds*:

- *the Oka property*;
- *the stratified Oka property*;
- *strong dominability*;
- *dominability*;
- *$\mathbb{C}$-connectedness*;
- *strong Liouvilleness.*

7.4 Oka Maps

In analogy to the class of Oka manifolds, we now introduce the class of *Oka maps*, following Lárusson [371, Sect. 16] and the author [201].

Let $\pi: E \to B$ be a holomorphic map of a complex space E onto a complex space B. (All complex spaces are assumed to be reduced.) Assume that X is a Stein space, $P_0 \subset P$ are compact Hausdorff spaces (the parameter spaces), and $f: P \times X \to B$ is a continuous X-holomorphic map, meaning that the map $f(p, \cdot): X \to B$ is holomorphic for every fixed $p \in P$. Assume that $F_0: P \times X \to E$ is continuous lifting of f, i.e., $\pi \circ F_0 = f$, which is X-holomorphic over P_0. We are looking for a homotopy of liftings $F_t: P \times X \to E$ of f $(t \in [0, 1])$ which is fixed on $P_0 \times X$ such that $F_1 = F$ is an X-holomorphic lifting. The situation is illustrated with the following diagram; note that F makes both triangles commute:

$$\begin{array}{ccc} P_0 \times X & \xrightarrow{F_0} & E \\ {\scriptstyle incl}\downarrow & \overset{F}{\nearrow} & \downarrow{\scriptstyle \pi} \\ P \times X & \xrightarrow{f} & B \end{array}$$

We say that the map $\pi: E \to B$ satisfies the *parametric Oka property* if such a homotopy F_t exists for any data (X, P_0, P, f, F_0). In the following definition, we also include approximation and interpolations conditions.

Definition 7.4.1 A holomorphic map $\pi: E \to B$ between reduced complex spaces enjoys the *Parametric Oka Property with Approximation and Interpolation (POPAI)* if for any collection $(X, X', K, P, P_0, f, F_0)$ where X is a reduced Stein space, X' is a closed complex subvariety of X, K is a compact $\mathcal{O}(X)$-convex subset of X, $P_0 \subset P$ are compact Hausdorff spaces, $f: P \times X \to B$ is a continuous X-holomorphic map, and $F_0: P \times X \to E$ is a continuous map such that $\pi \circ F = f$, the map $F_0(p, \cdot)$ is holomorphic on X for all $p \in P_0$ and is holomorphic on $K \cup X'$ for all $p \in P$, there exists a homotopy $F_t: P \times X \to E$ such that the following hold for all $t \in [0, 1]$:

(i) $\pi \circ F_t = f$,
(ii) $F_t = F_0$ on $(P_0 \times X) \cup (P \times X')$,
(iii) F_t is X-holomorphic on K and uniformly close to F_0 on $P \times K$, and
(iv) the map $F_1: P \times X \to E$ is X-holomorphic.

The map $\pi: E \to B$ enjoys the *Basic Oka Property with Approximation and Interpolation (BOPAI)* if the above holds when P is a singleton and $P_0 = \varnothing$.

Neglecting the approximation and the interpolation conditions, the parametric Oka property of a holomorphic map $\pi: E \to B$ can be illustrated by the following commutative diagram of maps:

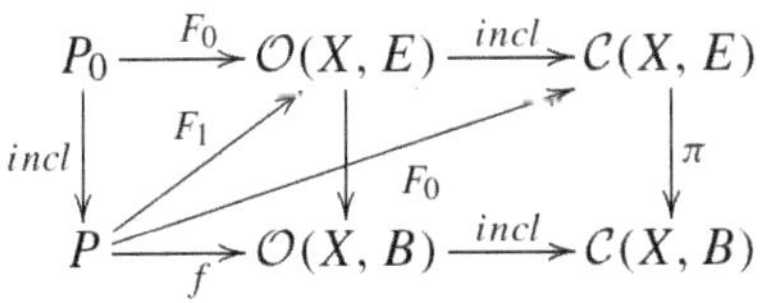

Note that a complex manifold E enjoys a certain Oka property if and only if the trivial map $\pi: E \to \textit{point}$ does. For a complex manifold Y, all Oka properties for parameter spaces $P_0 \subset P \subset \mathbb{R}^m$ are equivalent (see Proposition 5.15.1, p. 259), and such Y is an Oka manifold.

Note that the problem of lifting a holomorphic map $f: X \to B$ to a holomorphic map $F: X \to E$ reduces to the problem of finding a section of the pullback of $\pi: E \to B$ by f. As an example, we prove the following result.

Corollary 7.4.2 (i) *Every stratified subelliptic submersion enjoys* BOPAI.
(ii) *Every stratified holomorphic fibre bundle with Oka fibres enjoys* BOPAI.

Proof Let $\widetilde{\pi}: f^*E \to X$ denote the pullback of a holomorphic submersion $\pi: E \to B$ by a holomorphic map $f: X \to B$.

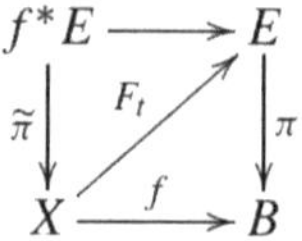

If $\pi: E \to B$ is a stratified subelliptic submersion (case (i)), then $\widetilde{\pi}: f^*E \to X$ is also a stratified subelliptic submersion. (Stratify X such that each stratum is mapped by f into a stratum of B over which the submersion π is subelliptic.) Then, liftings $X \to E$ of f are in bijective correspondence with sections of $\widetilde{\pi}: f^*E \to X$. Since the latter satisfy BOPAI by Theorem 6.2.2 (p. 265), π also satisfies BOPAI. The same proof applies in case (ii). □

The above argument no longer holds in the parametric case as there is no holomorphic dependence of the pullback $f^*E \to X$ on the map $f: X \to B$ in the given family. Nevertheless, BOPAI $\Longrightarrow$ POPAI still holds.

Theorem 7.4.3 ([201]) *For every holomorphic submersion $\pi: E \to B$ of reduced complex spaces we have the implication BOPAI $\Longrightarrow$ POPAI, where* POPAI *is restricted to parameter spaces $P_0 \subset P$ that are Euclidean compacts.*

Proof (sketch) In [201, Proposition 2.1] it is shown that BOPAI implies the *homotopy approximation property* (HAP) for liftings, similar to the one expressed for sections by Theorem 6.6.2. (Compare with Definition 6.6.5 on p. 287.) Next, one shows that if every point $b \in B$ admits an open neighborhood $U \subset B$ such that the restricted submersion $\pi: E|_U = \pi^{-1}(U) \to U$ satisfies HAP, then $\pi: E \to B$ satisfies POPAI (see [4, Theorem 4.2]). The main point is that, by using HAP over small open sets in B, we can assemble complexes of local holomorphic liftings over $\mathcal{C}$-strings into holomorphic liftings exactly as was done for sections in the proof of Theorem 6.2.2. □

The same proof shows that the following result (see [200, Theorem 4.7]) to the effect that POPAI is a local property of a map.

Theorem 7.4.4 (Localization principle for POPAI) *A holomorphic submersion $\pi: E \to B$ of a reduced complex space E onto a reduced complex space B satisfies* POPAI *if and only if every point $x \in B$ admits an open neighborhood $U_x \subset B$ such that the restricted submersion $\pi: E|_{U_x} \to U_x$ satisfies* POPAI.

The following is a consequence of Theorem 7.4.3 and Corollary 7.4.2.

Corollary 7.4.5 (i) *Every stratified subelliptic submersion enjoys* POPAI.
(ii) *Every stratified holomorphic fibre bundle with Oka fibres enjoys* POPAI.

Example 7.4.6 We say that f is a *stratified holomorphic function* on a complex space X if there is a stratification $X = X_0 \supset X_1 \supset \cdots \supset X_m = \varnothing$ such that f is defined over the union of some of the strata $S_j = X_j \setminus X_{j+1}$, it is holomorphic over every stratum S_j, and its graph $\Gamma_f \subset X \times \mathbb{C}$ is closed. Then, $Z = (X \times \mathbb{C}) \setminus \Gamma_f \to X$ is a submersion, and the restriction $Z|_S \to S$ to any stratum S in the above stratification is a holomorphic fibre bundle with fibre $\mathbb{C}$ or $\mathbb{C}^*$. Hence, $Z \to X$ is a stratified elliptic submersion and consequently it enjoys POPAI. In particular, if X is Stein, then sections $X \to Z$ satisfy the Oka property. (These are holomorphic functions $g \in \mathcal{O}(X)$ such that $g(x) \neq f(x)$ for each x.) A particular example of this type is a meromorphic function f on X with the polar locus $X_1 \subset X$ and without the indeterminacy set; $f|_{X_1}$ can again be a function of the same type, etc. □

We now introduce the class of *Oka maps*. For the definition of a Serre fibration, see e.g. [569].

Definition 7.4.7 A holomorphic map $\pi: E \to B$ of reduced complex spaces is an *Oka map* if it is a Serre fibration and it enjoys POPAI (see Definition 7.4.1).

A holomorphic map is an Oka map precisely when it is an *intermediate fibration* in Lárusson's model category [371, 372] (see Sect. 7.5 below).

Corollary 7.4.5 implies the following result.

Corollary 7.4.8 (i) *A holomorphic fibre bundle projection with Oka fibre is an Oka map.*

(ii) *A stratified subelliptic submersion, or a stratified holomorphic fibre bundle projection with Oka fibres, is an Oka map if and only if it is a Serre fibration.*

Assume that $\pi: E \to B$ is a holomorphic submersion. If π enjoys the (basic or the parametric) Oka property, then by considering liftings of constant maps $X \to b \in B$ we see that every fibre of π is an Oka manifold. If we allow noncompact fibres, then the converse fails as shown by the following result of Eremenko and its corollary.

Theorem 7.4.9 ([155, Theorem 2]) *Assume that D is a domain in $\mathbb{C}$ and $f: D \times \mathbb{C} \to \mathbb{C}$ a holomorphic function such that $\mathbb{C} \ni w \mapsto f(z, w)$ is nonconstant for all $z \in D$. Assume that for some function $g: D \to \mathbb{C}$ we have $f(z, w) \neq g(z)$ for all $w \in \mathbb{C}$. Then there exists a discrete set $\Lambda = \{z_j\} \subset D$ such that g is holomorphic in $D \setminus \Lambda$ and $g(z) \to \infty$ as $z \to z_j$ for every $z_j \in \Lambda$.*

The function g in Eremenko's theorem can take arbitrary values at the points $z_j \in \Lambda$, and hence it is stratified holomorphic (see Example 7.4.6).

Corollary 7.4.10 *Let $g: D \to \mathbb{C}$ be a continuous function on a domain $D \subset \mathbb{C}$, and denote by Γ_g its graph. Then, the projection $\pi: E_g = D \times \mathbb{C} \setminus \Gamma_g \to D$ is an Oka map if and only if g is holomorphic.*

Proof If g is holomorphic, then E_g is biholomorphic to $D \times \mathbb{C}^*$ by the map $(z, w) \mapsto (z, w - g(z))$, so $\pi: E_g \to D$ is an Oka map by Corollary 7.4.8(i). Conversely, assume that the projection $\pi: E_g \to D$ is an Oka map. Then, the continuous lifting $F_0: D \times \mathbb{C}^* \to E_g$, $(z, w) \mapsto (z, w + g(z))$ of the projection map $D \times \mathbb{C}^* \to D$ can be deformed to a holomorphic lifting $F_1(z, w) = (z, f(z, w))$. Note that f satisfies the hypothesis of Theorem 7.4.9, and hence g is holomorphic. □

In the case of compact fibres the following problem is open.

Problem 7.4.11 Assume that $\pi : E \to B$ is a family of compact complex manifolds such that each fibre E_b $(b \in B)$ is an Oka manifold. Does it follow that π is an Oka map?

Note that a family of compact complex manifolds is always a smooth fibre bundle and hence a Serre fibration. The following positive result in this direction was proved by Lárusson [375, Theorem 8].

Theorem 7.4.12 *The projection of a family of complex tori is an Oka map.*

Proof Let $\mathscr{B}$ be the space of $n \times n$ complex matrices T with $\det \Im T > 0$. Let $\mathscr{X}$ be the quotient of $\mathbb{C}^n \times \mathscr{B}$ by the free and properly discontinuous action of $\mathbb{Z}^{2n}$ given by the formula

$$m \cdot (z, T) = \left(z + m \cdot \begin{bmatrix} I \\ T \end{bmatrix}, T\right),$$

with the induced projection $p : \mathscr{X} \to \mathscr{B}$. (Here $m \in \mathbb{Z}^{2n}$ is a row vector multiplied by the $2n \times n$ matrix with blocks I and T.) It is well known that the family p is complete and effective and contains every n-dimensional torus [343]; that is, p is a minimal versal deformation of every n-dimensional torus.

Next, observe that p has a dominating fibre spray σ defined on the trivial vector bundle $\mathscr{X} \times \mathbb{C}^n$ over $\mathscr{X}$ by the formula $\sigma([z, T], w) = [z + w, T]$. Hence, p is elliptic and thus Oka.

Finally, let $\pi : E \to B$ be a family of n-dimensional tori. Let $t \in B$. By completeness of the family p, there is an open neighborhood U of t in B and a holomorphic map $f : U \to \mathscr{B}$ such that the restriction $\pi: \pi^{-1}(U) \to U$ is isomorphic to the pullback family f^*p. Since pullbacks of Oka maps are Oka, we conclude that π is Oka over a neighborhood of each point in B. Hence, π is Oka by Theorem 7.4.4. □

7.5 A Homotopy-Theoretic Viewpoint on Oka Theory

This section was contributed by Finnur Lárusson and is taken from [204, Appendix]. Its purpose is to explain how Oka manifolds and Oka maps naturally fit into an abstract homotopy-theoretic framework, not merely by way of analogy, but in precise, rigorous terms. The connection between Oka theory and abstract homotopy theory was developed in the papers [370–373].

Abstract homotopy theory, also known as *homotopical algebra*, was founded by D. Quillen in his 1967 monograph [465]. The fundamental notion of the theory is the concept of a *model category*, or a model structure on a category. A model structure is an abstraction of the essential features of the category of topological spaces that make ordinary homotopy theory possible. Model structures are good for many things. They have been introduced and applied in various areas of mathematics outside of homotopy theory, for example in homological algebra, algebraic geometry, category theory, and theoretical computer science. Here we view them as a tool for studying lifting and extension properties of holomorphic maps. Model structures provide a framework for investigating two classes of maps such that the first has the right lifting property with respect to the second and the second has the left lifting property with respect to the first in the absence of topological obstructions. It is more natural, in fact, to consider homotopy lifting properties rather than plain lifting properties, that is, liftings of families of maps varying continuously with respect to a parameter in a nice parameter space rather than liftings of individual maps.

One of the main results of Gromov in his seminal paper [266, Sect. 2.9] suggests a link with homotopical algebra. Let $T \hookrightarrow S$ be the inclusion into a Stein manifold S of a closed complex submanifold T (we call such an inclusion a *Stein inclusion*), and let $X \to Y$ be a holomorphic fibre bundle whose fibre is an elliptic manifold (let us call such a map an *elliptic bundle*). Consider a commuting square

where $T \to X$ and $S \to Y$ are otherwise arbitrary holomorphic maps. A basic version of Gromov's Oka principle states that every continuous lifting in the square, that is, every continuous map $S \to X$ such that the diagram (7.1) commutes, can be deformed through such liftings to a holomorphic lifting. Since $T \hookrightarrow S$ is a topological cofibration and $X \to Y$ a topological fibration, by elementary homotopy theory there is a continuous lifting in the square, and hence by Gromov's theorem a holomorphic lifting, if one of the two vertical maps is a homotopy equivalence (let us call a homotopy equivalence an *acyclic map*). Thus, elliptic bundles have the right lifting property with respect to acyclic Stein inclusions, and acyclic elliptic bundles have the right lifting property with respect to Stein inclusions.

Compare this with one of Quillen's axioms for a model category: A lifting $S \to X$ exists in every commuting square (7.1) in which $T \to S$ is a cofibration, $X \to Y$ is a fibration, and one of them is acyclic. It is now natural to ask whether there is a model category containing the category of complex manifolds (it is too small to carry a model structure itself) in which Stein inclusions are cofibrations, elliptic bundles are fibrations, and weak equivalences are defined topologically.

The answer is affirmative. There is a natural, explicit way to embed the category of complex manifolds into a model category such that Gromov's theorem becomes an instance of Quillen's axiom. In fact, a holomorphic map is a fibration in this model structure if and only if it is an Oka map [371, Corollary 20]. In particular, a complex manifold is fibrant as an object in the model category (meaning that the map to the terminal object is a fibration) if and only if it is Oka. Also, a complex manifold is cofibrant if and only if it is Stein [372, Theorem 6].

Model Categories and Simplicial Sets

A *model category* is a category with all small limits and colimits and three distinguished classes of maps, called weak equivalences or acyclic maps, fibrations, and cofibrations, such that the following axioms hold.

(A1) If f and g are composable maps, and two of $f, g, f \circ g$ are acyclic, then so is the third.

(A2) The classes of weak equivalences, fibrations, and cofibrations are closed under retraction. (Also, it follows from the axioms that the composition of fibrations is a fibration, and the pullback of a fibration by an arbitrary map is a fibration.)

(A3) A lifting $S \to X$ exists in every commuting square (7.1) in which $T \to S$ is a cofibration, $X \to Y$ is a fibration, and one of them is acyclic.

(A4) Every map can be functorially factored as

$$\text{acyclic fibration} \circ \text{cofibration}$$

and as

$$\text{fibration} \circ \text{acyclic cofibration}.$$

For the theory of model categories, we refer the reader to [139, 296, 303, 411].

There are many examples of model categories. A fundamental example, closely related to the category of topological spaces, is the category of simplicial sets. Simplicial sets are combinatorial objects that have a homotopy theory equivalent to that of topological spaces, but tend to be more useful or at least more convenient than topological spaces for various homotopy-theoretic purposes. In homotopy-theoretic parlance, the distinction between topological spaces and simplicial sets is blurred and the latter are often referred to as spaces. The prototypical example of a simplicial set is the singular set sX of a topological space X. It consists of a sequence $sX_0, sX_1, sX_2, \ldots$ of sets, where sX_n is the set of n-simplices in X, that is, the set of all continuous maps into X from the standard n-simplex

$$T_n = \{(t_0, \ldots, t_n) \in \mathbb{R}^{n+1} : t_0 + \cdots + t_n = 1,\ t_0, \ldots, t_n \geq 0\},$$

along with face maps $sX_n \to sX_{n-1}$ and degeneracy maps $sX_n \to sX_{n+1}$. The weak homotopy type of X is encoded in sX. For an introduction to simplicial sets, we refer the reader to [243, 410].

Complex Manifolds as Prestacks on the Stein Site

We have claimed that the embedding of the category of complex manifolds into a model category that realizes the Oka property as fibrancy is natural and explicit, but it is still quite technical. We shall give a sketch here; the details may be found in [371].

First of all, how could we expect to be able to do homotopy theory with complex manifolds in a way that takes not only their topology but also their complex structure into account? The answer lies in the following key observations, which vastly expand the scope of homotopical algebra.

- Not only can we do homotopy theory with individual spaces, but also with diagrams or sheaves of them.
- Manifolds and varieties can be thought of as sheaves of spaces, so we can do homotopy theory with them too. The general idea is known as the Yoneda lemma.

This line of thought has found a spectacular application in V. Voevodsky's homotopy theory of schemes and the resulting proof of the Milnor conjecture [562].

The gist of the Yoneda lemma, sometimes called the most basic theorem in mathematics, is that an object is determined up to isomorphism by its relationships with other objects, that is, by the system of arrows into it from all other objects. More precisely, there is a full embedding of each small category $\mathscr{M}$ into the category of presheaves of sets on itself, taking an object X to the presheaf $\mathscr{M}(\cdot, X)$.

In our case, Stein manifolds play a special role as sources of maps, so we think of a complex manifold X as defining a presheaf $\mathscr{O}(\cdot, X)$ on the full subcategory $\mathscr{S}$ of Stein manifolds of the category $\mathscr{M}$ of complex manifolds. The presheaf consists of the set $\mathscr{O}(S, X)$ of holomorphic maps $S \to X$ for each Stein manifold S, along with the precomposition map $\mathscr{O}(S_2, X) \to \mathscr{O}(S_1, X)$ induced by each holomorphic map $S_1 \to S_2$ between Stein manifolds. Even though $\mathscr{S}$ is much smaller than $\mathscr{M}$, it may be shown that the presheaf $\mathscr{O}(\cdot, X)$ determines X, so we have an embedding, in fact a full embedding, of $\mathscr{M}$ into the category of presheaves of sets on $\mathscr{S}$.

Each set $\mathscr{O}(S, X)$ carries the compact-open topology. A map between such sets defined by pre- or postcomposition by a holomorphic map is continuous. We may therefore consider a complex manifold X as a presheaf of topological spaces on $\mathscr{S}$. This presheaf has the property that as a holomorphic map $S_1 \to S_2$ between Stein manifolds is varied continuously in $\mathscr{O}(S_1, S_2)$, the induced precomposition map $\mathscr{O}(S_2, X) \to \mathscr{O}(S_1, X)$ varies continuously as well. We would like to do homotopy theory with complex manifolds viewed as presheaves with this property.

Somewhat unexpectedly, as explained in [371, Sect. 3], there are solid reasons, beyond mere convenience, to rephrase the above entirely in terms of simplicial sets. For the technical terms that follow, we refer the reader to [371] and the references cited there. To summarize, we turn $\mathscr{S}$ into a simplicial site and obtain an embedding of $\mathscr{M}$ into the category $\mathfrak{S}$ of prestacks on $\mathscr{S}$. The basic homotopy theory of prestacks on a simplicial site was developed by B. Toën and G. Vezzosi for use in algebraic geometry [551]. A new model structure on $\mathfrak{S}$, called the *intermediate structure* and based on ideas of J.F. Jardine, later published in [309], was constructed in [371]. It is in this model structure that Gromov's Oka principle finds a natural home.

The main results of [371] along with [372, Theorem 6] can be summarized as follows.

Theorem 7.5.1 *The category of complex manifolds and holomorphic maps can be embedded into a model category such that:*

- *a holomorphic map is acyclic when viewed as a map in the ambient model category if and only if it is a homotopy equivalence in the usual topological sense.*
- *a holomorphic map is a fibration if and only if it is an Oka map. In particular, a complex manifold is fibrant if and only if it is Oka.*
- *a complex manifold is cofibrant if and only if it is Stein.*
- *a Stein inclusion is a cofibration.*

A characterization of those holomorphic maps that are cofibrations is missing from this result. It may be that Stein inclusions and biholomorphisms are the only ones.

Knowing that Oka maps are fibrations in a model structure helps us understand and predict their behaviour. For example, it is immediate by abstract nonsense that the composition of Oka maps is Oka, that a retract of an Oka map is Oka, and that the pullback of an Oka map by an arbitrary holomorphic map is Oka (it is easily seen that the pullback exists in $\mathscr{M}$ and agrees with the pullback in $\mathfrak{S}$). Also, in any model category, the source of a fibration with a fibrant target is fibrant. It follows that the source of an Oka map with an Oka target is Oka. On the other hand, the fact that the image of an Oka map with an Oka source is Oka is a surprising feature of Oka theory not predicted by abstract nonsense, the reason being that the Oka property can be detected using Stein inclusions of the special kind $T \hookrightarrow \mathbb{C}^n$, where T is contractible.

Fibrant and Cofibrant Models

It is a familiar process in mathematics to associate to an object a closely related but better behaved object, with a good map between the two. For example, to a topological space we can associate a CW approximation, to a simplicial set a Kan complex, and to a module over a ring a projective resolution. These are examples of fibrant and cofibrant models, also known as approximations or resolutions. A fibrant model for an object X in a model category is a fibrant object Z with an acyclic cofibration (or sometimes only an acyclic map) $X \to Z$. Factoring the map from X to the terminal object as an acyclic cofibration followed by a fibration using axiom A4 above, we see that X has a fibrant model, and by axioms A1 and A3, any two fibrant models for X are weakly equivalent. The dual notion is that of a cofibrant model.

Thinking about fibrant and cofibrant models for complex manifolds in the model structure described above leads to interesting questions. As far as we know, these concepts had not been considered previously. A cofibrant model for a complex manifold X that lives in the category of complex manifolds (and not merely in the ambient model category) is a cofibrant complex manifold S, that is, a Stein manifold, with an acyclic fibration $S \to X$, that is, a surjective Oka map with contractible fibres. We call S a *Stein model* for X. Note that X is Oka if and only if S is Oka, and by Proposition 5.6.15 (p. 230) this holds if and only if S is elliptic.

It is an open question whether every complex manifold has a Stein model. Many complex manifolds X, for example all quasi-projective manifolds, are known to have a Stein model of a very special kind, namely an affine bundle $E \to X$ whose total space E is Stein (see Example 1.4.4 on p. 14 and [372], [211, Sect. 8G]). As mentioned above, such a manifold X is Oka if and only if E is elliptic. This is a purely geometric characterization of the Oka property, generalizing Gromov's result that a Stein manifold is Oka if and only if it is elliptic.

An example of a fibrant model for a Stein manifold S is an acyclic Stein inclusion $S \hookrightarrow X$ such that the Stein manifold X is Oka, that is, elliptic. It is an open question whether every Stein manifold S has such a fibrant model. We are asking for a variant of the Remmert-Bishop-Narasimhan embedding theorem with a proper holomorphic embedding that preserves the homotopy type of S and whose target retains the key properties of affine space of being Stein and elliptic. T. Ritter proved that every open Riemann surface acyclically embeds into an elliptic manifold [472, Theorem 4.2], and that when the surface is an annulus, the target may be taken to be $\mathbb{C} \times \mathbb{C}^*$ [473]. The latter statement is a special case of Ritter's result, proved using the embedding techniques of E. F. Wold, that every continuous map from a finitely connected circular domain in $\mathbb{C}$ to $\mathbb{C} \times \mathbb{C}^*$ is homotopic to an embedding. More general results, for finitely connected planar domains, are proved in [377]. The subject of embedding open Riemann surfaces into (domains in) $\mathbb{C}^2$ is treated in Sects. 9.10–9.11.

Affine Simplices in Oka Manifolds

Motivated by Gromov's comments in [266, Sects. 3.5.G, 3.5.G'], the affine singular set eX of a complex manifold X was defined in [373] as the simplicial set whose n-simplices for each $n \geq 0$ are the holomorphic maps into X from the affine n-simplex

$$A_n = \{(t_0, \dots, t_n) \in \mathbb{C}^{n+1} : t_0 + \cdots + t_n = 1\},$$

viewed as a complex manifold biholomorphic to $\mathbb{C}^n$, with the obvious face maps and degeneracy maps. If X is Brody hyperbolic, then eX is discrete and carries no topological information about X. On the other hand, when X is Oka, eX is "large".

A holomorphic map $A_n \to X$ is determined by its restriction to $T_n \subset A_n$, so we have a monomorphism, that is, a cofibration $eX \hookrightarrow sX$ of simplicial sets. When X is Oka, eX, which is of course much smaller than sX, carries the weak homotopy type of X. More precisely, the cofibration $eX \hookrightarrow sX$ is the inclusion of a strong deformation retract [373, Theorem 1]. Even for complex Lie groups, this result appears not to have been previously known.

7.6 Miscellanea and Open Problems

In this section we collect additional results, open problems and promising directions for future research in Oka theory.

Complements and Unions

Problem 7.6.1

(a) Let K be a compact convex set in $\mathbb{C}^n$ for some $n > 1$. Is $\mathbb{C}^n \setminus K$ an Oka manifold?
(b) In particular, is the complement of the closed ball $\overline{\mathbb{B}^n}$ an Oka manifold? Does there exist a surjective holomorphic map $\mathbb{C}^n \to \mathbb{C}^n \setminus \overline{\mathbb{B}^n}$?
(c) Is the complement $\mathbb{C}^n \setminus \mathbb{R}^k$ of a totally real affine subspace Oka?

In connection with (b), Dixon and Esterle [125, Theorem 8.13, p. 182] constructed for every $\epsilon > 0$ a finitely sheeted holomorphic map $f : \mathbb{C}^2 \to \mathbb{C}^2$ whose image avoids the closed unit ball $\overline{\mathbb{B}^2}$ but contains the complement of the ball of radius $1+\epsilon$:

$$\mathbb{C}^2 \setminus (1+\epsilon)\overline{\mathbb{B}^2} \subset f(\mathbb{C}^2) \subset \mathbb{C}^2 \setminus \overline{\mathbb{B}^2}.$$

Since every connected Oka manifold Y of dimenson n is the image of a surjective holomorphic map $\mathbb{C}^n \to Y$ (see Theorem 5.5.1, p. 219), a negative answer to the second question in (b) would imply that $\mathbb{C}^n \setminus \overline{\mathbb{B}^n}$ fails to be Oka. Since $\mathbb{C}^n \setminus \overline{\mathbb{B}^n}$ is a union of Fatou-Bieberbach domains, this would also provide an example of a strongly dominable manifold which is not Oka.

By Proposition 7.1.13 (p. 325), $\mathbb{C}^n \setminus \overline{\mathbb{B}^n}$ fails to be subelliptic if $n \geq 3$. On the other hand, it was shown in [220] that for any compact convex set K in $\mathbb{C}^n$ for $n > 1$ the complement $\mathbb{C}^n \setminus K$ has the Oka property for maps from all Stein manifolds of dimension $< n$ (see Theorem 9.8.8 on p. 441). The same holds for the complement $\mathbb{C}^n \setminus \mathbb{R}^k$ when $1 \leq k < n$, see [229].

Problem 7.6.2

(a) Assuming that Y is an Oka manifold of dimension $\dim Y > 1$, is the complement $Y \setminus \{p\}$ of an arbitrary point $p \in Y$ also an Oka manifold?
(b) Assume that Y is a quasi-projective Oka manifold and $A \subset Y$ is an algebraic subvariety of codimension ≥ 2; is $Y \setminus A$ an Oka manifold?

Problem 7.6.3 Is every stratified Oka manifold an Oka manifold?

An affirmative answer to the following problem would imply that every stratified Oka manifold is an Oka manifold (see Definition 7.1.7(a)). It would also follow that every Kummer surface is an Oka manifold (see Lemma 7.2.2).

Problem 7.6.4 (The union problem for Oka manifolds) Let Y be a complex manifold and $Y' \subset Y$ be a closed complex submanifold. If Y' and $Y \setminus Y'$ are Oka manifolds, does it follow that Y is also an Oka manifold?

Complements of Hypersurfaces

We mentioned in Sect. 5.6 that the complement of a hypersurface in an algebraic Oka manifold is not Oka in general. In fact, many interesting examples of hyperbolic manifolds arise as complements of smooth affine or projective hypersurfaces. Kobayashi asked [339, Problem 3, p. 132] whether the complement in $\mathbb{P}^n$ of a generic hypersurface of sufficiently high degree is hyperbolic. This was proved for curves in $\mathbb{P}^2$ by Siu and Yeung [508], and for any $n \geq 2$ by Siu [506, Theorem 2]. (For a recent survey on hyperbolicity, see the monograph by Diverio and Rousseau [123].) Hence, one expects that the complement of a projective hypersurface is Oka only in very special cases.

The case of the complement of a finite collection of hyperplanes is well understood; see e.g. Green [262] and Kobayashi [341, Sect. 3.10]. Let $f_1, \ldots, f_N$ be nonzero homogeneous linear forms in $n+1$ variables $z = (z_0, \ldots, z_n)$. We say that the hyperplanes in $\mathbb{P}^n$ defined by the equations $H_j = \{f_j = 0\}$, $j = 1, \ldots, N$, are in general position if every subset of $\{f_1, \ldots, f_N\}$ of size at most $n+1$ is linearly independent. If $N \leq n+1$, then a set of N hyperplanes is in general position if and only if coordinates can be chosen so that the given hyperplanes are the coordinate hyperplanes $z_j = 0$, $j = 0, \ldots, N-1$. The complement in $\mathbb{P}^n$ of at least $2n+1$ hyperplanes in general position is hyperbolic, and the complement of a collection of at most $2n$ hyperplanes is never hyperbolic. For hyperplanes not in general position, some necessary conditions for hyperbolicity are known.

Hanysz [278] determined which complements of hyperplane arrangements in a projective space are Oka.

Theorem 7.6.5 (Hanysz [278, Theorem 3.1]) *Let $H_1, \ldots, H_N$ be distinct hyperplanes in $\mathbb{P}^n$. Then the complement $X = \mathbb{P}^n \setminus \bigcup_{j=1}^N H_j$ is Oka if and only if the hyperplanes are in general position and $N \leq n+1$. Furthermore, if X is not Oka, then it is not dominable by $\mathbb{C}^n$ and not $\mathbb{C}$-connected.*

Another interesting result of Hanysz concerns complements of graphs of meromorphic functions.

Theorem 7.6.6 (Hanysz [278, Theorem 4.6]) *Let X be a complex manifold, and let $m\colon X \to \mathbb{P}^1$ be a meromorphic map with graph Γ. Suppose m can be written in the form $m = f + 1/g$ for holomorphic functions f and g. Then $(X \times \mathbb{C}) \setminus \Gamma$ is Oka if and only if X is Oka.*

Buzzard and Lu proved that the complement of a smooth cubic curve in $\mathbb{P}^2$ is dominable [80, Proposition 5.1]. Their analysis shows that there is a finite branched cover $X \to \mathbb{P}^2 \setminus \mathbb{C}$ where X is an Oka manifold, so the following problem is related to Problem 7.6.11.

Problem 7.6.7 Is the complement $\mathbb{P}^2 \setminus C$ of every smooth cubic curve Oka?

Siu and Yeung studied complements of hypersurfaces in abelian varieties [507, 509]. In [507] the following conjecture of Lang is proved.

Theorem 7.6.8 (Lang's conjecture) *Let A be an abelian variety and D be an ample divisor in A. Let $f:\mathbb{C} \to A$ be a nonconstant holomorphic map. Then the image of f must intersect D.*

The compact version of the above result, known as Bloch's theorem, says that the Zariski closure of the image of a holomorphic map from $\mathbb{C}$ to an abelian variety is precisely the translate of an abelian subvariety.

Problem 7.6.9 Classify hypersurfaces with Oka complements in abelian varieties.

Fibre Bundles, Branched Maps, Proper Modifications

Problem 7.6.10 Let $\pi: E \to B$ be a holomorphic fibre bundle with fibre Y. Assume that E is an Oka manifold; does it follow that the base B and the fibre Y are Oka manifolds? This obviously holds for the trivial bundle $E = B \times Y$.

The analysis in [80] shows that an affirmative answer to the following problem would imply an affirmative answer to Problem 7.6.7.

Problem 7.6.11 Let $\pi: X \to X'$ be a proper finite holomorphic map. If X is an Oka manifold, does it follow that X' is an Oka manifold?

The converse implication is false in general as shown by meromorphic functions on compact Riemann surfaces of genus $g > 1$; such a function defines a finite branched holomorphic map $X \to \mathbb{P}^1$ onto the Riemann sphere which is Oka, but X is not Oka since it is hyperbolic.

Problem 7.6.12 Suppose that Y is an Oka manifold with $\dim Y > 1$.

(a) Is the blowup of Y at a single point $p \in Y$ an Oka manifold?
(b) What about the blowup of Y along a closed complex submanifold?
(c) If Y is the blowup of a complex manifold X, is X necessarily Oka?

It would be interesting to know the answers to these questions at least in the class of compact projective manifolds. A positive answer to parts (a), (b), and (c) is known for algebraic manifolds that are Zariski locally Euclidean (see Definition 6.4.4 and Theorem 6.4.8 on p. 274). For a partial result on blowdowns (part (c)), see [212, Sect. 6].

K3 Surfaces, Enriques Surfaces, and Their Generalizations

The following problem is related to the discussion in Sect. 7.3.

Problem 7.6.13 Are any (or all) K3 surfaces and Enriques surfaces Oka?

Among the classes of complex manifolds generalizing the class of K3 surfaces that conjecturally are not volume hyperbolic (a necessary condition to be Oka) are the *Calabi-Yau manifolds* and the *hyper-Kähler manifolds*. A compact complex manifold Y is a Calabi-Yau manifold if it is simply connected, $c_1(Y)=0$, the canonical bundle K_Y is trivial, and $H^0(Y,\Omega^p)=\mathbb{C}$ in dimensions $p=0$ and $p=\dim Y$, and it equals zero otherwise. A Kähler manifold is hyper-Kähler if it is simply connected and carries a holomorphic symplectic form whose cohomology class spans $H^{2,0}(Y)$ (see O'Grady [438]). A hyper-Kähler manifold of dimension 2 is a K3 surface.

Oka Property in Families of Manifolds

We consider the variation of the Oka property in a holomorphic family of compact complex manifolds. Assume that $\pi: X\to B$ is a proper holomorphic submersion of a complex manifold X onto a complex manifold B, so each fibre $X_b=\pi^{-1}(b)$ is a compact complex manifold. Consider the set of Oka fibres:

$$\mathrm{Oka}(\pi)=\{b\in B: X_b \text{ is an Oka manifold}\}.$$

The following result was proved by Lárusson [375, Corollary 8].

Theorem 7.6.14 *In a holomorphic family $\pi: X\to B$ of compact complex manifolds, the set* $\mathrm{Oka}(\pi)$ *of Oka fibres is a G_δ subset of the base manifold B.*

Corollary 7.3.3 (p. 332) shows that the set of Oka fibres is not necessarily closed. In [375], Lárusson gave a necessary and sufficient condition for the limit fibre of a sequence of Oka fibres to be Oka in terms of a *uniform Oka property*. He also considered holomorphic submersions with noncompact fibres.

It is an open question whether $\mathrm{Oka}(\pi)$ is always open. Explicitly:

Problem 7.6.15 Assume that X is a complex manifold and $\pi: X\to\mathbb{D}$ is a proper holomorphic submersion onto the disc. Is there an example when the central fibre X_0 is Oka, but X_t is not Oka for some $t\neq 0$ arbitrarily near 0?

Topology and Geometry of Oka Manifolds

Problem 7.6.16

(a) Are there any restrictions on the homotopy type of a compact Oka manifold?
(b) Is there a contractible Stein Oka manifold that is not a Euclidean space?

Kaliman and Kutzschebauch [323] and Leuenberger [392] gave many examples of contractible algebraic hypersurfaces in $\mathbb{C}^n$ satisfying the density property (see Sect. 4.10). Every such manifold is Oka by Proposition 5.6.23. Are all these examples biholomorphic to Euclidean spaces? In particular:

Problem 7.6.17 Is the Koras-Russell cubic threefold (4.53) (p. 148) biholomorphic to $\mathbb{C}^3$?

The Koras-Russell cubic $C \subset \mathbb{C}^4$ is especially interesting since it is known to be diffeomorphic to $\mathbb{R}^6$, but is not algebraically equivalent to $\mathbb{C}^3$ since the algebraic automorphism group does not act transitively on C [138, 406],

Problem 7.6.18 Is there a geometric characterization of Oka manifolds in the spirit of Gromov's ellipticity (see Definition 5.6.13)? In particular, is every Oka manifold elliptic or subelliptic?

A partial answer is offered by the concept of a good manifold introduced by Lárusson [372]. Let us call a complex manifold X *good* if it is the image of an Oka map $\pi: E \to X$ (see Definition 7.4.7 on p. 335) from a Stein manifold E, and *very good* if X carries a holomorphic affine bundle $\pi: E \to X$ whose total space E is Stein. (These definitions are slightly different from those introduced in [372].) In such case, the manifold X is Oka if and only if E is Oka (see Theorem 5.6.5, p. 225). If this holds then E, being Stein, is elliptic by Proposition 5.6.15. Hence, in the class of good manifolds we have the following geometric characterization of the Oka property.

Proposition 7.6.19 *Let X be a good complex manifold. Then, X is an Oka manifold if and only if there is a surjective Oka map $\pi: E \to X$ (Definition 7.4.7) such that E is an elliptic Stein manifold.*

How big is the class of good (or very good) manifolds?

By definition, these classes include all Stein manifolds. By Example 1.4.3 on p. 14 there is an affine holomorphic bundle $E \to \mathbb{P}^n$ with fibre $\mathbb{C}^n$ whose total space E is Stein, so $\mathbb{P}^n$ is very good. By restricting the bundle $E \to \mathbb{P}^n$ to a closed complex submanifold $X \subset \mathbb{P}^n$ we see that every complex projective manifold is very good. (Note that $E|_X$, being a closed complex submanifold of a Stein manifold E, is Stein.) More generally, all quasi-projective algebraic manifolds are very good [372]. Lárusson showed that the classes of good, resp. very good, complex manifolds are closed under taking products, covering spaces, finite branched covering spaces, submanifolds, and complements of analytic hypersurfaces, so they seem to be quite large.

Oka Properties of Special Mapping Spaces

Assume that X and Y are compact complex manifolds. When is the space $\mathcal{O}(X, Y)$ of holomorphic maps $X \to Y$ an Oka manifold?

This question was studied by Hanysz [279] when $X = Y = \mathbb{P}^1$, and by Bowman [65] when X is a compact Riemann surface of genus 1 (a torus) and $Y = \mathbb{P}^1$. In both cases we have a decomposition in connected components $\mathcal{O}(X, Y) = \bigcup_{d=0}^{\infty} R_d$, where R_d is the complex manifold of maps $X \to Y$ of degree d, so $\mathcal{O}(X, Y)$ is Oka if and only if R_d is Oka for all d. If $X = \mathbb{P}^1$ then $R_1 = \mathrm{Aut}(\mathbb{P}^1)$ is the Möbius group of automorphism of $\mathbb{P}^1$ (a complex Lie group), while $R_1 = \varnothing$ if X is a torus. In both cases, the manifold R_2 of degree 2 maps is complex homogeneous and hence Oka, but the analysis is surprisingly difficult for $d \geq 3$. Here is a summary of their results.

Theorem 7.6.20 (Hanysz [279])

(a) [279, Theorem 1.1] *The space R_2 of rational maps $\mathbb{P} \to \mathbb{P}$ of degree 2 is complex homogeneous and hence Oka.*
(b) [279, Theorems 1.7 and 1.8] *The space R_3 of rational maps $\mathbb{P} \to \mathbb{P}$ of degree 3 is strongly dominable and strongly $\mathbb{C}$-connected.*

Problem 7.6.21 Is the manifold R_3 of degree 3 holomorphic maps $\mathbb{P}^1 \to \mathbb{P}^1$ Oka? What about R_d for $d > 3$?

Theorem 7.6.22 (Bowman [65]) *Let X be a compact Riemann surface of genus 1 (a complex torus).*

(a) [65, Theorem 24] *The complex manifold R_2 of holomorphic maps $X \to \mathbb{P}^1$ of degree 2 is biholomorphic to $(\mathrm{Aut}(\mathbb{P}) \times X)/(\mathbb{Z}_2 \times \mathbb{Z}_2)$. In particular, it is a homogeneous space and hence an Oka manifold.*
(b) [65, Theorem 34] *The space R_3 of holomorphic maps $X \to \mathbb{P}^1$ of degree 3 is a connected 6-dimensional complex manifold which admits a 6-sheeted connected branched covering space that is an Oka manifold.*

Problem 7.6.23 Let X be a compact Riemann surface of genus 1. Is the manifold R_3 of degree 3 holomorphic maps $X \to \mathbb{P}^1$ Oka? What about the manifold R_d of maps of degree $d > 3$?

Another mapping space that has been investigated from the perspective of Oka theory is the group $\mathrm{Aut}(\mathbb{C}^n)$ of holomorphic automorphisms of $\mathbb{C}^n$. We have see in Chap. 4 that for any $n > 1$ this is a huge and complicated group which does not have any natural structure of a (infinite dimensional) complex manifold. However, it makes sense to speak of holomorphic maps $f\colon X \to \mathrm{Aut}(\mathbb{C}^n)$ from a complex manifold X by asking that the associated map $X \times \mathbb{C}^n \to \mathbb{C}^n$, $(x, z) \mapsto f(x)(z)$, be holomorphic. With this in mind, Lárusson and the author showed in [213] that $\mathrm{Aut}(\mathbb{C}^n)$ and several of its subgroups enjoy the parametric Oka property with approximation and with interpolation on discrete subsets. (Interpolation on subvarieties of positive dimension remains an open problem.) The basic fact underlying their analysis is that every holomorphic automorphism $F \in \mathrm{Aut}(\mathbb{C}^n)$ can be written as

$$F(z) = a + AH(z), \quad a = F(0) \in \mathbb{C}^n, \ A = F'(0) \in GL_n(\mathbb{C}),$$

where the automorphism $H \in \mathrm{Aut}(\mathbb{C}^n)$ is of the form

$$H(z) = z + \sum_{|\alpha|=2}^{\infty} c_\alpha z^\alpha, \quad z \in \mathbb{C}^n, \ c_\alpha \in \mathbb{C}^n.$$

The subgroup $\mathcal{S}$ of $\mathrm{Aut}(\mathbb{C}^n)$ consisting of all automorphisms H normalized by the conditions $H(0) = 0$ and $H'(0) = \mathrm{Id}$ is clearly contractible, so the topology of $\mathrm{Aut}(\mathbb{C}^n)$ is carried by the general linear group $GL_n(\mathbb{C})$ to which the Oka-Grauert theorem applies. The main work in [213] is to show that $\mathcal{S}$ has the parametric Oka property with approximation and interpolation on discrete subsets. This boils down to a certain parametric version of the Andersén-Lempert theorem (see Theorem 4.9.1), similar to the analysis by Kutzschebauch and Ramos-Peon [366] who proved the infinite parametric transitivity property on Stein manifolds with the density property.

It is reasonable to expect that similar results hold for the holomorphic automorphism group $\mathrm{Aut}(S)$ of any Stein manifold S with the density property. However, the analysis in [213] does not carry over to this more general case in the absence of a decomposition into a (finite dimensional) complex Lie subgroup and a contractible (infinite dimensional) subgroup. The first open question along this line is the following.

Problem 7.6.24 Does the group $\mathrm{Aut}(\mathbb{C}^* \times \mathbb{C})$ enjoy any Oka properties?

Oka Theory for Singular Targets

It is natural to ask whether the Oka theory could also be developed for singular targets, that is, for maps $X \to Y$ from reduced Stein spaces to reduced complex spaces Y. The first, and so far the only attempt in this direction has been made by Lärkäng and Lárusson [369] who studied the case when Y is a singular affine toric variety. (By Theorem 5.6.12 on p. 228, smooth toric varieties are Oka manifolds.)

A complex affine toric variety Y is a subvariety of $\mathbb{C}^n$ for some $n > 1$ given by finitely many binomial equations $z^a = z^b$, where $a = (a_1, \ldots, a_n) \in \mathbb{Z}_+^n$ and $b = (b_1, \ldots, b_n) \in \mathbb{Z}_+^n$ are such that at least one of the numbers a_i and b_i equals zero for every $i = 1, \ldots, n$. The differences $a - b \in \mathbb{Z}^n$ span a lattice $\Lambda \subset \mathbb{Z}^n$, and hence affine toric varieties $Y \subset \mathbb{C}^n$ correspond to lattices in $\mathbb{Z}^n$. The ideal $I \subset \mathbb{C}[x_1, \ldots, x_n]$ generated by the defining functions $z^a - z^b$ is prime if and only if the quotient group $\mathbb{Z}^n/\Lambda$ is torsion-free or, equivalently, free; we assume this to be the case. We have that $0 \in Y$ if and only if there exist positive integers $k_1, \ldots, k_n \in \mathbb{N}$ such that Y is preserved under the homotopy $Y \times [0, 1] \to Y$,

$(y, t) \mapsto (t^{k_1} y_1, \ldots, t^{k_n} y_n)$, and hence is holomorphically contractible. In this case there is no obstruction to extending a continuous map into Y from a subcomplex of a CW complex to the whole complex. If S is a subvariety of a complex space X, there is a triangulation of X that restricts to a triangulation of S, and hence every continuous map $S \to Y$ has a continuous extension $X \to Y$.

A holomorphic map $f = (f_1, \ldots, f_n) \colon X \to Y \subset \mathbb{C}^n$ to a toric variety is said to be *nondegenerate* if no component f_i is identically zero on any irreducible component of X.

Lärkäng and Lárusson gave several positive and also some negative results concerning the possibility of extending a holomorphic map $S \to Y$ from a closed complex subvariety S of a Stein space X to a holomorphic map $X \to Y$. The following result combines Theorems 1.1 and 1.2 in [369].

Theorem 7.6.25 *Let Y be a not necessarily normal affine toric variety of dimension d in $\mathbb{C}^n$.*

(a) *If S is a factorial subvariety of a reduced Stein space X such that $H^p(X; \mathbb{Z}) \to H^p(S; \mathbb{Z})$ is surjective for $p = 0, 1, 2$, then every holomorphic map $S \to Y$ extends to a holomorphic map $X \to Y$.*
(b) *Assume that $0 \in Y$ and the normalization of Y is $\mathbb{C}^d$. If S is a normal subvariety of a reduced Stein space X, then every nondegenerate holomorphic map $S \to Y$ extends to a holomorphic map $X \to Y$.*
(c) *Assume that $0 \in Y$ and the normalization of Y is not $\mathbb{C}^d$. There is a smooth surface S in $\mathbb{C}^4$, biholomorphic to the product of two annuli, and a nondegenerate holomorphic map $S \to Y$ that does not extend to a holomorphic map $\mathbb{C}^4 \to Y$.*

The problem seems to be that singularities of Y may hide nontrivial topology which is revealed when passing to the desingularization. Hence, the existence of a continuous extension need not imply the existence of a holomorphic extension. This subject definitely deserves further investigation.

Part III
Applications

Chapter 8
Applications of Oka Theory and Its Methods

In this chapter we apply the results and methods of Oka theory from Part II to a variety of analytic and geometric problems on Stein manifolds.

We begin with applications of the Oka-Grauert principle (Sects. 8.1–8.5). Results of modern Oka theory (i.e., Theorems 5.4.4 and 6.2.2) are used in Sect. 8.6 (elimination of intersections of holomorphic maps with complex subvarieties), Sect. 8.7 (the holomorphic Vaserstein problem), and in the construction of proper holomorphic embeddings and immersions of Stein spaces to Euclidean spaces of minimal dimension (Sects. 9.3–9.5). In Sects. 8.8 and 8.9 we discuss transversality theorems for holomorphic and algebraic maps. In Sects. 8.10–8.13 we further develop and exploit the technique of local holomorphic sprays of maps and sections that was introduced in Sects. 5.8–5.9.

8.1 Principal Fibre Bundles

In this section we recall some basic notions on principal G-bundles in preparation for Grauert's classification theorem in the following section.

Definition 8.1.1 Let G be a (real or complex) Lie group. A *topological principal G-bundle* is a fibre bundle $\pi: E \to X$ with fibre $E_x \cong G$, endowed with a faithful right action of G on E whose orbits are the fibres $E_x = \pi^{-1}(x)$ $(x \in X)$ and which is locally trivial in the following sense: Every point $x_0 \in X$ admits an open neighborhood $U \subset X$ and a homeomorphism $\theta: E|_U \to U \times G$, $\theta(e) = (\pi(e), \vartheta(e))$, satisfying

$$\theta(eg) = \big(\pi(e), \vartheta(e)\, g\big), \quad e \in E,\ g \in G. \tag{8.1}$$

(Such θ is said to be *G-equivariant*.) If in addition X and E are complex spaces, the map $\pi: E \to X$ is a holomorphic fibre bundle projection whose fibre G is a complex Lie group acting holomorphically on E, and there exist G-equivariant local holomorphic trivializations θ as above, then we have a *holomorphic principal G-bundle structure* on $E \to X$.

F. Forstnerič, *Stein Manifolds and Holomorphic Mappings*, Ergebnisse der Mathematik und ihrer Grenzgebiete. 3. Folge / A Series of Modern Surveys in Mathematics 56, DOI 10.1007/978-3-319-61058-0_8

The same definition applies in the smooth and the real analytic category. It follows that every G-bundle $\pi: E \to X$ admits a G-atlas $\Theta = \{(U_j, \theta_j)\}_{j\in J}$ where $\mathcal{U} = \{U_j\}_{j\in J}$ is an open cover of X, every map

$$\theta_j: E|_{U_j} = \pi^{-1}(U_j) \to U_j \times G, \quad j \in J$$

is a homeomorphism (or a biholomorphism) as in (8.1), and the transition maps $\theta_{ij} = \theta_i \circ \theta_j^{-1}: U_{ij} \times G \to U_{ij} \times G$ (where $U_{ij} = U_i \cap U_j$) are of the form

$$\theta_{ij}(x, v) = \big(x, g_{ij}(x)\, v\big), \quad x \in U_{ij},\ v \in G, \tag{8.2}$$

where $g_{ij}: U_{ij} \to G$ are continuous (resp. holomorphic) maps to G. The collection $g = \{g_{ij}\}_{i,j\in J}$ clearly satisfies the 1-cocycle condition

$$g_{ii} = 1, \quad g_{ij}g_{ji} = 1, \quad g_{ij}g_{jk}g_{ki} = 1, \quad i, j, k \in J; \tag{8.3}$$

such g is a *1-cocycle* with values in the sheaf $\mathcal{C}^G = \mathcal{C}^G_X$ (resp. $\mathcal{O}^G = \mathcal{O}^G_X$) of germs of continuous (resp. holomorphic) maps $X \to G$. (The fibres E_x do not admit a natural Lie group structure since the transition maps are not Lie group automorphisms.) Conversely, every 1-cocycle (8.3) is the transition cocycle of a G-atlas on a principal G-bundle $E \to X$: Define E as the quotient of the disjoint union of the products $U_j \times G$ by the equivalence relation which identifies $(x, v) \in U_j \times G$ with $(x', v') \in U_i \times G$ if and only if $x = x' \in U_{ij}$ and $v' = g_{ij}(x)v$. The right multiplication of G on itself induces a right action of G on E, and the local trivializations are just the identifications of $U_j \times G$ with the corresponding subsets in E.

Two G-atlases $\Theta = \{(U_i, \theta_i)\}$ and $\Theta' = \{(U'_j, \theta'_j)\}$ on E are *equivalent* if their union $\Theta \cup \Theta'$ is again a G-atlas; this holds if and only if the transition maps $\theta'_j \circ \theta_i^{-1}$ are of the form (8.2). A *principal G-bundle structure* is determined by an equivalence class of G-atlases or, equivalently, by a maximal G-atlas obtained as the union of all atlases in an equivalence class.

If $\mathcal{V} = \{V_j\}_{j\in J'}$ is an open cover of X which is a refinement of $\mathcal{U} = \{U_j\}_{j\in J}$, with a refinement map $\alpha: J' \to J$ such that $V_j \subset U_{\alpha(j)}$ for every $j \in J'$, then any G-atlas on $\mathcal{U}$ induces a G-atlas on $\mathcal{V}$ by restriction. This allows us to consider any pair of G-atlases over a common refinement.

Suppose that $\Theta = \{(U_j, \theta_j)\}$ and $\Theta' = \{(U_j, \theta'_j)\}$ are equivalent G-atlases on an open cover $\mathcal{U} = \{U_j\}_{j\in J}$. For each index $j \in J$ the transition map $\theta'_j \circ \theta_j^{-1}: U_j \times G \to U_j \times G$ is given by $(x, v) \to (x, c_j(x)v)$ for some continuous map $c_j: U_j \to G$. If g_{ij} (resp. g'_{ij}) is the transition 1-cocycle of Θ (resp. of Θ') then $g'_{ij}c_j = c_i g_{ij}$ on U_{ij}; equivalently,

$$g'_{ij} = c_i g_{ij} c_j^{-1} \quad \text{on } U_{ij}. \tag{8.4}$$

Conversely, for every 1-cocycle $g = \{g_{ij}\}$ and every collection $c = \{c_j\}_{j\in J}$ of maps $c_j: U_j \to G$ (such c is called a *0-cochain with values in the sheaf* $\mathcal{C}^G$) we can define another 1-cocycle $g' = \{g'_{ij}\}$ by (8.4). This operation is called *twisting of the 1-cocycle g by the 0-cochain c* and is denoted by

$$g' = c \,\square\, g.$$

If $\Theta = \{(U_j, \theta_j)\}_{j\in J}$ is a G-bundle atlas on $E \to X$ with the transition cocycle g, and if we define $\theta_j' : E|_{U_j} \to U_j \times G$ by $\theta_j' = C_j \circ \theta_j$, where $C_j(x,v) = (x, c_j(x)v)$ for $x \in U_j$ and $v \in G$, then $\Theta' = \{(U_j, \theta_j')\}$ is a G-atlas on E with the transition cocycle $g' = c \,\square\, g$ which is equivalent to Θ.

Let $\pi : E \to X$ and $\pi' : E' \to X$ be principal G-bundles. A map $\Phi : E \to E'$ is a *G-bundle map* if $\pi' \circ \Phi = \pi$ and Φ is G-equivariant:

$$\Phi(eg) = \Phi(e)g, \quad e \in E,\ g \in G.$$

In any pair of G-bundle charts such Φ is given as left multiplication by a map from the base into the group G. More precisely, given G-atlases $\Theta = \{(U_j, \theta_j)\}$ for E and $\Theta' = \{(U_j, \theta_j')\}$ for E' over an open cover $\mathcal{U} = \{U_j\}_{j\in J}$ of X, each map $\Phi_j = \theta_j' \circ (\Phi|_{E|_{U_j}}) \circ \theta_j^{-1} : U_j \times G \to U_j \times G$ is of the form

$$\Phi_j(x, v) = \big(x, \phi_j(x)v\big), \quad x \in U_j,\ v \in G$$

for some continuous (resp. holomorphic) map $\phi_j : U_j \to G$. The 0-cochain $\phi = \{\phi_j\}_{j\in J}$ with values in the sheaf $\mathcal{C}^G$ (resp. $\mathcal{O}^G$) then represents Φ with respect to the given pair of G-atlases. Denoting the transition 1-cocycle of Θ, Θ' by (g_{ij}), (g'_{ij}), respectively, we have

$$g'_{ij}\phi_j = \phi_i\, g_{ij} \quad \text{on } U_{ij}. \tag{8.5}$$

Conversely, any 0-cochain $\phi = \{\phi_j\}$ satisfying (8.5) defines a G-bundle map $\Phi : E \to E'$. Such Φ is automatically invertible and hence a G-bundle isomorphism; its inverse given by the 0-cochain $\{\phi_j^{-1}\}$. Note that (8.5) means $g' = \phi \,\square\, g$ exactly as in (8.4). This is explained as follows: By Φ we pull the G-atlas Θ' back to the G-atlas $\widetilde{\Theta} = \Phi^*\Theta'$ (on E) with the bundle charts

$$\widetilde{\theta}_j = \theta_j' \circ \Phi|_{E_{U_j}} : E_j \to U_j \times G.$$

The transition cocycle of $\widetilde{\Theta}$ is g' (the same as for Θ'), and ϕ is now seen as the transition 0-cochain between Θ and $\widetilde{\Theta}$. Similarly we can use Φ to push forward a G-atlas Θ on E to a G-atlas $\Phi_*\Theta$ on E'.

Denote by $Z^1(\mathcal{U}, \mathcal{C}^G)$ the set of all 1-cocycles, and by $C^0(\mathcal{U}, \mathcal{C}^G)$ the set of all 0-cochains on a cover $\mathcal{U}$ with values in the sheaf $\mathcal{C}^G$. We declare 1-cocycles $g, g' \in Z^1(\mathcal{U}, \mathcal{C}^G)$ equivalent if and only if $g' = c \,\square\, g$ for some $c \in C^0(\mathcal{U}, \mathcal{C}^G)$. Denote by $H^1(\mathcal{U}; \mathcal{C}^G)$ the set of equivalence classes. Then

$$H^1\big(X; \mathcal{C}^G\big) = \varinjlim H^1\big(\mathcal{U}; \mathcal{C}^G\big)$$

(the direct limit over open covers, called *the first Čech cohomology group with coefficients in $\mathcal{C}^G$*) is precisely the set of equivalence classes of topological principal G-bundles on X. In the holomorphic category,

$$H^1\big(X; \mathcal{O}^G\big) = \varinjlim H^1\big(\mathcal{U}; \mathcal{O}^G\big)$$

is the set of all holomorphic equivalence classes of holomorphic principal G-bundles on X. The natural map $H^1(X; \mathcal{O}^G) \to H^1(X; \mathcal{C}^G)$ induced by the inclusion $\mathcal{O}^G \hookrightarrow \mathcal{C}^G$ is in general neither injective nor surjective as we have already seen in Sect. 5.3 in the case when $G = \mathbb{C}^*$.

8.2 The Oka-Grauert Principle for G-Bundles

In this section we prove the following theorem of Grauert.

Theorem 8.2.1 ([253, 254]) *Let X be a reduced Stein space. For every complex Lie group G the map $H^1(X;\mathcal{O}^G) \to H^1(X;\mathcal{C}^G)$ is bijective; that is, the holomorphic and the topological classes of principal G-bundles over X are in one to one correspondence. More precisely:*

(i) *For every topological G-bundle isomorphism $\Phi_0: E \to E'$ of holomorphic principal G-bundles over X there is a homotopy $\Phi_t: E \to E'$ ($t \in [0,1]$) of topological G-bundle isomorphisms such that Φ_1 is a holomorphic G-bundle isomorphism. If in addition Φ_0 is holomorphic on a closed complex subvariety $M \subset X$ and also in an open neighborhood of a compact $\mathcal{O}(X)$-convex subset K of X, then the homotopy $\{\Phi_t\}$ can be chosen fixed on $E|_M$ and uniformly close to Φ_0 on $E|_K$.*

(ii) *Every principal G-bundle $E \to X$ admits a holomorphic G-bundle structure. If K and M are as in* (i), *then a holomorphic G-bundle structure on $E|_{K\cup M}$ extends to a holomorphic G-bundle structure on E.*

(iii) *In particular, we have $H^1(X;\mathcal{O}^G) = 0$ for each 1-dimensional Stein space X and every connected complex Lie group G.*

Part (iii) follows from the main statement. Indeed, since $G \setminus \{1\}$ is connected and X is homotopy equivalent to a 1-dimensional CW complex by Corollary 3.12.2, every principal G-bundle $E \to X$ admits a section $f: X \to E$ such that $f(x) \neq 1 \in G$ for every $x \in X$, and hence is topologically trivial.

Parts (i) and (ii) in Theorem 8.2.1 (without the approximation and the interpolation condition) are equivalent to the following:

(i′) Given 1-cocycles $g, g' \in Z^1(\mathcal{U}, \mathcal{O}^G)$ and a 0-cochain $\phi^0 \in C^0(\mathcal{U}, \mathcal{C}^G)$ such that $\phi^0 \,\square\, g = g'$, there exists a homotopy of 0-cochains $\phi^t \in C^0(\mathcal{U}, \mathcal{C}^G)$ ($t \in [0,1]$) such that $\phi^t \,\square\, g = g'$ for all $t \in [0,1]$ and $\phi^1 \in C^0(\mathcal{U}, \mathcal{O}^G)$.

(ii′) Let $\mathcal{U}$ be a Stein open cover of X. For every 1-cocycle $g \in Z^1(\mathcal{U}, \mathcal{C}^G)$ there exists a 0-cochain $c \in C^0(\mathcal{U}, \mathcal{C}^G)$ such that $c \,\square\, g \in Z^1(\mathcal{U}, \mathcal{O}^G)$ is a holomorphic 1-cocycle.

Hence the natural map of Čech cohomology groups $H^1(\mathcal{U};\mathcal{O}^G) \to H^1(\mathcal{U};\mathcal{C}^G)$ is injective for every open cover by (i′), and is surjective for every Stein open cover by (ii′). The equivalence (i) $\Leftrightarrow$ (i′) follows from the discussion in Sect. 8.1. For (ii) $\Leftrightarrow$ (ii′) we need to observe that, by (i), a holomorphic G-bundle $E \to X$ which admits a topological atlas over a Stein open cover $\mathcal{U}$ of X also admits a holomorphic atlas over the same cover.

Proof of Theorem 8.2.1, part (i) Choose an open cover $\mathcal{U} = \{U_j\}_{j\in J}$ of X and holomorphic G-atlases $\Theta = \{(U_j, \theta_j)\}$ on E and $\Theta' = \{(U_j, \theta'_j)\}$ on E'. Let $g = \{g_{ij}\}$, $g' = \{g'_{ij}\}$ denote the holomorphic transition cocycles of Θ, Θ', respectively. The isomorphism $\Phi_0: E \to E'$ is then given by a 0-cochain $\phi^0 = \{\phi_j^0\}_{j\in J} \in C^0(\mathcal{U}, \mathcal{C}^G)$ satisfying $\phi^0 \,\square\, g = g'$ (8.5); if Φ_0 is holomorphic over $M \cup V$ then ϕ_j^0 is holomorphic on $U_j \cap (M \cup V)$ for every $j \in J$. We can view ϕ^0 as a section of the holomorphic fibre bundle $H \to X$ with fibre G which is defined over the cover $\mathcal{U}$ by the transition maps

$$\psi_{ij}(x,v) = \big(x, g'_{ij}(x)\, v\, g_{ij}^{-1}(x)\big), \quad x \in U_{ij},\ v \in G.$$

By Corollary 5.5.4 (p. 221) there exists a homotopy $\phi^t: X \to H$ ($t \in [0,1]$) of continuous sections such that ϕ^1 is a holomorphic section, the homotopy is fixed on M, and every ϕ^t is holomorphic on an open set $V' \supset K$ and is uniformly close to ϕ^0 on K (also uniformly in $t \in [0,1]$). Viewing ϕ^t again as a 1-cocycle on the cover $\mathcal{U}$ we have $\phi^t \,\square\, g = g'$, and hence ϕ^t determines a topological G-bundle isomorphism $\Phi_t: E \to E'$. At $t = 1$ we get a holomorphic G-bundle isomorphism $\Phi_1: E \to E'$. This proves part (i).

Applying the above argument with $E = E'$ and Φ_0 the identity map on E, considered as a topological equivalence between a pair of holomorphic atlases on E, we obtain the following corollary.

Corollary 8.2.2 *For any pair of topologically equivalent holomorphic G-atlases Θ_0 and Θ_1 on a principal G-bundle $E \to X$ there exists a homotopy of topological G-bundle automorphisms $\Phi_t: E \to E$ ($t \in [0,1]$) such that Φ_0 is the identity and $(\Phi_1)_*\Theta_0 = \Theta_1$. Furthermore, if K, V and M are as in Theorem* 8.2.1 *and if Θ_0 and Θ_1 define the same holomorphic structure on $E|_{M\cup V}$, then the homotopy Φ_t can be chosen fixed on M, holomorphic on an open set $V' \supset K$, and uniformly close to the identity on K.*

If $\{\Phi_t\}_{t\in[0,1]}$ is as in Corollary 8.2.2 then $\Theta_t = (\Phi_t)_*\Theta_0$ is a homotopy of holomorphic G-atlases on E connecting Θ_0 and Θ_1, and the holomorphic structure on $E|_{M\cup V'}$ defined by Θ_t does not depend on $t \in [0,1]$. □

Proof of Theorem 8.2.1, part (ii) We first consider the special case when X is a Stein manifold and $M = \varnothing$. Then $X = \bigcup_{k=0}^{\infty} A_k$ where each A_k is a compact strongly pseudoconvex domain and $A_{k+1} = A_k \cup B_k$, where B_k is a small compact strongly pseudoconvex domain such that E is trivial over an open neighborhood of B_k. If we are given a holomorphic structure on $E|_V$ over an open neighborhood V of a compact $\mathcal{O}(X)$-convex subset $K \subset X$, then A_0 may be chosen such that $K \subset A_0 \subset V$; otherwise take $A_0 = \varnothing$.

Assume inductively that we have an open set $V \supset A_k$ and a holomorphic G-bundle structure on $E|_V$. (This holds for $k = 0$.) By shrinking V around A_k we may assume that V is Stein. By the assumption there are an open Stein neighborhood $W \supset B_k$ and a trivialization $\theta: E|_W \to W \times G$; in particular, E is trivial over the Stein domain $U = V \cap W$. By part (i) there exists a trivialization $\theta': E|_U \to U \times G$ which is holomorphic with respect to the holomorphic principal bundle structure on $E|_V$. Then

$$\Phi(x,v) := \theta' \circ \theta^{-1}(x,v) = \big(x, g(x)v\big), \quad x \in U,\ v \in G$$

for some continuous map $g: U \to G$. Since U is Stein, Corollary 5.5.4 on p. 221 provides a homotopy $g_t: U \to G$ ($t \in [0,1]$) such that $g_0 = g$ and g_1 is holomorphic. Set

$$\Phi_t(x, v) = \big(x, g_t(x)v\big), \quad x \in U, \ v \in G, \ t \in [0,1];$$

thus $\Phi_0 = \Phi$. Choose a smooth function $\chi: X \to [0,1]$ such that $\operatorname{supp} \chi \subset V$ and $\chi = 1$ in an open set $V' \supset A_k$. Define $\Psi_t: W \times G \to W \times G$ by

$$\Psi_t(x, v) = \big(x, g_{t\chi(x)}^{-1} g_0(x)\, v\big), \quad x \in W, \ v \in G, \ t \in [0,1].$$

Although g_t is only defined on $U = V \cap W$, Ψ_t extends to $W \times G$ as the identity on $(W \setminus V) \times G$ since $\chi = 0$ on $W \setminus V$. Then $\{\Psi_t\}_{t \in [0,1]}$ is a homotopy of continuous automorphisms of the trivial G-bundle over W, with Ψ_0 the identity map. Over $U' = V' \cap W$ we have $\Phi_t \circ \Psi_t = \Phi_0$ since $\chi = 1$ there. Now define

$$\theta_t = \Psi_t \circ \theta : E|_W \to W \times G, \quad t \in [0,1];$$

this is a homotopy of G-bundle charts on $E|_W$. On $E|_{U'}$ we have

$$\theta' = \Phi_0 \circ \theta = \Phi_t \circ \Psi_t \circ \theta = \Phi_t \circ \theta_t, \quad t \in [0,1].$$

Since Φ_1 is holomorphic, the holomorphic structure on $E|_{V'}$ agrees with the holomorphic structure on $E|_W$ induced by the chart $\theta_1: E|_W \to W \times G$. Our change of the trivialization of $E|_W$ did not affect the G-bundle structure of E, and we obtained a holomorphic G-bundle structure on $E|_{V' \cup W}$ extending the structure on $E|_{V'}$. Since $V' \cup W \supset A_{k+1}$, the induction may proceed.

This completes the proof of part (ii) in the special case when X is smooth and $M = \emptyset$.

Note that the above procedure produces a holomorphic G-bundle structure by a stepwise extension that does not require any convergence arguments. Together with Corollary 8.2.2 we get the following lemma.

Lemma 8.2.3 (Gluing holomorphic G-bundle structures) *Let $E \to X$ be a topological principal G-bundle over a complex manifold X, and let M be a closed complex subvariety of X. Let $V, W \subset X$ be open sets such that $E|_V$ and $E|_W$ admit holomorphic G-bundle structures that coincide over $M \cap V \cap W$ and are topologically compatible with the structure on E. If $V \cap W$ is Stein, then for every open set $V' \subset V$ with $\overline{V}' \cap W \subset V$ there is a holomorphic G-bundle structure on $E|_{V' \cup W}$ that agrees with the initial structures on $E|_{V'}$ (inherited from $E|_V$) and on $E|_{M \cap (V' \cup W)}$.*

Proof Since $U = V \cap W$ is Stein, Corollary 8.2.2 furnishes a homotopy of topological G-bundle isomorphisms $\Phi_t: E|_U \to E|_U$ ($t \in [0,1]$) such that Φ_0 is the identity, the homotopy is fixed on $M \cap U$, and Φ_1 pushes forward to holomorphic structure on $E|_W$ to the holomorphic structure on $E|_V$ over U. The assumption $\overline{V}' \cap W \subset V$ implies that there is a smooth function $\chi: W \to [0,1]$ which equals one on $\overline{V}' \cap W$ and zero on $W \setminus V$. Then $\widetilde{\Phi}_t(x) = \Phi_{t\chi(x)}(x): E_x \to E_x$ ($x \in W$, $t \in [0,1]$) is a homotopy of topological G-bundle automorphisms of $E|_W$ such that $\widetilde{\Phi}_0$ is the identity, the homotopy is fixed on $M \cap W$, and $\widetilde{\Phi}_t = \Phi_t$ on $E|_{V' \cap W}$. The push-forward by $\widetilde{\Phi}_1$ of the holomorphic structure on $E|_W$ is a holomorphic structure on $E|_W$ that agrees with the structure on $E|_V$ over the set $V' \cap W$. □

The next lemma provides the inductive step for extending a holomorphic structure over a Stein stratum in a stratified complex space.

Lemma 8.2.4 *Assume that $\pi : E \to X$ and $M, K \subset X$ are as in Theorem* 8.2.1(i). *Let $M_1 \subset M_0$ be closed complex subvarieties of M such that $M_0 \setminus M_1$ is a Stein manifold. Given an open set $U_1 \supset M_1 \cup K$ and a holomorphic G-bundle structure on $E|_{M \cup U_1}$, there exist an open set $U_0 \supset M_0 \cup K$ and a holomorphic G-bundle structure on $E|_{M \cup U_0}$ which agrees with the given structure over $M \cup U_1'$ for an open neighborhood $U_1' \subset U_1$ of $K \cup M_1$.*

Proof The Stein manifold $S = M_0 \setminus M_1$ admits an open Stein neighborhood W in X and a holomorphic retraction $\tau : W \to S$ (Theorem 3.3.3 on p. 74; here we may consider a neighborhood of S in X as embedded in some $\mathbb{C}^N$). Using τ we extend the given holomorphic structure on $E|_{M_0 \cap W}$ to a holomorphic structure on $E|_W$. By Lemma 8.2.3 we can assume that this extension agrees with the given structure on $E|_M$ over the set $M \cap W$. (It suffices to push it forward by a G-bundle automorphism of $E|_W$ which is the identity on $E|_{M_0 \cap W}$.) By shrinking $U_1 \supset K \cup M_1$ we may assume that U_1 is Stein. Lemma 8.2.3 then shows that we can correct the structure on $E|_W$ (without changing it on $E|_{M \cap W}$) to make it agree with the structure on $E|_{U_1}$ over $W \cap U_1'$, where U_1' is an open set with $K \cup M_1 \subset U_1' \subset U_1$. Setting $U_0 = U_1' \cup W \supset K \cup M_0$ we get a holomorphic G-bundle structure on $E|_{M \cup U_0}$. □

Proof of Theorem 8.2.1(ii): the general case The first task is to find an open set $U_0 \supset M \cup K$ in X and a holomorphic structure on $E|_{U_0}$ which extends the given holomorphic structure on $E|_{K \cup M}$. Choose a stratification $M = M_0 \supset M_1 \supset \cdots \supset M_k = \varnothing$ with smooth Stein strata $S_j = M_j \setminus M_{j+1}$. (To find such stratification, choose $f \in \mathcal{O}(X)$ that vanishes on M_{sing} but does not vanish identically on any irreducible component of maximal dimension in M, take $M_1 = M \cap \{f = 0\}$ and $S_1 = M \setminus M_1$, and continue inductively.) Assume inductively that for some index $j \in \{1, \ldots, k\}$ we have an open set $U_j \subset X$ containing $K \cup M_j$ and a holomorphic extension of the given structure to $E|_{M \cup U_j}$. (By the hypothesis this holds for $j = k$ since $M_k = \varnothing$.) Since $S_{j-1} = M_{j-1} \setminus M_j$ is a Stein manifold, Lemma 8.2.4 provides an extension to a holomorphic structure on $E|_{M \cup U_{j-1}}$, where U_{j-1} is a neighborhood of $K \cup M_{j-1}$. At $j = 0$ we obtain a holomorphic structure on $E|_{U_0}$ over an open set $U_0 \supset M \cup K$ which extends the original structure.

It remains to extend this structure to all of E. Choose a stratification $X = X_0 \supset X_1 \supset \cdots \supset X_m = M$ whose strata $X_j \setminus X_{j+1}$ are Stein manifolds and proceed again by a downward induction on j, starting with $j = m$. In the j-step we have a holomorphic structure over an open set $W_j \supset X_j \cup K$. (For $j = m$ take $W_m = U_0$.) By (the proof of) Proposition 5.12.1 (p. 247), W_j can be chosen such that the subvariety X_{j-1} is obtained by attaching strongly pseudoconvex bumps to $W_j \cap X_{j-1}$. We extend the holomorphic structure across each bump exactly as in the special case when X is a Stein manifold. In this way we extend the holomorphic structure first to $E|_{X_{j-1}}$, and then to an open neighborhood $W_{j-1} \supset X_{j-1} \cup K$ by Lemma 8.2.4. The induction may proceed. At $j = 0$ we obtain a holomorphic G-bundle structure on E that extends the original structure over $M \cup K$. □

Remark 8.2.5 (A) In the case of principal $GL_r(\mathbb{C})$-bundles, Theorem 8.2.1 is equivalent to Theorem 5.3.1 (p. 356) since we can associate to every holomorphic vector bundle $E \to X$ with fibre $\mathbb{C}^r$ the principal $GL_r(\mathbb{C})$-bundle whose transition maps are precisely those of E, and vice versa.
(B) In our construction of a holomorphic G-structure, each step is accomplished by adapting the structure on the new set to the already given one. This scheme avoids any convergence process and is well adapted to the problem of extending the structure from a subvariety. For the classical approach with induction on analytic cubes, see the papers [88, 254, 386].
(C) Another construction of a holomorphic vector bundle structure on a complex vector bundle $E \to X$ over a Stein space was given by Cornalba and Griffiths [106, Chap. D]. A holomorphic structure is determined by its sheaf of holomorphic sections. The latter is chosen as the sheaf of sections which are annihilated by a (0, 1)-connection operator $\nabla = \nabla^{0,1}$ satisfying $\nabla^2 = 0$. Using such a flat (0, 1)-connection they define a complex structure operator on $\mathrm{T}E$ whose associated $\overline{\partial}$-operator on section $X \to E$ is just ∇. The condition $\nabla^2 = 0$ implies that this structure is involutive and hence integrable by the Newlander-Nirenberg theorem [432]. This procedure works well over the regular part of X, and the induction argument uses a suitable stratification of X in essentially the same way as above. □

8.3 Homomorphisms and Generators of Vector Bundles

Let $\pi: E \to X$ and $\pi': E' \to X$ be holomorphic vector bundles over a complex space X. A complex vector bundle map $\Phi: E \to E'$ is said to be of *maximal rank* if for every $x \in X$, $\Phi_x: E_x \to E'_x$ is a $\mathbb{C}$-linear map of maximal rank (injective if $\mathrm{rank}\, E \le \mathrm{rank}\, E'$, resp. surjective if $\mathrm{rank}\, E \ge \mathrm{rank}\, E'$). The following is an Oka principle for vector bundle maps of maximal rank.

Theorem 8.3.1 *Let $\pi: E \to X$ and $\pi': E' \to X$ be holomorphic vector bundles of rank r and r', respectively, over an n-dimensional reduced Stein space X. Then, the following hold:*

(a) *A topological complex vector bundle map $\Phi: E \to E'$ of maximal rank is homotopic, through complex vector bundle maps of maximal rank, to a holomorphic vector bundle map of maximal rank.*
(b) *If Φ in* (a) *is holomorphic over a closed complex subvariety X' of X and in a neighborhood of a compact $\mathcal{O}(X)$-convex subset $K \subset X$, then the homotopy in* (a) *can be chosen to be fixed on X', holomorphic on a neighborhood of K, and uniformly close to Φ on K.*

(c) *If* $|r - r'| \geq [\frac{n}{2}]$ *then there exists a holomorphic vector bundle map* $E \to E'$ *of maximal rank. Furthermore, given a closed complex subvariety* Y *of* X*, any holomorphic vector bundle map* $E|_Y \to E'|_Y$ *of maximal rank extends to a holomorphic vector bundle map* $E \to E'$ *of maximal rank.*

Proof We identify complex vector bundle maps $\Phi: E \to E'$ with sections of the holomorphic vector bundle $\mathrm{Hom}(E, E') = E^* \otimes_{\mathbb{C}} E'$ over the base X. The fibre of the latter bundle is $M_{r',r} = \mathbb{C}^{r'r}$, the set of all complex $r' \times r$ matrices. The set $\Sigma \subset M_{r',r}$ consisting of all matrices of less than maximal rank is an algebraic subvariety which is given locally at every point by at least $|r' - r| + 1$ independent equations (this is the number of independent minors of maximal size); hence Σ has complex codimension $q = |r' - r| + 1$ in $M_{r',r}$. (See Example 1.2.7 on p. 9 or Lemma 8.9.2 on p. 387.) The complement $V_{r',r} = M_{r',r} \setminus \Sigma$ (the Stiefel manifold) is the homogeneous space of the Lie group $GL_N(\mathbb{C})$, with $N = \max\{r, r'\}$, acting by matrix multiplication (on the left if $r \geq r'$, and on the right if $r \leq r'$).

Let A be the closed complex subvariety of the total space of $\mathrm{Hom}(E, E')$ whose fibre $A_x \subset \mathrm{Hom}(E_x, E'_x) = E^*_x \otimes E'_x$ at a point $x \in X$ corresponds to the set $\Sigma \subset M_{r',r}$ under an identification $M_{r',r} \cong E^*_x \otimes E'_x$. (We can use any pair of holomorphic vector bundle charts to make this identification; the set A_x will be independent of the choices.) Then a complex vector bundle map $\Phi: E \to E'$ is of maximal rank if and only $\Phi_x \in \mathrm{Hom}(E_x, E'_x) \setminus A_x$ for all $x \in X$. The conclusion (a) now follows from Corollary 5.5.4 (p. 221) which is a special case of Theorem 5.4.4 (p. 216). The latter result also contains the additions concerning the approximation and interpolation to get part (b).

Part (c) follows from Corollary 5.14.3 on p. 257; the condition $2\,\mathrm{codim}\, A_x > \dim X$, which ensures the existence of a continuous section of $\mathrm{Hom}(E, E') \setminus A$, reads $2(|r' - r| + 1) > n$ which is equivalent to $|r - r'| \geq [\frac{n}{2}]$. □

Applying Theorem 8.3.1(c) with E' a trivial bundle, we obtain

Corollary 8.3.2 *Let* $E \to X$ *be a holomorphic vector bundle of rank* r *over an* n*-dimensional reduced Stein space* X*.*

1. *There is a holomorphic vector bundle embedding* $E \hookrightarrow X \times \mathbb{C}^{r+[\frac{n}{2}]}$*.*
2. *If* $r = [\frac{n}{2}] + d$ *for some* $d > 0$ *then there exists a holomorphic vector bundle embedding* $X \times \mathbb{C}^d \hookrightarrow E$*. Hence,* E *admits* d *pointwise linearly independent holomorphic sections, and it splits into a holomorphic direct sum* $E = E' \oplus \mathbb{T}^d$ *of a holomorphic vector bundle* E' *of rank* $[\frac{n}{2}]$ *and a trivial rank* d *vector bundle* $\mathbb{T}^d$*.*
3. *There is a surjective holomorphic vector bundle map* $\mathbb{T}^{r+[\frac{n}{2}]} \to E$*, that is,* E *admits* $r + [\frac{n}{2}]$ *holomorphic sections spanning each fibre* E_x*.*

The following special case of Corollary 8.3.2, part 1, is of special interest.

Corollary 8.3.3 *Every holomorphic vector bundle over a* 1*-dimensional Stein space is holomorphically trivial.*

All statements in Corollary 8.3.2 also hold in the relative form: A holomorphic vector bundle map of the given type extends from a restricted bundle $E|_Y$ over a subvariety $Y \subset X$ to the whole bundle. The last statement in part 3 is a special case of Theorem 8.4.3, due to Forster and Ramspott [175], which gives optimal lower bound for the number of generators of a coherent analytic sheaf on a Stein space.

Example 8.3.4 In [178, Satz 3] the authors show that for every pair of integers $n, r \geq 1$ there exists a holomorphic vector bundle $E \to X = (\mathbb{C}^*)^n$ of rank r which is not generated by fewer than $r + [\frac{n}{2}]$ sections. □

In [178] it is also shown how dual Chern classes of a complex vector bundle can be used to estimate the number of generators. (Recall that the total dual Chern class $\tilde{c}(E)$ of a complex vector bundle E is defined by $c(E) \cdot \tilde{c}(E) = 1$, where $c(E) = 1 + c_1(E) + c_2(E) + \cdots$ is the total Chern class of E.) For example, we have the following result that we quote without proof.

Theorem 8.3.5 ([178, Satz 2]) *Let X be an n-dimensional reduced Stein space and $E \to X$ be a holomorphic vector bundle of rank r. Given an integer $q \geq \frac{n}{2}$, the module $H^0(X; E)$ can be generated by $r + q - 1$ holomorphic sections if and only if the q-th dual Chern class $\tilde{c}_q(E) = 0$ vanishes.*

We state the following special case of Theorem 8.3.1 for future applications.

Corollary 8.3.6 *Let E, E' be holomorphic vector bundles over a reduced Stein space X. Every topological complex vector bundle isomorphism $\Phi : E \to E'$ is homotopic (in the space of topological complex vector bundle isomorphisms) to a holomorphic one. If Φ is holomorphic over a closed complex subvariety X' of X and over a (neighborhood of a) $\mathcal{O}(X)$-convex subset K of X, then the homotopy can be chosen fixed on X' and holomorphic and close to Φ on K.*

The Oka principle in Corollary 8.3.6 has the following converse due to M. Putinar [464]. For references to earlier converse results, see the MathSciNet review MR1237106 of the paper [464]. We do not prove this result here.

Theorem 8.3.7 ([464]) *Let X be a Stein space of finite Zariski dimension d and let $r = \max\{\dim_{\mathbb{C}} X, d-1\}$. Then, an open subset Y of X is Stein if and only if every holomorphic vector bundle of rank r on Y which is topologically trivial is also holomorphically trivial.*

We denote by $\mathrm{Vect}^k_{\mathbb{R}}(X)$ (resp. $\mathrm{Vect}^k_{\mathbb{C}}(X)$) the topological isomorphism classes of vector bundles of rank k over a CW-complex X. By $\mathbb{T}^k_{\mathbb{R}}$ (resp. $\mathbb{T}^k_{\mathbb{C}}$) we denote the trivial vector bundle of rank k over a given base. Recall the following standard result (see Husemoller [305, Chap. 9]).

Theorem 8.3.8 *Let X be an n-dimensional CW complex.*

(a) *The map $\mathrm{Vect}^k_{\mathbb{R}}(X) \to \mathrm{Vect}^{k+r}_{\mathbb{R}}(X)$, $E \mapsto E \oplus \mathbb{T}^r_{\mathbb{R}}$ $(k, r \ge 1)$ is surjective if $k \ge n$, and is bijective if $k \ge n + 1$.*
(b) *The map $\mathrm{Vect}^k_{\mathbb{C}}(X) \to \mathrm{Vect}^{k+r}_{\mathbb{C}}(X)$, $E \mapsto E \oplus \mathbb{T}^r_{\mathbb{C}}$ $(k, r \ge 1)$ is surjective when $k \ge [\frac{n}{2}]$, and is bijective when $k \ge [\frac{n+1}{2}]$. In particular, if $E \to X$ is a nontrivial complex vector bundle of rank $k \ge [\frac{n+1}{2}]$ then the bundle $E \oplus \mathbb{T}^r_{\mathbb{C}}$ is also nontrivial for each $r \in \mathbb{N}$.*

This result applies in particular when X is an n-dimensional Stein space; in that case Corollary 8.3.6 gives $\mathrm{Vect}^r_{\mathrm{holo}}(X) = \mathrm{Vect}^r_{\mathbb{C}}(X)$. Thus, part (b) of Theorem 8.3.8 implies the following.

Corollary 8.3.9 ([175], [488, Satz 1.4]) *If X is an n-dimensional reduced Stein space, then the map*

$$\mathrm{Vect}^k_{\mathrm{holo}}(X) \to \mathrm{Vect}^{k+r}_{\mathrm{holo}}(X), \qquad E \mapsto E \oplus \mathbb{T}^r_{\mathbb{C}}, \quad k, r \ge 1$$

is surjective for $k \ge [\frac{n}{2}]$, and is bijective for $k \ge [\frac{n+1}{2}]$. In particular, if E, E' are holomorphic vector bundles of rank $k \ge [\frac{n+1}{2}]$ over X such that $E \oplus \mathbb{T}_{\mathbb{C}} \cong E' \oplus \mathbb{T}_{\mathbb{C}}$, then $E \cong E'$.

Corollary 8.3.10 *Let X be a reduced Stein space, $D \subset \mathbb{C}^n$ a contractible open set containing the origin, and $\rho: X \times D \to X \times \{0\}$ the map $\rho(x, \zeta) = (x, 0)$. A holomorphic vector bundle $E \to X \times D$ is holomorphically equivalent to the bundle $\widetilde{E} = \rho^*(E|_{X \times \{0\}}) \to X \times D$.*

Proof There is a topological isomorphism $\Phi: E \to \widetilde{E}$ of complex vector bundles which is the identity over $X \times \{0\}$ (where the two bundles coincide). Thus Φ may be considered as a continuous section, holomorphic over $X \times \{0\}$, of an associated fibre bundle $Z \to X \times D$ (see the proof of Theorem 5.3.1 on p. 213). By Corollary 5.5.4 (p. 221) we can deform Φ to a holomorphic section $\widetilde{\Phi}$ by a homotopy which is fixed on $X \times \{0\}$. Then $\widetilde{\Phi}$ defines a holomorphic isomorphism $E \to \widetilde{E}$ which equals the identity over $X \times \{0\}$. □

From Theorem 6.13.1 we obtain the following classification of vector bundles over 1-convex complex spaces, due to Henkin and Leiterer [287].

Corollary 8.3.11 *Let X be a reduced 1-convex space with the exceptional set S.*

(a) *If $E \to X$ is a topological complex vector bundle which is holomorphic over a neighborhood of S, then there exists a compatible holomorphic vector bundle structure on E that is also holomorphically compatible with the existing structure over a small neighborhood of S.*
(b) *Assume that E and E' are holomorphic vector bundles over X. If $\Phi_0: E \to E'$ is a topological complex vector bundle isomorphism that is holomorphic over*

a neighborhood of S in X, then Φ_0 can be deformed to a holomorphic vector bundle isomorphism $\Phi_1 \colon E \to E'$ by a homotopy of complex vector bundle isomorphisms $\Phi_t \colon E \to E'$ ($t \in [0, 1]$) that are holomorphic over a neighborhood of S and the homotopy is fixed over S.

In conclusion, we give two types of examples where the Oka-Grauert principle fails either because the structure group of the fibre is not a Lie group (it is infinite dimensional), or because the base space is non-Stein.

The following important problem in the development of Stein manifold theory was posed by J-P. Serre:

If $E \to X$ is a holomorphic fibre bundle whose base X and fibre Y are Stein manifolds, is the total space E also Stein?

The first counterexamples, with the fibre $\mathbb{C}^n$ ($n > 1$) and nonlinear transition automorphisms, were given by Skoda [510, 511]. In his first example [510, Theorem], the base is the unit disc with eight discs removed. Around each deleted disc the bundle is defined by a holonomy automorphism of the fibre which, up to a possible permutation of the coordinates, is a multiplicative shear of the form (4.2). In Skoda's second example [511, Theorem 1], the base is the disc with two removed discs; around one of the deleted discs the holonomy automorphism of the fibre $\mathbb{C}^2$ is the entire shear $(z_1, z_2) \mapsto (z_1, z_2 e^{z_1})$, and for the other deleted disc it is the affine map $(z_1, z_2) \mapsto (\mathrm{i}z_2, z_1)$.

Next, Demailly [111, 112] gave the following explicit example over an annulus with a polynomial transition automorphism.

Theorem 8.3.12 *Let $X = \{\zeta \in \mathbb{C} \colon 1 < |\zeta| < r\}$, and let $E \to X$ be a holomorphic fibre bundle with fibre $\mathbb{C}^2$ whose holonomy automorphism is the Hénon map $(z_1, z_2) \mapsto (z_2, -z_1 + z_2^k)$. If $k \in \mathbb{N}$ is big enough (for a fixed r), or if $r > 1$ is big enough (for a fixed $k \geq 2$), then E is non-Stein.*

Demailly showed that, under the stated conditions, E does not admit any plurisubharmonic functions other than those coming from the base X. He also gave counterexamples where the base is the disc or the plane $\mathbb{C}$, the fibre is $\mathbb{C}^2$, and the gluing automorphisms are entire.

It remained an open question for another twenty years whether there exists a non-Stein fibre bundle with fibre $\mathbb{C}^n$ and a contractible Stein base whose transition maps are polynomial automorphisms of the fibre. Such examples were constructed in 2007 by Rosay.

Theorem 8.3.13 (Rosay [476]) *There exists a holomorphic fibre bundle with fibre $\mathbb{C}^2$ over the disc $\mathbb{D}$ (or the plane $\mathbb{C}$) whose transition maps are polynomial automorphisms of $\mathbb{C}^2$ and whose total space is not Stein.*

A detailed presentation of Rosay's example can be found in Sect. 4.21 of the first edition of this book.

Note that, in these examples, the Oka principle holds for sections $X \to E$ since the fibre is a Euclidean space and the structure group does not play a role (see Theorem 5.4.4).

We now give a few examples illustrating the failure of the Oka-Grauert principle for complex line bundles over non-Stein manifolds.

Example 8.3.14 (A) Let $X = \mathbb{C}^2 \setminus \{0\}$. Then $H^1(X; \mathcal{O})$ is infinite dimensional. (In fact, for a domain $X \subset \mathbb{C}^n$, the group $H^p(X; \mathcal{O})$ is either zero or infinite dimensional for any $p \geq 1$ by Laufer [380].) Clearly, X is homotopic to the 3-sphere and hence $H^1(X; \mathbb{Z}) = H^2(X; \mathbb{Z}) = 0$. The top row in the exact sequence (5.2) (p. 212) shows that the exponential map $\mathcal{O} \to \mathcal{O}^*$ induces an isomorphism $H^1(X; \mathcal{O}) \cong H^1(X; \mathcal{O}^*)$. Thus, the trivial bundle $X \times \mathbb{C}$ supports an infinite dimensional family of holomorphic line bundle structures.
(B) Let X be a compact Riemann surface of genus g. Then $H^1(X; \mathcal{O}) = \mathbb{C}^g$, $H^1(X; \mathbb{Z}) = \mathbb{Z}^{2g}$, and the homomorphism $H^1(X; \mathbb{Z}) \to H^1(X; \mathcal{O})$ induced by the inclusion $\mathbb{Z} \hookrightarrow \mathcal{O}$ is an injective homomorphism $\mathbb{Z}^{2g} \hookrightarrow \mathbb{C}^g$ whose image is a lattice $\Gamma \subset \mathbb{C}^g$. Furthermore, $H^2(X; \mathbb{Z}) = \mathbb{Z}$ is generated by the fundamental class of X, and $H^2(X; \mathcal{O}) = 0$ by the Dolbeault isomorphism theorem. (See, e.g., [173] for these facts.) From the exponential sheaf sequence (see (5.2) on p. 212) we obtain an exact sequence

$$0 \to \mathbb{T}^g = \mathbb{C}^g/\Gamma \hookrightarrow H^1\big(X; \mathcal{O}^*\big) \xrightarrow{c_1} \mathbb{Z} \to 0.$$

The kernel $\mathbb{T}^g \subset H^1(X; \mathcal{O}^*)$ of the first Chern class map c_1 is the *Jacobi torus* of X which parametrizes the distinct classes of holomorphic line bundle structures on the topologically trivial bundle $X \times \mathbb{C}$.
(C) Example (A) above shows that the map $H^1(X; \mathcal{O}^*) \xrightarrow{c_1} H^2(X; \mathbb{Z})$ is non-injective on certain simple domains in $\mathbb{C}^2$. It fails to be surjective on any compact Kähler manifold X of dimension > 1 with $H^{(2,0)}(X) \neq 0$. This follows from the Hodge decomposition $H^2(X; \mathbb{C}) = H^{(2,0)}(X) \oplus H^{(1,1)}(X) \oplus H^{(0,2)}(X)$ since the class $c_1(E) \in H^2(X, \mathbb{Z})$ of any complex line bundle $E \to X$ is represented by the de Rham cohomology class $[\frac{\mathrm{i}}{2\pi}\Theta] \in H^{(1,1)}(X)$, where Θ is the Chern curvature form of any Hermitian metric on E.
(D) There also exist domains in $\mathbb{C}^n$ on which the first Chern class map c_1 fails to be surjective (J. Winkelmann, private communication). Let $S \subset \mathbb{P}^n$ be a compact projective surface with a nontrivial holomorphic 2-form, so $H^2(S; \mathbb{Z}) \to H^2(S; \mathcal{O})$ is not the zero map. (There are plenty of examples: any complex 2-torus, any K3 surface, any hypersurface in $\mathbb{P}^3$ of degree at least four, etc.) Consider the diagram

$$\begin{array}{ccc} Z & \hookrightarrow & \mathbb{C}^{n+1}_* \\ \pi \downarrow & & \downarrow \pi \\ S & \hookrightarrow & \mathbb{P}^n \end{array}$$

where $\mathbb{C}^{n+1}_* = \mathbb{C}^{n+1} \setminus \{0\}$, $\pi : \mathbb{C}^{n+1}_* \to \mathbb{P}^n$ is the universal $\mathbb{C}^*$-bundle over $\mathbb{P}^n$, and $Z = \pi^{-1}(S)$. Choose a domain $X \subset \mathbb{C}^{n+1}_*$ which admits a strong deformation retraction onto Z. Then, $c_1: H^1(X;\mathcal{O}^*) \to H^2(X;\mathbb{Z})$ is not surjective. To see this, consider the commutative diagram

$$\begin{array}{ccc} H^2(X;\mathbb{Z}) & \longrightarrow & H^2(X;\mathcal{O}) \\ \| & & \downarrow \\ H^2(Z;\mathbb{Z}) & \longrightarrow & H^2(Z;\mathcal{O}) \\ \uparrow & & \uparrow \\ H^2(S;\mathbb{Z}) & \longrightarrow & H^2(S;\mathcal{O}) \end{array}$$

By the choice of S, the map in the bottom row is not the zero map. Using that $Z \to S$ is a principal $\mathbb{C}^*$ bundle, it can be shown that the bottom right vertical map $H^2(S;\mathcal{O}) \to H^2(Z;\mathcal{O})$ is injective; we omit the details. Hence the composed map $H^2(S;\mathbb{Z}) \to H^2(S;\mathcal{O}) \to H^2(Z;\mathcal{O})$ is nonzero. It follows by diagram chasing that the map $H^2(X;\mathbb{Z}) \to H^2(X;\mathcal{O})$ in the top row is nonzero as well. From exactness of (5.2) we see that $c_1: H^1(X;\mathcal{O}^*) \to H^2(X;\mathbb{Z})$ is surjective if and only if the next map $H^2(X;\mathbb{Z}) \to H^2(X;\mathcal{O})$ in the long exact cohomology sequence is the zero map. It follows that the map $H^1(X;\mathcal{O}^*) \to H^2(X;\mathbb{Z})$ is not surjective.
(E) K. Stein [524] gave an example of a nontrivial complex line bundle over a domain of holomorphy X in $\mathbb{C}^n$ whose restriction to each relatively compact subset of X is trivial. □

8.4 Generators of Coherent Analytic Sheaves

In this section we use the Oka principle to find a minimal number of generators of a coherent analytic sheaf $\mathcal{F}$ over a finite dimensional Stein space X. For a locally trivial sheaf, the answer is given by Corollary 8.3.2. For a general coherent analytic sheaf this problem was treated by Forster and Ramspott [175, 177] using the Oka principle for *Oka pairs of sheaves*, established in [176]. As we shall see, the problem of decreasing the number of generators reduces to the problem of finding a holomorphic section of a certain stratified holomorphic fibre bundle with homogeneous fibres, so their results also follow from Theorems 5.4.4 and 5.14.1 in Chap. 5.

Let $(X,\mathcal{O})$ be a complex space and $\mathcal{F}$ a coherent analytic sheaf over X. Let $\mathfrak{m}_x$ denote the maximal ideal in the ring $\mathcal{O}_x$. Then $\mathfrak{m}_x\mathcal{F}_x$ is a submodule of $\mathcal{F}_x$ and the quotient $L_x(\mathcal{F}) = \mathcal{F}_x/\mathfrak{m}_x\mathcal{F}_x$ is a finite dimensional vector space over $\mathcal{O}_x/\mathfrak{m}_x\mathcal{O}_x = \mathbb{C}$. Consider the quotient projection

$$\lambda_x : \mathcal{F}_x \to L_x(\mathcal{F}) = \mathcal{F}_x/\mathfrak{m}_x\mathcal{F}_x \cong \mathbb{C}^{k_x}.$$

By the Nakayama lemma, a collection $f_1,\ldots,f_r \in \mathcal{F}_x$ generates the stalk $\mathcal{F}_x$ as an $\mathcal{O}_x$-module if and only if their images $\lambda_x(f_1),\ldots,\lambda_x(f_r) \in L_x(\mathcal{F})$ span the vector space $L_x(\mathcal{F})$ over $\mathbb{C}$. By choosing a basis for the vector space $L_x(\mathcal{F})$ we may represent $\lambda_x(f_j)$ by a row vector with k_x components. We assemble the germs $f_j \in \mathcal{F}_x$ into a vector $f = (f_1,\ldots,f_r)^t \in \mathcal{F}_x^r$ and consider $\lambda_x(f) \in M_{r,k_x}$ as a complex $r \times k_x$ matrix.

There exists a stratification $X = X_0 \supset X_1 \supset X_2 \supset \cdots$ (possibly infinite) by closed complex subvarieties

$$X_x = X_k(\mathcal{F}) = \{x \in X : \dim L_x(\mathcal{F}) \geq k\}. \tag{8.6}$$

If there exist finitely many global sections $f_1, \ldots, f_r \in \Gamma(X, \mathcal{F}) = \mathcal{F}(X)$ whose germs at any point $x \in X$ generate the stalk $\mathcal{F}_x$ as an $\mathcal{O}_x$-module, then clearly $r \geq \dim L_x(\mathcal{F})$ for every $x \in X$, and hence

$$k(\mathcal{F}) := \sup_{x \in X} \dim L_x(\mathcal{F}) < +\infty. \tag{8.7}$$

Conversely, if X is a Stein space, then $\mathcal{F}$ is finitely generated if and only if $k(\mathcal{F}) < +\infty$ (Forster [170, Corollary 4.4]). The proof uses Cartan's Theorem A and proceeds by induction on the dimension of the exceptional subvariety of X over which a given collection of sections of $\mathcal{F}$ fails to generate.

Assume now that X is a Stein space and the sheaf $\mathcal{F}$ is finitely generated, say, by the sections $f_1, \ldots, f_r \in \mathcal{F}(X)$. We have the short exact sequence

$$0 \longrightarrow \mathcal{R} \longrightarrow \mathcal{O}^r \stackrel{\sigma}{\longrightarrow} \mathcal{F} \longrightarrow 0$$

where σ is the sheaf epimorpism $\sigma(\alpha_1, \ldots, \alpha_r) = \sum_{j=1}^r \alpha_j f_j$ and $\mathcal{R} = \ker \sigma$ is the sheaf of relations (2.4). We have $\operatorname{rank} \lambda_x(f) = k_x = \dim L_x(\mathcal{F})$ for every $x \in X$, and the above stratification of X ends with $X_{r+1} = \varnothing$. We wish to know whether $\mathcal{F}$ is generated by a smaller number, say $s < r$, of sections $g_1, \ldots, g_s \in \mathcal{F}(X)$. (A necessary condition is that $s \geq k(\mathcal{F})$.) By Cartan's Theorem B, any s-tuple $g = (g_1, \ldots, g_s)^t \in \mathcal{F}(X)^s$ can be expressed as $g = Af$ for some holomorphic matrix function $A \colon X \to M_{s,r}$. We can easily see that

$$\lambda_x(g) = \lambda_x(Af) = A(x)\lambda_x(f), \quad x \in X.$$

Hence, the components of $g = Af \in \mathcal{F}(X)^s$ generate $\mathcal{F}_x$ if and only if

$$\operatorname{rank}\big(A(x)\lambda_x(f)\big) = \dim L_x(\mathcal{J}).$$

Thus, A must be a holomorphic section of the projection $\pi \colon E \to X$ where

$$E = E(\mathcal{F}, f, s) = \big\{(x, T) \in X \times M_{s,r} : \operatorname{rank}\big(T\lambda_x(f)\big) = \dim L_x(\mathcal{F})\big\}. \tag{8.8}$$

In [175] this is called the *Endromisbündel*.

Recall that $V_{s,k} \subset M_{s,k}$ is the Stiefel manifold of all $s \times k$ matrices of full rank k (assuming that $s \geq k$). Since $GL_s(\mathbb{C})$ acts transitively on $V_{s,k}$ by left multiplication, $V_{s,k}$ is an Oka manifold.

Proposition 8.4.1 ([175, Satz 1]) *Let $\mathcal{F}$ be a coherent analytic sheaf with r generators over a complex space X, and let $X_k \subset X$ be the subvariety defined by* (8.6). *The set E* (8.8) *is then an open subset of $X \times M_{s,r}$, and the restriction of the submersion $E \to X$ to the subvariety $Y_k = X_k \setminus X_{k+1}$ is a holomorphic fibre bundle with the fibre $F_{s,r,k} = V_{s,k} \times M_{s,r-k} \cong V_{s,k} \times \mathbb{C}^{s(r-k)}$.*

The proof uses elementary sheaf theory and linear algebra; we refer the reader to the original paper. This means that $E \to X$ is a stratified fibre bundle whose fibres $F_{s,r,k}$ are Oka manifolds, and hence the Oka principle applies to sections $X \to E$ in view of Theorem 5.4.4. In particular:

Corollary 8.4.2 ([175, Satz 6]) *Let X be a reduced Stein space and $\mathcal{F}$ a coherent analytic sheaf over X with generators $f = (f_1, \ldots, f_r)$. Then the $\mathcal{O}(X)$-module $\mathcal{F}(X) = H^0(X; \mathcal{F})$ admits s generators for some integer $s \in \mathbb{N}$ if and only if the bundle $E(\mathcal{F}, f, s)$ (8.8) admits a continuous section.*

By using elementary obstruction theory, we now find a condition for the existence of a continuous section $X \to E = E(\mathcal{F}, f, s)$ (compare with Corollary 5.14.3 on p. 257). Note that each variety $Y_k = X_k \setminus X_{k+1}$ can be substratified such that all strata are Stein manifolds. We need that for each integer $k = 1, \ldots, k(\mathcal{F})$ the homotopy groups $\pi_q(F_{s,r,k}) = \pi_q(V_{s,k})$ of the fibre over Y_k vanish for all $q = 0, 1, \ldots, \dim Y_k - 1$. Since $\pi_q(V_{s,k}) = 0$ for $q \le 2(s-k)$ (see Example 1.2.7 on p. 9), we see that E admits a continuous section when

$$\dim Y_k - 1 \le 2s - 2k, \quad \forall k = 1, \ldots, k(\mathcal{F}).$$

This gives the following result of Forster and Ramspott [175], [177, Satz 6].

Theorem 8.4.3 *Let $\mathcal{F}$ be a coherent analytic sheaf with finitely many generators over a finite dimensional reduced Stein space X. Set*

$$s(\mathcal{F}) = \max_{k \in \mathbb{N}} \left\{ k + \left[\frac{1}{2} \dim Y_k \right] \right\} \tag{8.9}$$

where $Y_k = \{x \in X : \dim L_x(\mathcal{F}) = k\}$. Then, $\mathcal{F}$ admits $s(\mathcal{F})$ generators.

If $\mathcal{F} = \mathcal{O}^k$ is a locally free sheaf or rank k, then $X = Y_k$ and we get $s = k + [\frac{n}{2}]$ generators; this coincides with part 2 in Corollary 8.3.2. This number can not be decreased in general (see Proposition 9.3.3).

In this connection we mention another interesting problem that was first considered by Cartan in his 1940 paper on holomorphic matrices [86]. Suppose that $\mathcal{F}$ is a coherent analytic sheaf on a Stein space X and $f = (f_1, \ldots, f_r)$, $g = (g_1, \ldots, g_r)$ are two sets of generators for the algebra of global sections $\mathcal{F}(X)$. It follows from Cartan's Theorem B that there exist holomorphic $r \times r$ matrix functions $A, B \colon X \to M_{r,r}(\mathbb{C}) \cong \mathbb{C}^{r^2}$ such that $Af = g$ and $Bg = f$. However, Cartan showed by a simple example over the polydisc in $\mathbb{C}^2$ that in general there does not exist an *invertible* holomorphic matrix $A \colon X \to GL_r(\mathbb{C})$ satisfying $Af = g$. The obstruction in his example is purely topological, and he conjectured that this problem has an analytic solution if it has a topological solution. The Oka principle for this problem was established in 1968 by Forster and Ramspott [177]. Define

$$E = \big\{(x, A) : x \in X,\ A \in GL_r(\mathbb{C}),\ A\lambda_x(f) = \lambda_x(g)\big\} \subset X \times GL_r(\mathbb{C})$$

and let $\pi \colon E \to X$ be the projection $\pi(x, A) = x$.

Theorem 8.4.4 ([177, Satz 7]) *If there exists a continuous section $X \to E$, then there also exists a holomorphic section $T = (\mathrm{Id}_X, A): X \to E$. In such case, the holomorphic matrix function $A: X \to GL_r(\mathbb{C})$ satisfies $Af = g$.*

With X_k defined by (8.6) and $Y_k = X_k \setminus X_{k+1}$, the restriction $E|_{Y_k}$ is a holomorphic fibre bundle with the fibre $F_{r,k}$ consisting of all matrices of the form $\left(\begin{smallmatrix} I & Q \\ 0 & R \end{smallmatrix}\right)$ with $R \in GL_{r-k}(\mathbb{C})$ and $Q \in M_{k,r-k}(\mathbb{C})$ [177, Hilfssatz 6]. Even though the projection $\pi: E \to X$ is not a submersion, and hence not a stratified fibre bundle in our sense, it is still possible to extend a holomorphic section $X_{k+1} \to E$ to an open neighborhood of X_{k+1} in X and then proceed in the usual way to extend it to a holomorphic section over X_k.

If the matrix function A in Theorem 8.4.4 is null-homotopic (this is always the case if X is contractible), then the homotopy $A_t: X \to GL_r(\mathbb{C})$ $(t \in [0,1])$ from $A = A_0$ to $A_1 = I$ provides an isotopy $A_t f$ of generators of $H^0(X; \mathcal{F})$ connecting f to g. This problem is treated in more detail in [177].

8.5 The Number of Equations Defining a Subvariety

Every algebraic subvariety Y of the affine complex space $\mathbb{A}^n$ is the common zero set of finitely many polynomials in n complex variables. In the holomorphic category, Cartan's Theorem A implies that every closed complex subvariety Y in a Stein manifold X (or in a finite dimensional Stein space) is the common zero set of finitely many global holomorphic functions on X.

Question 8.5.1 How many holomorphic functions are needed to define a closed complex subvariety Y in a Stein space X? What is the answer when $X = \mathbb{C}^n$?

We show how the Oka principle sheds light on this subject in the holomorphic category. The reader is referred to the papers [41, 61, 172, 488, 489] for additional results on complete intersections in the holomorphic, affine algebraic and differentiable categories.

There are two natural notions of defining a subvariety $Y \subset X$. In the set theoretic sense, we are looking for functions $f_1, \ldots, f_k \in \mathcal{O}(X)$ (or polynomials in $\mathbb{C}[z_1, \ldots, z_n]$ in the affine case) such that

$$Y = \{x \in X : f_1(x) = 0, \ldots, f_k(x) = 0\}. \tag{8.10}$$

If Y is of pure codimension k in X, then Y is said to be a *set theoretic complete intersection*. Replacing $\mathbb{C}^n$ by $\mathbb{P}^n$ and functions by projective hypersurfaces gives the corresponding notion in the projective category.

A theorem of Kronecker [350] asserts that any affine algebraic subvariety Y of an affine algebraic variety $X \subset \mathbb{C}^n$ is the common zero set of $n+1$ polynomials on $\mathbb{C}^n$ restricted to X; if $X = \mathbb{C}^n$ then n polynomials suffice [140]. Kronecker's problem, whether every affine algebraic curve is an algebraic set theoretic complete intersection, is apparently still open.

On the analytic side, it was proved by Grauert [251] that each complex subvariety in an n-dimensional Stein space is the common zero set of $n+1$ functions; this was improved to n by Forster and Ramspott [174] (see Corollary 8.5.5).

A stronger notion of defining a subvariety is the ideal theoretic one. Let $\mathcal{J} = \mathcal{J}_Y \subset \mathcal{O}_X$ denote the ideal sheaf of Y in X. In addition to (8.10) we now ask that $f_1, \ldots, f_k$ generate the ideal $\mathcal{J}_x \subset \mathcal{O}_x$ of Y at every point $x \in Y$. This is equivalent to asking that the functions $f_1, \ldots, f_k$ generate the algebra of global sections $\mathcal{J}(Y)$ of the sheaf $\mathcal{J}$. (Note that $\mathcal{J}(Y)$ is a closed ideal in the Fréchet algebra $\mathcal{O}(X)$. A collection of functions generating the ideal $\mathcal{J}(Y)$ also defines Y in the sense of (8.10), but the converse need not hold.) A subvariety Y in an n-dimensional Stein space X can be defined by finitely many functions in the ideal theoretic sense if and only if its ideal sheaf $\mathcal{J}$ satisfies the condition $k(\mathcal{J}) < +\infty$ (8.7); in this case Theorem 8.4.3 gives a bound on the number of generators.

Definition 8.5.2 Let Y be a complex subvariety of pure codimension k in a complex space X.

(a) Y is an *ideal theoretic complete intersection* if its ideal sheaf $\mathcal{J} = \mathcal{J}_Y$ is generated by k sections $f_1, \ldots, f_k \in \mathcal{J}(X)$.
(b) Y is an (ideal theoretic) *local complete intersection* if the ideal $\mathcal{J}_x$ at any point $x \in Y$ is generated by k elements.

Clearly, every complex submanifold Y of a complex manifold X is a local complete intersection. However, if Y is a global ideal theoretic complete intersection in X, then its normal bundle $\mathrm{N}_{Y/X} = \mathrm{T}X|_Y/\mathrm{T}Y$ is necessarily trivial. Indeed, if $Y = \{f_1 = 0, \ldots, f_k = 0\}$ and the differentials $\mathrm{d}f_1, \ldots, \mathrm{d}f_k$ are linearly independent at each point of Y, then the map $(\mathrm{d}f_1, \ldots, \mathrm{d}f_k)\colon \mathrm{T}X|_Y \to Y \times \mathbb{C}^k$ induces an isomorphism of the normal bundle $\mathrm{N}_{Y/X}$ onto $Y \times \mathbb{C}^k$. Conversely, the tubular neighborhood theorem for Stein manifolds (see Theorem 3.3.3 on p. 74) implies that any *Stein* submanifold Y in an arbitrary complex manifold X is an ideal theoretic complete intersection in an open neighborhood $U \subset X$ of Y.

The above observation provides a nontrivial topological restriction for global complete intersections as shown in the following proposition.

Proposition 8.5.3 *A complete intersection Stein submanifold Y in a parallelizable complex manifold X is itself parallelizable. In particular, any smooth complex hypersurface in $\mathbb{C}^n$ is parallelizable.*

Proof From $\mathrm{T}Y \oplus \mathrm{N}_{Y/X} = \mathrm{T}X|_Y \cong Y \times \mathbb{C}^n$ and $\mathrm{N}_{Y/X}$ being trivial, we infer by Corollary 8.3.9 that $\mathrm{T}Y$ is trivial. Assume now that $X = \mathbb{C}^n$. By the universal solvability of the second Cousin problem on $\mathbb{C}^n$ (see Sect. 5.2), every closed smooth complex hypersurface $Y \subset \mathbb{C}^n$ is the zero set of a holomorphic function $f \in \mathcal{O}(\mathbb{C}^n)$ with $\mathrm{d}f \neq 0$ on Y. Hence, Y is a complete intersection and therefore parallelizable. □

Assume now that Y is a local complete intersection of pure codimension k in an n-dimensional Stein space X. The sheaf $\mathcal{J}/\mathcal{J}^2$ is then locally free of rank k over Y, and hence it is the sheaf of sections of a holomorphic vector bundle over Y. The dual bundle of $\mathcal{J}/\mathcal{J}^2$ is called the *normal bundle of Y in X*, denoted $\mathrm{N}_{Y/X}$. (In the nonsingular case, this is the usual normal bundle of Y in X.) The sheaf $\mathcal{J}/\mathcal{J}^2$ is clearly trivial over the complement $X \setminus Y$. Recall from Sect. 8.4 that $L_x(\mathcal{J}) = \mathcal{J}_x/\mathfrak{m}_x\mathcal{J}_x$; in the present case, we have $\dim L_x(\mathcal{J}) = k$ if $x \in Y$ and $\dim L_x(\mathcal{J}) = 1$ if $x \in X \setminus Y$. Stratifying X by $X \supset Y \supset \varnothing$ we see that the number $s = s(\mathcal{J})$ (8.9) equals

$$s = \max\left\{k + \left[\frac{n-k}{2}\right], 1 + \left[\frac{n}{2}\right]\right\}.$$

Hence, Theorem 8.4.3 implies the following result of Forster and Ramspott [175, Satz 9].

Theorem 8.5.4 *Let Y be a local complete intersection subvariety of pure codimension k in an n-dimensional reduced Stein space X. Then, the ideal sheaf $\mathcal{J}_Y$ is generated by $[\frac{n+k}{2}]$ functions if $k \ge 2$, and by $1 + [\frac{n}{2}]$ functions if $k = 1$.*

Corollary 8.5.5

(a) *Every nonsingular analytic curve in a connected Stein manifold of dimension $n \ge 3$ is a complete intersection.*
(b) *Every discrete set in a Stein manifold X is defined by $n = \dim X$ functions.*

Assume that Y is a local complete intersection in X. When is Y also a global complete intersection? According to Boratyński, a local complete intersection in $\mathbb{C}^n$ is a global set theoretic complete intersection; see [172]. However, if we ask the question in the ideal theoretic sense, then an obvious necessary condition is that the normal bundle $\mathrm{N}_{Y/X}$ be trivial. Conversely, triviality of $\mathrm{N}_{Y/X}$ guarantees that Y is a complete intersection in an open neighborhood $U \subset X$ of Y, defined by a holomorphic map $f = (f_1, \ldots, f_k) \colon U \to \mathbb{C}^k$ with $Y = f^{-1}(0)$. If f can be chosen such that it extends continuously to all of X and maps $X \setminus Y$ to $\mathbb{C}^k \setminus \{0\}$, then the following result shows that we can deform f to a holomorphic map $\tilde{f} \colon X \to \mathbb{C}^k$ which defines Y as a complete intersection in X.

Theorem 8.5.6 (Oka principle for complete intersections) *Let X be a Stein manifold and $f = (f_1, \ldots, f_k) \colon X \to \mathbb{C}^k$ be a continuous map that is holomorphic in an open neighborhood $U \subset X$ of the subvariety $Y = f^{-1}(0)$. Given a coherent analytic sheaf of ideals $\mathcal{S} \subset \mathcal{O}_X$ supported on Y, there exists a holomorphic map $F \colon X \to \mathbb{C}^k$ which is homotopic to f relative to Y such that $F^{-1}(0) = Y$ and $F - f$ is a section of $\mathcal{S}^k$. In particular, if f defines Y as an ideal theoretic complete intersection in U, then F can be chosen to define Y as an ideal theoretic complete intersection in X.*

Proof Since the manifold $\mathbb{C}^k \setminus \{0\}$ is complex homogeneous and hence an Oka manifold, this follows from Theorem 8.6.1 in the following section. □

The problem of extending f continuously to X as a zero-free map on $X \setminus Y$ can be treated by obstruction theory as in the following result.

Theorem 8.5.7 ([175, Satz 10], [489, Theorem 2.5]) *Assume that X is a Stein manifold and $Y \subset X$ is a pure k-codimensional local complete intersection with trivial normal bundle $\mathrm{N}_{Y/X}$. Then, the ideal sheaf $\mathcal{J}_Y$ is generated by $k+1$ functions. If in addition $H^q(X,Y;\pi_q(S^{2k-1}))=0$ for all $q \ge 2k-1$, then Y is an ideal theoretic complete intersection in X.*

Proof The conditions imply that Y is a complete intersection in an open neighborhood $U \subset X$, defined by functions $f_1,\ldots,f_k \in \mathcal{O}(U)$. By Cartan's Theorem B we can find functions $g_j \in \mathcal{O}(X)$ such that $g_j - f_j$ is a section of the coherent sheaf of ideals $\mathcal{J}_Y^2$ in a neighborhood of Y for every $j=1,\ldots,k$. This implies that the map $g=(g_1,\ldots,g_k)\colon X \to \mathbb{C}^k$ defines Y as a complete intersection in an open neighborhood of Y. We have $g^{-1}(0)=Y \cup Y'$ where the subvariety Y' is disjoint from Y. Choose a function $g_{k+1} \in \mathcal{O}(X)$ that vanishes on Y and equals one on Y'; then the functions $g_1,\ldots,g_{k+1}$ generate the ideal sheaf $\mathcal{J}_Y$. This proves the first part of the theorem.

Assume now that $H^q(X,Y;\pi_q(S^{2k-1}))=0$ for all $q \ge 2k-1$. Since $\mathbb{C}^k \setminus \{0\}$ is homotopy equivalent to S^{2k-1}, this condition implies that, after shrinking U around Y if necessary, the map $f=(f_1,\ldots,f_k)\colon U \to \mathbb{C}^k$ extends to a continuous map $f\colon X \to \mathbb{C}^k$ such that $Y=f^{-1}(0)$. Applying Theorem 8.5.6 to f gives a holomorphic map $F\colon X \to \mathbb{C}^k$ which defines Y as an ideal theoretic complete intersection. □

Corollary 8.5.8 ([175, Satz 11, Satz 12]) *Let Y be a closed complex submanifold of a Stein manifold X. If $\dim Y < \frac{1}{2}\dim X$, then Y is an ideal theoretic complete intersection in X if and only if the normal bundle $\mathrm{N}_{Y/X}$ is trivial. If $X=\mathbb{C}^n$ then the same conclusion holds when $\dim Y \le \frac{2}{3}(n-1)$.*

Proof Let $n=\dim X$ and $k=n-\dim Y$. Then $2k>n$ and $H^q(Y;G)=H^{q+1}(X;G)=0$ for $q \ge n$ and for every Abelian group G (see Sect. 3.12). From the exact cohomology sequence

$$\cdots \longrightarrow H^q(Y;G) \longrightarrow H^{q+1}(X,Y;G) \longrightarrow H^{q+1}(X;G) \longrightarrow \cdots$$

we infer that $H^{q+1}(X,Y;G)=0$ for all $q \ge 2k-1$. Taking $G=\pi_q(S^{2k-1})$ we see that the condition in Theorem 8.5.7 holds, so Y is a complete intersection.

In the case $X=\mathbb{C}^n$ we have $H^{q+1}(\mathbb{C}^n;G)=0$ for all $q \ge 0$. The above exact cohomology sequence gives $H^q(Y;G)=H^{q+1}(\mathbb{C}^n,Y;G)$ for $q \ge 1$, and hence $H^{q+1}(\mathbb{C}^n,Y;G)=0$ for $q>d=\dim Y$. If $d \le \frac{2}{3}(n-1)$ then $d<2k-1$ and the conclusion follows from Theorem 8.5.7. □

Corollary 8.5.9 ([169, p. 713]) *Let X be a closed complex (hence Stein) submanifold of dimension n in $\mathbb{C}^N$.*

(i) *If X is an ideal theoretic complete intersection, then it is parallelizable.*
(ii) *If X is parallelizable and $N \ge \frac{3n}{2}+1$, then X is an ideal theoretic complete intersection.*

Proof If $X \subset \mathbb{C}^N$ is a complete intersection then its normal bundle $\mathrm{N}_{X/\mathbb{C}^N}$ is trivial, whence X is parallelizable by Proposition 8.5.3. Conversely, if X is parallelizable and $N \geq \frac{3n}{2} + 1$, then Corollary 8.3.9 implies that $\mathrm{N}_{X/\mathbb{C}^N}$ is trivial, hence X is a complete intersection by Corollary 8.5.8. □

Corollary 8.5.10 *A Stein manifold is parallelizable if and only if it can be embedded as an ideal theoretic complete intersection submanifold in some complex Euclidean space* $\mathbb{C}^N$.

Example 8.5.11 In [189], an example is given of a closed complex submanifold $X^3 \subset \mathbb{C}^5$ which is a differentiable complete intersection, but not a holomorphic complete intersection. In fact, for any compact orientable 2-surface M of genus $g \geq 2$ there exists a three dimensional Stein manifold X, homotopy equivalent to M, whose tangent bundle TX is trivial as a real vector bundle, but is nontrivial as a complex vector bundle. The normal bundle of any proper holomorphic embedding of such X in $\mathbb{C}^5$ or $\mathbb{C}^7$ is trivial as a real vector bundle, but nontrivial as a complex vector bundle. □

8.6 Elimination of Intersections

The Oka principle for complete intersections, given by Theorem 8.5.6, is a special case of Theorem 8.6.1 below. The main problem is the following.

Assume that X is a reduced Stein space, $h: Z \to X$ is a holomorphic submersion onto X, and $\Sigma \subset Z$ is a closed complex subvariety of Z whose fibre $\Sigma_x = \Sigma \cap Z_x$ is a proper subvariety of $Z_x = h^{-1}(x)$ for every $x \in X$. Suppose that $f: X \to Z$ is a holomorphic section and that

$$f^{-1}(\Sigma) = Y \cup Y', \quad Y \cap Y' = \varnothing,$$

where Y and Y' are unions of connected components of $f^{-1}(\Sigma)$. The problem is to find a homotopic deformation of f that is fixed on Y and that pushes Y' out of the intersection set with Σ.

Theorem 8.6.1 (The Oka principle for elimination of intersections) *Under the above assumptions, suppose that there exists a homotopy of continuous sections $g_t: X \to Z$ ($t \in [0, 1]$) which is fixed near Y such that $g_0 = f$ and $g_1^{-1}(\Sigma) = Y$. Let $\mathcal{S} \subset \mathcal{O}_X$ be a coherent sheaf of ideals supported on Y.*

(a) *If the submersion $h: Z \setminus \Sigma \to X$ is stratified subelliptic over $X \setminus Y$ (Definition 6.2.1 on p. 265), then there exists a holomorphic section $\tilde{f}: X \to Z$ such that $(\tilde{f})^{-1}(\Sigma) = Y$ and $\tilde{f}$ is $\mathcal{S}$-tangent to f along Y (Definition 1.3.2 on p. 12).*
(b) *If in addition the submersion $h: Z \to X$ is stratified subelliptic over $X \setminus Y$, then there is a homotopy of holomorphic sections $f_t: X \to Z$ such that $f_0 = f$, $f_1 = \tilde{f}$, and f_t is $\mathcal{S}$-tangent to f along Y for every $t \in [0, 1]$.*

The same conclusions hold if the submersions $Z \to X$ and $Z \setminus \Sigma \to X$ are stratified fibre bundles with Oka fibres.

Proof If $Y = \varnothing$, this is the Oka principle for sections $X \to Z \setminus \Sigma$ avoiding Σ (a special case of Theorem 6.2.2). The proof in the general case requires minor adjustments that we now explain. Let $\mathcal{J}_\Sigma$ denote the sheaf of ideals of the subvariety $\Sigma \subset Z$. Let $\mathcal{R} \subset \mathcal{O}_X$ be the coherent sheaf of ideals whose stalk $\mathcal{R}_x$ at each point $x \in Y$ is the pullback of $\mathcal{J}_\Sigma$ by f, while for $x \in X \setminus Y$ we take $\mathcal{S}_x = \mathcal{O}_{X,x}$. (More precisely, if $x \in Y$ and if $\mathcal{J}_{\Sigma, f(x)}$ is generated by the germs $h_1, \ldots, h_m$, we take the functions $h_j \circ f$ $(j = 1, \ldots, m)$ as the generators of $\mathcal{S}_x$.) It suffices to prove the theorem with $\mathcal{S}$ replaced by the product sheaf $\mathcal{S}\mathcal{R}\mathcal{J}_Y$, where $\mathcal{J}_Y$ is the ideal sheaf of Y in X. It is easily seen that for every section $f' \colon U \to Z|_U$ in an open set $U \supset Y$ such that f' is $\mathcal{S}$-tangent to f along Y there is a smaller open set $V \supset Y$ such that $(f')^{-1}(\Sigma) \cap V = Y$. (See the proof of Theorem 1.3 in [189] for the details when $\Sigma \subset \mathbb{C}^n$.)

Assume now that $g_1 \colon X \to Z$ is a continuous section which agrees with f near Y and satisfies $g_1^{-1}(\Sigma) = Y$. We apply to g_1 the same procedure as in the proof of Theorem 6.2.2 to deform it to a holomorphic section $\tilde{f} \colon X \to Z$ that is $\mathcal{S}$-tangential to f along Y and satisfies $\tilde{f}^{-1}(\Sigma) = Y$. The condition that the sections in the homotopy are holomorphic near Y and $\mathcal{S}$-tangential to f implies that no new intersections with Σ appear in a sufficiently small open neighborhood V of Y, while on $X \setminus V$ we can ensure this by the construction.

In a typical inductive step, we choose a relatively compact Runge domain $W \Subset X$ and stratify it by a descending chain of closed complex subvarieties $W = W_0 \supset W_1 \supset \cdots \supset W_m = Y \cap W$ such that the submersion $Z \setminus \Sigma \to X$ is subelliptic over each (smooth) stratum $W_j \setminus W_{j+1}$. This allows the construction of a holomorphic section with the stated properties by following the proof of Theorem 6.2.2, resp. of Theorem 5.4.4 in the case of a stratified fibre bundle with Oka fibres. (See Sects. 5.12 and 6.12.) The proof is concluded by another induction over a normal exhaustion of X by Runge domains.

By the construction, we have a continuous homotopy of sections $X \to Z$ connecting f to $\tilde{f}$ that are holomorphic near Y and $\mathcal{S}$-tangential to f along Y. If $Z \to X$ is a stratified subelliptic submersion over $X \setminus Y$, Theorem 6.2.2 gives a deformation (with fixed ends at $t = 0, 1$) to a homotopy of holomorphic sections that are $\mathcal{S}$-tangential to f along Y. □

Corollary 8.6.2 *Assume that $h \colon Z \to X$ is a holomorphic fibre bundle with fibre $\mathbb{C}^n$ and $\Sigma \subset Z$ is a closed complex subvariety that is locally uniformly tame (see Proposition* 6.4.14 *on p.* 278) *and satisfies* $\dim \Sigma_x \le n - 2$ *for every point $x \in X$. Then the conclusion of Theorem* 8.6.1 *holds. The same is true if the fibre is $\mathbb{P}^n$ and* $\dim \Sigma_x \le n - 2$ *for every $x \in X$.*

Proof In the first case with fibre $\mathbb{C}^n$, the submersion $Z \setminus \Sigma \to X$ is stratified elliptic by Proposition 6.4.14, and obviously $Z \to X$ is elliptic. In the second case with fibre $\mathbb{P}^n$, the subvariety Σ has algebraic fibres, and $E \setminus \Sigma \to X$ is a stratified subelliptic submersion by Proposition 6.4.15. □

The following is a special case of Corollary 8.6.2; complete intersections correspond to $\Sigma = \{0\} \subset \mathbb{C}^n$. It is easily seen that this fails if Σ is a nonavoidable discrete set in $\mathbb{C}^n$ (see Theorem 4.7.2 on p. 130).

Corollary 8.6.3 ([189, Theorem 1.3]) *If Σ is a closed complex analytic subvariety of $\mathbb{C}^n$ such that $\mathbb{C}^n \setminus \Sigma$ is an Oka manifold, then the conclusion of Theorem* 8.6.1 *holds for maps from any reduced Stein space X to $\mathbb{C}^n$.*

8.7 Holomorphic Vaserstein Problem

In this section we outline the solution to *Gromov's Vaserstein problem* obtained by Ivarsson and Kutzschebauch [306]. For this application, we use the Oka principle for sections of stratified elliptic submersions.

Let I_m denote the identity matrix of size m. It is well known that $SL_m(\mathbb{C})$ is generated by elementary matrices $I_m + \alpha e_{ij}$ $i \neq j$, i.e., matrices with 1's on the diagonal and all entries outside the diagonal except one equal to zero. Equivalently, every $A \in SL_m(\mathbb{C})$ can be written as a finite product of upper and lower diagonal unipotent matrices (in alternating order). The same question for matrices in $SL_m(R)$, where R is a commutative ring instead of the field $\mathbb{C}$, can be much more delicate. For example, if R is the ring of complex valued functions (continuous, smooth, algebraic or holomorphic) from a space X, the problem amounts to finding for a given map $f\colon X \to SL_m(\mathbb{C})$ a factorization as a product of upper and lower triangular unipotent matrices

$$f(x) = \begin{pmatrix} 1 & 0 \\ G_1(x) & 1 \end{pmatrix} \begin{pmatrix} 1 & G_2(x) \\ 0 & 1 \end{pmatrix} \cdots \begin{pmatrix} 1 & G_N(x) \\ 0 & 1 \end{pmatrix},$$

where the G_i are maps $G_i\colon X \to \mathbb{C}^{m(m-1)/2}$. Since any product of such matrices is homotopic to a constant map (multiplying each entry outside the diagonals by $t \in [0, 1]$ we get a homotopy to the identity matrix), one has to assume that the given map $f\colon X \to SL_m(\mathbb{C})$ is homotopic to a constant map.

This problem has been studied in the case of polynomials of n variables. For $n = 1$, i.e., $f\colon \mathbb{C} \to SL_m(\mathbb{C})$ a polynomial map (the ring R equals $\mathbb{C}[z]$), it is an easy consequence of the fact that $\mathbb{C}[z]$ is an Euclidean ring that such f factors into a product of upper and lower diagonal unipotent matrices. For $m = 2$ and $n = 2$ it was shown by Cohn [101] that the matrix

$$\begin{pmatrix} 1 - z_1 z_2 & z_1^2 \\ -z_2^2 & 1 + z_1 z_2 \end{pmatrix} \in SL_2(\mathbb{C}[z_1, z_2])$$

does not decompose as a finite product of unipotent matrices.

For $m \geq 3$ and any $n \in \mathbb{N}$ it is a result of Suslin [539] that any matrix in $SL_m(\mathbb{C}^{[n]})$ decomposes as a finite product of unipotent matrices. (Here, $\mathbb{C}^{[n]}$ is the polynomial ring in n complex variables.) Further results in the algebraic setting can be found in [539] and [271]. For a connection to the Jacobian conjecture see [587].

In the case of continuous complex valued functions on a topological space X the problem was partially solved by Thurston and Vaserstein [549], and completely solved by Vaserstein [561, Theorem 4].

It is natural to consider the Vaserstein problem for rings of holomorphic functions on Stein spaces, in particular, on $\mathbb{C}^n$. This problem was posed by Gromov [266, Sect. 3.5.G] and solved by Ivarsson and Kutzschebauch [306].

Theorem 8.7.1 ([306]) *Let X be a finite dimensional reduced Stein space and let $f: X \to SL_m(\mathbb{C})$ be a null-homotopic holomorphic mapping. There exist $k \in \mathbb{N}$ and holomorphic mappings $G_1, \ldots, G_k: X \to \mathbb{C}^{m(m-1)/2}$ such that*

$$f(x) = \begin{pmatrix} 1 & 0 \\ G_1(x) & 1 \end{pmatrix} \begin{pmatrix} 1 & G_2(x) \\ 0 & 1 \end{pmatrix} \cdots \begin{pmatrix} 1 & G_k(x) \\ 0 & 1 \end{pmatrix}$$

is a product of upper and lower diagonal unipotent matrices. In particular, if X is contractible then this holds for every holomorphic map $f: X \to SL_m(\mathbb{C})$.

An effective version of Theorem 8.7.1 with a bound on the number of factors was obtained in [307] for the group $SL_2(\mathbb{C})$. In the special case when X is an open Riemann surface, a more elementary solution was given already in 1988 by Klein and Ramspott [336], but it went unnoticed.

By the definition of the Whitehead K_1-group [478, p. 61], the above result has the following consequence.

Corollary 8.7.2 *Let $(X, \mathcal{O}_X)$ be a finite dimensional reduced Stein space that is topologically contractible. Then $SK_1(\mathcal{O}(X))$ is trivial, and the determinant induces an isomorphism* $\det: K_1(\mathcal{O}(X)) \to \mathcal{O}(X)^\star$.

Proof of Theorem 8.7.1 for maps $X \to SL_2(\mathbb{C})$ Define $\Psi_k: \mathbb{C}^k \to SL_2(\mathbb{C})$ by

$$\Psi_k(z_1, \ldots, z_k) = \begin{pmatrix} 1 & 0 \\ z_1 & 1 \end{pmatrix} \begin{pmatrix} 1 & z_2 \\ 0 & 1 \end{pmatrix} \cdots \begin{pmatrix} 1 & z_k \\ 0 & 1 \end{pmatrix}.$$

We must find a holomorphic map $G = (G_1, \ldots, G_k): X \to \mathbb{C}^k$ such that the following diagram commutes:

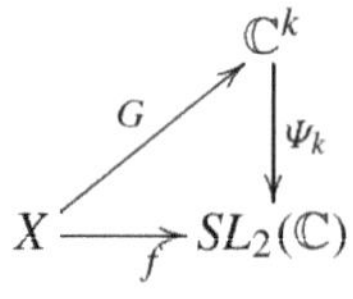

$$\begin{array}{ccc} & & \mathbb{C}^k \\ & \overset{G}{\nearrow} & \downarrow \Psi_k \\ X & \xrightarrow{f} & SL_2(\mathbb{C}) \end{array}$$

Vaserstein's result [561, Theorem 4] gives a continuous lifting of f. We shall deform it to a holomorphic lifting by applying the Oka principle to the diagram

$$\begin{array}{ccc} & & \mathbb{C}^k \\ & \overset{F}{\nearrow} & \Big\downarrow \pi_2\circ\Psi_k=\Phi_k \\ X & \xrightarrow[\pi_2\circ f]{} & \mathbb{C}^2\setminus\{0\} \end{array}$$

where $\pi_2\colon SL_2(\mathbb{C}) \to \mathbb{C}^2 \setminus \{0\}$ is the projection of a matrix to its second row.

The following lemma is easily verified.

Lemma 8.7.3 *The mapping $\Phi_k = \pi_2 \circ \Psi_k\colon \mathbb{C}^k \to \mathbb{C}^2 \setminus \{0\}$ is a holomorphic submersion exactly at points $\mathbb{C}^k \setminus S_k$, where for $k \ge 2$,*

$$S_k = \big\{(z_1, \ldots, z_k) \in \mathbb{C}^k : z_1 = \cdots = z_{k-1} = 0\big\}.$$

The submersion $\Phi_k = \pi_2 \circ \Psi_k\colon \mathbb{C}^k \setminus S_k \to \mathbb{C}^2 \setminus \{0\}$ is surjective when $k \ge 3$.

The following lemma enables us to apply the Oka principle.

Lemma 8.7.4 *If $k \ge 3$ then the map*

$$\Phi_k\colon \mathbb{C}^k \setminus S_k \to \mathbb{C}^2 \setminus \{0\} \tag{8.11}$$

is a stratified elliptic submersion.

Proof Write $\Phi_k(z_1, \ldots, z_k) = (P_k(z_1, \ldots, z_k), Q_k(z_1, \ldots, z_k))$. Note that

$$P_k(z_1, \ldots, z_k) = P_{k-1}(z_1, \ldots, z_{k-1}),$$
$$Q_k(z_1, \ldots, z_k) = Q_{k-1}(z_1, \ldots, z_{k-1}) + z_k P_{k-1}(z_1, \ldots, z_{k-1})$$

when k is even, and $P_k = z_k P_{k-1} + Q_{k-1}$, $Q_k = Q_{k-1}$ when k is odd. We concentrate on the case when k is even; the odd case is handled similarly. Let $(a, b) \in \mathbb{C}^2 \setminus \{0\}$ and study the fibre $P_k = a$, $Q_k = b$. When $a \ne 0$, we have $z_k = \frac{1}{a}(b - Q_{k-1})$, so the fibre is a graph in $\mathbb{C}^{k-1} \times \mathbb{C}_{z_k}$ over the fibre $\{P_{k-1} = a\} \subset \mathbb{C}^{k-1}$. When $a = 0$, the fibre is $\Phi_{k-1}^{-1}(0, b) \times \mathbb{C}_{z_k}$ and, since in this case $b \ne 0$, $\Phi_{k-1}^{-1}(0, b)$ is a graph in $\mathbb{C}^{k-2} \times \mathbb{C}_{z_{k-1}}$ over $\{Q_{k-2} = b\} \subset \mathbb{C}^{k-2}$. So in both cases the fibre is a graph over a hypersurface given by a single polynomial equation.

We now find globally integrable tangential holomorphic vector fields spanning the tangent space at each point of the fibre; a dominating spray is then obtained by composing their flows. We stratify $\mathbb{C}^2 \setminus \{0\}$ by

$$\mathbb{C}^2 \setminus \{0\} \supset \big\{(a, b) \in \mathbb{C}^2 \setminus \{0\} : a = 0\big\} \supset \varnothing.$$

We construct a spray over the highest dimensional stratum $\{(a, b) \in \mathbb{C}^2 \setminus \{0\}: a \ne 0\}$; the stratum $\{(a, b) \in \mathbb{C}^2 \setminus \{0\}: a = 0\}$ is handled similarly. We need to find complete holomorphic vector fields that are tangential to $\{P_{k-1} = a \ne 0\}$ and span the tangent space at each point. We claim that the vector fields

$$V_{ij} = \frac{\partial P_{k-1}}{\partial z_i}\frac{\partial}{\partial z_j} - \frac{\partial P_{k-1}}{\partial z_j}\frac{\partial}{\partial z_i}, \qquad 1 \le i < j \le k$$

have the required properties. They obviously are tangential, and one easily checks the spanning property. (It helps to observe that the nonsmooth points of $\{P_{k-1} = a\}$ are contained in S_k.) To see that they are globally integrable, note that P_{k-1} is a polynomial that is no more than linear in each variable separably, so $\partial P_{k-1}/\partial z_i$ is independent of z_i and $\partial P_{k-1}/\partial z_j$ is independent of z_j. Hence, the flow of V_{ij} is the solution of a system of two independent differential equations that are both globally integrable. $\square$

We can now conclude the proof. By [561, Theorem 4] we have a continuous map $F': X \to \mathbb{C}^{k'}$ for some $k' \in \mathbb{N}$ such that $f(x) = \Psi_{k'}(F'(x))$. By Lemma 8.7.3 we see that $F = (F_1, \ldots, F_{k'}, 1, 0, -1)$ gives a map from X into $\mathbb{C}^{k'+3} \setminus S_{k'+3}$. Putting $k = k' + 3$ we have $f(x) = \Psi_k(F(x))$, so $\Psi_k(F(x)) f(x)^{-1} = I_2$. Since (8.11) is a stratified elliptic submersion, Corollary 7.4.5 (p. 334) gives a holomorphic map $G: X \to \mathbb{C}^k \setminus S_k$ such that

$$\Phi_k\big(F(x)\big) = \pi_2\big(f(x)\big) = \Phi_k\big(G(x)\big);$$

that is, the matrices $\Psi_k(F(x))$ and $\Psi_k(G(x))$ have the same last row. Hence

$$\Psi_k\big(G(x)\big) f(x)^{-1} = \begin{pmatrix} a(x) & b(x) \\ 0 & 1 \end{pmatrix}$$

with a, b holomorphic. Moreover, $a = 1$ since the matrix is in $SL_2(\mathbb{C})$. Thus,

$$f(x) = \begin{pmatrix} 1 & -b(x) \\ 0 & 1 \end{pmatrix} \Psi_k\big(G(x)\big)$$

which solves the problem. $\square$

The proof for $m > 2$ is by induction on the size of the matrices. The difficult part is to show Lemma 8.7.4. One can reduce stratawise the m polynomial equations which define the fibres of Φ_k to a single equation. The special form of the occurring polynomials allows one to find finitely many complete holomorphic vector fields which span the tangent space of the fibre at each point. The details are considerable and can be found in [306].

Since a stratified elliptic submersion also enjoys the parametric Oka property (POP) according to Corollary 7.4.5 (p. 334), the same proof also gives a parametric version of Theorem 8.7.1.

8.8 Transversality Theorems for Holomorphic Maps

If X and Y are smooth manifolds and $k \in \mathbb{Z}_+$, then $\mathrm{J}^k(X, Y)$ denotes the manifold of k-jets of smooth maps $X \to Y$ (see Sect. 1.10). If Z is a smooth closed submanifold of $\mathrm{J}^k(X, Y)$ then, for a generically chosen smooth map $f: X \to Y$, the k-jet extension map $\mathrm{j}^k f: X \to \mathrm{J}^k(X, Y)$ is transverse to Z (see Thom [546, 547] and also [3, 250, 422, 555, 573]). This is referred to as the *jet transversality theorem*; the *basic transversality theorem* is the same statement for $k = 0$.

The analogous result holds only rarely for holomorphic maps between complex manifolds. For example, if holomorphic maps $\mathbb{C} \to Y$ satisfies the transversality theorem for 1-jets, then a generic holomorphic maps $\mathbb{C} \to Y$ has isolated critical points, and hence Y is not Brody hyperbolic. Similarly, the transversality theorem for 1-jets of holomorphic maps $\mathbb{C}^n \to Y$ with $n = \dim Y$ implies that Y is *dominable* by $\mathbb{C}^n$. Furthermore, if such Y is compact and connected, then Y is not of Kodaira general type according to Kobayashi and Ochiai [342].

In the positive direction, the basic transversality theorem holds for holomorphic maps to any complex manifold with a submersive family of holomorphic self-maps [3]; a classical example is *Bertini's theorem* that almost all projective hyperplanes in $\mathbb{P}^n$ intersect a given complex submanifold $Z \subset \mathbb{P}^n$ transversely [250, p. 150]. The jet transversality theorem holds for holomorphic maps of Stein manifolds to Euclidean spaces according to Forster [171].

We shall use the following notion of a *Whitney stratification*.

Theorem 8.8.1 ([573, Theorem 8.5]) *Every complex analytic subvariety A in a complex space X admits a stratification $\{A_\alpha\}$ satisfying the following Whitney's condition* (a):

(a) *Suppose that $A_\alpha \subset \overline{A_\beta}$, $x_i \in A_\beta$ $(i = 1, 2, \ldots)$ is a sequence converging to a point $x \in A_\alpha$, and in some local coordinates in a neighborhood of $x \in X$ the tangent spaces $\mathrm{T}_{x_i} A_\beta$ converge to a subspace $\tau \subset \mathrm{T}_x X$. Then $\mathrm{T}_x A_\alpha \subset \tau$.*

We recall how Whitney's condition (a) is used in transversality arguments. Given stratified subvarieties $A \subset X$ and $B \subset Y$, we let $\mathrm{NT}_{A,B} \subset \mathrm{J}^1(X,Y)$ denote the set of all 1-jets $(x, y, \lambda) \in \mathrm{J}^1(X,Y)$ such that, if x belongs to a stratum A_α of A and y belongs to a stratum B_β of B, then

$$\lambda(\mathrm{T}_x A_\alpha) + \mathrm{T}_y B_\beta \neq \mathrm{T}_y Y.$$

If $x \notin A$ or $y \notin B$ then $(x, y, \lambda) \notin \mathrm{NT}_{A,B}$.

The following elementary lemma is proved in [250, p. 38].

Lemma 8.8.2 *If $A \subset X$ and $B \subset Y$ are Whitney stratified complex subvarieties then $\mathrm{NT}_{A,B}$ is closed in $\mathrm{J}^1(X,Y)$. The set $\mathrm{NT}_{A,B}$ is also closed in $\mathrm{J}^1(X,Y)$ if B is a closed smooth submanifold of Y.*

Let $A \subset X$ and $B \subset Y$ be stratified complex subvarieties. Given a holomorphic map $f \in \mathcal{O}(X,Y)$, we say that $f|_A$ *is transverse to* B if the range of the 1-jet extension map $\mathrm{j}^1 f \colon X \to \mathrm{J}^1(X,Y)$ does not intersect $\mathrm{NT}_{A,B}$:

$$\big(x \in A_\alpha,\ f(x) \in B_\beta\big) \Longrightarrow \mathrm{d}f_x(\mathrm{T}_x A_\alpha) + \mathrm{T}_{f(x)}(B_\beta) = \mathrm{T}_{f(x)} Y.$$

The forthcoming discussion is based on the following condition introduced by Gromov [265, pp. 71–73]; see also [196].

Definition 8.8.3 Let X and Y be complex (resp. algebraic) manifolds. Holomorphic (resp. algebraic) maps $X \to Y$ satisfy *Condition* Ell_1 if for every holomorphic (resp. algebraic) map $f \colon X \to Y$ there is a holomorphic (resp. algebraic) map

$F: X \times \mathbb{C}^N \to Y$ for some $N \geq \dim Y$ such that $F(\cdot, 0) = f$ and $F(x, \cdot): \mathbb{C}^N \to Y$ has rank $\dim Y$ at $0 \in \mathbb{C}^N$ for every $x \in X$.

Such F is called a *dominating spray over f*.

Example 8.8.4 If $Y = \mathbb{C}^n$ then $F(x, t) = f(x) + t$ ($x \in X$, $t \in \mathbb{C}^n$) is a dominating spray over a map $f: X \to \mathbb{C}^n$. More generally, assume that a complex Lie group G acts holomorphically and transitively on a complex manifold Y. Let $\exp: \mathfrak{g} \to G$ denote the exponential map. Then, $F(x, v) = \exp v \cdot f(x)$ ($x \in X$, $v \in \mathfrak{g}$) is a dominating spray over the map $f: X \to Y$. □

The following two theorems are the main results of this section.

Theorem 8.8.5 *Let X be a Stein manifold and Y be a complex manifold. If holomorphic maps $X \to Y$ satisfy Condition* Ell_1, *then for every pair of closed, Whitney stratified complex analytic subvarieties $A \subset X$, $B \subset \mathrm{J}^k(X, Y)$ the set*

$$\big\{ f \in \mathcal{O}(X, Y) : \mathrm{j}^k f|_A \textit{ is transverse to } B \big\}$$

is residual in $\mathcal{O}(X, Y)$. The same holds if B is a smooth closed submanifold of $\mathrm{J}^k(X, Y)$. For $k = 0$ the conclusion holds even if X is not Stein.

For algebraic maps, we have the following transversality theorem in which $\mathcal{O}(X, Y)$ denotes the space of regular algebraic maps.

Theorem 8.8.6 *Let X be an affine algebraic manifold and Y be an algebraic manifold. If algebraic maps $X \to Y$ satisfy Condition* Ell_1, *then for every compact set $K \subset X$ and every pair of Whitney stratified complex subvarieties $A \subset X$, $B \subset \mathrm{J}^k(X, Y)$ the set*

$$\big\{ f \in \mathcal{O}(X, Y) : \mathrm{j}^k f|_A \textit{ is transverse to } B \textit{ on } A \cap K \big\}$$

is open and dense in $\mathcal{O}(X, Y)$. The same holds if B is a smooth closed submanifold of $\mathrm{J}^k(X, Y)$. For $k = 0$ this holds without assuming that X is affine.

Theorems 8.8.5 and 5.4.4 (p. 216) together imply the following result. The special case for $Y = \mathbb{C}^n$ is due to Forster [171].

Corollary 8.8.7 *Holomorphic maps from a Stein manifold X to an Oka manifold Y satisfy Condition* Ell_1, *and hence the jet transversality theorem.*

Proof Let $f: X \to Y$ be a holomorphic map. By Theorem 3.1.1 (p. 66), the graph $G = \{(x, f(x)) : x \in X\} \subset X \times Y$ has an open Stein neighborhood $\Omega \subset X \times Y$. Let $E \to \Omega$ denote the restriction to Ω of the vertical tangent bundle of the projection $\pi_X: X \times Y \to X$, $\pi_X(x, y) = x$. By Corollary 8.3.2, E has finitely many generators, i.e., holomorphic vector fields $V_1, \ldots, V_N$ tangent to the fibres of π_X. Let ϕ_t^i denote the local flow of V_i. Denote by $\pi_Y: X \times Y \to Y$ the projection $\pi_Y(x, y) = y$. Then, the map

$$F_0(x, t_1, \dots, t_N) = \pi_Y \circ \phi^1_{t_1} \circ \dots \circ \phi^N_{t_N}\big(x, f(x)\big) \in Y$$

is defined and holomorphic for (x, t) in an open neighborhood $U \subset X \times \mathbb{C}^N$ of $X \times \{0\}^N$. Clearly, $F_0(x, 0) = f(x)$ and $\partial_t|_{t=0} F_0(x, t): \mathbb{C}^N \to T_{f(x)}Y$ is surjective for every $x \in X$. Since Y is an Oka manifold, Theorem 5.4.4 (p. 216) gives a holomorphic map $F: X \times \mathbb{C}^N \to Y$ which agrees with F_0 to the second order along $X \times \{0\}^N$. Clearly, F is then a dominating spray over f. This shows that maps $X \to Y$ satisfy Condition Ell_1. □

Proof of Theorems 8.8.5 and 8.8.6 We follow Abraham's reduction [3] to Sard's theorem. The proofs are parallel up to the point where the Baire property of the space $\mathcal{O}(X, Y)$ is invoked; in the algebraic case this leaves us with the weaker statement.

Lemma 8.8.8 *Let X and Y be complex manifolds and let $A \subset X$ and $B \subset \mathrm{J}^k(X, Y)$ be closed Whitney stratified complex analytic subvarieties. For every compact subset K of X, the set*

$$\mathcal{T}_{A,B,K} = \big\{ f \in \mathcal{O}(X, Y) : \mathrm{j}^k f|_A \text{ is transverse to } B \text{ on } A \cap K \big\}$$

is open in $\mathcal{O}(X, Y)$. The same holds if A and B are smooth closed submanifolds of X and $J^k(X, Y)$, respectively.

Proof Consider the basic case with $B \subset Y$. Given a map $f: X \to Y$ and a compact set $K \subset X$, the restriction $f|_A: A \to B$ is transverse to B at each point of $A \cap K$ if and only if $(\mathrm{j}^1 f)(K) \cap \mathrm{NT}_{A,B} = \varnothing$. Assuming this to be the case, and taking into account that $\mathrm{NT}_{A,B}$ is closed in $\mathrm{J}^k(X, Y)$ by Whitney's condition, there is a compact set $L \subset X$, with $K \subset \mathring{L}$, such that $(\mathrm{j}^1 f)(L) \cap \mathrm{NT}_{A,B} = \varnothing$. If $g \in \mathcal{O}(X, Y)$ is sufficiently uniformly close to f on L, then $\mathrm{j}^1 g$ is close to $\mathrm{j}^1 f$ on K, and hence $(\mathrm{j}^1 g)(K) \cap \mathrm{NT}_{A,B} = \varnothing$. In the general case, one applies the same argument with f replaced by its k-jet extension $\mathrm{j}^k f: X \to \mathrm{J}^k(X, Y)$. □

To prove Theorem 8.8.5, it suffices to show that for every compact K in X the set $\mathcal{T}_{A,B,K} \subset \mathcal{O}(X, Y)$, which is open by Lemma 8.8.8, is everywhere dense in $\mathcal{O}(X, Y)$. Since $\mathcal{O}(X, Y)$ is a Baire space, the conclusion then follows by taking the intersection of such sets over a countable family of compacts exhausting X. In the algebraic case, the last step fails which leaves us with the weaker statement in Theorem 8.8.6.

Consider first the basic case with $A = X$ and $B \subset Y$. Let $f: X \to Y$ be a holomorphic (resp. an algebraic) map. Choose a map $F: X \times \mathbb{C}^N \to Y$ as in Definition 8.8.3. Let $\pi: X \times \mathbb{C}^N \to \mathbb{C}^N$ denote the projection $\pi(x, t) = t$. Fix a compact set K in X. Since $\partial_t F(x, 0): \mathbb{C}^N \to \mathrm{T}_{f(x)}Y$ is surjective for every $x \in X$, there are a ball $D \subset \mathbb{C}^N$ around the origin and an open set $U \subset X$ containing K such that $\partial_t F(x, t): \mathbb{C}^N \to \mathrm{T}_{F(x,t)}Y$ is surjective for every $(x, t) \in V = U \times D$. Hence,

$B' = F^{-1}(B) \cap V$ is a closed, Whitney stratified, complex analytic subvariety of V. (The strata B_β of B pull back by $F|_V$ to the strata B'_β of B').

Set $f_t = F(\cdot, t)\colon X \to Y$ for $t \in \mathbb{C}^N$. If $(x,t) \in B'_\beta$ then $y = f_t(x) \in B_\beta$, and by inspecting the definitions we see that the following are equivalent (compare with [250, p. 40]):

(a) $(\mathrm{d}f_t)_x(\mathrm{T}_x X) + \mathrm{T}_y B_\beta = \mathrm{T}_y Y$;
(b) (x,t) is a regular point of the restricted projection $\pi\colon B'_\beta \to D$.

By Sard's theorem, applied inductively to the components of a projection π in (b), we see that the set of regular values of all projections in (b) is residual in D. Choosing t in this set and close to 0 we get a map $f_t\colon X \to Y$ which is transverse to B on U and approximates $f = f_0$ uniformly on K. The same argument applies if B is a smooth closed submanifold of Y.

If A is a Whitney stratified complex subvariety (or a closed smooth submanifold) of X, one applies the above argument with U replaced by $U \cap A_\alpha$ for a fixed stratum A_α of A (f and F are still defined globally on X). This gives a residual set of points $t \in D$ for which $f_t|_{A_\alpha \cap U}$ is transverse to a stratum B_β of B. Since A and B have at most countably many strata, we find $t \in \mathbb{C}^N$ arbitrarily close to 0 such that $f_t|_{A\cap U}$ is transverse to B.

This proves the basic transversality theorem for holomorphic or algebraic maps $X \to Y$. In fact, the above proof gives the following result of independent interest which holds for any pair of complex manifolds.

Proposition 8.8.9 *Assume that X and Y are complex manifolds, $D \subset \mathbb{C}^N$ is a nonempty open set, and $F\colon X \times D \to Y$ is a holomorphic map such that*

$$\partial_t F(x,t) : \mathbb{C}^N \to T_{F(x,t)} Y \quad \text{is surjective for every } (x,t) \in X \times D.$$

Let $A \subset X$ and $B \subset Y$ be closed Whitney stratified complex analytic subvarieties. Then, there is a residual set of parameter values $t \in D$ for which the map $f_t = F(\cdot, t)\colon X \to Y$ is transverse to B on A.

Consider now the jet transversality theorem. Fix a map $f\colon X \to Y$ and a compact set $K \subset X$. The goal is to prove that f can be approximated uniformly on K by holomorphic (resp. algebraic) maps $X \to Y$ whose k-jet extension $\mathrm{j}^k f$ is such that $\mathrm{j}^k f|_A\colon A \to \mathrm{J}^k(X,Y)$ is transverse to the subvariety $B \subset \mathrm{J}^k(X,Y)$ at each point of $A \cap K$.

Recall that X is assumed to be Stein (in Theorem 8.8.5) or affine algebraic (in Theorem 8.8.6). Thus, we may assume that X is a closed holomorphic (resp. algebraic) submanifold of a Euclidean space $\mathbb{C}^n$. Let $\mathcal{W}$ denote the complex vector space of all holomorphic polynomial maps $P\colon \mathbb{C}^n \to \mathbb{C}^N$ of degree at most k. Let $F\colon X \times \mathbb{C}^N \to Y$ be a dominating spray over $F(\cdot, 0) = f$. Consider the holomorphic (resp. algebraic) map $G\colon X \times \mathcal{W} \to Y$ defined by

$$G(x,P) = F\big(x, P(x)\big), \quad x \in X,\ P \in \mathcal{W}.$$

Given $P \in \mathcal{W}$ set $G_P = G(\cdot, P)\colon X \to Y$; then $G_0(x) = F(x,0) = f(x)$.

Lemma 8.8.10 *The map $\Phi: X \times \mathcal{W} \to J^k(X,Y)$, defined by $\Phi(x,P) = \mathrm{j}^k_x G_P$, is a submersion in an open neighborhood of $X \times \{0\}$ in $X \times \mathcal{W}$.*

Proof The argument is local, and hence we may assume that $X = \mathbb{C}^n$. Write $P = (P_1, \dots, P_N) \in \mathcal{W}$, and let $t = (t_1, \dots, t_N)$ be the coordinates on $\mathbb{C}^N$. For every multi-index $I = (i_1, \dots, i_n)$ we have

$$\partial^I_x (G_P) = \sum_{j=1}^N \frac{\partial}{\partial t_j} F\big(x, P(x)\big)\, \partial^I_x P_j(x) + H_I(x),$$

where $H_I(x)$ contains only terms $\partial^J_x P$ with $|J| < |I|$, multiplied by various partial derivatives of F. Hence, the k-jet extension map $\mathrm{j}^k_x G_P$ is triangular with respect to the components of $\mathrm{j}^k_x P$, and the diagonal terms are nondegenerate at $P = 0$ since $G_0(x) = F(x,0)$ and $\partial_t F(x,0)$ is nondegenerate. □

Sard's theorem, applied to the map Φ in Lemma 8.8.10, shows that for most $P \in \mathcal{W}$ the map $\mathrm{j}^k G_P|_A$ is transverse to the subvariety $B \subset \mathrm{J}^k(X,Y)$ at every point of $A \cap K$. This proves Theorems 8.8.5 and 8.8.6. □

Combining Theorems 8.8.5 and 8.8.6 with the following proposition gives the transversality theorem in several geometrically interesting cases; compare with [265, p. 72]. Note that (b) is a special case of Corollary 8.8.7 (since every subelliptic manifold is an Oka manifold by Corollary 5.6.14); however, we give a more direct proof of Ell_1 for subelliptic manifolds.

Proposition 8.8.11 *Let X and Y be complex manifolds.*

(a) *If Y admits a dominating spray $s: Y \times \mathbb{C}^N \to Y$ defined on a trivial bundle over Y, then holomorphic maps $X \to Y$ satisfy Condition* Ell_1. *This holds in particular if Y is a complex homogeneous space.*
(b) *If Y is subelliptic (see Definition* 5.6.13 *on p.* 229*), then holomorphic maps from any Stein manifold X to Y satisfy* Ell_1.
(c) *If X and Y are algebraic manifolds and Y admits a dominating algebraic spray $s: Y \times \mathbb{C}^N \to Y$, then algebraic maps $X \to Y$ satisfy* Ell_1.
(d) *If X is an affine algebraic manifold and Y is algebraically subelliptic, then algebraic maps $X \to Y$ satisfy* Ell_1.

Proof Fix a holomorphic map $f: X \to Y$.

If (E, p, s) is a dominating spray on Y, then $f^*E \to X$ is a holomorphic vector bundle, and there is a fibrewise bijective holomorphic map $\iota: f^*E \to E$ covering f. The holomorphic map $F = s \circ \iota: f^*E \to Y$ satisfies Ell_1 for f, except that f^*E need not be a trivial bundle over X. In case (a), the bundle $E \to Y$ is assumed to be trivial and hence f^*E is also trivial, so (a) follows. An analogous argument proves (c).

If X is a Stein manifold, then by Corollary 8.3.2 there is a surjective holomorphic vector bundle map $\tau: X \times \mathbb{C}^N \to f^*E$ for some $N \in \mathbb{N}$. The map

$$F = s \circ \iota \circ \tau : X \times \mathbb{C}^N \to Y$$

satisfies Ell_1 with respect to $f = F(\cdot, 0)$. The analogous argument holds in the algebraic case by using Serre's Theorem A. This proves parts (b) and (d) for (algebraically) elliptic manifolds Y.

Assume now that X is Stein and Y is subelliptic. Let (E_j, p_j, s_j) for $j = 1, \ldots, k$ be a finite dominating family of holomorphic sprays on Y (see Definition 5.6.13, p. 229). Let $E_1' = f^*E_1 \to X$ be the pullback of $\pi_1: E_1 \to Y$ by the map $f: X \to Y$, and define $\sigma_1: E_1' \to Y$ by $\sigma_1(x, e) = s_1(f(x), e)$. As before, there is a surjective complex vector bundle map $X \times \mathbb{C}^{n_1} \to E_1'$ for some $n_1 \in \mathbb{N}$. By composing it with σ_1 we obtain a map $f_1: X_1 = X \times \mathbb{C}^{n_1} \to Y$ satisfying $f_1(x, 0) = f(x) = y \in Y$ and

$$\partial_t f_1(x, t)|_{t=0}\big(\mathrm{T}_0\mathbb{C}^{n_1}\big) = (\mathrm{d}s_1)_y(E_{1,y}) \subset \mathrm{T}_y Y.$$

Repeating the construction with $f_1: X_1 \to Y$ and the spray $s_2: E_2 \to Y$, we find $n_2 \in \mathbb{N}$ and a holomorphic map $f_2: X_2 = X_1 \times \mathbb{C}^{n_2} = X \times \mathbb{C}^{n_1} \times \mathbb{C}^{n_2} \to Y$ satisfying $f_2(x, t, 0) = f_1(x, t)$ (hence $f_2(x, 0, 0) = f(x) = y$) and

$$\partial_u f_2(x, 0, u)|_{u=0}\big(\mathrm{T}_0\mathbb{C}^{n_2}\big) = (\mathrm{d}s_2)_y(E_{2,y}) \subset \mathrm{T}_y Y.$$

After k steps we obtain a map $F: X \times \mathbb{C}^N \to Y$ $(N = n_1 + \cdots + n_k)$ satisfying the following for every $x \in X$ and $y = f(x) \in Y$:

$$F(x, 0) = f(x), \quad \partial_t F(x, t)|_{t=0}\big(\mathrm{T}_0\mathbb{C}^N\big) = \sum_{j=1}^{k} (\mathrm{d}s_j)_y(E_{j,y}) = \mathrm{T}_y Y.$$

This proves (b). The same proof applies in case (d) by appealing to Theorem A of Serre when passing at each step to a trivial bundle. □

Corollary 8.8.12 *Holomorphic maps from any complex manifold X to a complex homogeneous manifold Y satisfy the basic transversality theorem. The same holds if $Y = \mathbb{C}^n \setminus A$, where A is an algebraic subvariety of codimension ≥ 2; for such A, the basic transversality theorem also holds for algebraic maps from any algebraic manifold to $\mathbb{C}^n \setminus A$.*

In the case $Y = \mathbb{P}^n$ we recover Bertini's theorem [250, p. 150].

Kaliman and Zaidenberg proved the following transversality theorem for maps from Stein manifolds to arbitrary complex manifolds, but the domain of the map is allowed to shrink around a given compact set.

Theorem 8.8.13 ([328]) *Assume that X is a Stein manifold, Y is a complex manifold, and $A \subset X$, $B \subset \mathrm{J}^k(X, Y)$ are closed, Whitney stratified, complex analytic subvarieties. For every holomorphic map $f \in \mathcal{O}(X, Y)$ and compact set $K \subset X$ there are a neighborhood $U \subset X$ of K and a holomorphic map $g: U \to Y$ such that $\mathrm{j}^k g|_{A\cap U}$ is transverse to B, and g approximates f as closely as desired uniformly on K.*

Proof As in the proof of Corollary 8.8.7, we find $N \in \mathbb{N}$, an open Stein neighborhood $U \subset X$ of K, and a holomorphic map $F: U \times \mathbb{B}^N \to Y$ such that $F(\cdot, 0) = f$ and F is submersive in the second variable $t \in \mathbb{B}^N \subset \mathbb{C}^N$. For $k = 0$ the result then follows from Proposition 8.8.9. Furthermore, since the set U is relatively compact, the proof of Theorem 8.8.5 (precomposing F by polynomials) applies also for the k-jet extension for any $k \in \mathbb{N}$. □

Remark 8.8.14 Theorem 8.8.13 gives the following alternative proof of the jet transversality theorem for holomorphic maps from a Stein manifold X to an Oka manifold Y.

Let $f: X \to Y$ be a holomorphic map. Choose compact $\mathcal{O}(X)$-convex subsets $K, L \subset X$ with $K \subset \mathring{L}$. By Theorem 8.8.13 we can approximate f uniformly on L by a holomorphic map $g: U \to Y$ on an open set $U \supset L$ such that $\mathrm{j}^k g|_{U \cap A}$ is transverse to a given Whitney stratified subvariety B of $\mathrm{J}^k(X, Y)$. If the approximation is sufficiently close, there is a smooth map $\widetilde{g}: X \to Y$ which agrees with g in a neighborhood of L and with f on $X \backslash U$. Since Y is an Oka manifold, the map $\widetilde{g}$ can be approximated uniformly on L by holomorphic maps $\widetilde{f}: X \to Y$. If the approximation is sufficiently close, then $\widetilde{f}$ still satisfies the given transversality condition on K by Lemma 8.8.8. This shows the density of transverse maps on compacts in X, while openness is given by Lemma 8.8.8. □

We also have an interpolation version of Theorem 8.8.5. Given a closed complex subvariety X_0 of X, $f_0 \in \mathcal{O}(X, Y)$ and $r \in \{0, 1, \ldots\}$, the set

$$\mathcal{O}(X, Y; X_0, f_0, r) = \left\{ f \in \mathcal{O}(X, Y) : \mathrm{j}^r f|_{X_0} = \mathrm{j}^r f_0|_{X_0} \right\}$$

is a closed metric subspace of $\mathcal{O}(X, Y)$, hence a Baire space.

Theorem 8.8.15 *Assume that X is a Stein manifold and Y is an Oka manifold. Let $A \subset X$ and $B \subset \mathrm{J}^k(X, Y)$ be closed Whitney stratified subvarieties. If $f_0 \in \mathcal{O}(X, Y)$ and $\mathrm{j}^k f_0|_A$ is transverse to B at all points of $A \cap X_0$, then for every integer $r \geq k$ there is a residual set of $f \in \mathcal{O}(X, Y; X_0, f_0, r)$ for which $\mathrm{j}^k f|_A$ is transverse to B.*

Proof Since $r \geq k$, the set of all $f \in \mathcal{O}(X, Y; X_0, f_0, r)$ for which $\mathrm{j}^k f|_A$ is transverse to B at all points of $A \cap X_0$ is open in $\mathcal{O}(X, Y; X_0, f_0, r)$. It thus suffices to show that we can approximate the initial map f_0, uniformly on any compact $\mathcal{O}(X)$-convex subset $K \subset X$, by $f \in \mathcal{O}(X, Y; X_0, f_0, r)$ such that $\mathrm{j}^k f|_A$ is transverse to B at every point of $A \cap K$.

Since Y is an Oka manifold, holomorphic maps $X \to Y$ satisfy Condition Ell_1 by Corollary 8.8.7. Let $F: X \times \mathbb{C}^N \to Y$ be a dominating spray over f_0. Consider the basic case $k = 0$, $B \subset Y$. There exist functions $g_1, \ldots, g_l \in \mathcal{O}(X)$ which vanish to order $r + 1$ on the subvariety $X_0 = \{x \in X : g_j(x) = 0,\ j = 1, \ldots, l\}$. For every $x \in X$ let $\sigma_x: (\mathbb{C}^N)^l \to \mathbb{C}^N$ be defined by

$$\sigma_x(t_1, \ldots, t_l) = \sum_{j=1}^{l} t_j g_j(x), \quad t_j \in \mathbb{C}^N,\ j = 1, 2, \ldots, l.$$

Clearly, σ_x is surjective if $x \in X \setminus X_0$ and is the zero map if $x \in X_0$. The map $\widetilde{F}: X \times \mathbb{C}^{Nl} \to Y$, defined by

$$\widetilde{F}(x,t) = \widetilde{F}(x, t_1, \ldots, t_l) = F\big(x, \sigma_x(t_1, \ldots, t_l)\big),$$

is a submersion with respect to t (at $t = 0$) if $x \in X \backslash X_0$, and is degenerate (constant) if $x \in X_0$. Hence, the proof of Theorem 8.8.5 applies over $X \backslash X_0$.

Let $f_t = \widetilde{F}(\cdot, t): X \to Y$ for $t \in \mathbb{C}^{Nl}$. By the construction, $\mathrm{j}^r f_t|_{X_0} = \mathrm{j}^r f_0|_{X_0}$ for every t. Choose a compact set $K \subset X$. By the assumption, $f_0|_A$ is transverse to B on $A \cap X_0$. Hence, there is an open neighborhood $U \subset X$ of $A \cap X_0 \cap K$ such that $f_t|_{A \cap U}$ is transverse to B for every t close to 0 (see Lemma 8.8.8). The set $K' = K \backslash U \subset X \backslash X_0$ is compact, and hence for most values of t the map $f_t|_A$ is transverse to B on $A \cap K'$. Thus, $f_t|_A$ is transverse to B on $K \cap A$ for most t close to 0 which concludes the proof for $k = 0$. Similarly, one obtains the proof for $k > 0$ by following the arguments in the proof of Theorem 8.8.5. We also get a semiglobal version of Theorem 8.8.15, analogous to Theorem 8.8.13, without any restriction on the manifold Y. □

In the algebraic category, the global transversality theorem holds under the following stronger assumption on Y.

Proposition 8.8.16 ([196, Proposition 4.10]) *If Y is an algebraic manifold with a submersive algebraic spray $s: E \to Y$ (i.e., such that $s: E_y \to Y$ is a submersion for every $y \in Y$), then algebraic maps $X \to Y$ from any affine algebraic manifold X to Y satisfy the jet transversality theorem with respect to closed complex analytic subvarieties.*

Proof Let $f_0: X \to Y$ be an algebraic map. By pulling back the submersive algebraic spray $s: E \to Y$ by the map f_0 we obtain an algebraic submersion $F: X \times \mathbb{C}^N \to Y$ satisfying $f_0 = F(\cdot, 0)$ (compare with the proof of Proposition 8.8.11). Given closed complex subvarieties $A \subset X$ and $B \subset Y$, Sard's theorem shows that for a generic choice of $t \in \mathbb{C}^N$ the algebraic map $f_t|_A = F(\cdot, t)|_A$ is transverse to B. (See the proof of Theorem 8.8.5.) Similarly, we obtain the jet transversality theorem by considering maps $x \to F(x, P(x))$ for polynomials $P: \mathbb{C}^n \to \mathbb{C}^N$, where X is an affine subvariety in $\mathbb{C}^n$. □

8.9 Singularities of Holomorphic Maps

Let X and Y be complex manifolds of dimensions n, m, respectively. Consider the following *singularity sets* associated to a holomorphic map $f: X \to Y$:

$$\Delta_f^2 = \big\{(x_1, x_2) \in X \times X : x_1 \neq x_2,\ f(x_1) = f(x_2)\big\}, \tag{8.12}$$

$$\Sigma_f^i = \{x \in X : \dim \ker \mathrm{d} f_x = i\}, \tag{8.13}$$

$$\Lambda_f^i = \bigcup_{j \geq i} \Sigma_f^j. \tag{8.14}$$

Denoting by $\Delta_X = \{(x,x) : x \in X\}$ the diagonal of $X \times X$, we have

$$\Delta_f^2 = (f \times f)^{-1}(\Delta_Y) \setminus \Delta_Y,$$

and $\Delta_f^2 = \varnothing$ if and only if f is injective.

For every $i \in \mathbb{N}$ the set Λ_f^i is a complex subvariety consisting of all points $x \in X$ at which the kernel of the differential $\mathrm{d}f_x$ has dimension at least i.

If $n \le m$, we have $\Lambda_f^1 = \varnothing$ if and only if f is an immersion. Thus, f is an injective immersion if and only if $\Lambda_f^1 = \varnothing$ and $\Delta_f^2 = \varnothing$.

If $n > m$, then $\Sigma_f^i = \varnothing$ for $i = 1, \dots, n-m-1$, and f is a submersion if and only if $X = \Sigma_f^{n-m}$ (equivalently, $\Lambda_f^i = \varnothing$ for $i > n-m$).

Theorem 8.9.1 *Assume that X is an n-dimensional Stein manifold and Y is an m-dimensional Oka manifold. Then there is a residual set in $\mathcal{O}(X,Y)$ consisting of maps $f: X \to Y$ satisfying*

$$\dim \Delta_f^2 \le 2n - m \quad \textit{and} \quad \operatorname{codim} \Sigma_f^i = i(m-n+i), \quad i = 1, 2, \dots \tag{8.15}$$

This holds in particular for holomorphic maps $X \to \mathbb{C}^m$, and also for a generic polynomial map $\mathbb{C}^n \to \mathbb{C}^m$.

Proof By Corollary 8.8.7, holomorphic maps from a Stein manifold X to an Oka manifold Y satisfy the jet transversality theorem. In particular, the following conditions hold for a generic holomorphic map $f: X \to Y$:

- The map $f \times f: X \times X \to Y \times Y$, $(x_1, x_2) \to (f(x_1), f(x_2))$ is transverse to the diagonal Δ_Y of $Y \times Y$.
- For every $i = 1, 2, \dots$ the 1-jet extension $\mathrm{j}^1 f: X \to \mathrm{J}^1(X,Y)$ is transverse to the subvariety $\Lambda^i \subset \mathrm{J}^1(X,Y)$ consisting of all 1-jets $(x, y, \lambda) \in \mathrm{J}^1(X,Y)$ such that $\dim\ker\lambda \ge i$.

Since transverse maps preserve the codimension, we have $\operatorname{codim} \Delta_f^2 = \operatorname{codim} \Delta_Y = m$ and hence $\dim \Delta_f^2 = 2n - m$ at every point of Δ_f^2. This gives the first estimate in (8.15). To prove the second equality in (8.15), we need the following lemma.

Lemma 8.9.2 ([244, Proposition 5.3, p. 60]) *Let V and W be vector spaces of dimension n and m, respectively, over a field $F \in \{\mathbb{R}, \mathbb{C}\}$. The set $L^i(V,W) = \{S \in \mathrm{Hom}_F(V,W) : \dim_F \ker S = i\}$ is a (real resp. complex) submanifold of $\mathrm{Hom}_F(V,W)$ satisfying* $\operatorname{codim} L^i(V,W) = i(m-n+i)$.

The proof amounts to finding the number of independent minors of an $m \times n$ matrix A whose vanishing implies that A has i-dimensional kernel.

Let $\Sigma^i \subset \mathrm{J}^1(X,Y)$ be the set of all complex 1-jets with i-dimensional kernel. Then $\Sigma_f^i = (\mathrm{j}^1 f)^{-1}(\Sigma^i)$. Since the fibre $\Sigma^i_{(x,y)}$ of Σ^i over any point $(x,y) \in X \times Y$ is isomorphic to $L^i(\mathrm{T}_x X, \mathrm{T}_y Y)$, the second equality in (8.15) follows from Lemma 8.9.2 and the jet transversality theorem.

The same arguments apply to algebraic (polynomial) maps $\mathbb{C}^n \to \mathbb{C}^m$. $\square$

Corollary 8.9.3 *Assume that X is an n-dimensional Stein manifold and Y is an m-dimensional Oka manifold.*

(a) *If $2n \le m$, then a generic holomorphic map $X \to Y$ is an immersion.*
(b) *If $2n + 1 \le m$, then a generic holomorphic map $X \to Y$ is an injective immersion.*
(c) *If $n > m$, then a generic holomorphic map $f : X \to Y$ satisfies the condition*

$$\operatorname{codim}\{x \in X : \operatorname{rank}(\mathrm{d}f_x) < m\} \ge n - m + 1 \ge 2.$$

Proof Parts (a) and (b) follow from Theorem 8.9.1 applied with $i = 1$, while (c) follows by taking $i = n - m + 1$. □

8.10 Local Sprays of Class $\mathcal{A}(D)$

In this section we develop some technical tools concerning local holomorphic sprays of sections, refining and extending the results in Sects. 5.8–5.9. Corollary 8.10.4 is of central importance for subsequent applications. We also prove an up to the boundary version of the Docquier-Grauert tubular neighborhood theorem; see Corollary 8.10.5.

Definition 8.10.1 Assume that D is relatively compact domain with $\mathcal{C}^\ell$ boundary ($\ell \ge 2$) in a Stein manifold X, Y is a complex manifold, and $r \in \{0, 1, \ldots, \ell\}$. An *$\mathcal{A}^r_Y(D)$-bundle* with fibre Y is a fibre bundle $h : Z \to \bar{D}$ with $h^{-1}(x) \cong Y$ $(x \in \bar{D})$ that is of class $\mathcal{C}^r(\bar{D})$ and holomorphic over D.

We denote by $\Gamma^r(\bar{D}, Z)$ the set of all $\mathcal{C}^r$-sections $f : \bar{D} \to Z$ and

$$\Gamma^r_{\mathcal{A}}(D, Z) = \{f \in \Gamma^r(\bar{D}, Z) : f \text{ is holomorphic on } D\}. \tag{8.16}$$

The following result gives the existence of a local spray of class $\mathcal{A}(D)$ with a given core section and a prescribed normal 1-jet. This improves Lemma 5.10.4 on p. 243 where shrinking of the domain is allowed.

Proposition 8.10.2 ([136, Proposition 4.1]) *Assume that D is a relatively compact strongly pseudoconvex domain with $\mathcal{C}^\ell$ boundary ($\ell \ge 2$) in a Stein manifold X, $r \in \{0, 1, \ldots, \ell\}$, and $h : Z \to \bar{D}$ is either an $\mathcal{A}^r_Y(D)$-bundle (Definition 8.10.1) or the restriction to $\bar{D}$ of a holomorphic submersion $\tilde{h} : \tilde{Z} \to X$. Given a section $f_0 \in \Gamma^r_{\mathcal{A}}(D, Z)$ and a surjective complex vector bundle map $L : \bar{D} \times \mathbb{C}^N \to \mathrm{VT}Z|_{f_0(\bar{D})}$ of class $\mathcal{A}^r(D)$ over f_0, there exist a domain $P \subset \mathbb{C}^N$ containing the origin and a dominating fibre spray $f : \bar{D} \times P \to Z$ of class $\mathcal{A}^r(D)$ (see Definition 5.9.1) satisfying*

$$f(x, 0) = f_0(x), \qquad \partial_t|_{t=0} f(x, t) = L_x, \quad x \in \bar{D}. \tag{8.17}$$

Furthermore, given a homotopy of sections $f_0^s \in \Gamma^r_{\mathcal{A}}(D, Z)$ $(s \in [0, 1])$, covered by a homotopy of surjective complex vector bundle maps

$$L^s : \bar{D} \times \mathbb{C}^N \to \mathrm{VT}Z|_{f_0^s(\bar{D})}$$

which are holomorphic over D such that for $s = 0, 1$ the map f_0^s is the core of a spray f^s over $\bar{D} \times P$ satisfying $\partial_t|_{t=0} f^s(x,t) = L_x^s$ ($x \in \bar{D}$), there exist a domain $P_1 \subset P$ containing the origin and a 1-parametric spray $F : [0,1] \times \bar{D} \times P_1 \to Z$ of class $\mathcal{A}^r(D)$ such that $F^s = F(s,\cdot,\cdot)$ agrees with f^s on $\bar{D} \times P_1$ for $s = 0, 1$, and such that

$$F^s(x,0) = f_0^s(x), \qquad \partial_t|_{t=0} F^s(x,t) = L_x^s, \quad x \in \bar{D},\ s \in [0,1]. \tag{8.18}$$

If f_0^s and L^s are smooth of order k with respect to the parameter $s \in [0,1]$, then F^s may also be chosen smooth of order k in $s \in [0,1]$.

Proof Each point $z_0 \in Z$ admits a *special* open neighborhood $Z_0 \subset Z$, isomorphic to a product $Z_0 \cong U \times V$ of a relatively open subset U of $\bar{D}$ and an open subset V of a Euclidean space $\mathbb{C}^M$ such that, in the coordinates $z = (x, y)$ ($x \in U$, $y \in \mathbb{C}^M$), h is the projection $z = (x, y) \to x$.

By Proposition 5.10.3 (p. 242) there exist finitely many strongly pseudoconvex domains $D_0 \subset D_1 \subset \cdots \subset D_m = D$ with $\mathcal{C}^\ell$ boundaries such that $\bar{D}_0 \subset D$, and for every $j = 0, 1, \ldots, m-1$ we have $D_{j+1} = D_j \cup B_j$ where B_j is a special convex bump on D_j (see Definition 5.10.2, p. 242). Each set B_j may be chosen small enough such that $f_0(\bar{B}_j)$ is contained in a special coordinate neighborhood of Z.

Lemma 5.10.4 (p. 243) furnishes a dominating fibre spray f with core f_0 over a neighborhood of $\bar{D}_0$. We claim that f can be chosen such that $\partial_t|_{t=0} f(x,t) = L_x$ for each $x \in \bar{D}_0$. Indeed, let $\{e_j\}_{j=1}^N$ be the standard basis of $\mathbb{C}^N$. We define a holomorphic section L_j of the vertical tangent bundle $\mathrm{VT}Z$ on the complex submanifold $f_0(D) \subset Z$ by setting

$$L_j\big(f_0(x)\big) = L_x e_j \in \mathrm{VT}_{f_0(x)} Z, \quad x \in D,\ j = 1, \ldots, N.$$

Since $f_0(D)$ is a closed complex submanifold of $Z|_D = h^{-1}(D)$, it admits an open Stein neighborhood $\Omega \subset Z|_D$ by Siu's theorem (Theorem 3.1.1 on p. 66). By Cartan's Theorem B, the section L_j extends from the submanifold $f(D)$ of Ω to a holomorphic vector field on the domain Ω that is tangential to the fibres of h. Denote by θ_t^j its flow for small values of $t \in \mathbb{C}$. The map

$$f(x, t_1, \ldots, t_N) = \theta_{t_N}^N \circ \cdots \circ \theta_{t_2}^2 \circ \theta_{t_1}^1 \big(f_0(x)\big)$$

is a dominating spray over a neighborhood of $\bar{D}_0$ satisfying $\partial_t|_{t=0} f(x,t) = L_x$.

To find a desired spray on $\bar{D}$, we perform a stepwise extension of f over the convex bumps $B_0, \ldots, B_{m-1}$. At the j-th step we assume that we have a spray $\bar{D}_j \times P_j \to Z$ with the required properties, and we approximate it by a spray $\bar{D}_{j+1} \times P_{j+1} \to Z$ with a smaller parameter set $0 \in P_{j+1} \subset P_j$. Since all steps are of the same kind, it suffices to explain the initial step $j = 0$. We need the following local approximation lemma for sprays.

Lemma 8.10.3 *Assume that $C \subset B$ is a pair of bounded strongly convex domains with $\mathcal{C}^2$ boundaries in $\mathbb{C}^n$, $P \subset \mathbb{C}^N$ is an open set containing the origin, and Q is an open polydisc in $\mathbb{C}^M$. Given a spray of maps $f : \bar{C} \times P \to Q \subset \mathbb{C}^M$ of class $\mathcal{A}^r(C)$ ($r \in \mathbb{Z}_+$) whose core $f_0 = f(\cdot, 0)$ extends to a map $\bar{B} \to Q$ of class $\mathcal{A}^r(B)$, there* *exist*

(i) *open sets* $P_0 \supset P_1 \supset P_2 \supset \cdots$ *in* $\mathbb{C}^N$ *with* $P_0 \subset P$ *and* $0 \in P_j$ $(j \in \mathbb{Z}_+)$,
(ii) *relatively open sets* $\Omega_j \subset \bar{B} \times \mathbb{C}^N$ *satisfying*

$$(\bar{C} \times \bar{P}_0) \cup (\bar{B} \times \bar{P}_j) \subset \Omega_j, \quad j = 0, 1, 2, \ldots,$$

(iii) *a sequence of maps* $F_j : \Omega_j \to Q$ *of class* $\mathcal{A}^r(\Omega_j)$ $(j = 0, 1, 2, \ldots)$

such that $F_j(\cdot, 0) = f_0$ *on* $\bar{B}$ *for all* $j \in \mathbb{Z}_+$ *and the sequence* F_j *converges to* f *in* $\mathcal{C}^r(\bar{C} \times \bar{P}_0)$ *as* $j \to \infty$. *If in addition we are given a map* $L : \bar{B} \times \mathbb{C}^N \to \mathbb{C}^M$ *of class* $\mathcal{A}^r(B)$ *such that* $L_x : \mathbb{C}^N \to \mathbb{C}^M$ *is* $\mathbb{C}$*-linear for each* $x \in \bar{B}$ *and* $L_x = \partial_t f(x, t)|_{t=0}$ *for* $x \in \bar{C}$, *then the sequence* F_j *can be chosen such that, in addition to the above,* $\partial_t F_j(x, t)|_{t=0} = L_x$ *holds for every* $x \in \bar{B}$ *and* $j \in \mathbb{Z}_+$.

Proof It suffices to consider the case $M = 1$. We may assume that C contains the origin $0 \in \mathbb{C}^n$ in its interior. Taylor's expansion in the t variable gives

$$f(x, t) = f_0(x) + \sum_{j=1}^{N} g_j(x, t) t_j, \quad (x, t) \in \bar{C} \times P$$

with $g_j \in \mathcal{A}^r(C \times P)$ for $j = 1, \ldots, N$. For each number $s \in (0, 1)$ the function $g_j^s(z, t) = g_j(sz, t)$ is holomorphic in $(s^{-1}C) \times P \supset \bar{C} \times P$; choosing s close to 1 we ensure that the approximation is as close as desired in $\mathcal{C}^r(\bar{C} \times P)$. Fix an s, choose a polydisc $P_0 \Subset P$ containing $0 \in \mathbb{C}^N$ and apply Runge's theorem to approximate g_j^s in $\mathcal{C}^r(\bar{C} \times \bar{P}_0)$ by an entire function $\tilde{g}_j$ on $\mathbb{C}^n \times \mathbb{C}^N$. The function

$$F(x, t) = f_0(x) + \sum_{j=1}^{N} \tilde{g}_j(x, t) t_j, \quad (x, t) \in \bar{B} \times \mathbb{C}^N$$

then approximates f on $\bar{C} \times \bar{P}_0$ and it agrees with f_0 on $\bar{B} \times \{0\}$. Since f maps the compact set $(\bar{C} \times \bar{P}_0) \cup (\bar{B} \times \{0\})$ to the polydisc Q, F also maps a neighborhood of this set to Q provided that the approximations are close enough. This gives sequences F_j, P_j and Ω_j with the stated properties.

Assume in addition that $L : B \times \mathbb{C}^N \to \mathbb{C}$ is a map of class $\mathcal{A}^r(B \times \mathbb{C}^N)$ that is linear in the second variable and satisfies $L_x = \partial_t|_{t=0} f(x, t)$ for all $x \in \bar{C}$. By Taylor's formula we have

$$f(x, t) = f_0(x) + L_x(t) + \sum_{j,k=1}^{N} g_{j,k}(x, t) t_j t_k, \quad (x, t) \in \bar{C} \times P$$

for some $g_{j,k} \in \mathcal{A}^r(C \times P)$. Approximating each $g_{j,k}$ in $\mathcal{C}^r(\bar{C} \times \bar{P}_0)$ by an entire function $\tilde{g}_{j,k} : \mathbb{C}^n \times \mathbb{C}^N \to \mathbb{C}$ and setting

$$F(x, t) = f_0(x) + L_x(t) + \sum_{j,k=1}^{N} \tilde{g}_{j,k}(x, t) t_j t_k, \quad (x, t) \in \bar{B} \times \mathbb{C}^N$$

gives the desired approximation. □

We apply Lemma 8.10.3 (p. 389) with the sets $C = D_0 \cap B_0$, $B = B_0$ (that are strictly convex in local holomorphic coordinates), and with Y a polydisc $Q \subset \mathbb{C}^M$; this choice is permissible since $f_0(\bar{B}_0)$ is contained in a special coordinate chart of Z. This gives an open set $P' \subset P_0$ containing the origin of $\mathbb{C}^N$, a relatively open set $\Omega \subset \bar{B} \times \mathbb{C}^N$ containing $(\bar{C} \times \bar{P}_0) \cup (\bar{B} \times \bar{P}')$, and a map $f' \colon \Omega \to Q \subset \mathbb{C}^M$ of class $\mathcal{A}^r(\Omega)$ that approximates f in the $\mathcal{C}^r$ topology on $\bar{C} \times P_0$, such that $f(x,0) = f'(x,0)$ and $\partial_t|_{t=0} f(x,t) = \partial_t|_{t=0} f'(x,t)$ for $x \in \bar{C}$. If the approximation is sufficiently close, then Proposition 5.9.2 (p. 240) allows us to glue f and f' into a new spray over the union $\bar{D}_1 = \bar{D}_0 \cup \bar{B}_0$ that satisfies (8.17) on this set. After m steps of this kind we obtain the first part of Proposition 8.10.2.

It remains to prove the parametric case. Fix a number $s \in [0,1]$. By the first part there exists a fibre spray $f^s \colon \bar{D} \times P \to Z$ of class $\mathcal{A}^r(D)$ satisfying

$$f^s(x,0) = f_0^s(x), \qquad \partial_t|_{t=0} f^s(x,t) = L_x^s, \quad x \in \bar{D}.$$

For $s = 0, 1$ we use the already given sprays. We wish to choose these sprays to depend continuously (or smoothly) on the parameter s. To do this, we shall first use a fixed spray f^s to find a solution in an open interval $I_s \subset \mathbb{R}$ around s, and finally we patch these solutions together.

Fix a number $u \in [0,1]$. Since L^u is surjective, there is a direct sum splitting $\bar{D} \times \mathbb{C}^N = E \oplus G$, where E and G are vector bundles of class $\mathcal{A}^r(D)$ and $E_x = \ker L_x^u$ for each $x \in \bar{D}$. We split the fibre variable on $\{x\} \times \mathbb{C}^N$ accordingly as $t = t_x' \oplus t_x'' \in E_x \oplus G_x$. Note that $L^u \colon G \to \mathrm{VT}Z|_{f^u(\bar{D})}$ is a complex vector bundle isomorphism of class $\mathcal{A}^r(D)$. By the inverse function theorem there is an open interval $I_u = (u - \delta, u + \delta) \subset \mathbb{R}$ such that for each $s \in I_u \cap [0,1]$ there exists a unique section $g_s \colon \bar{D} \to G$ of class $\mathcal{A}^r(D)$ satisfying

$$f^u\big(x, 0_x' \oplus g_s(x)\big) = f_0^s(x), \quad x \in \bar{D}.$$

It follows that the map H^s, defined by

$$H^s(x,t) = f^u\big(x, t_x' \oplus \big(t_x'' + g_s(x)\big)\big),$$

is a dominating 1-parametric spray with the core f_0^s for $s \in I_u \cap [0,1]$. For $s = u$ we have $g_u = 0$ and $H^u = f^u$.

It remains to adjust the t-differential of H^s at $t = 0$. Elementary linear algebra shows that, after shrinking the interval I_u if necessary, there exist for every $s \in I_u$ a unique complex vector bundle automorphism $A^s \colon G \to G$ and a unique complex vector bundle map $B^s \colon E \to G$, both of class $\mathcal{A}^r(D)$ and continuous in s, such that the map

$$F^s(x,t) = H^s\big(x, t_x' \oplus \big(B_x^s t_x' + A_x^s t''\big)\big), \quad s \in I_u \cap [0,1]$$

is a 1-parametric spray satisfying $\partial_t|_{t=0} F^s(x,t) = L_x^s(t)$ for all $x \in \bar{D}$ and $s \in I_u \cap [0,1]$. This gives a finite cover of $[0,1]$ by intervals $I_j = [a_j, b_j]$ $(j = 0, 1, \ldots, m)$, where $a_0 = 0 < a_1 < b_0 < a_2 < b_1 < \cdots < b_m = 1$, and for each j a 1-parametric spray $F_j \colon I_j \times \bar{D} \times P \to Z$ satisfying the conclusion of the Proposition on I_j. It remains to patch the sprays $\{F_j\}_{j=0}^m$ into a 1-parametric spray F on $[0,1]$. Since each pair of adjacent intervals I_j, I_{j+1} intersect in the segment $[a_{j+1}, b_j]$ and each three intervals are disjoint, it suffices to explain how to patch F_j and F_{j+1}. Choose a point $u \in (a_{j+1}, b_j)$. Using a decomposition $\bar{D} \times \mathbb{C}^N = E \oplus G$ as above, with $E = \ker \partial_t|_{t=0} F_j(u, \cdot, t)$, the implicit function theorem gives a segment $J_j = [\alpha_j, \beta_j] \subset (a_{j+1}, b_j)$ with $\alpha_j < u < \beta_j$, a polydisc $P_0 \subset P$ containing $0 \in \mathbb{C}^N$, and for each $s \in J_j$ a unique map

$$\gamma(s, x, t) = t'_x \oplus \big(t''_x + c(s, x, t)\big), \quad (x, t) \in \bar{D} \times P_0$$

of class $\mathcal{A}^r(D \times P_0)$ whose derivatives of order $\leq r$ in (x, t) are continuous in all variables and $c(s, x, t)$ vanishes to second order at $t = 0$, such that

$$F_j\big(s, x, \gamma(s, x, t)\big) = F_{j+1}(s, x, t), \quad s \in I_j,\ x \in \bar{D},\ t \in P_0.$$

The special form of γ is ensured by the fact that the first order jets of the sprays F_j and F_{j+1} with respect to t agree at $t = 0$ for every $s \in I_j$. (For more details, see [136, Lemma 4.4].) Choose a smooth function $\chi: \mathbb{R} \to [0, 1]$ such that $\chi(s) = 0$ for $s \leq \alpha_j$ and $\chi(s) = 1$ for $s \geq \beta_j$. The map

$$(s, x, t) \mapsto F_j\big(s, x, t'_x \oplus \big(t''_x + \chi(s) c''(s, x, t)\big)\big)$$

is then a 1-parametric spray of class $\mathcal{A}^r(D)$ satisfying Proposition 8.10.2 on $I_j \cup I_{j+1}$. After m steps we obtain a solution on $[0, 1]$. □

A map L as in Proposition 8.10.2 always exists for a sufficiently big N as follows from Theorem A for coherent sheaves of $\mathcal{A}^r$ modules on strongly pseudoconvex domains (see [293, Theorem 6] and [385]). Hence, we get the following corollary on the existence of local sprays.

Corollary 8.10.4 ([136, Corollary 4.2]) *Let $h: Z \to \bar{D}$ and $f_0 \in \Gamma^r_{\mathcal{A}}(D, Z)$ be as in Proposition* 8.10.2. *Then, the following hold.*

(a) *There exist a domain $0 \in P \subset \mathbb{C}^N$ for some $N \in \mathbb{N}$ and a dominating fibre spray $f: \bar{D} \times P \to Z$ of class $\mathcal{A}^r(D)$ with the core section f_0.*
(b) *Given a homotopy of sections $f_0^s \in \Gamma^r_{\mathcal{A}}(D, Z)$ $(s \in [0, 1])$, there exists a homotopy of dominating fibre sprays $f^s: \bar{D} \times P \to Z$ of class $\mathcal{A}^r(D)$ such that the core of f^s equals f_0^s for every $s \in [0, 1]$.*

Corollary 8.10.4 implies the following up to the boundary version of the Docquier-Grauert tubular neighborhood theorem (see Theorem 3.3.3) which allows linearization of analytic problems near a given section.

Corollary 8.10.5 ([136, Corollary 4.3]) *Let $h: Z \to \bar{D}$ be as in Proposition* 8.10.2. *Given $f \in \Gamma^r_{\mathcal{A}}(D, Z)$ there exist a holomorphic vector bundle $\pi: E \to \widetilde{D}$ over an open neighborhood $\widetilde{D} \subset S$ of $\bar{D}$, a relatively open neighborhood Ω of the zero section of the restricted bundle $E|_{\bar{D}} := \pi^{-1}(\bar{D})$, and a fibre preserving $\mathcal{C}^r$ diffeomorphism $\Phi: \Omega \to \Phi(\Omega) \subset Z$ which is biholomorphic on $\Omega \cap \pi^{-1}(D)$ and maps the zero section of $E|_{\bar{D}}$ onto $f(\bar{D})$.*

Proof Let $F: \bar{D} \times P \to Z$ be a dominating fibre spray furnished by Corollary 8.10.4, with $F_0 = f$. There is a splitting $\bar{D} \times \mathbb{C}^N = E \oplus E'$ of $\mathcal{A}^r(D)$-vector bundles, where $E'_x = \ker \partial_t|_{t=0} F(x,t)$ and E is a complementary bundle (Theorem B for $\mathcal{A}^r$ bundles, [293]). The restriction of F to $\Omega = E \cap (\bar{D} \times P)$ fulfills Corollary 8.10.5. Note that E can be identified with the restriction of the vertical tangent bundle $\mathrm{VT}Z = \ker dh$ to $f(\bar{D})$. □

Corollary 8.10.6 *Assume that D is a relatively compact domain with strongly pseudoconvex boundary of class $\mathcal{C}^\ell$ $(\ell \geq 2)$ in a Stein manifold. If $r \in \{0, 1, \dots, \ell\}$ and Y is a complex manifold, then for any $f_0 \in \mathcal{A}^r(D, Y)$, the space of all $f \in \mathcal{A}^r(D, Y)$ which are sufficiently $\mathcal{C}^0$-close to f_0 is isomorphic to the space of $\mathcal{C}^0$-small sections in $\Gamma^r_{\mathcal{A}}(D, E)$, where $E = f_0^*(\mathrm{T}Y)$.*

Proof We may consider maps $f: \bar{D} \to Y$ as sections of the product fibration $h: Z = \bar{D} \times Y \to \bar{D}$. Fix $f_0 \in \Gamma^r_{\mathcal{A}}(D, Z)$. Let $\Phi: \Omega \to \Phi(\Omega) \subset Z$ be as in Corollary 8.10.5, where Ω is an open neighborhood of the zero section in the complex vector bundle $E = f_0^*(\mathrm{T}Y)$ and Φ maps the zero section of E onto $f_0(\bar{D})$. If $f: \bar{D} \to Z$ is a section in $\Gamma^r_{\mathcal{A}}(D, Z)$ which is sufficiently uniformly close to f_0, then $f(\bar{D}) \subset \Phi(\Omega)$ and hence $f = \Phi \circ f'$ for a unique $f' \in \Gamma^r_{\mathcal{A}}(D, E)$ with $f'(\bar{D}) \subset \Omega$. □

8.11 Stein Neighborhoods of $\mathcal{A}(D)$-Graphs

In this section we construct open Stein neighborhoods of graphs of $\mathcal{A}(D)$ maps over bounded strongly pseudoconvex domains, and we approximate manifold-valued maps of class $\mathcal{A}^r(D)$ by maps holomorphic in neighborhoods of $\bar{D}$.

Theorem 8.11.1 ([198, Theorem 1.2], [457]) *Assume that $h: Z \to X$ is a holomorphic submersion of a complex manifold Z onto a Stein manifold X, D is a relatively compact, strongly pseudoconvex domain with $\mathcal{C}^2$ boundary in X, and $f: \bar{D} \to Z$ is a continuous section that is holomorphic in D. There exists a holomorphic vector bundle $\pi: E \to U$ over an open set $U \subset X$ with $\bar{D} \subset U$, and for every open set $\Omega_0 \subset Z$ containing $f(\bar{D})$ there exist a Stein open set Ω in Z, with $f(\bar{D}) \subset \Omega \subset \Omega_0 \cap \pi^{-1}(U)$, and a biholomorphic map $\Theta: \Omega \to \Theta(\Omega) \subset E$ that maps each fibre $\Omega_x = h^{-1}(x) \cap \Omega$ $(x \in h(\Omega))$ biholomorphically onto an open convex set $\widetilde{\Omega}_x = \Theta(\Omega_x)$ in $E_x = \pi^{-1}(x)$.*

With Θ as in Theorem 8.11.1, the map $g \mapsto \Theta \circ g$ induces an isomorphism between the space of sections of the restricted submersion $Z_{\bar{D}} = h^{-1}(\bar{D}) \to \bar{D}$ which are sufficiently uniformly close to f and the space of sections of the complex vector bundle $E_{\bar{D}} \to \bar{D}$ which are close to the section $\Theta \circ f$.

The following is an immediate corollary to Theorem 8.11.1.

Corollary 8.11.2 *If $D \Subset X$ are as in Theorem 8.11.1, Y is a complex manifold, and $f: \bar{D} \to Y$ is a continuous map that is holomorphic in D, then the graph*

$G_f = \{(x, f(x)) : x \in \bar{D}\}$ admits a basis of Stein neighborhoods in $X \times Y$. In particular, if D is a smoothly bounded domain in $\mathbb{C}$, then the graph G_f of any map $f: \bar{D} \to Y$ of class $\mathcal{A}(D, Y)$ admits a basis of neighborhoods that are biholomorphic to Stein domains in $\mathbb{C}^{1+\dim Y}$.

We prove Theorem 8.11.1 by the method of sprays from Sect. 8.10, thereby explicitly producing a holomorphic vector bundle $\pi: E \to U$ whose total space E contains a Stein neighborhood of $f(\bar{D})$. Another proof of the existence of a Stein neighborhood basis of $f(\bar{D})$ was given by Poletsky [457] by gluing plurisubharmonic functions and appealing to Grauert's Theorem 2.5.2.

Proof of Theorem 8.11.1 Set $n = \dim X$ and $n + m = \dim Z$. Let $f: \bar{D} \to Z$ be as in the theorem and set $\Sigma = f(\bar{D})$. Corollary 8.10.4 furnishes a dominating fibre spray $F: \bar{D} \times r\mathbb{B}^N \to Z$ of class $\mathcal{A}(D)$ such that $F(x, 0) = f(x)$ for all $x \in \bar{D}$. The set $E'' \subset \mathbb{T}^N := \bar{D} \times \mathbb{C}^N$ with fibres

$$E''_x = \ker\left(\partial_t|_{t=0} F(x, t)\right), \quad x \in \bar{D}$$

is an $\mathcal{A}(D)$-subbundle of the trivial bundle $\mathbb{T}^N$. By [293, 385] there exists a complementary $\mathcal{A}(D)$-subbundle $E' \subset \mathbb{T}^N$ such that $\mathbb{T}^N = E' \oplus E''$. By [291] (see also [135, Appendix]) we can approximate $E'|_{\bar{D}}$ by a holomorphic vector subbundle $E \subset U \times \mathbb{C}^N$ over an open set $U \supset \bar{D}$ such that $\mathbb{T}^N = E_{\bar{D}} \oplus E''$. For every $r > 0$ and $x \in \bar{D}$ set

$$E_{x,r} = \left\{t = t' \oplus 0'' \in E_x : |t| < r\right\} = E_x \cap r\mathbb{B}^N.$$

Given a subset $K \subset \bar{D}$ and a number $r > 0$, we write

$$E_K = \bigcup_{x \in K} \{x\} \times E_x, \qquad E_{K,r} = \bigcup_{x \in K} \{x\} \times E_{x,r} = E_K \cap \left(K \times r\mathbb{B}^N\right). \tag{8.19}$$

Let G denote the restriction of F to $E_{\bar{D},r} = E_{\bar{D}} \cap (\bar{D} \times r\mathbb{B}^N)$ (8.19). Then,

$$\partial_{t'}|_{t'=0} G\left(x, t'\right) : E_x \to \mathrm{VT}_{f(x)} Z$$

is a linear isomorphism for every $x \in \bar{D}$. By decreasing $r > 0$ we can ensure that G is injective holomorphic on each fibre; such G will be called a *fibrewise biholomorphic spray*. Note that $E_{\bar{D}}$ is isomorphic to the bundle $\mathrm{VTZ}|_{\Sigma}$; when $Z = X \times Y$ and Σ is the graph of an $\mathcal{A}(D, Y)$-map $f: \bar{D} \to Y$, then the latter bundle is isomorphic to $f^*\mathrm{T}Y$. □

Lemma 8.11.3 *There exist a number $r' \in (0, r)$, a decreasing sequence of open sets $O_1 \supset O_2 \supset \cdots$ in X with $\bigcap_{s=1}^{\infty} O_s = \bar{D}$, and a sequence of fibrewise biholomorphic sprays $G_s: E_{O_s,r'} \to Z$ such that G_s converges to G uniformly on $E_{\bar{D},r'}$ as $s \to \infty$, and $\Sigma \subset G_s(E_{O_s,r'})$ for all $s = 1, 2, \ldots$.*

Assume Lemma 8.11.3 for the moment. Let $\Omega_0 \subset Z$ be an open neighborhood of $f(\bar{D})$. Choose an initial fibrewise biholomorphic spray $G: E_{\bar{D},r} \to Z$ as above such that $G(E_{\bar{D},r}) \subset \Omega_0$. Let $G_s: E_{O_s,r'} \to Z$ $(s = 1, 2, \ldots)$ be a sequence of sprays furnished by Lemma 8.11.3. Set

$$\widetilde{\Omega}_s = E_{O_s,r'} \subset E, \quad \Omega_s = G_s(E_{O_s,r'}) \subset Z, \quad \Theta_s = G_s^{-1}: \Omega_s \to \widetilde{\Omega}_s.$$

If $s \in \mathbb{N}$ is chosen sufficiently big, then $\Omega_s \subset \Omega_0$ and the map $\Theta_s: \Omega_s \to \widetilde{\Omega}_s$ satisfies the conclusion of Theorem 8.11.1. □

Proof of Lemma 8.11.3 Since $h: Z \to X$ is a holomorphic submersion, there exist for each point $z_0 \in Z$ open neighborhoods $z_0 \in W \subset Z$, $x_0 = h(z_0) \in V \subset X$, and biholomorphic maps $\phi: V \to \mathbb{B}^n \subset \mathbb{C}^n$, $\Phi: W \to \mathbb{B}^n \times \mathbb{B}^m \subset \mathbb{C}^n \times \mathbb{C}^m$ such that $\phi(h(z)) = pr_1(\Phi(z))$ for every $z \in W$. Note that

$$\Phi(z) = \big(\phi\big(h(z)\big), \phi'(z)\big) \in \mathbb{B}^n \times \mathbb{B}^m, \quad z \in W,$$

where $\phi' = pr_2 \circ \Phi$. We call such (W, V, Φ) a *special coordinate chart* on Z.

Recall that $G: E_{\bar{D},r} \to Z$ is fibrewise biholomorphic spray over $\bar{D}$ with the core f. Narasimhan's lemma (on local convexification of a strongly pseudoconvex hypersurface) gives finitely many special coordinate charts (W_j, V_j, Φ_j) on Z, with $\Phi_j = (\phi_j \circ h, \phi'_j)$, such that $bD \subset \bigcup_{j=1}^{j_0} V_j$ and the following hold for all $j = 1, \ldots, j_0$ (for (ii) and (iii) we may have to decrease $r > 0$):

(i) $\phi_j(bD \cap V_j)$ is a strongly convex hypersurface in the ball $\mathbb{B}^n$,
(ii) the spray G maps $E_{\bar{D}\cap V_j,r}$ into W_j, and
(iii) $\phi'_j \circ G(E_{\bar{D}\cap V_j,r}) \Subset \mathbb{B}^m$.

Fix an $r > 0$ such that the above properties hold and choose a number $r' \in (0, r)$. Also, choose a number $c \in (0, 1)$ sufficiently close to 1 such that the open sets $U_j = \phi_j^{-1}(c\mathbb{B}^n) \Subset V_j$ $(j = 1, \ldots, j_0)$ still cover bD. By a finite induction we shall find strongly pseudoconvex domains $D = D_0 \subset D_1 \subset \cdots \subset D_{j_0} \Subset U$, numbers $r = r_0 > r_1 > \cdots > r_{j_0} = r'$, and fibrewise biholomorphic sprays $G_k: E_{\bar{D}_k,r_k} \to Z$ of class $\mathcal{A}$ $(k = 0, 1, \ldots, j_0)$, with $G_0 = G$ such that for every $k = 1, \ldots, j_0$ the restriction of G_k to $E_{\bar{D}_{k-1},r_k}$ will approximate G_{k-1} in the sup-norm topology. The domain D_k will be chosen such that

$$D_{k-1} \subset D_k \subset D_{k-1} \cup V_k, \qquad bD_{k-1} \cap U_k \subset D_k$$

for $k = 1, \ldots, j_0$; that is, we enlarge (bump out) D_{k-1} inside V_k so that the part of bD_{k-1} which lies in the smaller set U_k is contained in the next domain D_k. As the U_j's cover bD, the final domain D_{j_0} will contain $\bar{D}$ in its interior, and the spray $\widetilde{G} = G_{j_0}: E_{\bar{D}_{j_0},r'} \to Z$ will approximate G as closely as desired on $E_{\bar{D},r'}$. In particular, we shall arrange that $\Sigma = f(\bar{D})$ is contained in $\widetilde{G}(E_{\bar{D}_{j_0},r'})$. To keep the induction going, we will also ensure at every step that conditions (ii) and (iii) remain valid with (D, G) replaced by (D_k, G_k) for all $k = 1, \ldots, j_0$. The restriction of $\widetilde{G}$ to the interior $E_{D_{j_0},r'}$ can be taken as one of the sprays in the conclusion of the lemma.

Since all steps are of the same kind, it suffices to explain how to get the pair (D_1, G_1) from $(D, G) = (D_0, G_0)$. We begin by finding a domain $D'_1 \subset S$ with $\mathcal{C}^2$

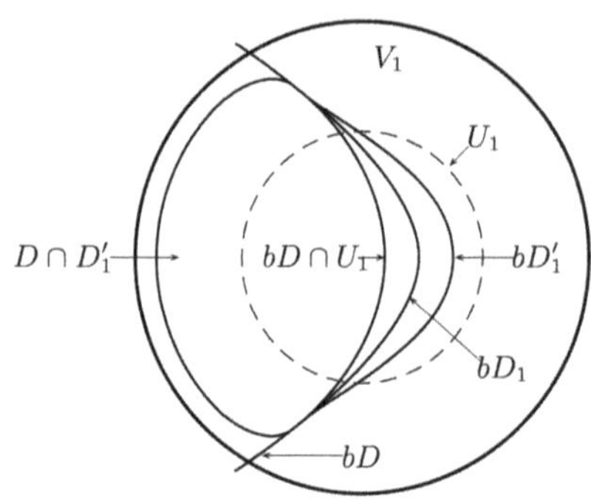

Fig. 8.1 The domains D_1' and D_1

boundary which is a convex bump on $D = D_0$ (see Definition 5.7.1, p. 234) and such that $\overline{U}_1 \cap \bar{D} \subset \bar{D}_1' \subset V_1$. To do this, we shall first find a set $\widetilde{D}_1' \subset \mathbb{B}^n$ with suitable properties and then take $D_1' = \phi_1^{-1}(\widetilde{D}_1')$.

Choose a smooth function $\chi \geq 0$ with compact support on $\mathbb{B}^n$ such that $\chi = 1$ on $c\mathbb{B}^n$. Recall that $U_1 = \phi_1^{-1}(c\mathbb{B}^n)$. Let $\tau : \mathbb{B}^n \to \mathbb{R}$ be a strongly convex defining function for the domain $\phi_1(D \cap V_1) \subset \mathbb{B}^n$. Choose $c' \in (c, 1)$ close to 1 such that the hypersurface $\phi_1(bD \cap V_1) = \{\tau = 0\}$ intersects the sphere $\{\zeta \in \mathbb{C}^n : |\zeta| = c'\}$ transversely. If $\delta > 0$ is small enough then the set

$$\{\zeta \in \mathbb{C}^n : |\zeta| < c',\ \tau(\zeta) < \delta\chi(\zeta)\}$$

could serve our purpose, except that it is not smooth along the intersection of the convex hypersurfaces $\{|\zeta| = c'\}$ and $\{\tau = \delta\chi\}$. By rounding off the corners we obtain a strongly convex set $\widetilde{D}_1' \subset \mathbb{B}^n$ such that $D_1' = \phi_1^{-1}(\widetilde{D}_1') \subset V_1$ satisfies the desired properties (see Fig. 8.1.)

Choose numbers r_1, r_1', r_1'' with $r' < r_1 < r_1' < r_1'' < r$. By using the special coordinate chart (W_1, V_1, Φ_1) we find an open set $V_1' \subset V_1$ containing $\bar{D} \cap V_1$ (V_1' will depend on the choice of G') and a fibrewise biholomorphic spray $G' : E_{\overline{V}_1', r_1''} \to Z$ with range in W_1 whose restriction to $E_{\bar{D} \cap \overline{V}_1, r_1''}$ approximates the initial spray $G = G_0$ as closely as desired in the uniform topology. If the approximation is sufficiently close on $E_{\overline{V}_1', r_1''}$, then there exists a unique fibrewise biholomorphic map $\gamma : E_{\bar{D} \cap \overline{V}_1, r_1'} \to E$ of class $\mathcal{A}$ which is close to the identity map and satisfies

$$G(x, t) = G'\big(\gamma(x, t)\big) = G'\big(x, \psi(x, t)\big), \quad x \in \bar{D} \cap \overline{V}_1,\ t \in E_{x, r_1'}.$$

Applying Propositions 5.8.1 (p. 235) on the Cartan pair (D, D_1'), we obtain

$$\gamma \circ \alpha = \beta \quad \text{on } E_{\bar{D} \cap \overline{V}_1, r_1},$$

where $\alpha : E_{\bar{D}, r_1} \to E$ and $\beta : E_{\bar{D}_1', r_1} \to E$ are injective holomorphic maps of the same form as γ which are close to the identity map on their respective domains. It follows that the fibrewise biholomorphic sprays

$$G \circ \alpha : E_{\bar{D}, r_1} \to Z, \qquad G' \circ \beta : E_{\bar{D}_1' \cap \overline{V}_1', r_1} \to Z$$

agree on the intersection of their domains, and hence they amalgamate into a fibrewise biholomorphic spray $G_1: E_{\bar{D}\cup(\bar{D}'_1\cap\overline{V}'_1),r_1} \to Z$ of class $\mathcal{A}$. Clearly, G_1 can be chosen to approximate G as closely as desired on $E_{\bar{D},r_1}$.

It remains to restrict G_1 to a suitably chosen strongly pseudoconvex domain $D_1 \Subset S$ contained in $D \cup (D'_1 \cap V'_1)$ and satisfying the other required properties. We choose D_1 such that it agrees with D outside of V_1, while

$$D \cap V_1 = \phi_1^{-1}\big(\{\zeta \in \mathbb{B}^n : \tau(\zeta) < \epsilon\chi(\zeta)\}\big)$$

for a small $\epsilon > 0$ (see Fig. 8.1). By choosing ϵ small enough (depending on G_1), we can ensure that conditions (i)–(iii) are satisfied by the pair (D_1, G_1).

Applying the same procedure to (D_1, G_1), using the special coordinate chart (W_2, V_2, Φ_2), we get the next pair (D_2, G_2). After j_0 steps we find a domain $D_{j_0} \subset S$ containing $\bar{D}$ and a fibrewise biholomorphic spray $\widetilde{G} = G_{j_0}: E_{\bar{D}_{j_0},r'} \to Z$ which approximates G as closely as desired uniformly on $E_{\bar{D},r'}$. If the approximation is close enough, then the range of $\widetilde{G}$ contains Σ.

The sequence G_s in Lemma 8.11.3 is chosen to consist of sprays $\widetilde{G}$ obtained as above, approximating G ever more closely on $E_{\bar{D},r'}$. □

Theorem 8.11.1, together with classical results on approximation of functions, gives the following approximation result for maps from compact strongly pseudoconvex Stein domains to an arbitrary complex manifold.

Theorem 8.11.4 ([136, Theorem 1.2]) *Assume that D is a bounded strongly pseudoconvex domain with $\mathcal{C}^\ell$ boundary ($\ell \ge 2$) in a Stein manifold X, Y is a complex manifold, and $r \in \{0, 1, \dots, \ell\}$. Then, every map $f: \bar{D} \to Y$ of class $\mathcal{A}^r(D, Y)$ can be approximated in the $\mathcal{C}^r(\bar{D}, Y)$ topology by holomorphic maps $U \to Y$ in open neighborhoods U of $\bar{D}$ in X. If in addition Y is an Oka manifold and f extends to a continuous map $X \to Y$, then f can be approximated in $\mathcal{C}^r(\bar{D}, Y)$ by holomorphic maps $X \to Y$. The analogous approximation result holds for sections as in Theorem* 8.11.1.

Proof The case of functions (i.e., when $Y = \mathbb{C}$) is well known, see e.g. [289]. The general case can be reduced to this special case as follows.

Denote by $\pi_Y: X \times Y \to Y$ the projection onto the second factor. Choose a Stein neighborhood $\Omega \subset X \times Y$ of the graph G_f (see Theorem 8.11.1) and a holomorphic embedding $\Theta: \Omega \hookrightarrow \mathbb{C}^N$. Next, we apply the approximation theorem for the component functions of $\tilde{f} = \Theta \circ f: \bar{D} \to \mathbb{C}^N$ to get a holomorphic map $\tilde{g}: U \to \mathbb{C}^N$ from an open set $U \supset \bar{D}$ that approximates $\tilde{f}$ in $\mathcal{C}^r(\bar{D})$. Let ι be a holomorphic retraction from a neighborhood $V \subset \mathbb{C}^N$ of the submanifold $\Theta(\Omega)$ onto $\Theta(\Omega)$ (see Theorem 3.3.3 on p. 74). Then,

$$g = \pi_Y \circ \Theta^{-1} \circ \iota \circ \tilde{g} : U' = \{x \in U : \tilde{g}(x) \in V\} \to Y$$

is a holomorphic map. If $\tilde{g}$ is sufficiently close to $\tilde{f}$ on $\bar{D}$, then $\bar{D}$ is contained in U' and g approximates f in $\mathcal{C}^r(\bar{D})$. □

Remark 8.11.5 The above proof of Theorem 8.11.4, by reducing it to Theorem 8.11.1, is rather convenient, but in fact is not necessary. Indeed, an inspection of the proof of Theorem 8.11.1 shows that the approximation comes for free as a result of the stepwise extension (by approximation) of the map across finitely many bumps which together give a neighborhood of $\bar{D}$ in the ambient manifold X. By using this argument and some additional techniques, Stopar [531, Theorem 3.1] proved that Theorem 8.11.4 also holds when D is a 1-convex domain (see Definition 3.1.3) with strongly pseudoconvex boundary of class $\mathcal{C}^\ell$ ($\ell \geq 2$). (This means that D contains a certain maximal compact complex subvariety E of positive dimension.) The condition in his theorem, that the projection $\pi_X: X \times Y \to X$ satisfies the technical condition $\mathcal{E}$ introduced in the paper, is always satisfied according to the new results of Prezelj [462] (see Sect. 6.13).

8.12 Oka Principle on Strongly Pseudoconvex Domains

The main result of this section is the following Oka principle for sections of $\mathcal{A}^r_Y(D)$-bundles with Oka fibres over compact strongly pseudoconvex Stein domains. These are fibre bundles that are smooth of order r over $\bar{D}$ and holomorphic in D. (Compare with Theorem 5.4.4 concerning Stein manifolds without boundary.) We use the terminology and notation from Sect. 8.10.

Theorem 8.12.1 ([136, Theorem 6.1]) *Assume that X is a Stein manifold, D is a relatively compact strongly pseudoconvex domain in X with boundary of class $\mathcal{C}^\ell$ ($\ell \geq 2$), and $h: Z \to \bar{D}$ is an $\mathcal{A}^r_Y(D)$-bundle with fibre Y (see Definition* 8.10.1*). Choose $r \in \{0, 1, \ldots, \ell\}$ and let* dist *denote a distance function on the manifold $\mathrm{J}^r(\bar{D}, Z)$ of all r-jets of sections $\bar{D} \to Z$ of h. Let K be a compact $\mathcal{A}(D)$-convex subset of $\bar{D}$ and let $U \subset X$ be an open set containing K. If Y is an Oka manifold (see Definition* 5.4.1*), then sections $\bar{D} \to Z$ satisfy the following*:

(i) (Basic Oka principle.) *Given a continuous section $f_0: \bar{D} \to Z$ that is of class $\mathcal{C}^r$ on $U \cap \bar{D}$ and holomorphic on $U \cap D$, there exist for every $\epsilon > 0$ an open neighborhood $V \subset U$ of K and a homotopy of sections $f_t: \bar{D} \to Z$ ($t \in [0, 1]$) which are of class $\mathcal{C}^r$ on $V \cap \bar{D}$ and holomorphic on $V \cap D$ such that $f_1 \in \Gamma^r_{\mathcal{A}}(\bar{D}, Z)$* (8.16) *and* $\sup_{x \in K} \mathrm{dist}(j^r_x f_t, j^r_x f_0) < \epsilon$, $t \in [0, 1]$.

(ii) (1-parametric Oka principle.) *Given a homotopy of continuous sections $f_t: \bar{D} \to Z$ ($t \in [0, 1]$) such that $f_0, f_1 \in \Gamma^r_{\mathcal{A}}(\bar{D}, Z)$, f_t is holomorphic in $U \cap D$ and of class $\mathcal{C}^r$ on $U \cap \bar{D}$ for each $t \in [0, 1]$, with continuous dependence of $j^r f_t|_{U \cap \bar{D}}$ on t, there exist an open neighborhood $V \subset U$ of K and a homotopy of sections $g_{t,s}: \bar{D} \to Z$ ($t, s \in [0, 1]$) satisfying*

1. $g_{t,0} = f_t$, $g_{0,s} = f_0$, $g_{1,s} = f_1$ *for all* $t, s \in [0, 1]$,
2. $g_{t,1} \in \Gamma^r_{\mathcal{A}}(\bar{D}, Z)$ *for all* $t \in [0, 1]$, *and*
3. $g_{t,s}$ *is of class* $\mathcal{C}^r$ *on* $V \cap \bar{D}$, *holomorphic on* $V \cap D$, *and satisfies*

$$\sup_{x \in K} \mathrm{dist}\big(j^r_x g_{t,s}, j^r_x f_t\big) < \epsilon, \quad s, t \in [0, 1].$$

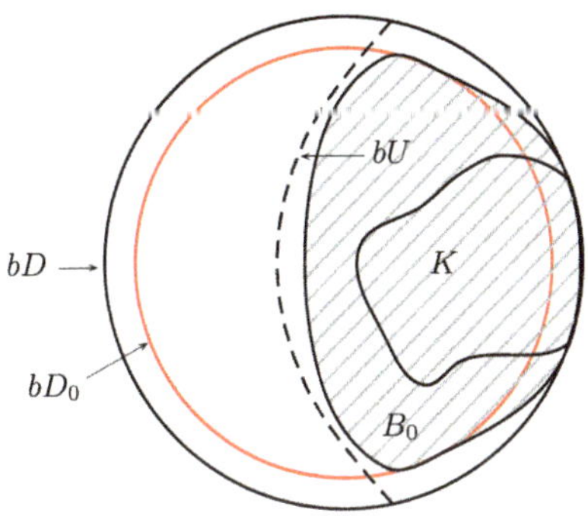

Fig. 8.2 The sets K, D_0, B_0 and U

Proof The proof of (i) will consist of two parts. In the first part, we find a section g_1 of Z satisfying (i) on a strongly pseudoconvex domain $\bar{D}_1 \subset \bar{D}$ containing K and such that $\bar{D}$ is obtained from $\bar{D}_1$ by attaching finitely many convex bumps. In the second part, we show how to approximately extend the solution over each bump to get a solution f_1 on all of $\bar{D}$.

Part 1. We have $D = \{\rho < 0\}$ where ρ is a $\mathcal{C}^\ell$ strongly plurisubharmonic function in a neighborhood of $\bar{D}$, with $d\rho \neq 0$ on $bD = \rho^{-1}(0)$.

Consider first the case when $K \subset D$. Choose a constant $c_1 < 0$ close to 0 such that $K \subset D_1 := \{x \in D : \rho(x) < c_1\}$ and $d\rho \neq 0$ on $\{c_1 \le \rho \le 0\} = \bar{D} \setminus D_1$. Since the fibre Y is an Oka manifold, the Oka principle on open Stein manifolds (see Theorem 5.4.4, p. 216) gives a holomorphic section g_1 of Z over an open neighborhood of $\bar{D}_1$ which satisfies the conclusion (i) over $\bar{D}_1$ (with f_1 replaced by g_1). Now proceed directly to part 2.

The situation is more complicated when $K \cap bD \neq \varnothing$. Let $D_0 = \{\rho < c_0\}$ for some $c_0 < 0$ close to 0 such that $d\rho \neq 0$ on $\{c_0 \le \rho \le 0\}$. If $c_0 < 0$ is chosen sufficiently close to 0, there is a strongly pseudoconvex domain $B_0 \subset D$ such that $K \subset \bar{B}_0$, (B_0, D_0) is a Cartan pair of class $\mathcal{C}^\ell$ (see Definition 5.7.1, p. 234), and $\bar{B}_0 \cap \bar{D}_0$ is holomorphically convex in D. (We used the assumption that K, and hence $K \cap \bar{D}_0$, is $\mathcal{A}(D)$-convex.) Set $D_1 - D_0 \cup B_0$ (see Fig. 8.2). Suitable choices of D_0 and B_0 ensure that $\bar{D}$ is obtained from $\bar{D}_1$ by attaching finitely many special convex bumps (see Definition 5.10.2 and Lemma 5.10.3, p. 242).

Let U be as in the statement of the theorem. Choose a strongly pseudoconvex domain $D' \subset D \cap U$ such that $\bar{B}_0 \subset D' \cup bD$. By Corollary 8.10.4 (p. 392) there exist a domain $P_0 \subset \mathbb{C}^N$ containing the origin and a dominating fibre spray $f: \bar{D}' \times P_0 \to Z$ of class $\mathcal{A}^r(D')$ with the core section $f_0|_{\bar{D}'}$.

Let $P \Subset P_0$ be a ball centered at the origin in $\mathbb{C}^N$. Since the set $(\bar{B}_0 \cap \bar{D}_0) \times \bar{P}$ is holomorphically convex in $D \times P_0$ and Y is an Oka manifold, Theorem 5.4.4 (p. 216) gives an open neighborhood $V \subset D$ of $\bar{D}_0$ and a holomorphic spray of sections $f': V \times P_0 \to Z$ that approximates f uniformly on a neighborhood of $(\bar{B}_0 \cap \bar{D}_0) \times \bar{P}$; by the Cauchy estimates, the approximation is then $\mathcal{C}^r$ close on the latter set. Applying Proposition 5.9.2 (p. 240) we glue the sprays f (over $\bar{B}_0$) and f' (over $\bar{D}_0$) to a new spray $g: \bar{D}_1 \times P' \to Z$ of class $\mathcal{A}^r(D_1)$ which approximates f in the $\mathcal{C}^r$ topology on $(\bar{B}_0 \cap \bar{D}_0) \times P'$. Its core section $g_1 = g(\cdot, 0): \bar{D}_1 \to Z$ then satisfies the condition (i) over $\bar{D}_1$.

Part 2: By Proposition 5.10.3 (p. 242) there are strongly pseudoconvex domains $D_2 \subset \cdots \subset D_m = D$ with $\mathcal{C}^\ell$ boundaries such that for every $j = 1, \ldots, m-1$ we have $D_{j+1} = D_j \cup B_j$, where B_j is a convex bump on D_j and $Z|_{\bar{B}_j} \to \bar{B}_j$ is a trivial bundle. To find a solution $f_1 \colon \bar{D} \to Z$, we successively extend the section $g_1 \colon \bar{D}_1 \to Z$ obtained in part 1 over each bump; this will give sections $g_j \in \Gamma^r_{\mathcal{A}}(D_j, Z)$ $(j = 2, 3, \ldots, m)$ such that the restriction of g_j to $\bar{D}_{j-1}$ is homotopic to g_{j-1} and close to g_{j-1} in $\Gamma^r_{\mathcal{A}}(\bar{D}_{j-1}, Z)$. All steps are of the same kind, and we explain how to obtain g_2.

Corollary 8.10.4 (p. 392) furnishes a ball $P \subset \mathbb{C}^N$ containing the origin and a dominating fibre spray $G_1 \colon \bar{D}_1 \times P \to Z$ of class $\mathcal{A}^r(D_1)$ with the core section g_1. Let $P' \Subset P$ be a smaller ball around the origin. Since the fibre bundle Z is trivial over $\bar{B}_1$, we can identify sections of Z over (subsets of) $\bar{B}_1$ with maps to the fibre Y. Also, using local holomorphic coordinates in a neighborhood of $\bar{B}_1$ in X, we identify $\bar{B}_1 \cap \bar{D}_1 \subset \bar{B}_1$ with compact convex sets in $\mathbb{C}^n$. Using these identifications, we first approximate G_1 in the $\mathcal{C}^r$ topology on $(\bar{B}_1 \cap \bar{D}_1) \times \bar{P}'$ by a holomorphic map from an open neighborhood of this set (in $\mathbb{C}^n \times \mathbb{C}^N$) to Y (just precompose G_1 with linear contractions to an interior point of its domain). Since Y is an Oka manifold, we can approximate this map by an entire map $\mathbb{C}^n \times \mathbb{C}^N \to Y$. Passing back to subsets of X, the above procedure gives a spray $G_1' \colon \bar{B}_1 \times P' \to Z$ of class $\mathcal{A}^r(B_1)$ which approximates G_1 as closely as desired in the $\mathcal{C}^r$ topology on $(\bar{D}_1 \cap \bar{B}_1) \times P'$. Finally, we glue G_1 and G_1' into a spray $G_2 \colon \bar{D}_2 \times P'' \to X$ of class $\mathcal{A}^r(D_2)$ by appealing to Proposition 5.9.2 (p. 240). Letting $g_2 = G_2(\cdot, 0) \colon \bar{D}_2 \to Z$ be the core section of G_2 completes the induction step.

After m steps of this kind we obtain a section $g_m \in \Gamma^r_{\mathcal{A}}(D, Z)$ (which we now call f_1) satisfying condition (i) in Theorem 8.12.1. A homotopy from f_0 to f_1 with the stated properties is obtained by combining the homotopies obtained in the individual steps of the proof.

Part (ii) is proved in the same way by using the 1-parametric Oka property of Y. A 1-parametric spray with the given core is furnished by Corollary 8.10.4 (p. 392), and we glue sprays by Proposition 5.9.2 (p. 240). □

Remark 8.12.2 Theorem 8.12.1 was extended by Stopar [531, Theorem 3.1] to the case when D is a 1-convex domain (see Definition 3.1.3) with strongly pseudoconvex boundary of class $\mathcal{C}^\ell$ $(\ell \geq 2)$. As mentioned in Remark 8.11.5, the additional technical assumption regarding Condition $\mathcal{E}$ in his paper is always satisfied due to the new results of Prezelj in [462]. □

8.13 Banach Manifolds of Holomorphic Mappings

In this section we show that certain classes of maps from a compact strongly pseudoconvex Stein domain to an arbitrary complex manifold admit the structure of a Banach (resp. Hilbert or Fréchet) manifold.

Let D be a relatively compact domain with piecewise $\mathcal{C}^1$ boundary in $\mathbb{C}^n$. We consider the following function spaces:

- For $k \in \mathbb{Z}_+$ and $0 \le \alpha < 1$, $\mathcal{A}^{k,\alpha}(D)$ is the Banach space of all functions $\bar{D} \to \mathbb{C}$ in the Hölder class $\mathcal{C}^{k,\alpha}(\bar{D})$ which are holomorphic in D. When $\alpha = 0$, we write $\mathcal{A}^{k,0} = \mathcal{A}^k$ and $\mathcal{A}^0 = \mathcal{A}$.
- $\mathcal{A}^\infty(D) = \bigcap_{k=0}^\infty \mathcal{A}^k(D)$ is the Fréchet space consisting of all $\mathcal{C}^\infty$ function $\bar{D} \to \mathbb{C}$ that are holomorphic in D.
- For $k \in \mathbb{Z}_+$ and $p \ge 1$, $L^{k,p}_{\mathcal{O}}(D)$ is the Banach space (Hilbert if $p = 2$) consisting of all holomorphic functions $D \to \mathbb{C}$ whose partial derivatives of order $\le k$ belong to $L^p(D)$ (with respect to the Lebesgue measure). These are Sobolev spaces of holomorphic functions on D.

If $L(D)$ is any of the above function spaces, we denote by $L(D, \mathbb{C}^m)$ the locally convex topological vector space consisting of maps whose components belong to $L(D)$. In the case (iii) we assume that $kp > 2n$, so the Sobolev embedding theorem provides a continuous (compact) inclusion map $L^{k,p}_{\mathcal{O}}(D) \hookrightarrow \mathcal{A}(D)$. (See Palais [448, 449].)

Assume now that D is a relatively compact domain with piecewise $\mathcal{C}^1$ boundary in an n-dimensional complex manifold X. Given a complex manifold Y of dimension m without boundary, one defines the mapping space $L(D, Y)$ as follows. (See Lempert [391, Sect. 2] for the case when $\bar{D}$ is a compact smooth manifold and we are considering the space $\mathcal{C}^k(\bar{D}, Y)$, $k \in \mathbb{Z}_+ \cup \{\infty\}$.) Fix a continuous map $f : \bar{D} \to Y$. Choose finitely many holomorphic coordinates systems $\phi_j : U_j \to \widetilde{U}_j \subset \mathbb{C}^n$ on X, and $\psi_j : W_j \to \widetilde{W}_j \subset \mathbb{C}^m$ on Y, such that $f(\bar{D} \cap U_j) \subset W_j$ for all j. Also, choose open subsets $V_j \Subset U_j$ such that $\bar{D} \subset \bigcup_j V_j$ and $V_j \cap D$ has piecewise $\mathcal{C}^1$ boundary for each j. Then, $f \in L(D, Y)$ precisely when for each j the restriction f_j of the map $\psi_j \circ f \circ \phi_j^{-1}$ to the set $\phi_j(\overline{D \cap V_j}) \Subset \widetilde{U}_j$ belongs to $L(\phi_j(D \cap V_j), \mathbb{C}^m)$; the definition is independent of the choices of charts. Further, given an open neighborhood $\mathcal{U}_j \subset L(\phi_j(D \cap V_j), \mathbb{C}^m)$ of f_j for every j, the corresponding neighborhood of f in $L(D, Y)$ consists of all maps $g : \bar{D} \to Y$ such that $g(\overline{D \cap V_j}) \subset W_j$ and the restriction g_j of $\psi_j \circ g \circ \phi_j^{-1}$ to $\phi_j(\overline{D \cap V_j})$ belongs to $\mathcal{U}_j$ for all j.

Theorem 8.13.1 ([198]) *Let D be a relatively compact strongly pseudoconvex domain in a Stein manifold, and let Y be a complex manifold.*

(i) *$\mathcal{A}^{k,\alpha}(D, Y) = \mathcal{C}^{k,\alpha}(\bar{D}, Y) \cap \mathcal{O}(D, Y)$ is a complex Banach manifold for every $k \in \mathbb{Z}_+$ and $0 \le \alpha < 1$.*
(ii) *$\mathcal{A}^\infty(D, Y) = \mathcal{C}^\infty(\bar{D}, Y) \cap \mathcal{O}(D, Y)$ is a complex Fréchet manifold.*
(iii) *For each $k \in \mathbb{N}$ and $p \ge 1$ such that $kp > \dim_{\mathbb{R}} D$, the Sobolev space $L^{k,p}_{\mathcal{O}}(D, Y) = L^{k,p}(\bar{D}, Y) \cap \mathcal{O}(D, Y)$ is a complex Banach manifold (a complex Hilbert manifold if $p = 2$).*

If $L(D, Y)$ denotes any of the above manifolds of maps then the tangent space $\mathrm{T}_f L(D, Y)$ at a point $f \in L(D, Y)$ is $L_h(D, f^*\mathrm{T}Y)$, the space of sections of *class* $L(D)$ *of the complex vector bundle* $h: f^*\mathrm{T}Y \to \bar{D}$. *If* D *is contractible, or if* $\dim D = 1$, *then* $\mathrm{T}_f L(D,Y) \cong L(D, \mathbb{C}^m)$ *with* $m = \dim Y$.

Theorem 8.13.1 also holds if $\bar{D}$ is a compact complex manifold with Stein interior and smooth strongly pseudoconvex boundary; according to [292, 440] such $\bar{D}$ embeds as a smoothly bounded strongly pseudoconvex domain in a Stein manifold.

The following is an immediate corollary to Theorem 8.13.1. (Compare with the results in Sect. 8.8.) The analogous result for pseudoholomorphic discs in almost complex manifolds was obtained by Sukhov and Tumanov [538].

Corollary 8.13.2 *Let* D *be a relatively compact strongly pseudoconvex domain in a Stein manifold, and let* Y *be a complex manifold. Then, maps of class* $\mathcal{A}^\infty(D,Y)$ *satisfy the jet transversality theorem.*

Proof of Theorem 8.13.1 Let $L(D,Y)$ denote any one of the above spaces; note that it is a subset of $\mathcal{A}(D,Y)$. We need to construct holomorphic charts in $L(D,Y)$. Given a holomorphic vector bundle $\pi: E \to U$ over an open set $U \subset X$ containing $\bar{D}$, we denote by $L_h(D,E)$ the space of all section $\bar{D} \to E_{\bar{D}}$ of h over $\bar{D}$ which belong to $L(D,E)$. This is a locally convex topological vector space; Banach for $\mathcal{A}^{k,\alpha}$ or $L^{k,p}_{\mathcal{O}}$, Hilbert for $L^{k,2}_{\mathcal{O}}$, and Fréchet for $\mathcal{A}^\infty$. Fix a map $f \in L(D,Y)$. Theorem 8.11.1 (p. 393) furnishes an open Stein neighborhood $\Omega \subset X \times Y$ of the graph $G_f = \{(x, f(x)) : x \in \bar{D}\}$ and a biholomorphic map $\Theta: \Omega \to \widetilde{\Omega} \subset E$ onto an open set $\widetilde{\Omega}$ in the total space of a holomorphic vector bundle $\pi: E \to U$ such that $\bar{D} \subset U$ and $\pi \circ \Theta: \Omega \to X$ is the restriction to Ω of the base projection $(x,y) \to x$. Since Θ is holomorphic in a neighborhood of G_f, the map $\theta(f): \bar{D} \to E_{\bar{D}}$ defined by

$$\bar{D} \ni x \to \theta(f)(x) = \Theta\big(x, f(x)\big) \in E_x$$

is a section of the restricted bundle $E_{\bar{D}} \to \bar{D}$ which belongs to the space $L_h(D,E)$. The graph G_g of any $g \in L(D,Y)$ sufficiently near f is also contained in Ω, and the composition with Θ defines an isomorphism $g \mapsto \theta(g) = \Theta(\cdot\,, g)$ between an open neighborhood of f in $L(D,Y)$ and an open neighborhood of $\theta(f)$ in $L_h(D,E)$; we take θ as a Banach (or Fréchet) chart on $L(D,Y)$. It is easily verified that the transition map between any such pair of charts is holomorphic; for the Sobolev classes, see Palais [449, Theorem 9.10]. The collection of all such charts defines a holomorphic Banach (Hilbert, Fréchet) manifold structure on $L(D,Y)$.

The above also shows that the tangent space to the manifold $L(D,Y)$ at a point $f \in L(D,Y)$ can be identified with the space of sections of class $L(D)$ of the complex vector bundle $f^*\mathrm{T}Y \to \bar{D}$ (the pullback to $\bar{D}$ of the tangent bundle $\mathrm{T}Y$ by the map f). By the Oka-Grauert principle (see Theorem 8.12.1), homotopic maps induce isomorphic pullback bundles, and hence $\mathrm{T}_f L(D,Y)$ is independent of a point f in a connected component of $L(D,Y)$ up to a complex Banach space isomorphism. If D is contractible, or if $\dim D = 1$, then every $\mathcal{A}(D)$-vector bundle over $\bar{D}$ is trivial; in this case $\mathrm{T}_f L(D,Y) \cong L(D,\mathbb{C}^m)$ with $m = \dim Y$ for every $f \in L(D,Y)$. □

Chapter 9
Embeddings, Immersions and Submersions

In this chapter, Oka theory is applied to the study of some of the most challenging problems of Stein geometry concerning *regular* holomorphic maps—immersions, embeddings, and submersions—of Stein manifolds and Stein spaces to complex Euclidean spaces and other complex manifolds.

We begin in Sect. 9.1 by considering totally real immersions, and complex submersions, of smooth manifolds into complex manifolds. In Sect. 9.2 we construct almost proper holomorphic maps $X \to \mathbb{C}^n$ from any Stein manifold X of dimension n, following Bishop [58]. In Sects. 9.3–9.5 we present results on proper holomorphic embeddings and immersions of Stein spaces of dimension > 1 to Euclidean spaces of minimal dimension, due to Eliashberg and Gromov (with an improvement by Schürmann); see Theorem 9.3.1. The main ingredient is the Relative Embedding Theorem (see Theorem 9.3.7) whose proof in Sect. 9.4 uses the Oka principle for sections of holomorphic vector bundles that avoid certain complex subvarieties of the total space. In Sect. 9.6 we outline the proof of the Eliashberg-Gromov homotopy principle for holomorphic immersions of Stein manifolds to Euclidean spaces.

In Sect. 9.7 we prove a Cartan type splitting lemma for biholomorphic maps close to the identity on a Cartan pair. This result is essential in subsequent constructions in the chapter. In Sect. 9.8 we look at the Oka principle for proper holomorphic maps to q-convex manifolds, and to Stein manifolds with the density property. In Sects. 9.9–9.11 we consider the classical problem of embedding open Riemann surfaces into $\mathbb{C}^2$. In Sect. 9.12 we use holomorphic automorphisms of $\mathbb{C}^n$ to prove approximation theorems for holomorphic submersions $\mathbb{C}^n \to \mathbb{C}^q$ with $1 \le q < n$. By using these results and the gluing technique from Sect. 9.7, we prove the homotopy principle for holomorphic submersions of Stein manifolds X to Euclidean spaces $\mathbb{C}^q$ for $q < \dim X$ (see Sect. 9.13) and to other complex manifolds satisfying a suitable flexibility property (see Sect. 9.14). In Sect. 9.15 we show that the cohomology group $H^1(X;\mathbb{C})$ of a Stein manifold is represented by nowhere vanishing closed holomorphic 1-forms. In Sect. 9.16 we give applications to holomorphic foliations on Stein manifolds.

F. Forstnerič, *Stein Manifolds and Holomorphic Mappings*, Ergebnisse der Mathematik und ihrer Grenzgebiete. 3. Folge / A Series of Modern Surveys in Mathematics 56,
DOI 10.1007/978-3-319-61058-0_9

9.1 The H-Principle for Totally Real Immersions and for Complex Submersions

In this section we consider two classes of maps from smooth real manifolds to complex manifolds—totally real immersions and complex submersions (see Definition 9.1.1). Totally real immersions and embeddings have been studied extensively in the literature. The second class of maps, which we call complex submersions, is less familiar, but is natural when considering holomorphic submersions from complexifications of real analytic manifolds. These two classes of maps are important in proofs of the main results of this chapter, namely, the construction of holomorphic immersions, embeddings and submersions of Stein manifolds to Euclidean spaces and to other complex manifolds. Both classes satisfy the h-principle (see Theorem 9.1.4), and by using the jet transversality theorem it is easy to establish the universal dimensions for their existence (see Theorem 9.1.2). Examples provided by Theorem 9.1.3 show that the universal holomorphic immersion and submersion dimensions for all Stein manifolds, which are established later in the chapter, are optimal already for Stein complexifications of compact real analytic manifolds.

Definition 9.1.1 Let M be a smooth manifold, X be a complex manifold, and $f\colon M \to X$ be a map of class $\mathcal{C}^1$.

(a) The map f is a *totally real immersion* at a point $p \in M$ if the complexified differential $\mathrm{d}f_p\colon \mathbb{C}\mathrm{T}_pM \to \mathrm{T}_{f(p)}X$ (see Sect. 1.6) is injective; it is a totally real immersion if this holds at every point $p \in M$.
(b) The map f is *a complex submersion* at $p \in M$ if $\mathrm{d}f_p\colon \mathbb{C}\mathrm{T}_pM \to \mathrm{T}_{f(p)}X$ is surjective; it is a complex submersion if this holds at every point $p \in M$.

Note that these two classes of maps coincide when $\dim_{\mathbb{R}} M = \dim_{\mathbb{C}} X$. For example, an immersed real curve in $\mathbb{C}$ is a totally real immersion and a complex submersion.

By elementary linear algebra, a map $f\colon M \to X$ is a totally real immersion at $p \in M$ if and only if is an immersion and the image $\mathrm{d}f_p(\mathrm{T}_pM)$ is a real subspace of $\mathrm{T}_{f(p)}X$ which does not contain any complex line, i.e.,

$$\mathrm{d}f_p(\mathrm{T}_pM) \cap J\big(\mathrm{d}f_p(\mathrm{T}_pM)\big) = \{0\} \in \mathrm{T}_{f(p)}X$$

where J denotes the almost complex structure operator on $\mathrm{T}X$ (see (1.19)). Such a point $p \in M$ has a neighborhood $U \subset M$ such that $f(U)$ is a totally real submanifold of X (see Sect. 3.5). In local holomorphic coordinates $z = (z_1, \dots, z_N)$ on X around $f(p)$ and writing $f = (f_1, \dots, f_N)$ accordingly, we have that

- f is a totally real immersion at $p \in M$ if and only if the differentials $\mathrm{d}f_1, \dots, \mathrm{d}f_N$ span the complexified cotangent space $\mathbb{C}\mathrm{T}^*_pM$;
- f is a complex submersion at p if and only if the differentials $\mathrm{d}f_1, \dots, \mathrm{d}f_N$ are $\mathbb{C}$-linearly independent elements of $\mathbb{C}\mathrm{T}^*_pM$.

The geometric significance of these classes of maps is seen by the following observation. Assume that M is real analytic; recall that every $\mathscr{C}^1$ manifold admits a compatible real analytic structure [571]. Then M admits a complexification (see [575]), i.e., a complex manifold $M^{\mathbb{C}}$ of complex dimension equal to $\dim_{\mathbb{R}} M$ which contains M as a totally real submanifold; furthermore, M has a basis of tubular Stein neighborhoods in $M^{\mathbb{C}}$ (see Grauert [255]). The holomorphic tangent bundle $TM^{\mathbb{C}}$, restricted to M, is isomorphic to the complexified tangent bundle $\mathbb{C}TM$ of M. Every real analytic map $f\colon M \to X$ into a complex manifold X admits a complexification, i.e., a holomorphic map $f^{\mathbb{C}}\colon U \to X$ from an open neighborhood $U \subset M^{\mathbb{C}}$ of M such that $f^{\mathbb{C}}|_M = f$. The germs of $M^{\mathbb{C}}$ and $f^{\mathbb{C}}$ are unique along M. It follows that

(a) $f\colon M \to X$ is a totally real immersion if and only if its complexification $f^{\mathbb{C}}\colon U \to X$ is a holomorphic immersion in a neighborhood $U \subset M^{\mathbb{C}}$ of M;
(b) $f\colon M \to X$ is a complex submersion if and only if its complexification $f^{\mathbb{C}}\colon U \to X$ is a holomorphic submersion in a neighborhood of M.

Hence, if we wish to prove that every Stein manifold X^n of dimension n admits a holomorphic immersion to $\mathbb{C}^N$ for a certain integer $N > n$, then we must at least be able to show that every smooth n-manifold admits a totally real immersion into $\mathbb{C}^N$. Similarly, if every Stein manifold X^n admits a holomorphic submersion to $\mathbb{C}^q$ for some $q < n$, then every smooth n-manifold admits a complex submersion to $\mathbb{C}^q$. The optimal dimensions $N = N(n)$ and $q = q(n)$ in these problems are provided by the following result, together with the examples in Theorem 9.1.3. It is interesting that these necessary conditions are also sufficient for all Stein manifolds, and not only for complexifications of real manifolds; see Theorem 9.6.3 (p. 430) for immersions and Corollary 9.13.2 (p. 462) for submersions.

Theorem 9.1.2 *Let M be a smooth n-dimensional manifold and X be a complex N-dimensional manifold.*

(a) *If $N \ge [\frac{3n}{2}]$ then a generic smooth map $M \to X$ is a totally real embedding. In particular, there exists a totally real embedding $M \hookrightarrow \mathbb{C}^{[3n/2]}$.*
(b) *If $N \le [\frac{n+1}{2}]$ then a generic smooth map $M \to X$ is a complex submersion. In particular, there exists a complex submersion $M \to \mathbb{C}^{[(n+1)/2]}$.*

By an *embedding*, we mean an injective immersion. The word *generic* is understood in the fine $\mathscr{C}^1$ topology on the space of maps $M \to X$. When $X = \mathbb{C}^N$, part (a) can be found in [221, Lemma 5.3].

Note that the totally real immersion dimension $N(n)$, and the complex submersion dimensions $q(n)$, are symmetric with respect to n:

$$N(n) = n + \left[\frac{n}{2}\right], \qquad q(n) = n - \left[\frac{n}{2}\right]. \tag{9.1}$$

We shall see from the proof that this is not a coincidence.

Proof Let us first consider part (a), so $N \ge n$. In local real coordinates $u = (u_1, \ldots, u_n)$ on M and complex coordinates $z = (z_1, \ldots, z_N)$ on X, a map

$f = (f_1, \dots, f_N)$ from M to X is a totally real immersion if and only if its Jacobian matrix $f'(u) = (\partial f_i/\partial u_j)$ (of size $N \times n$) has maximal rank n at each point. The number of independent minors in this matrix is $N - n + 1$, and hence the subset Σ of the space of complex $N \times n$ matrices (which equals $\mathbb{C}^{nN}$), consisting of matrices of less than maximal rank, is a complex subvariety of codimension $N - n + 1$, hence of real codimension $2(N - n + 1)$. (See Lemma 8.9.2 on p. 387. Note that the set of $N \times n$ matrices of maximal rank is the Stiefel manifold $V_{n,N}$; see Example 1.2.7 on p. 9.) If $n = \dim M < 2(N - n + 1)$, then by the jet transversality theorem the differential of a generic map $M \to X$ misses Σ, and hence is a totally real immersion. It is easy to check that the above inequality is equivalent to $N \geq [3n/2]$. The fact that a generic totally real immersion is injective when $N > n$ follows from the standard result that a generic map $M \to X$ is an injective immersion if $\dim_{\mathbb{R}} X \geq 2\dim_{\mathbb{R}} M + 1$, together with the observation that totally real immersions form an open set in the fine $\mathcal{C}^1$ topology on the space of maps $M \to X$.

In case (b) we have $N \leq n$, and the variety of matrices of less than maximal rank has complex codimension $n - N + 1$. This leads to the same conclusion when $n < 2(n - N + 1)$ which is equivalent to $N \leq [\frac{n+1}{2}]$. □

Theorems obtained using transversality, such as this one, are often, but not always, optimal in the sense that the target dimension cannot be decreased (in case (a)) or increased (in case (b)). For example, a transversality argument implies that every n-dimensional manifold M admits an immersion into $\mathbb{R}^{2n}$ [571], but a more delicate argument due to Whitney [572] decreases $2n$ to $2n - 1$ for $n > 1$. However, the totally real immersion dimension $N(n)$ and the complex submersion dimension $q(n)$ (see (9.1)) are optimal for every n. A simple family of examples was described by Ho, Jacobowitz and Landweber [298]. For every $k \in \mathbb{Z}_+$ let

$$M^{4k} = \overbrace{\mathbb{CP}^2 \times \cdots \times \mathbb{CP}^2}^{k} = (\mathbb{CP}^2)^k,$$

the product of k copies of the complex projective plane. We also let

$$M^{4k+1} = M^{4k} \times S^1, \qquad M^{4k+2} = M^{4k} \times \mathbb{RP}^2, \qquad M^{4k+3} = M^{4k} \times \mathbb{RP}^2 \times S^1.$$

Theorem 9.1.3 ([298, Theorems 2.1 and 3.1]) *For every integer $n \in \mathbb{N}$, the manifold M^n defined above does not admit a totally real immersion in $\mathbb{C}^{N(n)-1}$, and it does not admit a complex submersion to $\mathbb{C}^{q(n)+1}$. (Here, the numbers $q(n)$ and $N(n)$ are given by* (9.1).)

The proof, which we do not include here, is based on the observation that the existence of a totally real immersion $M^n \to \mathbb{C}^N$ induces a decomposition $\mathbb{T}^N = M \times \mathbb{C}^N = \mathbb{C}TM \oplus E$ for some complex vector bundle $E \to M$ of rank $N - n$. Similarly, a complex submersions $M \to \mathbb{C}^q$ induces a decomposition $\mathbb{C}TM = \mathbb{T}^q \oplus E$ where $\mathbb{T}^q$ is a trivial bundle of rank q and E is a complementary subbundle. By using Stiefel-Whitney classes one shows that, in the first case, $N - n$ cannot be too small, and in the second case q cannot be too big, provided M has sufficiently complicated topology. (Compare with Corollary 8.3.2 and see Example 9.3.2 and Proposition 9.3.3, due to Forster.)

Note that the manifold M^n in the above list is orientable when the residue of n modulo 4 is either 0 or 1, and is nonorientable when the residue is 2 or 3. This is not a coincidence—orientable manifold are simpler and have better properties in such problems. For example, every connected open orientable 2-manifold is parallelizable and admits a totally real immersion into $\mathbb{C}^2$. More generally, by [298, Theorem 4.1] every orientable $(4k+2)$-manifold M admits a totally real immersion into $\mathbb{C}^{6k+2}$, and this is optimal in general. By the same theorem, if an orientable $4k$-dimensional manifold has vanishing top dual Pontryagin class, then it admits a totally real immersion into $\mathbb{C}^{6k-1}$.

The following h-principle for totally real immersions is due to Gromov [268, p. 332], [265, p. 192]. The same proof applies to complex submersions.

Theorem 9.1.4 *Totally real immersion and complex submersions satisfy the h-principle.*

Proof As in the proof of Theorem 9.1.2, the same idea applies to both cases, so we focus on totally real imersions. Let $m = \dim_{\mathbb{R}} M$ and $n = \dim_{\mathbb{C}} X$. Given points $p \in M$ and $x \in X$, we denote by $\mathrm{Hom}_{\mathrm{tr}}(\mathrm{T}_p M, \mathrm{T}_x X)$ the set of all $\mathbb{R}$-linear homomorphisms $\mathrm{T}_p M \to \mathrm{T}_x X$ which are totally real, i.e., whose complexifications are injective maps $\mathbb{C}\mathrm{T}_p M \to \mathrm{T}_x X$. We have seen in the proof of Theorem 9.1.2 that $\mathrm{Hom}_{\mathrm{tr}}(\mathrm{T}_p M, \mathrm{T}_x X)$ is isomorphic to the Stiefel variety $V_{n,N}$ of n-frames in $\mathbb{C}^N$. Consider the following open subset of the 1-jet manifold $\mathrm{J}^1(M, X)$ (see Sect. 1.10):

$$\Omega = \big\{(p, x, \lambda) : p \in M,\ x \in X,\ \lambda \in \mathrm{Hom}_{\mathrm{tr}}(\mathrm{T}_p M, \mathrm{T}_x X)\big\}. \tag{9.2}$$

Note that a $\mathscr{C}^1$ map $f\colon M \to X$ is a totally real immersion if and only if its 1-jet extension $\mathrm{j}^1 f$ is a section of Ω. Theorem 9.1.6 then follows immediately from Theorem 1.10.5 on p. 42 (Gromov's h-principle for ample differential relations) in view of the following lemma.

Lemma 9.1.5 *The differential relation* Ω (9.2) *is ample.*

Proof In any pair of local smooth coordinates $u = (u_1, \ldots, u_m)$ on M and local holomorphic coordinates $z = (z_1, \ldots, z_n)$ on X, the 1-jet of a $\mathscr{C}^1$ map $f\colon M \to X$ with $f(u) = z$ is represented by

$$z^{(1)} = (u, z; v_1, \ldots, v_m) \in \mathrm{J}^1(M, X),$$

where $v_k = \frac{\partial f}{\partial u_k}(u) \in \mathbb{C}^n$ for $k = 1, \ldots, m$. Clearly, $z^{(1)}$ belongs to Ω if and only if the vectors $v_1, \ldots, v_m \in \mathbb{C}^n$ are $\mathbb{C}$-linearly independent. Consider a restricted 1-jet $z_k^{(1)} = (u, z; v_1, \ldots, \hat{v}_k, \ldots, v_m)$. Clearly, the subset

$$\big\{v \in \mathbb{C}^n : (u, z; v_1, \ldots, v_{k-1}, v, v_{k+1}, \ldots, v_m) \in \Omega\big\}$$

is either empty or the complement of a complex subspace of $\mathbb{C}^n$; in the latter cases, Ω is connected and its convex hull equals $\mathbb{C}^n$. □

We introduce the following terminology. Assume that $E \to M$ is a real vector bundle and $F \to X$ is a complex vector bundle. A vector bundle map $\Phi\colon E \to F$ (over some base map $f\colon M \to X$) is said to be *totally real* if for every $p \in M$ the fibre map $\Phi_p\colon E_p \to F_{f(p)}$ is injective and the image $\Phi_p(E_p) \subset F_{f(p)}$ is a totally real subspace of the complex vector space $F_{f(p)}$. Equivalently, the complexified map $\Phi^{\mathbb{C}}\colon E \otimes_{\mathbb{R}} \mathbb{C} \to F$ is a complex vector bundle injection. Let

$$\mathrm{Hom}_{\mathrm{tr}}(E, F)$$

denote the set of all totally real vector bundle maps $E \to F$. Thus, a $\mathcal{C}^1$ map $f\colon M \to X$ is a totally real immersion if and only if its tangent map $\mathrm{T}f\colon \mathrm{T}M \to \mathrm{T}X$ is a totally real vector bundle map.

The following result is a corollary to Theorem 9.1.4.

Theorem 9.1.6 *Let M be a smooth manifold and X be a complex manifold.*

(a) *For every $\Phi \in \mathrm{Hom}_{\mathrm{tr}}(\mathrm{T}M, \mathrm{T}X)$ there exists a totally real immersion $f\colon M \to X$ whose tangent map $\mathrm{T}f$ is homotopic to Φ in $\mathrm{Hom}_{\mathrm{tr}}(\mathrm{T}M, \mathrm{T}X)$.*
(b) *If $f_0, f_1\colon M \to X$ are totally real immersions whose tangent maps are homotopic in $\mathrm{Hom}_{\mathrm{tr}}(\mathrm{T}M, \mathrm{T}X)$, then f_0 and f_1 are regularly homotopic through totally real immersions.*

Let us now combine Theorem 9.1.6 with the h-principle for ordinary immersions. Denote by $\mathrm{Hom}_{\mathrm{inj}}(E, F)$ the set of all fibrewise injective vector bundle maps $E \to F$. Consider the open differential relation

$$\Omega_0 = \big\{(p, x, \lambda) : p \in M,\ x \in X,\ \lambda \in \mathrm{Hom}_{\mathrm{inj}}(\mathrm{T}_p M, \mathrm{T}_x X)\big\} \subset \mathrm{J}^1(M, X).$$

Clearly, a $\mathcal{C}^1$ map $f\colon M \to X$ is an immersion if and only if $\mathrm{j}^1 f\colon M \to \mathrm{J}^1(M, X)$ is a section of Ω_0. If $\dim_{\mathbb{R}} M = m$ and $\dim_{\mathbb{R}} X = N$, then the fibre of the projection $\Omega_0 \to M \times X$ is the Stiefel manifold $V^{\mathbb{R}}_{m,N}$ of real m-frames in $\mathbb{R}^N$ (see Example 1.2.7). It is easily seen that Ω_0 is ample (see Definition 1.10.4) if and only if $m < N$. Indeed, the set of vectors $v \in \mathbb{R}^N$ that complete a restricted 1-jet $z^{(1)}_k$ to a 1-jet $z^{(1)} \in \Omega_0$ is either empty or the complement of a real subspace of codimension $N - m + 1$ in $\mathbb{R}^N$. Theorem 1.10.5 applied to Ω_0 yields the Hirsch-Smale h-principle for immersions when $\dim M < \dim X$ [295, 518]. If $\dim M = \dim X = n$ and M is open and connected, we triangulate M and retract it by a diffeotopy into a neighborhood of its $(n-1)$-skeleton, so the h-principle for immersions $M \to X$ still holds.

Corollary 9.1.7 *Let M be a smooth manifold and X be a complex manifold.*

(a) *An immersion $f\colon M \to X$ is regularly homotopic to a totally real immersion if and only if its tangent map $\mathrm{T}f\colon \mathrm{T}M \to \mathrm{T}X$ is homotopic in $\mathrm{Hom}_{\mathrm{inj}}(\mathrm{T}M, \mathrm{T}X)$ to a totally real vector bundle map $\mathrm{T}M \to \mathrm{T}X$.*

(b) *A regular homotopy of immersions* $f_t\colon M \to X$ $(t \in [0,1])$ *such that* f_0 *and* f_1 *are totally real can be deformed (rel.* $\{0,1\}$*) to a regular homotopy of totally real immersions if and only if the path* $\mathrm{T}f_t \in \mathrm{Hom}_{\mathrm{inj}}(\mathrm{T}M, \mathrm{T}X)$ *can be deformed (rel.* $\{0,1\}$*) to a path in* $\mathrm{Hom}_{\mathrm{tr}}(\mathrm{T}M, \mathrm{T}X)$.

Proof The latter condition in (a) is clearly necessary. Conversely, assume that $\tau_t \in \mathrm{Hom}_{\mathrm{inj}}(\mathrm{T}M, \mathrm{T}X)$ $(t \in [0,1])$ is a homotopy with $\tau_0 = \mathrm{T}f$ and $\tau_1 \in \mathrm{Hom}_{\mathrm{tr}}(\mathrm{T}M, \mathrm{T}X)$. By Theorem 9.1.6(i) there is a totally real immersion $f_1\colon M \to X$ such that $\mathrm{T}f_1$ is homotopic in $\mathrm{Hom}_{\mathrm{tr}}(\mathrm{T}M, \mathrm{T}X)$ to τ_1. Coupling this homotopy from $\mathrm{T}f_1$ to τ_1 with the homotopy $\{\tau_t\}_{t\in[0,1]}$ gives a path in $\mathrm{Hom}_{\mathrm{inj}}(\mathrm{T}M, \mathrm{T}X)$ connecting $\mathrm{T}f_0$ to $\mathrm{T}f_1$. By the h-principle for immersions $M \to X$, we infer that f_0 and f_1 are regularly homotopic through immersions. A similar argument proves part (b). □

Consider now totally real immersions to $\mathbb{C}^n$. A totally real vector bundle map $\Phi\colon \mathrm{T}M \to \mathrm{T}\mathbb{C}^n$ over a map $f\colon M \to \mathbb{C}^n$ induces a complex vector bundle injection $\Phi^{\mathbb{C}}\colon \mathbb{C}\mathrm{T}M \hookrightarrow f^*\mathrm{T}\mathbb{C}^n \cong M \times \mathbb{C}^n$. If $\dim_{\mathbb{R}} M = n$, then this map is a complex vector bundle isomorphism $\mathbb{C}\mathrm{T}M \xrightarrow{\cong} M \times \mathbb{C}^n$.

Proposition 9.1.8 *Let* M *be a smooth manifold.*

(a) *Regular homotopy classes of totally real immersions* $M \to \mathbb{C}^n$ *are in one-to-one correspondence with the homotopy classes of injective complex vector bundle maps* $\mathbb{C}\mathrm{T}M \hookrightarrow M \times \mathbb{C}^n$.
(b) M *admits a totally real immersion into* $\mathbb{C}^n$ *with* $n = \dim_{\mathbb{R}} M$ *if and only if the complexified tangent bundle* $\mathbb{C}\mathrm{T}M$ *is trivial. If this holds then the regular homotopy classes of totally real immersions* $M \to X$ *are in bijective correspondence with the homotopy classes of maps* $M \to GL_n(\mathbb{C})$*, that is, with the elements of the complex* K*-group* $K^1(M)$.
(c) *A compact smooth orientable manifold* M^n *which admits a totally real embedding in* $\mathbb{C}^n$ *has Euler number* $\chi(M) = 0$.

Proof Part (a) follows from Corollary 9.1.7, and (b) is a special case of (a). The last statement in (b) follows by observing that any two complex vector bundle trivializations $\mathbb{C}\mathrm{T}M \xrightarrow{\cong} M \times \mathbb{C}^n$ differ by a map $M \to GL_n(\mathbb{C})$.

Part (c), due to Wells [567], is seen as follows. Given a totally real immersion $f\colon M^n \to X^n$, the complex structure operator J on X induces an isomorphism between the tangent bundle $\mathrm{T}M$ and the normal bundle N_f of the immersion, so we have $\mathbb{C}\mathrm{T}M \cong \mathrm{T}M \oplus \mathrm{T}M \cong \mathrm{T}M \oplus \mathrm{N}_f \cong f^*\mathrm{T}X$. Hence, the Euler number $\chi(M)$ equals the normal Euler number $\chi(\mathrm{N}_f)$ of the immersion. If f is an embedding, then the normal Euler number is the self-intersection number of $f(M)$ in X. When $X = \mathbb{C}^n$, this number is zero since we can move the image $f(M)$ off itself; hence $\chi(M) = 0$. □

An especially interesting subclass of totally real immersions are *Lagrangian immersions*. Let ω be a *symplectic form* on a smooth manifold X^{2n}, that is, a closed

2-form with $\omega^n \neq 0$. (A Kähler form (p. 31) on a complex manifold is also a symplectic form.) An immersion $f\colon M \to X$ of a real manifold M into X is *ω-isotropic* if $f^*\omega = 0$; this requires that $\dim_{\mathbb{R}} M \le n$. If $\dim M = n$ then an isotropic immersion $M \to X$ is called an *ω-Lagrangian immersion*. The standard symplectic (Kähler) form on $\mathbb{R}^{2n} = \mathbb{C}^n$ is

$$\omega = \sum_{j=1}^{n} \mathrm{d}x_j \wedge \mathrm{d}y_j = \mathrm{d}\alpha, \quad \alpha = \sum_{j=1}^{n} x_j\, \mathrm{d}y_j.$$

If $f\colon M^n \to \mathbb{C}^n$ is a Lagrangian immersion, then $0 = f^*\omega = f^*\mathrm{d}\alpha = \mathrm{d}f^*\alpha$, so $f^*\alpha$ is a closed 1-form on M. The immersion f is called *exact Lagrangian* if $f^*\alpha$ is an *exact* 1-form, $f^*\alpha = \mathrm{d}\beta$ for some $\beta\colon M \to \mathbb{R}$. We have the following result (see Lees [381] and Gromov [265, pp. 60–61]).

Theorem 9.1.9 *Let M be a smooth compact manifold of real dimension n, and let ω be the standard symplectic form on $\mathbb{C}^n$.*

(a) *A totally real immersion $f\colon M \to \mathbb{C}^n$ with $[f^*\omega] = 0 \in H^2(M;\mathbb{R})$ is regularly homotopic through totally real immersions to a Lagrangian immersion.*
(b) *If M is stably parallelizable (i.e., $\mathrm{T}M \oplus \mathbb{R} \cong M \times \mathbb{R}^{n+1}$ is a trivial bundle), then M admits a smooth exact Lagrangian immersion in $\mathbb{C}^n$.*
(c) *The n-sphere S^n admits an exact smooth Lagrangian immersion $S^n \to \mathbb{C}^n$ for each $n \in \mathbb{N}$.*

Explicit examples of immersed Lagrangian spheres $S^n \to \mathbb{C}^n$ in part (c) were given by Weinstein [565, p. 26].

By a more refined version of Gromov's h-principle one also obtains the following h-principle for totally real embeddings. (This is a nontrivial addition only when $\dim_{\mathbb{R}} M = \dim_{\mathbb{C}} X$.) For real surfaces in complex surfaces, this is also seen from Corollary 10.5.1 on p. 492.

Theorem 9.1.10 ([265, p. 192], [180]) *Let $f_0\colon M \hookrightarrow X$ be a smooth embedding into a complex manifold X. If the tangent map $\mathrm{T}f_0\colon \mathrm{T}M \to \mathrm{T}X$ is homotopic in $\mathrm{Hom}_{\mathrm{inj}}(\mathrm{T}M, \mathrm{T}X)$ to a totally real map $\tau_1 \in \mathrm{Hom}_{\mathrm{tr}}(\mathrm{T}M, \mathrm{T}X)$, then f_0 is isotopic to a totally real embedding $f_1\colon M \hookrightarrow X$.*

Results on classification of totally real immersions and embeddings can be found in Gromov's monograph [265] and in the papers [39, 118, 180, 513, 536], among others. The paper [118] by Derdzinski and Januszkiewicz contains a good set of references. Here we mention without proofs some of the main results on totally real immersions and embeddings of compact n-manifolds into $\mathbb{C}^n$, with emphasis on spheres.

By part (c) in Theorem 9.1.9, the sphere S^n admits a totally real immersion in $\mathbb{C}^n$ for each $n \in \mathbb{N}$. The question arises for which values of n does there exist a totally real embedding $S^n \hookrightarrow \mathbb{C}^n$. According to Kervaire [335], every embedding of the

n-sphere into $\mathbb{C}^n$ has trivial normal bundle. If the embedding is totally real, then the normal bundle is isomorphic to the tangent bundle TS^n and hence the latter must be trivial; this holds precisely when $n \in \{1, 3, 7\}$ (see Bott and Milnor [64], Kervaire [334], and Adams and Atiyah [7]). For $n = 1$ the standard inclusion $S^1 \hookrightarrow \mathbb{C}$ is totally real. Explicit totally real embeddings $S^3 \hookrightarrow \mathbb{C}^3$ were found by Ahern and Rudin [11]. (For generalizations to certain quotients of S^3 see [181].) The 7-sphere does not admit a totally real embedding into $\mathbb{C}^7$ (see Gromov [265, p. 193] and Stout and Zame [536]). In fact, we have the following more precise result.

Theorem 9.1.11 ([536]) *The 7-dimensional Stein manifold*

$$\Sigma^7 = \left\{ (z_1, \ldots, z_8) \in \mathbb{C}^8 : \sum_{j=1}^{8} z_j^2 = 1 \right\}$$

(a complexification of the real 7-sphere S^7) is real analytically equivalent to a domain in $\mathbb{C}^7$, but is not holomorphically equivalent to a domain in $\mathbb{C}^7$. In particular, S^7 does not admit a totally real embedding in $\mathbb{C}^7$.

Summarizing the above discussion, we have the following result.

Corollary 9.1.12 *The n-sphere S^n admits a totally real embedding in $\mathbb{C}^n$ if and only if $n = 1$ or $n = 3$.*

Further results concerning totally real embeddings $S^k \hookrightarrow \mathbb{C}^n$ of spheres for $k < n$ were obtained by Gong [249].

One also knows that every closed orientable 3-manifold is parallelizable (Stiefel's theorem). Furthermore, every open connected orientable 3-manifold admits an immersion into $\mathbb{R}^3$ and therefore is parallelizable, which was proved by J.H.C. Whitehead [570]. Hence every orientable 3-manifold admits a totally real immersion into $\mathbb{C}^3$. We have the following more precise result.

Theorem 9.1.13 ([180, Theorem 1.4]) *Every immersion of a compact smooth orientable 3-manifold M into $\mathbb{C}^3$ is regularly homotopic to a totally real immersion. Furthermore, such M admits a totally real embedding in $\mathbb{C}^3$.*

The last statement follows from Theorem 9.1.10.

9.2 (Almost) Proper Maps to Euclidean Spaces

This section contains some preparations for the proof of the proper embedding and immersion theorems for Stein manifolds.

Let X and Y be locally compact topological spaces. Recall that a continuous map $f : X \to Y$ is *proper* if for every compact set K in Y the inverse image $f^{-1}(K)$ is compact, and is *almost proper* if each connected component of $f^{-1}(K)$ is compact. A proper map with finite fibres is called a *finite map*. If X is Stein then any proper holomorphic map $f\colon X\to Y$ is finite, and any almost proper holomorphic map has discrete fibres.

Recall that a subset of a Fréchet space is *residual* if it is the intersection of at most countably many dense open subsets; such a set is everywhere dense. Given a map $f\colon X\to Y$ we write

$$\Delta_f^2=\big\{(x_1,x_2)\in X\times X: x_1\neq x_2,\ f(x_1)=f(x_2)\big\}.$$

Theorem 9.2.1 ([58], [274, Theorem 2, p. 220]) *Let X be a Stein manifold of dimension n. There is a residual set $\mathcal{A}\subset\mathcal{O}(X)^n$ such that for every $f\in\mathcal{A}$ the map $f=(f_1,\ldots,f_n)\colon X\to\mathbb{C}^n$ is almost proper and satisfies*

$$\dim\Delta_f^2\le n,\qquad \dim\{x\in X:\dim\ker \mathrm{d}f_x\ge i\}\le n-i^2\quad (i\in\mathbb{N}). \tag{9.3}$$

We have seen in Sect. 8.9 that the sets in (9.3) are complex subvarieties of X.

Proof An *analytic polyhedron* in a complex manifold X is a compact set of the form $P=\{x\in U: |f_k(x)|\le 1,\ k=1,\ldots,r\}$, where f_k are holomorphic functions in an open set U containing P. Since $f=(f_1,\ldots,f_r)$ maps the interior of P properly onto the polydisc $\mathbb{D}^r\subset\mathbb{C}^r$, we have $r\ge n=\dim X$. An analytic polyhedron is said to be *special* if it is defined by $n=\dim X$ functions. Bishop proved that every analytic polyhedron can be approximated from the outside by special analytic polyhedra (see [58, p. 691] or [274, Corollary 4, p. 218]). Hence every Stein manifold X can be exhausted by an increasing sequence $P_1\subset P_2\subset\cdots\subset\bigcup_{j=1}^\infty P_j=X$ of special analytic polyhedra

$$P_j=\big\{x\in U_j:\ \big|f_{j,k}(x)\big|\le 1,\ k=1,\ldots,n\big\},$$

where $f_{j,k}\in\mathcal{O}(X)$ for all $j\in\mathbb{N}$ and $k=1,\ldots,n$.

Let $h\in\mathcal{O}(X)^n$, and let K be a compact set in X. Choose $j_0\in\mathbb{N}$ big enough such that K is contained in the interior of P_{j_0}, and pick an increasing sequence $\alpha_1<\alpha_2<\cdots$ with $\lim_{j\to\infty}\alpha_j=+\infty$. Given $\epsilon>0$, we can find a holomorphic map $f=(f_1,\ldots,f_n)\colon X\to\mathbb{C}^n$ of the form

$$f_k(x)=\sum_{j\ge j_0}a_j\big(f_{j,k}(x)\big)^{t_j}+h_k(x),\quad k=1,\ldots,n$$

such that $|f-h|<\epsilon$ on K and $|f|>\alpha_j$ on bP_j for all $j\ge j_0$. These conditions are easily achieved by an inductive choice of the constants $a_j>0$ and exponents $t_j\in\mathbb{N}$. (See [274, Theorem 2, p. 220] for the details.)

For every $j\in\mathbb{N}$ let $\mathcal{H}(j)$ denote the set of all $f\in\mathcal{O}(X)^n$ such that there exist j different integers $k_1,\ldots,k_j\in\mathbb{N}$ with the property that $|f(x)|>k_i$ for all $x\in bP_{k_i}$ and $i=1,\ldots,j$. Clearly, $\mathcal{H}(j)$ is open in $\mathcal{O}(X)^n$ and, by the first part of the proof, it is also dense. Hence, $\mathcal{H}=\bigcap_{j=1}^\infty\mathcal{H}(j)$ is residual in $\mathcal{O}(X)^n$. Every $f\in\mathcal{H}$

determines an almost proper map $f: X \to \mathbb{C}^n$ since we get an infinite sequence of integers $k \in \mathbb{N}$ such that $|f(x)| > k$ holds for all $x \in bP_k$.

By Theorem 8.9.1 on p. 387 there is a residual set $\mathcal{H}' \subset \mathcal{O}(X)^n$ such that every $f \in \mathcal{H}'$ satisfies (9.3). The set $\mathcal{H} \cap \mathcal{H}'$ is still residual in $\mathcal{O}(X)^n$, and every $f \in \mathcal{H} \cap \mathcal{H}'$ satisfies Theorem 9.2.1. □

Assume that X is an n-dimensional Stein manifold and $f: X \to \mathbb{C}^n$ is an almost proper holomorphic map. Then, there is a dense set of functions $f_{n+1} \in \mathcal{O}(X)$ such that $(f, f_{n+1}): X \to \mathbb{C}^{n+1}$ is proper [274, Theorem 4, p. 221]. (It suffices to choose f_{n+1} to be big enough on the compact connected components of any sublevel set $\{|f| \le k\}$, $k \in \mathbb{N}$.) An argument similar to the one in the proof of Theorem 9.2.1 gives the following result.

Theorem 9.2.2 *For every n-dimensional Stein manifold X there is a dense subset $\mathcal{P} \subset \mathcal{O}(X)^{n+1}$ such that every $f \in \mathcal{P}$ is a proper holomorphic map $f: X \to \mathbb{C}^{n+1}$ satisfying* $\dim \Delta_f^2 \le n-1$ *and*

$$\dim\{x \in X : \dim \ker \mathrm{d} f_x \ge i\} \le n - i(i+1), \quad i = 1, 2 \ldots \tag{9.4}$$

Variants of Theorems 9.2.1 and 9.2.2, with interpolation on a discrete set in X, were proved by Prezelj in [459, Sect. 2]. A different construction of proper holomorphic maps $X^n \to \mathbb{C}^{n+1}$ which does not depend on Theorem 9.2.1 was given by Ritter and the author in [220]; see Theorem 9.8.8 on p. 441.

Assume now that $(X, \mathcal{O}_X)$ is a (not necessarily reduced) Stein space of dimension n and of finite embedding dimension:

$$n_0 = \mathrm{embdim} X = \sup_{x \in X} \mathrm{embdim}_x X < \infty. \tag{9.5}$$

According to Theorem 2.4.1(c) (p. 52), such X is biholomorphic to a closed complex subspace of $\mathbb{C}^m$ with $m = \max\{n + n_0, 2n + 1\}$. We consider the more general situation when X is a closed complex subspace of an m-dimensional Stein manifold $\widetilde{X}$. Choose a stratification

$$X = X_0 \supset X_1 \supset X_2 \supset \cdots \supset X_l = \varnothing$$

by closed complex subvarieties of X such that $\mathrm{embdim}_x X$ is constant on every (smooth) stratum $X_r^* = X_r \setminus X_{r+1}$. After refining the stratification, we may assume that it is a Whitney stratification (see Theorem 8.8.1, p. 379) and hence the jet transversality theorem applies (see Theorem 8.8.5, p. 380). For every $k = 0, 1, \ldots, n_0 = \mathrm{embdim} X$ let $n(k)$ denote the number

$$n(k) = \dim\{x \in X : \mathrm{embdim}_x X = k\} \le k. \tag{9.6}$$

Note that $n(k)$ is the maximal complex dimension of the strata Z in the above stratification of X such that $\mathrm{embdim}_x X = k$ for all $x \in Z$. We also introduce the *Schür-*

mann invariants (see [492]):

$$b_X = \max\left\{k + \left[\frac{n(k)}{2}\right] : k = 0, 1, \ldots, n_0\right\}, \tag{9.7}$$

$$N(n) = n + q(n) = \max\left\{\left[\frac{3n}{2}\right] + 1, b_X, 3\right\}. \tag{9.8}$$

Note that the number b_X (9.7) is denoted by $b'(X)$ in [492, (0.2)].

Let $\mathcal{J}_X$ denote the sheaf of ideals of the subvariety X in $\widetilde{X}$. Then, $\mathcal{J}(X) = \Gamma(\widetilde{X}, \mathcal{J}_X)$ is a closed subspace of $\mathcal{O}(X)$ and the set

$$\mathcal{H} = \mathcal{O}(\widetilde{X})^n \times \mathcal{J}(X)^{m-n} \subset \mathcal{O}(\widetilde{X})^m \tag{9.9}$$

is a Fréchet space.

For a holomorphic map between complex spaces, $\mathrm{d}f_x : \mathrm{T}_x X \to \mathrm{T}_{f(x)} Y$ denotes its differential on the Zariski tangent spaces (see Sect. 1.3).

Theorem 9.2.3 ([492, Theorem 1.1]) *If X is a closed n-dimensional complex subspace of an m-dimensional Stein manifold $\widetilde{X}$, then the set of all $H = (h, h') \in \mathcal{H}$ (9.9) satisfying the following two conditions is residual in $\mathcal{H}$:*

(i) *$H : \widetilde{X} \to \mathbb{C}^m$ is almost proper (and hence so is $h|_X : X \to \mathbb{C}^n$), and*
(ii) *$\dim \Delta_h^2 \le n$ and $\dim\{x \in X : \dim \ker(\mathrm{d}h|_X)_x \ge i\} < 2(q(n) - i + 1)$ for $i = 1, \ldots, n_0$, where $q(n) = N(n) - n$ is defined by* (9.8).

Proof The proof of (i) is similar to the first part of proof of Theorem 9.2.1; the main addition is that we can exhaust the manifold $\widetilde{X}$ by special analytic polyhedra defined by maps from $\mathcal{H}$. (See [58] or [274] for the details.)

We now turn to the proof of (ii). As in Sect. 8.9, we consider the varieties

$$\Sigma_h^i = \left\{x \in X : \dim \ker(\mathrm{d}h|_X)_x = i\right\}, \qquad \Lambda_h^i = \bigcup_{j \ge i} \Sigma_h^j.$$

As before, let $X = X_0 \supset X_1 \supset \cdots \supset X_l = \varnothing$ be a stratification by closed complex subvarieties such that $\mathrm{embdim}_x X$ is constant on every difference $X_r^* = X_r \setminus X_{r+1}$. Assume that $Z \subset X_r^*$ is a stratum with $\dim \mathrm{T}_x X = k$ for all $x \in Z$; then $\dim Z \le n(k) \le k$ by (9.6). By Theorem 8.9.1 (p. 387) there is a residual set of holomorphic maps $h : \widetilde{X} \to \mathbb{C}^n$ that satisfy

$$\dim \Lambda_h^i \cap Z \le n(k) - i(n + i - k) \tag{9.10}$$

for $k = 0, \ldots, n_0$ and $\max\{1, k - n\} \le i \le k$. (This estimate is trivial for the other values of i.) From the definition of $q(n)$ (9.8) it follows by an elementary computation (see [492, Appendix A]) that the inequality

$$n(k) - i(n + i - k) < 2\big(q(n) - i + 1\big)$$

holds for the indicated ranges of i and k. Since a generic map $h\colon \widetilde{X} \to \mathbb{C}^n$ satisfies condition (9.10) on every stratum Z in a given stratification of X, there is a residual set of maps $h \in \mathcal{O}(\widetilde{X})^n$ that satisfy Theorem 9.2.3(ii). The product of this set with $\mathcal{J}(X)^{m-n}$ is clearly residual in $\mathcal{H}$. □

9.3 Embedding and Immersing Stein Manifolds into Euclidean Spaces of Minimal Dimension

In this and the following two sections we prove results on the existence of proper holomorphic immersions and embeddings of Stein manifolds of dimension $n > 1$ to Euclidean spaces of minimal dimension. These constructions depend in an essential way on the Oka principle from Chaps. 5 and 6.

The following theorem was proved by Eliashberg and Gromov [150]; the improvement of the embedding dimension by one for odd values of $n = \dim X$ is due to Schürmann [492]. A slightly weaker version of this theorem was announced by Eliashberg and Gromov [269] in 1971; however, some of the key analytic ingredients in the proof were not available at that time.

Theorem 9.3.1 ([150, 492]) *Every Stein manifold X of dimension n immerses properly holomorphically into $\mathbb{C}^M$ with $M = [\frac{3n+1}{2}]$, and if $n > 1$ then X embeds properly holomorphically into $\mathbb{C}^N$ with $N = [\frac{3n}{2}] + 1$.*

The proof for embeddings fails in dimension $n = 1$, that is, for open Riemann surfaces, due to a hyperbolicity obstruction. By the Bishop-Narasimhan theorem (see Theorem 2.4.1, p. 52), every open Riemann surface admits a proper holomorphic embedding in $\mathbb{C}^3$, but only a few are known to embed in $\mathbb{C}^2$. See Sects. 9.10–9.11 for a discussion of this classical problem.

Theorem 9.3.1 was conjectured by Forster in 1970 (see [171, p. 183]). He gave the following example which shows that the proper embedding dimension $N = [\frac{3n}{2}] + 1$ is optimal for every $n > 1$, and the proper immersion dimension $M = [\frac{3n+1}{2}]$ is optimal if n is even, while for n odd it may be off by at most one. As we shall see, the underlying reasons are purely topological. Deleting the properness conditions, Theorem 9.6.3 says that every n-dimensional Stein manifold admits a holomorphic immersion to $\mathbb{C}^{[3n/2]}$, and this result is optimal in view of Theorem 9.1.3.

Example 9.3.2 Let Y be the Stein surface

$$Y = \{[x:y:z] \in \mathbb{P}^2 : x^2 + y^2 + z^2 \neq 0\}.$$

Given an integer $n \geq 2$, let X^n denote the n-dimensional Stein manifold

$$X^n = \begin{cases} Y^m, & \text{if } n = 2m; \\ Y^m \times \mathbb{C}, & \text{if } n = 2m+1. \end{cases} \tag{9.11}$$

Proposition 9.3.3 ([171, Proposition 3, p. 183]) *For each $n \geq 2$ the Stein manifold X^n (9.11) does not admit a proper holomorphic embedding into $\mathbb{C}^{[3n/2]}$, and it does not admit a holomorphic immersion into $\mathbb{C}^{[3n/2]-1}$.*

Proof We begin by showing that the Stein surface Y is not parallelizable.

It is easily seen that the smooth totally real embedded submanifold M of Y, consisting of all points $[x : y : z] \in \mathbb{P}^2$ with $x, y, z \in \mathbb{R}$, is a deformation retract of Y [169, p. 714]. Note that M is the real projective plane $\mathbb{RP}^2$. The total Stiefel-Whitney class of $\mathbb{RP}^2$ equals $(1+\alpha)^3$, where $\alpha \in H^1(\mathbb{RP}^2; \mathbb{Z}_2) \cong \mathbb{Z}_2$ is the generator of the cohomology ring $H^*(\mathbb{RP}^2; \mathbb{Z}_2)$. Since $TY|_M \cong TM \oplus TM$, the Stiefel-Whitney class $w(TY) = w(TY|_M)$ equals

$$w(TM \oplus TM) = (1+\alpha)^6 = 1 + \alpha^2,$$

so $w_2(TY) = \alpha^2 \neq 0$. Since w_2 is the mod 2 reduction of the first Chern class, we see that $c_1(TY)$ is the nonzero element of $H^2(Y; \mathbb{Z}) \cong H^2(\mathbb{RP}^2; \mathbb{Z}) \cong \mathbb{Z}_2$.

This proves that Y is not parallelizable. Hence, Y does not admit a proper holomorphic embedding into $\mathbb{C}^3$ (since every smooth complex hypersurface in $\mathbb{C}^n$ is parallelizable by Proposition 8.5.3), and it does not admit a holomorphic immersion to $\mathbb{C}^2$ (its differential would induce a trivialization of the tangent bundle TY). This proves the proposition when $n = 2$.

Assume now that $n > 2$ and let $X = X^n$ be the manifold (9.11). A calculation shows that the dual Stiefel-Whitney class $\widetilde{w}_{2m}(TX)$ is the nonzero element of the group $H^{2m}(X; \mathbb{Z}_2) = H^{2m}((\mathbb{RP}^2)^m; \mathbb{Z}_2) = \mathbb{Z}_2$, and hence the dual Chern class $\tilde{c}_m(TX)$ is the nonzero element of $H^{2m}(X; \mathbb{Z}) = \mathbb{Z}_2$ (see [169, p. 715]). It follows that X does not embed properly holomorphically into $\mathbb{C}^{[3n/2]}$, and it does not immerse holomorphically into $\mathbb{C}^{[3n/2]-1}$ (see [305, Theorem 10.2] or [418, Corollary 11.4, p. 120]). □

On the other hand, by the Eliashberg-Gromov Oka principle for holomorphic immersions (see Theorem 9.6.3 on p. 430), every n-dimensional Stein manifold admits a (not necessarily proper) holomorphic immersion into $\mathbb{C}^{[3n/2]}$. (For $n = 1$ this is due to Gunning and Narasimhan [273].) Hence, Theorem 9.3.1 gives the optimal proper immersion dimension if n is even (in this case $[\frac{3n+1}{2}] = \frac{3n}{2}$), but there are two possibilities if n is odd. □

The topological obstruction in Example 9.3.2 is the nonvanishing of the dual Chern class $\tilde{c}_m(TX)$ for $m = [\frac{n}{2}]$. For Stein manifolds with trivial Chern classes one may expect that they embed in lower dimension. In particular, the following is a reasonable problem.

Problem 9.3.4 Let X be a Stein manifold of dimension n with trivial tangent bundle. Does there exist a proper holomorphic embedding $X \hookrightarrow \mathbb{C}^{n+1}$? Is this true for strongly pseudoconvex domains in $\mathbb{C}^n$?

Little is known about this problem when $n > 1$. A result in this direction, due to Globevnik [237], is that there exist arbitrarily small perturbations of the polydisc in $\mathbb{C}^n$ which embed properly holomorphically in $\mathbb{C}^{n+1}$. Globevnik constructed a sequence of compositions of shear automorphisms of $\mathbb{C}^{n+1}$ in coordinate directions whose domain of convergence (see Corollary 4.4.2 on p. 125) is a Fatou-Bieberbach domain $\Omega \subset \mathbb{C}^{n+1}$ such that one of the connected components of $\Omega \cap (\mathbb{C}^n \times \{0\})$, call it X, is a domain as close as desired to the polydisc $\mathbb{D}^n$. Since Ω is biholomorphic to $\mathbb{C}^{n+1}$, this gives a proper holomorphic embedding $X \hookrightarrow \mathbb{C}^{n+1}$.

A little more is known about Problem 9.3.4 when $n = 1$. For example, a special case of the results of Forstnerič and Wold [228] is that every planar domain $X \subset \mathbb{C}$ with at most countably many boundary components, at most finitely many of which are points, admits a proper holomorphic embedding into $\mathbb{C}^2$. (See Sects. 9.10–9.11 and in particular Theorem 9.11.4.)

Every real analytic manifold M of dimension n embeds as a closed real analytic and totally real submanifold of an n-dimensional Stein manifold (Grauert [255]). Hence, Theorem 9.3.1 implies the following result.

Corollary 9.3.5 *Every real analytic manifold M of dimension n admits a proper real analytic embedding onto a totally real submanifold of $\mathbb{C}^{[3n/2]+1}$.*

It does not seem to be known what is the minimal dimension N for the existence of proper real analytic embeddings $M \hookrightarrow \mathbb{R}^N$ of all real analytic manifolds M of a given dimension n.

Schürmann also proved the following embedding theorem for Stein spaces of finite embedding dimension. This includes Theorem 9.3.1 since $b_X = [\frac{3n}{2}]$ when X is a Stein manifold of dimension n.

Theorem 9.3.6 ([492, Theorem 0.5]) *An n-dimensional Stein space X with $n_0 = \mathrm{embdim} X < +\infty$ (9.5) embeds properly holomorphically into $\mathbb{C}^N$ with $N = \max\{[\frac{3n}{2}] + 1, b_X, 3\}$, where $b_X = \max\{k + [\frac{n(k)}{2}] : k = 1, \ldots, n_0\}$ and $n(k) = \dim\{x \in X : \mathrm{embdim}_x X \geq k\}$. Furthermore, such X immerses properly holomorphically to $\mathbb{C}^M$ with $M = \max\{[\frac{3n+1}{2}], b_X\}$.*

In this theorem, the space X needs not be reduced since one can always pass to the reduction. If X does not have bounded embedding dimension, then Theorem 2.4.1(b) still seems the best known result. However, for Stein spaces with isolated singular points, see Theorem 9.5.3 (p. 427) due to Prezelj.

Let $h\colon X \to Y$ be a holomorphic map. We say that a holomorphic map $f\colon X \to \mathbb{C}^q$ is a *relative embedding* (resp. a *relative immersion*) over h if $(h, f)\colon X \to Y \times \mathbb{C}^q$ is an embedding (resp. an immersion). Let

$$\Delta_h^2 = \big\{(x_1, x_2) \in X \times X : x_1 \neq x_2,\ h(x_1) = h(x_2)\big\},$$

$$\Sigma_h^i = \{x \in X : \dim \ker dh_x = i\}, \quad i = 1, 2, \ldots.$$

The main ingredient in the proof of Theorem 9.3.1 is the following *Relative Embedding Theorem* [150, 2.A., p. 126] proved in the next section.

Theorem 9.3.7 *Let X be a Stein space with* $\dim X = n$ *and* $\operatorname{embdim} X = n_0 < \infty$. *Assume that Y is a Stein manifold and $h\colon X \to Y$ is a proper holomorphic map satisfying the following conditions for some integer $q \ge 2$:*

(i) $\dim \Delta_h^2 < 2q$, *and*
(ii) $\dim \Sigma_h^i < 2(q-i+1)$ *for* $i = 1, 2, \ldots, n_0$.

Let K be a compact $\mathcal{O}(Y)$-convex set in Y, and assume that $h^{-1}(K) = L \cup Q$ where L and Q are compact and $L \cap Q = \varnothing$. Given a neighborhood $U \subset X$ of L and a relative embedding $f_0\colon U \to \mathbb{C}^q$ over h, there exists for every $\epsilon > 0$ a relative embedding $f\colon X \to \mathbb{C}^q$ over h such that $\operatorname{dist}(f(x), f_0(x)) < \epsilon$ *for all $x \in L$. If $q = 1$ then this holds if the set of triple points of h is 0-dimensional.*

If h satisfies only condition (ii) *but not necessarily* (i), *then there exists a relative immersion $f\colon X \to \mathbb{C}^q$ over h.*

Assuming Theorem 9.3.7, we now prove Theorems 9.3.1 and 9.3.6.

Proof of Theorem 9.3.6 Assume first that X is smooth, i.e., a Stein manifold of dimension $n \ge 2$. We will prove that X embeds properly holomorphically into $\mathbb{C}^N$ with $N = n + q$, where $q = [\frac{n}{2}] + 1$.

Let $h = (h_1, \ldots, h_n)\colon X \to \mathbb{C}^n$ be an almost proper map furnished by Theorem 9.2.1 (p. 412). It is easily seen that $n < 2q$ and $n - i^2 < 2(q - i + 1)$ for $i = 1, \ldots, n$, and hence h satisfies conditions (i) and (ii) in Theorem 9.3.7.

We begin by constructing an exhaustion $L_1 \subset L_2 \subset \cdots \subset \bigcup_{j=1}^\infty L_j = X$ by a sequence $L_j \subset \mathring{L}_{j+1}$ of compact $\mathcal{O}(X)$-convex sets to which we shall inductively apply Theorem 9.3.7. Let P be the closed unit polydisc in $\mathbb{C}^n$. Choose an increasing sequence $1 \le r_1 < r_2 < \cdots$ with $\lim_{j\to\infty} r_j = \infty$ such that every r_j is a regular value of the function $|h_i|\colon X \to \mathbb{R}_+$ for $i = 1, \ldots, n$. Let $K_j = r_j P$. Then, $h^{-1}(K_j) \subset X$ is a union of at most countably many compact analytic polyhedra (the compact connected components of $h^{-1}(\mathring{K}_j)$) which do not cluster on each other; in particular, each compact set in X intersects at most finitely many of them. Furthermore, $h^{-1}(K_j)$ is contained in the interior of $h^{-1}(K_{j+1})$ for every $j \in \mathbb{N}$. Let L_j be a union of finitely many connected components of $h^{-1}(K_j)$ such that $L_j \subset \mathring{L}_{j+1}$ for every $j \in \mathbb{N}$ and $\bigcup_{j=1}^\infty L_j = X$. Note that $h\colon \mathring{L}_j \to \mathring{K}_j$ is a proper holomorphic map for every $j \in \mathbb{N}$. The set $L'_j = h^{-1}(K_j) \setminus L_j$ is also a union of connected components of $h^{-1}(K_j)$ and is closed. Furthermore, every connected component of L'_j is either contained in L_{j+1} or else is disjoint from it. The intersection $Q_j = L'_j \cap L_{j+1}$ is compact and $\mathcal{O}(X)$-convex. Set $K_0 = \varnothing$, $L_0 = Q_0 = \varnothing$.

We now construct a sequence of holomorphic maps $f_j : U_j \to \mathbb{C}^q$ on open neighborhoods $U_j \supset L_j$ and a sequence of positive numbers $1 = \epsilon_0 > \epsilon_1 > \epsilon_2 > \cdots$ such that the following conditions hold for $j = 1, 2, \ldots$:

(a) the map $(h, f_j)\colon U_j \to \mathbb{C}^n \times \mathbb{C}^q = \mathbb{C}^N$ is an embedding on L_j,
(b) $|f_j(x) - f_{j-1}(x)| < 2^{-j}\epsilon_{j-1}$ for all $x \in L_{j-1}$,
(c) $|f_j| > j$ on Q_{j-1}, and

(d) for every holomorphic map $\widetilde{f}\colon U_j \to \mathbb{C}^q$ satisfying $|\widetilde{f} - f_j|_{L_j} < \epsilon_j$ the map $(h, \widetilde{f})$ is an embedding on L_{j-1}.

Choose a number $\epsilon = \epsilon_1 \in (0, 1)$ and let $f_1 = f$; hence (a) holds on L_1 while properties (b)–(d) are vacuous for $j = 1$. Assume inductively that we have found maps $f_1, \ldots, f_j$ and numbers $\epsilon_1, \ldots, \epsilon_j$ such that conditions (a)–(d) hold up to index j. Since Q_j is $\mathcal{O}(X)$-convex, Theorem 9.3.7 furnishes a holomorphic map $g_j\colon V_j \to \mathbb{C}^q$ on a neighborhood $V_j \supset Q_j$ such that (h, g_j) is an embedding on Q_j. By adding a big constant to the first component of g_j we ensure that $|g_j| > j + 2$ on Q_j. We extend f_j to a neighborhood of Q_j by setting $f_j = g_j$ there. Another application of Theorem 9.3.7 gives a holomorphic map $f_{j+1}\colon U_{j+1} \to \mathbb{C}^q$ on a neighborhood U_{j+1} of L_{j+1} such that (h, f_{j+1}) is an embedding on L_{j+1} and $|f_{j+1} - f_j| < 2^{-j-1}\epsilon_j$ on $L_j \cup Q_j$. It follows that conditions (a)–(c) hold for the index $j + 1$. Since embeddings are stable under perturbations, we can choose a number $\epsilon_{j+1} \in (0, \epsilon_j)$ such that (d) holds for $j + 1$. This completes the induction step.

By the construction, the limit $f = \lim_{j\to\infty} f_j\colon X \to \mathbb{C}^q$ exists and is holomorphic on X. Indeed, on L_j we have $f = f_j + \sum_{l=j}^{\infty}(f_{l+1} - f_l)$, and hence $|f - f_j| \le \sum_{l=j}^{\infty} 2^{-l-1}\epsilon_l < 2^{-j}\epsilon_j$ on L_j for $j = 1, 2, \ldots$. Condition (d) implies that (h, f) is an embedding on L_{j-1}. Since this holds for every j, f is a relative embedding over h. Finally, by (c) we have $|f| > j$ on Q_j for all $j \in \mathbb{N}$. It follows that $(h, f)^{-1}(K_j \times j\overline{\mathbb{B}}) \subset L_j$ for every j, and hence this set is compact. (Here, $\mathbb{B}$ is the unit ball in $\mathbb{C}^q$.) Since the sets $K_j \times j\overline{\mathbb{B}}$ exhaust $\mathbb{C}^N$, the map $(h, f)\colon X \to \mathbb{C}^N$ is proper.

This completes the proof of the proper embedding theorem if X is smooth.

Assume now that X is a closed n-dimensional complex subspace of $\mathbb{C}^{n_0}$. Let $H = (h, h')\colon \mathbb{C}^{n_0} \to \mathbb{C}^{n_0}$ be an almost proper map furnished by Theorem 9.2.3. In particular, h' vanishes on X and $h\colon X \to \mathbb{C}^n$ is an almost proper map satisfying $\dim\{x \in X : \dim\ker(dh|_X)_x \ge i\} < 2(q(n) - i + 1)$ for $i = 1, \ldots, n_0$, where $q(n) = N(n) - n$ is defined by (9.8). As before, we can see that h satisfies conditions (i) and (ii) in Theorem 9.3.7.

Let $L_1 \subset L_2 \subset \cdots \subset \bigcup_{j=1}^{\infty} L_j = \mathbb{C}^{n_0}$ be an exhaustion associated to the almost proper map $H\colon \mathbb{C}^{n_0} \to \mathbb{C}^{n_0}$ as explained above. Then the compact sets $L_j \cap X$ are $\mathcal{O}(X)$-convex and provide an exhaustion of X to which Theorem 9.3.7 can be applied by the same inductive scheme as in the special case.

Replacing N by $M = \max\{[\frac{3n+1}{2}], b_X\} = n + q$, condition (ii) in Theorem 9.3.7 still holds (since if follows from $n + q \le b_X$) and we get a proper immersion $(h, f)\colon X \to \mathbb{C}^M$. (See also Remark 9.4.5 on p. 425.) □

Remark 9.3.8 Eliashberg and Gromov [150] begin instead with a generic *proper* background map $h\colon X^n \to \mathbb{C}^{n+1}$; this gives the same result for even values of n, but their result is off by one for odd values of n. Almost proper maps were introduced to this proof by Schürmann [492]. □

9.4 Proof of the Relative Embedding Theorem

We now prove Theorem 9.3.7, following Eliashberg and Gromov [150].

We begin with a few reductions. We may assume that the proper holomorphic map $h\colon X \to Y$ is surjective, and hence $\dim Y = n$. (In the general case, we replace Y by the closed complex subvariety $f(X) \subset Y$; singularities do not present any difficulties in the proof.) When X is singular, we may assume that it is embedded as a closed n-dimensional reduced complex subspace of $\mathbb{C}^{n_0}$ and that h extends to a (non-proper) holomorphic map from a Stein open neighborhood of X in $\mathbb{C}^{n_0}$ to Y. Next, it suffices to prove the approximation result in the special case when $L = h^{-1}(K)$. Indeed, since $h^{-1}(K) = L \cup Q$ with L, Q compact and $L \cap Q = \varnothing$, both L and Q are $\mathcal{O}(X)$-convex. By localization of proper maps, h maps suitably chosen small open neighborhoods of Q in X properly onto small neighborhoods of K in Y. The absolute version of the theorem, applied to this restricted map, gives a relative embedding f_0' on a neighborhood of Q. By taking f_0 near L and f_0' near Q we get a relative embedding of a neighborhood $h^{-1}(K)$, and it suffices to prove the approximation statement for this new map.

We stratify X and Y so that the basic topological characteristics of the map $h\colon X \to Y$ are constant along each stratum in the following sense.

Definition 9.4.1 Let S be a locally closed connected complex submanifold of Y and set $\widetilde{S} = h^{-1}(S) \subset X$. We say that S is *equisingular* with respect to a proper holomorphic map $h\colon X \to Y$ if the following conditions hold:

(i) the map $h|_{\widetilde{S}}\colon \widetilde{S} \to S$ is a (necessarily proper) immersion, and
(ii) $\dim \ker dh_x$ is constant on every connected component of $\widetilde{S}$.

It follows from (i) that $h\colon \widetilde{S} \to S$ is a finitely sheeted holomorphic covering projection. Condition (ii) means that each connected component of $\widetilde{S}$ is contained in one of the sets $\Sigma_h^i \setminus \Sigma_h^{i+1}$, where $\Sigma_h^i = \{x \in X\colon \dim \ker dh_x = i\}$.

The proof of the following lemma is straightforward.

Lemma 9.4.2 ([150, Proposition 3.A$_2$], [557, Proposition 2.1]) *Every subvariety $Y_0 \subset Y$ admits an equisingular stratification, that is, a descending sequence of closed complex subvarieties $Y_0 \supset Y_1 \supset \cdots \supset Y_m$, where the strata (the connected components of differences $S_i = Y_i \setminus Y_{i+1}$) are smooth locally closed subvarieties of Y that are equisingular with respect to h.*

The proof of Theorem 9.3.7 consists of a repeated application of two basic operations: (A) separations of points over a stratum, and (B) elimination of the kernel of the differential over a stratum.

(A) *Separation of points over a stratum.* Let $Y_1 \subset Y_0$ be closed complex subvarieties of Y such that $S = Y_0 \setminus Y_1$ is a connected complex submanifold that is equisingular with respect to h. (In a typical situation, Y_0 and Y_1 are two subvarieties from an equisingular stratification of Y.) Let $X_j = h^{-1}(Y_j)$ for $j = 0, 1$; these are closed complex subvarieties of X such that $\widetilde{S} = X_0 \setminus X_1$ is smooth and $h\colon \widetilde{S} \to S$ is a finite holomorphic covering map. Under these assumptions we have the following lemma.

Lemma 9.4.3 ([150, Lemma 3.B$_1$]) *Let $K \subset Y$ and $L \subset X$ be as in Theorem 9.3.7. Assume that $f\colon X \to \mathbb{C}^q$ is a holomorphic map such that $(h, f)\colon X \to Y \times \mathbb{C}^q$ is an embedding on a neighborhood of $X_1 \cup L$. If $q \ge 2$ and $2q > \dim Y_0$, then for a suitably small neighborhood L' of L and for every $\epsilon > 0$ there exists a holomorphic map $f'\colon X \to \mathbb{C}^q$ such that*

(i) *$(h, f')\colon X \to Y \times \mathbb{C}^q$ is injective on $\widetilde{S}$,*
(ii) *$f - f'$ vanishes to the second order along the subvariety X_1, and*
(iii) *$|f(x) - f'(x)| < \epsilon$ for $x \in L'$.*

If $q = 1$ and $\dim Y_0 = 1$, then a map f' satisfying these conditions exists provided that h has no triple points over $S = Y_0 \setminus Y_1$.

Proof The assumptions imply that $f(x_1) \ne f(x_2)$ for any pair of distinct points $x_1, x_2 \in X$ such that $h(x_1) = h(x_2) \in Y_1 \cup K$, and $\mathrm{d}f_x \cdot v \ne 0$ for every vector $0 \ne v \in \ker \mathrm{d}h_x$, $x \in X_1 \cup L$. The same is then true if we replace K and L by a pair of slightly bigger $\mathcal{O}(Y)$-convex sets $K_1 \subset Y$, resp. $L_1 \subset X$.

We seek a map f' with the desired properties to be of the form

$$f'(x) = f_\alpha(x) = f(x) + \sum_{j=1}^{N} \alpha_j\big(h(x)\big)\psi_j(x), \quad x \in X, \tag{9.12}$$

where $\alpha_j \in \mathcal{O}(Y)$ are holomorphic functions on Y (to be determined), and $\psi_1, \ldots, \psi_N\colon X \to \mathbb{C}^q$ (with $N \ge dq$) are holomorphic maps chosen such that

(a) each ψ_j vanishes to the second order along the subvariety X_1, and
(b) for every point $y \subset S$ let $h^{-1}(y) = \{x_1, \ldots, x_d\} \in \widetilde{S}$ (distinct points); then the following $dq \times N$ matrix has maximal rank dq:

$$\begin{pmatrix} \psi_1(x_1) & \psi_2(x_1) & \ldots & \psi_N(x_1) \\ \psi_1(x_2) & \psi_2(x_2) & \ldots & \psi_N(x_2) \\ \cdots & \cdots & \cdots & \cdots \\ \psi_1(x_d) & \psi_2(x_d) & \ldots & \psi_N(x_d) \end{pmatrix}$$

The existence of such maps $\psi_1, \ldots, \psi_N$ follows from Cartan's Theorem A. It is trivial to achieve this rank condition at one point $y \in S$, and hence the same holds outside a proper subvariety. Adding more functions, we complete the proof by induction on the dimension of the exceptional set.

Property (b) of the collection $\{\psi_j\}$ implies that for every fixed $y \in S$ and $h^{-1}(y) = \{x_1, \ldots, x_d\} \in \widetilde{S}$, the linear map $\Phi_y\colon \mathbb{C}^N \to \mathbb{C}^{dq}$ given by

$$\alpha = (\alpha_1, \ldots, \alpha_N) \stackrel{\Phi_y}{\longmapsto} \left(f(x_1) + \sum_{j=1}^{N} \alpha_j \psi_j(x_1), \ldots, f(x_d) + \sum_{j=1}^{N} \alpha_j \psi_j(x_d) \right)$$

is surjective. Consider the trivial bundle $\pi\colon E = Y \times \mathbb{C}^N \to Y$. Let $\Sigma \subset E$ be the subset of the total space whose fibre Σ_y over any point $y \in S$ consists of all

$\alpha = (\alpha_1, \ldots, \alpha_N) \in \mathbb{C}^N$ such that the equality

$$f(x_i) + \sum_{j=1}^{N} \alpha_j \psi_j(x_i) = f(x_k) + \sum_{j=1}^{N} \alpha_j \psi_j(x_k)$$

holds for at least one pair of indexes $1 \le i \neq k \le d$. (For $y \in Y \setminus S$ set $\Sigma_y = \varnothing$.) By introducing the codimension q linear subspaces

$$\Lambda_{i,k} = \left\{(b_1, \ldots, b_d) \in \left(\mathbb{C}^q\right)^d : b_i = b_k\right\} \subset \mathbb{C}^{dq}$$

we see that

$$\Sigma_y = \bigcup_{1 \le i \neq k \le d} \Phi_y^{-1}(\Lambda_{ik}) \subset \mathbb{C}^N.$$

Since the map $\Phi_y \colon \mathbb{C}^N \to \mathbb{C}^{dq}$ is affine linear and surjective, Σ_y is a union of $\binom{d}{2}$ affine linear subspaces of codimension q in $\mathbb{C}^N$. Furthermore, the points $x_i(y)$ in the fibre $h^{-1}(y)$ depend locally holomorphically on $y \in S$, and hence Σ_y are the fibres of a complex subvariety $\Sigma \subset E|_S$.

To complete the proof, it suffices to find a holomorphic section $A(y) = (y, \alpha(y))$ of the trivial vector bundle $Y \times \mathbb{C}^N \to Y$ whose image avoids Σ. Indeed, for such α the map $f' = f_\alpha \colon X \to \mathbb{C}^q$ (9.12) satisfies the separation property over the stratum S, and the remaining conditions can also be arranged. We are thus looking for a holomorphic map $\alpha \colon Y \to \mathbb{C}^N$ such that

$$\alpha(y) \notin \Sigma_y \quad \text{for all } y \in Y. \tag{9.13}$$

For points y in a neighborhood of $Y_1 \cup K$ this holds for the constant map $\alpha = 0$ since f is a relative embedding there. The assumption $2q > \dim Y_0$ ensures the existence of a continuous extension α_0 to $Y_0 \cup K$ satisfying (9.13) and such that $\alpha_0 = 0$ on $Y_1 \cup K$.

Clearly, the subvariety $\Sigma \subset E|_S$ is locally uniformly tame in the sense of Proposition 6.4.14 on p. 278. If $q \ge 2$, the cited proposition implies that the restricted submersion $\pi \colon E|_S \setminus \Sigma \to S$ is elliptic. If $q = 1$ and $d = 2$ then $\binom{d}{2} = 1$, and hence Σ_y consists of a single affine hyperplane in $\mathbb{C}^N$ depending holomorphically on $y \in S$; hence $\mathbb{C}^N \setminus \Sigma_y \cong \mathbb{C}^{N-1} \times \mathbb{C}^*$ is complex homogeneous, and thus elliptic. In each of these two case, Theorem 6.2.2 (the Oka principle for sections of stratified elliptic submersions) shows that α_0 is homotopic to a holomorphic section α avoiding Σ such that $\alpha|_{Y_1} = 0$ and α approximates $\alpha_0 = 0$ uniformly on a neighborhood of K. (See also Theorem 6.2.3 that pertains to this situation.) The corresponding map $f' = f_\alpha \colon X \to \mathbb{C}^q$ (9.12) then satisfies Lemma 9.4.3. Since $f' - f$ vanishes to the second order on X_1 by the choise of the maps ψ_j, f' is still a relative embedding on X_1, and on L this holds if the approximation is close enough over a neighborhood of L. □

(B) *Elimination of the kernel.* Let $Y_1 \subset Y_0 \subset Y$, $S = Y_0 \setminus Y_1$, $X_1 \subset X_0 \subset X$ and $\widetilde{S} = h^{-1}(S)$ be as in (A) above, and let $f' \colon X \to \mathbb{C}^q$ be furnished by Lemma 9.4.3. Thus, (h, f') is an embedding over a neighborhood of $X_1 \cup L$ and it is injective on $\widetilde{S} = h^{-1}(S)$, but it need not be an embedding on X_0 since its differential may have nontrivial kernel transverse to $T\widetilde{S}$ (recall that $h \colon \widetilde{S} \to S$ is an immersion). We now explain how to change f' to a immersion over X_0.

Lemma 9.4.4 ([150, Lemma 3.C_1]) *If* $\dim \Sigma_h^i < 2(q-i+1)$ *holds for some integer* $q \in \mathbb{N}$ *and for all* $i = 1, 2, \ldots, n_0$, *then for a suitably small neighborhood* $L' \subset X$ *of* L *and for every* $\epsilon > 0$ *there exists a holomorphic map* $f'' \colon X \to \mathbb{C}^q$ *satisfying the following conditions*:

(i) $f'' = f'$ *on* X_0,
(ii) $f'' - f'$ *vanishes to second order along* X_1,
(iii) $|f''(x) - f'(x)| < \epsilon$ *for all* $x \in L'$, *and*
(iv) $\ker dh_x \cap \ker df''_x = \{0\}$ *for every* $x \in X_0$.

Condition (i) shows that (h, f'') is injective on X_0 (as it equals (h, f') there), (ii) shows that (h, f'') is an embedding on X_1 (since it agrees along X_1 to the second order with (h, f') which is an embedding on X_1), (iii) shows that (h, f'') remains an embedding on a neighborhood of L if the approximation is close enough, and (iv) ensures that (h, f'') is and immersion, and hence an embedding, also on the bigger subvariety X_0.

Proof of Lemma 9.4.4 The submanifold $\widetilde{S} = h^{-1}(S) = X_0 \setminus X_1$ of X is mapped by h properly onto the connected manifold $S = Y_0 \setminus Y_1$, and hence $\widetilde{S}$ is the union of at most finitely many connected manifold $\widetilde{S}_1, \ldots, \widetilde{S}_m$. Since S is equisingular with respect to h, every component $\widetilde{S}_k$ is contained in $\Sigma_h^i \setminus \Sigma_h^{i+1}$ for some $i = i(k)$, that is, $\dim \ker dh_x = i$ for every point $x \in \widetilde{S}_k$. The kernels $\ker dh_x$ form a holomorphic vector subbundle $E^k \subset TX|_{\widetilde{S}_k}$ of rank i.

By Cartan's Theorem A, there exist finitely many holomorphic maps $\psi_1, \ldots, \psi_N \colon X \to \mathbb{C}^q$ satisfying the following conditions:

- ψ_j vanishes on the subvariety X_0,
- ψ_j vanishes to the second order on the subvariety X_1, and
- $(\bigcap_{j=1}^N \ker(d\psi_j)_x) \cap \ker dh_x = \{0\}$ for every $x \in \widetilde{S}$.

The last property means that, setting $\psi = (\psi_1, \ldots, \psi_N) \colon X \to (\mathbb{C}^q)^N$, the differential $d\psi_x$ is injective on $E_x^k = \ker dh_x$ for every $x \in \widetilde{S}_k$ and $k = 1, \ldots, m$. This is satisfied for a generically chosen ψ provided that N is big enough. We now seek a solution $f'' \colon X \to \mathbb{C}^q$ in the form

$$f''(x) = f'(x) + \sum_{j=1}^N \alpha_j(x)\psi_j(x), \quad x \in X \tag{9.14}$$

where $\alpha_j \in \mathcal{O}(X)$ must be chosen. Since ψ vanishes on X_0, we have

$$\mathrm{d}f''_x = \mathrm{d}f'_x + \sum_{j=1}^{N} \alpha_j(x)\,(\mathrm{d}\psi_j)_x, \quad x \in X_0.$$

Hence the injectivity of $\mathrm{d}f''_x$ on $\ker \mathrm{d}h_x$ for $x \in \widetilde{S}$ leads to the algebraic condition that the graph of the map $\alpha = (\alpha_1, \dots, \alpha_N)\colon X \to \mathbb{C}^N$ avoids a certain subvariety $\Sigma \subset X \times \mathbb{C}^N$ which projects to $X_0 \setminus X_1$. Indeed, given a point $x \in \widetilde{S}_k \subset \widetilde{S}$, the 'bad' set $\Sigma_x \subset \mathbb{C}^N$ consists of those N-tuples $\alpha = (\alpha_1, \dots, \alpha_N) \in \mathbb{C}^N$ for which the $\mathbb{C}$-linear map

$$\theta_x(\alpha) = \mathrm{d}f'_x + \sum_{j=1}^{N} \alpha_j\,(\mathrm{d}\psi_j)_x : \mathrm{T}_x X \to \mathrm{T}_{f'(x)}\mathbb{C}^q$$

fails to be injective on the i-dimensional subspace $E^k_x \subset \mathrm{T}_x X$ (where $i = i(k)$).

Choose a local holomorphic trivialization of the holomorphic vector bundle $\mathrm{T}X|_{\widetilde{S}_k}$, and of its rank i holomorphic vector subbundle $E^k = \ker \mathrm{d}h \subset \mathrm{T}X|_{\widetilde{S}_k}$, near the point x. For every $\alpha \in \mathbb{C}^N$ let

$$\Theta_x(\alpha) \in M_{q,i} \cong \mathbb{C}^q \times \mathbb{C}^i$$

denote the $q \times i$ matrix representing $\theta_x(\alpha)|_{E^k_x} \in \mathrm{Hom}(E^k_x, \mathrm{T}_{f'(x)}\mathbb{C}^q)$. Our choice of the maps $\psi_j : X \to \mathbb{C}^q$ ensures that the affine map

$$\mathbb{C}^N \ni \alpha \mapsto \theta_x(\alpha)|_{E^k_x} \in \mathrm{Hom}\big(E^k_x, \mathrm{T}_{f'(x)}\mathbb{C}^q\big) \cong \mathrm{Hom}\big(\mathbb{C}^i, \mathbb{C}^q\big)$$

is surjective for every $x \in \widetilde{S}$ and it depends holomorphically on x. Hence,

$$\Sigma_x = \Theta_x^{-1}(\Lambda) \subset \mathbb{C}^N, \quad \Lambda = \big\{A \in M_{q,i} : \mathrm{rank}\, A < \min\{i, q\}\big\}.$$

By Lemma 8.9.2 (p. 387), Λ is a subvariety of codimension $|q - i| + 1 \ge 1$ whose complement $M_{q,i} \setminus \Lambda$ is $\mathbb{C}$-homogeneous. Hence, Σ_x is an algebraic subvariety of $\mathbb{C}^N$ with homogeneous complement, and $\widetilde{S} \times \mathbb{C}^N \setminus \Sigma \to \widetilde{S}$ is a holomorphic fibre bundle with homogeneous fibres. Thus, the Oka principle (Theorem 6.2.3 on p. 265) applies to sections $\alpha\colon X \to X \times \mathbb{C}^N$ whose graphs avoid Σ.

The properties of f' ensure that the zero section $\alpha_0 = 0$ avoids Σ for all points in a neighborhood of $X_1 \cup L$. By the assumption, the estimate

$$\dim \widetilde{S}_k \le \dim \Sigma^i_h < 2(q - i + 1)$$

holds for every connected component $\widetilde{S}_k$ of $\widetilde{S}$. If $q < i$ then $\widetilde{S}_k = \varnothing$ and hence this case need not be considered. If $q \ge i$ then the above condition reads

$$\dim \widetilde{S}_k < 2\,\mathrm{codim}\, \Sigma_x, \quad x \in \widetilde{S}.$$

Since $\widetilde{S} = X_0 \setminus X_1$ and X_0 is Stein, this conditions suffices to extend the constant map $\alpha_0 = 0$ from a neighborhood of $X_1 \cup L$ to a continuous map $\alpha_0 : X \to \mathbb{C}^N$ such that $\alpha_0(x) \notin \Sigma_x$ for all $x \in X_0$. (See Theorem 5.14.1 and Corollary 5.14.2 on p. 257 for the details.) By the Oka principle (see Theorem 6.2.3, p. 265), the map α_0 is homotopic to a holomorphic map $\alpha : X \to \mathbb{C}^q$ such that $\alpha = 0$ on X_1, α is arbitrary close to 0 on a neighborhood of L, and $\alpha(x) \notin \Sigma_x$ for all $x \in X_0$. (Theorem 6.2.3 shows that we can make α holomorphic over X_0, and we can then extend it to X in an arbitrary way since the subvariety Σ projects to X_0.) The corresponding holomorphic map f'' given by (9.14) then satisfies Lemma 9.4.4, and hence $(h, f'') : X \to Y \times \mathbb{C}^q$ is an embedding over a neighborhood of $X_0 \cup L$. □

Proof of Theorem 9.3.7 Let K be a compact $\mathcal{O}(Y)$-convex set in Y and $L = f^{-1}(K)$. Assume that $f : U \to \mathbb{C}^q$ is a holomorphic map on an open set $U \supset L$ such that f is a relative embedding over h. Choose another compact $\mathcal{O}(Y)$-convex set $K_1 \subset Y$ containing K and set $L_1 = f^{-1}(K_1) \subset X$. It suffices to show that f can be approximated uniformly on L by a holomorphic map $f_1 : U_1 \to \mathbb{C}^q$ on a neighborhood $U_1 \supset L_1$ such that (h, f_1) is an embedding over L_1; the proof is then completed by induction over a suitable exhaustion of X by compact holomorphically convex sets.

We stratify Y as in Lemma 9.4.2 above, and let $X_j = h^{-1}(Y_j)$ be the corresponding stratification of X. Although every difference $Y_k \setminus Y_{k+1}$ may have infinitely many connected components (strata), only finitely many of them intersect the compact set K_1. Replacing Y and X by a neighborhood of K_1 and L_1, respectively, and refining the stratifications if necessary, we may assume that every difference $Y_k \setminus Y_{k+1}$ is connected and there are only finitely many strata. Hence, the induction described below ends in finitely many steps.

We begin by extending f to a map $f : X \to \mathbb{C}^q$ that is a relative embedding over $L \cup X_m$ where $X_m = h^{-1}(Y_m)$ is zero-dimensional (discrete). We then proceed to change f in finitely many steps so as to make it a relative embedding over the successive strata in X. In a typical step, we have a pair of closed complex subvarieties $Y_{k+1} \subset Y_k$ of Y whose difference $S = Y_k \setminus Y_{k+1}$ is equisingular with respect to h, and a holomorphic map $f : X \to \mathbb{C}^q$ such that (h, f) is an embedding over a neighborhood of the set $X_{k+1} \cup L$ in X. We now change f so that we get an embedding over $X_k \cup L$. The situation is exactly as described above for the pair $Y_1 \subset Y_0$: First, use Lemma 9.4.3 to separate points over $X_k = h^{-1}(Y_k)$, and then apply Lemma 9.4.4 to eliminate the kernel of the differential over X_k. The dimension estimates in Theorem 9.3.7 ensure that each of these steps can be carried out. □

Remark 9.4.5 In the proof of Theorem 9.3.6, with $M = \max\{[\frac{3n+1}{2}], b_X\} = n + q$, condition (i) in Theorem 9.3.7 holds for all strata S of dimension $< n$ since $2q \geq 2[\frac{n+1}{2}] \geq n$. Hence we obtain a holomorphic map $f : X \to \mathbb{C}^q$ such that $(h, f) : X \to \mathbb{C}^M$ is a proper immersion that is injective on all strata of dimension $< n$ in a given stratification of X. However, if $2q = n$ then we might not be able to separate points over the n-dimensional stratum.

If we take $q = n+1$, then the separation of points and the elimination of the kernel (in the proof of Theorem 9.3.7) can be achieved by the transversality theorem. This gives a proper holomorphic embedding $X^n \hookrightarrow \mathbb{C}^{2n+1}$ and a proper holomorphic immersion $X^n \to \mathbb{C}^{2n}$ (cf. Theorem 2.4.1 on p. 52). □

9.5 Weakly Regular Embeddings and Interpolation

In this section we present some further results on embedding Stein spaces in complex Euclidean spaces. We begin with a theorem of Prezelj [460] on *weakly regular embeddings* of Stein spaces with isolated singular points. Before stating her result, we must recall the following notions.

Given a point x in a complex space X, let $C_4(X,x) \subset C_5(X,x)$ denote the Whitney cones of X at x (see [574] or [98, p. 91]). Embedding a neighborhood of x in X as a complex subvariety of some $\mathbb{C}^m$, we have

$$C_4(X,x) = \big\{v \in \mathbb{C}^m : \text{there exist a sequence } x_j \in X_{\rm reg} \text{ such that } x_j \to x, \\ \text{and a sequence } v_j \in \mathrm{T}_{x_j}X \text{ such that } v_j \to v\big\};$$

$$C_5(X,x) = \big\{v \in \mathbb{C}^m : \text{there exist sequences } x_j, y_j \in X_{\rm reg} \text{ with } x_j, y_j \to x, \\ \text{and a sequence } \lambda_j \in \mathbb{C} \text{ with } \lambda_j(x_j - y_j) \to v\big\}.$$

The cones $C_4(X,x) \subset C_5(X,x)$ are closed algebraic sets in $\mathbb{C}^m$ contained in the Zariski tangent space T_xX, and we have

$$n \le \dim C_4(X,x) \le \dim C_5(X,x) \le 2n, \quad n = \dim_x X.$$

If $x \in X_{\rm reg}$ then clearly $\dim C_4(X,x) = \dim C_5(X,x) = n$; conversely, $\dim C_5(X,x) = n$ implies that $x \in X_{\rm reg}$.

Definition 9.5.1 ([459, Definition 2.11]) A holomorphic map $f\colon X \to \mathbb{C}^N$ is *weakly regular* at a point $x \in X$ if $\ker \mathrm{d}f_x \cap C_5(X,x) = \{0\}$; f is weakly regular on X if this holds at every point $x \in X$.

The notions of regularity and weak regularity coincide at a regular point $x \in X_{\rm reg}$. The following lemma explains the significance of this condition.

Lemma 9.5.2 ([459, Corollary 2.10]) *If a holomorphic map $f\colon X \to \mathbb{C}^N$ is weakly regular at a point $x \in X$, then there is a neighborhood $U \subset X$ of x such that $f|_U\colon U \to \mathbb{C}^N$ is an injective map which is an immersion on $U \cap X_{\rm reg}$.*

Proof We first show that f is regular on $X_{\rm reg}$ near the point x. If not, there exists a sequence $x_j \in X_{\rm reg}$ converging to x and tangent vectors $v_j \in \mathrm{T}_{x_j}X$ of length $|v_j| = 1$ such that $\mathrm{d}f_{x_j} \cdot v_j = 0$ for all $j = 1, 2, \ldots$. Passing to a subsequence, we have $\lim_{j\to\infty} v_j = v \in \mathrm{T}_x X$. Then, $0 \neq v \in C_4(X,x)$ and $\mathrm{d}f_x \cdot v = 0$, contradicting the assumption that f is weakly regular at x.

If f fails to be injective in every neighborhood of x, there exist sequences $x_j \neq y_j$ in X converging to x such that $f(x_j) - f(y_j) = 0$ for $j = 1, 2, \ldots$. Let $v_j = \frac{x_j - y_j}{|x_j - y_j|}$. From $0 = f(x_j) - f(y_j) = \mathrm{d}f_{x_j} \cdot (x_j - y_j) + o(|x_j - y_j|)$ we see that $\mathrm{d}f_{x_j} \cdot v_j \to 0$ as $j \to \infty$. By passing to a subsequence we may assume that $v_j \to v$, and hence $\mathrm{d}f_x \cdot v = 0$. Since $v \in C_5(X,x)$ and $|v| = 1$, this again contradicts the assumption that f is weakly regular at x. □

Theorem 9.5.3 (Prezelj [459, Theorem 1.1]) *Assume that X is a Stein space of dimension n with isolated singular points. Let*

$$N(n) = \max\left\{\left[\frac{3n}{2}\right] + 1, \max_{x\in X} \dim C_5(X,x), 3\right\}. \tag{9.15}$$

Then, there exists an injective proper holomorphic map $f\colon X \hookrightarrow \mathbb{C}^{N(n)}$ which is regular (an immersion) on the regular locus X_{reg}.

Since $\dim C_5(X,x) \le 2\dim_x X$, we have $N(n) \le 2n$ for $n > 1$, and hence this result improves Theorem 2.4.1(b). Note that $f\colon X \to f(X) \subset \mathbb{C}^{N(n)}$ is a homeomorphism of X onto its image.

Proof Without loss of generality we may assume that the space X is reduced. By the assumption, the singular locus $X_{\mathrm{sing}} = \{x_1, x_2, \ldots\}$ is discrete. Let $N = N(n) = n + q$. The condition $N \ge \dim C_5(X,x)$ ensures the existence of a holomorphic map $g_0 = (h_0, f_0)\colon X \to \mathbb{C}^{n+q} = \mathbb{C}^N$ such that

$$\ker(\mathrm{d}g_0)_{x_j} \cap C_5(X,x_j) = \{0\}, \quad j = 1, 2, \ldots.$$

In view of Lemma 9.5.2 it follows that, locally near every singular point x_j, g_0 is regular on X_{reg} and injective. By a minor improvement of Theorem 9.2.3 (see [459, Proposition 3.1]) we find an almost proper map $h\colon X \to \mathbb{C}^n$ such that $h - h_0$ vanishes to a given finite order $m_j \in \mathbb{N}$ at each of the singular points x_j, and $\ker \mathrm{d}h$ satisfies the estimates in Theorem 9.2.3(ii) on the regular locus X_{reg}. We then find a holomorphic map $f\colon X \to \mathbb{C}^q$ such that $f - f_0$ vanishes to a high order at each point $x_j \in X_{\mathrm{sing}}$ and the map $(h, f)\colon X \to \mathbb{C}^N$ is injective, proper, and regular on X_{reg}. This is done as in the proof of Theorem 9.3.6. If $g = (h, f)$ agrees with $g_0 = (h_0, f_0)$ to a sufficiently high order at each point $x_j \in X_{\mathrm{sing}}$ (which can be ensured by the construction), then we have

$$\ker \mathrm{d}g_{x_j} \cap C_5(X,x_j) = \{0\}, \quad x_j \in X_{\mathrm{sing}}.$$

Hence, g is injective and regular on X_{reg} near x_j. Elsewhere, we ensure injectivity, and regularity on X_{reg}, as in the proof of Theorem 9.3.6. □

Remark 9.5.4 As pointed out by Prezelj in [459], Theorem 9.5.3 gives embedding results for smooth, compact, strongly pseudoconvex, integrable CR manifolds of dimension $2n-1 \geq 5$ and CR dimension $n-1 \geq 2$. Any such M bounds a relatively compact domain D in a pure n-dimensional Stein space X with finitely many isolated singularities; hence, the problem of finding a CR embedding $M \hookrightarrow \mathbb{C}^N$ with minimal N reduces to the problem of finding a weakly regular embedding $X \hookrightarrow \mathbb{C}^N$. □

We now consider the extension problem for proper regular maps and embeddings. Given a Stein space X, a closed complex subspace X' of X, and a proper holomorphic embedding $\varphi\colon X' \hookrightarrow \mathbb{C}^N$, the problem is to find a holomorphic map $f\colon X \to \mathbb{C}^N$ that extends φ and satisfies some other regularity properties (such as being proper, injective, regular on X_{reg}, or an embedding). The following result of Acquistapace, Broglia and Tognoli generalizes the theorem of Bishop and Narasimhan (see Theorem 2.4.1 on p. 52).

Theorem 9.5.5 ([5, Theorem 1]) *Assume that X is a reduced Stein space of dimension n, X' is a closed complex subspace of X, and $\varphi\colon X' \hookrightarrow \mathbb{C}^N$ is a proper holomorphic embedding for some $N \geq 2n+1$. Then the set of all holomorphic maps $f\colon X \to \mathbb{C}^N$ that extend φ and are proper, injective and regular on X_{reg}, is dense in the space of all holomorphic maps $X \to \mathbb{C}^N$ extending φ. In particular, if X is nonsingular (a Stein manifold), then the above holds for proper holomorphic embeddings $X \hookrightarrow \mathbb{C}^N$ extending φ.*

The following example shows that Theorem 9.5.5 is close to optimal—interpolation fails in general for proper holomorphic embeddings $X^n \hookrightarrow \mathbb{C}^N$ when $N < 2n$. We identify $\mathbb{C}^{n-1}$ with the hyperplane $\mathbb{C}^{n-1} \times \{0\} \subset \mathbb{C}^n$.

Proposition 9.5.6 *For every integer $n > 1$ there exists a proper holomorphic embedding $\varphi\colon \mathbb{C}^{n-1} \hookrightarrow \mathbb{C}^{2n-1}$ which cannot be extended to an injective holomorphic map $f\colon \mathbb{C}^n \hookrightarrow \mathbb{C}^{2n-1}$.*

Proof Theorem 4.18.9 on p. 188 furnishes a proper holomorphic embedding $\varphi\colon \mathbb{C}^{n-1} \hookrightarrow \mathbb{C}^{2n-1}$ such that $\Omega = \mathbb{C}^{2n-1} \setminus \varphi(\mathbb{C}^{n-1})$ is Eisenman n-hyperbolic. Write $z = (z', z_n) \in \mathbb{C}^n$ and identify $\mathbb{C}^{n-1}$ with the subspace $\{z_n = 0\} \subset \mathbb{C}^n$. Assume that $f\colon \mathbb{C}^n \to \mathbb{C}^{2n-1}$ is an injective holomorphic map extending φ; then f has rank n at a generic point. Note that $f(z', 0) = \varphi(z')$ and $f(z', z_n) \notin \varphi(\mathbb{C}^{n-1})$ if $z_n \neq 0$. The entire map $F(z', z_n) = f(z', \mathrm{e}^{z_n})$ maps $\mathbb{C}^n$ to Ω and has rank n at a generic point; a contradiction. □

A similar construction gives for every pair of integers $1 \leq k < n$ a proper holomorphic embedding $\varphi\colon \mathbb{C}^k \hookrightarrow \mathbb{C}^{n+k}$ that cannot be extended from $\mathbb{C}^k = \mathbb{C}^k \times \{0\}^{n-k} \subset \mathbb{C}^n$ to an injective holomorphic map $\mathbb{C}^n \to \mathbb{C}^{n+k}$.

In light of Theorem 9.3.1 and Example 9.5.6, the best possible result concerning the interpolation of holomorphic embeddings would be a positive answer to the following question.

Problem 9.5.7 Let X be an n-dimensional Stein manifold and X' be a k-dimensional complex submanifold of X. Let

$$N(n,k) = \max\left\{\left[\frac{3n}{2}\right]+1, n+k+1, 3\right\}.$$

Does every proper holomorphic embedding $X' \hookrightarrow \mathbb{C}^N$ for $N \geq N(n,k)$ extend to a proper holomorphic embedding $X \hookrightarrow \mathbb{C}^N$? □

In the special case when X' is discrete, we have the following analogue of Theorem 9.3.1. Let b_X denote the number (9.7) on p. 414.

Theorem 9.5.8 ([208, 459]) *If X is an n-dimensional Stein space with finite embedding dimension and $N \geq \max\{[\frac{3n}{2}]+1, b_X, 3\}$, then for every pair of discrete sequences $\{a_j\} \subset X$, $\{b_j\} \subset \mathbb{C}^N$ without repetitions there is a proper holomorphic embedding $f\colon X \hookrightarrow \mathbb{C}^N$ satisfying $f(a_j) = b_j$ for $j = 1, 2, \ldots$. If X is a Stein manifold of dimension $n > 1$ then this holds for $N \geq [\frac{3n}{2}]+1$.*

Proof Write $N = n+q$. The proof of Theorem 9.3.1 furnishes an almost proper map $h\colon X \to \mathbb{C}^n$ and a proper holomorphic embedding $G = (h, g)\colon X \hookrightarrow \mathbb{C}^{n+q} = \mathbb{C}^N$ such that the discrete sequence $\{G(a_j)\}_{j\in\mathbb{N}}$ is tame in $\mathbb{C}^N$ (see Definition 4.6.1, p. 128). In fact, it suffices to ensure in the construction of the map $g\colon X \to \mathbb{C}^q$ that the sequence $|g(a_j)|$ tends to infinity fast enough so that the projection of the set $\{G(a_j)\}_{j\in\mathbb{N}} \subset \mathbb{C}^{n+q}$ onto $\{0\}^n \times \mathbb{C}^q$ is proper; tameness then follows from Theorem 4.6.2(a) on p. 128. By Theorem 4.17.1 (p. 181), there exist a Fatou-Bieberbach domain $\Omega \subset \mathbb{C}^N$, containing the complex subvariety $G(X)$ of $\mathbb{C}^N$, and a biholomorphic map $\Phi\colon \Omega \to \mathbb{C}^N$ such that $\Phi(G(a_j)) = b_j$ for $j = 1, 2, \ldots$. The map $f = \Phi \circ G\colon X \hookrightarrow \mathbb{C}^N$ then satisfies the conclusion of the theorem. □

9.6 The Oka Principle for Holomorphic Immersions

In the proof of Lemma 9.4.4 we did not use the hypothesis that the manifold Y is Stein or that the background map $h\colon X \to Y$ is proper. Essentially the same proof gives the following *relative Oka principle for immersions*, due to Eliashberg and Gromov [150], for more general background maps.

Given a holomorphic map $h\colon X \to Y$, we define

$$\dim_h X = \sup_{y\in Y} \dim h^{-1}(y).$$

Thus, $\dim_h X = 0$ precisely when the fibre $h^{-1}(y)$ over any point $y \in Y$ is 0-dimensional, and hence a discrete subset of X.

Theorem 9.6.1 ([150, 2.E$_1$, p. 128]) *Let X be a Stein space and $h\colon X \to Y$ be a holomorphic map. If $\dim_h X = 0$, then relative immersions $f\colon X \to \mathbb{C}^q$ over h satisfy the following homotopy principle: There exists a holomorphic map $f\colon X \to \mathbb{C}^q$ such that $(h, f)\colon X \to Y \times \mathbb{C}^q$ is an immersion if and only if there exists a continuous complex vector bundle map $\theta\colon \mathrm{T}X \to \mathrm{T}\mathbb{C}^q$ such that $\mathrm{d}h \oplus \theta\colon \mathrm{T}X \hookrightarrow \mathrm{T}(Y \times \mathbb{C}^q)$ is injective.*

Proof (sketch) We stratify X such that for every stratum S, $h|_S\colon S \to Y$ is an immersion and $\dim\ker \mathrm{d}h_x$ is independent of the point $x \in S$. It follows that $\ker \mathrm{d}h|_S$ is a holomorphic vector subbundle of $\mathrm{T}X|_S$ which is transverse to S. We then construct a relative immersion $f\colon X \to \mathbb{C}^q$ by passing from one stratum to the next one as in the proof of Theorem 9.3.7, using Lemma 9.4.4 at every step. The necessary topological condition to apply the Oka principle is the existence of a map θ as in the statement of the theorem. □

Theorem 9.6.1 also holds with approximation on a compact $\mathcal{O}(X)$-convex subset K in X: If $f_0\colon X \to \mathbb{C}^q$ is a continuous map that is holomorphic on a neighborhood of K and there exists a vector bundle map $\theta\colon \mathrm{T}X \to \mathrm{T}\mathbb{C}^q$ as in the theorem such that $\theta = \mathrm{d}f_0$ on a neighborhood of K, then we can deform f_0 to a holomorphic map $f\colon X \to \mathbb{C}^q$ which approximates f_0 as closely as desired on K and such that $(h, f)\colon X \to Y \times \mathbb{C}^q$ is an immersion.

The proof of Lemma 9.4.4 shows that the dimension estimates on the subvarieties $\Sigma_h^i = \{x \in X : \dim\ker \mathrm{d}h_x = i\}$ in the following corollary imply the existence of a map $\theta\colon \mathrm{T}X \to \mathrm{T}\mathbb{C}^q$ in Theorem 9.6.1.

Corollary 9.6.2 ([150, 2.E$_2$.]) *Let X be a Stein space of finite embedding dimension, and let $h\colon X \to Y$ be a holomorphic map with $\dim_h X = 0$. If for some $q \in \mathbb{N}$ and for all $i = 1, 2, \ldots, \mathrm{embdim} X$ we have*

$$\dim\{x \in X : \dim\ker \mathrm{d}h_x = i\} \le 2(q - i + 1),$$

then there exists a holomorphic immersion $(h, f)\colon X \to Y \times \mathbb{C}^q$.

It is not known whether *proper* holomorphic immersions or embeddings of Stein manifolds to Euclidean spaces satisfy the Oka principle. However, without insisting on properness, holomorphic immersions of Stein manifolds satisfy the following Oka principle due to Eliashberg and Gromov [269]; see also [265, Sect. 2.1.5].

Theorem 9.6.3 *If X is a Stein manifold whose cotangent bundle $\mathrm{T}^*X \cong \mathrm{T}^{*1,0}X$ is generated by $(1,0)$-forms $\theta_1, \ldots, \theta_q$ for some $q > \dim X$, then there exists a holomorphic immersion $X \to \mathbb{C}^q$. More precisely, every such q-tuple $\theta = (\theta_1, \ldots, \theta_q)$ can be deformed through a homotopy of q-tuples generating T^*X to a q-tuple $\mathrm{d}f = (\mathrm{d}f_1, \ldots, \mathrm{d}f_q)$, where $f = (f_1, \ldots, f_q)\colon X \to \mathbb{C}^q$ is a holomorphic immersion. In particular, every n-dimensional Stein manifold admits a holomorphic immersion into $\mathbb{C}^{[3n/2]}$.*

Proof (sketch) By the Oka-Grauert principle (see Sect. 8.2) we may assume that $\theta_1, \ldots, \theta_q$ are holomorphic 1-forms. In the first step, the last form θ_q is replaced by the differential $\mathrm{d} f_q$ of a holomorphic function on X such that $\theta_1, \ldots, \theta_{q-1}, \mathrm{d} f_q$ still generate $\mathrm{T}^* X$ pointwise. Since $q > \dim X$, we may assume that the forms $\theta_1, \ldots, \theta_{q-1}$ generate $\mathrm{T}^* X$ outside a proper complex subvariety $\Sigma \subset X$, and hence f_q must satisfy an algebraic condition on its 1-jet along Σ provided that the forms $\theta_1, \ldots, \theta_{q-1}$ are chosen generically (this requires a small perturbation). Once f_q has been chosen, one proceeds in the same way with the forms $\theta_1, \ldots, \theta_{q-2}, \mathrm{d} f_q$; after a small perturbation of these forms, we can replace θ_{q-1} with an exact differential $\mathrm{d} f_{q-1}$ such that $\theta_1, \ldots, \theta_{q-2}, \mathrm{d} f_{q-1}, \mathrm{d} f_q$ span $\mathrm{T}^* X$. In q steps of this kind all forms θ_j are replaced by holomorphic differentials.

We refer to the paper by Kolarič [349] for a detailed exposition of this proof and some generalizations, including approximation on compact $\mathcal{O}(X)$-convex subsets and the 1-parametric Oka principle. □

The fully parametric Oka principle for holomorphic immersions of Stein manifolds to Euclidean spaces is not known, with the exception of the following result for open Riemann surfaces, due to Forstnerič and Lárusson.

Theorem 9.6.4 ([210, Theorem 1.4]) *Let X be an open Riemann surface. For every $n \in \mathbb{N}$ the space $\mathfrak{I}(X, \mathbb{C}^n)$ of all holomorphic immersions $X \to \mathbb{C}^n$ is weakly homotopy equivalent to the space $\mathcal{C}(X, S^{2n-1})$ of continuous maps from X to the sphere S^{2n-1}. The two spaces are homotopy equivalent when X has finite topological type.*

More precisely, if θ is any holomorphic 1-form without zeros on X, then the map $\mathfrak{I}(X, \mathbb{C}^n) \to \mathcal{C}(X, S^{2n-1})$, which associates to an immersion $f\colon X \to \mathbb{C}^n$ the radial projection of the map $\mathrm{d} f/\theta\colon X \to \mathbb{C}^n_*$ to the unit sphere, is a weak homotopy equivalence. It follows in particular that path components of $\mathfrak{I}(X, \mathbb{C})$ are in bijective correspondence with the elements of $H_1(X; \mathbb{Z})$, and $\mathfrak{I}(X, \mathbb{C}^n)$ is $(2n-3)$-connected if $n > 1$ (see [210, Corollary 1.5]).

9.7 A Splitting Lemma for Biholomorphic Maps

Let dist be a Riemannian distance function on a complex manifold X. Given a subset $U \subset X$ and a map $\gamma\colon U \to X$, we set

$$\mathrm{dist}_U(\gamma, \mathrm{Id}) = \sup_{x \in U} \mathrm{dist}\big(\gamma(x), x\big).$$

We prove the following result on decomposition of a biholomorphic map close to the identity. This is used in a number of subsequent constructions.

Theorem 9.7.1 *Let A and B be compact sets in a complex manifold X such that $D = A \cup B$ is a Stein compact and $\overline{A \setminus B} \cap \overline{B \setminus A} = \varnothing$. Given an open set $\widetilde{C} \subset X$ containing $C = A \cap B$, there exist open sets $A' \supset A$, $B' \supset B$, $C' \supset C$ such that $C' \subset A' \cap B' \subset \widetilde{C}$, and constants $\epsilon_0 > 0$ and $c > 0$, such that the following holds. For every injective holomorphic map $\gamma \colon \widetilde{C} \to X$ with $\epsilon = \mathrm{dist}_{\widetilde{C}}(\gamma, \mathrm{Id}) < \epsilon_0$ there exist injective holomorphic maps $\alpha \colon A' \to X$, $\beta \colon B' \to X$, depending continuously on γ and satisfying*

$$\gamma \circ \alpha = \beta \text{ on } C', \quad \mathrm{dist}_{A'}(\alpha, \mathrm{Id}) < c\epsilon, \quad \mathrm{dist}_{B'}(\beta, \mathrm{Id}) < c\epsilon. \tag{9.16}$$

If in addition C is a Stein compact and X' is a closed complex subvariety of X such that $C \cap X' = \varnothing$, then α and β can be chosen tangent to the identity map to any given finite order along X'.

This is [191, Theorem 4.1], except that we now have linear estimates (9.16). In [191] (and in the first edition of the book) this result was proved by a rapidly convergent iteration scheme. Here we give a simpler proof by using the implicit function theorem in Banach spaces. A minor modification of the construction also provides the foliated version that was given in both mentioned sources. Theorem 9.7.4 below, which is taken from [206, Theorem 3.2], gives the analogous result on complex spaces with singularities.

Proof By following the proof of Proposition 5.7.3 on p. 234, we can slightly enlarge the compact sets A and B so that they are smoothly bounded, their union $D = A \cup B$ is smoothly bounded strongly pseudoconvex, and $C = A \cap B$ is contained in the given open set $\widetilde{C}$. If the original set C is also a Stein compact, then we can choose the new sets such that C is a compact smoothly bounded strongly pseudoconvex domain satisfying $C \cap X' = \varnothing$. Clearly it suffices to find maps $\alpha \in \mathcal{A}(A, X)$ and $\beta \in \mathcal{A}(B, X)$ satisfying (9.16); if $\epsilon_0 > 0$ is small enough then these maps are automatically injective in smaller neighborhoods of the original sets A and B in the theorem.

Replacing X by a Stein neighborhood of $D = A \cup B$, we may assume that it is a Stein manifold. Choose holomorphic vector fields $V_1, \ldots, V_N$ on X which generate TX on a Stein neighborhood U of D. Let ϕ_t^i denote the flow of V_i for complex time t, and let Φ be the composition of these flows:

$$\Phi(x, z) = \phi_{z_1}^1 \circ \cdots \circ \phi_{z_N}^N(x) \in X, \quad x \in X. \tag{9.17}$$

Here, $z = (z_1, \ldots, z_N)$ are coordinates on $\mathbb{C}^N$. The map Φ is defined and holomorphic in an open neighborhood $\Omega \subset X \times \mathbb{C}^N$ of $X \times \{0\}$. Let $\Omega_x \subset \mathbb{C}^N$ denote the fibre of Ω over x. Since the vector fields V_i span $\mathrm{T}_x X$ for all $x \in U$, we may assume (after shrinking Ω around $X \times \{0\}$ if necessary) that the map $\Phi(x, \cdot) \colon \Omega_x \to X$ is a submersion for every $x \in U$. Let $\Theta_x \colon \mathbb{C}^N \to T_x X$ denote the differential of the map $z \mapsto \Phi(x, z) \in X$ at $z = 0$. This gives a holomorphic vector bundle splitting $U \times \mathbb{C}^N = E \oplus E'$, where $E'_x = \ker \Theta_x$ and E is some complementary subbundle. Set $W_x = E_x \cap \Omega_x$ for $x \in U$. After shrinking Ω around $X \times \{0\}$ and U around D, it follows that W_x contains a ball of a fixed size around the origin $0_x \in E_x$ independent of $x \in U$, and the map $\Phi(x, \cdot)\colon W_x \to X$ takes W_x biholomorphically onto an open neighborhood of x in X. This obviously gives the following lemma.

Lemma 9.7.2 (Notation as above) *There are constants $\eta_0 > 0$ and $M_0 \ge 1$ such that for every complex manifold S and pair of maps $f, g\colon S \to U$ with $\sup_{s\in S} \mathrm{dist}(f(s), g(s)) < \eta_0$ there exists a unique map $c\colon S \to \mathbb{C}^N$ such that for every $s \in S$ we have $c(s) \in W_{f(s)}$, $\Phi(f(s), c(s)) = g(s)$, and*

$$M_0^{-1}|c(s)| \le \mathrm{dist}\big(f(s), g(s)\big) \le M_0|c(s)|.$$

If f, g are continuous or holomorphic, then so is c.

Choose a compact smoothly bounded set C_0 in X such that $C \subset \mathring{C}_0 \subset C_0 \subset \widetilde{C}$. We shall consider maps $\gamma \in \mathcal{A}(C_0, X)$ (i.e., $\gamma\colon C_0 \to X$ is continuous and holomorphic in $\mathring{C}_0$) close to the identity Id on C_0. Denote by $\mathscr{C}_0$ the Banach space of maps $c\colon C_0 \to \mathbb{C}^N$ of class $\mathcal{A}(C_0)$ such that $c(x) \in E_x$ for every $x \in C_0$, endowed with the sup-norm. Lemma 9.7.2 applied with $S = C_0$ furnishes an isomorphism between a neighborhood of the identity map in $\mathcal{A}(C_0, X)$ and a neighborhood of the zero map in the Banach space $\mathscr{C}_0$.

The following lemma amounts to the solution of the Cousin-I problem with bounds; its proof is the same as that of Lemma 5.8.2 on p. 236.

Lemma 9.7.3 (Notation as above) *There exist bounded linear operators $\mathscr{A}_0\colon \mathcal{A}(C, \mathbb{C}^N) \to \mathcal{A}(A, \mathbb{C}^N)$ and $\mathscr{B}_0\colon \mathcal{A}(C, \mathbb{C}^N) \to \mathcal{A}(B, \mathbb{C}^N)$ such that*

$$c = \mathscr{B}_0 c - \mathscr{A}_0 c \quad \textit{holds for every } c \in \mathcal{A}\big(C, \mathbb{C}^N\big). \tag{9.18}$$

If in addition C is a Stein compact, then we may choose $\mathscr{A}_0 c$ and $\mathscr{B}_0 c$ to vanish to a given finite order on $X' \cap A$ and $X' \cap B$, respectively.

Denote by $\mathscr{C}$ the Banach space of maps $c\colon C \to \mathbb{C}^N$ of class $\mathcal{A}(C, \mathbb{C}^N)$ such that $c(x) \in E_x$ for every $x \in C$, endowed with the sup-norm. If $c \in \mathscr{C}$ is sufficiently close to the zero map, then $c(x) \in W_x$ for all $x \in C$. For such c, we define maps $\alpha_c \in \mathcal{A}(A, X)$ and $\beta_c \in \mathcal{A}(B, X)$ by

$$\alpha_c(x) = \Phi\big(x, (\mathscr{A}_0 c)(x)\big) \quad (x \in A), \qquad \beta_c(x) = \Phi\big(x, (\mathscr{B}_0 c)(x)\big) \quad (x \in B).$$

Here, $\mathscr{A}_0$ and $\mathscr{B}_0$ are the linear operators furnished by Lemma 9.7.3. If $c \in \mathscr{C}$ is close to the zero map, then α_c and β_c are well defined and close to the identity maps on A and B, respectively. In particular, we also have $\alpha_c(C) \subset \mathring{C}_0$, and hence $\gamma \circ \alpha_c\colon C \to X$ is a map in $\mathcal{A}(C, X)$ close to Id_C. Note that $\alpha_0 = \mathrm{Id}_A$ and $\beta_0 = \mathrm{Id}_B$. Assuming that $\gamma \in \mathcal{A}(C_0, X)$ is close enough to Id and $c \in \mathscr{C}$ is close enough to 0, Lemma 9.7.2 gives a unique map $\tilde{c} \in \mathscr{C}$ such that

$$\Phi\big(\gamma \circ \alpha_c(x), \tilde{c}(x)\big) = \beta_c(x), \quad x \in C. \tag{9.19}$$

This defines a map $\phi(\gamma, c) = \tilde{c}$ from a neighborhood of $(\mathrm{Id}, 0) \in \mathcal{A}(C_0, X) \times \mathscr{C}$ to $\mathscr{C}$ such that $\phi(\mathrm{Id}, 0) = 0$. It is obvious from the construction that the map ϕ is smooth. Furthermore, we claim that the partial differential of ϕ with respect to the second variable at $\gamma = \mathrm{Id}$ and $c = 0$ is the identity:

$$\partial_c \phi(\mathrm{Id}, 0) = \mathrm{Id}. \tag{9.20}$$

This is seen as follows. Recall that $\Theta_x : \mathbb{C}^N \to T_x X$ is the differential of $z \mapsto \Phi(x, z) \in X$ at $z = 0$. Note that $\Theta_x : E_x \to \mathrm{T}_x X$ is a linear isomorphism for every $x \in U$. In any local coordinate x on X we have the Taylor expansion

$$\Phi(x, z) = x + \Theta_x(z) + O\big(|z|^2\big), \quad z \to 0.$$

Let $\|c\|$ denote the sup-norm of $c \in \mathscr{C}$. Since the maps $a = \mathscr{A}_0 c \in \mathcal{A}(A, \mathbb{C}^N)$ and $b = \mathscr{B}_0 c \in \mathcal{A}(B, \mathbb{C}^N)$ are given by bounded linear operators, we have

$$\big|a(x)\big| = O\big(\|c\|\big) \quad \text{for } x \in A, \qquad \big|b(x)\big| = O\big(\|c\|\big) \quad \text{for } x \in B.$$

This gives

$$\alpha_c(x) = x + \Theta_x\big(a(x)\big) + O\big(\|c\|^2\big), \qquad \beta_c(x) = x + \Theta_x\big(b(x)\big) + O\big(\|c\|^2\big). \tag{9.21}$$

Therefore, we have

$$\begin{aligned} \Phi\big(\alpha_c(x), \tilde{c}(x)\big) &= \alpha_c(x) + \Theta_{\alpha_c(x)}\big(\tilde{c}(x)\big) + \big|\tilde{c}(x)\big|^2 \\ &= x + \Theta_x\big(a(x)\big) + \Theta_x\big(\tilde{c}(x)\big) + O\big(\|c\|^2\big), \end{aligned}$$

where we used that $\tilde{c}(x) = O(\|c\|)$ and

$$\Theta_{\alpha_c(x)}(z) = \Theta_x(z) + |z| O\big(\big|a(x)\big|\big) = \Theta_x(z) + |z| O\big(\|c\|\big)$$

Writing $\tilde{c} = c + c'$ and using that $a + c = b$ in view of (9.18), we obtain

$$\Phi\big(\alpha_c(x), \tilde{c}(x)\big) = x + \Theta_x\big(b(x)\big) + \Theta_x\big(c'(x)\big) + O\big(\|c\|^2\big).$$

Comparing with the expansion of $\beta_c(x)$ (9.21), we see that $\Theta_x(c'(x)) = O(\|c\|^2)$ for $x \in C$. Since $c'(x) \in E_x$ and $\Theta_x : E_x \to T_x X$ is a linear isomorphism, it follows that $|c'(x)| = O(\|c\|^2)$ and hence $\tilde{c} = c + O(\|c\|^2)$. This proves (9.20).

From $\phi(\mathrm{Id}, 0) = 0$ and (9.20) it follows by the implicit function theorem that the equation $\phi(\gamma, c) = 0$ has a unique local solution $c = \psi(\gamma) \in \mathscr{C}$ for $\gamma \in \mathcal{A}(C_0, X)$ near Id, with $\psi(\mathrm{Id}) = 0$. It follows from (9.19) that the maps

$$\alpha = \alpha_{\psi(\gamma)} \in \mathcal{A}(A, X), \qquad \beta = \beta_{\psi(\gamma)} \in \mathcal{A}(B, X)$$

satisfy

$$\gamma \circ \alpha_{\psi(\gamma)} = \Phi(\gamma \circ \alpha_{\psi(\gamma)}, 0) = \beta_{\psi(\gamma)} \quad \text{on } C,$$

so (9.16) holds. The linear estimates in (9.16) follow by observing that $\gamma \mapsto \alpha_{\psi(\gamma)}$ and $\gamma \mapsto \beta_{\psi(\gamma)}$ are given by smooth (nonlinear) operators. This completes the proof of Theorem 9.7.1. □

The following result generalizes Theorem 9.7.1 to complex spaces. This is used in the proof of Theorem 9.13.13 concerning the existence of holomorphic functions without critical points on reduced Stein spaces.

Theorem 9.7.4 ([206, Theorem 3.2]) *Assume that X is a reduced complex space and X' is a closed complex subvariety of X containing X_{sing}. Let (A, B) be a Cartan pair in X (see Definition 5.7.1) such that $C = A \cap B \subset X \setminus X'$. Then the conclusion of Theorem 9.7.1 holds for (A, B).*

The individual sets A and B need not be Stein compacts in this result.

Proof We adapt the proof of Theorem 9.7.1 to this situation, following the notation established in the earlier theorem as much as possible.

We replace X by a relatively compact Stein neighborhood of the Stein compact $D = A \cup B$ and embed it properly holomorphically into a Euclidean space $\mathbb{C}^n$. Then, (A, B) is also a Cartan pair in $\mathbb{C}^n$ by Lemma 5.7.2. Since $C = A \cap B$ is a Stein compact contained in $X \setminus X' \subset X_{\mathrm{reg}}$, there exist a Stein neighborhood $U_0 \Subset \mathbb{C}^n$ of C, with $U = U_0 \cap X \subset X_{\mathrm{reg}}$, and a holomorphic retraction $\rho\colon U_0 \to U$ (see Theorem 3.3.3 on p. 74).

By Proposition 5.7.3 there is a smooth compact strongly pseudoconvex Cartan pair (A_0, B_0) in $\mathbb{C}^n$ such that A_0, B_0 contain the sets A, B in their respective interiors and approximate them as closely as desired. It is easy to arrange that the boundary of the set $C_0 = A_0 \cap B_0 \subset U_0$ intersects the submanifold X_{reg} transversely and we have $\rho(C_0) = C_0 \cap X$. (See [206, proof of Theorem 3.2] for the details.)

Recall that $U = U_0 \cap X \subset X_{\mathrm{reg}}$ is a relatively compact Stein neighborhood of C. Choose holomorphic vector fields $V_1, \ldots, V_N$ on X which generate the tangent sheaf $\mathcal{T}X$ over $\overline{U}$ and vanish to a given order on the subvariety X'. Let $\Phi\colon \Omega \to X$ be the induced map (9.17) defined on a neighborhood $\Omega \subset X \times \mathbb{C}^N$ of $X \times \{0\}$. As in the proof of Theorem 9.7.1, we have a holomorphic vector bundle decomposition $U \times \mathbb{C}^N = E \oplus E'$ and neighborhoods $0_x \in W_x \subset E_x$ of uniform size for all $x \in U$ such that W_x is mapped by $\Phi(x, \cdot)$ bijectively onto a neighborhood of x in X_{reg}. It follows that Lemma 9.7.2 holds.

Let $A' = A_0 \cap X$, $B' = B_0 \cap X$, and $C' = C_0 \cap X \subset X_{\mathrm{reg}}$; then, (A', B') is a Cartan pair in X which contains (A, B) in the interior.

Denote by $\mathscr{C}$ the Banach space of maps $c\colon C' \to \mathbb{C}^N$ of class $\mathcal{A}(C', \mathbb{C}^N)$ such that $c(x) \in E_x$ for every $x \in C'$, endowed by the sup-norm. Recall that $\rho(C_0) = C'$. Hence, the map $\mathscr{C} \ni c \mapsto c \circ \rho \in \mathcal{A}(C_0, \mathbb{C}^N)$ is a bounded linear extension operator. By using this operator and Lemma 9.7.3 on the Cartan pair (A_0, B_0), we obtain bounded linear operators $\mathscr{A}_0\colon \mathscr{C} \to \mathcal{A}(A', \mathbb{C}^N)$ and $\mathscr{B}_0\colon \mathscr{C} \to \mathcal{A}(B', \mathbb{C}^N)$ such that (9.18) holds:

$$c = \mathscr{B}_0 c - \mathscr{A}_0 c \quad \text{for every } c \in \mathscr{C}.$$

From this point, the proof follows that of Theorem 9.7.1 essentially verbatim. Note that the local analysis concerning the Taylor expansion of the maps in the proof takes place in a neighborhood of $C' \subset X_{\mathrm{reg}}$, so we need not deal with the singularities of X. Assuming that the map $\gamma: \widetilde{C} \to X$ is sufficiently uniformly close to the identity, we thus get holomorphic maps $\alpha: A' \to X$ and $\beta: B' \to X$ satisfying

$$\gamma \circ \alpha = \beta \quad \text{on } C'$$

and the linear estimates (9.16), thereby improving those in [206, Theorem 3.2]. If these maps are sufficiently close to the identity on their respective domains, then it is easily seen that their restrictions to a pair of smaller open neighborhoods of A and B, respectively, are injective holomorphic maps (see [206, Lemma 2.13]). □

9.8 The Oka Principle for Proper Holomorphic Maps

In this section we consider the Oka principle for proper holomorphic maps, immersions and embeddings of Stein manifolds to noncompact complex manifolds more general than Euclidean spaces. We discuss two types of results. In the first one, the source manifold is a relatively compact, smoothly bounded, strongly pseudoconvex domain D in a Stein manifold S. In this case we have an essentially optimal result under a purely geometric condition on X, namely q-convexity for a suitable value of q depending on $\dim D$; see Theorem 9.8.1 and Corollaries 9.8.2 and 9.8.4. Results of the second type pertain to the case when the source S is an arbitrary Stein manifold and the target X is a Stein manifold with the holomorphic (volume) density property; see Theorems 9.8.6 and 9.8.8. Such X is an Oka manifold by Proposition 5.6.23 (p. 233).

Theorem 9.8.1 (Drinovec Drnovšek and Forstnerič [137, Corollary 1.2]) *Let X be an n-dimensional complex manifold, and let D be a relatively compact, smoothly bounded, strongly pseudoconvex domain in a Stein manifold S of dimension d. Assume that $2d \le n$ and $q \in \{1, \ldots, n-2d+1\}$. If $\rho: X \to \mathbb{R}$ is an exhaustion function which is q-convex on $\{\rho > c\}$ for some $c \in \mathbb{R}$, then every holomorphic map $f: \bar{D} \to X$ satisfying $\rho \circ f > c$ on bD can be approximated uniformly on compacts in D by proper holomorphic immersions $D \to X$ (embeddings if $2d+1 \le n$).*

The following case of Theorem 9.8.1 with $d = 1$ is of particular interest. This is also a special case of [135, Theorem 1.1].

Corollary 9.8.2 *Assume that X is an n-dimensional complex manifold with a smooth exhaustion function $\rho: X \to \mathbb{R}$ whose Levi form has at least two positive eigenvalues at every point. Let $\bar{D}$ be a compact bordered Riemann surface. Then, every continuous map $\bar{D} \to X$ which is holomorphic in the interior D can be approximated, uniformly on compacts in D, by proper holomorphic immersions $D \to X$ (embeddings if $n \ge 3$).*

These results present a culmination of a long series of developments by many authors; we refer to the papers [135, 137] for the discussion, history and references. (See also Sect. 8.6 in the first edition of this book.)

The following example shows that the assumptions in Theorem 9.8.1 cannot be weakened for any pair of dimensions $1 \le d < n$.

Example 9.8.3 Given integers $1 \le d < n$, set $m = n - d + 1 \in \{2, \dots, n\}$. Let $\mathbb{T}^m = \mathbb{C}^m / \Gamma$ be a complex torus that does not contain any closed complex curves. (Most tori of dimension > 1 are such; an explicit example with $m = 2$ can be found in [568, p. 222].) Set

$$X = (\mathbb{T}^m \setminus \{p\}) \times \mathbb{C}^{n-m} = (\mathbb{T}^m \setminus \{p\}) \times \mathbb{C}^{d-1}. \tag{9.22}$$

Note that X is an Oka manifold (see Corollary 5.6.18, p. 231). Choose a function $\tau \colon \mathbb{T}^m \setminus \{p\} \to \mathbb{R}$ that equals $|y - y(p)|^{-2}$ in some local holomorphic coordinates y on $\mathbb{T}^m$ near p. The exhaustion function $\rho(y, w) = \tau(y) + |w|^2$ on X has no critical points in a deleted neighborhood of p, and its Levi form has $1 + n - m = d$ positive eigenvalues near $\{p\} \times \mathbb{C}^{d-1}$. Thus, X satisfies the assumptions of Theorem 9.8.1 for domains D of dimension $< d$, but not for domains of dimension $\ge d$.

We claim that no d-dimensional Stein manifold S admits a proper holomorphic map to the manifold X (9.22). Indeed, suppose that $f \colon S \to X$ is a proper holomorphic map. Let $\pi \colon X \to \mathbb{C}^{d-1}$ denote the projection $\pi(y, w) = w$. For reasons of dimension, there is a point $w \in \mathbb{C}^{d-1}$ for which $\Sigma = \{z \in S : \pi(f(z)) = w\}$ is a subvariety of positive dimension in S. Since S is Stein, Σ contains a one dimensional subvariety C, and $f(C)$ is then a closed complex curve in $\mathbb{T}^m \setminus \{p\} \times \{w\}$. Since a point is a removable singularity for positive dimensional analytic subvarieties, $\overline{f(C)}$ is a nontrivial closed complex curve in $\mathbb{T}^m \times \{w\}$, a contradiction to the choice of $\mathbb{T}^m$. □

Natural examples of q-convex and q-complete manifolds are complements of closed complex submanifolds in projective spaces and other compact manifolds. If X is a complex manifold and $A \subset X$ is a compact complex submanifold of codimension q with Griffiths positive normal bundle, then $X \setminus A$ admits a function $\rho \colon X \setminus A \to \mathbb{R}$ which tends to $+\infty$ along A and is q-convex and without critical points in a deleted neighborhood of A (see Schneider [486]). This holds in particular for every projective submanifold $A \subset \mathbb{P}^n$ (see Barth [44]); the manifold $\mathbb{P}^n \setminus \mathbb{P}^{n-q}$ is even q-complete. Thus, Theorem 9.8.1 applies when $n - q + 1 > \dim D$ or, equivalently, when $\dim D \le \dim A$. This gives the following corollary on the existence of proper holomorphic maps $D \to \mathbb{C}^n = \mathbb{P}^n \setminus \mathbb{P}^{n-1}$ for $n > \dim D$.

Corollary 9.8.4 ([137, Corollary 1.4]) *Let A be a compact complex submanifold of $\mathbb{P}^n$. Every smoothly bounded, relatively compact, strongly pseudoconvex Stein domain D with* $\dim D \le \dim A$ *admits a proper holomorphic map $D \to \mathbb{P}^n \setminus A$. In particular, if* $\dim D < n$ *then D admits a proper holomorphic map into $\mathbb{P}^n \setminus A$ for any smooth complex hypersurface A in $\mathbb{P}^n$. The analogous conclusion holds for* *maps $D \to X \setminus A$, where A is a compact complex submanifold with Griffiths positive normal bundle in a complex manifold X.*

The proof of Theorem 9.8.1 combines three analytic techniques. The first one is a method of lifting small pieces of the boundary of $\bar{D}$ (considered as a subset of X) to higher levels of the exhaustion function ρ. Here, the Levi geometry plays a key role, and we use q-convexity of X for suitable q. This technique is fairly elementary when $\dim D = 1$, but is rather delicate in higher dimensions. The main lifting lemma (see [137, Lemma 5.3]) employs special holomorphic peak functions that reach their maximum along certain complex tangential submanifolds of maximal real dimension in the boundary of D. Its proof relies on the work of Dor [132–134], building upon earlier results of many authors. Each local modification is patched with the global map $\bar{D} \to X$ by the method of gluing holomorphic sprays, explained in Sect. 5.9. To pass a critical level of the exhaustion function ρ, one applies Lemma 3.11.1 (p. 98) and the general position argument. For further details, see [135, 137].

By using Theorem 9.8.1, Forstnerič, Smrekar and Sukhov [224] gave the following affirmative solution to the Hodge conjecture for the top dimensional a priori nontrivial cohomology group of a q-complete manifold. (We state a special case of their result and refer to [224] for references on this subject.)

Theorem 9.8.5 (Hodge theorem for q-complete manifolds) *Let X be a complex manifold of dimension $n > 1$ with a q-convex Morse exhaustion function with only finitely many critical points for some $q \in \{1, \dots, n-1\}$. If the number $n + q - 1 = 2p$ is even, then every cohomology class in $H^{2p}(X; \mathbb{Z})$ is Poincaré dual to an analytic cycle $Z = \sum_j n_j Z_j$ of complex dimension p with integer coefficients, where each Z_j is an embedded complex submanifold (immersed with normal crossings if $q = 1$) biholomorphic to the ball in $\mathbb{C}^p$.*

A version of this result also holds if X admits a q-convex Morse exhaustion function $\rho\colon X \to \mathbb{R}$ with infinitely many critical points. In this case we take an exhaustion $M_1 \subset M_2 \subset \cdots \subset \bigcup_{j=1}^{\infty} M_j = X$, where each M_j is a regular sublevel set of ρ. The inclusion $M_j \hookrightarrow M_{j+1}$ induces a homomorphism $H^k(M_{j+1}; \mathbb{Z}) \to H^k(M_j; \mathbb{Z})$, and we have a well defined inverse limit $\mathcal{H}^k(X; \mathbb{Z}) = \lim_j H^k(M_j; \mathbb{Z})$. Then, every class in $\mathcal{H}^{n+q-1}(X; \mathbb{Z})$ can be represented by an analytic chain as in Theorem 9.8.5.

We now discuss results of the second type where X is a Stein manifold with the (volume) density property (see Definition 4.10.1 on p. 142). The following result is due to Andrist and Wold [36] in the special case when S is an open Riemann surface, to Andrist et al. [29, Theorems 1.1, 1.2] for embeddings when $\dim X > 2\dim S$, and to the author [230, Theorem 1.1] for immersions when $\dim X = 2\dim S$.

Theorem 9.8.6 *Let X be a Stein manifold with the density property or the volume density property (see Definition 4.10.1, p. 142). Let S be a Stein manifold and $f\colon S \to X$ a continuous map. If $\dim X > 2\dim S$ then f is homotopic to a proper* *holomorphic embedding $S \hookrightarrow X$, and if* $\dim X = 2\dim S$ *then it is homotopic to a proper holomorphic immersion $S \to X$ with simple double points.*

Theorem 9.8.6 generalizes results of Bishop, Narasimhan and Remmert concerning the existence of proper holomorphic immersions and embedding of Stein manifolds to Euclidean spaces (see Theorem 2.4.1(a) on p. 52).

Proof The scheme is similar as in the proof of Theorem 5.4.4 which gives holomorphic maps of Stein manifolds to Oka manifolds. The nontrivial addition where the density property is used comes in the following situation which describes the main induction step.

Set $n = \dim X$ and $d = \dim S$. Let $L \subset X$ be a compact $\mathcal{O}(X)$-convex set. Assume that $A \subset S$ is a compact strongly pseudoconvex domain and $B \subset X$ is a convex bump attached to A such that (A, B) is a special convex pair (see Definition 5.10.2). In particular, there is a holomorphic coordinate on a neighborhood of B in which B and $C = A \cap B$ are strongly convex sets. (In the paper [230], even more special Cartan pairs are used in which the attaching set C is a thin slab of B; this is necessary in the limit case when $n = 2d$.) Furthermore, we are given a compact set $K \subset \mathring{A} \setminus C$ such that $K \cup C$ is $\mathcal{O}(A)$-convex, an open set $W \subset S$ containing A, and a holomorphic immersion $f\colon W \to X$ (which can be made injective when $n > 2d$) such that $f^{-1}(L) \subset \mathring{K}$. We must approximate the restricted map $f|_C\colon C \to X \setminus L$ as closely as desired by a holomorphic map $g\colon B \to X$ whose image lies in $X \setminus L$; this is used to control properness. This is accomplished by [29, Lemma 2.2] (or [230, Lemma 3.1] when $n = 2d$); we describe it below. By gluing f and g we get a new holomorphic map $\tilde{f}\colon A \cup B \to X$ with $\tilde{f}(B) \subset X \setminus L$. We complete the proof by an inductive application of this procedure with respect to an exhaustion of X by a sequence $L_1 \subset L_2 \subset \cdots \subset \bigcup_{i=1}^{\infty} L_i = X$ of compact $\mathcal{O}(X)$-convex sets, together with the method of attaching handles to domains in S as in the proof of Theorem 5.4.4, and using also the general position argument at every step to ensure that we have embeddings if $n > 2d$ and immersions with simple double points when $n = 2d$.

In order to solve the local extension-approximation problem described above, we proceed as follows (see [29, Lemma 2.2] and [230, Lemma 3.1] for the details). We focus on the case $n > 2d$ when f is an embedding. By a small perturbation of f we can achieve that the set $L \cup f(K) \cup f(C)$ is $\mathcal{O}(X)$-convex; this uses Theorem 9.8.1. Setting $L' = L \cup f(K)$, it follows that $L' \cap f(C) = \varnothing$ and the sets L', $f(C)$ and $L' \cup f(C)$ are $\mathcal{O}(X)$-convex. Pick a compact set $P \subset X \setminus L'$ such that $f(C) \subset \mathring{P}$ and $L' \cup P$ is also $\mathcal{O}(X)$-convex. By the hypotheses on $C \subset B$, there is a holomorphic coordinate map $z\colon V_0 \to \mathbb{C}^d$ on a neighborhood $V_0 \subset S$ of B which makes B and C geometrically convex. Choose open convex neighborhoods $U, V \subset S$ of C and B, respectively, such that $U \subset V \cap W$ and $V \subset V_0$. We can find an isotopy $r_t\colon V \to V$ of injective holomorphic self-maps, depending smoothly on $t \in [0, 1]$, such that r_0 is the identity map on V, $r_t(U) \subset U$ for all $t \in [0, 1]$, and $r_1(V) \subset U$. In the coordinates z on V_0 we can simply choose r_t to be a family of linear contractions towards a point in U.

Since U is convex, the normal bundle of the embedding $f\colon W \hookrightarrow X$ is holomorphically trivial over U by the Oka-Grauert principle (see Theorem 5.3.1). Hence, after shrinking W around A and U around C if necessary, there is a holomorphic map $F\colon W \times \mathbb{D}^{n-d} \to X$, where $\mathbb{D}^{n-d}$ denotes the polydisc in $\mathbb{C}^{n-d}$, which is injective on $U \times \mathbb{D}^{n-d}$ and satisfies $F(z,0) = f(z)$ for all $z \in W$. In the case $n = 2d$ the same can be achieved provided the C is a very thin slab of B on which f is an embedding. By a further shrinking of $U \supset C$ and a rescaling of $\mathbb{D}^{n-d}$ we may also assume that the Stein domain $\Omega = F(U \times \mathbb{D}^{n-d}) \subset P \subset X \setminus L'$ is Runge in $\mathring{P}$ and its closure $\overline{\Omega}$ is $\mathcal{O}(P)$-convex. Since $L' \cup P$ is $\mathcal{O}(X)$-convex, it follows that $L' \cup \overline{\Omega}$ is also $\mathcal{O}(X)$-convex. Hence there is a Stein neighborhood $\Omega' \subset X$ of L' such that $\overline{\Omega} \cap \overline{\Omega}' = \varnothing$ and the union $\Omega_0 := \Omega \cup \Omega'$ is a Stein Runge domain in X. Consider the isotopy of biholomorphic maps $\phi_t\colon V \times \mathbb{D}^{n-d} \to V \times \mathbb{D}^{n-d}$ given by

$$\phi_t(z,w) = \big(r_t(z), w\big), \quad z \in V,\ w \in \mathbb{D}^{n-d},\ t \in [0,1]. \tag{9.23}$$

We define a smooth isotopy of injective holomorphic maps $\psi_t\colon \Omega_0 \to X$ by

$$\psi_t = F \circ \phi_t \circ F^{-1} \quad \text{on } \Omega, \qquad \psi_t = \mathrm{Id} \quad \text{on } \Omega', \quad t \in [0,1]. \tag{9.24}$$

The map ψ_t is defined on Ω since $r_t(U) \subset U$ for all $t \in [0,1]$. Note that ψ_0 is the identity on Ω_0 and the domain $\psi_t(\Omega_0)$ is Runge in X for all $t \in [0,1]$. Since X has the density property, Theorem 4.10.5 shows that we can approximate the map $\psi_1 = F \circ \phi_1 \circ F^{-1}\colon \Omega_0 \to X$ uniformly on compacts in Ω_0 by holomorphic automorphisms $\Psi \in \mathrm{Aut}(X)$. Fix such Ψ and consider the injective holomorphic map

$$G = \Psi^{-1} \circ F \circ \phi_1\colon V \times \mathbb{D}^{n-d} \to X.$$

Note that G is indeed defined on $V \times \mathbb{D}^{n-d}$ since $\phi_1(z,w) = (r_1(z), w)$ and $r_1(V) \subset U$, so $\phi_1(V \times \mathbb{D}^{n-d}) \subset U \times \mathbb{D}^{n-d}$. Since ψ_1 equals the identity map on $\Omega' \supset L'$ by (9.24), Ψ can be chosen to approximate the identity as closely as desired on a neighborhood of L', so we may assume that $G(V \times \mathbb{D}^{n-d}) \subset X \setminus L'$. From the first equation in (9.24) we see that

$$G = \Psi^{-1} \circ \psi_1 \circ F \quad \text{on } U \times \mathbb{D}^{n-d}.$$

Since $\Psi^{-1} \circ \psi_1$ is close to the identity map on $F(U \times \mathbb{D}^{n-d})$ by the choice of Ψ_1, G is close to F on $U \times \mathbb{D}^{n-d}$. More precisely, the above argument shows that for any given compact subset M of $U \times \mathbb{D}^{n-d}$ we can choose the automorphism $\Psi \in \mathrm{Aut}(X)$ such that the map G is as close as desired to F on M.

Assuming that the approximation of F by G is close enough, we can use Theorem 9.7.1 to glue them into a holomorphic map $\widetilde{F}\colon (A \cup B) \times \rho\mathbb{D}^{n-d} \to X$ for some $0 < \rho < 1$ such that $\widetilde{F}$ is close to F on $A \times \rho\mathbb{D}^{n-d}$, and is close to G on $B \times \rho\mathbb{D}^{n-d}$. The holomorphic map $\tilde{f} = \widetilde{F}(\cdot\,,0)\colon A \cup B = D \to X$ then satisfies the desired condition. By a small perturbation we can assume that it is an injective embedding if $n > 2d$ and an immersion if $n \geq 2d$. If the approximations are close enough, then we have $\tilde{f}^{-1}(L) \subset \mathring{K}$.

This solves the approximation problem if X has the density property. A similar construction applies in the case when X enjoys the volume density property; we refer to [29] for the details. □

Problem 9.8.7 Assume that S and X are Stein manifolds and X has the density property. If $\dim S < \dim X$, does there exist a proper holomorphic map $S \to X$?

By Theorem 2.4.1 this question has an affirmative answer when $X = \mathbb{C}^n$. The following result of Forstnerič and Ritter [220] provides a considerably more precise result and gives a new construction of proper holomorphic maps $S \to \mathbb{C}^n$ for Stein manifolds S of dimension $< n$.

Theorem 9.8.8 ([220]) *Let L be a compact convex set in $\mathbb{C}^n$. Given a Stein manifold S with $\dim S < n$, a compact $\mathcal{O}(S)$-convex set $K \subset S$, a closed complex subvariety S' of S, and holomorphic map $f\colon K \cup S' \to \mathbb{C}^n$ such that $f(bK \cup (S' \setminus K)) \cap L = \varnothing$, there exists for every $\epsilon > 0$ a holomorphic map $F\colon S \to \mathbb{C}^n$ satisfying the following conditions:*

$$\text{(a) } F(S \setminus K) \subset \mathbb{C}^n \setminus L, \qquad \text{(b) } \|F - f\|_K < \epsilon, \qquad \text{(c) } F|_{S'} = f|_{S'}.$$

If the map $f|_{S'}\colon X' \to \mathbb{C}^n$ is proper, then F can also be chosen proper. If $2\dim S \le n$ then the same holds when the set L is polynomially convex. In particular, if $f(K \cup S') \cap L = \varnothing$, then there exists a holomorphic map $F\colon S \to \mathbb{C}^n \setminus L$ satisfying the above properties. If in addition $f|_{S'}\colon X' \to \mathbb{C}^n$ is proper, then F can also be chosen proper as a map to $\mathbb{C}^n$.

The proof of Theorem 9.8.8 follows the same path as that of Theorem 9.8.6 outlined above. It says in particular that the complement $\mathbb{C}^n \setminus K$ of a compact convex set K enjoys the Oka property for maps from Stein manifolds of dimension $< n$. It is an open problem whether $\mathbb{C}^n \setminus K$ is an Oka manifold.

Another result of this type is due to Forstnerič and Wold [229] who showed that for any pair of integers $1 \le k < n$, the domain $\mathbb{C}^n \setminus \mathbb{R}^k$ enjoys the Oka property for maps from Stein manifolds of dimension $< n$. Its proof uses Theorem 4.14.7 (p. 169) by Kutzschebauch and Wold on Carleman approximation by holomorphic automorphisms of $\mathbb{C}^n$.

9.9 Exposing Points of Bordered Riemann Surfaces

In this section we prove the following result on exposing boundary points of smoothly bounded domains in Riemann surfaces.

Theorem 9.9.1 ([226, Theorem 2.3]) *Assume that D is a relatively compact smoothly bounded domain in a Riemann surface R. Choose finitely many pairwise distinct points $a_1, \ldots, a_k \in bD$, $b_1, \ldots, b_k \in R \setminus D$, and $c_1, \ldots, c_l \in D \setminus \{a_1, \ldots, a_k\}$ such that for each $j = 1, \ldots, k$ the points a_j and b_j are connected by a Jordan arc $\gamma_j \subset R \setminus D$ with $\gamma_j \cap \bar{D} = \{a_j\}$. Suppose that for every $j = 1, \ldots, k$ we are given neighborhoods $U'_j \Subset U_j$ of the point a_j and a neighborhood $V_j \subset R$ of the arc γ_j. Then, for every integer $N \in \mathbb{N}$ there exists a smooth diffeomorphism $\phi \colon \bar{D} \to \bar{D}'$ onto a compact smoothly bounded domain $\bar{D}' \subset R$ such that the following conditions hold:*

1. *$\phi \colon D \to D'$ is biholomorphic,*
2. *ϕ is as close as desired to the identity map in $\mathscr{C}^\infty(\bar{D} \setminus \bigcup_{j=1}^k U'_j)$,*
3. *ϕ is tangent to the identity map to order N at each point c_j, and*
4. *$\phi(a_j) = b_j$ and $\phi(\bar{D} \cap U'_j) \subset U_j \cup V_j$ for $j = 1, \ldots, k$.*

These conditions mean that $\bar{D}' = \phi(\bar{D})$ is contained in a small neighborhood of $\bar{D} \cup (\bigcup_{j=1}^k \gamma_j)$, it is close to $\bar{D}$ away from a small neighborhood of each point $a_j \in bD$, and at a_j it includes a spike reaching out to $b_j \in bD'$.

We need the following lemma. Denote by $\mathbb{D}$ the open unit disc in $\mathbb{C}$ and by $r\mathbb{D}$ the disc of radius $r > 0$.

Lemma 9.9.2 *Assume that R is a connected open Riemann surface, $G \Subset R$ is an open simply connected domain with smooth boundary, $V' \Subset V'' \subset R$ are small neighborhoods of a boundary point $a \in bG$, b is a point in $R \setminus \overline{G}$, γ is a smooth Jordan arc with endpoints a and b such that $\gamma \cap \overline{G} = \{a\}$ and the tangent lines to γ and bG at the point a are transverse, and V is a neighborhood of γ. Then, there exists a sequence of smooth diffeomorphisms $\psi_n \colon \overline{G} \to \psi_n(\overline{G}) \subset R$ that are holomorphic on G and satisfy the following:*

(i) $\psi_n \to \mathrm{Id}$ *uniformly on $\overline{G} \setminus V'$ as $n \to \infty$,*
(ii) *$\psi_n(a) = b$ for $n = 1, 2 \ldots$, and*
(iii) *$\psi_n(\overline{V}' \cap \overline{G}) \subset V'' \cup V$ for $n = 1, 2 \ldots$.*

Proof Since $\overline{G} \cup \gamma$ admits a simply connected neighborhood in R and we are going to construct maps with images near $\overline{G} \cup \gamma$, we may assume that we are working in the complex plane, $a \in bG$ is the origin, and the positive real axis lies in the complement of $\overline{G}$ near the origin. For each $n \in \mathbb{N}$, let l_n denote the line segment between 0 and $\frac{1}{n}$ in $\mathbb{R} \subset \mathbb{C}$. Let $\widetilde{V}$ be a neighborhood of the origin with $V' \Subset \widetilde{V} \Subset V''$. Choose a diffeomorphism $h_n \colon \overline{G} \cup l_n \to \overline{G} \cup \gamma$ which is the identity on $\overline{G}$ and maps l_n diffeomorphically onto γ. By Mergelyan's theorem [413] there are neighborhoods U_n of $\overline{G} \cup l_n$ and holomorphic injections $f_n \colon U_n \hookrightarrow \mathbb{C}$ such that the following hold for all $n \in \mathbb{N}$:

(1) $f_n \to \mathrm{Id}$ uniformly on $\overline{G}$ as $n \to \infty$,
(2) $f_n(l_n)$ approximates γ, with $f_n(\frac{1}{n}) = b$ and $f_n(l_n) \subset V$, and
(3) $f_n(\overline{G} \cap \widetilde{V}) \subset V''$. (This is a consequence of (1) for big enough n.)

For small $\epsilon > 0$ let Ω_ϵ denote a smoothly bounded domain obtained by adding to G an ϵ-strip around l_n, containing the point $\frac{1}{n}$ in the boundary $b\Omega_\epsilon$ and smoothing the corners. Set $R_\epsilon = \Omega_\epsilon \setminus G$. Choose a sequence $\epsilon_n \to 0$ such that $\overline{\Omega}_{\epsilon_n} \subset U_n$ for each $n \in \mathbb{N}$. Write $\Omega_n = \Omega_{\epsilon_n}$ and $R_n = R_{\epsilon_n}$. By choosing the ϵ_n's small enough we get that

(4) $f_n(\overline{R}_n) \subset V$ for each $n \in \mathbb{N}$.

Choose a point $p \in G$ and a sequence of biholomorphic maps $g_n\colon G \to \Omega_n$ such that $g_n(p) = p$ and $g_n'(p) > 0$ for $n = 1, 2, \ldots$. Since our domains are smoothly bounded, the map g_n extends to a smooth diffeomorphism of $\overline{G}$ onto $\overline{\Omega}_n$. Furthermore, since the domains $\overline{\Omega}_n$ converge to $\overline{G}$ as $n \to \infty$, we conclude by Rado's theorem (see [458, Corollary 2.4, p. 22]) that

(5) $g_n \to \mathrm{Id}$ uniformly on $\overline{G}$ as $n \to \infty$.

Hence, for n big enough we have that $g_n(\overline{V'} \cap \overline{G}) \subset (\widetilde{V} \cap \overline{G}) \cup R_n$ (keeping in mind that g_n is injective and close to the identity). Combining this with (3) and (4), we see that $(f_n \circ g_n)(\overline{V'} \cap \overline{G}) \subset V'' \cup V$ if n is big enough. Hence, the map $\psi_n = f_n \circ g_n$ satisfies conditions (i) and (iii) for all big $n \in \mathbb{N}$.

To see that condition (ii) holds, let $a_n \in bG$ denote the point that g_n sends to $\frac{1}{n} \in b\Omega_n$. By (5), the sequence a_n converges to the origin, and hence there is a sequence of conformal automorphisms φ_n of G fixing the point p, sending the origin to a_n, with $\varphi_n \to \mathrm{Id}$ uniformly on $\overline{G}$. Replacing the maps g_n by $g_n \circ \varphi_n$ in the above argument also gives (ii). □

The next lemma provides the main step in the proof of Theorem 9.9.1.

Lemma 9.9.3 *There is a smooth diffeomorphism $\phi\colon \bar{D} \to \bar{D}' \subset R$ such that $\phi\colon D \to D'$ is biholomorphic, $\phi(a_1) = b_1$, ϕ is tangent to the identity map to order N at each of the points $a_2, \ldots, a_k$ and $c_1, \ldots, c_l$, ϕ is as close as desired to the identity in the smooth topology on $\bar{D} \setminus U_1'$, and $\phi(\bar{D} \cap U_1') \subset U_1 \cup V_1$.*

Proof We may assume that $N > 2$. Write $\gamma = \gamma_1$. We may choose γ such that it intersects bD transversely at a_1. Then, γ has an open, connected and simply connected neighborhood $W \subset R$ which is conformally equivalent to the disc. Let z be a holomorphic coordinate on W with $z(a_1) = 0$. By shrinking the neighborhood $U' = U_1'$ of a_1 we may assume that $\overline{U}' \subset W$, that $\overline{U}'$ does not contain any of the points $a_2, \ldots, a_k, c_1, \ldots, c_l$, and that $z(U') = r'\mathbb{D} \subset \mathbb{C}$ for some $r' > 0$. Choose $r'' \in (0, r')$ and let $U'' \subset U'$ be such that $z(U'') = r''\mathbb{D}$. Pick a connected and simply connected domain $G \subset W$ with smooth boundary, with a defining function ρ such that $G = \{\rho < 0\}$ and $d\rho \neq 0$ on bG, satisfying the following conditions (see Fig. 9.1):

(i) $\bar{D} \cap U' \subset G \cup \{a_1\}$,
(ii) $-\rho(z) \geq \mathrm{const}\cdot \mathrm{dist}(z, a_1)^2$ for points $z \in bD$ close to a_1, and
(iii) $\gamma \cap \overline{G} = \{a_1\}$.

Condition (iii) can be achieved since the arc γ is transverse to bD at a_1.

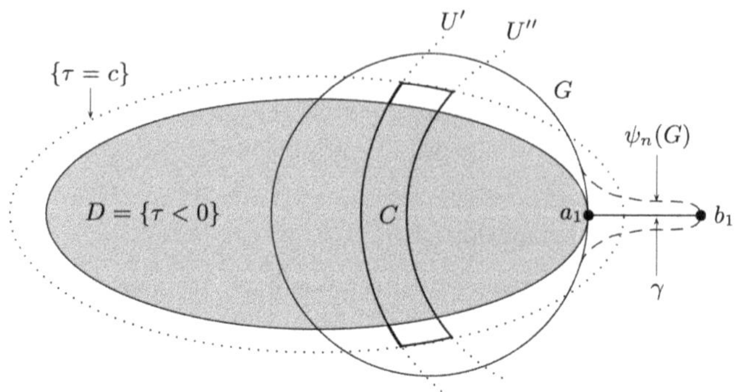

Fig. 9.1 The domains D and G. ([226, p. 104, Fig. 1])

Choose a smooth function τ on R such that $D = \{\tau < 0\}$ and $d\tau \neq 0$ on $bD = \{\tau = 0\}$. Pick a small number $c > 0$ and let

$$A = \{\tau \le c\} \setminus U'', \qquad B = \{\tau \le c\} \cap \overline{U}', \qquad C = \{\tau \le c\} \cap \left(\overline{U}' \setminus U''\right).$$

By taking $c > 0$ small enough, we ensure that C is a compact set contained in G (see Fig. 9.1) and we have

$$A \cup B = \{\tau \le c\}, \qquad A \cap B = C.$$

On Fig. 9.1, the set C is bounded by the two circular arcs (left and right) and by the two arcs in the dotted ellipse representing the level set $\{\tau = c\}$. The set A is the part of the dotted ellipse lying on the left hand side of the right boundary arc of C, and B is the part of the dotted ellipse on the right hand side of the left boundary arc of C.

Choose open neighborhoods $V' \Subset V''$ of the point a_1 such that $\overline{V}''$ is contained in the interior of $B \setminus A$. By shrinking the neighborhood $V = V_1$ around the arc γ, we may assume that $\overline{V} \cap \overline{A \setminus B} = \varnothing$. Let $\psi_n : \overline{G} \to \psi_n(\overline{G})$ be a sequence of conformal diffeomorphisms satisfying Lemma 9.9.2 with respect to the sets V, V', V''. Choose an open set $C' \Subset G$ containing the compact set $C = A \cap B$. As $n \to \infty$, the sequence ψ_n converges to the identity in the smooth topology on C'. For every sufficiently big $n \in \mathbb{N}$, Theorem 9.7.1 on p. 432 furnishes a splitting

$$\alpha_n = \psi_n \circ \beta_n \quad \text{on a neighborhood of } C, \tag{9.25}$$

where α_n is a small holomorphic perturbation of the identity map on a neighborhood of A (independent of n) that is tangent to the identity to order N at each of the points $a_2, \ldots, a_k, c_1, \ldots, c_l$, and β_n is a small holomorphic perturbation of the identity map on a neighborhood of B that is tangent to the identity to order N at the point a_1. The closeness of α_n and β_n to the identity in any C^r norm on A and B, respectively, can be estimated by the closeness of ψ_n to the identity on C'. If the approximations are sufficiently close (which holds for n big enough), then the two sides in (9.25), restricted to $A \cap \bar{D}$ and $B \cap \bar{D}$, respectively, define a diffeomorphism $\phi_n : \bar{D} \to \phi_n(\bar{D}) \subset R$ that is holomorphic in D and such that $\phi_n(a_1) = b_1$,

ϕ_n is tangent to the identity to order N at each of the points $a_2, \dots, a_k, c_1, \dots, c_l$, ϕ_n converges to the identity in the smooth topology on $\bar{D} \setminus U'$ as $n \to +\infty$, while $\psi_n(\bar{D} \cap U'') \subset U' \cup V$ for all big n. Indeed, both sides α_n and $\psi_n \circ \beta_n$ satisfy the stated properties on their respective domains. For α_n, this is clear from the construction. For β_n, we need a more precise argument to see that it maps $B \cap \bar{D}$ into $G \cup \{a_1\}$ for sufficiently big $n \in \mathbb{N}$. By the construction, its Taylor expansion in a local holomorphic coordinate z near a_1, with $z(a_1) = 0$, equals

$$\beta_n(z) = z + M_n z^N + O(z^{N+1}).$$

By using the Cauchy estimates, the size of the constant M_n and the remainder term can be estimated by $\operatorname{dist}_B(\beta_n, \mathrm{Id})$, and hence by $\operatorname{dist}_{C'}(\psi_n, \mathrm{Id})$. Since G osculates D from the outside to the second order at the point a_1 by property (ii), it follows that for a sufficiently small neighborhood U' of a_1 and for all big enough $n \in \mathbb{N}$ we have

$$\beta_n(\bar{D} \cap U'') \subset (G \cup \{a_1\}) \cap V'. \tag{9.26}$$

On $(B \cap \bar{D}) \setminus U''$ the map β_n is close to the identity for big n, and hence it takes this set into a fixed compact set in G independent of n. Thus, the composition $\psi_n \circ \beta_n$ is well defined on $B \cap \bar{D}$ and it satisfies the stated conditions.

It is also easily seen that ϕ_n is injective if n is big enough. Indeed, each of the two maps defining ϕ_n on $A \cap \bar{D}$ and $B \cap \bar{D}$, respectively, is injective by the construction. Hence, it suffices to verify that no point from $(A \setminus B) \cap \bar{D}$ gets identified with a point from $(B \setminus A) \cap \bar{D}$ under the map ϕ_n for big n. Note that the points from the first set remain nearby since α_n is close to the identity. Consider now points $x \in (B \setminus A) \cap \bar{D}$. If $x \in U''$, then $\beta_n(x) \in (G \cup \{a_1\}) \cap V'$ by (9.26), and hence $\psi_n \circ \beta_n(x) \in V'' \cup V$ by condition (iii) in Lemma 9.9.2. Since the set $V'' \cup V$ is at a positive distance from $(A \setminus B) \cap \bar{D}$, we see that $\psi_n \circ \beta_n(x) \neq \alpha_n(x')$ for any point $x' \in (A \setminus B) \cap \bar{D}$ provided that n is big enough. The remaining set $(B \setminus A) \cap \bar{D} \setminus U''$ is compactly contained in $B \cap \bar{D} \cap G$ where $\psi_n \circ \beta_n$ is close to the identity for big n, and hence no point from this set can get identified with a point from $(A \setminus B) \cap \bar{D}$. □

Proof of Theorem 9.9.1 We may assume that $N \geq 2$ and the neighborhoods $V_j \subset R$ of the arcs γ_j (with the endpoints a_j and b_j) for $j = 1, \dots, k$ have pairwise disjoint closures. We seek ϕ of the form

$$\phi = \phi_k \circ \phi_{k-1} \circ \cdots \circ \phi_1 \colon \bar{D} \to \bar{D}'.$$

Lemma 9.9.3 furnishes a conformal diffeomorphism $\phi_1 \colon \bar{D} \to \bar{D}_1 = \phi_1(\bar{D}) \subset R$ such that $\phi_1(a_1) = b_1$, ϕ_1 is tangent to the identity to order N at each of the points $a_2, \dots, a_k, c_1, \dots, c_l$, ϕ_1 is close to the identity map on $\bar{D} \setminus U_1'$, and $\phi_1(\bar{D} \cap U_1') \subset U_1 \cup V_1$. Hence, the points $b_1 = \phi_1(a_1), a_2, \dots, a_k$ lie on bD_1, while $c_1, \dots, c_l \in D_1$.

Next, we apply Lemma 9.9.3, with $\bar{D}$ replaced by $\bar{D}_1 = \phi_1(\bar{D})$, to find a conformal diffeomorphism $\phi_2 \colon \bar{D}_1 \to \phi_2(\bar{D}_1) = \bar{D}_2$ such that $\phi_2(a_2) = b_2$, ϕ_2 is tangent to the identity to order N at the points $b_1, a_3, \dots, a_k, c_1, \dots, c_l$, ϕ_2 is close to the identity map on $\bar{D}_1 \setminus U_2'$, and $\phi_2(\bar{D}_1 \cap U_2') \subset U_2 \cup V_2$.

Continuing inductively, we obtain after k steps a map ϕ satisfying the conclusion of Theorem 9.9.1 with $D' = D_k$. □

9.10 Embedding Bordered Riemann Surfaces in $\mathbb{C}^2$

One of the oldest open problems in classical function theory is whether every open Riemann surface, R, is biholomorphically equivalent to a closed smooth embedded complex curve in $\mathbb{C}^2$. Equivalently,

Problem 9.10.1 Does every open Riemann surface embed properly holomorphically in $\mathbb{C}^2$? (See Bell and Narasimhan [53, Conjecture 3.7, p. 20].)

Recall that every open Riemann surface embeds properly holomorphically in $\mathbb{C}^3$ by the Remmert-Bishop-Narasimhan theorem (see Theorem 2.4.1 on p. 52). The proof of the optimal embedding theorem for higher dimensional Stein manifolds (see Theorem 9.3.1) breaks down for Riemann surfaces.

The first positive results were obtained in the 1970's: for the unit disc (Kasahara and Nishino [523]), annuli (Laufer [379]), and the punctured disc (Alexander [21]). These were essentially the only known results at the time of the survey [53] by Bell and Narasimhan. In 1995, Globevnik and Stensønes proved the following major result.

Theorem 9.10.2 ([242]) *Every finitely connected domain in $\mathbb{C}$ without isolated boundary points embeds properly holomorphically into $\mathbb{C}^2$.*

The main idea of their proof is to embed the closure $\bar{D}$ of the given domain $D \subset \mathbb{C}$ into $\overline{\mathbb{D}} \times \mathbb{C}$, choosing the first coordinate to be an inner function that is continuous on $\bar{D}$. The boundary of D is then pushed to infinity by the method explained in Sect. 4.4, using compositions of shears in coordinate directions. For technical reasons, they make small modifications of the conformal structure on D due to cutting away small pieces near the boundary; hence, an additional argument using uniformization theory is needed to see that every planar domain of the indicated type can be embedded in this way.

Small improvements and extensions of Theorem 9.10.2, using solutions of a Riemann-Hilbert boundary value problem, can be found in [92, 93]. By [92, Theorem 1.1], every bordered orientable surface has a complex structure in which it admits a proper holomorphic embedding into $\mathbb{C}^2$. Alarcón and Lopez showed in [19] that the same is true for every open orientable surface, so there are no topological obstructions to the Bell-Narasimhan conjecture.

In 2013, Forstnerič and Wold [228] generalized Theorem 9.10.2 to all planar domains with at most countably many boundary components, none of which are points; see Theorem 9.11.4. Their proof gives the analogous result for domains in tori. They also constructed embeddings of certain domains with punctures. (We discuss surfaces of infinite connectivity in Sect. 9.11.) These developments were made possible by using a conceptually new method, introduced by Wold [581–583], which we now describe.

In Wold's first paper [582], he extended the Globevnik-Stensønes result (Theorem 9.10.2) to all finitely connected and some infinitely connected planar domains. In subsequent papers he developed a method that, together with another key addition from [226] which is given by Theorem 9.9.1 in the previous section, culminated in the proof of Theorem 9.10.3 below.

Recall that a *bordered Riemann surface* is the interior D of a compact one dimensional complex manifold, $\bar{D}$, not necessarily connected, with smooth boundary bD consisting of finitely many closed Jordan curves. Every such surface is conformally equivalent to a smoothly bounded domain D' in another Riemann surface R. More precisely, if bD is of class $\mathscr{C}^r$ and $r > 1$ is a noninteger, then any biholomorphic map $D \to D'$ extends to a $\mathscr{C}^r$ diffeomorphism $\bar{D} \to \bar{D}'$. (See the discussion in [226, Sect. 6].) The problem of finding a proper holomorphic embedding of such D into $\mathbb{C}^2$ naturally decouples into the following two problems:

(A) find a holomorphic embedding $f\colon \bar{D} \hookrightarrow \mathbb{C}^2$ onto a smooth compact complex curve with boundary $\overline{\Sigma} = f(\bar{D}) \subset \mathbb{C}^2$;
(B) given a compact complex curve $\overline{\Sigma} \subset \mathbb{C}^2$ with smooth boundary, find a proper holomorphic embedding of its interior Σ to $\mathbb{C}^2$.

The following result gives a complete answer to problem (B).

Theorem 9.10.3 ([226, Corollary 1.2]) *Assume that $\bar{D}$ is a compact bordered Riemann surface with $\mathscr{C}^r$ boundary for some $r > 1$ and $f\colon \bar{D} \hookrightarrow \mathbb{C}^2$ is a $\mathscr{C}^1$ embedding that is holomorphic in the interior D. Then, f can be approximated, uniformly on compacts in D, by proper holomorphic embeddings $D \hookrightarrow \mathbb{C}^2$.*

More generally, given a compact complex curve $\overline{\Sigma} \subset \mathbb{C}^2$ with smooth boundary $b\Sigma$, possibly with (finitely many) interior singularities, it was shown that the inclusion $\Sigma \hookrightarrow \mathbb{C}^2$ can be approximated uniformly on compacts in Σ by proper holomorphic embeddings $\Sigma \to \mathbb{C}^2$ (see [226, Theorem 1.1]).

Not much is known about problem (A). In view of the solution of (B), it would suffice to solve problem (A) affirmatively for complements of small discs in compact Riemann surfaces; this would imply that every bordered Riemann surface embeds properly in $\mathbb{C}^2$.

Theorem 9.10.3, together with [365], implies the following result on embeddings with interpolation. For the disc $\mathbb{D}$, see also [239].

Corollary 9.10.4 *Let D be as in Theorem 9.10.3. Given discrete sequences of points $\{a_j\} \subset D$ and $\{b_j\} \subset \mathbb{C}^2$ without repetitions, there is a proper holomorphic embedding $f\colon D \hookrightarrow \mathbb{C}^2$ such that $f(a_j) = b_j$ for $j = 1, 2, \ldots$.*

Example 9.10.5 ([583], [226, Example 1.4]) Let R be a smooth closed algebraic curve in $\mathbb{P}^2$. If $U_1, \ldots, U_k$ are pairwise disjoint smoothly bounded discs in R whose union contains the intersection of R with a projective line $\mathbb{P}^1 \subset \mathbb{P}^2$, then the bordered Riemann surface $D = R \setminus \bigcup_{i=1}^k \overline{U}_i \subset \mathbb{P}^2 \setminus \mathbb{P}^1$ embeds properly holomorphically into $\mathbb{C}^2$ according to Theorem 9.10.3. Since every one dimensional complex torus embeds as a smooth cubic curve in $\mathbb{P}^2$, we obtain:

Corollary 9.10.6 *Every finitely connected subset without isolated boundary points in a torus embeds properly holomorphically into $\mathbb{C}^2$.*

Example 9.10.7 ([92], [226, Example 1.5]) A compact Riemann surface R is called *hyperelliptic* if it admits a meromorphic function of degree two, i.e., a two-sheeted branched holomorphic covering $R \to \mathbb{P}^1$ (see [264, p. 247]). Such R is the normalization of a complex curve in $\mathbb{P}^2$ given by $w^2 = \Pi_{j=1}^k (z - z_j)$ for some choice of points $z_1, \ldots, z_k \in \mathbb{C}$. A bordered Riemann surface $\bar{D}$ is said to be hyperelliptic if its double is hyperelliptic. (The double of D is obtained by gluing two copies of $\bar{D}$, the second one with the conjugate conformal structure, along their boundaries.) A hyperelliptic bordered Riemann surface $\bar{D}$ admits a holomorphic embedding into the closed bidisc $\overline{\mathbb{D}}^2 \subset \mathbb{C}^2$ by a pair of inner functions mapping bD to the torus $(b\mathbb{D})^2$ [480]. Hence, Theorem 9.10.3 implies

Corollary 9.10.8 *Every smoothly bounded domain in a hyperelliptic bordered Riemann surface admits a proper holomorphic embedding in $\mathbb{C}^2$.*

In the proof of Theorem 9.10.3, we follow [226, proof of Theorem 1.1]. Another approach using the uniformization theory is outlined in [226, Sect. 6]. The following notion from [581] plays the key role.

Definition 9.10.9 Let $\pi : \mathbb{C}^2 \to \mathbb{C}$ be the projection $\pi(z_1, z_2) = z_1$. Assume that $\Sigma \subset \mathbb{C}^2$ is a locally closed complex curve, possibly with boundary. A point $p = (p_1, p_2) \in \Sigma$ is *exposed* (with respect to π) if the complex line

$$\Lambda_p = \pi^{-1}\big(\pi(p)\big) = \big\{(p_1, \zeta) : \zeta \in \mathbb{C}\big\}$$

intersects Σ precisely at p and the intersection is transverse. If $\Sigma = f(R)$, where R is a Riemann surface and $f : R \to \mathbb{C}^2$ is a holomorphic map, then a point $a \in R$ is *f-exposed* if the point $p = f(a) \in \Sigma$ is exposed.

Theorem 9.10.3 follows immediately from the following two results.

Theorem 9.10.10 ([226, Theorem 4.2]) *Let $f : \bar{D} \hookrightarrow \mathbb{C}^2$ be an embedding as in Theorem* 9.10.3. *Pick a point a_j in each connected component C_j of $bD = \bigcup_{j=1}^m C_j$ and let U_j be a neighborhood of a_j in $\bar{D}$. Then, f can be approximated in the $\mathscr{C}^1$ topology on $\bar{D} \setminus \bigcup_{j=1}^m U_j$ by holomorphic embeddings $F : \bar{D} \hookrightarrow \mathbb{C}^2$ such that $F(a_j)$ is an exposed point of $F(\bar{D})$ for every $j = 1, \ldots, m$, and F agrees with f to a given finite order at any given finite set of interior points of D.*

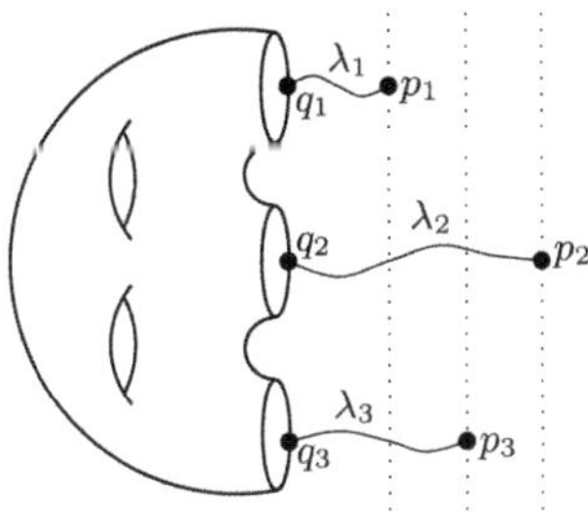

Fig. 9.2 A complex curve with exposed tails. ([226, p. 109, Fig. 2])

Theorem 9.10.11 ([581, Theorem 1.1]) *Let $f\colon \bar{D} \hookrightarrow \mathbb{C}^2$ be as in Theorem 9.10.3. If every boundary component of $f(\bar{D})$ contains an exposed point then the conclusion of Theorem 9.10.3 holds, i.e., f can be approximated uniformly on compacts in D by proper holomorphic embeddings $D \hookrightarrow \mathbb{C}^2$.*

Proof of Theorem 9.10.10 We realize $\bar{D}$ as a domain with smooth boundary in an open Riemann surface R; the corresponding biholomorphic map is of class $\mathscr{C}^1$ up to the boundary. By the Bishop-Mergelyan theorem [57, 413], we can approximate f in the $\mathscr{C}^1(\bar{D})$ topology by a holomorphic map $g\colon U \to \mathbb{C}^2$ on an open neighborhood $U \subset R$ of $\bar{D}$. Replacing f by g and R by a sufficiently small open neighborhood of $\bar{D}$ in R, we may therefore assume that $f\colon R \hookrightarrow \mathbb{C}^2$ is a (not necessarily proper) holomorphic embedding.

The boundary $bD = \bigcup_{j=1}^m C_j$ consists of finitely many smooth Jordan curves C_j. For every $j = 1, \ldots, m$ we choose a point $a_j \in C_j$ and a smooth embedded arc $\gamma_j \subset R$ that is transversely attached to $\bar{D}$ at its endpoint $a_j \in bD$. The rest of the arc, $\gamma_j \setminus \{a_j\}$, is contained in $R \setminus \bar{D}$. Let b_j denote the other endpoint of γ_j. Choose an open set $V \Subset R \setminus \{b_1, \ldots, b_m\}$ that contains $\bar{D}$ and such that $\gamma_j \cap V = \widetilde{\gamma}_j$ is an arc with an endpoint a_j.

For every $j = 1, \ldots, m$ we choose a smooth embedded arc λ_j in $\mathbb{C}^2$ that agrees with the arc $f(\widetilde{\gamma}_j)$ near the endpoint $q_j = f(a_j)$, while $\lambda_j \setminus f(\widetilde{\gamma}_j)$ does not intersect $f(V)$. We also ensure that the arcs $\lambda_1, \ldots, \lambda_m$ are pairwise disjoint and that the other endpoint p_j of λ_j is an exposed point for the set $f(\bar{D}) \cup (\bigcup_{j=1}^m \lambda_j) \subset \mathbb{C}^2$ (see Fig. 9.2).

Consider the compact set $K = \bar{D} \cup (\bigcup_{j=1}^m \gamma_j)$ in the Riemann surface R. Let $f'\colon V \cup (\bigcup_{j=1}^m \gamma_j) \to \mathbb{C}^2$ be a smooth map that agrees with f on V and maps each arc $\gamma_j \subset R$ diffeomorphically onto the corresponding arc $\lambda_j \subset \mathbb{C}^2$. In particular, the endpoint b_j of γ_j is mapped by f' to the exposed endpoint p_j of λ_j. By the Bishop-Mergelyan theorem, we can approximate f', uniformly on a neighborhood of $\bar{D}$ in R and in the $\mathscr{C}^1$ topology on each of the arcs γ_j, by a holomorphic map $\widetilde{f}\colon V \to \mathbb{C}^2$ from an open neighborhood of K in R. At the same time we ensure that $\widetilde{f}$ agrees with f' to a high order at each of the points $a_1, \ldots, a_m, b_1, \ldots, b_m$. If the approximation is close enough, the neighborhood $V \supset K$ is chosen small enough, and the interpolation at the indicated points is to a sufficiently high order, then $\widetilde{f}\colon V \to \mathbb{C}^2$ is a (non-proper) holomorphic embedding, the complex curve

$\widetilde{\Sigma} = \widetilde{f}(D) \subset \mathbb{C}^2$ is biholomorphic to the curve $\Sigma = f(D)$, and the points $p_j = \widetilde{f}(b_j)$ are exposed in $\widetilde{f}(V)$.

Theorem 9.9.1 on p. 442 furnishes a diffeomorphism $\phi\colon \bar{D} \to \phi(\bar{D}) \subset V$ that is holomorphic in D, that sends the point $a_j \in bD$ to the point b_j for every $j = 1, \ldots, m$, and that is close to the identity map outside a small neighborhood of $\{a_1, \ldots, a_m\}$ in $\bar{D}$. The composition $F = \widetilde{f} \circ \phi\colon \bar{D} \to \mathbb{C}^2$ maps $\bar{D}$ onto the domain $F(\bar{D})$ in the complex curve $\widetilde{f}(V) \subset \mathbb{C}^2$ such that each point $p_j = F(a_j)$ for $j = 1, \ldots, m$ is an exposed boundary point of $F(\bar{D})$. In addition, by the properties of ϕ given by Theorem 9.9.1, we can ensure that $F(\bar{D})$ lies in an arbitrary given open neighborhood of the set $f(\bar{D}) \cup (\bigcup_{j=1}^k \lambda_j)$. Finally, we can approximate F in the smooth topology on $\bar{D}$ by an embedding that is holomorphic in a neighborhood of $\bar{D}$. □

Proof of Theorem 9.10.11 Let $bD = \bigcup_{j=1}^m C_j$, and assume that $a_j \in C_j$ is an f-exposed point for each $j = 1, \ldots, m$ with respect to the first coordinate projection $\pi(z, w) = z$. Let $\pi_2\colon \mathbb{C}^2 \to \mathbb{C}$ be the second coordinate projection: $\pi_2(z, w) = w$. Define a rational shear

$$g(z, w) = \left(z, w + \sum_{j=1}^m \frac{\alpha_j}{z - \pi(f(a_j))} \right). \tag{9.27}$$

The numbers $\alpha_j \in \mathbb{C} \setminus \{0\}$ can be chosen such that π_2 maps the unbounded curves $\lambda_j = (g \circ f)(C_j \setminus \{a_j\}) \subset \mathbb{C}^2$ properly to pairwise disjoint unbounded curves $\gamma_j = \pi_2(\lambda_j) \subset \mathbb{C}$, and $\pi_2\colon \lambda_j \to \gamma_j$ is a diffeomorphism near infinity for every j. Thus, $\bigcup_{j=1}^m \lambda_j$ has the nice projection property with respect to π_2 (see Definition 4.14.2, p. 165). At the same time, the absolute values $|\alpha_j|$ can be chosen as small as desired, thereby ensuring that the embedding $g \circ f\colon \bar{D} \setminus \{a_j\}_{j=1}^m \hookrightarrow \mathbb{C}^2$ is close to f on a given compact set in D. The embedded complex curve $X = (g \circ f)(D) \subset \mathbb{C}^2$ with boundary $bX = \bigcup_{j=1}^m \lambda_j$ is then biholomorphic to $\Sigma = f(D)$. Since bX has the nice projection property, the conclusion follows from Theorem 4.14.6 (p. 168). □

The techniques of this section were used by Ritter [473] to show that if $D \subset \mathbb{C}$ is a finitely connected domain without punctures, then every continuous map $f\colon D \to \mathbb{C} \times \mathbb{C}^*$ is homotopic to a proper holomorphic embedding $D \hookrightarrow \mathbb{C} \times \mathbb{C}^*$. In the presence of finitely many punctures, Lárusson and Ritter [377] showed that every such map f is homotopic to a proper holomorphic immersion $D \to \mathbb{C} \times \mathbb{C}^*$ with at most finitely many normal crossings.

9.11 Infinitely Connected Complex Curves in $\mathbb{C}^2$

Theorem 9.10.3 was extended to certain infinitely connected Riemann surfaces by Majcen [404] in 2009. The following is her main result.

Theorem 9.11.1 ([404, Theorem 1]) *Assume that $\bar{R}$ is a compact bordered Riemann surface and $f\colon \bar{R} \hookrightarrow \mathbb{C}^2$ is an embedding which is holomorphic in R. Let $C_1, \ldots, C_m$ be the boundary components of R. Let D be an open set in R obtained by removing countably many closed pairwise disjoint discs $\overline{\Delta}_i$. Assume that for every $j = 1, \ldots, m$ there is a point $a_j \in C_j$ such that any point in $(bD \setminus bR) \cup \{a_1, \ldots, a_m\}$ has a neighborhood in $\bar{R}$ meeting only finitely many of the discs $\overline{\Delta}_i$. Given a compact set K in D and an $\epsilon > 0$, there exists a proper holomorphic embedding $F\colon D \hookrightarrow \mathbb{C}^2$ such that $\|F - f\|_K < \epsilon$.*

The discs $\overline{\Delta}_i$ are clustering only on the boundary curves $C_1, \ldots, C_m$ of R, and they stay away from the points $a_j \in C_j$ $(j = 1, \ldots, m)$. It is not clear whether the latter condition is really necessary, but the induction procedure used in the proof cannot be controlled without it.

Proof We use the same techniques as in the proof of Theorem 9.10.3, but the induction scheme and the technical details are substantially more involved.

We denote by $r\mathbb{B} \subset \mathbb{C}^2$ the ball of radius r centered at the origin. Choose an exhaustion $K \Subset K_1 \Subset K_2 \Subset \cdots \Subset \bigcup_{j=1}^{\infty} K_j = D$ by compact $\mathcal{O}(D)$-convex sets with smooth boundaries. Let $z = (z_1, z_2)$ be coordinates on $\mathbb{C}^2$ with the corresponding projections $\pi_i(z_1, z_2) = z_i$ for $i = 1, 2$. At the n-th step of the inductive construction we shall obtain the following:

- integers $m_n, k_n \in \mathbb{N}$,
- a number $\epsilon_n > 0$ such that $2\epsilon_n \le \epsilon_{n-1}$,
- curves $\Gamma_j = b\Delta_j$ $(j = 1, \ldots, k_n)$ which are the intersection of $bD \setminus bR$ with the $\mathcal{O}(R)$-convex hull of the set K_{m_n},
- the domain $D_n = R \setminus \bigcup_{j=1}^{k_n} \overline{\Delta}_j$ with boundary $bD_n = bR \cup (\bigcup_{j=1}^{k_n} \Gamma_j)$,
- points $c_j \in \Gamma_j$ for $j = 1, \ldots, k_n$,
- a holomorphic embedding $f_n\colon \bar{D}_n \hookrightarrow \mathbb{C}^2$ that exposes the points $a_1, \ldots, a_m$ and $c_1, \ldots, c_{k_n}$ with respect to the projection π_1 (see Theorem 9.10.10),
- a rational shear

$$g_n(z_1, z_2) = \left(z_1, z_2 + \sum_{j=1}^{m} \frac{\alpha_j}{z_1 - \pi_1(f_n(a_j))} + \sum_{j=1}^{k_n} \frac{\beta_j}{z_1 - \pi_1(f_n(c_j))} \right)$$

  with poles at the exposed points $f(a_j)$ and $f(c_j)$ of $f_n(\bar{D}_n)$, and
- an automorphism ϕ_n of $\mathbb{C}^2$

such that, setting

$$\Phi_n = \phi_n \circ \Phi_{n-1} = \phi_n \circ \phi_{n-1} \cdots \circ \phi_1, \qquad F_{n-1} = \Phi_{n-1} \circ g_n \circ f_n,$$

the following conditions hold:

$$\big|g_n \circ f_n(x) - g_{n-1} \circ f_{n-1}(x)\big| < \epsilon_n, \quad x \in K_{m_n}, \tag{9.28}$$

$$(j-1)\overline{\mathbb{B}} \cap F_{n-1}(D) \subset F_{n-1}(\mathring{K}_{m_j}), \quad j = 1, \ldots, n, \tag{9.29}$$

$$\big|\phi_n(z) - z\big| < \frac{\epsilon}{2^{n+1}}, \quad z \in (n-1)\overline{\mathbb{B}} \cup F_{n-1}(K_{m_n}), \tag{9.30}$$

$$\big|\phi_n(z)\big| > n, \quad z \in F_{n-1}(bD). \tag{9.31}$$

Note that $R \supset D_1 \supset D_2 \supset \cdots \supset \bigcap_{j=1}^{\infty} D_j = D$.

We begin the induction with $n = 0$, $m_0 = k_0 = 0$, $\epsilon_0 = \epsilon$, $K_0 = \varnothing$, $f_0 = f$, $g_0 = \mathrm{Id}$, and $\Phi_0 = \mathrm{Id}$. In this case, (9.28)–(9.30) are vacuous while (9.31) holds if $(0,0) \notin f(bR)$; this can be arranged by a small translation.

We now explain the induction step $n \to n+1$. By (9.31) there exists an integer $m_{n+1} > m_n$ such that

$$n\overline{\mathbb{B}} \cap \big(\Phi_n \circ g_n \circ f_n(D)\big) \subset \Phi_n \circ g_n \circ f_n(\mathring{K}_{m_{n+1}}).$$

The $\mathcal{O}(R)$-convex hull of $K_{m_{n+1}}$ contains the boundary curves $\Gamma_1, \ldots, \Gamma_{k_n}$ of D and at most finitely many other boundary curves which we label $\Gamma_{k_n+1}, \ldots, \Gamma_{k_{n+1}}$. Let $D_{n+1} = R \setminus \bigcup_{j=1}^{k_{n+1}} \overline{\Delta}_j$. Choose a point $c_j \in \Gamma_j$ for each $j = k_n + 1, \ldots, k_{n+1}$. Also, choose a number $\epsilon_{n+1} \in (0, \epsilon_n/2)$ such that any holomorphic map $h\colon D \to \mathbb{C}^2$ satisfying $\|h - g_n \circ f_n\|_{K_{m_{n+1}}} < 2\epsilon_{n+1}$ is an embedding on the smaller set K_{m_n}. Theorem 9.10.10 furnishes a holomorphic embedding $f_{n+1}\colon \bar{D}_{n+1} \hookrightarrow \mathbb{C}^2$ that agrees with f_n at the points $a_1, \ldots, a_m, c_1, \ldots, c_{k_n}$, it exposes the boundary points $a_1, \ldots, a_m, c_1, \ldots, c_{k_{n+1}}$, and it approximates f_n as closely as desired outside small neighborhoods of these points. The image $f_{n+1}(\bar{D}_{n+1})$ stays as close as desired to the union of $f_n(\bar{D}_{n+1}) \subset \mathbb{C}^2$ with the family of arcs that were attached to this domain in order to expose the points $c_{k_n+1}, \ldots, c_{k_{n+1}}$. In particular, we ensure that none of the complex lines $z_1 = \pi_1(f_{n+1}(c_j))$ for $j = 1, \ldots, k_{n+1}$ intersects the set $\Phi_n^{-1}(n\overline{\mathbb{B}})$. The rational shear

$$g_{n+1}(z_1, z_2) = g_n(z_1, z_2) + \left(0, \sum_{j=k_n+1}^{k_{n+1}} \frac{\beta_j}{z_1 - \pi_1(f_{n+1}(c_j))}\right)$$

sends the exposed points $f_{n+1}(a_1), \ldots, f_{n+1}(a_m), f_{n+1}(c_1), \ldots, f_{n+1}(c_{k_{n+1}})$ to infinity. A suitable choice of the arguments of the numbers $\beta_j \in \mathbb{C}^*$ for $j = k_n + 1, \ldots, k_{n+1}$ ensures that the unbounded curve $g_{n+1} \circ f_{n+1}(bD_{n+1})$ has the nice projection property with respect to π_2; at the same time we can choose their absolute values $|\beta_j|$ arbitrarily small. Set $F_n = \Phi_n \circ g_{n+1} \circ f_{n+1}$. If the approximations of f_n, g_n by f_{n+1}, g_{n+1}, respectively, were close enough then the conditions (9.28) and (9.29) hold with n replaced by $n+1$.

Note that $X_n = F_n(D_{n+1})$ is a complex curve in $\mathbb{C}^2$ satisfying the hypotheses of Theorem 4.14.6 (p. 168). In particular, the boundary bX_n consists of unbounded arcs diffeomorphic to $\mathbb{R}$ and $\Phi_n^{-1}(bX_n) = g_{n+1} \circ f_{n+1}(bD_{n+1})$ has the nice projection property with respect to the projection π_2.

Consider the discs $\overline{\Delta}_j$ that are contained in D_{n+1} (this holds for all but finitely many of them), and let $\Lambda_n \subset X_n$ denote the union of their images with respect to the map F_n. Set

$$L_n = n\overline{\mathbb{B}} \cup F_n(K_{m_{n+1}}). \tag{9.32}$$

By (the proof of) Theorem 4.14.6, there exists $\phi_{n+1} \in \mathrm{Aut}(\mathbb{C}^2)$ satisfying

$$\left|\phi_{n+1}(z) - z\right| < \frac{\epsilon}{2^{n+2}}, \quad z \in L_n, \tag{9.33}$$

$$\left|\phi_{n+1}(z)\right| > n+1, \quad z \in bX_n \cup \Lambda_n. \tag{9.34}$$

Such ϕ_{n+1} is found by a modification in the proof of Lemma 4.14.4 (p. 166). By that argument, it suffices to explain the construction when bX_n has the nice projection property with respect to π_2; the general case is obtained by conjugating with $\alpha = \Phi_n^{-1} \in \mathrm{Aut}(\mathbb{C}^2)$. We look for a composition

$$\phi_{n+1} = \varphi_2 \circ \varphi_1 \circ \psi$$

of three automorphisms. In the first step, we choose a long arc λ_j on each of the boundary curves of X_n and apply Corollary 4.13.5 (p. 163) to find $\varphi_1 \in \mathrm{Aut}(\mathbb{C}^2)$ that it is close to the identity near L_n and sends each of the arcs λ_j out of the ball $(n+1)\overline{\mathbb{B}}$. We may assume that the first m arcs $\lambda_1, \ldots, \lambda_m$ belong to those boundary curves in bX_n that are F_n-images of the boundary curves C_j of R. Choosing the arcs λ_j sufficiently long, the condition in the theorem implies that the discs in Λ_n cluster only on $\lambda = \bigcup_{j=1}^m \lambda_j$.

We now explain the construction of φ_2 and ψ. Choose a smoothly bounded, simply connected closed domain $E_n \subset \overline{X}_n$ that is a relative neighborhood of λ in $\overline{X}_n$, is disjoint from the set L_n (9.32), and such that every disc $\overline{\Delta}_i \subset \Lambda_n$ is either entirely contained in E_n or else is disjoint from E_n. Such E_n is found by entirely elementary geometric considerations. Let Λ'_n be the union of those finitely many discs in Λ_n that remained out of E_n. Since L_n is polynomially convex and $E_n \cup \Lambda'_n$ is a union of finitely many pairwise disjoint smoothly bounded holomorphic discs contained in the complex curve $\overline{X}_n$, the sets $E_n \cup L_n$ and $E_n \cup L_n \cup \Lambda'_n$ are also polynomially convex by [530]. (See the argument in the proof of Theorem 4.14.6, property 1.)

The set Λ'_n can be moved out of the ball $(n+1)\overline{\mathbb{B}}$ by an isotopy of biholomorphic maps that are close to identity near the compact polynomially convex set $E_n \cup L_n$ and such that the trace of this isotopy is polynomially convex in the extended phase space. (We start by contracting the holomorphic discs belonging to Λ'_n to make them very small, and then we drag them out of $(n+1)\overline{\mathbb{B}}$ without doing anything near $E_n \cup L_n$.) By Theorem 4.12.1 on p. 157 (which uses Andersén-Lempert theory) we can therefore find $\varphi_2 \in \mathrm{Aut}(\mathbb{C}^2)$ that is close to the identity near $E_n \cup L_n$ and that maps Λ_n' out of $(n+1)\overline{\mathbb{B}}$. It follows that

$$\left\{z \in bX_n : \left|\varphi_2 \circ \varphi_1(z)\right| \le n+1\right\} \subset \bigcup_{j=1}^{m} F_n\left(C_j \setminus \{a_j\}\right) \setminus \lambda.$$

If the arcs $\lambda_j \subset \lambda$ have been chosen long enough, then this remaining intersection of $\varphi_2 \circ \varphi_1(bX_n)$ with the ball $(n+1)\overline{\mathbb{B}}$ can be removed by precomposing $\varphi_2 \circ \varphi_1$ by a shear $\psi(z_1, z_2) = (z_1 + h(z_2), z_2)$. (See (4.60) in the proof of Lemma 4.14.4 for the construction of such a shear ψ.)

This completes the construction of $\phi_{n+1} = \varphi_2 \circ \varphi_1 \circ \psi$ satisfying (9.33) and (9.34). Hence, the map $\Phi_{n+1} = \phi_{n+1} \circ \Phi_n$ satisfies

$$\left|\Phi_{n+1}(z)\right| > n+1, \quad z \in g_{n+1} \circ f_{n+1}(bD).$$

Thus, condition (9.31) holds for $n+1$ which completes the induction step.

By (9.28) and the choice of ϵ_n we see that $G = \lim_{n\to\infty} g_n \circ f_n : D \hookrightarrow \mathbb{C}^2$ is a holomorphic embedding. Condition (9.30) and Corollary 4.4.2 (p. 125) imply that the sequence $\Phi_n \in \mathrm{Aut}(\mathbb{C}^2)$ converges on $\Omega = \bigcup_{n=2}^{\infty} \Phi_n^{-1}((n-1)\overline{\mathbb{B}})$ to a Fatou-Bieberbach map $\Phi = \lim_{n\to\infty} \Phi_n : \Omega \to \mathbb{C}^2$. From (9.29) and (9.31) we see that G embeds D properly into Ω. Hence, $F = \Phi \circ G : D \hookrightarrow \mathbb{C}^2$ is a proper holomorphic embedding satisfying $\|F - f\|_K < \epsilon$. □

For planar domains, Wold proved the following stronger result which generalizes Theorem 9.10.2 of Globevnik and Stensønes.

Theorem 9.11.2 ([582]) *Any finitely connected domain D in $\mathbb{C}$ can be embedded properly holomorphically into $\mathbb{C}^2$. Moreover, let $\{p_j\} \subset D$ be a sequence converging to a point p in the boundary bD (we allow $p = \infty$), and assume there exists a continuous curve $\gamma : [0,1] \to \bar{D}$ such that $\gamma([0,1)) \subset D$, $\gamma(1) = p$, and $\{p_j\} \subset \gamma$. Then, $D \setminus \{p_j\}_{j=1}^{\infty}$ embeds properly holomorphically in $\mathbb{C}^2$.*

Proper holomorphic embeddings of bordered Riemann surfaces can be almost prescribed on certain smooth curves approaching the boundary. A result of this type is Theorem 4.14.1 on p. 165. In the following result of Majcen, $\mathbb{C}$ is replaced by a finitely connected domain $D \subset \mathbb{C}$ and $\mathbb{R}$ is replaced by a union of arcs which may terminate at boundary points of D (see Fig. 9.3).

Theorem 9.11.3 ([405]) *Let $n \ge 2$ and $r \ge 1$ be integers. Let D be a finitely connected domain in $\mathbb{C}$. Let $\ell_i = \{\gamma_i(t) : t \in [0,1]\}$ $(i = 1, 2, \ldots, k)$ be smooth embedded curves in $\mathbb{C}$ such that $\gamma_i((0,1)) \subset D$ and $\ell_i \cap \ell_j \cap D = \varnothing$ for all $1 \le i < j \le k$. Set $\ell = (\bigcup_{i=1}^{k} \ell_i) \setminus bD$. Given a proper C^r embedding $\lambda : \ell \hookrightarrow \mathbb{C}^n$ and a positive continuous function $\eta : \ell \to (0,\infty)$, there exists a proper holomorphic embedding $f : D \hookrightarrow \mathbb{C}^n$ such that $|f^{(s)}(x) - \lambda^{(s)}(x)| < \eta(x)$ holds for all $x \in \ell$ and $s = 0, 1, \ldots, r$.*

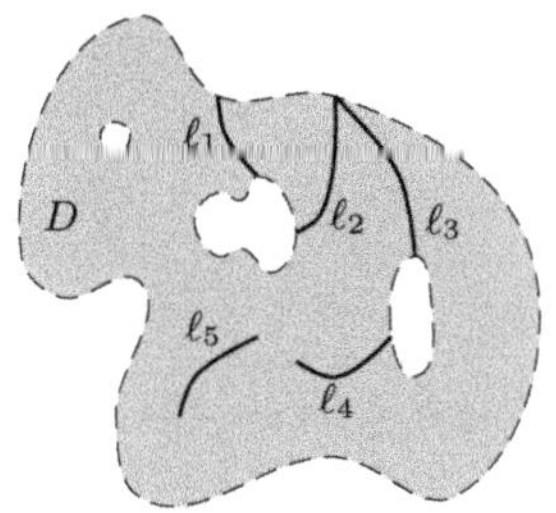

Fig. 9.3 Planar domain D and curves ℓ_i in Theorem 9.11.3

Note that some arcs $\ell_i = \gamma_i([0,1])$ may be entirely contained in D, and in such case every embedding $\ell_i \hookrightarrow \mathbb{C}^2$ is proper. On the other hand, an endpoint of ℓ_i that happens to belong to bD is sent by a proper embedding of $\ell_i \setminus bD$ to infinity. Theorem 9.11.3 is obtained by following the proof of Theorem 9.10.3 but using Lemma 4.14.4 on p. 166 (in place of Corollary 4.14.5) when dealing with the curves in ℓ.

We now present without proof the most general known embedding result for planar domains, due to Forstnerič and Wold [228]. It concerns domains in $\mathbb{P}^1$ with at most countably many boundary components. By the uniformization theorem of He and Schramm [284, 285], such a domain is conformally equivalent to a *generalized circled domain*, i.e. a domain $\Omega \subset \mathbb{P}^1$ such that every connected component of $\mathbb{P}^1 \setminus \Omega$ is a round disc or a point (puncture).

Theorem 9.11.4 ([228, Theorem 5.1]) *Let Ω be a generalized circled domain in $\mathbb{P}^1$. If all but finitely many punctures in $\mathbb{P}^1 \setminus \Omega$ are limit points of discs in $\mathbb{P}^1 \setminus \Omega$, then Ω embeds properly holomorphically in $\mathbb{C}^2$.*

The proof in [228] is conceptually similar to the proof of Theorem 9.11.1, but it uses an improved induction scheme for dealing with an infinite sequence of boundary components, clustering them together into suitable subsets to which the analytic methods described above can be applied. The details are considerable. By using essentially the same proof, one can show the analogous result for domains in tori.

We conclude with an example of a rather different type.

Theorem 9.11.5 (Orevkov [445]) *There exists a Cantor set K in $\mathbb{P}^1$ whose complement $\mathbb{P}^1 \setminus K$ admits a proper holomorphic embedding into $\mathbb{C}^2$.*

Proof We construct such a curve in $\mathbb{C}^2$ as a limit of algebraic curves A_n obtained from each other by a birational transformation $F_n\colon \mathbb{C}^2 \to \mathbb{C}^2$. For some exhaustion of $\mathbb{C}^2$ by nested bidiscs $B_1 \subset B_2 \subset \cdots$ the topological type of $A_n \cap B_n$ will not change under further transformations.

Let (x, y) be complex coordinates on $\mathbb{C}^2$. Choose a sequence of complex numbers $a_1, a_2, \ldots$ whose absolute values are strictly increasing and tend to infinity. We inductively define a sequence of birational mappings $F_n\colon \mathbb{C}^2 \to \mathbb{C}^2$ by setting F_0 to be the identity mapping and by setting $F_n = f_n \circ F_{n-1}$, where

$$f_n(x, y) = \begin{cases} (x, y + g_n(x)), & \text{if } n \text{ is odd}, \\ (x + g_n(y), y), & \text{if } n \text{ is even}, \end{cases}$$

where $g_n(t) = \frac{\epsilon_n}{t - a_n}$ and the sequence $0 < \epsilon_n \ll \epsilon_{n-1}$ tend to zero sufficiently fast (to be specified below). Let $\overline{\mathbb{C}}^2 = \mathbb{C}^2 \cup \{\infty\}$ be the one point compactification of $\mathbb{C}^2$. Let $\gamma_n : \mathbb{P}^1 \to \overline{\mathbb{C}}^2$ be defined by $\gamma_n(z) = F_n(z, 0)$. Then, for a suitable choice of ϵ_n, the limit $\gamma = \lim_{n\to\infty} \gamma_n$ is a continuous mapping $\gamma : \mathbb{P}^1 \to \overline{\mathbb{C}}^2$ such that $K = \gamma^{-1}(\infty)$ is a Cantor set, and the restriction of γ to $\mathbb{P}^1 \setminus K$ is a proper embedding of the open Riemann surface $\mathbb{P}^1 \setminus K$ into $\mathbb{C}^2$.

Let us describe the choice of the parameters ϵ_n and explain why γ satisfies the stated properties. Choose numbers $R_n > 0$ such that $|a_n| < R_n < |a_{n+1}|$ for every $n \in \mathbb{N}$. Let $A_n = F_n(\mathbb{P}^1)$ and $D_n = \{z \in \mathbb{C} : |z| < R_n\}$. Denote the projection $(z_1, z_2) \to z_i$ by π_i $(i = 1, 2)$. Set

$$C_n^{(i)} = \pi_i^{-1}(D_n), \qquad B_n = C_n^{(1)} \cap C_n^{(2)} = D_n \times D_n, \qquad C_n = C_n^{(1)} \cup C_n^{(2)}.$$

Then $B_1 \subset B_2 \subset \cdots \subset \bigcup_{n=1}^\infty B_n = \mathbb{C}^2$. We define the sequence $\epsilon_n > 0$ inductively so that the following conditions hold for every $n \in \mathbb{N}$:

1. $A_n \subset C_n$,
2. $A_n \cap (C_n^{(i)} \setminus B_n)$ $(i = 1, 2)$ has a finite number of connected components each being mapped by π_i biholomorphically onto $\mathbb{C} \setminus D_n$,
3. all curves $A_p \cap B_n$ for $p \geq n$ are isotopic to each other in B_n and they converge in the $\mathcal{C}^\infty$ topology to an analytic curve which is also isotopic to all of them, and
4. $\lim_{n\to\infty} d_n = 0$, where d_n is the maximum of diameters (with respect to some fixed metric on $\overline{C}$) of the connected components of $F_n^{-1}(A_n \setminus B_n)$.

Let us call a boundary component of $A_n \cap B_n$ horizontal if it is contained in $bD_n \times D_n$, and vertical if it is contained in $D_n \times bD_n$. (Condition 1 shows that there are no other components.) If the numbers ϵ_n are chosen as described then, up to a small perturbation, $A_{2n+1} \cap B_{2n+1}$ is obtained from $A_{2n} \cap B_{2n}$ by attaching an annulus to each vertical boundary component, and by attaching a pair of pants (an annulus with a hole) to each horizontal boundary component. So each vertical component at the $2n$-th step provides a single vertical component at the next step, but each horizontal component provides one horizontal and one vertical component at the next step. When passing from $A_{2n+1} \cap B_{2n+1}$ to $A_{2n+2} \cap B_{2n+2}$, the roles of vertical and horizontal boundary components are exchanged. □

Problem 9.11.6 Does the complement $\mathbb{P}^1 \setminus K$ of every Cantor set $K \subset \mathbb{P}^1$ admit a proper holomorphic embedding into $\mathbb{C}^2$? □

9.12 Approximation of Holomorphic Submersions

In this section we prove results on approximation of holomorphic submersions from domains in $\mathbb{C}^n$ to $\mathbb{C}^q$ for $1 \le q < n$. These results are used in subsequent sections. We begin with the simplest case of approximating noncritical holomorphic functions on polynomially convex sets in $\mathbb{C}^n$.

Theorem 9.12.1 ([191, Theorem 3.1]) *Let K be a compact polynomially convex subset of $\mathbb{C}^n$ for $n \ge 1$. If f is a holomorphic function in an open set $U \supset K$ such that $\mathrm{d}f_z \ne 0$ for every $z \in U$, then for every $\epsilon > 0$ there exists an entire function $g \in \mathcal{O}(\mathbb{C}^n)$ satisfying $\mathrm{d}g \ne 0$ on $\mathbb{C}^n$ and $\sup_K |f - g| < \epsilon$.*

Proof Choose a point $z_0 \in K$ and a compact polynomially convex set $L \subset U$ with smooth boundary and containing K in the interior.

Consider first the case $n = 1$. Then, L is the union of finitely many compact, connected and simply connected sets (discs). Let P be a holomorphic polynomial on $\mathbb{C}$ such that $\sup_L |f - P| < \frac{\epsilon}{2}$ and $P' \ne 0$ on L. The critical set $\Sigma = \{P' = 0\}$ consists of finitely many points in $\mathbb{C} \setminus L$. Hence, there is a connected and simply connected domain $V \subset \mathbb{C} \setminus \Sigma$ with $L \subset V$. Choose a smoothly bounded, connected and simply connected compact set $M \subset V$ containing L. Since $P' \ne 0$ on V, we have $P'|_V = \mathrm{e}^h$ for some $h \in \mathcal{O}(V)$. We approximate h uniformly on M by a holomorphic polynomial η and set

$$g(z) = g(z_0) + \int_{z_0}^{z} \mathrm{e}^{\eta(\zeta)}\, \mathrm{d}\zeta, \quad z \in \mathbb{C}.$$

The integral is clearly independent of the choice of a path. We then have $g' = \mathrm{e}^\eta \ne 0$ on $\mathbb{C}$, and $|g' - P'| = |\mathrm{e}^\eta - \mathrm{e}^h|$ is small on M. For every point $z \in M$ the path of integration can be chosen in M, with length bounded by $C|z - z_0|$ for some constant $C \ge 1$ independent of the point $z \in M$. If the approximation of h by η is close enough on M then $|g - P| < \frac{\epsilon}{2}$ on M, and hence $|g - f| < \epsilon$ holds on K. This proves the case $n = 1$.

Assume now that $n \ge 2$. Since L is polynomially convex, there is a holomorphic polynomial h on $\mathbb{C}^n$ satisfying $\sup_L |f - h| < \frac{\epsilon}{2}$. If $\epsilon > 0$ is chosen sufficiently small, then $\mathrm{d}h \ne 0$ on K. For a generic choice of h, its critical set $\Sigma = \{z \in \mathbb{C}^n : \mathrm{d}h_z = 0\} \subset \mathbb{C}^n \setminus K$ is finite. Let $c = \sup_{z\in L} |\mathrm{d}h_z|$. Choose $\delta \in \mathbb{R}$ such that

$$0 < \delta < \min\left\{\mathrm{dist}\big(K, \mathbb{C}^n \setminus L\big), \frac{\epsilon}{2c}\right\}.$$

By Corollary 4.12.7 (p. 160) there exists a biholomorphic map ϕ of $\mathbb{C}^n$ onto a subset $\Omega \subset \mathbb{C}^n \setminus \Sigma$ such that $\sup_{z\in K} |\phi(z) - z| < \delta$. The entire function $g = h \circ \phi$ then satisfies $\mathrm{d}g_z = \mathrm{d}h_{\phi(z)} \cdot \mathrm{d}\phi_z \ne 0$ for every $z \in \mathbb{C}^n$ (since $\phi(z) \in \Omega \subset \mathbb{C}^n \setminus \Sigma$ and $\mathrm{d}h \ne 0$ on $\mathbb{C}^n \setminus \Sigma$). Our choice of δ ensures that for every point $z \in K$ the line segment connecting z and $\phi(z)$ is contained in L. Hence,

$$\big|g(z) - h(z)\big| = \big|h\big(\phi(z)\big) - h(z)\big| \le c\big|\phi(z) - z\big| < c\delta < \frac{\epsilon}{2}$$

and therefore $|g(z) - f(z)| < \epsilon$ for every $z \in K$. □

We also have the following approximation result for holomorphic submersions $\mathbb{C}^n \to \mathbb{C}^q$, $1 \le q < n$ on compact *convex* sets in $\mathbb{C}^n$.

Theorem 9.12.2 *Let K be a compact convex set in $\mathbb{C}^n$. Given a holomorphic submersion $f\colon K \to \mathbb{C}^q$ $(1 \le q < n)$ and an $\epsilon > 0$, there exists a holomorphic submersion $g\colon \mathbb{C}^n \to \mathbb{C}^q$ such that $\sup_K |f - g| < \epsilon$.*

Proof By the assumption, $f\colon U \to \mathbb{C}^q$ is a holomorphic submersion on an open set $U \supset K$. Choose a compact convex set $L \subset U$ containing K in the interior. By the Oka-Weil theorem there is a polynomial map $h\colon \mathbb{C}^n \to \mathbb{C}^q$ satisfying $\sup_L |f - h| < \frac{\epsilon}{2}$. A generic choice of h ensures that the set

$$\Sigma = \big\{z \in \mathbb{C}^n : \operatorname{rank} dh_x < q\big\}$$

is an algebraic subvariety of dimension $q - 1 \le n - 2$ disjoint from L (see Theorem 8.9.1, p. 387). By Corollary 4.12.2 (p. 157) there exists for every $\delta > 0$ a biholomorphic map ϕ of $\mathbb{C}^n$ onto a subset $\Omega \subset \mathbb{C}^n \setminus \Sigma$ such that $\sup_{z\in K} |\phi(z) - z| < \delta$. The entire map $g = h \circ \phi\colon \mathbb{C}^n \to \mathbb{C}^q$ is then a submersion at each point, and if $\delta > 0$ is chosen small enough then $\sup_K |g - h| < \epsilon/2$. It follows that $\sup_K |g - f| < \epsilon$. □

The proof of Theorem 9.12.2 breaks down when $q = n$: In this case the subvariety $\Sigma = \{z \in \mathbb{C}^n : \operatorname{rank} dh_z < n\}$ (the zero locus of the Jacobian determinant $\det(\partial h_j/\partial z_k)$) is a hypersurface in $\mathbb{C}^n$, and in general there do not exist any nondegenerate holomorphic maps $\mathbb{C}^n \to \mathbb{C}^n \setminus \Sigma$.

Problem 9.12.3 Let $n > 1$. Is it possible to approximate every locally biholomorphic map $\mathbb{C}^n \supset U \to \mathbb{C}^n$ on a neighborhood of a compact convex set $K \subset \mathbb{C}^n$, uniformly on K, by entire locally biholomorphic maps $\mathbb{C}^n \to \mathbb{C}^n$?

Theorem 9.12.2 has been extended to holomorphic maps $f\colon \mathbb{C}^n \to Y$ to manifolds Y of Class $\mathcal{A}$ (see Definition 6.4.4, p. 273) such that the rank of f satisfies a pointwise estimate from below. The following result generalizes [196, Proposition 5.4] and also the main result of [347].

Theorem 9.12.4 *Let n, q, r be integers satisfying $(n - r + 1)(q - r + 1) \ge 2$, and let Y be a manifold of Class $\mathcal{A}$ and of dimension q. If K is a compact convex set in $\mathbb{C}^n$ and $f\colon K \to Y$ is a holomorphic map such that $\operatorname{rank} df_z \ge r$ for every point $z \in K$, then f can be approximated uniformly on K by entire maps $g\colon \mathbb{C}^n \to Y$ such that $\operatorname{rank} dg_z \ge r$ for every point $z \in \mathbb{C}^n$.*

If $Y = \mathbb{C}^q$, then by taking $n > q = r$ we get Theorem 9.12.2, while for $n = r < q$ we get the following result for immersions $\mathbb{C}^n \to \mathbb{C}^q$.

Corollary 9.12.5 *Let K be a compact convex set in $\mathbb{C}^n$. Every holomorphic immersion $K \to \mathbb{C}^q$ for $q > n$ can be approximated, uniformly on K, by entire immersions $\mathbb{C}^n \to \mathbb{C}^q$.*

Proof of Theorem 9.12.4 The proof for $Y = \mathbb{C}^q$ is a straightforward generalization of the proof of Theorem 9.12.2. The condition $(n - r + 1)(q - r + 1) \geq 2$ ensures that for a generic polynomial map $h \colon \mathbb{C}^n \to \mathbb{C}^q$, the algebraic subvariety $\{z \in \mathbb{C}^n : \operatorname{rank} dh_z < r\}$ has codimension at least two in $\mathbb{C}^n$ (see Theorem 8.9.1 on p. 387). The rest of the proof is exactly as before.

If Y is a manifold of Class $\mathcal{A}$, we need a more careful argument since we do not have a global jet transversality theorem for algebraic maps $\mathbb{C}^n \to Y$. We give the details for the case of a submersion, i.e., $n > q = r$.

We may assume that Y is connected. Let $f \colon K \to Y$ be a holomorphic submersion from an open neighborhood of compact convex set $K \subset \mathbb{C}^n$. By Proposition 6.4.5 (p. 274) every manifold Y of Class $\mathcal{A}$ is algebraically subelliptic. By Corollary 6.15.2 (p. 313) it follows that f can be approximated uniformly on K by algebraic maps (morphisms) $\mathbb{C}^n \to Y$. We may thus assume that $f \colon \mathbb{C}^n \to Y$ is an algebraic map which is a submersion on K.

The set $\Sigma = \{z \in \mathbb{C}^n : \operatorname{rank} df_z < q = \dim Y\}$ is an algebraic subvariety of $\mathbb{C}^n$ which does not intersect K. We will show that a small perturbation of f gives an algebraic map for which $\dim \Sigma \leq n - 2$; the proof is then concluded exactly as in Theorem 9.12.2.

Assume that $\dim \Sigma = n - 1$. We shall inductively remove all $(n-1)$-dimensional irreducible components from Σ, changing the map only a little at every step. Choose an irreducible component $\Sigma' \subset \Sigma$ of dimension $n - 1$ and a point $z_0 \in \Sigma'$ that does not belong to any other irreducible component of Σ. By the definition of Class $\mathcal{A}$ (see Definition 6.4.4) we have $Y = \widehat{Y} \backslash A$, where $\widehat{Y}$ is a connected manifold of Class $\mathcal{A}_0$ (i.e., locally Zariski affine) and A an algebraic subvariety of $\widehat{Y}$ of codimension at least two. Let $U \subset \widehat{Y}$ be a Zariski open set isomorphic to $\mathbb{C}^q$ and containing the point $y_0 = f(z_0)$. Let $s_0 \colon U \times \mathbb{C}^q \to U \cong \mathbb{C}^q$ denote the spray $s_0(y,t) = y + t$. Choose a closed algebraic subvariety Y_0 of $\widehat{Y}$ of pure dimension $q - 1$ (a hypersurface) such that $\widehat{Y} = Y_0 \cup U$ and $y_0 \notin Y_0$. Let $[Y_0] \to \widehat{Y}$ denote the holomorphic line bundle defined by the divisor of Y_0, and let $L = [Y_0]^{-1}$. Let $\mathbb{T}^q = \widehat{Y} \times \mathbb{C}^q$ denote the trivial bundle of rank q over $\widehat{Y}$.

By Proposition 6.4.2 (p. 273) there are an integer $m \in \mathbb{N}$ and an algebraic spray $s \colon E = \mathbb{T}^q \otimes L^{\otimes m} \to \widehat{Y}$ such that $s(y,t) = y$ for all $y \in Y_0$ and $t \in E_y$, and such that s equals s_0 over the open set $\widehat{Y} \backslash Y_0 \subset U$ (using an obvious identification of $E|_U$ with $\mathbb{T}^q|_U$). Note that L is trivial over $\widehat{Y} \setminus Y_0$.

By Serre's Theorem A, the algebraic vector bundle $f^*(E)$ over $\mathbb{C}^n$ is generated by finitely many (say p) algebraic sections, and hence there is a surjective algebraic vector bundle map $\rho \colon \mathbb{C}^n \times \mathbb{C}^p \to f^*E$. Let $\iota \colon f^*E \to E$ be the natural vector bundle map covering the base map $f \colon \mathbb{C}^n \to Y$. Set $Z = f^{-1}(Y_0) \subset \mathbb{C}^n$. The algebraic map

$$F = s \circ \iota \circ \rho : \mathbb{C}^n \times \mathbb{C}^p \to \widehat{Y}$$

then satisfies the following conditions:

(a) $F(z,0) = f(z)$ for $z \in \mathbb{C}^n$,
(b) $F(z,t) = f(z)$ for $z \in Z$ and $t \in \mathbb{C}^p$, and
(c) $F(z,\cdot)\colon \mathbb{C}^p \to \widehat{Y}$ is a submersion for every point $z \in \mathbb{C}^n \setminus Z$.

The proof of Theorem 8.8.5 (see especially Lemma 8.8.10 and Proposition 8.8.16) gives a polynomial map $P\colon \mathbb{C}^n \to \mathbb{C}^p$ such that the algebraic map

$$f_1\colon \mathbb{C}^n \to \widehat{Y}, \qquad f_1(z) = F\big(z, P(z)\big) \quad \big(z \in \mathbb{C}^n\big)$$

satisfies the following conditions:

(i) f_1 approximates f as closely as desired on a neighborhood of K,
(ii) $\mathrm{j}^1_z f = \mathrm{j}^1_z f_1$ for every $z \in Z = f^{-1}(Y_0)$,
(iii) $f_1|_{\mathbb{C}^n \setminus Z}$ is transverse to the subvariety $A \subset \widehat{Y}$, and
(iv) the branch locus of $f_1|_{\mathbb{C}^n \setminus Z}$ has dimension $\le n-2$.

To obtain (iv), choose P such that $\mathrm{j}^1 f_1\colon X \to J^1(\mathbb{C}^n, Y)$ is transverse to the subvariety of $J^1(\mathbb{C}^n, Y)$ of all 1-jets of rank $< q = \dim Y$. By Lemma 8.9.2 (p. 387) this subvariety has codimension $n - q + 1 \ge 2$ which implies (iv). Let $C \subset \mathbb{C}^n$ be the branch locus of $f_1\colon \mathbb{C}^n \to \widehat{Y}$; thus, $\dim C \setminus Z \le n-2$ by (iv). The set

$$\Sigma_1 = (\Sigma \cap Z) \cup f_1^{-1}(A) \cup C$$

is an algebraic subvariety of $\mathbb{C}^n$ which does not intersect K, provided that the approximation of f by f_1 is sufficiently close near K. The restriction of f_1 to $\mathbb{C}^n \setminus \Sigma_1$ maps the latter set submersively to $Y = \widehat{Y} \setminus A$. We claim that Σ_1 has fewer $(n-1)$-dimensional irreducible components than Σ. Note that $\dim(\Sigma_1 \setminus Z) \le n-2$ by properties (iii) and (iv) of f_1. Next, we show that

$$\dim\big[(Z \setminus \Sigma) \cap \big(f_1^{-1}(A) \cup C\big)\big] \le n-2.$$

If $z \in Z \setminus \Sigma$, then f is unbranched at z by the definition of Σ. Furthermore, $\mathrm{j}^1_z f_1 = \mathrm{j}^1_z f$ by property (ii), and hence f_1 is also unbranched at such point z, thereby showing that $(Z \setminus \Sigma) \cap C = \varnothing$. This also implies that $f_1|_{Z \setminus \Sigma}$ is transverse to A and hence $\dim f_1^{-1}(A) \cap (Z \setminus \Sigma) \le n-2$. It follows that the $(n-1)$-dimensional irreducible components of Z_1 are the same as those of $\Sigma \cap Z$. Since $z_0 \in \Sigma' \setminus Z$, the component Σ' of Σ is not among them, thereby proving the claim.

Repeating this argument with the pair (f_1, Σ_1) gives an algebraic map $f_2\colon \mathbb{C}^n \to \widehat{Y}$ and an algebraic subvariety $\Sigma_2 \subset \mathbb{C}^n$ with fewer $(n-1)$-dimensional components than Σ_1 such that $f_2\colon \mathbb{C}^n \setminus \Sigma_2 \to Y$ is a submersion which approximates f_1 (and hence f) uniformly on K. In finitely many steps of this kind we obtain an algebraic map $\widetilde{f}\colon \mathbb{C}^n \to Y$ that approximates f on K and whose branch locus $\widetilde{\Sigma}$ satisfies $\dim \widetilde{\Sigma} \le n-2$. □

In the sequel, we shall need the following result on the existence of a holomorphic transition map between a pair of nearby holomorphic submersions.

Lemma 9.12.6 ([191, Lemma 5.1]) *Let K be a Stein compact in a complex manifold X. Assume that $U \subset X$ is an open set containing K and $f\colon U \to \mathbb{C}^q$ is a holomorphic submersion for some $q \le \dim X$. There exist constants $\epsilon_0 > 0$, $M > 0$, and an open set $V \subset X$ with $K \subset V \Subset U$, satisfying the following property. Given a number $\epsilon \in (0, \epsilon_0)$ and a holomorphic submersion $g\colon U \to \mathbb{C}^q$ with $\sup_{x\in U} |f(x) - g(x)| < \epsilon$, there is an injective holomorphic map $\gamma\colon V \to U$ satisfying $f = g \circ \gamma$ on V and $\sup_{x\in V} \mathrm{dist}(\gamma(x), x) < M\epsilon$.*

Proof We may assume that U is Stein and hence $\mathrm{T}X|_U = \ker \mathrm{d}f \oplus E$ for some trivial rank q holomorphic subbundle $E \subset \mathrm{T}X|_U$. Thus, E is spanned by q pointwise independent holomorphic vector fields on U. Denote by $\phi(x, t_1, \ldots, t_q) = \phi^1_{t_1} \circ \cdots \circ \phi^q_{t_q}(x)$ the composition of their local flows. Then ϕ is defined in an open set $\Omega \subset U \times \mathbb{C}^q$ containing $U \times \{0\}^q$. For $x \in U$ write $\Omega_x = \{t \in \mathbb{C}^q : (x,t) \in \Omega\}$. After shrinking Ω we may assume that for each $x \in U$ the fibre Ω_x is connected and that $F_x = \{\phi(x,t) : t \in \Omega_x\} \subset X$ is a local complex submanifold of X which intersects the level set $\{f = f(x)\}$ transversely at x (since $\mathrm{T}_x F_x = E_x$ is complementary to the kernel of $\mathrm{d}f_x$). The implicit function theorem shows that, after shrinking $\Omega \supset U \times \{0\}^q$, the map $\Omega_x \ni t \to f(\phi(x,t)) \in \mathbb{C}^q$ maps Ω_x biholomorphically onto a neighborhood of the point $f(x)$ in $\mathbb{C}^q$. If $g\colon U \to \mathbb{C}^q$ is sufficiently uniformly close to f, then the same holds for the map $\Omega_x \ni t \mapsto g(\phi(x,t))$, provided that we restrict x to a compact subset of U. It follows that, if $K \subset V \Subset U$ and if g is sufficiently close to f on U, there exists for every $x \in V$ a unique point $c(x) \in \Omega_x$ such that $g(\phi(x, c(x))) = f(x)$. Clearly $c\colon V \to \mathbb{C}^q$ is holomorphic, and the map $\gamma(x) = \phi(x, c(x)) \in X$ $(x \in V)$ satisfies Lemma 9.12.6. □

9.13 Noncritical Holomorphic Functions

Let X be a complex manifold. A holomorphic map $f = (f_1, \ldots, f_q)\colon X \to \mathbb{C}^q$ is a submersion if its differential $\mathrm{d}f_x\colon \mathrm{T}_x X \to \mathrm{T}_{f(x)}\mathbb{C}^q \cong \mathbb{C}^q$ is surjective at every point $x \in X$; equivalently,

$$\mathrm{d}f_1 \wedge \mathrm{d}f_2 \wedge \cdots \wedge \mathrm{d}f_q \neq 0$$

holds at every point of X. Thus, the tangent map of a holomorphic submersion $X \to \mathbb{C}^q$ is surjective complex vector bundle map $\mathrm{T}X \to X \times \mathbb{C}^q$. On a Stein manifold, this necessary condition for the existence of a holomorphic submersion $X \to \mathbb{C}^q$ is also sufficient, except perhaps in the equidimensional case $\dim X = q > 1$. The following result is [191, Theorem II], except the case $n = q = 1$ which is due to Gunning and Narasimhan [273].

Theorem 9.13.1 (Oka principle for holomorphic submersions to Euclidean spaces) *If X is a Stein manifold of dimension n and if $1 \le q < n$ or $q = n = 1$, then every surjective complex vector bundle map $\mathrm{T}X \to X \times \mathbb{C}^q$ is homotopic (through surjective complex vector bundle maps $\mathrm{T}X \to X \times \mathbb{C}^q$) to the tangent map of a holomorphic submersion $X \to \mathbb{C}^q$. In particular, every open Riemann surface can be represented as a Riemann domain over $\mathbb{C}$.*

By Theorem 8.3.1(c), the tangent bundle of an n-dimensional Stein manifold X^n admits a surjective complex vector bundle map onto the trivial bundle $X \times \mathbb{C}^{[(n+1)/2]}$. Hence, Theorem 9.13.1 implies the following result.

Corollary 9.13.2 ([191]) *Every Stein manifold admits a holomorphic function without critical points. More precisely, an n-dimensional Stein manifold admits $[\frac{n+1}{2}]$ holomorphic functions with pointwise independent differentials, and this number is maximal for every $n \in \mathbb{N}$.*

That the number $q(n) = [\frac{n+1}{2}]$ is maximal in Corollary 9.13.2 follows from Theorem 9.1.3 due to Ho et al. [298], and it can also be seen using the manifolds in Forster's Example 9.3.2 (p. 415).

Theorem 9.13.1 is a holomorphic analogue of the h-principle for submersions of smooth open manifolds [267, 451]. It is not known whether it is valid also for $q = n > 1$. In this equidimensional case, submersions and immersions coincide with locally biholomorphic maps.

Problem 9.13.3 ([265, p. 70]) Assume that X is a Stein manifold of dimension $n > 1$ with trivial holomorphic tangent bundle $\mathrm{T}X$. Does X admit a holomorphic immersion $X \to \mathbb{C}^n$? Equivalently, is every such X a Riemann domain over $\mathbb{C}^n$? Is every complex vector bundle trivialization $\mathrm{T}X \xrightarrow{\cong} X \times \mathbb{C}^n$ homotopic to the differential of a holomorphic immersion $X \to \mathbb{C}^n$?

Our proof of Theorem 9.13.1 implies the following result.

Theorem 9.13.4 *If Problem 9.12.3 (p. 458) has a positive answer, then for every n-dimensional Stein manifold X with trivial complex tangent bundle $\mathrm{T}X$, each complex vector bundle trivialization $\mathrm{T}X \xrightarrow{\cong} X \times \mathbb{C}^n$ is homotopic to the differential of a locally biholomorphic map $X \to \mathbb{C}^n$.*

Example 9.13.5 (Parallelizable Stein manifolds) A complex manifold X whose complex tangent bundle $\mathrm{T}X$ is holomorphically trivial is said to be (holomorphically) parallelizable. Examples are holomorphic quotients of $\mathbb{C}^n$ and its subdomains. If X is Stein then, by the Oka-Grauert principle (Corollary 8.3.6), $\mathrm{T}X$ is topologically trivial if and only if it is holomorphically trivial.

Every closed smooth complex hypersurface $X \subset \mathbb{C}^{n+1}$ is parallelizable [169]. To see this, note that the normal bundle N of X in $\mathbb{C}^{n+1}$ is trivial by the solution of the

Cousin II problem; since $TX \oplus N = T\mathbb{C}^{n+1}|_X \cong X \times \mathbb{C}^{n+1}$ is trivial, Corollary 8.3.9 (p. 363) implies that TX is also trivial. The same argument shows that a closed complex submanifold $X^n \subset \mathbb{C}^m$, $n \geq \lfloor \frac{m}{2} \rfloor$, with trivial normal bundle is parallelizable. Only a few of these examples are known to immerse holomorphically to $\mathbb{C}^n$. Loeb found explicit holomorphic immersions $X \to \mathbb{C}^n$ of algebraic hypersurfaces of the form

$$X = \left\{(z_0, z_1, \ldots, z_k) : z_0^d + P_1(z_1) + \cdots + P_k(z_k) = 1\right\} \subset \mathbb{C}^{n+1},$$

where $z_0 \in \mathbb{C}$, $z_j \in \mathbb{C}^{n_j}$, P_j is a homogeneous polynomial of some degree d_j on $\mathbb{C}^{n_j}$ for every $j = 1, \ldots, k$, and $n_1 + \cdots + n_k = n$ (see [53, p. 19]). These manifolds are even algebraically parallelizable, but they do not admit any algebraic immersions to $\mathbb{C}^n$. An example of this type is the complex n-sphere $\Sigma^n = \{z \in \mathbb{C}^{n+1} : \sum z_j^2 = 1\}$. In another direction, Nishimura found explicit holomorphic immersions $\mathbb{P}^2 \setminus C \to \mathbb{C}^2$, where C is an irreducible cuspidal cubic curve in $\mathbb{P}^2$ [433]. Further examples and remarks on parallelizable Stein manifolds can be found in [169]. □

Example 9.13.6 There exist Stein Riemann domains $X \to \mathbb{C}^n$ such that X is not biholomorphic (and not even homeomorphic) to a domain in $\mathbb{C}^n$. For example, if X is a complex torus $\mathbb{C}/\Gamma$ with one point removed, then X contains a pair of embedded closed real curves intersecting transversely at one point. Such X is a Riemann domain over $\mathbb{C}$ by the Gunning-Narasimhan theorem [273], but is not homeomorphic to a domain in $\mathbb{C}$ since a pair of closed curves in $\mathbb{C}$ intersecting transversely meet at an even number of points. (This is standard in the smooth intersection theory; for the relevant topological intersection theory see [129, Corollary 4.9, p. 199].) □

We now turn to the proof of Theorem 9.13.1, beginning with preliminaries.

Every complex vector bundle map $\Theta : TX \to X \times \mathbb{C}^q$ is of the form $\Theta(x, v) = (x, \theta v)$, where $\theta = (\theta_1, \ldots, \theta_q)$ is a q-tuple of differential $(1,0)$-forms on X; that is, a q-tuple of sections of the holomorphic cotangent bundle $T^{*1,0}X$. Furthermore, Θ is surjective if and only if the $(1,0)$-forms $\theta_1, \ldots, \theta_q$ are $\mathbb{C}$-linearly independent at every point $x \in X$:

$$\theta_{1,x} \wedge \cdots \wedge \theta_{q,x} \neq 0, \quad x \in X. \tag{9.35}$$

A q-tuple satisfying (9.35) will be called a *q-coframe* on X. Such a q-coframe spans a trivial rank q complex subbundle E^* of the complex cotangent bundle $T^{*1,0}X \cong T^*X$. Conversely, every q-coframe induces a surjective complex vector bundle map $\Theta : TX \to X \times \mathbb{C}^q$. Clearly, we may speak of holomorphic q-coframes, homotopies of q-coframes, etc. If every component θ_j of a q-coframe θ is exact holomorphic, $\theta_j = df_j$ for some $f_j \in \mathcal{O}(X)$, then we write $\theta = df$ and say that θ is an *exact holomorphic q-coframe*.

We may view q-coframes as sections of the holomorphic fibre bundle $V^q(T^*X) \to X$ whose fibre V_x^q is the Stiefel manifold of all q-tuples of $\mathbb{C}$-independent elements in T_x^*X. (see Example 1.2.7 on p. 9). Since the fibres are $GL_n(\mathbb{C})$-homogeneous ($n = \dim X$), the Oka-Grauert principle applies to sections of $V^q(T^*X)$; hence every q-coframe is homotopic to a holomorphic one.

Theorem 9.13.1 is an immediate consequence of the following result that includes the approximation and interpolation conditions.

Theorem 9.13.7 ([191, Theorem 2.5]) *Assume that X is a Stein manifold of dimension n, X' is a closed complex subvariety of X, and K is a compact $\mathscr{O}(X)$-convex subset of X. Assume that $q \in \{1,\dots,n-1\}$ or $q=n=1$. Let $\epsilon>0$ and $r\in\mathbb{N}$. Given a q-coframe θ^0 on X such that $\theta^0 = \mathrm{d}f^0$ is exact holomorphic in an open set $U\supset X'\cup K$, there exists a homotopy of q-coframes θ^t $(t\in[0,1])$ such that $\theta^1=\mathrm{d}f$ is the differential of a holomorphic submersion $f=(f_1,\dots,f_q)\colon X\to\mathbb{C}^q$ and the following hold for every $t\in[0,1]$:*

(i) $\theta^t = \mathrm{d}f^t$ *is exact holomorphic in a neighborhood of* $X'\cup K$,
(ii) $\sup_{x\in K}|f^t(x)-f^0(x)|<\epsilon$, *and*
(iii) f^t-f^0 *vanishes to order r on the subvariety* X'.

Proof We follow the scheme of proof of Theorem 5.4.4 (see Sect. 5.11). Consider first the case $X'=\varnothing$. Choose a smooth strongly plurisubharmonic Morse exhaustion function $\rho\colon X\to\mathbb{R}$ with nice critical points (Definition 3.10.2, p. 95) such that $\rho<0$ on K and $\rho>0$ on $X\setminus U$. We shall inductively modify the given q-coframe θ such that, at every step, we make it exact holomorphic on a somewhat bigger set, taking care to satisfy conditions (i)–(iii).

There are two main cases to consider.

The noncritical case: This amounts to passing from $\{\rho\le c\}$ to $\{\rho\le c'\}$ when ρ has no critical values on the interval $[c,c']$. By Lemma 5.10.3 (p. 242) this is accomplished in finitely many basic steps, each consisting of attaching a special convex bump B to a compact strongly pseudoconvex domain A; i.e., such that (A,B) is a special Cartan pair in X (see Definition 5.10.2, p. 242).

We explain the procedure for such a pair (A,B). By the assumption, we have a q-coframe θ on X that is exact holomorphic in an open neighborhood U_A of A, i.e., $\theta=\mathrm{d}f$ where $f\colon U_A\to\mathbb{C}^q$ is a holomorphic submersion. The goal is to approximate f uniformly on A by a holomorphic submersion $f'\colon U'_D\to\mathbb{C}^q$ over a neighborhood U'_D of $D=A\cup B$. Then, $\theta'=\mathrm{d}f'$ is exact holomorphic near D and is homotopic to θ there. Outside of a bigger neighborhood of D we patch θ' with θ by using a cut-off function in the parameter of the homotopy (see the arguments in Sect. 5.11).

In suitable local holomorphic coordinates on X in a neighborhood of B, B and $C=A\cap B$ are compact convex subsets of $\mathbb{C}^n$ (see Fig. 5.2 on p. 242). By Theorem 9.12.2 on p. 458 (if $q<n$) or Theorem 9.12.1 (if $q=1\le n$) we can approximate f uniformly on a compact neighborhood C' of C by a holomorphic submersion $g\colon U_B\to\mathbb{C}^n$ on neighborhood of B. If g is sufficiently close to f, then Lemma 9.12.6 furnishes an injective holomorphic map $\gamma\colon U_C\to X$ in a neighborhood of C such that γ is uniformly close to the identity map (depending on $\mathrm{dist}_{C'}(f,g)$) and $f=g\circ\gamma$ holds on a neighborhood of C. If g is close enough to f on C', then Theorem 9.7.1 (p. 432) gives injective holomorphic maps α and β, close to the identity map on a neighborhood of A and B, respectively, such that

$\gamma \circ \alpha = \beta$ holds on a neighborhood of C. It follows that $f \circ \alpha = g \circ \beta$ holds near C, and hence these two maps define a holomorphic submersion f' from a neighborhood of $D = A \cup B$ to $\mathbb{C}^q$ which approximates f uniformly on A. This concludes the noncritical case.

The critical case: We use the local normal form of ρ at a nice critical point p (see Sects. 3.10–3.11). We may assume that $\rho(p) = 0$. Let $k \in \{0, 1, \dots, n\}$ denote the Morse index of ρ at p. If $k = 0$, a new connected component of the sublevel set appears at p, and we can take an arbitrary submersion in this component. Assume now that $k \geq 1$. In suitable local holomorphic coordinates $z = (z_1, \dots, z_n) = (z', z'') \in \mathbb{C}^k \times \mathbb{C}^{n-k}$ around the critical point p, $\rho(z)$ is a normal form given by Lemma 3.10.1. In particular, the stable manifold of p,

$$E = \{(x' + iy', z'') : y' = 0,\ z'' = 0,\ |x'|^2 \leq |c|\},$$

is a k-dimensional totally real disc, attached from the outside to the sublevel set $\{\rho \leq c\}$ along the $(k-1)$-sphere $bE \subset \{\rho = c\}$. Choose numbers $c < r < 0$ close enough to 0 such that $\theta = \mathrm{d}f$ is exact holomorphic on $\{\rho < r\}$ and ρ has no critical values on $[c, 0)$. The extension across the critical level is accomplished in three steps (see the *critical case* in Sect. 5.11):

1. Smooth extension of f to a neighborhood of E such that $\overline{\partial} f|_E = 0$ and $\mathrm{d}f|_E = \partial f|_E$ has maximal rank q along E.
2. Approximation of the map f from Step 1 by a holomorphic submersion g defined in a neighborhood of the handlebody $\{\rho \leq c\} \cup E$.
3. Applying the noncritical case with another strongly plurisubharmonic function τ to extend g across the critical level $\{\rho = 0\}$.

Let us explain Step 1. The components θ_j of the q-coframe θ are expressed in the z-coordinates by $\theta_j(z) = \sum_{l=1}^{n} \theta_{j,l}(z)\,\mathrm{d}z_l$ $(j = 1, \dots, q)$ where the continuous $q \times n$ matrix $(\theta_{j,l})$ has complex rank q at each point. For $z \in E$ near bE we have $\theta = \mathrm{d}f$, and hence $\theta_{j,l} = \frac{\partial f_j}{\partial z_l} = \frac{\partial f_j}{\partial x_l}$.

Let $M_{q,n} \cong \mathbb{C}^{q \times n}$ denote the set of all complex $q \times n$ matrices, and let $V_{q,n}$ consist of all matrices in $M_{q,n}$ of maximal rank q (the Stiefel manifold).

Lemma 9.13.8 *The map f extends smoothly to a neighborhood of the set $\{\rho \leq c\} \cup E$ such that the following conditions hold:*

(i) *the extension satisfies $\overline{\partial} f = 0$ at every point of E,*
(ii) *the Jacobian matrix $f' = (\frac{\partial f_j}{\partial z_l})$ has complex rank q along E, and*
(iii) *f' is homotopic to $(\theta_{j,l})$ by a homotopy of maps $E \to V_{q,n}$ that is fixed near $bE = \{\rho \leq c\} \cap E$.*

Proof We identify $x \in \mathbb{R}^n$ with $x + \mathrm{i}0 \in \mathbb{C}^n$. Choose a small $\delta > 0$. Let $A \subset D \subset \mathbb{R}^n$ be subsets of X given in the local coordinates by

$$D = \{(x', x'') : |x'|^2 \le |c|,\ |x''| \le \delta\},$$
$$A = \{(x', x'') : |r| \le |x'|^2 \le |c|,\ |x''| \le \delta\}.$$

Choosing $\delta > 0$ small enough, f is a holomorphic submersion to $\mathbb{C}^q$ in a neighborhood of the annular set $A \subset \mathbb{R}^n$, and hence its Jacobian matrix defines a map $f' \colon A \to V_{q,n}$. By the assumption we have $f' = \theta$ on A.

Lemma 9.13.9 *There exists a smooth map $h = (h_1, \ldots, h_q) \colon D \to \mathbb{C}^q$ such that $h = f$ on A, $h' = (\frac{\partial h_j}{\partial x_l}) \colon D \to V_{q,n}$, and the map $\theta \colon D \to V_{q,n}$ is homotopic to h' (relative to A) in the space of maps $D \to V_{q,n}$.*

Proof Note that a map $h \colon D \to \mathbb{C}^q$ satisfies the conclusion of the lemma if and only if it is a complex submersion (see Definition 9.1.1(b)). Hence the conclusion follows from the h-principle for complex submersions (see Theorem 9.1.4). The case when $k < 2(n - q + 1)$ (which is equivalent to $q \le n - [\frac{k}{2}]$) follows also from Theorem 9.1.2 which only uses the jet transversality theorem. □

Since D is contained in the real subspace $\mathbb{R}^n$ of $\mathbb{C}^n$, we can apply Lemma 3.5.5 (p. 81) to extend the map h from Lemma 9.13.9 smoothly to a neighborhood of D in $\mathbb{C}^n$ (considered as a subset of X) such that $\overline{\partial} h = 0$ on D and the extension agrees with f near A. By setting $f = h$ near D we thus obtain an extension of f satisfying Lemma 9.13.8. □

Clearly, the homotopy of the q-coframe θ on D can be extended to a homotopy on X such that $\theta = \mathrm{d}f$ holds on $\{\rho \le c\} \cup D$. This completes Step 1 in the critical case. Now apply Theorem 3.8.1 (p. 88) to approximate f in the $\mathscr{C}^1$ topology on $\{\rho \le c\} \cup E$ by a holomorphic map g from a neighborhood of this set to $\mathbb{C}^q$. If the approximation is sufficiently close, then g is a submersion, and this completes Step 2. Step 3 is accomplished by using Lemma 3.11.1 (p. 98) just as in Sect. 5.11.

This proves Theorem 9.13.7 when $X' = \varnothing$. The general case with interpolation is obtained exactly as in the proof of Theorem 5.4.4 (see Sect. 5.12). □

The above construction allows improvements in several directions. First, we can add the interpolation condition for finite order jets at a discrete set of points, obtaining holomorphic functions on a Stein manifold with a prescribed critical set. Furthermore, we can add jet interpolation along a subvariety.

Theorem 9.13.10 ([191, Theorem 2.1]) *Let X be a Stein manifold, X' be a closed complex subvariety of X, and K be a compact $\mathcal{O}(X)$-convex subset of X. Let $U \subset X$ be an open set containing $X' \cup K$ and let $f \in \mathcal{O}(U)$ be a holomorphic function with discrete critical set $P = \{p_1, p_2, \ldots\} \subset X' \cup K$. Given numbers $\epsilon > 0$ and $r, n_1, n_2, \ldots \in \mathbb{N}$, there exists $g \in \mathcal{O}(X)$ with the critical set P and satisfying the following additional properties:*

(i) $|f(x) - g(x)| < \epsilon$ *for all* $x \in K$,
(ii) $\mathrm{j}_x^r f = \mathrm{j}_x^r g$ *for all* $x \in X'$, *and*
(iii) $\mathrm{j}_{p_k}^{n_k} f = \mathrm{j}_{p_k}^{n_k} g = 0$ *for* $k = 1, 2, \ldots$.

In particular, for any discrete set $P = \{p_j\} \subset X$ there exists a holomorphic function $f \in \mathcal{O}(X)$ whose critical locus equals P.

In the proof of Theorem 9.13.7, the topological assumption concerning the existence of a q-coframe θ on X was used only in the critical case—Step 1 (extension of a submersion across the stable manifold of a critical point p of ρ). If the Morse index k of p satisfies $k < 2(n - q + 1)$ (equivalently, $q \le n - [\frac{k}{2}]$), then this maximal rank condition holds for a generic map on the handle (see the proof of Lemma 9.13.9). This gives the following corollary.

Corollary 9.13.11 *Let X be a Stein manifold of dimension n. Assume that $\rho\colon X \to \mathbb{R}$ is a strongly plurisubharmonic Morse exhaustion function, c is a regular value of ρ, and every critical point of ρ in $\{x \in X : \rho(x) > c\}$ has Morse index $\le k$. If*

$$q \le q(k, n) = \min\left\{n - \left[\frac{k}{2}\right], n - 1\right\}$$

then every holomorphic submersion $\{x \in X : \rho(x) < c\} \to \mathbb{C}^q$ can be approximated uniformly on compacts by holomorphic submersions $X \to \mathbb{C}^q$.

The theorem on the existence of noncritical functions on Stein manifolds was extended to (paracompact, reduced) Stein spaces in [206]. We begin by recalling the relevant notions.

Let X be a complex space. Denote by $\mathcal{O}_{X,x}$ the ring of germs of holomorphic function at a point $x \in X$ and by $\mathfrak{m}_x$ the maximal ideal of $\mathcal{O}_{X,x}$, so $\mathcal{O}_{X,x}/\mathfrak{m}_x \cong \mathbb{C}$. Given $f \in \mathcal{O}_{X,x}$ we denote by $f - f(x) \in \mathfrak{m}_x$ the germ obtained by subtracting from f its value $f(x) \in \mathbb{C}$ at the point x.

Definition 9.13.12 Let x be a nonisolated point of a complex space X.

(a) A germ $f \in \mathcal{O}_{X,x}$ at x is *critical* (and x is a *critical point* of f) if $f - f(x) \in \mathfrak{m}_x^2$ (the square of the maximal ideal $\mathfrak{m}_x$), and is *noncritical* if $f - f(x) \in \mathfrak{m}_x \setminus \mathfrak{m}_x^2$.
(b) A germ $f \in \mathcal{O}_{X,x}$ is *strongly noncritical* if the germ at x of the restriction $f|_V$ to any local irreducible component V of X is noncritical.

Any function is considered (strongly) noncritical at an isolated point of X.

One can characterize these notions by the (non-) vanishing of the differential $\mathrm{d}f_x$ on the *Zariski tangent space* T_xX. Recall that T_xX is isomorphic to $(\mathfrak{m}_x/\mathfrak{m}_x^2)^*$, the dual of $\mathfrak{m}_x/\mathfrak{m}_x^2$, the latter being the cotangent space T_x^*X (see [158, p. 78]). The number $\dim_{\mathbb{C}} \mathrm{T}_xX$ is the embedding dimension of the germ X_x of X at x. The differential $\mathrm{d}f_x \colon \mathrm{T}_xX \to \mathbb{C}$ of $f \in \mathcal{O}_{X,x}$ is determined by the class $f - f(x) \in \mathfrak{m}_x/\mathfrak{m}_x^2 = \mathrm{T}_x^*X$, so f is critical at x if and only if $df_x = 0$. If $X_x = \bigcup_{j=1}^k V_j$ is a decomposition into local irreducible components, then f is strongly noncritical at x if $\mathrm{d}f_x \colon \mathrm{T}_xV_j \to \mathbb{C}$ is nonvanishing for every $j = 1, \ldots, k$. At a regular point $x \in X_{\mathrm{reg}}$ these notions coincide with the usual one: x is a critical point of f if and only if in some (hence in any) local holomorphic coordinates $z = (z_1, \ldots, z_n)$ on a neighborhood of x, with $z(x) = 0$ and $n = \dim_x X$, we have $\frac{\partial f}{\partial z_j}(0) = 0$ for $j = 1, \ldots, n$.

Theorem 9.13.13 ([206, Theorem 1.3]) *Every reduced Stein space admits a holomorphic function which is strongly noncritical at every point.*

Furthermore, given a closed discrete set $P = \{p_1, p_2, \ldots\}$ in X, germs $f_k \in \mathcal{O}_{X,p_k}$ and integers $n_k \in \mathbb{N}$, there exists a holomorphic function $f \in \mathcal{O}(X)$ which is strongly noncritical on $X \setminus P$ and agrees with the germ f_k to order n_k at the point $p_k \in P$ (i.e., $f_{p_k} - f_k \in \mathfrak{m}_{p_k}^{n_k}$) for every $k = 1, 2, \ldots$.

The proof follows the same pattern as that of Theorem 9.13.10, but is more involved from the technical point of view. The main gluing lemma used in the proof is given by Theorem 9.7.4.

In conclusion, we mention that Forstnerič and Ohsawa [216] constructed a noncritical holomorphic function of finite order on any compact Riemann surface with a puncture. Let X be a compact Riemann surface, let p be a point of X, and set $R = X \setminus \{p\}$. Pick a local coordinate z on a neighborhood $U \subset X$ of p with $z(p) = 0$. A holomorphic function $f \in \mathcal{O}(R)$ is said to be *of finite order* if there exist positive numbers λ and μ such that

$$\left|f(z)\right| \le \lambda \exp|z|^{-\mu} \quad \text{holds on } U \setminus \{p\}. \tag{9.36}$$

Theorem 9.13.14 ([216]) *If X is a compact Riemann surface and $p \in X$, then the punctured Riemann surface $R = X \setminus \{p\}$ admits a holomorphic function of finite order without critical points, i.e., such that $\mathrm{d}f_x \neq 0$ for every $x \in R$.*

Theorem 9.13.14 is optimal in the sense that, unless $X = \mathbb{P}^1$ is the Riemann sphere and $R = \mathbb{C}$, there do not exist any noncritical *algebraic* functions on R. Indeed, an algebraic function on R extends to a holomorphic map $f \colon X \to \mathbb{P}^1$. Denoting by $g \ge 1$ the genus of X and by $d \ge 1$ the local degree of f at the point p, the Riemann-Hurwitz formula shows that f has $(d+1) - (2-2g) = d + 2g - 1 \ge 2$ branch points in R when counted with algebraic multiplicities (see [216, Proposition 1.2]).

Problem 9.13.15 Does every affine algebraic manifold $X \subset \mathbb{C}^N$ admit a noncritical holomorphic function of finite order? Does Corollary 9.13.2 hold in the finite order category?

9.14 The Oka Principle for Holomorphic Submersions

The Oka principle in Theorem 9.13.1 can be extended to holomorphic submersions of Stein manifolds to a certain class of complex manifolds that satisfy the following approximation property analogous to CAP.

Definition 9.14.1 A complex manifold Y satisfies *Property* S_n for some integer $n \geq \dim Y$ if every holomorphic submersion $K \to Y$ from a compact convex set $K \subset \mathbb{C}^n$ can be approximated by holomorphic submersions $\mathbb{C}^n \to Y$.

Note that a manifold of Class $\mathcal{A}$ (see Definition 6.4.4) satisfies Property S_n for any $n > \dim Y$ according to Theorem 9.12.4 (p. 458). We have the following Oka principle which extends the results for smooth submersions, due to Phillips [451] and Gromov [265, 267].

Theorem 9.14.2 ([194, Theorem 1.1]) *Assume that Y is a complex manifold of dimension q satisfying Property S_n for some $n \geq q$. Let X be a Stein manifold of dimension n. A continuous map $f: X \to Y$ is homotopic to a holomorphic submersion $X \to Y$ if and only if there exists a surjective complex vector bundle map $\iota: TX \to f^*TY$.*

If Y satisfies a 1-parametric version of Property S_n (see [194, Definition 2.1(b)]), then we also have the following

One-parametric Oka principle for holomorphic submersions: *Regular homotopy classes of holomorphic submersions $X \to Y$ from n-dimensional Stein manifolds X to Y are in one-to-one correspondence with homotopy classes of fibrewise surjective complex vector bundle maps $TX \to TY$.*

The proof of Theorem 9.14.2 is similar to that of Theorem 9.13.7. The only analytic property of the target manifold $\mathbb{C}^q$ that was used in the proof is the approximation property for holomorphic submersions $\mathbb{C}^n \to \mathbb{C}^q$ on compact convex subsets of $\mathbb{C}^n$ (see Theorem 9.12.2, p. 458); this is precisely Property S_n of $\mathbb{C}^q$ for $n > q$. The last statement follows from Theorem 9.12.4.

Property S_n obviously passes up and down in an unbranched holomorphic covering $\widetilde{Y} \to Y$. For further functorial properties of S_n, see [194]. While its relationship with CAP (characterizing the class of Oka manifolds) is unclear, S_n has been verified in many cases when CAP holds and $n > \dim Y$.

Problem 9.14.3 Does every Oka manifold Y enjoy Property S_n for each integer $n > \dim Y$? Does $\mathbb{C}^n$ for $n > 1$ enjoy Property S_n? (Note that this is a rephrasing of Problem 9.12.3.)

Problem 9.14.4 Assume that X is a Stein manifold of dimension $n > 1$ and $f_1, \ldots, f_q$ are holomorphic functions on X satisfying $\mathrm{d}f_1 \wedge \cdots \wedge \mathrm{d}f_q \neq 0$ where $1 \le q < n$. Assume that θ is a continuous $(1,0)$-form on X such that

$$\mathrm{d}f_1 \wedge \cdots \wedge \mathrm{d}f_q \wedge \theta \neq 0. \tag{9.37}$$

Is θ homotopic to an exact holomorphic differential $\theta_1 = \mathrm{d}g$ in the space of $(1,0)$-forms satisfying the independence condition (9.37)?

That is, we are looking for a function $g \in \mathcal{O}(X)$ whose restriction to every fibre $f^{-1}(z)$ $(z \in \mathbb{C}^q)$ of the holomorphic submersion $f = (f_1, \ldots, f_q)\colon X \to \mathbb{C}^q$ is noncritical on that fibre. At the moment this is open even for $q = 1$, i.e., when the fibres are hypersurfaces. A positive solution for $q = n - 1$ (when the fibres are complex curves) would clearly imply that every Stein manifold X with trivial tangent bundle admits a locally biholomorphic map to $\mathbb{C}^n$ with $n = \dim X$ (see Problem 9.13.3 on p. 462). When $X = \mathbb{C}^n$, the tangent bundle of a nonsingular foliation by curves is a line bundle and hence trivial, so there is a nowhere vanishing holomorphic vector field L on $\mathbb{C}^n$ tangent to the leaves. This brings us to the following problem (see [265, p. 70]).

Problem 9.14.5 Let L be a nowhere vanishing holomorphic vector field on a Stein manifold X. Does there exist a holomorphic function $g \in \mathcal{O}(X)$ such that $L(g) \neq 0$ at every point of X?

This problem is open even for polynomial vector fields on $\mathbb{C}^n$. Simple examples show that in general one cannot solve the equation $L(g) = h$ for a given $h \in \mathcal{O}(X)$, not even for the constant function $h = 1$.

9.15 Closed Holomorphic 1-Forms Without Zeros

By a theorem of Serre [496, Theorem 1], each element of the de Rham cohomology group $H^1(X;\mathbb{C})$ of a Stein manifold X is represented by a closed holomorphic 1-form. By Corollary 9.13.2 on p. 462, the zero class in $H^1(X;\mathbb{C})$ is also represented by a closed holomorphic 1-form *without zeros*, namely, the differential $\mathrm{d}f$ a noncritical holomorphic function $f\colon X \to \mathbb{C}$. The following result of I. Majcen shows that every cohomology class in $H^1(X;\mathbb{C})$ is represented by a closed holomorphic 1-form without zeros.

Theorem 9.15.1 (Majcen [403]) *Let X be a Stein manifold of dimension n whose holomorphic cotangent bundle T^*X admits a trivial complex subbundle of rank q for some $1 \le q < n$. Given closed 1-forms $\theta_1, \ldots, \theta_q$ on X, there exist closed holomorphic 1-forms $\omega_1, \ldots, \omega_q$ on X satisfying*

$$[\omega_j] = [\theta_j] \in H^1(X;\mathbb{C}) \ (j = 1, \ldots, q), \qquad \omega_1 \wedge \omega_2 \wedge \cdots \wedge \omega_q \neq 0.$$

In particular, every cohomology class in $H^1(X;\mathbb{C})$ is represented by a closed holomorphic 1-form without zeros.

On open Riemann surfaces this was shown by Kusunoki and Sainouchi [355]. Theorem 9.15.1 fails on general complex manifolds. For example, if X is a compact Riemann surface of genus $g > 1$, the Riemann-Roch theorem shows that each nonconstant holomorphic 1-form has $2g - 2 \geq 2$ zeros.

Proof We explain the main idea in the case of a single closed 1-form θ.

By Serre's theorem (see [496, Theorem 1]) we may assume that θ is a closed holomorphic 1-form. If ω is another such form, de Rham's theorem implies that $[\theta] = [\omega] \in H^1(X;\mathbb{C})$ if and only if $\int_C \omega = \int_C \theta$ for every closed path C in X. Thus, we need to find a closed holomorphic 1-form without zeros and with prescribed periods on a set of curves forming a basis of the free part of $H_1(X;\mathbb{Z})$. We follow the proof of Theorem 9.13.7 and explain the necessary modifications.

Choose a strongly plurisubharmonic exhaustion function $\rho\colon X \to \mathbb{R}$ with nice critical points (see Definition 3.10.2). Let $c < c'$ be regular values of ρ. Assume that ω is a closed holomorphic 1-form without zeros on a neighborhood of $\{\rho \leq c\}$ whose periods over loops in $\{\rho \leq c\}$ agree with those of θ. We must approximate ω by a 1-form $\widetilde{\omega}$ satisfying the same conditions in a neighborhood of $\{\rho \leq c'\}$; the proof is then finished by an obvious induction.

Consider first the case when ρ has no critical values in $[c, c']$. The topology of the sublevel sets does not change from c to c', and no new closed homologically nontrivial curves appear. It suffices to explain how to extend ω across a convex bump B attached to a certain set A such that (A, B) is a special convex pair (Definition 5.4.2, p. 215). In a neighborhood of $A \cap B$ (which is a compact convex set in $\mathbb{C}^n$ in some local coordinates) the form ω is exact holomorphic, $0 \neq \omega = \mathrm{d}f$. Now, approximate f by a noncritical holomorphic function g in a neighborhood of B, find a holomorphic transition map γ close to the identity near $A \cap B$ such that $f = g \circ \gamma$, and decompose $\gamma = \beta \circ \alpha^{-1}$ where α and β are small perturbations of the identity map near A and B, respectively. (See the proof of Theorem 9.13.7, p. 464). Thus, $f \circ \alpha = g \circ \beta$ holds near $A \cap B$. Define a holomorphic 1-form ω' in a neighborhood of $A \cup B$ by taking $\omega' = \alpha^*\omega$ near A and $\omega' = \beta^*\mathrm{d}g$ near B. On the intersection $A \cap B$ these two expressions agree:

$$\beta^*\mathrm{d}g = \mathrm{d}(g \circ \beta) = \mathrm{d}(f \circ \alpha) = \alpha^*\mathrm{d}f = \alpha^*\omega.$$

In finitely many such steps we accomplish the task. The closed holomorphic 1-form $\widetilde{\omega}$ without zeros on $\{\rho \leq c'\}$, obtained in this way, is conjugate to ω on $\{\rho \leq c\}$ by a biholomorphism close to the identity map, and hence the periods over the closed curves in $\{\rho \leq c\}$ remain the same.

It remains to explain how to extend ω across a critical level of ρ. Suppose that p is a critical point of ρ and let k be its Morse index. Let E be the local stable manifold of p (an embedded totally real disc of dimension k) with $bE \subset \{\rho = c\}$, where $c < \rho(p)$ is close to $\rho(p)$. (See Sect. 3.11.)

If $k = 0$, a new connected component of the sublevel set appears at p, and it is trivial to extend ω to this component.

If $k \in \{2, \ldots, n\}$ then no new curves appear in the homology basis when passing the critical level at p. Furthermore, we claim that ω is exact holomorphic in a neighborhood of the $(k-1)$-sphere bE. This is obvious if $k > 2$, while for $k = 2$ (when $bE \cong S^1$) we have $\int_{bE} \omega = \int_{bE} \theta = \int_E \mathrm{d}\theta = 0$ by Stokes' theorem; the claim follows. Hence, $\omega = \mathrm{d}f$ in a neighborhood of bE for some noncritical holomorphic function f. By Theorem 9.12.1 we can approximate f in a neighborhood of bE by a noncritical holomorphic function g defined in a neighborhood of E. If the approximation is close enough, we can proceed as before to glue ω and $\mathrm{d}g$ into a closed holomorphic 1-form ω' without zeros in a neighborhood of $\{\rho \le c\} \cup E$. We complete this step by applying the noncritical case with a different strongly plurisubharmonic function (see the proof of Theorem 9.13.7, critical case, Step 3).

It remains to consider the case $k = 1$. There are two possibilities:

- The arc E joins two distinct connected components of $\{\rho \le c\}$. In this case, no new curve appears in the homology basis and we proceed as before.
- E is attached with both endpoints to the same component of $\{\rho \le c\}$.

In the latter case, we add to E another embedded arc $E' \subset \{\rho \le c\}$ such that $C = E \cup E'$ is a closed loop in X. This loop is a nontrivial new element of the first homology group when passing the critical level at p. We proceed as before, but taking care to ensure that $\int_C \omega = \int_C \theta$. This can be arranged by a suitable choice of the noncritical holomorphic function f along E such that $\mathrm{d}f \approx \omega$ near bE. Since the value of the integral changes a little when performing approximation and patching, we choose a submersive family of functions and use the implicit function theorem to obtain an extension with the correct period within this family. For further details, see [403]. □

9.16 Holomorphic Foliations on Stein Manifolds

The results on holomorphic submersions in the previous two sections give interesting consequences concerning the existence and approximation of nonsingular holomorphic foliations on Stein manifolds. The results in this section are taken from [191]. We begin with two immediate consequences of the existence of noncritical functions.

Corollary 9.16.1 *Every Stein manifold X admits a nonsingular holomorphic foliation by closed complex hypersurfaces; in addition, such a foliation may be chosen transverse to a given closed complex submanifold of X.*

Proof A closed complex submanifold V of a Stein manifold X is itself a Stein manifold, and hence it admits a noncritical function $f \in \mathcal{O}(V)$ by Corollary 9.13.2. By Cartan's Theorem B, f extends to a holomorphic function on X. Since the extension remains noncritical on V, Theorem 9.13.7 (p. 464), applied with $q = 1$ and $X' = V$, gives a noncritical function $g \in \mathcal{O}(X)$ such that $g|_V = f$. The levels sets $\{g = c\}_{c \in \mathbb{C}}$ form a foliation of X by closed smooth complex hypersurfaces that are transverse to V. □

Corollary 9.16.2 *If V is a smooth closed complex hypersurface with trivial normal bundle in a Stein manifold X, then V is a union of leaves in a nonsingular holomorphic foliation of X by closed complex hypersurfaces. This holds in particular if $H^2(V;\mathbb{Z})=0$, or if $X=\mathbb{C}^n$.*

In particular, every smooth connected complex curve in a Stein surface is a leaf in a nonsingular holomorphic foliation.

Proof Choose a holomorphic trivialization $N = \mathrm{T}X|_V/\mathrm{T}V \xrightarrow{\cong} V\times\mathbb{C}$ of the normal bundle $\mathcal{N}$ of the submanifold V in X. The projection $h\colon \mathcal{N}\to\mathbb{C}$ of this trivialization onto the second factor is a noncritical holomorphic function on $\mathcal{N}$. We identify the zero section $\{h=0\}$ of $\mathcal{N}$ with V. By Theorem 3.3.3 (the Docquier-Grauert theorem), there are an open neighborhood Ω of V in X and an injective holomorphic map $\phi\colon \Omega\to\mathcal{N}$ which equals the identity on V. Then, $f=h\circ\phi$ is a noncritical holomorphic function on Ω such that $\{f=0\}=V$. Applying Theorem 9.13.7 (p. 464) with $q=1$ and $X'=V$, we obtain a noncritical function $g\in\mathcal{O}(X)$ which vanishes on V. The foliation $\{g=c\}_{c\in\mathbb{C}}$ clearly satisfies Corollary 9.16.2.

The second statement follows from $\mathrm{Pic}(V)=H^1(V;\mathcal{O}^*)\cong H^2(V;\mathbb{Z})$ (see Theorem 5.2.1 on p. 211); the latter group vanishes if V is an open Riemann surface. Since every divisor on $\mathbb{C}^n$ is a principal divisor, the normal bundle of any closed complex hypersurface $V\subset\mathbb{C}^n$ is trivial. □

The Oka principle for holomorphic submersions $X\to\mathbb{C}^q$ (Theorem 9.13.1 and Corollary 9.13.11, p. 467) implies the following corollary; the last statement follows from the Oka-Grauert principle in Sect. 8.2.

Corollary 9.16.3 *Let X be a Stein manifold. Every trivial complex subbundle $\Theta\subset\mathrm{T}^*X$ of rank $q<\dim X$ is homotopic to a subbundle generated by independent holomorphic differentials $\mathrm{d}f_1,\ldots,\mathrm{d}f_q$. If Θ is holomorphic, then the homotopy can be chosen through holomorphic subbundles of T^*X.*

Corollary 9.16.3 admits the following dual formulation in terms of subbundles of the tangent bundle $\mathrm{T}X$. See also Theorem 9.16.7.

Corollary 9.16.4 *Let X be a Stein manifold of dimension n. Every complex subbundle $E\subset\mathrm{T}X$ of rank $k\ge 1$ with trivial quotient bundle $\mathrm{T}X/E$ is homotopic to an integrable holomorphic subbundle of the form $\ker\mathrm{d}f\subset\mathrm{T}X$, where $f\colon X\to\mathbb{C}^{n-k}$ is a holomorphic submersion.*

Proof The complex subbundle $\Theta=E^{\perp}\subset\mathrm{T}^*X$ with fibres

$$\Theta_x=\left\{\lambda\in\mathrm{T}^*_xX:\lambda(v)=0 \text{ for all } v\in E_x\right\}$$

(the complex conormal bundle of E) satisfies $\Theta\simeq(\mathrm{T}X/E)^*$, and hence is trivial. Corollary 9.16.3 gives a homotopy of complex subbundles $\Theta^t\subset\mathrm{T}^*X$ ($t\in[0,1]$)

from $\Theta^0 = \Theta$ to a subbundle $\Theta^1 \subset \mathrm{T}^*X$ that is spanned by $n-k$ independent holomorphic differentials $\mathrm{d}f_1, \ldots, \mathrm{d}f_{n-k}$. The homotopy of subbundles $E^t = (\Theta^t)^\perp \subset \mathrm{T}X$ then satisfies Corollary 9.16.4. □

We now mention a couple of results on the existence of submersion foliations that either contain a given submanifold as a union of leaves, or else are transverse to it. These results generalize Corollaries 9.16.1 and 9.16.2 to foliations of lower dimension. The proofs are similar but slightly more involved, depending on Theorem 9.13.7; see [191, p. 181] for the details.

Corollary 9.16.5 *Let X be an n-dimensional Stein manifold and $V \subset X$ be a closed complex submanifold. If the tangent bundle $\mathrm{T}X$ admits a trivial complex subbundle $\mathcal{N}$ satisfying $\mathrm{T}X|_V = \mathrm{T}V \oplus \mathcal{N}|_V$, then there is a holomorphic submersion $f\colon X \to \mathbb{C}^q$ ($q = n - \dim V$) such that V is a union of connected components of the fibre $f^{-1}(0)$. If $\dim V \geq [\frac{n}{2}]$, then the above conclusion holds if the normal bundle $\mathcal{N}_V = \mathrm{T}X|_V/TV$ is trivial.*

Corollary 9.16.6 *Assume that X is a Stein manifold of dimension n, $\iota\colon V \hookrightarrow X$ is a closed complex submanifold, and $f = (f_1, \ldots, f_q)\colon V \to \mathbb{C}^q$ is a holomorphic submersion. If there is a q-coframe $\theta = (\theta_1, \ldots, \theta_q)$ on X satisfying $\iota^*\theta_j = \mathrm{d}f_j$ for $j = 1, \ldots, q$, then there exists a holomorphic submersion $F\colon X \to \mathbb{C}^q$ with $F|_V = f$. Such F always exists if $q \leq [\frac{n+1}{2}]$.*

All results mentioned so far concern submersion foliations. Corollary 9.16.4 generalizes to subbundles $E \subset \mathrm{T}X$ with flat quotients $\mathcal{N} = \mathrm{T}X/E$, furnishing foliations that are not necessarily given by submersions. Recall that a complex vector bundle $\mathcal{N} \to X$ is *flat* if it admits a vector bundle atlas with locally constant transition functions (see Example 1.4.8 on p. 15).

Theorem 9.16.7 ([191, Theorem 7.1]) *Let X be a Stein manifold. If E is a complex subbundle of the tangent bundle $\mathrm{T}X$ such that the quotient bundle $\mathcal{N} = \mathrm{T}X/E$ is flat, then E is homotopic (through complex subbundles of $\mathrm{T}X$) to the tangent bundle of a nonsingular holomorphic foliation of X.*

Analogous results for smooth foliations on open manifolds were proved by Gromov [267] and Phillips [452–455], and on closed manifolds by Thurston [548, 550].

Proof We may assume that X is connected. Assume that the bundle $\mathcal{N} = \mathrm{T}X/E$ has rank q. Let $\pi : \mathcal{N} \to X$ denote the vector bundle projection. By flatness of $\mathcal{N}$ there is an open cover $\mathcal{U} = \{U_i\}_{i\in\mathbb{N}}$ of X and vector bundle charts $\phi_i : \mathcal{N}|_{U_i} = \pi^{-1}(U_i) \to U_i \times \mathbb{C}^q$ with transition maps of the form

$$\phi_{ij}(x,z) = \phi_i \circ \phi_j^{-1}(x,z) = \big(x, h_{ij}(z)\big), \quad x \in U_i \cap U_j,\ z \in \mathbb{C}^q,$$

where $h_{ij} \in GL_q(\mathbb{C})$ is independent of the base point $x \in U_i \cap U_j$. The latter property ensures that the product foliations of $\mathcal{N}|_{U_i} \cong U_i \times \mathbb{C}^q$ over the sets $U_j \in \mathcal{U}$

amalgamate into a global holomorphic foliation $\mathcal{H}$ of $\mathcal{N}$, with the zero section of $\mathcal{N}$ as a leaf. Precisely, if $U_i \cap U_j \neq \varnothing$ and $z \in \mathbb{C}^q$, then $\phi_i^{-1}(U_i \times \{h_{ij}(z)\})$ and $\phi_j^{-1}(U_j \times \{z\})$ belong to the same leaf of $\mathcal{H}$.

The tangent bundle of $\mathcal{N}$ decomposes as $\mathrm{T}\mathcal{N} = H \oplus V$, where the horizontal component $H = \mathrm{T}\mathcal{H}$ is the tangent bundle of the foliation $\mathcal{H}$, while the vertical component V is the tangent bundle of the vertical foliation $\pi^{-1}(x)$ $(x \in X)$. Denote by $\tau\colon \mathrm{T}\mathcal{N} \to V$ the projection onto V with $\ker \tau = H$. Note that $V = \pi^*\mathcal{N}$ is the pullback of the vector bundle $\mathcal{N} \to X$ to the total space $\mathcal{N}$ by the projection map $\pi\colon \mathcal{N} \to X$, and for every section $f\colon X \to \mathcal{N}$ of π we have $f^*V = \mathcal{N}$. If $f\colon X \to \mathcal{N}$ is a holomorphic section transverse to the horizontal foliation $\mathcal{H}$, then the intersection of its graph $f(X) \subset \mathcal{N}$ with $\mathcal{H}$ defines a holomorphic foliation $\mathcal{H}_f$ on X, of dimension $k = n - q$, whose tangent bundle $\mathrm{T}\mathcal{H}_f$ has fibres

$$\mathrm{T}_x\mathcal{H}_f = \big\{\xi \in \mathrm{T}_x X : \mathrm{d}f_x(\xi) \in H_{f(x)}\big\} = \big\{\xi \in \mathrm{T}_x X : \tau \circ \mathrm{d}f_x(\xi) = 0\big\}.$$

Transversality of f to the foliation $\mathcal{H}$ means that the vector bundle map

$$f' = f^* \circ \tau \circ \mathrm{d}f : \mathrm{T}X \to f^*V = \mathcal{N} \tag{9.38}$$

is surjective, and hence it induces an isomorphism $\mathrm{T}X/\mathrm{T}\mathcal{H}_f \xrightarrow{\cong} \mathcal{N}$. In particular, $\mathcal{N}$ is the normal bundle of any foliation $\mathcal{H}_f$ obtained in this way.

To prove the theorem, we construct a holomorphic section $f\colon X \to \mathcal{N}$ transverse to $\mathcal{H}$ and a continuous complex vector bundle injection $\iota\colon \mathcal{N} \to \mathrm{T}X$ such that the subbundle $\mathrm{T}\mathcal{H}_f \subset \mathrm{T}X$ is homotopic to E, and $f' \circ \iota\colon \mathcal{N} \to \mathcal{N}$ is a complex vector bundle automorphism homotopic to the identity through complex vector bundle automorphisms of $\mathcal{N}$.

On every sufficiently small connected open set $U \subset X$ we have $\mathcal{N}|_U \cong U \times \mathbb{C}^q$ and the plaques of $\mathcal{H}$ are $U \times \{z\}$ $(z \in \mathbb{C}^q)$. A section of $\mathcal{N}|_U$ is of the form $f(x) = (x, \tilde{f}(x))$ where $\tilde{f}\colon U \to \mathbb{C}^q$, and f is transverse to $\mathcal{H}$ if and only if $\tilde{f}$ is a submersion. This reduces every local problem in the construction of a transverse section to the corresponding problem for submersions.

From this point on, the construction follows the proof of Theorem 9.13.7. Choose a strongly plurisubharmonic Morse exhaustion function $\rho\colon X \to \mathbb{R}$ and a holomorphic vector bundle embedding $\tau\colon \mathcal{N} \hookrightarrow \mathrm{T}X$ such that $\mathrm{T}X = E \oplus \iota(\mathcal{N})$. Suppose f is a transverse holomorphic section, defined on a sublevel set $\{\rho \le c\}$, such that $\ker(\tau \circ \mathrm{d}f)$ is complementary to $\iota(\mathcal{N})$ and $f' \circ \iota$ is homotopic to the identity over the domain of f. (Here, f' is defined by (9.38).) We now inductively enlarge the domain of f as in the proof of Theorem 9.13.7. Whenever we change f, the injection ι is changed accordingly (by a homotopy of vector bundle injections $\mathcal{N} \to \mathrm{T}X$) so that the composition $f' \circ \iota$ remains homotopic to the identity on $\mathcal{N}$. The noncritical case is essentially the same as before. In the critical case we extend f across a totally real handle by using the h-principle given by Theorem 1.10.5 (p. 42). The necessary topological condition for the existence of such an extension is that $f' \circ \iota\colon \mathcal{N} \to \mathcal{N}$ be a complex vector bundle automorphism homotopic to the identity. The remaining steps are just as before. □

Holomorphic foliations considered in this section are rather special, given by holomorphic submersions or sections of flat bundles. Such foliations satisfy the Runge approximation theorem; see Theorem 9.13.7. No such result seems known in general. Specifically, we ask the following questions.

Problem 9.16.8 Let $\mathcal{F}$ be a nonsingular holomorphic foliation in a neighborhood of a compact $\mathcal{O}(X)$-convex set K in a Stein manifold X. Is it possible to approximate $\mathcal{F}$ on K by global holomorphic foliations of X (possibly with singularities on $X \setminus K$)?

Furthermore, assuming that the holomorphic tangent bundle $\mathrm{T}\mathcal{F}$ extends to a complex subbundle E of $\mathrm{T}X$, is it possible to deform E to an integrable holomorphic subbundle of $\mathrm{T}X$ (the tangent bundle of a nonsingular foliation) by a deformation that is holomorphic and close to the identity in a neighborhood of K?

Chapter 10
Topological Methods in Stein Geometry

In this chapter we apply some of the more advanced topological tools to the study of Stein manifolds.

We begin by considering closed real surfaces S embedded or immersed in a complex surface X. A generic embedding has isolated complex points; their algebraic number $I(S)$, and the numbers $I_\pm(S)$ if S is oriented, are isotopy invariants called *Lai indexes* (see Sects. 10.1–10.3). Every surface is isotopic to one with minimal number of complex points (see Sect. 10.4). The *Seiberg-Witten theory* shows that every closed oriented real surface S, smoothly embedded in a compact Kähler surface X with $b^+(X) > 1$ and such that S is not a homologically trivial sphere, satisfies the *adjunction inequality* $I_\pm(S) \le 0$; similar arguments yield the *generalized Thom conjecture* (see Sects. 10.6–10.7). On the other hand, the inequalities $I_\pm(S) \le 0$ imply that S is isotopic to a surface with a basis of open tubular Stein neighborhoods in X; this gives Stein domains of a given diffeomorphism type in compact Kähler surfaces.

Another interesting direction is the Eliashberg-Gompf construction of integrable Stein structures on almost complex manifolds with a suitable handlebody decomposition (see Sects. 10.8–10.11). On 4-manifolds, these Stein structures are typically *exotic* due to framing obstructions coming from symplectic geometry and the adjunction inequality in Stein surfaces. In Gompf's words [245], the classification theory for smooth structures on a fixed topological manifold is entirely anomalous in dimension four. One of the main results proved in this chapter is the following version of the Oka principle (see Sect. 10.9):

The Soft Oka principle: Any continuous map $X \to Y$ from a Stein manifold (X, J) to a complex manifold Y is homotopic to a holomorphic map in some Stein structure $\widetilde{J}$ on X that is homotopic to J. If $\dim_{\mathbb{R}} X = 4$ then this holds after a change of the smooth structure on X.

There are a number of interesting open problems in this interface between Stein geometry, symplectic geometry and 4-dimensional topology. A more complete exposition of this subject can be found in the monograph by Cieliebak and Eliashberg [99].

F. Forstnerič, *Stein Manifolds and Holomorphic Mappings*, Ergebnisse der Mathematik und ihrer Grenzgebiete. 3. Folge / A Series of Modern Surveys in Mathematics 56, DOI 10.1007/978-3-319-61058-0_10

10.1 Real Surfaces in Complex Surfaces

In this section, S denotes a real surface smoothly embedded (of class at least $\mathcal{C}^2$) in a complex surface (X, J). Given a point $p \in S$, the maximal complex tangent space $\Lambda_p = \mathrm{T}_p S \cap J(\mathrm{T}_p S)$ is a complex subspace of the tangent space $\mathrm{T}_p X$ that is contained in $\mathrm{T}_p S$. We have precisely two possibilities:

- $\Lambda_p = \{0\}$, in which case p is a totally real point of S, or
- $\Lambda_p = \mathrm{T}_p S$ is a complex line in $\mathrm{T}_p X$. Such p is a *complex point* of S.

Locally near any point $p \in S$ the surface S can be represented in suitably chosen local holomorphic coordinates (z, w) as the graph $\{w = f(z)\} \subset \mathbb{C}^2$ of a smooth function over a domain in $\mathbb{C}$. It is easily seen that $p = (z, f(z))$ is a complex point of S if and only if the Cauchy-Riemann equation holds at z:

$$2\frac{\partial f}{\partial \bar z}(z) = \frac{\partial f}{\partial x}(z) + \mathrm{i}\frac{\partial f}{\partial y}(z) = 0.$$

We may further assume that p corresponds to the origin $(0, 0) \in \mathbb{C}^2$ and that $\mathrm{T}_0 S = \{w = 0\}$ (equivalently, $\mathrm{d}f_0 = 0$). The complex point is said to be *nondegenerate* if the second order Taylor polynomial of f at 0 does not vanish identically; assume that this holds. If $\frac{\partial^2 f}{\partial z \partial \bar z}(0) \neq 0$, then by Bishop [59] there exist local holomorphic coordinates at $(0, 0)$ in which S is given by

$$w = |z|^2 + \lambda\big(z^2 + \bar z^2\big) + o\big(|z|^2\big) = (1 + 2\lambda)x^2 + (1 - 2\lambda)y^2 + o\big(|z|^2\big) \quad (10.1)$$

for some real number $\lambda \geq 0$ that is a biholomorphic invariant of the complex point. Note that $1 \pm 2\lambda$ are the eigenvalues of the real Hessian of f at 0. In particular, up to an error term $o(|z|^2)$, S is locally contained in the real hyperplane $\{\Im w = 0\} \cong \mathbb{C} \times \mathbb{R}$. If $\frac{\partial^2 f}{\partial z \partial \bar z}(0) = 0$ but the second order Taylor polynomial of f at 0 is nonzero, then we can simplify S to the form

$$w = z^2 + \bar z^2 + o\big(|z|^2\big). \quad (10.2)$$

This case corresponds to $\lambda = +\infty$ in (10.1).

A complex point $p \in S$ is said to be *quadratic* if, in some local holomorphic coordinates near p, S is of the form (10.1) or (10.2) without the remainder term. The point p is *flat* if S is locally near p of the form (10.1) or (10.2) with $\Im o(|z|^2) = 0$; that is, if S is locally contained in $\mathbb{C} \times \mathbb{R}$.

By using the invariant $\lambda = \lambda(p) \geq 0$ we classify nondegenerate complex points as follows:

- *elliptic* if $0 \leq \lambda < \frac{1}{2}$ (the real Hessian of f at 0 is positive definite),
- *parabolic* if $\lambda = \frac{1}{2}$ (the Hessian of f is degenerate), or
- *hyperbolic* if $\lambda > \frac{1}{2}$ (the eigenvalues of the Hessian have opposite signs); this includes the case $\lambda = \infty$ when S is given by (10.2).

Elliptic and hyperbolic complex points are isolated and stable under small $\mathscr{C}^2$ deformations of S in X. Parabolic and degenerate complex points need not be isolated. For example, taking $\lambda = \frac{1}{2}$ and $o(|z|^2) = 0$ in (10.1) gives the surface $w = 2x^2$ with the line of parabolic complex points $\{x = 0, \ w = 0\}$. A generic compact real surface in a complex surface has at most finitely many complex points, each of them either elliptic or hyperbolic. Parabolic complex points may occur in a generic homotopy of real surfaces in a complex surface.

A more precise normalization near an elliptic complex point was obtained by Moser and Webster [421] when the surface S is real analytic. In this case there exist local holomorphic coordinates $z = x + \mathrm{i}y$, $w = u + \mathrm{i}v$ on X at p in which S is given by

$$u = |z|^2 + \Gamma(u)\big(z^2 + \bar{z}^2\big), \qquad v = 0,$$

where $\Gamma(u) = \lambda + \delta u^s$, $\lambda \in (0, \frac{1}{2})$ is as above, $\delta = \pm 1$, and $s \in \mathbb{Z}_+$, or $\Gamma = \lambda$ ($s = \infty$). In particular, an elliptic point in a real analytic surface is flat. The triple (λ, δ, s) completely determines the germ of S at p up to a local biholomorphic equivalence.

Complex analytic properties of elliptic points are very different from those of hyperbolic points. At an elliptic point $p \in S$ there exists a one parameter family of closed, embedded, pairwise disjoint holomorphic discs $\{\Delta_t\}_{t \in (0,c]}$ in X, with boundaries $b\Delta_t \subset S$ covering a deleted neighborhood of p in S, such that Δ_t shrinks to the point p as $t \to 0$. The union $\Sigma = \bigcup_{t \in (0,c]} \Delta_t$ of these discs is a Levi flat hypersurface with boundary $b\Sigma \subset S$ that is contained in the local rational hull of S. (This is so since every circle $b\Delta_t$ also bounds a disc $D_t \subset S$ around p, and hence any complex curve intersecting Δ_t must also intersect D_t since the total intersection number with $\Delta_t \cup D_t$ equals zero.) Furthermore, $\Sigma \cup \{p\}$ is the local envelope of holomorphy of S at p. In particular, S does not admit a basis of Stein neighborhoods at an elliptic complex point. This is a special case of the results obtained by Bishop in his seminal paper [59]; for related works see König and Webster [333] and Hill and Taiani [294]. Bishop's work had a profound impact on the study of envelopes of holomorphy of real surfaces in complex surfaces and on related questions concerning extension and approximation of holomorphic functions near real surfaces. Bishop's theorem was extended to real surfaces in almost complex surfaces by Sukhov and Tumanov [537].

In contrast to the above, a real surface is locally holomorphically convex at a hyperbolic point (see [225]). This is seen by considering the preimage of S (10.1) by the proper quadratic map $\Phi \colon \mathbb{C}^2 \to \mathbb{C}^2$ given by

$$\Phi(z, w) = \big(z, zw + \lambda\big(z^2 + w^2\big)\big).$$

A calculation shows that $\Phi^{-1}(S) = S_1 \cup S_2$, where S_1 and S_2 are small discs in totally real surfaces that are tangent at the origin to the totally real 2-planes

$$\Sigma_1 = \big\{(\zeta, \bar{\zeta}) : \zeta \in \mathbb{C}\big\}, \qquad \Sigma_2 = \left\{\left(\zeta, -\frac{1}{\lambda}\zeta - \bar{\zeta}\right) : \zeta \in \mathbb{C}\right\}.$$

Assuming that $\lambda > \frac{1}{2}$, one can see that for sufficiently small $\epsilon > 0$ (and taking S small around the origin) the polynomial $\psi(z, w) = \frac{1}{4}(z^2 - w^2) + \epsilon zw$ maps S_1 and S_2 to cones in $\mathbb{C}$ whose only common point is their vertex at the origin. By Kallin's lemma ([329], [534, p. 62]) it follows that $S_1 \cup S_2$ is polynomially convex. Hence, $S = \Phi(S_1 \cup S_2)$ is also polynomially convex.

If S is smooth and a hyperbolic complex point $p \in S$ is flat, then locally near p the surface S admits a basis of tubular strongly pseudoconvex Stein neighborhoods in X (see Slapar [512]).

Example 10.1.1 A complex point $p \in S$ is *special hyperbolic* if there exist local holomorphic coordinates (z, w) on X near p in which S is given by $w = z^2 + \bar{z}^2$. The nonnegative function

$$\rho(z, w) = \left|w - z^2 - \bar{z}^2\right|^2 = \left|w - z^2\right|^2 + |z|^4 - 2\Re\big((w - z^2)z^2\big)$$

is clearly plurisubharmonic and it vanishes quadratically on S. The sublevel sets $\{\rho < t\}$ form a basis of tubular Stein neighborhoods of S. □

This simple idea does not work directly with any other type of hyperbolic points, not even for the quadric $w = |z|^2 + \lambda(z^2 + \bar{z}^2)$ with $\frac{1}{2} < \lambda < \infty$, since the corresponding function ρ fails to be plurisubharmonic in a full neighborhood of the origin. The construction of Stein neighborhoods in [512] uses certain nonhomogeneous squared distance functions and is rather delicate. From the existence of local Stein neighborhoods at hyperbolic points one easily obtains global Stein neighborhoods of S as in the following result.

Theorem 10.1.2 (Slapar [512, Theorem 2]) *Let $S \hookrightarrow X$ be a compact real surface smoothly embedded in a complex surface X and having only flat hyperbolic points $\{p_1, \ldots, p_k\}$. Then, there exists a $\mathcal{C}^\infty$ function ρ in a neighborhood $U \subset X$ of S such that $S = \{\rho = 0\} = \{\mathrm{d}\rho = 0\}$ and ρ is strictly plurisubharmonic on $U \setminus \{p_1, \ldots, p_k\}$. The sublevel sets $\{\rho < t\}$ for small $t > 0$ define a basis of tubular strongly pseudoconvex Stein neighborhoods of S in X.*

It remains to consider degenerate complex point. Before proceeding, we define the *index* of an isolated complex point $p \in S$. Assume as before that in some local holomorphic coordinates near p we have

$$p = 0, \qquad S = \big\{w = f(z)\big\} \subset \mathbb{C}^2. \tag{10.3}$$

Then, $z = 0$ is an isolated zero of the function $\frac{\partial f}{\partial \bar{z}}$.

Definition 10.1.3 (Notation as above) The index $I(p) = I(p; S)$ of an isolated complex point is the winding number of the function $\frac{\partial f}{\partial \bar{z}}$ in (10.3) along any small enough positively oriented circle $|z| = \epsilon$, $\epsilon > 0$. The index of a totally real point equals 0.

It is easily seen that the index is a local holomorphic invariant of S at p (see e.g. [182]). If S is given by (10.1), then

$$\frac{\partial f}{\partial \bar{z}} = z + 2\lambda\bar{z} + o(|z|).$$

By looking at the winding number along small circles $|z| = \epsilon$ we conclude that

$$I(p) = +1 \quad \text{if } p \text{ is an elliptic point,} \qquad I(p) = -1 \quad \text{if } p \text{ is a hyperbolic point.}$$

Another characterization of the index was given by Webster [564]. Let $\tau\colon \mathrm{T}X|_S \to \mathrm{N}_{S/X}$ be the projection onto the normal bundle of S in X with $\ker\tau = \mathrm{T}S$. Choose a local orientation on $\mathrm{T}S$ and coorient $\mathrm{N}_{S/X}$ so that the two orientations add up to the standard orientation of the complex bundle $\mathrm{T}X|_S = \mathrm{T}S \oplus \mathrm{N}_{S/X}$. Let J denote the complex structure operator on X. Assume that p is an isolated complex point of S. Choose a tangent vector field V on S such that $V_p \neq 0$. Then, $\widetilde{V} = \tau(JV)$ is a normal vector field (a section of $\mathrm{N}_{S/X}$) with an isolated zero at p.

Lemma 10.1.4 (Notation as above) *The index $I(p; S)$ of an isolated complex point $p \in S$ equals the index of the normal vector field $\widetilde{V}$ at p (i.e., the winding number of the fibre coordinate of the map $S \ni x \to \widetilde{V}_x$ around p).*

Proof Choose local coordinates (z, w) near $p = 0$ in which $S = \{w = f(z)\}$. We can take $\mathrm{N}_{S/X} = S \times (\{0\} \oplus \mathbb{C})$. The vector fields

$$V = \frac{\partial}{\partial z} + \frac{\partial f}{\partial x}(z)\frac{\partial}{\partial w}, \qquad W = \mathrm{i}\frac{\partial}{\partial z} + \frac{\partial f}{\partial y}(z)\frac{\partial}{\partial w} \tag{10.4}$$

form a real basis of the tangent space $\mathrm{T}_{(z,f(z))}S$ for all points z near $0 \in \mathbb{C}$. Then, $JV = \mathrm{i}\frac{\partial}{\partial z} + \mathrm{i}\frac{\partial f}{\partial x}(z)\frac{\partial}{\partial w}$, and by subtracting W we get a vector field whose first component equals zero (a section of the normal bundle $\mathrm{N}_{S/X}$). Hence, the latter vector field is precisely $\widetilde{V} = \tau(JV)$:

$$\widetilde{V} = JV - W = \left(\mathrm{i}\frac{\partial f}{\partial x}(z) - \frac{\partial f}{\partial y}\right)\frac{\partial}{\partial w} = 2\mathrm{i}\,\frac{\partial f}{\partial \bar{z}}(z)\frac{\partial}{\partial w}.$$

The index of $\widetilde{V}$ at 0 equals the winding number of $\frac{\partial f}{\partial \bar{z}}$ around 0, and this coincides with our definition of the index $I(p)$. We leave it to the reader to verify that the index of $\widetilde{V}$ at p does not depend on the choice of V. □

Example 10.1.5 For $k \in \mathbb{N}$ the surface $w = z^k\bar{z}$ in $\mathbb{C}^2$ has an isolated complex point of index k at $z = 0$, while the surface $w = \bar{z}^{k+1}$ has an isolated complex point of index $-k$ at $z = 0$. Both surfaces are degenerate if $k > 1$. □

Remark 10.1.6 By Chern and Spanier [97], one can interpret $I(p; S)$ as a local intersection index in the following way. The Grassmann manifold $G_2(X)$ of oriented real 2-planes in the complex tangent bundle (TX, J) has two disjoint oriented submanifolds $\Sigma_\pm$ consisting of all complex lines with positive and negative orientation, respectively. Let $\tau: S \to G_2(X)$ denote the Gauss map of an oriented embedded surface $S \hookrightarrow X$, sending a point $p \in S$ to its oriented tangent plane $T_pS \in G_2(X)$. Then, $p \in S$ is a complex point if and only if $\tau(p) \in \Sigma = \Sigma_+ \cup \Sigma_-$, and $I(p; S)$ equals the local intersection number of $\tau(S) \subset G_2(X)$ with Σ at p. □

The last two characterizations of the index $I(p; S)$ also apply to real surfaces in almost complex (not necessarily integrable) surfaces.

A thorough investigation of the local hull at an isolated parabolic point $p \in S$ ($\lambda = \frac{1}{2}$) was made by Jöricke [313]. Further results concerning local polynomial convexity of a real surface in $\mathbb{C}^2$ were obtained by Bharali [54, 55].

Most results mentioned so far concern the local behavior of a surface near a complex tangent. There also exist a number of results concerning global holomorphic, polynomial and rational convexity of smooth real surfaces in $\mathbb{C}^2$ (see [534], especially Sect. 5.3 and Chap. 6). One of the most interesting global results in this subject is due to Bedford and Klingenberg [47] and Kruzhilin [353]. They proved that the envelope of holomorphy of a generic smooth 2-sphere S, contained in the boundary of a strongly pseudoconvex domain $\Omega \Subset \mathbb{C}^2$, is a real three dimensional Levi flat hypersurface contained in $\overline{\Omega}$ and foliated by analytic discs. This fails in general for spheres that are not contained in a strongly pseudoconvex boundary. For topological applications of the *filling by discs* methods, see Eliashberg [147].

10.2 Invariants of Smooth 4-Manifolds

The geometry of a smooth oriented 4-manifold X is to a large extent determined by the smooth real surfaces that it contains; in particular, by the intersection form on the second homology group $H_2(X; \mathbb{Z})$. In this section we briefly recall the relevant topological invariants of compact smooth 4-manifolds that are used in the sequel; the details can be found in any of the standard sources such as [234, 235, 264, 281].

Let X be an oriented smooth 4-manifold. An embedded oriented closed surface S in X determines an integral homology class $[S] \in H_2(X; \mathbb{Z})$; if X is closed (compact and without boundary) then every element of $H_2(X; \mathbb{Z})$ is of this form. To a pair of compact smooth oriented surfaces $S, S' \subset X$ we associate their intersection number $S \cdot S' \in \mathbb{Z}$. Assuming that S and S' intersect transversely, $S \cdot S'$ is the number of intersection points counted with their orientation signs. The intersection number $S \cdot S'$ only depends on the respective homology classes $\alpha = [S], \alpha' = [S'] \in H_2(X; \mathbb{Z})$, and we write $\alpha \cdot \alpha'$ accordingly. We denote by $S \cdot S = [S]^2$ the self-intersection number of S in X, obtained by deforming one copy of S into a generic position S' and taking $S \cdot S'$. (We avoid writing S^2 to avoid confusion with the 2-sphere.)

Assume now that X is a closed oriented 4-manifold. Then, the intersection pairing $H_2(X; \mathbb{Z}) \times H_2(X; \mathbb{Z}) \to \mathbb{Z}$ is a nondegenerate bilinear form on the free part $\widetilde{H}_2(X; \mathbb{Z})$ of $H_2(X; \mathbb{Z})$ (the quotient of $H_2(X; \mathbb{Z})$ modulo torsion) and it vanishes identically on the torsion part. Considering $\widetilde{H}_2(X;\mathbb{Z})$ as a lattice in the real homology group $H_2(X;\mathbb{R}) = H_2(X;\mathbb{Z}) \otimes \mathbb{R}$, the intersection form extends to a nondegenerate real-valued bilinear form on the finite dimensional vector space $H_2(X;\mathbb{R})$. Denote by $b_2^+(X)$ and $b_2^-(X)$ the number of positive and negative eigenvalues, respectively. Then,

$$b_2^+(X) + b_2^-(X) = b_2(X) = \dim H_2(X;\mathbb{R}), \quad b_2^+(X) - b_2^-(X) = \sigma(X)$$

are the *second Betti number* and the *signature* of X, respectively.

A nondegenerate intersection pairing on the second cohomology group $H^2(X;\mathbb{R})$ is induced by the cup product evaluated on the fundamental class of X. Consider $H^2(X;\mathbb{R})$ as the second de Rham group of X. Given a pair of classes $[\alpha], [\beta] \in H^2(X;\mathbb{R})$, their intersection product is defined by

$$[\alpha] \cdot [\beta] = \int_X \alpha \wedge \beta \in \mathbb{R}. \tag{10.5}$$

Considering $H^2(X;\mathbb{R})$ as the dual of $H_2(X;\mathbb{R})$, this pairing gives rise to the isomorphism of vector spaces $PD\colon H_2(X;\mathbb{R}) \xrightarrow{\cong} H^2(X;\mathbb{R})$ called *Poincaré duality*. Explicitly, given a smooth oriented closed real surface $S \subset X$, its Poincaré dual $PD([S]) = [\beta_S] \in H^2(X;\mathbb{R})$ is represented by a closed 2-form β_S on X such that for every $[\alpha] \in H^2(X;\mathbb{R})$ we have

$$\big\langle [\alpha], [S] \big\rangle = \int_S \alpha = \int_X \alpha \wedge \beta_S. \tag{10.6}$$

We often write $PD(S)$ instead of $PD([S])$. It is customary to denote the inverse of PD by the same sign, thereby writing $PD([\alpha]) = PD(\alpha) \in H_2(X;\mathbb{R})$ for a class $[\alpha] \in H^2(X;\mathbb{R})$.

Poincaré duality maps integral elements of $H_2(X;\mathbb{R})$ (those coming from $H_2(X;\mathbb{Z})$) to integral elements of $H^2(X;\mathbb{R})$. The intersection pairing (10.5) on $H^2(X;\mathbb{R})$ is then Poincaré dual to the intersection pairing of homology classes in $H_2(X;\mathbb{Z})$, in the sense that $[S]\cdot[S'] = PD(S)\cdot PD(S')$.

Assume now that X is a complex surface. The first Chern class of its complex tangent bundle $\mathrm{T}X$ will be denoted $c_1(X) \in H^2(X;\mathbb{Z})$. Recall that $c_1(X) = c_1(\Lambda^2\mathrm{T}X)$ where $\Lambda^2\mathrm{T}X$ is the determinant bundle of $\mathrm{T}X$ (a holomorphic line bundle over X). The standard definition of the Chern class $c_1(L) \in H^2(X;\mathbb{Z})$ of a complex line bundle $L \to X$ is via the exact exponential sheaf sequence

$$0 \longrightarrow \mathbb{Z} \longrightarrow \mathcal{C} \xrightarrow{e^{2\pi i \cdot}} \mathcal{C}^* \longrightarrow 0$$

which induces an isomorphism $c_1\colon H^1(X;\mathcal{C}^*) \xrightarrow{\cong} H^2(X;\mathbb{Z})$. (See (5.1) on p. 212.) The dual bundle $(\Lambda^2\mathrm{T}X)^* = \Lambda^2\mathrm{T}^*X = K_X$ is the *canonical bundle* of X, and its Poincaré dual $PD(K_X) = [K_X] \in H_2(X;\mathbb{Z})$ is the *canonical class* of X. We follow the custom of writing K_X both for the line bundle and for the corresponding homology class $[K_X] \in H_2(X;\mathbb{Z})$.

Given a smooth embedded surface $S \subset X$, the integer $c_1(X) \cdot S = -c_1(K_X) \cdot S$ is the degree of the restricted bundle $TX|_S$, which is just the oriented self-intersection number of the zero section in $TX|_S$.

It is customary to identify a second homology or cohomology class with a particular representative, or with its Poincaré dual. In the same vein, a complex line bundle L on X is often identified with its Chern class $c_1(L) \in H^2(X; \mathbb{Z})$, and also with its Poincaré dual $PD(c_1(L)) \in H_2(X; \mathbb{Z})$. If X is a complex surface and L is a line bundle determined by the divisor of a complex curve $C \subset X$, then $PD(c_1(L)) = [C] \in H_2(X; \mathbb{Z})$ is the homology class determined by C. If L, L' are line bundles over X determined by the divisors of C, C', respectively, then $L \cdot L' = c_1(L) \cdot c_1(L') = C \cdot C'$. Further, if S is a smooth closed real surface in X, then $L \cdot S$ stands for $c_1(L) \cdot [S] = C \cdot S$. In the same spirit we identify the canonical bundle K_X with its Poincaré dual $[K_X] \in H_2(X; \mathbb{Z})$ and write $c_1(X) \cdot S = -[K_X] \cdot [S] = -K_X \cdot S$.

10.3 Lai Indexes and Index Formulas

Recall that an orientable compact real surface S of genus $g \geq 0$ is the two-sphere if $g = 0$, and is a connected sum of g tori if $g > 0$. Its Euler number equals $\chi(S) = 2 - 2g$. A nonorientable surface of genus $g > 0$ is a connected sum of g copies of the real projective plane; its Euler number is $\chi(S) = 2 - g$. Assume now that S is smoothly embedded in a complex surface X. After a generic small perturbation of S in X, we may assume that it has only isolated complex points. If S is compact, then it has only finitely many complex points $p_1, \ldots, p_k \in S$, and we define the *index* of $S \hookrightarrow X$ as the integer

$$I(S) = \sum_{p \in S} I(p; S) = \sum_{j=1}^{k} I(p_j; S), \tag{10.7}$$

where $I(p_j; S)$ is the index of the complex point p_j in S (see Definition 10.1.3). Similarly we define the index $I(\pi)$ of an immersion $\pi : S \to X$.

If S is oriented, a complex point $p \in S$ is either *positive* or *negative*, depending on whether the orientation on T_pS, induced by the complex structure, agrees or disagrees with the orientation on S. Let $I_\pm(S)$ denote the sum of indexes over positive and negative complex points of S, respectively; hence $I(S) = I_+(S) + I_-(S)$. If S has only elliptic and hyperbolic points then, denoting by $e_\pm(S)$, $h_\pm(S)$ their respective numbers, we have

$$I_+(S) = e_+(S) - h_+(S), \qquad I_-(S) = e_-(S) - h_-(S).$$

These numbers are called *Lai indexes* (Lai [368]), although they were already present in the work of Chern and Spanier [97] and Bishop [59] who studied surfaces in $\mathbb{C}^2$. In view of Remark 10.1.6, the Lai index $I_\pm(S)$ equals the intersection index of the Gauss map $\tau\colon S \to G_2(X,J)$ with the submanifold $\Sigma_\pm$ of $G_2(X,J)$ consisting of all complex lines with positive or negative orientation, respectively. Lai [368] studied the relationship between these numbers and the topological invariants of immersions and embeddings of compact real k-dimensional manifolds into complex n-dimensional manifolds for $k \ge n$.

Suppose now that $\pi : S \to X$ is an immersion with only simple (transverse) double points such that both tangent planes at any double point are totally real. At each double point, π has self-intersection index ± 1 which is independent of the choice of the orientation on S. Double points with index $+1$ are called positive, and those with index -1 are negative. If π has δ_+ positive and δ_- negative double points, then $\delta(\pi) = \delta_+ - \delta_-$ is the (geometric) *self-intersection index* of π which only depends on its regular homotopy class.

The following result summarizes most of the known index theorems concerning complex points of immersed real surfaces in complex surfaces (see [151, 182, 193, 248, 368, 431, 564]). The notation is explained in the previous section. In particular, $[\pi(S)]^2 \in \mathbb{Z}$ is the self-intersection number of the homology class $[\pi(S)] \in H_2(X;\mathbb{Z})$, and $c_1(X)\cdot\pi(S)$ is the value of the first Chern class $c_1(X) = c_1(\mathrm{T}X) \in H^2(X;\mathbb{Z})$ on the homology class $[\pi(S)]$.

Theorem 10.3.1 *Let S be a closed connected real surface with genus $g(S)$ and Euler number $\chi(S)$. For every immersion $\pi : S \to X$ into a complex surface X we have*

$$I(\pi) = \chi(S) + \chi_n(\pi), \tag{10.8}$$

where $\chi_n(\pi)$ is the normal Euler number of π. If S is oriented, then

$$I(\pi) = 2 - 2g(S) - 2\delta(\pi) + \big[\pi(S)\big]^2, \tag{10.9}$$

$$I_\pm(\pi) = 1 - g(S) - \delta(\pi) + \frac{1}{2}\Big(\big[\pi(S)\big]^2 \pm c_1(X)\cdot\pi(S)\Big). \tag{10.10}$$

Proof Let J denote the almost complex structure operator on X. Choose a tangent vector field V to S with isolated zeros that do not coincide with any of the complex points of S. Let $\tau\colon \mathrm{T}X|_S \to \mathrm{N}_{S/X} = \mathrm{T}X|_S/\mathrm{T}S$ denote the projection onto the normal bundle of S in X. The vector field $W = \tau(JV)$ is then a section of $\mathrm{N}_{S/X}$ whose zeros are the zeros of V and the complex points of S. A choice of local orientation of S coorients the normal bundle $\mathrm{N}_{S/X}$ by the complex orientation of $\mathrm{T}X|_S = \mathrm{T}S \oplus \mathrm{N}_{S/X}$. The map $\tau \circ J\colon \mathrm{T}S \to \mathrm{N}_{S/X}$ is orientation reversing over the totally real part of S. It follows that at every zero of V the vector field W has zero of index opposite to that of V. At a complex point p of S, the index of W equals $I(p;S)$ by Lemma 10.1.4. Since the sum of indexes of V is the Euler number $\chi(S)$ and the sum of indexes of W is the normal Euler number of S in X, (10.8) follows. The same argument applies to immersions if V is chosen to be nonzero at any double point.

Suppose now that $\pi : S \to X$ is an immersed oriented surface with simple double point. Let $\mathrm{N}_\pi \to S$ denote the normal bundle of the immersion; this is an oriented

2-plane bundle whose fibre over a point $p \in S$ equals $\mathrm{T}_{\pi(p)}X/d\pi_p(\mathrm{T}_pS)$. Note that π extends to an immersion (still denoted π) of a neighborhood of the zero section of N_π (which we identify with S) onto a neighborhood of $\pi(S)$ in X; this map is an embedding if $\pi : S \hookrightarrow X$ is an embedding. The normal Euler number $\chi_n(\pi)$ is the oriented self-intersection number of the zero section S in N_π. Choosing a small generic deformation $S' \subset \mathrm{N}_\pi$ of S that does not intersect S at any of the double points of π, we have $\chi_n(\pi) = S \cdot S'$. Every intersection point of S and S' in N_π gives an intersection point of the image surfaces $\pi(S), \pi(S')$ in X with the same intersection index. In addition, each simple double point of π creates two nearby intersection point of $\pi(S)$ and $\pi(S')$ of the same sign in X. (Look at the cross $(\mathbb{R}^2 \times \{0\}^2) \cup (\{0\}^2 \times \mathbb{R}^2) \subset \mathbb{R}^4$ and translate it off itself.) This shows that

$$\chi_n(\pi) + 2\delta(\pi) = \left[\pi(S)\right]^2. \tag{10.11}$$

Substituting in (10.8) yields (10.9). To prove (10.10) it remains to show that

$$I_+(\pi) - I_-(\pi) = c_1(X) \cdot \pi(S). \tag{10.12}$$

Assume first that S is embedded. Choose a nonvanishing 2-form ω on S (such exists if S is oriented). For a local frame (V, W) of $\mathrm{T}S$ the expression $\omega(V, W)^{-1} V \wedge W$ is independent of the choice of the frame (V, W), and hence it gives a well defined global section ξ of the determinant bundle $\Lambda^2_{\mathbb{C}}(\mathrm{T}X)|_S$. The zeros of ξ are precisely the complex points of S. Locally near a complex point we represent S as the graph $\{w = f(z)\} \subset \mathbb{C}^2$ and use the local frame (V, W) for $\mathrm{T}S$ given by (10.4) on p. 481. Then,

$$V \wedge W = \left(\frac{\partial f}{\partial y}(z) - \mathrm{i}\frac{\partial f}{\partial x}(z)\right)\frac{\partial}{\partial z} \wedge \frac{\partial}{\partial w} = -2\mathrm{i}\,\frac{\partial f}{\partial \bar z}(z) \cdot \frac{\partial}{\partial z} \wedge \frac{\partial}{\partial w}.$$

This shows that the index of ξ at the complex point at $0 \in \mathbb{C}^2$ equals $I(0; S)$ if the point is positive, and it equals $-I(0; S)$ if the point is negative. Hence, the algebraic number of zeros of ξ is $I_+(S) - I_-(S)$. But this number also equals $c_1(X) \cdot S$ since ξ is a section of $\Lambda^2_{\mathbb{C}}(\mathrm{T}X)|_S$. This proves (10.12) and hence (10.10). The same argument applies in the immersed case. □

Remark 10.3.2 We give another derivation of the index formula (10.10) for an immersed surface $\pi(S) \subset X$ from the corresponding formula for embedded surfaces by using surgery that replaces every double point by a handle. This argument, which is also used in the proof of the adjunction inequality for immersed surfaces in Sect. 10.6, exposes the different roles played by the positive and the negative double points.

Assume that $\pi(S)$ has only simple double points, say δ_+ positive ones and δ_- negative ones, so $\delta_+ - \delta_- = \delta(\pi)$. After a regular homotopy we may assume that in certain local holomorphic coordinates $(z, w) = (x + \mathrm{i}y, u + \mathrm{i}v)$ on X near a double point $p \in \pi(S)$, with p corresponding to the origin $(0, 0) \in \mathbb{C}^2$, the immersed surface $\pi(S)$ intersects a small ball $\mathbb{B} \subset \mathbb{C}^2$ in the union $D_1 \cup D_2$ of discs contained in the Lagrangian planes $\Lambda_1 = \{y=0,\ v=0\} \subset \mathbb{C}^2$, $\Lambda_2 = \{x=0,\ u=0\} \subset \mathbb{C}^2$, respectively. We orient Λ_1 by $\partial_x \wedge \partial_u$ and Λ_2 by $\kappa(p)\,\partial_v \wedge \partial_y$, where $\kappa(p) = \pm 1$ is the self-intersection index of $\pi(S)$ at p. We now remove from $\pi(S)$ the union of intersecting discs $D_1 \cup D_2$ and glue in a smooth annular handle Σ diffeomorphic to $S^1 \times [0, 1]$. In topological language, the pair of oriented circles $bD_1 \cup bD_2$ is a *Hopf link* in the 3-sphere $b\mathbb{B} \subset \mathbb{R}^4$, and Σ is a *Seifert surface* for this link. This surgery removes the double point at p and increases the genus of the surface by one.

If $\kappa(p) = +1$, then an appropriate handle matching the orientation on $bD_1 \cup bD_2$ is $\Sigma_+ = \{(x+\mathrm{i}u)(y-\mathrm{i}v) = \epsilon\} \cap \mathbb{B}$ for a small $\epsilon \neq 0$. (This is a totally real analogue of the modification $zw = \epsilon$ of the complex double point of $zw = 0$ at $(0, 0)$.) Outside of $\mathbb{B}$, we can smoothly patch Σ_+ with $(\Lambda_1 \setminus D_1) \cup (\Lambda_2 \setminus D_2)$ without introducing any new complex points provided that $\epsilon > 0$ is small enough. A calculation shows that Σ_+ is totally real for every $\epsilon \neq 0$, so this surgery does not change the indexes $I_\pm$.

If $\kappa(p) = -1$, an appropriate handle is $\Sigma_- = \{(x+\mathrm{i}u)(y+\mathrm{i}v) = \epsilon\} \cap \mathbb{B}$ for small $\epsilon \neq 0$. It is easily seen that Σ_- has four complex points, located at $x = y = \pm\sqrt{\frac{\epsilon}{2}}$, $v = -u = \pm\sqrt{\frac{\epsilon}{2}}$, two positive ones and two negative ones, all of them hyperbolic. Hence $I_\pm(\Sigma_-) = -2$, so this surgery decreases each of the indexes $I_\pm$ by two.

After replacing all double points of $\pi(S)$ by handles, we obtain a smooth embedded surface $S' \subset X$ of genus $g(S') = g(S) + \delta_+ + \delta_-$, homology class $[S'] = [\pi(S)] \in H_2(X;\mathbb{Z})$, and Lai indexes $I_\pm(S') = I_\pm(S) - 2\delta_-$. Inserting these data in $I_\pm(S') = 1 - g(S') + \frac{1}{2}(S' \cdot S' \pm c_1(X) \cdot S')$ yields (10.10). □

Corollary 10.3.3 *Let $\pi : S \to X$ be a smooth immersion of a compact oriented real surface S in a complex surface X. If $[\pi(S)] = 0 \in H_2(X;\mathbb{Z})$, then*

$$I_\pm(\pi) = 1 - g(S) - \delta(\pi). \tag{10.13}$$

This holds in particular if $X = \mathbb{C}^2$.

Proof This follows from (10.9) since $[\pi(S)] = 0$. For $X = \mathbb{C}^2$ this is due to Chern and Spanier [97] (for the 2-sphere) and Bishop [59]. □

If C is a complex curve in X with its natural orientation induced by the complex structure, then a generic small perturbation of C in X yields a surface with only positive complex points, so in this case we have

$$I_-(C) = 0, \qquad I_+(C) = c_1(X) \cdot C = -K_X \cdot C.$$

The same holds if C is a symplectic curve for a symplectic form ω on X that is tamed by J, in the sense that $\langle \omega, v \wedge Jv \rangle > 0$ for every $0 \neq v \in TX$. (Recall that a curve C is ω-symplectic if $\omega > 0$ on the tangent bundle TC.) From the expression (10.10) for $I_-(C)$ we get the following:

Corollary 10.3.4 (Genus formula) *If C is a smoothly embedded complex or symplectic curve in a complex (resp. symplectic) surface X, then*

$$2g(C) - 2 = [C]^2 - c_1(X)\cdot C = [C]^2 + K_X \cdot C. \tag{10.14}$$

The standard proof of (10.14) uses the normal bundle exact sequence

$$0 \longrightarrow \mathcal{T}_C \longrightarrow \mathcal{T}|_C \longrightarrow \mathcal{N}_{C/X} \longrightarrow 0$$

which gives $\Lambda^2 \mathrm{T}X|_C = \mathrm{T}C \otimes \mathrm{N}_{C/X}$. Here, $\mathcal{T}$ and $\mathcal{N}$ denote the tangent and the normal sheaf, respectively. Since $\Lambda^2 \mathrm{T}X|_C = (K_X|_C)^{-1}$, $\mathrm{T}C = (K_C)^{-1}$ and $\mathcal{N}_{C/X} = \mathcal{O}_X(C)|_C$, where $\mathcal{O}_X(C)$ is the line bundle on X determined by the divisor of C, we obtain $K_C = K_X|_C \otimes \mathcal{O}_X(C)$ as line bundles over C. Taking the degrees and noting that $\deg K_C = 2g(C) - 2$ gives (10.14).

Example 10.3.5 The group $H_2(\mathbb{P}^2;\mathbb{Z}) = \mathbb{Z}$ is generated by the homology class $[H]$ of the projective line $H = \mathbb{P}^1 \hookrightarrow \mathbb{P}^2$. If $C \hookrightarrow \mathbb{P}^2$ is a compact complex or symplectic curve of degree $d > 0$, then $[C] = d[H]$ and hence $[C]^2 = d^2$. Furthermore, we have $c_1(\mathbb{P}^2)\cdot H = 3$ and hence $c_1(\mathbb{P}^2)\cdot C = c_1(\mathbb{P}^2)\cdot 3H = 3d$. Thus, for $X = \mathbb{P}^2$ the genus formula (10.14) reads

$$g(C) = \frac{1}{2}(d-1)(d-1). \tag{10.15}$$

10.4 Cancelling Pairs of Complex Points

In this section we prove a theorem of Eliashberg and Kharlamov [151] on cancelling a pair of an elliptic and a hyperbolic point of the same sign by a $\mathscr{C}^0$-small deformation of the given real surface in a complex surface. This shows that Lai indexes $I_\pm$ (see p. 484) are the only topological invariants of an embedded or immersed real surface with respect to the complex structure of the ambient complex surface (a form of the h-principle).

Assume that $S \hookrightarrow X$ is a real surface smoothly embedded in a complex surface X. Let $\gamma \subset S$ be an arc connecting a pair of complex points $p, q \in S$. We say that γ is orientation preserving if the orientation of the tangent plane $\mathrm{T}_q S$, obtained by transporting along γ the standard orientation of $\mathrm{T}_p S$ induced by the complex structure, agrees with the standard orientation on $\mathrm{T}_q S$; in the opposite case γ is orientation reversing. (This does not require S to be oriented.) The same definition applies to immersions $S \to X$. If S is oriented, then any arc in S connecting a pair of complex points is orientation preserving if its endpoints are of the same sign, and is orientation reversing if they are of opposite signs. If S is connected and nonorientable, then any pair of complex points can be connected by an orientation preserving arc.

Theorem 10.4.1 ([151]) *Let $p, q \in S$ be isolated complex points of an immersion $\pi_0 \colon S \to X$ of a real surface S to a complex surface X, and let $\gamma \subset S$ be an orientation preserving simple smooth arc connecting p to q. Assume that there is a neighborhood $U \subset S$ of γ such that $\overline{U}$ does not contain any other complex point or double point of π_0. Given $\epsilon > 0$, there is a regular homotopy of immersions $\pi_t \colon S \to X$ $(t \in [0,1])$ satisfying the following conditions:*

(a) *$\pi_t = \pi_0$ on $S \setminus U$ for all $t \in [0,1]$,*
(b) *$\sup_{x \in S} \operatorname{dist}(\pi_t(x), \pi_0(x)) < \epsilon$ for all $t \in [0,1]$,*
(c) *every π_t has the same number of double points as π_0, and*
(d) *$\pi_1|_U$ has a single complex point of index $I(s;\pi_1) = I(p;\pi_0) + I(q;\pi_0)$.*

In particular, if $I(p;\pi_0) + I(q;\pi_0) = 0$ then π_1 can be chosen totally real on U. If π_0 is an embedding, then $\{\pi_t\}_{t\in[0,1]}$ can be chosen an isotopy of embeddings.

This formulation of the cancellation theorem is taken from [182, Theorem 1.1]. The same result holds if (X, J) is an almost complex surface.

Corollary 10.4.2 *Let S be an immersed connected real surface with isolated complex points in a complex surface X. If S is oriented then every pair of an elliptic and a hyperbolic point of the same orientation class can be cancelled by a $\mathscr{C}^0$-small smooth deformation of S in X which does not create any new complex points or double points. If S is nonorientable then any pair of an elliptic and a hyperbolic complex point can be cancelled in this way.*

The key step in the proof of Theorem 10.4.1 is the following $\overline\partial$ modification lemma on the closed disc $\overline{\mathbb{D}} = \{z \in \mathbb{C} : |z| \le 1\}$.

Lemma 10.4.3 *Let f be a smooth complex function on the closed unit disc $\overline{\mathbb{D}}$ such that $\overline\partial f \neq 0$ on the circle $b\mathbb{D}$ and the winding number of $\overline\partial f$ along $b\mathbb{D}$ equals zero. Given an $\epsilon > 0$, there exists a smooth function g on $\overline{\mathbb{D}}$ that equals f in a neighborhood of $b\mathbb{D}$ such that $\overline\partial g \neq 0$ on $\overline{\mathbb{D}}$ and $|g - f| < \epsilon$ on $\overline{\mathbb{D}}$.*

Proof We give a complex analytic proof due to Nemirovski [431, p. 735]. A similar result holds for a nonintegrable almost complex structure on $\mathbb{C}^2$ that is close to the standard structure along the disc $\overline{\mathbb{D}} \times \{0\}$; the proof of [182, Lemma 4.1] can be adapted to this situation.

Recall that if ψ is a smooth function supported in the disc $\mathbb{D}$, then its Cauchy-Green transform

$$u(z) = \frac{1}{2\pi \mathrm{i}} \int_{\mathbb{C}} \frac{\psi(\zeta + z)}{\zeta} \, d\zeta \wedge d\bar\zeta \tag{10.16}$$

solves the equation $\overline\partial u = \psi$ and satisfies the estimate

$$\|u\|_{\mathscr{C}^0(\mathbb{D})} \le \text{const} \, \|\psi\|_{L^p(\mathbb{D})}, \quad p > 2.$$

(Here $\overline{\partial} u = \frac{\partial u}{\partial \bar{z}}$.) This well known fact follows from Hölder's inequality since the Cauchy kernel $1/\zeta$ belongs to L^q_{loc} for all $q < 2$. Fix a number $p > 2$.

By perturbing f away from $b\mathbb{D}$ we ensure that $\overline{\partial} f$ has only finitely many zeros $z_1, \ldots, z_k$ in $\mathbb{D}$. Choose a simple smooth arc γ in $\mathbb{D}$ connecting all these zeros. Pick a number $r \in (0,1)$ such that $\gamma \subset r\mathbb{D}$. Let $m = \inf\{|\overline{\partial} f(z)| : r \le |z| \le 1\} > 0$. Choose a smooth function $\chi \ge 0$ with support in $\mathbb{D}$ that equals 1 on $r\mathbb{D}$. Decrease the number $\epsilon > 0$ if necessary so that $m - \epsilon |\overline{\partial}\chi| > 0$. Note that $\overline{\partial} f$ has winding number zero around the arc γ. Hence, there is a smooth function ψ, supported in any given neighborhood $V \subset r\mathbb{D}$ of γ, such that $\overline{\partial} f + \psi \ne 0$ on $\overline{\mathbb{D}}$ and $\|\psi\|_{\mathcal{C}^0(\mathbb{D})} \le 2\|\overline{\partial} f\|_{\mathcal{C}^0(\mathbb{D})}$. By taking the area of V small enough, we can ensure that $\|\psi\|_{L^p}$ is so small that the corresponding function u (10.16) satisfies $\|u\|_{\mathcal{C}^0(\mathbb{D})} < \epsilon$. Set $g = f + \chi u$. Then, $g = f$ near $b\mathbb{D}$ and $\|f - g\|_{\mathcal{C}^0(\mathbb{D})} = \|\chi u\|_{\mathcal{C}^0(\mathbb{D})} \le \|u\|_{\mathcal{C}^0(\mathbb{D})} < \epsilon$. We claim that $\overline{\partial} g \ne 0$ on $\overline{\mathbb{D}}$. This is clear on $r\mathbb{D}$ where $\overline{\partial} g = \overline{\partial} f + \overline{\partial} u = \overline{\partial} f + \psi$. Outside of $r\mathbb{D}$ we have $\overline{\partial} u = \psi = 0$ and hence $|\overline{\partial} g| = |\overline{\partial} f + u\overline{\partial}\chi| \ge |\overline{\partial} f| - \epsilon|\overline{\partial}\chi| \ge m - \epsilon|\overline{\partial}\chi| > 0$. □

Proof of Theorem 10.4.1 We begin by showing that a neighborhood of the arc γ in S can be holomorphically spread as a graph of a complex function over a simply connected domain in $\mathbb{C}$; this will enable us to apply Lemma 10.4.3. Choose a pair of smooth $\mathbb{R}$-independent $(1,0)$ vector fields V and W in a neighborhood of γ in X that are tangential to S, with V tangential to the arc γ. The assumptions imply that V and W are $\mathbb{C}$-independent near γ, except at the two endpoints p and q. Pick another vector field V' near γ that is $\mathbb{C}$-linearly independent of V. Then, $W = aV + bV'$ for some smooth complex valued functions a and b, and b vanishes precisely at the endpoints p, q of γ. For any function c, the vector field $Z = cV + V'$ is $\mathbb{C}$-linearly independent of V and we have $W = (a - bc)V + bZ$. Write $a - bc = \alpha + \mathrm{i}\beta$ with α, β real. We wish to choose c such that $\beta(x) \ne 0$ for all $x \in \gamma$. Since γ is orientation preserving, $\Im a$ is of the same sign at both endpoints of γ, say positive. Taking $c(x) = -\mathrm{i}C\overline{b(x)}$ for a sufficiently large $C > 0$ gives $\beta(x) = \Im a(x) + C|b(x)|^2 > 0$ for $x \in \gamma$, and hence the requirement is fulfilled.

Let (z, w) be complex coordinates on $\mathbb{C}^2$. There is a neighborhood Ω of γ in X, with $\Omega \cap S \subset U$, and a smooth diffeomorphism $\Phi \colon \Omega \to \Omega' \subset \mathbb{C}^2$ satisfying the following conditions:

(i) $\Phi(\gamma) = [0, T] \times \{0\}$ for some $T > 0$,
(ii) the differential $\mathrm{d}\Phi = \Phi_*$ is $\mathbb{C}$-linear at each point of γ, and
(iii) $\Phi_* V = \frac{\partial}{\partial z}$ and $\Phi_* Z = \frac{\partial}{\partial w}$ at every point of γ.

It follows that along the arc γ we have

$$\Phi_* W_x = \big(\alpha(x) + \mathrm{i}\beta(x)\big)\frac{\partial}{\partial z} + b(x)\frac{\partial}{\partial w}, \qquad x \in \gamma.$$

Note that the tangent space $\Phi_*(\mathrm{T}_x S) \subset \mathrm{T}_{\Phi(x)}\mathbb{C}^2 \cong \mathbb{C}^2$ to $\Phi(S \cap \Omega)$ is spanned over $\mathbb{R}$ by the vectors $\Phi_* V_x$ and $\Phi_* W_x$. Since $\beta > 0$ on γ, we see that $\Phi_*(\mathrm{T}_x S)$ projects isomorphically onto $\mathbb{C} \times \{0\}$ for each point $x \in \gamma$. The implicit function theorem now shows that for a small enough neighborhood $\Omega \subset X$ of γ the set $\Phi(S \cap \Omega)$ is a smooth graph over a domain in $\mathbb{C}$.

The smooth arc γ has a basis of Stein neighborhoods in X by Corollary 3.5.2. Since the differential of Φ is $\mathbb{C}$-linear along γ, we can approximate Φ in the $\mathcal{C}^1$ topology on γ by a holomorphic mapping $\Psi = (\psi_1, \psi_2)$ on a neighborhood of γ in X (see Theorem 3.5.4 on p. 81). Then, Ψ is biholomorphic on a neighborhood Ω_1 of γ in X, and the image $\Psi(S \cap \Omega_1)$ is a graph

$$S' = \Psi(S \cap \Omega_1) = \big\{\big(z, f(z)\big) : z \in D\big\} \subset \mathbb{C}^2$$

of a smooth function f on a domain $D \subset \mathbb{C}$. By the construction, the only complex points of S' are the endpoints $p' = \Psi(p)$ and $q' = \Psi(q)$ of the arc $\gamma' = \Psi(\gamma)$. Since the index of a complex point is preserved under a biholomorphism, we have that $I(p; S) = I(p'; S')$ and $I(q; S) = I(q'; S')$. Write $p' = (z_0, f(z_0))$ and $q' = (z_1, f(z_1))$. The first coordinate projection $\psi_1(\gamma)$ of the arc $\Psi(\gamma)$ is a simple smooth arc in D with the endpoints z_0 and z_1.

The conclusion is now immediate if $k = 0$ (when the pair of complex point p, q should be cancelled). Choose a simply connected domain $D_0 \Subset D$ containing the arc $\psi_1(\gamma)$. The winding number $\mathcal{W}(\bar\partial f; bD_0)$ equals the sum of winding numbers of $\bar\partial f$ along small circles around the points z_0 and z_1, and the latter equals $I(p'; S') + I(q', S') = 0$. By Lemma 10.4.3, we can approximate f by a smooth function F on D_0 such that $F = f$ near bD_0 and $\bar\partial F \neq 0$ on D_0. The graph of $f_t = (1-t)f + tF$ over D_0 for $t \in [0, 1]$ then provides an isotopy $S'_t \subset \mathbb{C}^2$, fixed near the boundary, from the initial surface $S'_0 = S'$ to a totally real surface S'_1. The isotopy $\Psi^{-1}(S'_t) \subset X$ extends to the rest of S as the identity and makes the surface totally real near the arc γ.

In the general case, we let $s \in \gamma$ be the point at which we wish to produce a complex point of index $k = I(p; S) + I(q; S)$. Let $\Psi(s) = s' = (z_2, f(z_2)) \in \gamma'$. Pick a number $\eta > 0$. Choose a pair of smoothly bounded simply connected domains D_0, D_1 in $\mathbb{C}$ such that $\psi_1(\gamma) \subset D_1 \Subset D_0 \Subset D$. Choose a smooth function g on $\mathbb{C}$ whose graph has an isolated complex point of index k at z_2 and $g(z_2) = f(z_2)$. Let $D_2 \Subset D_1$ be a small disc centered at z_2 such that $|f - g| < \eta$ on $\bar D_2$. Choose a thin strip G connecting the disc D_2 with the annular region $D_0 \setminus D_1$ and satisfying $G \cap \psi_1(\gamma) = \varnothing$. We can find a smooth function F on the closed annular region $\bar A_* = (\bar D_0 \setminus D_1) \cup \bar G \cup \bar D_2$ such that $F = f$ on $\bar D_0 \setminus D_1$, $F = g$ on $\bar D_2$, $|F - f| < \eta$ on $\bar A_*$, and $\bar\partial F \neq 0$ on $\bar A_* \setminus \{z_2\}$. The complement $D_* = D_0 \setminus \bar A_* = D_1 \setminus (\bar G \cup \bar D_2) \Subset D_0$ is a disc whose boundary curve τ_* is homologous in $\bar A_* \setminus \{z_2\}$ to the cycle $bD_1 - bD_2$. We then have

$$\begin{aligned}\mathcal{W}(\bar\partial F; \tau_*) &= \mathcal{W}(\bar\partial F; bD_1) - \mathcal{W}(\bar\partial F; bD_2) = \mathcal{W}(\bar\partial f; bD_1) - \mathcal{W}(\bar\partial g; bD_2) \\ &= I\big(p'; S'\big) + I\big(q'; S'\big) - k\end{aligned}$$

which equals zero by the choice of k. By Lemma 10.4.3 we can smoothly extend F from $\bar A_*$ to $\bar D_0$ such that $\bar\partial F \neq 0$ on $\bar D_*$ and $|F - f| < \eta$ on $\bar D_0$. Thus, F is a smooth function on $\bar D_0$ such that $F = f$ near bD_0, $F = g$ near the point z_2, $\bar\partial F \neq 0$

on $\bar{D}_0 \setminus \{z_2\}$, and $\sup_{D_0} |F - f| < \eta$. We now conclude the proof exactly as before by letting $S'_t \subset D_0 \times \mathbb{C}$ be the graph of $F_t = (1-t)f + tF$ and considering the isotopy $\Psi^{-1}(S'_t) \subset X$. □

Remark 10.4.4 Theorem 10.4.1 also holds for real surfaces in almost complex surfaces. The reduction to the case of a graph in $\mathbb{R}^4$ is essentially the same as above. A suitable version of Lemma 10.4.3 for a nonintegrable almost complex structure on $\mathbb{R}^4$ can be proved by using Gromov's h-principle for ample differential relations (see Theorem 1.10.5 on p. 42). □

10.5 Applications of the Cancellation Theorem

The following is an immediate corollary to Theorem 10.4.1.

Corollary 10.5.1 *Let S be a compact connected real surface smoothly embedded in a complex surface X.*

(a) *If S is oriented and $I_+(S) = I_-(S) = 0$, or if S is nonorientable and $I(S) = 0$, then S is isotopic to an embedded totally real surface $S' \subset X$ by an isotopy that is $\mathcal{C}^0$-close to S.*
(b) *If S is oriented and $I_\pm(S) \le 0$, or if S is nonorientable and $I(S) \le 0$, then S is isotopic to an embedded surface with only hyperbolic complex points.*

The analogous conclusions hold for immersed surfaces.

Note that the pair of inequalities $I_\pm(S) \le 0$ for an embedded oriented surface $S \subset X$ is equivalent to

$$2g(S) - 2 \ge [S]^2 + \left|c_1(X) \cdot S\right| = [S]^2 + |K_X \cdot S|. \tag{10.17}$$

Since the right hand side only depends on the homology class $[S] \in H_2(X; \mathbb{Z})$, this *adjunction inequality* imposes a lower bound on the genus of an embedded surface $S \subset X$ representing a given homology class in $H_2(X; \mathbb{Z})$. The Seiberg-Witten theory shows that the inequality (10.17) holds for most real surfaces in compact Kähler surfaces (see Theorem 10.6.1, p. 498).

Part (b) of Corollary 10.5.1, together with Theorem 10.1.2 (p. 480), implies the following corollary.

Corollary 10.5.2 *Let S be a closed oriented real surface smoothly embedded in a complex surface X. If (10.17) holds, then S can be $\mathcal{C}^0$ approximated by an isotopic surface $S' \subset X$ with a basis of tubular Stein neighborhoods.*

A very useful operation for finding explicit examples is the *connected sum*; let us recall this briefly. Let X_1 and X_2 be connected smooth n-manifolds. Choose embedded n-discs $D_j \subset X_j$ for $j = 1, 2$ and let $\phi\colon bD_1 \to bD_2$ be a smooth diffeomorphism of their boundaries. The identification space

$$X_1 \# X_2 = (X_1 \setminus \operatorname{int} D_1) \bigcup_{\phi} (X_2 \setminus \operatorname{int} D_2)$$

has the structure of a smooth manifold called the connected sum of X_1 and X_2. The smoothing process can be visualized by connecting the two pieces $X_j \setminus \operatorname{int} D_j$ by a cylinder $\Sigma = [-1, 1] \times S^{n-1}$ glued along its boundary spheres $(\{-1\} \times S^{n-1}) \cup (\{+1\} \times S^{n-1})$ to the spheres bD_1, bD_2, respectively. If both X_1 and X_2 are oriented, then $X_1 \# X_2$ is also oriented provided that the gluing diffeomorphism $\phi\colon bD_1 \to bD_2$ is orientation reversing. If $X_1 = X_2 = X$ and ϕ is the identity map on $bD \subset X$ then we must reverse the orientation on one of the copies of X; in this case we write $X \# \overline{X}$. For example, if X is a complex n-manifold, then the connected sum $X \# \overline{\mathbb{P}^n}$ of X and the complex projective space $\mathbb{P}^n$ with the reversed orientation is the complex manifold obtained by blowing up a point in X. The connected sum operation extends to several terms; one writes $X_1 \# k X_2$ for the connected sum of X_1 with k copies of X_2.

Suppose now that $S_1, S_2 \subset X$ are embedded or immersed real surfaces in general position in a complex surface X. We can realize their *interior connected sum* $S_1 \# S_2$ as an immersed surface of genus $g(S_1 \# S_2) = g(S_1) + g(S_2)$ in X whose double points are the double points of S_1, S_2 and the intersection points $S_1 \cap S_2$. We obtain $S_1 \# S_2$ by removing a pair of totally real discs $D_j \subset S_j$ $(j = 1, 2)$ and gluing in a tube $\Sigma \cong [-1, 1] \times S^1$, identifying its boundary circles with bD_1 and bD_2. After an isotopy we may assume that we are working in a local chart in $\mathbb{C}^2$, with D_1 and D_2 being discs of center $(0, 0)$ and radius $\epsilon > 0$ in the totally real planes $\Lambda^{\pm} = \{(\pm 1 + \mathrm{i}y, u) : (u, y) \in \mathbb{R}^2\} \subset \mathbb{C}^2$. A standard tube used for gluing is

$$\Sigma = \left\{(x + \mathrm{i}y, u + \mathrm{i}0) \in \mathbb{C}^2 : -1 \le x \le 1,\ y^2 + u^2 = h(x)^2\right\},$$

where $h\colon [-1, 1] \to (0, \epsilon]$ is an even continuous function which is smooth and strictly convex on $(-1, +1)$, strictly increasing on $(0, 1]$ and satisfies $h(1) = \epsilon$ and $\lim_{t \to 1} h'(t) = +\infty$. These conditions imply that Σ has a pair of hyperbolic points of different orientation class at $(0, \pm h(0))$ and no elliptic points, so $I_{\pm}(\Sigma) = 1$. Assuming that $S_1 \cap S_2 = \varnothing$, we thus have

$$I(S_1 \# S_2) = I(S_1) + I(S_2) - 2. \tag{10.18}$$

If both surfaces S_1 and S_2 are oriented and the gluing was done by an orientation reversing diffeomorphism of the circles $bD_1 \to bD_2$, then we also have $[S_1 \# S_2] = [S_1] + [S_2] \in H_2(X; \mathbb{Z})$ and

$$I_+(S_1 \# S_2) = I_+(S_1) + I_+(S_2) - 1, \qquad I_-(S_1 \# S_2) = I_-(S_1) + I_-(S_2) - 1.$$

By taking the connected sum of any real surface $S \subset X$ with a homologically trivial torus $T \subset X$, we increase the genus of the surface by one and decrease the index by two. (In the oriented case, each of the indexes $I_\pm$ decreases by one.) After attaching k trivial handles, each of the indexes $I_\pm$ decreases by k. From Theorem 10.4.1 (p. 489) we get the following.

Corollary 10.5.3 *Let $S \subset X$ be a closed, connected, oriented real surface smoothly embedded in a complex surface X. Set $k = \max\{I_+(S), I_-(S), 0\}$. Then, S is homologous to an embedded surface $S' \subset X$ of genus $g(S') = g(S) + k$ with a basis of tubular Stein neighborhoods.*

We now consider surfaces in $\mathbb{C}^2$. The index formulas simplify to

$$I(\pi) = \chi(S) + \chi_n(\pi); \qquad I_\pm(\pi) = 1 - g(S) - \delta(\pi) \quad \text{if } S \text{ is orientable.} \tag{10.19}$$

For an oriented embedded surface $S \hookrightarrow \mathbb{C}^2$ we thus have $I_\pm(S) = 1 - g(S)$ which is ≤ 0 unless S is the 2-sphere. The standard 2-sphere

$$S = \left\{(x + \mathrm{i}y, u + \mathrm{i}v) \in \mathbb{C}^2 : x^2 + y^2 + u^2 = 1,\ v = 0\right\} \tag{10.20}$$

has two complex points $(0, \pm 1)$ which are both elliptic, so $I_\pm(S) = 1$.

Assume now that S is a nonorientable surface of genus $g(S)$ and Euler number $h = \chi(S) = 2 - g(S) \leq 1$. By Whitney and Massey [409], the set of normal Euler numbers of embeddings $S \hookrightarrow \mathbb{C}^2$ is $\{2h - 4, 2h, 2h + 4, \ldots, 4 - 2h\}$, and hence the set of indexes of embeddings $S \hookrightarrow \mathbb{C}^2$ is $\{3h - 4, 3h, \ldots, 4 - h\}$. This set always contains a nonpositive number. In summary:

Corollary 10.5.4 ([182, Theorem 1.8]) *Every closed real surface S other than the 2-sphere admits a smooth embedding in $\mathbb{C}^2$ with only hyperbolic complex points and with a basis of tubular strongly pseudoconvex Stein neighborhoods. Moreover, if S is orientable and $g(S) > 0$, then every embedding $S \hookrightarrow \mathbb{C}^2$ is isotopic to an embedding with this property.*

An explicit construction in [182] uses the connected sum of a suitable number of copies of the torus and the real projective plane. Similarly we obtain the following result concerning totally real embeddings into $\mathbb{C}^2$.

Corollary 10.5.5 ([182, Corollary 1.3]) *A closed real surface S of genus $g \geq 0$ admits a totally real embedding into $\mathbb{C}^2$ if and only if S is orientable and $g = 1$ (S is the 2-torus), or S is nonorientable and $g = 2 \pmod 4$, i.e., S is the connected sum of an odd number $\frac{g}{2}$ of Klein bottles.*

Proof If S is orientable, then (10.9) gives $I_\pm(S) = 1 - g$ which equal zero precisely when $g = 1$. The standard embedding of the torus $S^1 \times S^1 \hookrightarrow \mathbb{C}^2$ as the product of unit circles in both copies of $\mathbb{C}$ is Lagrangian.

If S is a nonorientable surface with the Euler number $h = 2 - g$, then the set of Lai indexes of embeddings $S \hookrightarrow \mathbb{C}^2$ is $\{3h - 4, 3h, \ldots, 4 - h\}$ (see above). Since this set includes the number 0 if and only if $2 - g = h = 0 \pmod 4$, S admits a totally real embedding into $\mathbb{C}^2$ if and only if it is the connected sum of an odd number $\frac{g}{?}$ of Klein bottles. An explicit totally real embedding of the Klein bottle into $\mathbb{C}^2$ was given by Rudin [481] (see Example 10.5.6). Givental [236] found a Lagrangian embedding of any nonorientable surface with Euler number $h \le -4$, $h \equiv 0 \pmod 4$. The Klein bottle does not admit a Lagrangian embedding into $\mathbb{C}^2$ (see Shevchishin [501]). □

Example 10.5.6 We recall Rudin's construction [481] of an embedded totally real Klein bottle in $\mathbb{C}^2$.

Let (x, y) be coordinates on $\mathbb{R}^2$. Pick numbers $a > b > 0$, put

$$g(x, y) = (a + b\cos y)e^{ix}, \qquad h(y) = \sin y + i\sin 2y,$$

and define the map $F(x, y) = (z, w)$ by

$$z = g(x, y)^2, \qquad w = g(x, y)h(y).$$

Then, $F(x, y + 2\pi) = F(x, y) = F(x + \pi, -y)$, so F identifies the opposite sides of the rectangle $Q = \{(x, y) \in \mathbb{R}^2 : 0 \le x \le \pi,\ -\pi \le y \le \pi\}$ consistently with the Klein bottle construction. It is easily verified that no other identifications in Q are made. Its Jacobian equals

$$z_x w_y - z_y w_x = 2g^2 g_x h' = 2ig^3 h' \ne 0.$$

Hence, $F(\mathbb{R}^2) = K$ is an embedded totally real Klein bottle in $\mathbb{C}^2$. □

It is not surprising that there are fewer restrictions on immersed surfaces in $\mathbb{C}^2$. We give the following result in the oriented case.

Corollary 10.5.7 *Let S be a closed orientable surface of genus g. Every totally real immersion $\pi : S \to \mathbb{C}^2$ satisfies $\delta(\pi) = 1 - g$. There exists a totally real immersion of S into $\mathbb{C}^2$ with precisely $|1 - g|$ simple double points.*

Proof The first statement follows from (10.31) since $[\pi(S)] = 0$. To find a totally real immersion with $|1 - g|$ normal crossings, we begin with the case $g = 0$. Consider the immersion $\pi : S \to \mathbb{C}^2$ given by

$$\pi(x, y, u) = \big(x(1 + 2iu), y(1 + 2iu)\big), \quad x^2 + y^2 + u^2 = 1. \tag{10.21}$$

Its image $S^+ = \pi(S) \subset \mathbb{C}^2$ is a Lagrangian immersed sphere with one double point; this double point is positive and is located at the origin $(0, 0) \in \mathbb{C}^2$. We have $\delta(S^+) = +1$ and $\chi_n(S^+) = -2$. (A 'figure eight' immersion $S^n \to \mathbb{C}^n$ with these properties exists for every n; see Weinstein [565, p. 26].)

Let S^- be the image of S^+ under the map $(z, w) \mapsto (z, \overline{w})$. This is an immersed sphere with a negative double point at the origin and $I(S^-) = 2 - 2\delta(S^-) = 4$. Let $T = S^1 \times S^1 = \{(e^{i\theta}, e^{i\phi}) : \theta, \phi \in \mathbb{R}\} \subset \mathbb{C}^2$ be the standard Lagrangian torus. The connected sum $T_1 = T\#S^-$ is an immersed torus with one negative double point and $I(T_1) = I(T) + I(S^-) - 2 = 2$ in view of (10.18). For any integer $g > 0$, the orientable immersed surface of genus g given by

$$M_g = T\#\overbrace{T_1\#\cdots\#T_1}^{(g-1)\text{ times}} \subset \mathbb{C}^2$$

has $g-1$ (negative) double points. We see from (10.18) that $I(M_g) = 0$, and hence it can be deformed into a totally real immersion. □

Example 10.5.8 (Real surfaces in $\mathbb{P}^2$) The group $H_2(\mathbb{P}^2;\mathbb{Z}) = \mathbb{Z}$ is generated by the homology class $h = [H]$ of the projective line $H = \mathbb{P}^1 \hookrightarrow \mathbb{P}^2$. We have $c_1(\mathbb{P}^2) = 3PD(h)$, where $PD(h) \in H^2(\mathbb{P}^2;\mathbb{Z})$ is the Poincaré dual of h. Let $S \subset \mathbb{P}^2$ be an oriented embedded surface of genus $g(S)$ and of degree $d \neq 0$, meaning that $[S] = d[H] \in H_2(\mathbb{P}^2;\mathbb{Z})$. After reversing the orientation of S if needed, it suffices to consider the case $d > 0$. We have $[S]^2 = [dH]\cdot[dH] = d^2$ and $c_1(\mathbb{P}^2)\cdot S = [3H]\cdot[S] = 3d$. Hence, the index formula (10.10) gives

$$I_\pm(S) = 1 - g(S) + \frac{d^2 \pm 3d}{2}.$$

In particular, the condition $I_\pm(S) \le 0$ is equivalent to

$$g(S) \ge \frac{(d+1)(d+2)}{2}.$$

Comparing with the genus formula $g(C) = \frac{1}{2}(d-1)(d-1)$ that holds for a smooth complex curve $C \subset \mathbb{P}^2$ of degree d (see (10.15) on p. 488), we see that one must attach at least $3d$ homologically trivial torus handles to C to obtain a surface $S \subset \mathbb{P}^2$ of degree d satisfying $I_\pm(S) \le 0$.

From $I_+(S) - I_-(S) = c_1(\mathbb{P}^2)\cdot S = 3d$ we also infer the following.

Corollary 10.5.9 *A homologically nontrivial embedded oriented surface in $\mathbb{P}^2$ is never totally real.*

On the other hand, $\mathbb{P}^2$ contains many totally real embedded nonorientable surfaces with nontrivial homology class in $H_2(\mathbb{P}^2;\mathbb{Z}_2) = \mathbb{Z}_2$. An obvious example is the real projective plane $\mathbb{RP}^2 \subset \mathbb{P}^2 = \mathbb{CP}^2$. □

It is a natural question whether there exist complex surfaces into which every closed connected real surface embeds as a totally real surface. This question was studied by Slapar in [513]; here is a summary of his results. For a discussion of elliptic and K3 surfaces we refer to [45].

- Every closed, connected real surface S admits a totally real embedding into a $K3$ surface blown up at a single point. If S is orientable, then the blowup is not necessary.

- Every closed, connected real surface S admits a totally real embedding into any elliptic surface of type $E(3)$.
- Let $D(n, g)$ be an open disc bundle with Euler number n over a closed, connected, orientable real surface of genus g. If $n \le 2g - 2$, then every elliptic surface of type $E(-n + 2g)$ contains a Stein domain diffeomorphic to $D(n, g)$.
- Let $\widetilde{D}(n, \chi)$ be the disc bundle with Euler number n over a closed, connected, nonorientable surface with Euler number χ. If $n + \chi \le 0$, then there is a Stein domain diffeomorphic to $\widetilde{D}(n, \chi)$ inside a complex surface obtained by blowing up $\mathbb{P}^2$ at finitely many points.

In this direction, see also the paper by Nemirovski and Siegel [430] where the authors classified disc bundles over smooth surfaces whose total spaces embed as strongly pseudoconvex rationally convex domains in $\mathbb{C}^2$.

Derdzinski and Januszkiewicz [118] studied totally real immersions of a real surface S into an almost complex surface X. They established a one-to-one correspondence between the set of regular homotopy classes of such immersions and the set of homotopy classes of maps $S \to E(X)$ subject to a simple cohomological condition, where $E(X)$ is an $\mathbb{RP}^1$-bundle over X naturally associated with the almost complex structure. For the complex surfaces $\mathbb{C}^2$, $\mathbb{P}^1 \times \mathbb{P}^1$, $\mathbb{P}^2$ and $\mathbb{P}^2 \# m\overline{\mathbb{P}^2}$, $1 \le m \le 7$, they provided explicit examples of immersions and embeddings which exhaust all possible equivalence classes.

Complex points of a real surface S in a complex surface (X, J) are the simplest examples of CR-singularities, i.e., points $p \in S$ where the dimension of the maximal complex subspace $\mathrm{T}_p^{\mathbb{C}} S = \mathrm{T}_p S \cap J(\mathrm{T}_p S)$ of $T_p S$ is bigger than at a generic point of S. If S is a smooth submanifold of real codimension two in a complex manifold X, then after a small generic perturbation the CR-singularities form a discrete subset of S, and at such points the tangent space $\mathrm{T}_p S$ is complex. Many of the results in this and the previous two sections extend to this situation. In particular, Slapar [514] obtained an analogue of the cancellation theorem (Theorem 10.4.1) for CR singular points of real codimension 2 submanifolds in complex manifolds. See also his papers [515–517] concerning the classification of such points.

Another case that has been studied intensively are smooth 3-manifolds in $\mathbb{C}^3$. A generic smooth embedding of a real 3-manifold M into a complex 3-manifold X has a curve of complex tangents, and it is natural to ask which curves arise in this way. Kasuya and Takase [330] showed that a one-dimensional submanifold C of a closed orientable 3-manifold M can be realized as the set of complex tangents of a smooth embedding $M \hookrightarrow \mathbb{C}^3$ if and only if $[C] = 0 \in H_1(M; \mathbb{Z})$. In particular, every knot or link in S^3 is the set of complex tangents of a smooth embedding $S^3 \hookrightarrow \mathbb{C}^3$. For earlier results on this subject see Elgindi [143, 145]. In [144], Elgindi described a topological obstruction to the removal of an isolated degenerate complex tangent of a real 3-manifold in a complex 3-fold.

10.6 The Adjunction Inequality in Kähler Surfaces

In this and the following section we explore applications of the Seiberg-Witten theory to the geometry of complex surfaces, and we explain the connection between the adjunction inequality and the *generalized Thom conjecture*.

Let S be a compact oriented surface smoothly embedded in a complex surface X. Recall that $I_\pm(S)$ is the number of complex points of S in the given orientation class (see Sect. 10.1). By Theorem 10.3.1 (p. 485) we have

$$2I_\pm(S) = 2 - 2g(S) + [S]^2 \pm c_1(X)\cdot S = 2 - 2g(S) + [S]^2 \mp K_X\cdot S,$$

where $K_X = -c_1(X)$ denotes the canonical class of X. If S has no negative complex points, then $I_-(S) = 0$ which is equivalent to the genus formula

$$2g(S) - 2 = [S]^2 - c_1(X)\cdot S = [S]^2 + K_X\cdot S. \qquad (10.22)$$

This holds in particular if S is a smooth complex curve in a complex surface X, or a symplectic surface embedded in a symplectic 4-manifold (X,ω).

The next theorem summarizes results of Fintushel and Stern [157], Kronheimer and Mrowka [351], [352, Theorem 1.7], Morgan, Szabó and Taubes [420], and Ozsváth and Szabó [447]. We restrict attention to Kähler surfaces; a similar result holds in many other 4-manifolds [248, Theorem 2.4.8]. Recall that $b_2^+(X)$ is the dimension of a maximal linear subspace $H^+(X;\mathbb{R})$ of $H^2(X;\mathbb{R})$ on which the intersection form is positive definite (see Sect. 10.2).

Theorem 10.6.1 *Assume that X is a compact Kähler surface with $b_2^+(X) > 1$ and $S \hookrightarrow X$ is a smoothly embedded closed oriented real surface of genus $g(S)$ which is not a 2-sphere with $[S] = 0 \in H_2(X;\mathbb{R})$. If $[S]^2 \geq 0$, or if $[S]^2 < 0$ and $g(S) > 0$, then S satisfies the adjunction inequality*

$$[S]^2 + |K_X\cdot S| \leq 2g(S) - 2. \qquad (10.23)$$

If the canonical bundle K_X is ample, then (10.23) *also holds if S is an embedded sphere with $[S]^2 < 0$. If $\pi\colon S \to X$ is an immersion with $\delta_+(\pi)$ positive (simple) double points and if $g(S) + \delta_+(\pi) > 0$, then*

$$\big[\pi(S)\big]^2 + \big|K_X\cdot\pi(S)\big| \leq 2g(S) + 2\delta_+(\pi) - 2. \qquad (10.24)$$

A sketch of proof can be found at the end of the section.

The main interest of the adjunction inequality (10.23) to Stein geometry is that by (10.10) it is equivalent to $I_\pm(S) \leq 0$, and by Corollary 10.5.1 (p. 492) such S can be $\mathcal{C}^0$ approximated by an isotopic embedding with tubular Stein neighborhoods. Conversely, Theorem 10.7.2 below shows that the adjunction inequality also holds for embedded surfaces in a Stein surface thanks to the Kähler embedding theorem (Theorem 10.7.1, p. 506). This connection, which was observed by Lisca and Matić [398] and Nemirovski [431], has interesting applications and gives nontrivial obstructions to the existence of Stein structures on smooth 4-manifolds. For immersed surfaces, the situation is less clear since $I_\pm(\pi) \leq 0$ is equivalent to the stronger inequality

$$[\pi(S)]^2 + \left|K_X \cdot \pi(S)\right| + 2\delta_-(\pi) \le 2g(S) + 2\delta_+(\pi) - 2 \qquad (10.25)$$

obtained by adding the contribution of negative double points. This is false in general since the number $\delta_-(\pi)$ can be increased at will by attaching to $\pi(S)$ sufficiently many homologically trivial immersed spheres with a negative double point (see the proof of Corollary 10.5.7, p. 495). However, see Theorem 10.7.7 below for immersed surfaces in Stein surfaces.

Remark 10.6.2 (A) If the homology class of S is a torsion element of $H_2(X;\mathbb{Z})$, then the left hand side of (10.23) clearly vanishes; in this case, the adjunction inequality trivially holds when $g(S) > 0$ and it fails for the sphere. Thus, it suffices to consider surfaces which are *essential* in the sense that $[S]$ is an element of infinite order in $H_2(X;\mathbb{Z})$; equivalently, $[S] \neq 0 \in H_2(X;\mathbb{R})$.

(B) Comparing (10.23) with the genus formula (10.22), we see that a smooth complex curve C in a complex surface X satisfies the adjunction inequality (10.23) if and only if $K_X \cdot C \ge 0$. If this holds for all complex curves in X, then the canonical class K_X is said to be *numerically effective* (nef). If K_X is ample, then $[K_X]^2 > 0$ and $K_X \cdot C > 0$ for every complex curve $C \subset X$; a surface with ample canonical bundle is minimal and of general type.

If on the other hand we have $K_X \cdot C < 0$ for a complex curve $C \subset X$, then (10.23) clearly fails for C, and possibly also for some surfaces of higher genus representing the homology class $[C]$. This happens for example in the projective plane $X = \mathbb{P}^2$ as was seen in the previous section, so the hypothesis $b_2^+(X) > 1$ cannot be removed in general. However, by Proposition 10.6.6 below this cannot happen if X is a compact Kähler surface with $b_2^+(X) > 1$ and $g(C) > 0$. Here is an alternative argument. On a Kähler surface X, the condition $b_2^+(X) > 1$ is equivalent to $h^{2,0}(X) > 0$ (the Morse index theorem says that there can be only one positive eigenvalue of the intersection form on $H^{1,1}(X)$). This implies that X cannot be uniruled. Hence, it has a minimal model Y whose canonical class K_Y is nef, and $K_X = \mu^* K_Y + \sum PD(E_i)$ where $\mu\colon X \to Y$ is the projection and $E_i \subset X$ are the exceptional curves. The only curves in X with $K_X \cdot C < 0$ are the exceptional rational curves E_i. All these things can be found in [45]. □

The adjunction inequality (10.23) is closely related to the *Thom conjecture* stating that a smooth embedded complex curve C in a complex surface X has the smallest genus among all oriented smoothly embedded surfaces $S \hookrightarrow X$ with $[C] = [S] \in H_2(X;\mathbb{Z})$. The following theorem summarizes results of Fintushel and Stern [157], Kronheimer and Mrowka [351, 352], Morgan, Szabó and Taubes [420], and Ozsváth and Szabó [447].

Theorem 10.6.3 (Generalized Thom Conjecture) *A smooth embedded complex curve in a compact Kähler surface is genus minimizing in its homology class. More generally, an embedded symplectic surface in a closed symplectic 4-manifold is genus minimizing in its homology class.*

To see the connection with the adjunction inequality (10.23), assume that C is a smooth complex or symplectic curve in X and $S \subset X$ is an embedded surface with $[S] = [C] \in H_2(X;\mathbb{Z})$. We have

$$[S]^2 + |c_1(X)\cdot S| = C\cdot C + |c_1(X)\cdot C| \geq 2g(C) - 2,$$

where the last inequality follows from the genus formula (10.22). If (10.23) is valid for any such S, it follows that $g(S) \geq g(C)$ and hence the Thom conjecture holds for the class $[C] \in H_2(X;\mathbb{Z})$. Conversely, if C satisfies the Thom conjecture and $K_X\cdot C \geq 0$, then $K_X\cdot C = |K_X\cdot C| = |K_X\cdot S|$; thus

$$g(S) \geq g(C) = 1 + \frac{1}{2}\big([C]^2 + K_X\cdot C\big) = 1 + \frac{1}{2}\big([S]^2 + |K_X\cdot S|\big)$$

and hence S satisfies (10.23). Although Theorem 10.6.3 does not formally imply Theorem 10.6.1 (since not all homology classes in $H_2(X;\mathbb{Z})$ are represented by complex curves), the proofs of both results follow the same pattern.

Before proceeding, we recall the basic notions and results of Seiberg-Witten theory [495]. We refer to Morgan's monograph [419] for a more complete exposition. An excellent introductory survey is [248, Appendix 2.4].

If X is a closed, oriented, smooth 4-manifold with a Riemannian metric g, then the *Seiberg-Witten invariant* of X is an integer valued function

$$SW_{X,g} : Spin^c(X) \to \mathbb{Z} \tag{10.26}$$

defined on the set of all spin^c structures on X. We explain what is a spin^c structure. Recall that the special orthogonal group $SO(4)$ has the fundamental group equal to $\mathbb{Z}_2 = \mathbb{Z}/2\mathbb{Z}$. The universal (double) covering group of $SO(4)$ is $Spin(4) = SO(4) \times \{\pm 1\}/\{\pm(I,1)\}$. The Lie group $Spin^c(4)$ is defined by

$$Spin^c(4) = Spin(4) \times S^1/\{\pm(I,1)\},$$

where $S^1 = U(1)$ is the unit circle in $\mathbb{C}$. By projecting $Spin^c(4)$ onto the first factor $Spin(4)$ and noting that, by doing so, we actually pass to the quotient $Spin(4)/\{\pm I\} = SO(4)$, we obtain an exact sequence (an S^1-fibration)

$$1 \longrightarrow S^1 \longrightarrow Spin^c(4) \stackrel{\tau}{\longrightarrow} SO(4) \longrightarrow 1.$$

Let $P \to X$ denote the principal $SO(4)$-bundle consisting of oriented g-orthonormal frames of the tangent bundle TX. A spin^c *structure* on X is a lifting of P to a principal $Spin^c(4)$-bundle $\widetilde{P} \to X$, together with a bundle map $\tau\colon \widetilde{P} \to P$ over X which equals the map $\tau : Spin^c(4) \to SO(4)$ fibrewise, so it gives an identification $\widetilde{P} \times_\tau SO(4) = P$. More precisely, there is an open cover $\{U_j\}$ of X and a 1-cocycle (g_{ij}) with values in the sheaf $\mathcal{C}_X^{Spin^c(4)}$ of smooth maps $X \to Spin^c(4)$ defining the principal $Spin^c(4)$-bundle $\widetilde{P} \to X$ such that the corresponding 1-cocycle $(\tau \circ g_{ij})$ with values in the sheaf $\mathcal{C}_X^{SO(4)}$ defines the bundle P. Such a lifting of the principal $SO(4)$-bundle $P \to X$ to a principal $Spin^c(4)$-bundle $\widetilde{P} \to X$ always exists.

The group $Spin^c(4)$ also admits the presentation

$$Spin^c(4) = \{(A, B) \in U(2) \times U(2) : \det A = \det B\}. \tag{10.27}$$

This gives projections $\mu^{\pm}: Spin^c(4) \to U(2)$ onto the first and the second factor. Using $\mu^{\pm}$ we associate to a spin^c structure $(\widetilde{P}, \tau) \in Spin^c(X)$ a pair of unitary vector bundles with fibre $\mathbb{C}^2$,

$$W^+ = \widetilde{P} \times_{\mu^+} \mathbb{C}^2 \to X, \qquad W^- = \widetilde{P} \times_{\mu^-} \mathbb{C}^2 \to X,$$

called the bundles of *positive and negative spinors*, respectively. This gives a presentation of a spin^c structure on X as a triple

$$\mathfrak{s} = (\rho, W^+, W^-), \tag{10.28}$$

where W^+, W^- are smooth $U(2)$-bundles over X with isomorphic $U(1)$ determinant bundles $\det W^+ = \det W^-$ and

$$\rho: \mathrm{T}^*X \otimes \Gamma(X; W^+) \to \Gamma(X; W^-) \tag{10.29}$$

is the *Clifford multiplication*, a purely algebraic operation (see e.g. [419]). The bundles $W^{\pm}$, and the Clifford multiplication ρ, can also be explained by the following representations of $Spin^c(4)$ on the field of quaternions $\mathbb{H}$ (see [248, p. 57]). Let us use the presentation

$$Spin^c(4) = S^1 \times SU(2) \times SU(2)/\{\pm(1, I, I)\}.$$

Identifying $SU(2)$ with the group of unit quaternions in $\mathbb{H}$, an element of $a = (\lambda, q_1, q_2) \in Spin^c(4)$ is given by a unit complex number $|\lambda| = 1$ and two unit quaternions q_1, q_2. The representations $\rho_0(\lambda, q_1, q_2)(h) = q_1 h \bar{q}_2$ ($h \in \mathbb{H}$, quaternionic multiplication) results in $\widetilde{P}' \times_{\rho_0} \mathbb{H} = \mathrm{T}X$. The representations $\rho_+(\lambda, q_1, q_2)(h) = q_1 h \lambda$ and $\rho_-(\lambda, q_1, q_2)(h) = q_2 h \lambda$ result in $\widetilde{P} \times_{\rho_\pm} \mathbb{H} = W^{\pm}$. It follows easily that $\mathrm{T}X \otimes \mathbb{C} \cong \mathrm{Hom}_{\mathbb{C}}(W^+, W^-)$, and this isomorphism gives the Clifford multiplication $\Gamma(X; \mathrm{T}X \otimes W^+) \to \Gamma(X; W^-)$. By using the Riemannian metric g on X we identify $\mathrm{T}X$ with T^*X, and this gives the Clifford product (10.29).

The Seiberg-Witten monopole equation is an elliptic system of two nonlinear partial differential equations that involve a smooth section $\psi: X \to W^+$ of the positive spinor bundle and a smooth $U(1)$-connection A on the unitary line bundle $L = \det W^+ = \det W^-$ over X. The connection A on L, together with the Riemannian connection on $\mathrm{T}X$ determined by the metric g, determine a covariant derivative ∇_A on any spin^c structure $\mathfrak{s}$ on X (10.28). This derivative maps smooth sections of W^+ to smooth sections of $\mathrm{T}^*X \otimes W^+$. Composing ∇_A with the Clifford product (10.29) gives the *Dirac operator*

$$\partial\!\!\!/_A = \rho \circ \nabla_A : \Gamma(X, W^+) \to \Gamma(X, W^-).$$

Given a pair (ψ, A), where A is a unitary connection on $L = \det W^+$ and ψ is a smooth section of W^+, the *unperturbed Seiberg-Witten equations* are

$$\partial\!\!\!/_A \psi = 0, \qquad F_A^+ = \mathrm{i}q(\psi) = \mathrm{i}\left(\psi \otimes \psi^* - \frac{|\psi|^2}{2}\mathrm{Id}\right)$$

where F_A^+ is the self-dual part (with respect to the Hodge operator $*_g$ associated to the metric g) of the curvature form F_A of the connection A, and $q(\psi)$ is naturally identified with a self-dual 2-form. (See [419, Sect. 4.1].) For technical reasons one replaces the second equation by a perturbed equation

$$F_A^+ = \mathrm{i}q(\psi) + \mathrm{i}\eta$$

for some self-dual 2-form η on X. The gauge transformations $X \to U(1) = S^1$ act on the space of solutions (A, ψ) of this system. For a generic choice of the metric g on X and of the perturbation η, the moduli space $\mathcal{M}_X(\mathfrak{s})$ (where $\mathfrak{s}$ is the spin^c structure (10.28)) of solutions (A, ψ) of the perturbed Seiberg-Witten equations $\partial\!\!\!/_A \psi = 0$, $F_A^+ = \mathrm{i}q(\psi) + \mathrm{i}\eta$ is a compact smooth oriented manifold of dimension

$$d(\mathfrak{s}) = \dim_{\mathbb{R}} \mathcal{M}_X(\mathfrak{s}) = \frac{c_1(\mathfrak{s})^2 - 2\chi(X) - 3\sigma(X)}{4}. \qquad (10.30)$$

Here, $c_1(\mathfrak{s})^2 \in \mathbb{Z}$ is the square of the Chern class $c_1(\mathfrak{s}) \in H^2(X;\mathbb{Z})$ (an element of $H^4(X;\mathbb{Z})$) evaluated on the fundamental class $[X] \in H_4(X;\mathbb{Z})$, $\chi(X)$ is the Euler number of X, and $\sigma(X) = b_2^+(X) - b_2^-(X)$ is the signature of the intersection form on $H_2(X;\mathbb{Z})$. In the most interesting case when $d(\mathfrak{s}) = 0$, $SW_{X,g}(\mathfrak{s})$ is simply the number of solutions of the Seiberg-Witten equation up to gauge transformations $X \to U(1)$ and counted with orientation signs. The definition of $SW_{X,g}$ is somewhat more involved when $d(\mathfrak{s}) > 0$.

Remark 10.6.4 The case of interest to us is when X is a compact Kähler surface. Recall that a cohomology class $c \in H^2(X;\mathbb{Z})$ is a *characteristic class* if c is an integral lift of the second Stiefel-Whitney class $w_2(X) \in H^2(X;\mathbb{Z}_2)$, that is, if $c \equiv w_2(X)$ (mod 2). We denote the set of all characteristic classes by $\mathcal{C}_X$. By a theorem of Hirzebruch and Hopf [297], a smooth, compact, oriented 4-manifold X admits an almost complex structure J if and only if there is a characteristic class $c \in \mathcal{C}_X \subset H^2(X;\mathbb{Z})$ satisfying

$$\langle c^2, [X]\rangle - 2\chi(X) - 3\sigma(X) = 0;$$

in this case there exists an almost complex structure J on X with $c_1(X, J) = c$. Thus, $d(\mathfrak{s}) = 0$ for any spin^c structure $\mathfrak{s}$ with $c_1(\mathfrak{s}) = c_1(X, J) = -K_X$. □

The hypothesis $b_2^+(X) > 1$ ensures that the Seiberg-Witten invariant (10.26) is independent of the generic choice of the metric g on X, so it is an invariant SW_X of the $\mathcal{C}^\infty$ structure on X. Furthermore, for any smooth map $f: X \to Y$ to a compact

Kähler surface Y and spin^c structure $\mathfrak{s} \in Spin^c(Y)$ we have $SW_X(f^*\mathfrak{s}) = \pm SW_Y(\mathfrak{s})$. Assume this to be the case from now on.

There is a finite-to-one map $\iota\colon Spin^c(X) \to \mathcal{C}_X \subset H^2(X;\mathbb{Z})$ of $Spin^c(X)$ onto the set $\mathcal{C}_X$ of characteristic classes of X. The map ι associates to any spin^c structure $\mathfrak{s} = (\rho, W^+, W^-)$ the first Chern class $c_1(W^+)$ of the positive spinor bundle W^+. We write $\iota(\mathfrak{s}) = c_1(\mathfrak{s}) = c_1(W^+)$. One also defines a function $SW_X\colon \mathcal{C}_X \to \mathbb{Z}$ by $SW_X(c) = \sum_{\iota(\mathfrak{s})=c} SW_X(\mathfrak{s})$. If the group $H_2(X;\mathbb{Z})$ has no 2-torsion (which holds in particular if X is simply connected), then ι is bijective and hence the above distinction becomes unnecessary.

A spin^c structure $\mathfrak{s}$ on X is said to be a *Seiberg-Witten structure* if $SW_X(\mathfrak{s}) \neq 0$; a characteristic class $c \in \mathcal{C}_X$ is a *Seiberg-Witten class* if $SW_X(\mathfrak{s}) \neq 0$ for some $\mathfrak{s} \in Spin^c(X)$ with $c_1(\mathfrak{s}) = c$. At most finitely many characteristic classes (and the associated spin^c structures) are Seiberg-Witten classes. A smooth 4-manifold X is said to be of *Seiberg-Witten simple type* if for any $\mathfrak{s} \in Spin^c(X)$ with $SW_X(\mathfrak{s}) \neq 0$ we have $d(\mathfrak{s}) = 0$; that is, for every $\mathfrak{s} \in Spin^c(X)$ with $SW_X(\mathfrak{s}) \neq 0$ the associated Seiberg-Witten equations have finitely many solutions up to gauge equivalence. So far there seem to be no known examples of 4-manifolds that are not of Seiberg-Witten simple type.

We quote the following result of Witten and Taubes.

Theorem 10.6.5 ([543, 544, 579]) *Every compact Kähler surface X satisfying $b_2^+(X) > 1$ is of Seiberg-Witten simple type and satisfies $SW_X(\pm K_X) \neq 0$, where K_X is the canonical class. If such X is also minimal and of general type, then $\pm K_X$ are the only Seiberg-Witten classes of X.*

Proof of Theorem 10.6.1 Assume first that S is embedded in X.

Case 1: $[S]^2 \geq 0$. In this case, S cannot be a sphere. Indeed, the existence of an embedded homologically essential sphere S with $[S]^2 \geq 0$ in a compact 4-manifold X implies that the Seiberg-Witten invariant $SW_X : Spin^c(X) \to \mathbb{Z}$ is the zero function [157, Lemma 5.1, p. 154]. (The proof amounts to showing that the existence of a Seiberg-Witten class would produce infinitely many Seiberg-Witten classes in X, a contradiction.) In our case this is not so according to Theorem 10.6.5. Assume now that $g(S) > 0$. Then Theorem 10.6.1 is a consequence of Proposition 4.2 in [420, p. 714] (the key step in the proof of the Thom conjecture). We state the special case of interest to us.

Proposition 10.6.6 ([420, Proposition 4.2]) *Assume that X is a compact Kähler surface with $b_2^+(X) > 1$ and $S \subset X$ is a smooth embedded real surface with $[S]^2 \geq 0$ and $g(S) > 1$. If for some Seiberg-Witten class $K \in \mathcal{C}_X$ we have $K \cdot S + [S]^2 = 2g(S) - 2$, then S has the smallest genus among all smooth embedded real surfaces $S' \subset X$ with $[S'] = [S] \in H_2(X;\mathbb{Z})$.*

For the canonical class $K = K_X$, the hypothesis in Proposition 10.6.6 is precisely that S satisfies the genus formula for complex curves. The proof reduces to the case $[S]^2 = 0$ as follows. Let $\widetilde{X} = X\#k\overline{\mathbb{P}^2} \xrightarrow{\mu} X$ be obtained by blowing up X at

$k = [S]^2 \geq 0$ distinct points of S. Let $E_1, \ldots, E_k \subset \widetilde{X}$ be the exceptional fibres of the blown-up points; so $E_j \cong \mathbb{P}^1$ and $[E_j]^2 = -1$. Let $S' \subset \widetilde{X}$ be the connected sum of S with $E_1, \ldots, E_k$. Note that $g(S') = g(S)$ and that S' is homologous to the total transform $\mu^{-1}(S) \subset \widetilde{X}$ of S. Since K is a Seiberg-Witten class in X, $K' = \mu^* K + \sum_{j=1}^k PD(E_j) \in H^2(\widetilde{X}; \mathbb{Z})$ is a Seiberg-Witten class in $\widetilde{X}$ [157]. It is easily seen that $S' \cdot S' = [S]^2 - k = 0$ and $K' \cdot S' = K \cdot S + k$. It now suffices to prove Proposition 10.6.6 for the pair (K', S') in $\widetilde{X}$. This is done by using a product formula for Seiberg-Witten invariants [420, Theorem 3.1].

To prove Theorem 10.6.1 in the case $[S]^2 \geq 0$, assume that the adjunction inequality (10.23) fails for a smooth embedded surface $S \subset X$ of genus $g(S) > 0$. Then $[S]^2 + |K_X \cdot S| = 2g(S) - 2 + k$ for a positive integer $k > 0$. Choose $\epsilon = \pm 1$ so that $\epsilon K_X \cdot S = |K_X \cdot S|$. Let $C \subset X$ be a surface obtained by adding k homologically trivial handles to S. Then, $[C] = [S] \in H_2(X; \mathbb{Z})$ and $g(C) = g(S) + k > 1$, so $C \cdot C + \epsilon K_X \cdot C = g(C) - 2$. Since $SW_X(\epsilon K_X) = \pm 1$, Proposition 10.6.6 implies that C has the smallest genus in its homology class, a contradiction to $g(S) < g(C)$.

Case 2: $[S]^2 < 0$. If $g(S) > 0$, one gets the adjunction inequality by applying the following result with the canonical class $K = K_X$.

Theorem 10.6.7 (Ozsváth and Szabó [447, Corollary 1.7]) *Let X be a compact oriented 4-manifold of Seiberg-Witten simple type with $b_2^+(X) > 1$, and let $S \subset X$ be a smoothly embedded, oriented closed real surface of genus $g(S) > 0$ with $[S]^2 < 0$. Then, for all Seiberg-Witten classes $K \in \mathcal{C}_X$ we have*

$$[S]^2 + |K \cdot S| \leq 2g(S) - 2. \tag{10.31}$$

This is an immediate consequence of [447, Theorem 1.3, p. 96]. The latter asserts that, if the adjunction inequality (10.31) fails for a certain Seiberg-Witten class $K \in \mathcal{C}_X$, then $K' = K + 2\epsilon PD(S) \in \mathcal{C}_X$ is also a Seiberg-Witten class. (Here, $PD(S)$ is the Poincaré dual of $[S]$ and $\epsilon = \pm 1$ is the sign of $K \cdot S$. Remark 1.6 in [447, p. 97] explains the presence of the coefficient 2, as opposed to [447, Theorem 1.3] which pertains to a spin^c structure $\mathfrak{s}$ with $c_1(\mathfrak{s}) = K$.) From (10.30) we get

$$d(K') = \frac{1}{4}\big[\big(K + 2\epsilon PD(S)\big)^2 - 2\chi(X) - 3\sigma(X)\big] = d(K) + \epsilon K \cdot S + PD(S)^2.$$

If (10.31) fails, then $\epsilon K \cdot S + PD(S)^2 = |K \cdot S| + [S]^2 \geq 2g \geq 2$. This implies $d(K') \geq 2$, contradicting the assumption that X is of simple type.

The last conclusion fails if S is the sphere ($g = 0$). However, under the stated conditions on X, its only Seiberg-Witten classes are $\pm K_X$ by Theorem 10.6.5. Applying the above argument with $K = K_X$, we see as before that $K' = K_X + 2\epsilon PD(S)$ is a Seiberg-Witten class. Hence, either $K' = K_X$ which gives $PD(S) = 0$, a contradiction since S is essential, or $K' = -K_X$ which gives $PD(S) = \pm K_X$. In the latter case we have $[S]^2 = PD(S)^2 = K_X^2 > 0$ since K_X is assumed to be ample. This contradicts the assumption $[S]^2 < 0$ and thus proves the adjunction inequality (10.31) also for spheres.

It remains to consider immersed surfaces. Assume that $\pi : S \to X$ is an immersion with simple double points, say δ_+ positive ones and δ_- negative ones. We replace each of the positive double points by an embedded totally real handle in X as described in Remark 10.3.2 on p. 486. This replaces $\Sigma = \pi(S) \subset X$ by a homologically equivalent immersed surface $\Sigma' = \pi'(S') \subset X$ of genus $g(S') = g(S) + \delta_+$. Let Y be the complex surface obtained by blowing up X at each of the remaining (negative) double points p_j of Σ'. Let $\iota \colon Y \to X$ denote the projection and $E_j = \iota^{-1}(p_j) \cong \mathbb{P}^1 \subset Y$ the exceptional fibres. Then, Y is a compact Kähler surface with $b_2^+(Y) > 1$ containing an embedded copy of S', namely the proper transform A of $\Sigma' \subset X$. We claim that $[A] = [\iota^{-1}(\Sigma')] \in H_2(\widetilde{X}; \mathbb{Z})$. To see this, note that a local model for Σ' at any negative double point p_j is $L_1 \cup \bar{L}_2 \subset \mathbb{C}^2$, where $L_1 \neq L_2$ is a pair of complex lines through $p = 0 \in \mathbb{C}^2$ and the bar on L_2 means the reversed orientation. The class of the total transform $[\iota^{-1}(\Sigma')]$ is obtained from the proper transform by adding a copy of $[E_j]$ (the contribution of L_1) and also a copy of $-[E_j]$ (the contribution of $\bar{L}_2$), so $[E_j]$ cancels out. This proves the claim. It follows that $[A]^2 = [\Sigma']^2$ and $K_Y \cdot A = K_X \cdot \Sigma'$. The first property is obvious, while the second follows from $K_Y = \iota^*(K_X) + \sum_j PD(E_j)$ and $PD(E_j) \cdot [A] = [E_j] \cdot [A] = [E_j] \cdot [\iota^{-1}(\Sigma')] = 0$. Hence, the left hand side of the inequality (10.24) remains unchanged when passing from $\Sigma' \subset X$ to $A \subset Y$, and the right hand side equals $2g(S') - 2$. Since $g(S') = g(S) + \delta_+ > 0$, the adjunction inequality (10.23) for embedded surfaces gives (10.24).

This leaves the case when $\Sigma \subset X$ is an immersed sphere with only negative simple double points. We still get the same conclusion if either $[\Sigma]^2 \geq 0$, or if $[\Sigma]^2 < 0$ and K_X is ample (so $\pm K_X$ are the only Seiberg-Witten classes of X). To see this, let $\iota \colon Y \to X$ and $A \subset Y$ be as above, so A is an embedded homologically nontrivial sphere. The case $[A]^2 \geq 0$ can be ruled out as before. Assume now that $[A]^2 < 0$ and K_X is ample. Then, the Seiberg-Witten classes of Y are $\pm\iota^*(K_X) + \sum_j \pm PD(E_j)$ (see [248, p. 54]). If (10.24) fails, we conclude as in the embedded case that $K_Y + 2\epsilon PD(A) \in H^2(Y; \mathbb{Z})$ is a Seiberg-Witten class for at least one of the choices of $\epsilon = \pm 1$. Hence, $PD(A) = \pm K_Y$ modulo the exceptional classes $PD(E_j)$. Dualizing gives $[A] = \pm[K_Y]$ in $H^2(Y; \mathbb{Z})$ modulo the exceptional curves $[E_j]$. Pushing down to X by the map $\iota \colon Y \to X$, we get $[\Sigma] = \pm[K_X]$ and hence $[A]^2 = [\Sigma]^2 = [K_X]^2 > 0$, a contradiction to the assumption $[A]^2 < 0$. □

10.7 The Adjunction Inequality in Stein Surfaces

It was observed by Nemirovski [431] and Lisca and Matić [398] that Theorem 10.6.1 (the adjunction inequality for compact Kähler surfaces) also implies the adjunction inequality for compact real surfaces in Stein surfaces. The main point of the reduction is the following embedding theorem which we quote here without proof (see Theorem 3.2 and Corollary 1.4 in [398]). Similar results without the Kähler condition (which is actually not used here) were obtained by Stout [533] and by Demailly, Lempert and Schiffman [117, 390].

Theorem 10.7.1 *Let X be a Stein manifold and $\rho\colon X \to \mathbb{R}$ be a smooth strongly plurisubharmonic exhaustion function. Let $r \in \mathbb{R}$ be a regular value of ρ. Then there exists a biholomorphic map $\phi\colon X_r = \{x \in X : \rho(x) < r\} \to \Omega$ onto a domain in a smooth minimal projective variety Y with ample canonical bundle K_Y, having a Kähler form ω whose pullback to X_r equals the Kähler form $\omega_\rho = d \circ J^* \circ d\rho$ determined by ρ. If $\dim_{\mathbb{C}} X = 2$, then Y may be chosen such that $b_2^+(Y) > 1$. Furthermore, if $S \subset X_r$ is an embedded real surface with $[S] \neq 0 \in H_2(X;\mathbb{Z})$, then we can arrange that $\phi(S)$ is a homologically essential surface in Y.*

Corollary 10.7.2 (Lisca and Matić [398, Theorem 5.2], Nemirovski [431]) *If S is a closed, connected, oriented real surface smoothly embedded in a Stein surface X such that S is not a homologically trivial two-sphere, then*

$$[S]^2 + |K_X \cdot S| \leq 2g(S) - 2.$$

In particular, a homologically nontrivial embedded sphere in a Stein surface satisfies $[S]^2 \leq -2$. If $\pi : S \to X$ is an immersion and $g(S) + \delta_+(\pi) > 0$, then

$$\pi(S)^2 + \left|K_X \cdot \pi(S)\right| \leq 2g(S) + 2\delta_+(S) - 2.$$

Proof A Stein surface has no homology above dimension 2 and the group $H_2(X;\mathbb{Z})$ is without torsion (see Sect. 3.12). Hence, $[S] \neq 0 \in H_2(X;\mathbb{Z})$ implies that S is essential in X. By Theorem 10.7.1, there exists a domain $\Omega \subset X$ containing S and a biholomorphic map $\phi\colon \Omega \to \Omega'$ onto a domain in a minimal projective (hence Kähler) surface Y of general type, with ample canonical bundle K_Y and satisfying the condition $b_2^+(Y) > 1$. The construction in [398] also ensures that $\phi(S)$ is essential in Y. The right hand side of the adjunction inequality (10.23) does not change if we replace $S \subset X$ with its image $\phi(S) \subset Y$ since the numbers $[S]^2$ and $K_X \cdot S$ only depend on the neighborhood of S in the ambient surface. Therefore, the corollary follows from Theorem 10.6.1 applied to $\phi(S) \subset Y$. The same argument holds in the immersed case. □

Example 10.7.3 The smooth 4-manifold $S^2 \times \mathbb{R}^2$ does not admit any Stein structure in view Corollary 10.7.2 since it contains the homologically essential embedded sphere $S^2 \times \{0\}$ with self-intersection number zero. □

Example 10.7.4 If X is a Stein surface, then a surface Y obtained by blowing up finitely many points in X does not admit any Stein structure. In fact, the exceptional fibre of a blown up point is an embedded rational curve with self-intersection number -1, so the claim follows from Corollary 10.7.2. □

We have seen in Sect. 10.5 that the adjunction inequality holds for an embedded oriented surface S of degree $d > 0$ in $\mathbb{P}^2$ if and only if $2g(S) \geq \mathrm{d}^2 + 3d + 2 \geq 6$. This fails if S is obtained by adding at most $3d - 1$ homologically trivial handles to a smooth complex curve of degree $d > 0$; hence, such a surface does not admit any Stein neighborhoods. Going a step further, S. Nemirovski proved the following result that in particular answers a question of Vitushkin.

Theorem 10.7.5 ([431, Theorem 10]) *Let $S \subset \mathbb{P}^2$ be a smoothly embedded oriented surface of genus g and of degree $d > 0$. If there exists a nonconstant holomorphic function in a neighborhood of S, then*

$$g(S) \geq \frac{1}{2}\left(d^2 + 3d + 2\right) \geq 3.$$

In particular, if S is a homologically nontrivial embedded sphere in $\mathbb{P}^2$, then every holomorphic function in a neighborhood of S is constant.

Proof Assume that there exists a nonconstant holomorphic function in a neighborhood $U \subset \mathbb{P}^2$ of the embedded surface $S \subset \mathbb{P}^2$. Then, the envelope of holomorphy of U is a Riemann domain $\pi : \Omega \to \mathbb{P}^2$ whose total space Ω is a Stein surface. (The envelope of holomorphy exists for every domain $U \subset \mathbb{P}^n$; see the references in [431, Sect. 1.4]. If U does not admit any nonconstant holomorphic functions, then its envelope equals $\mathbb{P}^n$; otherwise it is a Stein manifold.) Note that Ω contains a domain $\widetilde{U}$ that projects by π bijectively onto U. The surface $S' = (\pi|_{\widetilde{U}})^{-1}(S) \subset \Omega$ is homologically nontrivial in Ω (otherwise $S = \pi(S')$ would be homologically trivial in $\mathbb{P}^2$), and it clearly satisfies

$$\left[S'\right]^2 + \left|K_\Omega \cdot S'\right| = [S]^2 + |K_{\mathbb{P}^2} \cdot S| = d^2 + 3d.$$

By Corollary 10.7.2, the surface $S' \subset \Omega$ satisfies the adjunction inequality:

$$2g\left(S'\right) - 2 \geq \left[S'\right]^2 + \left|K_\Omega \cdot S'\right|.$$

Since $g(S) = g(S')$, the result follows. □

Theorem 10.7.5 is false under the weaker hypothesis that there exists a strongly plurisubharmonic function in a neighborhood of an embedded surface $S \subset \mathbb{P}^2$. Indeed, if S is a real submanifold of class $\mathcal{C}^1$ embedded in a complex manifold X with only isolated complex points, then there exists a strongly plurisubharmonic function in an open neighborhood of S in X (see [205, Proposition 1.3]). In particular, a generic smooth embedding of the 2-sphere (a surface of genus 0) into $\mathbb{P}^2$ satisfies this condition. This gives the following corollary (see [205, Theorem 1.1]).

Corollary 10.7.6 *There exists a complex surface X without nonconstant holomorphic functions, but with a strongly plurisubharmonic function. Such X can be chosen as a tubular neighborhood of an embedded sphere S^2 in $\mathbb{P}^2$.*

We have seen that the stronger inequality (10.25) on p. 499 is false in general for an immersed surface $\pi : S \to X$ in a Kähler or Stein surface X. (Recall that (10.25) is equivalent to the condition $I_\pm(\pi) \leq 0$ implying the existence of tubular Stein neighborhoods of $\pi(S)$ in X.) However, Nemirovski proved in [429] that the *essential negative double points* do contribute to (10.25). Let us explain this notion. A double point $x \in \pi(S) \subset X$ is the image of two different points $p \neq q \in S$. Let us connect p and q by an arc in S. Its image in X is a loop at x. The double point x is called *essential* if the homotopy class of any such 'double point loop' at x does not belong to the image $\pi_*(\pi_1(S))$ of the fundamental group of S in the fundamental group $\pi_1(X, x)$.

Theorem 10.7.7 (Nemirovski [429, Theorem 4.1]) *Let S be a closed oriented surface of genus g and $\pi: S \to X$ be an immersion with simple double points into a Stein surface X. Assume that π has δ_+ positive double points and δ_-^{ess} essential negative double points. If $\Sigma = \pi(S)$ is not a homotopically trivial 2-sphere in X, then*

$$[\Sigma]^2 + |K_X \cdot \Sigma| + 2\delta_-^{\mathrm{ess}} \leq 2g + 2\delta_+ - 2.$$

In particular, a smoothly embedded homotopically nontrivial 2-sphere $S \hookrightarrow X$ satisfies $[S]^2 + |K_X \cdot S| \leq -2$.

Note that a homotopically trivial surface is also homologically trivial, so Theorem 10.7.7 is stronger than Corollary 10.7.2.

Proof We first consider embedded spheres. Let $h: \widetilde{X} \to X$ be the universal covering of X. Then, $\widetilde{X}$ is a simply connected Stein surface, and h induces an isomorphism $\pi_2(\widetilde{X}) \to \pi_2(X)$. (By a theorem of Stein [525], any covering space of a Stein manifold is Stein.) Assume that $\pi : S \hookrightarrow X$ is a smooth embedded 2-sphere in X whose image $\pi(S)$ is homotopically nontrivial in X. Since S is simply connected, π lifts to an embedding $\tilde{\pi}: S \hookrightarrow \widetilde{X}$, and $\widetilde{S} = \tilde{\pi}(S) \subset \widetilde{X}$ is homotopically nontrivial in $\widetilde{X}$. By the Hurewicz theorem we have $\pi_2(\widetilde{X}) = H_2(\widetilde{X}; \mathbb{Z})$, and hence $\widetilde{S}$ is also homologically nontrivial in $\widetilde{X}$. It is easily seen that $[S]^2 = [\widetilde{S}]^2$ and $K_X \cdot S = K_{\widetilde{X}} \cdot \widetilde{S}$. Thus

$$[S]^2 + |K_X \cdot S| = [\widetilde{S}]^2 + K_{\widetilde{X}} \cdot \widetilde{S} \leq -2,$$

where the last inequality holds by Corollary 10.7.2.

Assume now that $\pi : S \to X$ is an immersed surface with δ_+ positive double points and δ_-^{ess} essential negative double points. We replace each positive double point by a handle as before; this does not change the number $g + \delta_+$ or the homology class of the image $\Sigma = \pi(S)$ in X. Let $h: \widetilde{X} \to X$ be a covering space determined by the condition that $h_*(\pi_1(\widetilde{X})) = \pi_*(\pi_1(S))$. This means that every loop in X which does not come (up to homotopy) from a nontrivial loop in S opens up in $\widetilde{X}$. Then, the immersion $\pi : S \to X$ lifts to an immersion $\tilde{\pi}: S \to \widetilde{X}$ which removes all essential negative double points (since the preimages $p \neq q \in S$ of any such point are mapped by $\tilde{\pi}$ to different levels in $\widetilde{X}$). One can easily see that $\widetilde{\Sigma} = \tilde{\pi}(S) \subset \widetilde{X}$ satisfies $[\widetilde{\Sigma}]^2 = [\Sigma]^2 + 2\delta_-^{\mathrm{ess}}$. Applying Corollary 10.7.2 to $\widetilde{\Sigma} \subset \widetilde{X}$ gives the conclusion. □

Theorem 10.7.7 has several applications. For example, if Y is a smooth open 3-manifold then $X = Y \times \mathbb{R}$ admits a handlebody decomposition without handles of index > 2, and hence X is *homeomorphic* to a Stein surface according to Theorem 10.9.1(b) on p. 517. However, such X is in general not *diffeomorphic* to a Stein surface as shown by the following result.

Corollary 10.7.8 ([429, Corollary 2.1]) *If Y is a smooth open 3-manifold such that $Y \times \mathbb{R}$ is diffeomorphic to a Stein surface, then every smoothly embedded 2-sphere in Y bounds a 3-ball.*

Proof Assume that $Y \times \mathbb{R}$ admits the structure of a Stein surface. Let $S \subset Y$ be an embedded 2-sphere. Then, $S \times \{0\} \subset Y \times \{0\} \subset Y \times \mathbb{R}$ is an embedded 2-sphere with the self-intersection number zero; hence, its homotopy class in $Y \times \mathbb{R}$ is trivial by Theorem 10.7.7. By composing with the projection $Y \times \mathbb{R} \to Y$ we see that S is homotopically trivial in Y. A null-homotopic embedded 2-sphere S in a 3-manifold Y bounds a homotopy 3-ball B in Y (see [283, Proposition 3.10]). Gluing to B a 3-ball along $bB = S$, we get a homotopy 3-sphere Σ. By Perelman's solution of the Poincaré Conjecture (*a simply connected closed 3-manifold is homeomorphic to the 3-sphere*), Σ is homeomorphic to the 3-sphere. Since S is smoothly embedded in Y, S is a locally flat 2-sphere in the 3-sphere Σ, and hence each of the two sides determined by S is homeomorphic to a 3-ball. □

Another application is the following result which is related to the Bedford-Gaveau-Klingenberg-Kruzhilin theorem concerning the envelope of holomorphy of certain embedded 2-spheres in $\mathbb{C}^2$ [47, 353].

Corollary 10.7.9 (Nemirovski [429]) *A smoothly embedded 2-sphere S in a domain $U \subset \mathbb{C}^2$ is homotopically trivial in the envelope of holomorphy of U.*

Proof The envelope of holomorphy of $U \subset \mathbb{C}^2$ is a Riemann domain $\widetilde{U} \to \mathbb{C}^2$ containing a biholomorphic copy of U; this embeds S in $\widetilde{U}$. Note that $[S]^2 = 0$ and hence the adjunction inequality fails (in U, and also in $\widetilde{U}$). Since $\widetilde{U}$ is a Stein surface, Theorem 10.7.7 implies that S is homotopically trivial in $\widetilde{U}$. □

10.8 Well Attached Handles

In this section we prove a key lemma that will be used in the Eliashberg-Gompf construction of integrable Stein manifold structures on a smooth almost complex manifold with a correct handlebody decomposition (see Sect. 10.9). We follow the exposition in the papers [222, 223]. Analogous results in symplectic geometry were obtained earlier by Weinstein [566]. For a more complete exposition of this subject we refer to the monograph by Cieliebak and Eliashberg [99].

Let J_{st} denote the standard complex structure on $\mathbb{C}^n$. For a fixed $k \in \{1, \dots, n\}$ let $z = (z_1, \dots, z_n) = (x' + \mathrm{i}y', x'' + \mathrm{i}y'')$, with $z_j = x_j + \mathrm{i}y_j$, denote the coordinates on $\mathbb{C}^n$ corresponding to the decomposition

$$\mathbb{C}^n = \mathbb{C}^k \oplus \mathbb{C}^{n-k} = \mathbb{R}^k \oplus \mathrm{i}\mathbb{R}^k \oplus \mathbb{R}^{n-k} \oplus \mathrm{i}\mathbb{R}^{n-k}.$$

Let $D = D^k \subset \mathbb{R}^k$ be the closed unit ball in $\mathbb{R}^k$ and $S = S^{k-1} = bD$ its boundary $(k-1)$-sphere. We shall often identify D with its image in $\mathbb{R}^k \oplus \{0\}^{2n-k} \subset \mathbb{C}^n$. We introduce the following (trivial) bundles over D:

$$\nu' = \mathrm{Span}\left\{\frac{\partial}{\partial y_1}, \ldots, \frac{\partial}{\partial y_k}\right\}\bigg|_D = D \times \left(\{0\}^k \oplus \mathbb{R}^k \oplus \{0\}^{2n-2k}\right),$$
$$\nu'' = \mathrm{Span}\left\{\frac{\partial}{\partial x_j}, \frac{\partial}{\partial y_j} : j = k+1, \ldots, n\right\}\bigg|_D = D \times \left(\{0\}^{2k} \oplus \mathbb{R}^{2n-2k}\right),$$
$$\nu = \nu' \oplus \nu'' = D \times \left(\{0\}^k \oplus \mathbb{R}^{2n-k}\right).$$

Clearly, ν is the real normal bundle to D in $\mathbb{C}^n$, and we have

$$\nu' = J_{st}(\mathrm{T}D), \qquad \mathrm{T}^{\mathbb{C}} D = \mathrm{T}D \oplus \nu', \qquad \mathrm{T}\mathbb{C}^n|_D = \mathrm{T}D \oplus \nu = \mathrm{T}^{\mathbb{C}} D \oplus \nu''.$$

Let $\upsilon \to S = bD$ denote the (trivial) real line bundle over S spanned by the radial vector field $\sum_{j=1}^k x_j \frac{\partial}{\partial x_j}$. Over S, we have further decompositions

$$\mathrm{T}D|_S = \upsilon \oplus \mathrm{T}S, \qquad \nu'|_S = J_{st}(\upsilon) \oplus J_{st}(\mathrm{T}S), \qquad \mathrm{T}D|_S \oplus \nu'|_S \simeq \upsilon^{\mathbb{C}} \oplus \mathrm{T}^{\mathbb{C}} S.$$

Note that $\mathrm{T}^{\mathbb{C}} S$ is a trivial complex vector bundle over S.

Let X be a smooth $2n$-dimensional manifold. Given a smooth embedding (or immersion) $G\colon D \to X$ of the disc $D = D^k \subset \mathbb{C}^n$, a *normal framing* over G is a homomorphism $\beta\colon \nu \to \mathrm{T}X|_{G(D)}$ such that

$$dG_x \oplus \beta_x : \mathrm{T}_x D \oplus \nu_x = \mathrm{T}_x \mathbb{C}^n \to \mathrm{T}_{G(x)} X$$

is a linear isomorphism for every $x \in D$.

Assume now that (X, J) is an almost complex manifold of real dimension $2n$. Let W be an open, relatively compact domain with smooth strongly J-pseudoconvex boundary $\Sigma = bW$ in X. Choose a real line subbundle $w \subset \mathrm{T}X|_\Sigma$ such that $Jw \subset \mathrm{T}\Sigma$; this implies

$$\mathrm{T}X|_\Sigma = w \oplus \mathrm{T}\Sigma = w \oplus Jw \oplus \xi, \quad \xi = \mathrm{T}\Sigma \cap J(\mathrm{T}\Sigma).$$

Let $D = D^k$, $S = S^{k-1} = bD$ and υ be as above. An *embedding of a pair* $G\colon (D, S) \hookrightarrow (X \setminus W, \Sigma)$ is a smooth embedding $G\colon D \hookrightarrow X \setminus W$ such that $G(S) = G(D) \cap \Sigma$ and G is transverse to Σ along $G(S)$. Such G is said to be *normal to* Σ if $dG_x(\upsilon_x) = w_{G(x)}$ holds for every $x \in S$, where w is as above. The analogous definition applies to immersions.

The following lemma of Eliashberg [148] is a key result in Stein geometry. This precise formulation is taken from [223, Lemma 3.1].

Lemma 10.8.1 (Existence of well attached handles) *Let W be an open relatively compact domain with smooth strongly J-pseudoconvex boundary $\Sigma = bW$ in an almost complex manifold (X, J). Let $1 \le k \le n = \dim_{\mathbb{C}} X$, $D = D^k \subset \mathbb{R}^k$ be the*

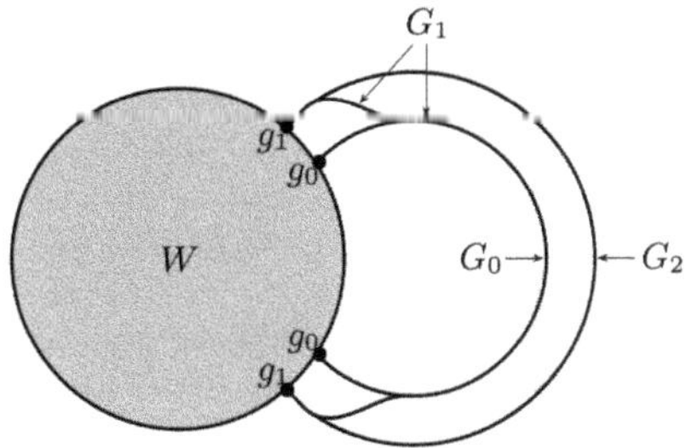

Fig. 10.1 Deformations of an attached disc. ([223, p. 622, Fig. 2])

closed k-disc, and $S = bD \cong S^{k-1}$. Given a smooth embedding $G_0\colon (D, S) \hookrightarrow (X \setminus W, \Sigma)$, there exists a regular homotopy of immersions $G_t\colon (D, S) \to (X \setminus W, \Sigma)$ ($t \in [0, 1]$) that is $\mathcal{C}^0$ close to G_0 such that the immersion $G_1\colon D \to X \setminus W$ is J-totally real and normal to Σ, and $g_1 = G_1|_S\colon S \hookrightarrow \Sigma$ is a Legendrian (complex tangential) embedding. If $k < n$, or if $k = n \neq 2$, then there exists an isotopy of embeddings G_t with these properties. If J is integrable in a neighborhood of $\overline{W}$ and $\Sigma = bW$ is real analytic, then G_1 can be chosen real analytic.

Lemma 10.8.1 shows in particular that one can attach an embedded totally real disc of any dimension $\le n = \dim X \neq 2$ to a strongly pseudoconvex domain $\overline{W}$ in a Stein manifold X along a complex tangential sphere in bW. When coupled with Theorem 3.9.2, this lemma is a strong tool for constructing Stein structures and will be exploited in Sect. 10.9. The main point is that, after a small perturbation of the set $\overline{W} \cup G_1(D)$ and straightening the disc $G_1(D)$, we can obtain the standard situation described in Lemma 3.9.3; hence, the Stein structure on a neighborhood of $\overline{W}$ extends to a Stein structure on a strongly pseudoconvex handlebody around $\overline{W} \cup M$.

Remark 10.8.2 As pointed out by Eliashberg [148], the framing obstruction in dimension $n = 2$ is essential. In particular, it is impossible to attach an embedded totally real 2-disc $D \subset \mathbb{C}^2 \setminus \mathbb{B}$ to the ball $\mathbb{B} \subset \mathbb{C}^2$ along a complex tangential curve $C \subset b\mathbb{B}$. Indeed, if this were possible, we would obtain a strongly pseudoconvex domain in $\mathbb{C}^2$ containing an embedded homologically essential sphere with self-intersection number zero (glue a 2-disc $D' \subset \mathbb{B}$ smoothly to D along $bD' = C = bD$). This contradicts Corollary 10.7.2. □

Proof We follow the construction in [148, Sect. 2] of a *special handle attaching triple*, with an essential additional argument in the critical case $k = n \neq 2$. (The details are taken from [223, Lemma 3.1].) First, we find a regular homotopy from the initial disc $G_0\colon D \hookrightarrow X \setminus W$ to an immersed disc $G_1\colon D \to X \setminus W$ which is attached to W with a correct normal framing along an embedded Legendrian sphere in bW. Next, we apply the h-principle for totally real immersions to deform G_1 by a regular homotopy that is fixed near the boundary to a totally real immersed disc G_2 (see Fig. 10.1). Finally, we show that, unless $k = n = 2$, the construction can be done by isotopies of embeddings.

Set $g_0 = G_0|_S : S \hookrightarrow bW$. By a correction of G_0 along S (keeping g_0 fixed) we may assume that it is normal to Σ, i.e., such that $\ell_0 := \mathrm{d}G_0|_v$ maps v to $w|_{g_0(S)}$. Choose a complex vector bundle isomorphism

$$\phi_0 : \mathrm{T}\mathbb{C}^n|_D = D \times \mathbb{C}^n \to \mathrm{T}X|_{G_0(D)}, \qquad \phi_0 \circ J_{st} = J \circ \phi_0$$

covering G_0. We use the coordinates on $\mathbb{C}^n$ introduced at the beginning of the section. The vector field $\tau = \sum_{j=1}^k x_j \frac{\partial}{\partial x_j}$ is the outer radial field to the sphere $S = bD$ in $\mathbb{R}^k \times \{0\}^{2n-k}$. Let $\widetilde{\tau}$ be the unique nonvanishing vector field on $\mathbb{C}^n$ over S satisfying $\phi_0(\widetilde{\tau}_x) = \ell_0(\tau_x)$ for every $x \in S$. For dimensional reasons there exists a map $A : D \to GL_n(\mathbb{C})$ satisfying $A_x \tau_x = \widetilde{\tau}_x$ for all $x \in S$. Replacing ϕ_0 by $\phi_0 \circ A$, we may assume from now on that $\phi_0|_v = \ell_0$. A further homotopic correction of ϕ_0 ensures that

$$\phi_0(\mathrm{T}^{\mathbb{C}} S \oplus \nu''|_S) = \xi|_{g_0(S)},$$

thereby providing a trivialization of the latter bundle.

Write $\phi_0 = \phi_0' \oplus \phi_0''$, where $\phi_0' = \phi_0|_{\mathrm{T}^{\mathbb{C}} D}$ and $\phi_0'' = \phi_0|_{\nu''}$. (We are using the notation introduced before Lemma 3.9.3.) Setting $\psi_0 := \phi_0|_{\mathrm{T}^{\mathbb{C}} S}$, we have

$$\phi_0'|_{\mathrm{T}^{\mathbb{C}} D|_S} = \ell_0^{\mathbb{C}} \oplus \psi_0 : v^{\mathbb{C}} \oplus \mathrm{T}^{\mathbb{C}} S \longrightarrow w^{\mathbb{C}} \oplus \xi|_{g_0(S)} = \mathrm{T}X|_{g_0(S)}.$$

Note that $\psi_0 \oplus \phi_0'' : \mathrm{T}^{\mathbb{C}} S \oplus \nu''|_S \to \xi|_{g_0(S)}$ is a complex vector bundle isomorphism. Furthermore, there is a homotopy of real vector bundle monomorphisms $\iota_s : \mathrm{T}D \hookrightarrow \mathrm{T}X|_{G_0(D)}$ $(s \in [0, 1])$ satisfying

$$\iota_0 = \mathrm{d}G_0, \qquad \iota_1 = \phi_0|_{\mathrm{T}D}, \qquad \iota_s|_v = \ell_0 : v \to w|_{g_0(S)} \quad \big(s \in [0, 1]\big).$$

Consider the pair (g_0, ψ_0) consisting of the embedding $g_0 : S \hookrightarrow \Sigma$ and the $\mathbb{C}$-linear embedding $\psi_0 : \mathrm{T}^{\mathbb{C}} S \hookrightarrow \xi|_{g_0(S)}$ over g_0 of the complexified tangent bundle of S (a trivial complex vector bundle of rank $k - 1$) into the contact subbundle $\xi \subset \mathrm{T}\Sigma$. By Gromov's Legendrization theorem [265, p. 339, (B')], there exists a Legendrian embedding $g_1 : S \hookrightarrow \Sigma$ whose complexified differential $\psi_1 := \mathrm{d}^{\mathbb{C}} g_1$ is homotopic to ψ_0 by a homotopy of $\mathbb{C}$-linear vector bundle embeddings $\psi_t : \mathrm{T}^{\mathbb{C}} S \hookrightarrow \xi$ $(t \in [0, 1])$.

Let $\mathrm{Hom}_{\mathrm{inj}}(\mathrm{T}S, \mathrm{T}\Sigma)$ denote the space of fibrewise injective real vector bundle maps $\mathrm{T}S \to \mathrm{T}\Sigma$. Consider the path in $\mathrm{Hom}_{\mathrm{inj}}(\mathrm{T}S, \mathrm{T}\Sigma)$ from $\mathrm{d}g_0$ to $\mathrm{d}g_1$ consisting of the homotopy $\iota_s|_{\mathrm{T}S}$ $(s \in [0, 1])$, followed by the homotopy $\psi_t|_{\mathrm{T}S}$ $(t \in [0, 1])$ (left and top side of the square on Fig. 10.2). By Hirsch's one-parametric h-principle for immersions [265, 295], this path can be deformed in $\mathrm{Hom}_{\mathrm{inj}}(\mathrm{T}S, \mathrm{T}\Sigma)$ (with fixed ends) to a path of differentials $\mathrm{d}g_t : \mathrm{T}S \hookrightarrow \mathrm{T}\Sigma|_{g_t(S)}$, where $g_t : S \to \Sigma$ $(t \in [0, 1])$ is a regular homotopy of immersions from g_0 to g_1. We can ensure that ψ_t covers the base map g_t for all $t \in [0, 1]$. This gives a homotopy $\theta_{t,s} \in \mathrm{Hom}_{\mathrm{inj}}(\mathrm{T}S, \mathrm{T}\Sigma)$ for $(t, s) \in [0, 1]^2$ satisfying the following conditions (see Fig. 10.2):

(i) $\theta_{t,0} = \mathrm{d}g_t$ (bottom side),
(ii) $\theta_{t,1} = \psi_t|_{\mathrm{T}S}$ (top side),

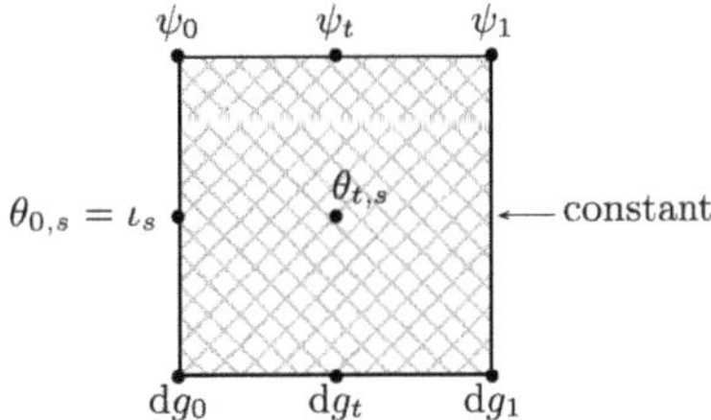

Fig. 10.2 The homotopy $\theta_{t,s}$. ([223, p. 623, Fig. 3])

(iii) $\theta_{0,s} = \iota_s|_{TS}$ (left side; hence $\theta_{0,0} = dg_0$ and $\theta_{0,1} = \psi_0|_{TS}$),
(iv) $\theta_{1,s} = dg_1$ (right side), and
(v) $\theta_{t,s}$ covers g_t for every $t, s \in [0, 1]$.

We can extend g_t to a regular homotopy $G_t\colon (D, S) \to (X \setminus W, \Sigma)$ $(t \in [0, 1])$ consisting of immersions normal to Σ, beginning at $t = 0$ with the map G_0. Let $\ell_t := dG_t|_v\colon v \to w|_{g_t(S)}$. By the homotopy lifting theorem there exists a homotopy of $\mathbb{C}$-linear complex vector bundle isomorphisms ϕ_t covering G_t,

$$\phi_t = \phi_t' \oplus \phi_t'' : T\mathbb{C}^n|_D = T^{\mathbb{C}}D \oplus \nu'' \longrightarrow TX|_{G_t(D)}, \quad t \in [0, 1],$$

beginning at $t = 0$ with the map ϕ_0, such that over $S = bD$ we have

$$\phi_t' = \ell_t^{\mathbb{C}} \oplus \psi_t, \quad t \in [0, 1],$$

and $dG_1 = \phi_1$ on $TD|_S$. Set $\widetilde{\theta}_{t,s} = \ell_t \oplus \theta_{t,s}\colon TD|_S \hookrightarrow TX|_{g_t(S)}$ for $t, s \in [0, 1]$; this is a real vector bundle monomorphism over the map g_t. From the above properties (i)–(v) of $\theta_{t,s}$ we obtain the following:

(i′) $\widetilde{\theta}_{t,0} = \ell_t \oplus dg_t = dG_t|_{TD|_S}$ (bottom side),
(ii′) $\widetilde{\theta}_{t,1} = \ell_t \oplus \psi_t|_{TS} = \phi_t|_{TD|_S}$ (top side),
(iii′) $\widetilde{\theta}_{0,s} = \iota_s|_{TD|_S}$ (left side),
(iv′) $\widetilde{\theta}_{1,s} = \ell_1 \oplus dg_1 = dG_1|_{TD|_S}$ (right side), and
(v′) $\widetilde{\theta}_{t,s}$ covers g_t for every $t, s \in [0, 1]$.

We wish to extend the monomorphisms $\widetilde{\theta}_{t,s}\colon TD|_S \hookrightarrow TX|_{g_t(S)}$ to real vector bundle monomorphisms $\Theta_{t,s}\colon TD \to TX$ $(t, s \in [0, 1])$ covering the immersions $G_t\colon D \hookrightarrow X$. Such an extension already exists for (t, s) in the bottom, top and left face of the parameter square $[0, 1]^2$ where we take dG_t, $\phi_t|_{TD}$ and ι_s, respectively (see properties (i′), (ii′) and (iii′)). The homotopy lifting property provides an extension $\Theta_{t,s}$ for all $(t, s) \in [0, 1]^2$ with the given boundary values on the bottom, top and left side of $[0, 1]^2$. (See Fig. 10.3. The front and the back face belong to the homotopy $\widetilde{\theta}_{t,s}$ over $S = bD$; compare also with Fig. 10.2.) Over the right face $\{t = 1\}$ we thus obtain a homotopy $\Theta_{1,s} \in \mathrm{Hom_{inj}}(TD, TX|_{G_1(D)})$ $(s \in [0, 1])$ satisfying

$$\Theta_{1,0} = dG_1 : TD \to TX|_{G_1(D)}, \qquad \Theta_{1,1} = \phi_1|_{TD} : TD \to TX|_{G_1(D)}.$$

The homotopy $\Theta_{1,s}$ is fixed over S where it coincides with $\widetilde{\theta}_{1,s} = dG_1|_{TD|_S}$ by property (iv′). (In Fig. 10.3, the map $\Theta_{1,s}$ appears on the right face of the cube, with bold vertical sides indicating that it is constant on $TD|_S$ where it equals $\ell_1 \oplus dg_1$.)

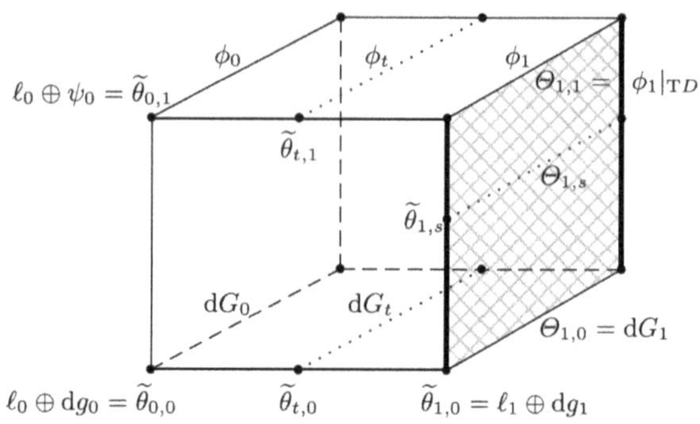

Fig. 10.3 The homotopy $\Theta_{t,s}$. ([223, p. 624, Fig. 4])

Since $\phi_1 \colon T\mathbb{C}^n|_D \to TX|_{G_1(D)}$ is a $\mathbb{C}$-linear vector bundle isomorphism, the h-principle for totally real immersions (see Sect. 9.1) provides a regular homotopy of immersions $G_t \colon D \to X \setminus W$ $(t \in [1, 2])$ that is fixed near S such that G_2 is J-totally real and its complexified differential $d^{\mathbb{C}} G_2$ is homotopic to ϕ_1 in the space of $\mathbb{C}$-linear maps $T\mathbb{C}^n|_D \to TX$ of maximal rank. If in addition G_1 is an embedding, then we can deform it to a totally real embedding G_2 by an isotopy that is fixed near S. This follows from the fact that totally real embeddings also satisfy the h-principle (see Sect. 9.1). For $k = n = 2$ this also follows from Theorem 10.4.1 (p. 489) on cancellation of pairs of complex points of real surfaces in complex surfaces.

Finally, we reparametrize the family $\{G_t : t \in [0, 2]\}$ back to the parameter interval $[0, 1]$ and obtain a regular homotopy with the required properties.

It remains to show that, unless $k = n = 2$, there also exists an *isotopy of embeddings* $\{G_t\}$ with these properties. If $k < n$, a small perturbation of $\{g_t\}$ with fixed ends at $t = 0, 1$ gives an isotopy which can be realized by an ambient diffeotopy, and we get an isotopy of embedded discs $G_t \colon D \hookrightarrow X \setminus W$ with $G_t|_S = g_t$. For $k = n = 1$ the conclusion of Lemma 10.8.1 obviously holds for any attached 1-disc (segment).

It remains to consider the critical case $k = n > 2$.

A generic choice of the isotopy $g_t \colon S \cong S^{n-1} \to \Sigma$ ensures that g_t is an embedding for all but finitely many values $t \in [0, 1]$, and it has a simple double point at each of the exceptional values of the parameter. We wish to change the Legendrian embedding g_1 by a regular homotopy of Legendrian immersions $g_t \colon S \hookrightarrow \Sigma$ $(t \in [1, 2])$ to another Legendrian embedding g_2 so that the resulting regular homotopy $\{g_t : t \in [0, 2]\}$ will have self-intersection index zero. More precisely, the map

$$\widetilde{g} \colon \widetilde{S} = S \times [0, 2] \to \widetilde{\Sigma} = \Sigma \times [0, 2], \quad \widetilde{g}(x, t) = \big(g_t(x), t\big)$$

is an immersion of the n-dimensional oriented manifold $\widetilde{S}$ into the $2n$-dimensional oriented manifold $\widetilde{\Sigma}$ such that the double points of $\widetilde{g}$ correspond to the double points of the regular homotopy $\{g_t\}$. We define the *index* $i(\{g_t\})$ as the number of double points of $\widetilde{g}$ counted with the orientation signs.

If this index equals zero, then a foliated version of the Whitney trick [572] allows us to deform $\{g_t\}_{t\in[0,2]}$ with fixed ends to an isotopy of embeddings. This is done by connecting a chosen pair of double points $q_0, q_1 \in \widetilde{g}(\widetilde{S})$ of opposite signs, lying over two different values $t_0 < t_1$ of the parameter, by a pair of curves

$$\lambda_j(t) = \widetilde{g}\big(c_j(t), t\big), \quad t \in [t_0, t_1], \ j = 1, 2$$

which together bound an embedded Whitney disc $D^2 \subset \widetilde{\Sigma}$ such that $D^2 \cap (\Sigma_t \times \{t\})$ is an arc connecting $\lambda_1(t)$ to $\lambda_2(t)$ for every $t \in [t_0, t_1]$, and it degenerates to q_0 and q_1 over the endpoints t_0, t_1, respectively. The rest of the procedure, removing this pair of double points by pulling $\widetilde{g}(\widetilde{S})$ across D^2, is standard, see e.g. [572]. Performing this operation finitely many times, one can remove all double points and change $\{g_t\}$ to an isotopy of embeddings. The rest of the proof can be completed exactly as before: We extend g_t to an isotopy of embedded discs $G_t \colon D \hookrightarrow X \setminus W$, with $G_t|_S = g_t$, covered by a homotopy of $\mathbb{C}$-linear isomorphisms $\phi_t \colon T\mathbb{C}^n|_D \to TX|_{G_t(D)}$. Note that $\{dg_t\}$ still has the correct homotopy property so that the final embedding G_2 can be deformed (with fixed boundary) to a totally real embedding.

To complete the proof, we now show that the index $i(\{g_t\})$ can be changed to zero by a Legendrian isotopy in Σ which introduces the correct number of double points. (A similar deformation is used in [148, Sect. 2.4]. It is here that the hypothesis $n \neq 2$ is used.)

The set $L = g_1(S)$ is an embedded Legendrian sphere in Σ. Choose a point $a \in L$. In suitable local coordinates $(z, q, p) \in \mathbb{R}^{2n-1}$ on Σ, with a corresponding to $0 \in \mathbb{R}^{2n-1}$, the contact form is $\eta = dz - \sum_{j=1}^{n-1} p_j dq_j$ and

$$L = \left\{ (z, q, p) \in \mathbb{R}^{2n-1} : z^2 = q_1^3, \ p_1^2 = \frac{9}{4} q_1, \ p_2 = \cdots = p_{n-1} = 0 \right\}$$

(see [148, Sect. 2.4]). Let $\pi \colon \mathbb{R}^{2n-1} \to \mathbb{R}^{n-1}$ denote the projection $\pi(z, q, p) = q$. Choose a small number $q_1^0 > 0$ and let $\Delta \subset \mathbb{R}^{n-1}$ denote the closed ball of radius $q_1^0/2$ centered at $(q_1^0, 0, \ldots, 0)$. Let $\phi \colon \Delta \to \mathbb{R}$ be a smooth function that equals 0 near $b\Delta$. Set

$$h_t(q) = q_1^{3/2}\big(1 + (t-1)\phi(q)\big), \quad t \in [1, 2].$$

Let L_t equal L outside of $\pi^{-1}(\Delta)$ and equal

$$\left\{ (z, q, p) : z = h_t(q), \ p = \frac{\partial h_t}{\partial q}(q) \right\} \cup \left\{ (z, q, p) : z = -h_t(q), \ p = -\frac{\partial h_t}{\partial q}(q) \right\}$$

over Δ. We choose ϕ with sufficiently small derivative to ensure that we remain in the given coordinate patch; this can be done if $q_1^0 > 0$ is chosen small enough. Let $g_t \colon S \to \Sigma$ $(t \in [1, 2])$ be the regular homotopy such that $g_t(S) = L_t$. The deformation is illustrated by Fig. 10.4. The top diagrams show the projection onto the (z, q)-plane at three typical stages, with the cusp at $(z, q) = (0, 0)$ and with a self-intersection shown in the middle figure.

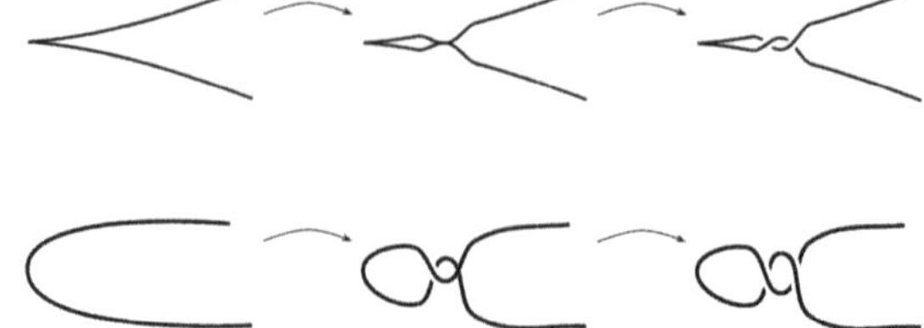

Fig. 10.4 Changing the index of a regular homotopy by +1. ([223, p. 626, Fig. 5])

The index of $\{g_t\}_{t\in[1,2]}$ equals the intersection number between the manifolds $M_\pm$ defined by

$$M_\pm = \left\{(z,q,p,t) \in \mathbb{R}^{2n} : z = \pm h_t(q),\ p = \pm\frac{\partial h_t}{\partial q}(q),\ q \in \Delta,\ t \in [1,2]\right\}.$$

The intersection points of M_+ and M_- are solutions of the equations

$$1 + (t-1)\phi = 0, \qquad \frac{\partial \phi}{\partial q} = 0, \quad t \in [1,2].$$

This is precisely the set of all critical points of ϕ with the critical values belonging to $(-\infty, -1]$. By a generic choice of ϕ we can ensure that -1 is not a critical value. A computation shows that each point (q,t) satisfying the above equations adds ± 1 to the index $i(\{g_t\})$, depending on the sign of the determinant of the Hessian of ϕ at q; hence we get $+1$ at a critical point of even Morse index and -1 at a critical point of odd Morse index. Similarly, as we increase $c \in \mathbb{R}$, the Euler characteristic of the sublevel set $\{\phi \le c\}$ increases by one at every critical point of ϕ of even Morse index, and it decreases by one at every critical point of odd Morse index.

We conclude that $i = i(\{g_t\}_{t\in[1,2]})$ equals the Euler number of the set $\{q \in \Delta : \phi(q) \le -1\}$. If $n > 2$, this can be arranged to equal any preassigned integer by a suitable choice of ϕ, and hence we can arrange the index i to equal zero. If $n = 2$ then the index i can be arranged to be any nonnegative number (since $\{\phi \le -1\}$ is a union of segments), but it cannot be negative. This completes the proof of Lemma 10.8.1 in the smooth case.

Assume now that J is integrable in a neighborhood of $\overline{W}$ in X and the hypersurface $\Sigma = bW$ is real analytic near the attaching sphere $G_0(S) \subset \Sigma$. We wish to find a real analytic disc satisfying Lemma 10.8.1. Since the embedded disc G_1 constructed above can be chosen arbitrarily $\mathcal{C}^0$-close to G_0, we may assume that the same conditions on J and Σ also hold near $G_1(D)$. By Gray's theorem on real analytic approximation of Legendrian embeddings (see [148, Lemma 2.5.1]) it is possible to approximate G_1 in the $\mathcal{C}^1$ topology by an embedded disc $G_1' \colon (D,S) \hookrightarrow (X \setminus W, \Sigma)$ which is real analytic near S such that $G_1'|_S \colon S \hookrightarrow \Sigma$ is Legendrian and G_1' is normal to Σ along S. It remains to perturb G_1' to a nearby real analytic embedding $G_1'' \colon D \hookrightarrow X$ that agrees with G_1' to the second order along S and to replace G_1 by G_1''. □

10.9 Stein Structures and the Soft Oka Principle

Recall from Sect. 3.12 that every n-dimensional Stein manifold admits a handlebody decomposition without handles of index $> n$, induced by a strongly plurisubharmonic Morse exhaustion function. The following seminal result in the opposite direction is due to Eliashberg (for $n \neq 2$) and Gompf (for $n = 2$).

Theorem 10.9.1 *Assume that (X, J) is a smooth almost complex manifold of real dimension $2n$ admitting a Morse exhaustion function $\rho \colon X \to \mathbb{R}$ without critical points of index $> n$.*

(a) [148] *If $n \neq 2$, then J is homotopic to an integrable Stein structure.*
(b) [246] *If $n = 2$, then there exists a Stein surface $(\widetilde{X}, \widetilde{J})$ and an orientation preserving homeomorphism $h \colon X \to \widetilde{X}$ such that $h^*\widetilde{J}$ is homotopic to J (i.e., (a) holds for a possibly exotic Stein structure on X).*

More precisely, (a) asserts that there exists a homotopy $\{J_t\}_{t\in[0,1]}$ of smooth almost complex structure on X connecting the given structure $J_0 = J$ to an integrable Stein structure $J_1 = \widetilde{J}$. The existence of a function ρ in the theorem is equivalent to asking that X admit a smooth handlebody decomposition with handles of index at most $n = \dim_{\mathbb{C}} X$ (i.e., half of the real dimension). Eliashberg also showed that $\widetilde{J}$ can be chosen such that the sublevel sets $\{x \in X : \rho(x) < c\}$ are strongly $\widetilde{J}$-pseudoconvex. Regarding part (b), the pullback $h^*\widetilde{J}$ of an almost complex structure by a homeomorphism $h \colon X \to \widetilde{X}$ gives a well-defined homotopy class of almost complex structures on X [246, p. 645]. A change of the smooth structure on X is necessary in general as shown by Examples 10.7.3 and 10.7.4 on p. 506.

Theorem 10.9.1 is proved in Sect. 10.10 for $n \neq 2$, or if $n = 2$ and X is a handlebody without handles of index > 1. The case $\dim_{\mathbb{R}} X = 4$ is discussed in Sect. 10.11. For $n \neq 2$ the result follows easily from Lemma 10.8.1 (providing well attached totally real handles) and Theorem 3.9.2 (p. 91) which enables one to extend the Stein structure across the handle.

When $\dim_{\mathbb{R}} X = 4$, the framing obstruction in Lemma 10.8.1 cannot be avoided in general, and one must use *kinky handles*; these are immersed 2-handles with a transverse double point (see Fig. 10.5). The superfluous double point loops are cancelled by again attaching kinky handles, and so on. The resulting procedure converges to an attached *Casson handle* which is homeomorphic, but not diffeomorphic, to the standard 4-handle $D^2 \times \mathrm{Int} D^2$, thereby changing the underlying $\mathcal{C}^\infty$ structure (see Freedman and Quinn [231]).

We now describe a certain soft version of the Oka principle which fits in the context of Theorem 10.9.1.

We have seen in Chap. 5 that it is only rarely possible to find a holomorphic representative in a given homotopy class of maps $X \to Y$ from a Stein manifold X to a complex manifold Y. The main problem is holomorphic rigidity of Y; the manifold X may be holomorphically 'too big' to fit into Y, at least in some homotopy classes of maps. To some extent, this obstruction can be explained by the distance decreasing property of holomorphic maps in standard biholomorphically invariant metrics (such as the Kobayashi metric). We now show that the source Stein manifold X can always be made holomorphically 'smaller' so as to accommodate a holomorphic map $X \to Y$ in any given homotopy class, while still retaining its essential topological (and smooth in dimension $\neq 2$) characteristics. We give two different formulations; the first one fits in the context of Theorem 10.9.1. There is no restriction whatsoever on the target manifold Y in Theorems 10.9.2 and 10.9.3.

Theorem 10.9.2 (The soft Oka principle, [223, Theorem 1.1]) *Let (X, J) be a smooth almost complex manifold of real dimension $2n$ admitting a Morse exhaustion function $\rho\colon X \to \mathbb{R}$ without critical points of index $> n$. Let $f\colon X \to Y$ be a continuous map to a complex manifold (Y, J_Y).*

(i) *If $\dim_{\mathbb{C}} X \neq 2$, there exist an integrable Stein structure $\widetilde{J}$ on X, homotopic to J, and a $(\widetilde{J}, J_Y)$-holomorphic map $\widetilde{f}\colon X \to Y$ homotopic to f.*
(ii) *If $\dim_{\mathbb{C}} X = 2$, then there exist an orientation preserving homeomorphism $h\colon X \to X'$ onto a Stein surface X' and a holomorphic map $f'\colon X' \to Y$ such that the map $\widetilde{f} = f' \circ h\colon X \to Y$ is homotopic to f.*

In each of these two cases, a family of maps $f_p\colon X \to Y$ depending continuously on the parameter p in a compact Hausdorff space can be deformed to a family of holomorphic maps with respect to some Stein structure $\widetilde{J}$ on X that is homotopic to J.

If (X, J) is an integrable complex manifold with a suitable handlebody structure, we can obtain a holomorphic map in a given homotopy class on a Stein domain $\Omega \subset X$ that is diffeotopic (homeotopic if $\dim_{\mathbb{R}} X = 4$) to X:

Theorem 10.9.3 ([223, Theorem 1.2]) *Let X be a complex manifold with a Morse exhaustion function without critical points of index $> n = \dim_{\mathbb{C}} X$. Let P be a compact Hausdorff space and $f\colon X \times P \to Y$ be a continuous map to a complex manifold Y.*

(i) *If $n \neq 2$, or if $n = 2$ and ρ has no critical points of index > 1, there is a Stein domain Ω in X, a diffeomorphism $h\colon X \to \Omega$ diffeotopic to Id_X, and a map $f'\colon \Omega \times P \to Y$ such that $f'_p = f'(\cdot, p)\colon \Omega \to Y$ is holomorphic for every $p \in P$ and the map $f' \circ h\colon X \times P \to Y$ is homotopic to f.*
(ii) *If $n = 2$, then the conclusion in* (i) *holds for a homeomorphism $h\colon X \to \Omega$ that is homeotopic to the identity map on X.*

Let us show that Theorem 10.9.2 follows immediately from Theorems 10.9.1 and 10.9.3. By Theorem 10.9.1 we can change J to an integrable Stein structure on X. Let $h_t\colon X \to h_t(X) \subset X$ $(t \in [0, 1])$ be a smooth diffeotopy from $h_0 = \mathrm{Id}_X$ to $h_1 = h\colon X \to \Omega$, furnished by Theorem 10.9.3. Then, $J_t = h_t^*(J)$ is a homotopy of integrable complex structures on X connecting $J = J_0$ to the Stein structure $J_1 = h^*(J|_{T\Omega})$, and $f' \circ h\colon X \to Y$ is a (J_1, J_Y)-holomorphic map that is homotopic to f.

Remark 10.9.4 Theorems 10.9.2 and 10.9.3 were proved by Slapar and the author in 2007 [223]; the proofs given here follow the original ones and are the same as in the first edition of this book. Meanwhile, Cieliebak and Eliashberg showed in their monograph [99] (2012) that the constraint $n \neq 2$ in these theorems is unnecessary provided that (X, J) is an integrable Stein manifold. In this case, there exists a homotopy J_t $(t \in [0,1])$ of integrable Stein structures on X, with $J_0 = J$, and a homotopy $f_t \colon X \to Y$ $(t \in [0,1])$ of continuous maps with $f_0 = f$, such that f_1 is (J_1, J_Y)-holomorphic. (See [99, Theorem 8.43 and Remark 8.44].) The same remark pertains to the remaining results in this section. This shows that Problems 9.10.5 and 9.10.6 in the first edition of this book have affirmative answers. These improvements depend on a stronger technical result given by [99, Theorem 8.5, p. 157]. □

The following example illustrates the failure of Theorems 10.9.2 and 10.9.3 for noncompact families of maps $X \to Y$.

Example 10.9.5 (Mappings of annuli) Let $X = A_r = \{z \in \mathbb{C} : \frac{1}{r} < |z| < r\}$ for some $r > 1$, and let $Y = A_R$ for another $R > 1$. We have $[X, Y] = \pi_1(S^1) = \mathbb{Z}$, and the homotopy class of a map $f \colon X \to Y$ is determined by its winding number. A homotopy class represented by $k \in \mathbb{Z}$ admits a holomorphic representative if and only if $r^{|k|} \le R$, and in this case a representative is $z \to z^k$. Since every complex structure on an annulus is biholomorphic to A_r for some $r > 1$, we see that at most finitely many homotopy classes of maps between a pair of annuli contain a holomorphic map. The conclusion of Theorem 10.9.2 is obtained in this example by decreasing the value of $r > 1$ to another value satisfying $r^{|k|} \le R$; this amounts to a smooth deformation of the complex structure on X. This allows us to simultaneously deform any compact family of maps $X \to Y$ to a family of holomorphic maps, but it is impossible to do the same for a sequence of maps in different homotopy classes.

The same phenomenon occurs whenever $\pi_1(Y)$ contains an element $[\alpha]$ of infinite order such that the minimal Kobayashi length l_N of loops in Y representing the class $N[\alpha] \in \pi_1(Y)$ tends to $+\infty$ as $N \to +\infty$. □

For a Stein source manifold we have the following improvement of Theorem 10.9.3 which includes the approximation and interpolation.

Theorem 10.9.6 ([222, Theorems 1.2 and 4.1]) *Assume that X is a Stein manifold, A is a closed complex subvariety of X, and $K \subset X$ is a compact $\mathcal{O}(X)$-convex subset. Let Y be a complex manifold and $f \colon X \to Y$ be a continuous map that is holomorphic in a neighborhood of $A \cup K$. If $\dim_{\mathbb{C}} X \neq 2$, then there exist a Stein domain $\Omega \subset X$ containing $A \cup K$, a holomorphic map $\widetilde{f} \colon \Omega \to Y$, and a diffeomorphism $h \colon X \to \Omega$ which is diffeotopic to Id_X by a diffeotopy that is fixed on a neighborhood $A \cup K$ such that $\widetilde{f}|_A = f|_A$, $\widetilde{f}$ approximates f uniformly on K as closely as desired, and the map $\widetilde{f} \circ h \colon X \to Y$ is homotopic to f relative to $A \cup K$. If $\dim_{\mathbb{C}} X = 2$, then the same conclusion holds for a homeomorphism h that is homeotopic to Id_X.*

Our construction also gives holomorphic maps of maximal rank (immersions or submersions) in the absence of topological obstructions. The following result is a simplified version of [223, Theorem 6.3].

Theorem 10.9.7 (Soft Oka principle for maps of pointwise maximal rank) *Let X be a Stein manifold with $\dim_{\mathbb{C}} X \neq 2$ and $f\colon X \to Y$ be a continuous map to a complex manifold Y. If there is a complex vector bundle map $\iota\colon \mathrm{T}X \to f^*(\mathrm{T}Y)$ of fibrewise maximal rank, then there exist a Stein structure $\widetilde{J}$ on X, homotopic to J, and a $\widetilde{J}$-holomorphic map $\widetilde{f}\colon X \to Y$ of pointwise maximal rank which is homotopic to f. The same holds if $\dim X = 2$ and X admits a Morse exhaustion function without critical points of index > 1.*

This shows that Theorem 9.13.1 on p. 462 (the h-principle for holomorphic submersions) holds without any assumptions on the target manifold Y if we allow a homotopic change of the Stein structure on the source manifold X. In particular, taking $Y = \mathbb{C}^n$ with $n = \dim_{\mathbb{C}} X$ we obtain the following corollary which shows that every parallelizable Stein manifold becomes a Riemann domain over $\mathbb{C}^n$ after a homotopic change of its Stein structure. (Compare with Problem 9.13.3 on p. 462.)

Corollary 10.9.8 *If (X, J) is a Stein manifold of dimension $n \neq 2$ whose holomorphic tangent bundle $\mathrm{T}X$ is trivial, then there are a Stein structure $\widetilde{J}$ on X, homotopic to J, and a $\widetilde{J}$-holomorphic immersion $\pi\colon X \to \mathbb{C}^n$.*

Theorem 10.9.6 was extended by Prezelj and Slapar [463] to 1-convex source manifolds X, provided that the initial map $f\colon X \to Y$ is holomorphic in a neighborhood of the maximal compact subvariety of X.

The results stated in this section are proved in the following two sections. Theorem 3.8.1 is used to obtain a holomorphic extension of the map across the handle at every inductive step. The essential difference from the standard Oka principle in Chaps. 5 and 6 is that one cannot enlarge the domain of existence of a holomorphic map in the absence of a Runge approximation property for maps $\mathbb{C}^n \to Y$, so we only get a holomorphic map on a thin Stein handlebody $\Omega \subset X$ that is diffeotopic (resp. homeotopic) to X.

10.10 The Case $\dim_{\mathbb{R}} X \neq 4$

In this section we prove the results stated in the previous section when $\dim_{\mathbb{R}} X \neq 4$, or when X is a 4-manifold without handles of index > 1.

Let P denote a compact Hausdorff space. A *P-map* $X \to Y$ is a continuous map $f\colon X \times P \to Y$. If X and Y are complex manifolds, then f is a *holomorphic P-map* if $f_p = f(\cdot, p)\colon X \to Y$ is holomorphic for every $p \in P$. (Compare with Definition 6.6.3.) The following result includes Theorem 10.9.1(a) and Theorem 10.9.2(i) as special cases. Except for the fact that we also find a holomorphic map, the construction is essentially the one of Eliashberg [148].

Theorem 10.10.1 ([223, Theorem 6.1]) *Let (X, J) be a smooth almost complex manifold of real dimension $2n$ endowed with a Morse exhaustion function $\rho: X \to \mathbb{R}$ without critical points of index $> n$. Assume that for some $c \in \mathbb{R}$ the structure J is integrable in $X_c = \{x \in X : \rho(x) < c\}$ and ρ is strongly J-plurisubharmonic in X_c. Let Y be a complex manifold with a complete distance function* dist_Y *and $f: X \times P \to Y$ be a P-map which is J-holomorphic in X_c. If $n \neq 2$, or if $n = 2$ and ρ has no critical points of index > 1 in the set $\{x \in X : \rho(x) \geq c\}$, then for every compact set $K \subset X_c$ and $\epsilon > 0$ there exist a Stein structure $\widetilde{J}$ on X and a homotopy of P-maps $f^t: X \times P \to Y$ $(t \in [0, 1])$ such that $f^0 = f$ and the following hold:*

(a) *the P-map $\widetilde{f} := f^1$ is $\widetilde{J}$-holomorphic on X;*
(b) *there is a homotopy J_t of almost complex structures on X which is fixed in a neighborhood of K such that $J_0 = J$ and $J_1 = \widetilde{J}$. If J is integrable on X, then J_t can be chosen integrable on X for all $t \in [0, 1]$;*
(c) *for every $t \in [0, 1]$ the P-map f^t is J-holomorphic in a neighborhood of K and satisfies* $\sup\{\mathrm{dist}_Y(f^t(x, p), f(x, p)) : x \in K,\ p \in P\} < \epsilon$.

Proof We give the proof for the nonparametric case; the parametric case is essentially the same. Pick a compact set $K \subset X_c$ and a regular value $c_0 \in \mathbb{R}$ of ρ such that $K \subset X_{c_0} \Subset X_c$. Let $p_1, p_2, \ldots$ be the critical points of ρ in the set $\{x \in X : \rho(x) > c_0\}$, ordered so that $\rho(p_j) < \rho(p_{j+1})$ for $j = 1, 2, \ldots$. Choose numbers c_j satisfying $c_{-1} = -\infty < c_0 < \rho(p_1) < c_1 < \rho(p_2) < c_2 < \cdots$. Let k_j denote the Morse index of ρ at p_j. For each $j = 0, 1, \ldots$ we set

$$X_j = \{x \in X : \rho(x) < c_j\}, \qquad \Sigma_j = bX_j = \{x \in X : \rho(x) = c_j\}.$$

Set $J_0 = J$. We construct a sequence of almost complex structures J_j on X and maps $f_j: X \to Y$ satisfying the following conditions for $j = 0, 1, 2, \ldots$:

(i) J_j is integrable in a neighborhood of $\overline{X}_j$ and (X_j, J_j) is a Stein manifold with strongly pseudoconvex boundary,
(ii) $J_j = J_{j-1}$ in a neighborhood of $\overline{X}_{j-1}$,
(iii) the set $\overline{X}_{j-1}$ is $\mathcal{O}(X_j, J_j)$-convex,
(iv) the map f_j is J_j-holomorphic in a neighborhood of $\overline{X}_j$,
(v) $\sup_{x \in X_{j-1}} \mathrm{dist}_Y(f_j(x), f_{j-1}(x)) < \epsilon 2^{-j-1}$, and
(vi) there is a homotopy from f_{j-1} to f_j which is J_j-holomorphic and uniformly close to f_{j-1} in a neighborhood of $\overline{X}_{j-1}$ (satisfying (v)).

These conditions hold for $j = 0$; in this case, (ii), (iii), (v) and (vi) are vacuous. Assume inductively that we have found a pair (J_l, f_l) for all indexes up to $j - 1 \geq 0$. If $k_j = 0$ then a new connected component of the sublevel set $\{\rho < t\}$ appears at the critical point p_j when t passes the value $\rho(p_j)$, and it is trivial to find the next pair (J_j, f_j). Assume now that $k_j > 0$. By Morse theory X_j is diffeotopic to a handlebody obtained by smoothly thickening $\overline{X}_{j-1} \cup M_j$ for an embedded disc $M_j \subset X_j \setminus X_{j-1}$ of dimension k_j (the stable manifold for the gradient flow of ρ at p_j). The manifold X admits a real analytic structure that is equivalent to the underlying smooth structure and is induced near $\overline{X}_{j-1}$ by the complex structure J_{j-1}. By perturbing $\Sigma_{j-1} = bX_{j-1}$ we may assume that it is real analytic. Applying Lemma 3.9.3 (p. 92) we can isotopically deform M_j to an embedded real analytic disc that is well attached to $\overline{X}_{j-1}$ along $bM_j \subset \Sigma_{j-1}$. Theorem 3.9.2 (p. 91) now furnishes a strongly pseudoconvex Stein handlebody $(\widetilde{W}_j, \widetilde{J}_j)$ in X such that

(a) $\overline{X}_{j-1} \subset \widetilde{W}_j \subset X_j$,
(b) X_j is diffeotopic to $\widetilde{W}_j$ by a deformation supported in $X_j \setminus X_{j-1}$,
(c) $\widetilde{J}_j$ coincides with J_{j-1} in a neighborhood of $\overline{X}_{j-1}$, and
(d) the sets $\overline{X}_{j-1}$ and $\overline{X}_{j-1} \cup M_j$ are $\mathcal{O}(\widetilde{W}_j, \widetilde{J}_j)$-convex.

By choosing $\widetilde{W}_j$ thin enough around $\overline{X}_{j-1} \cup M_j$, Theorem 3.8.1 (p. 88) furnishes a map $\widetilde{f}_j \colon X \to Y$ that is homotopic to f_{j-1}, it is $\widetilde{J}_j$-holomorphic on $\widetilde{W}_j$, and it approximates f_{j-1} uniformly on X_{j-1} so that Condition (v) holds.

There is a smooth diffeotopy $h_t \colon X \to X$ ($t \in [0,1]$) that is fixed in a neighborhood of $\overline{X}_{j-1}$ such that h_0 is the identity map on X and $h_1(X_j) = \widetilde{W}_j$. Taking $J_j = h_1^*(\widetilde{J}_j)$ and $f_j = \widetilde{f} \circ h_1$ completes the inductive step. A homotopy from f_{j-1} to f_j is obtained by composing the homotopy from f_{j-1} to $\widetilde{f}$ by the map h_1. The induction may proceed.

By conditions (i) and (ii) there is a unique integrable complex structure $\widetilde{J}$ on X which agrees with J_j on X_j for every j. Note that $\widetilde{J}$ is homotopic to $J = J_0$ since the structure J_j at the j-th stage of the construction was chosen homotopic to J_{j-1} by a homotopy that is fixed near $\overline{X}_{j-1}$. The complex manifold $(X, \widetilde{J})$ is exhausted by the increasing sequence of Stein domain X_j, and the Runge property (iii) implies that $(X, \widetilde{J})$ is Stein. Conditions (iv) and (v) ensure that the sequence $f_j \colon X \to Y$ converges uniformly on compacts in X to the $\widetilde{J}$-holomorphic map $\widetilde{f} = \lim_{j\to\infty} f_j \colon X \to Y$ satisfying $\sup_{x \in X_0} \mathrm{dist}_Y(\widetilde{f}(x), f_0(x)) < \epsilon$. Condition (vi) ensures that the homotopies from f_{j-1} to f_j also converge uniformly on compacts in X to a homotopy $f_t \colon X \to Y$ from the initial map f_0 to $f_1 = \widetilde{f}$. If the initial structure J on X is integrable, then clearly all intermediate structures are also integrable. □

Proof of Theorem 10.9.3 This requires a few minor modifications in the proof of Theorem 10.10.1. We do not change the complex structure J on X during the construction at the cost of remaining on subsets of X which are only diffeomorphic to sublevel sets of ρ. We get approximation at no additional cost, but for the interpolation condition in Theorem 10.9.6 one must employ the exhaustion method that has been used in Sect. 5.12.

We use the same notation as above. Thus, $W_0 = X_0$ is a sublevel set of a Morse exhaustion function $\rho \colon X \to \mathbb{R}$ without critical points of index $> n$ in $X \setminus W_0$, ρ is strongly plurisubharmonic in a neighborhood of $\overline{W}_0$, and $f_0 = f \colon X \to Y$ is holomorphic in a neighborhood of $\overline{W}_0$. Let $X_j = \{\rho < c_j\}$ where the constants c_j are chosen such that ρ has a unique critical point p_j in $X_j \setminus X_{j-1}$. Pick $\epsilon > 0$. Assume that $n \neq 2$. We inductively construct a sequence of relatively compact, strongly pseudoconvex domains $W_1 \subset W_2 \subset \cdots \subset X$, maps $f_j \colon X \to Y$, and diffeomorphisms $h_j \colon X \to X$ such that the following conditions hold for all $j = 1, 2, \ldots$:

(i) W_{j-1} is $\mathcal{O}(W_j)$-convex,
(ii) f_j is holomorphic in a neighborhood of $\overline{W}_j$ and is homotopic to f_{j-1} by a homotopy $f_{j,t} \colon X \to Y$ $(t \in [0,1])$ such that each $f_{j,t}$ is holomorphic near $\overline{W}_{j-1}$ and satisfies $\sup_{x \in W_{j-1}} \mathrm{dist}_Y(f_{j,t}(x), f_{j-1}(x)) < \epsilon 2^{-j}$,
(iii) $h_j(X_j) = W_j$, and
(iv) $h_j = g_j \circ h_{j-1}$, where $g_j \colon X \to X$ is a diffeomorphism of X which is diffeotopic to Id_X by a diffeotopy that is fixed in a neighborhood of $\overline{W}_{j-1}$. (In particular, h_j agrees with h_{j-1} near $\overline{W}_{j-1}$.)

It is easily seen that the map $\widetilde{f} = \lim_{j\to\infty} f_j \colon \Omega = \bigcup_{j=1}^{\infty} W_j \to Y$ and the diffeomorphism $h = \lim_{j\to\infty} h_j \colon X \to \Omega$ satisfy Theorem 10.9.3.

To prove the inductive step, we begin by attaching to $W_{j-1} = h_{j-1}(X_{j-1})$ the disc $M_j = h_{j-1}(D_j)$, where $D_j \subset X_j \setminus X_{j-1}$ (with $bD_j \subset bX_{j-1}$) is the stable manifold for the gradient flow of ρ at the unique critical point $p_j \in X_j \setminus X_{j-1}$. By Lemma 3.9.3 (p. 92) we can isotope M_j to a real analytic totally real disc in $X \setminus W_{j-1}$ that is well attached to the domain $\overline{W}_{j-1}$ along a Legendrian sphere $bM_j \subset bW_{j-1}$. Applying Theorem 3.8.1 (p. 88), we find the next map $f_j \colon X \to Y$ that is holomorphic in a thin strongly pseudoconvex handlebody $W_j \supset \overline{W}_{j-1} \cup M_j$. The next diffeomorphism h_j is furnished by Morse theory. Hence, the induction proceeds and the proof is concluded as before. With a bit more care, one can also ensure that $b\Omega$ is smoothly bounded and strongly pseudoconvex, but in general we cannot choose Ω to be relatively compact in X, unless X admits an exhaustion function with at most finitely many critical points. □

10.11 Exotic Stein Structures on Smooth 4-Manifolds

In this section we discuss the construction of Stein surface structures on a smooth oriented 4-manifold X with a handlebody decomposition without handles of index > 2. In particular, we prove Theorems 10.9.1 and 10.9.2 (p. 517) in the case $\dim_{\mathbb{R}} X = 4$. We first indicate Gompf's original construction [246], but for the actual proof we use the modification from [223] which is better suited to the task of finding a holomorphic map in the given homotopy class. We also give a construction inside a given integrable complex surface as in [247], thereby proving Theorem 10.9.3 for $\dim_{\mathbb{R}} X = 4$.

We begin with the main point of the construction—extending a Stein structure across a handle.

Assume that X is a smooth oriented 4-manifold, $W \Subset X$ is a smoothly bounded domain in X, and J is an integrable Stein structure on a neighborhood of $\overline{W}$ such that the boundary $\Sigma = bW$ is strongly J-pseudoconvex. The problem is to extend J to a Stein structure over a suitably chosen handle attached to W. Handles of index 0 or 1 do not present any problem (see the previous section), so we focus on handles of index 2.

Let $D = D^2 \subset \mathbb{R}^2 \times \{i0\}^2 \subset \mathbb{C}^2$ be the standard 2-disc and $S = bD$ its boundary circle. Let $G\colon (D, S) \hookrightarrow (X \setminus W, \Sigma)$ be an embedded disc with boundary $G(S) \subset \Sigma$. After an isotopy given by Lemma 3.9.3 (p. 92) we can assume that $C = G(S)$ is a Legendrian knot in Σ (i.e., tangent to the contact subbundle $\xi = \mathrm{T}\Sigma \cap J(\mathrm{T}\Sigma)$ of $\mathrm{T}\Sigma$), and the embedded disc $M = G(D) \subset X \setminus W$ is J-normal to Σ along C in the sense that $J\mathrm{T}_x M \in \mathrm{T}_x \Sigma$ for all $x \in C$. This implies that the embedding G is J-real near S. Let $\nu_D = D \times \mathbb{R}^2$ denote the normal bundle of D in $\mathbb{C}^2$. The normal bundle ν_M of M in X is a trivial 2-plane bundle; a trivialization $\beta\colon \nu_D \to \nu_M$ covering G is called a *normal framing* of M. If β can be chosen such that

$$\beta \circ J_{st} = J \circ \mathrm{d}G \quad \text{on } \mathrm{T}_x D, \quad x \in S, \tag{10.32}$$

then (after perturbing M slightly) we can extend J to a Stein structure in a neighborhood of $\overline{W} \cup M$ exactly as in the proof of Theorem 10.9.1(a).

The key problem is that, when $n = 2$, the framing condition (10.32) cannot be achieved in general by an isotopy of the disc G. Let us analyze the underlying geometry more carefully. Assume that C is an oriented Legendrian knot in the contact 3-manifold (Σ, ξ). (For the moment, C need not bound a disc.) We may assume that the global orientation of X agrees with the orientation induced by J near $\overline{W}$. We coorient the boundary $\Sigma = bW$ by the outer normal vector field w, chosen such that $Jw \in \mathrm{T}\Sigma$. These choices coorient the normal bundle ν_C of C in Σ so that the orientations on $\mathrm{T}\Sigma|_C = \mathrm{T}C \oplus \nu_C$ add up correctly. Let τ be a nonvanishing vector field tangent to C in the chosen orientation class. Then, $J\tau$ is tangent to ξ since C is Legendrian, and $(\tau, J\tau)$ is a framing of the 2-plane bundle $\xi|_C$. The pair of vector fields $(J\tau, Jw)$ spans the normal bundle $\nu_C \subset \mathrm{T}\Sigma|_C$ of the knot C in Σ. This is a *canonical framing*, also called a *Thurston-Bennequin (TB) framing* of the normal bundle ν_C of the Legendrian knot C in (Σ, ξ). Any pair of TB framings are homotopic to each other as framings of ν_C. Note that the triple $(\tau, J\tau, Jw)$ is a positive framing of $\mathrm{T}\Sigma|_C$. Indeed, w is the outer normal to Σ and $(w, \tau, J\tau, Jw)$ defines the same orientation of $\mathrm{T}X$ as $(\tau, J\tau, w, Jw)$, hence positive. Thus, any TB framing $(J\tau, Jw)$ defines a positive orientation of the normal bundle ν_C.

Assume now that C bounds an embedded 2-disc $M = G(D) \subset X \setminus W$ that is transverse to Σ along C. By an isotopic correction, keeping the boundary curve $C = bM$ fixed, we ensure that M is J-orthogonal to Σ along C. Let (x_1, x_2) be coordinates on $\mathbb{R}^2$. The vector fields

$$w_0 = -x_1 \partial_{x_1} - x_2 \partial_{x_2}, \qquad \tau_0 = -x_2 \partial_{x_1} + x_1 \partial_{x_2}$$

are the inner normal and the tangential field to $S = \{x_1^2 + x_2^2 = 1\} = bD$, respectively. Their images $w = G_* w_0$, $\tau = G_* \tau_0$ are vector field on X along $C = G(S)$ spanning the tangent bundle $\mathrm{T}M|_C$. Since τ is tangent to C and w is outer normal to $\Sigma = bW$ along C, the pair $(J\tau, Jw)$ is a TB framing of ν_C. Choose a normal framing $\beta\colon \nu_D \to \nu_M$ over $G\colon D \to M$ such that $\beta(\nu|_S) = \nu_C$ and the orientations match up. We thus have two framings of ν_C, namely β and the TB framing. Since ν_C is a trivial oriented 2-plane bundle over the circle C, any two framings in the same orientation class differ up to homotopy by a map $C \cong S^1 \to SO(2) = S^1$, hence by an integer. We can thus write

$$[\beta] = TB + k(\beta), \quad k(\beta) \in \mathbb{Z}.$$

The integer $k = k(\beta)$ is called the *framing index* of β.

The framing condition (10.32) that we wish to achieve is

$$\beta(\partial_{y_1}) = J(G_*\partial_{x_1}), \qquad \beta(\partial_{y_2}) = J(G_*\partial_{x_2})$$

along the circle S. In terms of the vector fields $w = G_* w_0$ and $\tau = G_*\tau_0$ introduced above, this is equivalent to

$$\beta(-x_2\partial_{y_1} + x_1\partial_{y_2}) = J\tau, \qquad \beta(-x_1\partial_{y_1} - x_2\partial_{y_2}) = Jw.$$

Since the framing $(-x_2\partial_{y_1} + x_1\partial_{y_2}, -x_1\partial_{y_1} - x_2\partial_{y_2})$ makes one positive twist against the framing $(\partial_{y_1}, \partial_{y_2})$ of ν_D as the point (x_1, x_2) traces the circle S, we see that β is obtained by one negative twist from the TB framing $(J\tau, Jw)$. Hence, the framing condition (10.32) is equivalent to

$$k(\beta) = -1 \quad \Longleftrightarrow \quad [\beta] = TB - 1. \tag{10.33}$$

(This holds in the model case when $M = D^2 \subset \mathbb{R}^2 \times \{i0\}^2 \subset \mathbb{C}^2$ and the domain W is tangent along $C = bD^2$ to a quadric domain $Q_\lambda \subset \mathbb{C}^2$ (3.10).) If the normal framing β of M satisfies (10.33), then J extends to an integrable complex structure in a neighborhood of $\overline{W} \cup M$ in X such that M is J-real. If this ideal situation occurs for all 2-handles in $X \setminus W$, then the construction of a Stein structure on X, and of a holomorphic map $X \to Y$ in a given homotopy class, can be completed exactly as in the previous section.

Suppose now that $k = [\beta] - TB \neq -1$ for some 2-handle. A basic fact [146, 149] is that for any Legendrian knot C in a contact 3-manifold there is a $\mathcal{C}^0$-small isotopy preserving the knot type, but changing its Legendrian knot type, which adds a desired number of negative twists to the TB framing. (One adds small spirals to the knot C.) Since the homotopy class $[\beta]$ of the framing is preserved under an isotopy of C in Σ, we see that $k = [\beta] - TB$ can be increased by any number of units. If $k < -1$, it is therefore possible to obtain an isotopic embedding $(D, S) \hookrightarrow (X \setminus W, \Sigma)$ satisfying (10.33), thereby reducing the problem to the previous case.

The problem becomes nontrivial when $k \geq 0$ since it is in general impossible to add positive twists to the TB framing (equivalently, to decrease the framing index k). This is only possible in a contact structure which is *overtwisted*, in the sense that it contains a topologically unknotted Legendrian knot K with the Thurston-Bennequin index $tb(K) = 0$; adding such a knot to a Legendrian knot adds a positive twist to the TB framing, making it possible to decrease $k = [\beta] - TB$ and hence reach $k = -1$.

However, Eliashberg proved in [146] that contact structures arising as boundaries of strongly pseudoconvex Stein manifolds are never overtwisted (they are *tight*). A 2-handle for which we cannot find an isotopy of the boundary circle to a Legendrian knot such that (10.33) holds will be called in the sequel *a wrongly attached handle*.

Gompf [246] circumvented the problem by replacing a wrongly attached 2-handle by a handle with the correct framing index -1 and then introducing sufficiently many positive self-plumbings on the core disc. These self-plumbings restore the second homology group of the manifold to the correct group of the initial manifold, at the cost of introducing new superfluous generators to its fundamental group. We end up with an immersed correctly attached 2-handle to which the Stein structure can be extended. In order to get the topologically correct manifold, one must then trivialize all double point loops coming from self-plumbings (double points) on the core disc. This is again done by correctly attaching an immersed 2-disc along each double point loop, etc. This procedure does not terminate in a finite number of steps, and it converges to an attached Casson handle that is homeomorphic, but not necessarily diffeomorphic, to the standard 4-handle. For a summary of Gompf's construction, see [223, p. 639].

We now recall a slightly different construction from [223] whose advantage is that we remain inside the given manifold X at all steps.

As said before, the TB invariant of a Legendrian knot can be increased by an arbitrary integer. Since the homotopy class $[\beta]$ of the normal framing does not change under such isotopy, we can assume that the framing coefficient is odd, $k(\beta) = [\beta] - TB = -1 + 2k$ for some $k \in \mathbb{Z}$. If $k < 0$, we add negative twists to the TB framing to get (10.33) and we are done. If $k > 0$, we glue onto the core disc M k copies of homologically trivial Weinstein immersed sphere $K = F(S^2)$ (10.21), given as the image of the map

$$F(x, y, z) = \big(x(1 + 2iz), y(1 + 2iz)\big) \in \mathbb{C}^2 \tag{10.34}$$

where $(x, y, z) \in \mathbb{R}^3$, $x^2 + y^2 + z^2 = 1$. (Precisely, we take the connected sum of M with k copies of K.) Recall (p. 495) that the normal bundle ν_K of $K \subset \mathbb{C}^2$ has Euler number $\chi(\nu_K) = -2$; hence, a copy of K glued to M reduces the framing coefficient of M by two units. Adding k copies of K will therefore reduce the framing coefficient to -1.

With F as in (10.34), let

$$\Delta = \big\{F(0, y, z) : y \geq 0,\ y^2 + z^2 \leq 1\big\} \subset \mathbb{C}^2.$$

This 2-disc is embedded, except along the side $\{y = 0\}$ which gets pinched to the point $0 \in \mathbb{C}^2$. Note that $b\Delta \subset F(S^2)$ and the union $F(S^2) \cup \Delta$ has a tubular neighborhood diffeomorphic to $S^2 \times \mathbb{R}^2$.

In order to make a self-intersection at a point p in the core disc M of a handle h in our 4-manifold X, we replace a small disc in M around p by a copy of the standard kink K. (See Fig. 10.5; we removed the small dotted disc and smoothly attached a kinky disc along its boundary.) This surgery reduces the relative Euler number over the immersed disc M by 2 for each kink. Adding k kinks on M inside X and

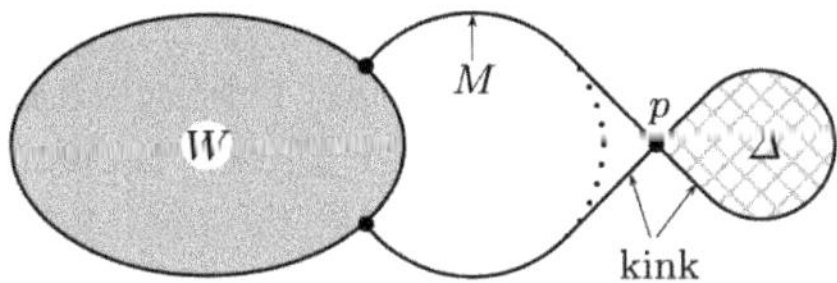

Fig. 10.5 A kinky disc M with a trivializing 2-cell Δ. ([223, p. 640, Fig. 8] and also [222, p. 356, Fig. 4])

then taking a tubular neighborhood has the same effect as first removing the handle h from X, reattaching it with a framing of the boundary reduced by $2k$, and then performing k self-plumbings on h (as was done by Gompf in [246]). This shows that the manifold constructed in Gompf's proof can be seen as a submanifold of the original manifold X, changed only by a surgery in a small coordinate neighborhood of each of the kinked points on the core disc of the handle h. We can also explicitly see the trivializing 2-cell Δ that needs to be added to each of the kinks in order to reconstruct the desired manifold. In the next stage of the construction, every such disc Δ will also have to receive a kink in order to correct its framing coefficient. This begins the *Casson tower* procedure which will converge to a Casson handle in place of the original removed disc in M.

We are now ready to prove the following more precise version of Theorem 10.9.2 when $\dim_{\mathbb{R}} X = 4$. Unlike in the case $n > 2$, we need not assume that the almost complex structure J is defined on all of X since the obstruction to extending J only appears for handles of index > 2. However, if J is already given on all of X, then one can choose (X', J') in the theorem such that the almost complex structure $h^*(J')$ on X is homotopic to J; this notion makes sense for orientation preserving homeomorphisms [246, p. 645].

Theorem 10.11.1 *Let X be a smooth oriented 4-manifold exhausted by a Morse function $\rho\colon X \to \mathbb{R}$ without critical points of index > 2. Assume that for some $c \in \mathbb{R}$ there is a Stein structure J on $X_c = \{x \in X : \rho(x) < c\}$ such that $\rho|_{X_c}$ is strongly J-plurisubharmonic. Let Y be a complex manifold, P be a compact Hausdorff space, and $f\colon X \times P \to Y$ be a P-map which is J-holomorphic in X_c. Given a compact set $K \subset X_c$ and a number $\epsilon > 0$, there exist a Stein surface (X', J'), an orientation preserving homeomorphism $h\colon X \to X'$ which is biholomorphic in a neighborhood of K, and a holomorphic P-map $f'\colon X' \times P \to Y$ such that the P-map $\widetilde{f}\colon X \times P \to Y$, defined by $\widetilde{f}(x, p) = f'(h(x), p)$, is homotopic to f and satisfies*

$$\sup_{x \in K,\ p \in P} \operatorname{dist}_Y\big(f(x, p), \widetilde{f}(x, p)\big) < \epsilon.$$

Proof Assume that our smooth 4-manifold X is constructed by successively attaching handles $h_1, h_2, h_3, \ldots$ of index ≤ 2, beginning with the compact domain $\overline{W} \subset X$ with smooth boundary $\Sigma = bW$. By the assumption, we also have an integrable complex structure J in a neighborhood of $\overline{W}$ such that W is Stein and its boundary Σ is strongly pseudoconvex.

Let $M_1, M_2, \ldots$ be the cores of the handles $h_1, h_2, \ldots$, chosen such that their union is a smoothly embedded CW complex inside X. Since we have not assumed that our handlebody is finite, we cannot ask for the ordering of the handles with regard to their indexes. However, due to local compactness we can ask that when a handle h_j with core M_j is being attached, all handles whose core discs intersect the boundary bM_j have already been attached in earlier steps. We can also assume that bM_j consists only of the core discs of handles of lower indexes. We can now proceed as in the proof of Theorem 10.10.1, but with the following modifications:

(1) When a 2-handle is attached with a wrong framing, we insert the right number of kinks to its core disc in order to change the framing coefficient to -1. This ensures that we can extend J to a Stein structure in a tubular strongly pseudoconvex neighborhood of the immersed disc. (The disc is totally real in this structure and has a special double point at each kink that corresponds in local coordinates to the union of totally real coordinate planes $\mathbb{R}^2 \times \{i0\}^2 \cup \{0\}^2 \times i\mathbb{R}^2$ in $\mathbb{C}^2$.)
(2) Each time before proceeding to the next handle h_{j+1} we perform one more step on each of the kinked discs appearing in the sequence before. More precisely, we add a new kinked disc which cancels the double point loop at the self-intersection point introduced in the previous step.

The first condition is essential since we need to build a manifold that is Stein. The second condition ensures that each handle is properly worked upon, thereby producing a Casson tower at every place where a kink was made in the initial 2-disc. At every step we also approximate the given map, which has already been made holomorphic in a tubular strongly pseudoconvex neighborhood of our partial (finite) subcomplex, by a map holomorphic in a tubular neighborhood of the previous domain with all core discs that have been added at the given step.

Note that whenever a handle is wrongly attached, the above process is never finite. The reason is that, in the standard kink K, the disc Δ needed to be added to reconstruct the original manifold requires exactly one positive kink in order to be able to extend the Stein structure to its neighborhood.

The proof can now be concluded as in Theorem 10.10.1. We construct an increasing sequence of Stein domains $X_1 \subset X_2 \subset \cdots$ inside the smooth 4-manifold X such that X_j is Runge in X_{j+1} for every j, and a sequence of maps $f_j : X \to Y$ $(j = 1, 2, \ldots)$ such that f_j is holomorphic on X_j, it approximates f_{j-1} uniformly on X_{j-1}, and is homotopic to f_{j-1} by a homotopy which is holomorphic and uniformly close to f_{j-1} on X_{j-1}. The complex structure on X_j does not change from the j-th step of the construction on, so we get a limit complex structure J' on the union $X' = \bigcup_j X_j \subset X$. The Runge property of each pair $X_j \subset X_{j+1}$ ensures that (X', J') is a Stein manifold. By the construction, X' is homeomorphic to X. (It is also diffeomorphic to X provided that no Casson handles were used in the construction.) A small ambient topological deformation moves the initial CW complex into X'; see [247] for the details. By the construction, the limit map $f' = \lim_{j\to\infty} f_j : X' \to Y$ is holomorphic, and the map $f' \circ h : X \to Y$ is homotopic to f.

The same proof applies to any smoothly embedded 2-complex M inside X—after a small ambient topological deformation we find a new embedding $M \hookrightarrow X$

with a Stein thickening $M \subset X' \subset X$ such that a given continuous map $M \to Y$ admits a holomorphic representative $X' \to Y$. □

Proof of Theorem 10.9.3 for $n = 2$ The proof, taken from [223], is similar to Gompf's construction in [247].

Assume that (X, J) is a complex surface, not necessarily Stein, such that the underlying smooth manifold X has a handlebody decomposition without handles of index > 2. As in the case $n > 2$, we find an increasing sequence of open Stein domains $X_1 \subset X_2 \subset \cdots$ in X and J-holomorphic maps $f_j \colon X_j \to Y$ such that each X_j is Runge in X_{j+1} and the other properties are as before, the only difference being that their union $\Omega = \bigcup_{j=1}^{\infty} X_j$ is only homeotopic (and not necessarily diffeotopic) to X.

The main inductive step requires certain *kinky handles*. One can use totally real immersed discs furnished by Lemma 3.9.3 (p. 92). We give another construction which depends on the cancellation theorem for complex points of real surface in complex surfaces (see Theorem 10.4.1 on p. 489).

Let K be the immersed Lagrangian sphere in $\mathbb{C}^2$ (10.34) with one positive double point. From the index formula (10.10) (p. 485) we see that the connected sum of an oriented real surface M in a complex surface X with a homologically trivial copy of K decreases each of the Lai indexes $I_\pm(M)$ by one. (Lai indexes were introduced in Sect. 10.3. They are well defined for surfaces that are totally real along the boundary curves, and are invariant under isotopies preserving this condition.) As before, we say that we added a positive kink to the surface. Adding a sufficient number of positive kinks to M we get a homologically equivalent immersed surface M' satisfying $I_\pm(M') \le 0$ that agrees with M near bM. By Theorem 10.4.1 (p. 489) we can deform M' to a surface with only special hyperbolic complex points, keeping the boundary fixed. Hence, the new surface has a basis of tubular Stein neighborhoods by Theorem 10.1.2 (p. 480). Theorem 3.8.1 on p. 88 (concerning holomorphic approximation on handlebodies) easily adapts to immersed totally real surfaces with isolated special double points (see [223, Theorem 4.1] for the details).

We now complete the proof of Theorem 10.9.3 in the case $\dim_{\mathbb{R}} X = 4$.

Assume that X is obtained from a strongly pseudoconvex domain $W \subset X$ by successively attaching handles $h_1, h_2, \ldots$ with core discs $M_1, M_2, \ldots$, where the ordering of these handles satisfies the same condition as in the proof of Theorem 10.11.1. We use the same notation as in the proof of Theorem 10.9.3 for $\dim_{\mathbb{R}} X \neq 4$, beginning with $W_0 = W$ and $f_0 = f$.

In the inductive step, we have a smoothly bounded, strongly pseudoconvex domain $W_j \subset X$ and a map $f_j \colon X \to Y$ that is holomorphic in a neighborhood of $\overline{W}_j$. The set W_j is a tubular neighborhood of the union of $\overline{W}$ with the cores of handles attached in the earlier steps. Since these cores may have received kinks, W_j does not have the correct homeomorphic type, but this will be corrected in the limit by the Casson handles resulting from the construction.

We now attach to $\overline{W}_j$ the next handle in the sequence. As before, 0- and 1-handles do not pose any problem. For a 2-handle h we first apply Gromov's Legendrization theorem [265, p. 339, (B')] to make sure that the boundary of its core disc M is a

Legendrian curve in bW_j and the attachment is J-orthogonal. We then add enough positive kinks to the core disc M in order to cancel all elliptic points as explained above. The new immersed disc M' obtained in this way is made real analytic by a small deformation that does not destroy any of the above properties. It follows from Theorem 3.9.2 (p. 91) that $\overline{W}_j \cup M'$ has a basis of strongly pseudoconvex tubular Stein neighborhoods in X. In the same way we add to $\overline{W}_j$ a new kinky disc at each of the kinks from the earlier stages of the construction (trivializing the double point loop), making sure that the conditions (1) and (2) in the proof of Theorem 10.11.1 are satisfied. These additional kinky discs $\Delta_1, \ldots, \Delta_k$ can be chosen such that the set $L_j = \overline{W}_j \cup M' \cup (\bigcup_{l=1}^{k} \Delta_l)$ admits a basis of tubular, strongly pseudoconvex Stein neighborhoods in X. We then approximate the map f_j from the previous step (which is holomorphic near $\overline{W}_j$) uniformly on L_j by a map $f_{j+1}\colon X \to Y$ that is holomorphic in a neighborhood U_j of L_j. By Theorem 3.9.2 (p. 91) there is a strongly pseudoconvex tubular neighborhood W_{j+1} of L_j, contained in U_j, such that W_j is Runge in W_{j+1}. This completes the induction step. In the limit we obtain a holomorphic map $f'\colon \Omega \to Y$ on the Stein domain $\Omega = \bigcup_j W_j \subset X$ with the stated properties. □

The following corollary is obtained by combining Theorem 10.11.1 (p. 527) with Corollary 3.2 and Theorem 3.3 in [246, p. 648].

Corollary 10.11.2 *Let X be a smooth closed oriented 4-manifold. There exists a smooth finite wedge of circles $\Gamma \subset X$ such that for every continuous map $f\colon X \setminus \Gamma \to Y$ to a complex manifold Y there is a (possibly exotic) Stein structure on $X \setminus \Gamma$ and a holomorphic map $\widetilde{f}\colon X \setminus \Gamma \to Y$ homotopic to f. In $X = \mathbb{P}^2$, this holds after removing a single point, and in this case any Stein structure on $\mathbb{P}^2 \setminus \{p\}$ is exotic. The analogous result holds for every open oriented smooth 4-manifold after removing a suitably chosen smooth 1-complex.*

The point is that there is a wedge of circles Γ in X such that $X \setminus \Gamma$ admits a handle decomposition without any 3-handles or 4-handles. The complex projective plane $\mathbb{P}^2$ has a single 4-cell (and no 3-cells) in its handlebody decomposition, hence removing a point leaves only cells of index ≤ 2.

Here is another result obtained by combining Gompf's methods from [247] with the proof of Theorem 10.9.2.

Corollary 10.11.3 *Let M be a tame, topologically embedded CW n-complex in a complex manifold X^n, and let U be an open neighborhood of M in X. For every continuous map $f\colon M \to Y$ to a complex manifold Y there exist a topological isotopy $h_t\colon X \to X$ satisfying $h_0 = \mathrm{Id}_X$ and $h_t(M) \subset U$ for all $t \in [0,1]$, a Stein thickening $\Omega \subset U$ of the CW complex $h_1(M)$, and a holomorphic map $\widetilde{f}\colon \Omega \to Y$ such that $\widetilde{f} \circ h_1\colon M \to Y$ is homotopic to f.*

Gompf showed that the necessary adjustment of the initial CW complex $M \subset X$ is quite mild from the topological point of view. When $n = 2$, the Stein domains Ω

obtained in this way will typically have nonsmooth boundaries and may be chosen to realize uncountably many distinct diffeomorphism types.

This approach has been developed further in Gompf's recent paper [245] from 2017. He showed in particular that if a smooth open 4-manifold X admits a Morse exhaustion function with indexes at most 2, then X admits an exotic smooth structure; it admits infinitely many diffeomorphism types of smoothings if $H_2(X;\mathbb{Z}) \neq 0$, and uncountably many types if $H_2(X;\mathbb{Z})$ has infinite rank. Furthermore, for every domain of holomorphy U in $\mathbb{C}^2$, the inclusion map is topologically isotopic to other embeddings, whose images are also domains of holomorphy and represent uncountably many diffeomorphism types of smoothings of U, while having the same genus function.

These results have interesting consequences concerning envelopes of holomorphy of real submanifolds of dimension $\leq n$ in $\mathbb{C}^n$. For example, assume that $M \subset \mathbb{C}^n$ is an embedded n-dimensional submanifold and $U \subset \mathbb{C}^n$ is an open set containing M. If $n \neq 2$ then there exist an isotopic deformation $M' \subset U$ of M and a Stein domain $\Omega \subset U$ containing M'. In particular, the envelope of holomorphy of the deformation M' remains close to M, something which definitely fails for the polynomial hull (see [22, 184]).

References

1. Abhyankar, S.S.: On the semigroup of a meromorphic curve. I. In: Proceedings of the International Symposium on Algebraic Geometry, Kyoto Univ., Kyoto, 1977, pp. 249–414. Kinokuniya Book Store, Tokyo (1978)
2. Abhyankar, S.S., Moh, T.T.: Embeddings of the line in the plane. J. Reine Angew. Math. **276**, 148–166 (1975)
3. Abraham, R.: Transversality in manifolds of mappings. Bull. Am. Math. Soc. **69**, 470–474 (1963)
4. Abraham, R., Marsden, J.E., Ratiu, T.: Manifolds, Tensor Analysis, and Applications, 2nd edn. Applied Mathematical Sciences, vol. 75. Springer, New York (1988)
5. Acquistapace, F., Broglia, F., Tognoli, A.: A relative embedding theorem for Stein spaces. Ann. Sc. Norm. Super. Pisa, Cl. Sci. (4) **2**(4), 507–522 (1975)
6. Adams, J.F.: Lectures on Lie Groups. W. A. Benjamin, New York (1969)
7. Adams, J.F., Atiyah, M.F.: K-theory and the Hopf invariant. Q. J. Math. Oxf. Ser. (2) **17**, 31–38 (1966)
8. Ahern, P., Flores, M., Rosay, J.-P.: On $\mathbb{R}^+$ and $\mathbb{C}$ complete holomorphic vector fields. Proc. Am. Math. Soc. **128**(10), 3107–3113 (2000)
9. Ahern, P., Forstnerič, F.: One parameter automorphism groups on $\mathbb{C}^2$. Complex Var. Theory Appl. **27**(3), 245–268 (1995)
10. Ahern, P., Forstnerič, F., Varolin, D.: Flows on $\mathbb{C}^2$ with polynomial time one map. Complex Var. Theory Appl. **29**(4), 363–366 (1996)
11. Ahern, P., Rudin, W.: Totally real embeddings of S^3 in $\mathbb{C}^3$. Proc. Am. Math. Soc. **94**(3), 460–462 (1985)
12. Ahern, P., Rudin, W.: Periodic automorphisms of $\mathbb{C}^n$. Indiana Univ. Math. J. **44**(1), 287–303 (1995)
13. Alarcón, A., Drinovec Drnovšek, B., Forstnerič, F., López, F.J.: Every bordered Riemann surface is a complete conformal minimal surface bounded by Jordan curves. Proc. Lond. Math. Soc. (3) **111**(4), 851–886 (2015)
14. Alarcón, A., Forstnerič, F.: Every bordered Riemann surface is a complete proper curve in a ball. Math. Ann. **357**(3), 1049–1070 (2013)
15. Alarcón, A., Forstnerič, F.: Null curves and directed immersions of open Riemann surfaces. Invent. Math. **196**(3), 733–771 (2014)
16. Alarcón, A., Forstnerič, F.: The Calabi-Yau problem, null curves, and Bryant surfaces. Math. Ann. **363**(3–4), 913–951 (2015)
17. Alarcon, A., Globevnik, J.: Complete embedded complex curves in the ball of $\mathbb{C}^2$ can have any topology. Anal. PDE **10**(8), 1987–1999 (2017). doi:10.2140/apde.2017.10.1987

F. Forstnerič, *Stein Manifolds and Holomorphic Mappings*, Ergebnisse der Mathematik und ihrer Grenzgebiete. 3. Folge / A Series of Modern Surveys in Mathematics 56, DOI 10.1007/978-3-319-61058-0

18. Alarcón, A., Globevnik, J., López, F.J.: A construction of complete complex hypersurfaces in the ball with control on the topology. J. Reine Angew. Math. (2016). doi:10.1515/crelle-2016-0061
19. Alarcón, A., López, F.J.: Proper holomorphic embeddings of Riemann surfaces with arbitrary topology into $\mathbb{C}^2$. J. Geom. Anal. **23**(4), 1794–1805 (2013)
20. Alexander, H.: Polynomial approximation and hulls in sets of finite linear measure in $\mathbb{C}^n$. Am. J. Math. **93**, 65–74 (1971)
21. Alexander, H.: Explicit imbedding of the (punctured) disc into $\mathbb{C}^2$. Comment. Math. Helv. **52**(4), 539–544 (1977)
22. Alexander, H.: Linking and holomorphic hulls. J. Differ. Geom. **38**(1), 151–160 (1993)
23. Andersén, E.: Volume-preserving automorphisms of $\mathbb{C}^n$. Complex Var. Theory Appl. **14**(1–4), 223–235 (1990)
24. Andersén, E.: Complete vector fields on $(\mathbb{C}^*)^n$. Proc. Am. Math. Soc. **128**(4), 1079–1085 (2000)
25. Andersén, E., Lempert, L.: On the group of holomorphic automorphisms of $\mathbb{C}^n$. Invent. Math. **110**(2), 371–388 (1992)
26. Andreotti, A., Frankel, T.: The Lefschetz theorem on hyperplane sections. Ann. Math. (2) **69**, 713–717 (1959)
27. Andreotti, A., Grauert, H.: Théorème de finitude pour la cohomologie des espaces complexes. Bull. Soc. Math. Fr. **90**, 193–259 (1962)
28. Andreotti, A., Vesentini, E.: Carleman estimates for the Laplace-Beltrami equation on complex manifolds. Publ. Math. IHÉS **25**, 81–130 (1965)
29. Andrist, R., Forstnerič, F., Ritter, T., Wold, E.F.: Proper holomorphic embeddings into Stein manifolds with the density property. J. Anal. Math. **130**, 135–150 (2016)
30. Andrist, R.B.: Lifting to the spectral ball with interpolation. J. Math. Anal. Appl. **435**(1), 315–320 (2016)
31. Andrist, R.B.: The density property for Gizatullin surfaces with reduced degenerate fibre. J. Geom. Anal. To appear. ArXiv e-prints (2017). arXiv:1701.03373
32. Andrist, R.B., Kutzschebauch, F.: The fibred density property and the automorphism group of the spectral ball. Math. Ann. (2017). doi:10.1007/s00208-017-1520-8
33. Andrist, R.B., Kutzschebauch, F., Lind, A.: Holomorphic automorphisms of Danielewski surfaces II: structure of the overshear group. J. Geom. Anal. **25**(3), 1859–1889 (2015)
34. Andrist, R.B., Kutzschebauch, F., Poloni, P.-M.: The density property for Gizatullin surfaces completed by four rational curves. Proc. Am. Math. Soc. In press. doi:10.1090/proc/13665
35. Andrist, R.B., Shcherbina, N., Wold, E.F.: The Hartogs extension theorem for holomorphic vector bundles and sprays. Ark. Mat. **54**(2), 299–319 (2016)
36. Andrist, R.B., Wold, E.F.: Riemann surfaces in Stein manifolds with the density property. Ann. Inst. Fourier (Grenoble) **64**(2), 681–697 (2014)
37. Arosio, L., Fornaess Wold, E.: Totally real embeddings with prescribed polynomial hulls. ArXiv e-prints (2017). arXiv:1702.01002
38. Arzhantsev, I., Flenner, H., Kaliman, S., Kutzschebauch, F., Zaidenberg, M.: Flexible varieties and automorphism groups. Duke Math. J. **162**(4), 767–823 (2013)
39. Audin, M.: Fibrés normaux d'immersions en dimension double, points doubles d'immersions lagragiennes et plongements totalement réels. Comment. Math. Helv. **63**(4), 593–623 (1988)
40. Baader, S., Kutzschebauch, F., Wold, E.F.: Knotted holomorphic discs in $\mathbb{C}^2$. J. Reine Angew. Math. **648**, 69–73 (2010)
41. Bănică, C., Forster, O.: Complete intersections in Stein manifolds. Manuscr. Math. **37**(3), 343–356 (1982)
42. Baouendi, M.S., Ebenfelt, P., Rothschild, L.P.: Real Submanifolds in Complex Space and Their Mappings. Princeton Mathematical Series, vol. 47. Princeton University Press, Princeton (1999)
43. Baouendi, M.S., Trèves, F.: A property of the functions and distributions annihilated by a locally integrable system of complex vector fields. Ann. Math. (2) **113**(2), 387–421 (1981)

44. Barth, W.: Der Abstand von einer algebraischen Mannigfaltigkeit im komplex-projektiven Raum. Math. Ann. **187**, 150–162 (1970)
45. Barth, W.P., Hulek, K., Peters, C.A.M., Van de Ven, A.: Compact Complex Surfaces, 2nd edn. Ergebnisse der Mathematik und ihrer Grenzgebiete, 3. Folge, vol. 4. Springer, Berlin (2004)
46. Bass, H., Meisters, G.: Polynomial flows in the plane. Adv. Math. **55**(2), 173–208 (1985)
47. Bedford, E., Klingenberg, W.: On the envelope of holomorphy of a 2-sphere in $\mathbf{C}^2$. J. Am. Math. Soc. **4**(3), 623–646 (1991)
48. Bedford, E., Smillie, J.: Fatou-Bieberbach domains arising from polynomial automorphisms. Indiana Univ. Math. J. **40**(2), 789–792 (1991)
49. Bedford, E., Smillie, J.: Polynomial diffeomorphisms of $\mathbb{C}^2$. II. Stable manifolds and recurrence. J. Am. Math. Soc. **4**(4), 657–679 (1991)
50. Behnke, H., Stein, K.: Entwicklung analytischer Funktionen auf Riemannschen Flächen. Math. Ann. **120**, 430–461 (1949)
51. Behnke, H., Stein, K.: Elementarfunktionen auf Riemannschen Flächen als Hilfsmittel für die Funktionentheorie mehrerer Veränderlichen. Can. J. Math. **2**, 152–165 (1950)
52. Behnke, H., Stein, K.: Modifikation komplexer Mannigfaltigkeiten und Riemannscher Gebiete. Math. Ann. **124**, 1–16 (1951)
53. Bell, S.R., Narasimhan, R.: Proper holomorphic mappings of complex spaces. In: Several Complex Variables, VI. Encyclopaedia Math. Sci., vol. 69, pp. 1–38. Springer, Berlin (1990)
54. Bharali, G.: Surfaces with degenerate CR singularities that are locally polynomially convex. Mich. Math. J. **53**(2), 429–445 (2005)
55. Bharali, G.: Polynomial approximation, local polynomial convexity, and degenerate CR singularities. J. Funct. Anal. **236**(1), 351–368 (2006)
56. Bieberbach, L.: Beispiel zweier ganzer Funktionen zweier komplexer Variablen, welche eine schlichte volumtreue Abbildung des $\mathbb{R}_4$ auf einen Teil seiner selbst vermitteln. Sitzungsber. Preuß. Akad. Wiss., Phys.-Math. Kl. **1933**, 476–479 (1933)
57. Bishop, E.: Subalgebras of functions on a Riemann surface. Pac. J. Math. **8**, 29–50 (1958)
58. Bishop, E.: Mappings of partially analytic spaces. Am. J. Math. **83**, 209–242 (1961)
59. Bishop, E.: Differentiable manifolds in complex Euclidean space. Duke Math. J. **32**, 1–21 (1965)
60. Boc Thaler, L., Forstnerič, F.: A long $\mathbb{C}^2$ without holomorphic functions. Anal. PDE **9**(8), 2031–2050 (2016)
61. Bochnak, J., Kucharz, W.: Complete intersections in differential topology and analytic geometry. Boll. Unione Mat. Ital., B (7) **10**(4), 1019–1041 (1996)
62. Borell, S., Kutzschebauch, F.: Non-equivalent embeddings into complex Euclidean spaces. Int. J. Math. **17**(9), 1033–1046 (2006)
63. Borelli, M.: Divisorial varieties. Pac. J. Math. **13**, 375–388 (1963)
64. Bott, R., Milnor, J.: On the parallelizability of the spheres. Bull. Am. Math. Soc. **64**, 87–89 (1958)
65. Bowman, D.: Holomorphic flexibility properties of spaces of elliptic functions. ArXiv e-prints (2016). arXiv:1609.07184
66. Bracci, F., Raissy, J., Stensønes, B.: Automorphisms of $\mathbb{C}^2$ with an invariant non-recurrent attracting Fatou component biholomorphic to $\mathbb{C} \times \mathbb{C}^*$. ArXiv e-prints (2017). arXiv:1703.08423
67. Bremermann, H.J.: Über die Äquivalenz der pseudokonvexen Gebiete und der Holomorphiegebiete im Raum von n komplexen Veränderlichen. Math. Ann. **128**, 63–91 (1954)
68. Bröcker, T., tom Dieck, T.: Representations of Compact Lie Groups. Graduate Texts in Mathematics, vol. 98. Springer, New York (1995). Corrected reprint of the 1985 translation
69. Brody, R.: Compact manifolds and hyperbolicity. Trans. Am. Math. Soc. **235**, 213–219 (1978)
70. Browder, A.: Cohomology of maximal ideal spaces. Bull. Am. Math. Soc. **67**, 515–516 (1961)

71. Brunella, M.: Complete polynomial vector fields on the complex plane. Topology **43**(2), 433–445 (2004)
72. Bungart, L.: On analytic fiber bundles. I. Holomorphic fiber bundles with infinite dimensional fibers. Topology **7**, 55–68 (1967)
73. Buzzard, G.T.: Kupka-Smale theorem for automorphisms of $\mathbb{C}^n$. Duke Math. J. **93**(3), 487–503 (1998)
74. Buzzard, G.T.: Tame sets, dominating maps, and complex tori. Trans. Am. Math. Soc. **355**(6), 2557–2568 (electronic) (2003)
75. Buzzard, G.T., Fornæss, J.E.: Complete holomorphic vector fields and time-1 maps. Indiana Univ. Math. J. **44**(4), 1175–1182 (1995)
76. Buzzard, G.T., Fornæss, J.E.: An embedding of $\mathbb{C}$ in $\mathbb{C}^2$ with hyperbolic complement. Math. Ann. **306**(3), 539–546 (1996)
77. Buzzard, G.T., Forstnerič, F.: A Carleman type theorem for proper holomorphic embeddings. Ark. Mat. **35**(1), 157–169 (1997)
78. Buzzard, G.T., Forstnerič, F.: An interpolation theorem for holomorphic automorphisms of $\mathbb{C}^n$. J. Geom. Anal. **10**(1), 101–108 (2000)
79. Buzzard, G.T., Hubbard, J.H.: A Fatou-Bieberbach domain avoiding a neighborhood of a variety of codimension 2. Math. Ann. **316**(4), 699–702 (2000)
80. Buzzard, G.T., Lu, S.S.Y.: Algebraic surfaces holomorphically dominable by $\mathbb{C}^2$. Invent. Math. **139**(3), 617–659 (2000)
81. Campana, F.: Orbifolds, special varieties and classification theory. Ann. Inst. Fourier (Grenoble) **54**(3), 499–630 (2004)
82. Campana, F.: Orbifolds, special varieties and classification theory: an appendix. Ann. Inst. Fourier (Grenoble) **54**(3), 631–665 (2004)
83. Campana, F., Winkelmann, J.: On the h-principle and specialness for complex projective manifolds. Algebraic Geom. **2**(3), 298–314 (2015)
84. Cartan, H.: Sur les fonctions de deux variables complexes. Les transformations d'un domaine borné D en un domaine intérieur à D. Bull. Soc. Math. Fr. **58**, 199–219 (1930)
85. Cartan, H.: Les problèmes de Poincaré et de Cousin pour les fonctions de plusieurs variables complexes, vol. 199, pp. 1284–1287. C. R. Acad. Sci., Paris (1934)
86. Cartan, H.: Sur les matrices holomorphes de n variables complexes. J. Math. Pures Appl. **19**, 1–26 (1940)
87. Cartan, H.: Variétés analytiques complexes et cohomologie. In: Colloque sur les fonctions de plusieurs variables, tenu à Bruxelles, 1953, pp. 41–55. Georges Thone/Masson & Cie, Liège/Paris (1953)
88. Cartan, H.: Espaces fibrés analytiques. In: Symposium internacional de topología algebraica (International Symposium on Algebraic Topology), pp. 97–121. Universidad Nacional Autónoma de México and UNESCO, Mexico City (1958)
89. Cartan, H.: Sur les fonctions de plusieurs variables complexes: les espaces analytiques. In: Proc. Internat. Congress Math., vol. 1958, pp. 33–52. Cambridge Univ. Press, New York (1960)
90. Cartan, H.: In: Remmert, R., Serre, J-P. (eds.) Œuvres, vols. I, II, III. Springer, Berlin (1979)
91. Cartan, H., Thullen, P.: Zur Theorie der Singularitäten der Funktionen mehrerer komplexen Veränderlichen. Regularitäts- und Konvergenzbereiche. Math. Ann. **106**, 617–647 (1932)
92. Černe, M., Forstnerič, F.: Embedding some bordered Riemann surfaces in the affine plane. Math. Res. Lett. **9**(5–6), 683–696 (2002)
93. Černe, M., Globevnik, J.: On holomorphic embedding of planar domains into $\mathbb{C}^2$. J. Anal. Math. **81**, 269–282 (2000)
94. Chakrabarti, D.: Coordinate neighborhoods of arcs and the approximation of maps into (almost) complex manifolds. Mich. Math. J. **55**(2), 299–333 (2007)
95. Chen, S.-C., Shaw, M.-C.: Partial Differential Equations in Several Complex Variables. AMS/IP Studies in Advanced Mathematics, vol. 19. Am. Math. Soc./International Press, Providence/Boston (2001)

96. Chern, S.S., Moser, J.K.: Real hypersurfaces in complex manifolds. Acta Math. **133**, 219–271 (1974)
97. Chern, S.S., Spanier, E.H.: A theorem on orientable surfaces in four dimensional space. Comment. Math. Helv. **25**, 205–209 (1951)
98. Chirka, E.: Complex Analytic Sets. Kluwer Academic Publishers, Dordrecht (1989). Translated from the Russian by R.A.M. Hoksbergen
99. Cieliebak, K., Eliashberg, Y.: From Stein to Weinstein and Back. Symplectic Geometry of Affine Complex Manifolds. Am. Math. Soc. Colloquium Publications, vol. 59. Am. Math. Soc., Providence (2012)
100. Cieliebak, K., Eliashberg, Y.: The topology of rationally and polynomially convex domains. Invent. Math. **199**(1), 215–238 (2015)
101. Cohn, P.M.: On the structure of the GL_2 of a ring. Publ. Math. IHÉS **30**, 5–53 (1966)
102. Colţoiu, M.: Complete locally pluripolar sets. J. Reine Angew. Math. **412**, 108–112 (1990)
103. Colţoiu, M.: q-convexity. A survey. In: Complex Analysis and Geometry, Trento, 1995. Pitman Res. Notes Math. Ser., vol. 366, pp. 83–93. Longman, Harlow (1997)
104. Colţoiu, M.: On the Oka-Grauert principle for 1-convex manifolds. Math. Ann. **310**(3), 561–569 (1998)
105. Colţoiu, M., Mihalache, N.: On the homology groups of Stein spaces and Runge pairs. J. Reine Angew. Math. **371**, 216–220 (1986)
106. Cornalba, M., Griffiths, P.: Analytic cycles and vector bundles on non-compact algebraic varieties. Invent. Math. **28**, 1–106 (1975)
107. Cousin, P.: Sur les fonctions de n variables complexes. Acta Math. **19**(1), 1–61 (1895)
108. Cox, D.A., Little, J.B., Schenck, H.K.: Toric Varieties. Graduate Studies in Mathematics, vol. 124. Am. Math. Soc., Providence (2011)
109. De Fabritiis, C.: On continuous dynamics of automorphisms of $\mathbb{C}^2$. Manuscr. Math. **77**(4), 337–359 (1992)
110. Demailly, J.-P.: Complex analytic and differential geometry. https://www-fourier.ujf-grenoble.fr/~demailly/manuscripts/agbook.pdf
111. Demailly, J.-P.: Différents exemples de fibrés holomorphes non de Stein. In: Séminaire Pierre Lelong-Henri Skoda (Analyse), Année 1976/77. Lecture Notes in Math., vol. 694, pp. 15–41. Springer, Berlin (1978)
112. Demailly, J.-P.: Un exemple de fibré holomorphe non de Stein à fibre $\mathbb{C}^2$ ayant pour base le disque ou le plan. Invent. Math. **48**(3), 293–302 (1978)
113. Demailly, J.-P.: Cohomology of q-convex spaces in top degrees. Math. Z. **204**(2), 283–295 (1990)
114. Demailly, J.-P.: Algebraic criteria for Kobayashi hyperbolic projective varieties and jet differentials. In: Algebraic Geometry—Santa Cruz 1995. Proc. Sympos. Pure Math., vol. 62, pp. 285–360. Am. Math. Soc., Providence, RI (1997)
115. Demailly, J.-P.: Holomorphic Morse inequalities and the Green-Griffiths-Lang conjecture. Pure Appl. Math. Q. **7**(4), 1165–1207 (2011). Special Issue: In memory of Eckart Viehweg
116. Demailly, J.-P., El Goul, J.: Hyperbolicity of generic surfaces of high degree in projective 3-space. Am. J. Math. **122**(3), 515–546 (2000)
117. Demailly, J.-P., Lempert, L., Shiffman, B.: Algebraic approximations of holomorphic maps from Stein domains to projective manifolds. Duke Math. J. **76**(2), 333–363 (1994)
118. Derdzinski, A., Januszkiewicz, T.: Totally real immersions of surfaces. Trans. Am. Math. Soc. **362**(1), 53–115 (2010)
119. Derksen, H., Kutzschebauch, F.: Nonlinearizable holomorphic group actions. Math. Ann. **311**(1), 41–53 (1998)
120. Diederich, K., Fornaess, J.E.: Pseudoconvex domains: an example with nontrivial Nebenhülle. Math. Ann. **225**(3), 275–292 (1977)
121. Diederich, K., Fornaess, J.E.: Pseudoconvex domains: bounded strictly plurisubharmonic exhaustion functions. Invent. Math. **39**(2), 129–141 (1977)
122. Diverio, S., Merker, J., Rousseau, E.: Effective algebraic degeneracy. Invent. Math. **180**(1), 161–223 (2010)

123. Diverio, S., Rousseau, E.: Hyperbolicity of Projective Hypersurfaces, 2nd edn. IMPA Monographs, vol. 5. Springer, Cham (2016)
124. Diverio, S., Trapani, S.: Quasi-negative holomorphic sectional curvature and positivity of the canonical bundle. ArXiv e-prints (2016). arXiv:1606.01381
125. Dixon, P.G., Esterle, J.: Michael's problem and the Poincaré-Fatou-Bieberbach phenomenon. Bull., New Ser., Am. Math. Soc. **15**(2), 127–187 (1986)
126. Dloussky, G.: From non-Kählerian surfaces to Cremona group of $\mathbb{P}^2(\mathbb{C})$. Complex Manifolds **1**, 1–33 (2014)
127. Dloussky, G., Oeljeklaus, K.: Vector fields and foliations on compact surfaces of class VII_0. Ann. Inst. Fourier (Grenoble) **49**(5), 1503–1545 (1999)
128. Docquier, F., Grauert, H.: Levisches Problem und Rungescher Satz für Teilgebiete Steinscher Mannigfaltigkeiten. Math. Ann. **140**, 94–123 (1960)
129. Dold, A.: Lectures on Algebraic Topology. Classics in Mathematics. Springer, Berlin (1995). Reprint of the 1972 edition
130. Donaldson, S.: Riemann Surfaces. Oxford Graduate Texts in Mathematics, vol. 22. Oxford University Press, Oxford (2011)
131. Donaldson, S.K.: Symplectic submanifolds and almost-complex geometry. J. Differ. Geom. **44**(4), 666–705 (1996)
132. Dor, A.: Proper holomorphic maps between balls in one co-dimension. Ark. Mat. **28**(1), 49–100 (1990)
133. Dor, A.: Approximation by proper holomorphic maps into convex domains. Ann. Sc. Norm. Super. Pisa, Cl. Sci. (4) **20**(1), 147–162 (1993)
134. Dor, A.: Immersions and embeddings in domains of holomorphy. Trans. Am. Math. Soc. **347**(8), 2813–2849 (1995)
135. Drinovec Drnovšek, B., Forstnerič, F.: Holomorphic curves in complex spaces. Duke Math. J. **139**(2), 203–253 (2007)
136. Drinovec-Drnovšek, B., Forstnerič, F.: Approximation of holomorphic mappings on strongly pseudoconvex domains. Forum Math. **20**(5), 817–840 (2008)
137. Drinovec Drnovšek, B., Forstnerič, F.: Strongly pseudoconvex domains as subvarieties of complex manifolds. Am. J. Math. **132**(2), 331–360 (2010)
138. Dubouloz, A., Moser-Jauslin, L., Poloni, P.-M.: Inequivalent embeddings of the Koras-Russell cubic 3-fold. Mich. Math. J. **59**(3), 679–694 (2010)
139. Dwyer, W.G., Spaliński, J.: Homotopy theories and model categories. In: Handbook of Algebraic Topology, pp. 73–126. North-Holland, Amsterdam (1995)
140. Eisenbud, D., Evans, E.G. Jr.: Every algebraic set in n-space is the intersection of n hypersurfaces. Invent. Math. **19**, 107–112 (1973)
141. Eisenman, D.A.: Intrinsic Measures on Complex Manifolds and Holomorphic Mappings. Memoirs of the Am. Math. Soc., vol. 96. Am. Math. Soc., Providence (1970)
142. El Kasimi, A.: Approximation polynômiale dans les domaines étoilés de $\mathbb{C}^n$. Complex Var. Theory Appl. **10**(2–3), 179–182 (1988)
143. Elgindi, A.M.: On the topological structure of complex tangencies to embeddings of S^3 into $\mathbb{C}^3$. N.Y. J. Math. **18**, 295–313 (2012)
144. Elgindi, A.M.: A topological obstruction to the removal of a degenerate complex tangent and some related homotopy and homology groups. Int. J. Math. **26**(5), 16 (2015)
145. Elgindi, A.M.: Totally real perturbations and nondegenerate embeddings of S^3. N.Y. J. Math. **21**, 1283–1293 (2015)
146. Eliashberg, Y.: Classification of overtwisted contact structures on 3-manifolds. Invent. Math. **98**(3), 623–637 (1989)
147. Eliashberg, Y.: Filling by holomorphic discs and its applications. In: Geometry of Low-Dimensional Manifolds, 2, Durham, 1989. London Math. Soc. Lecture Note Ser., vol. 151, pp. 45–67. Cambridge Univ. Press, Cambridge (1990)
148. Eliashberg, Y.: Topological characterization of Stein manifolds of dimension > 2. Int. J. Math. **1**(1), 29–46 (1990)

149. Eliashberg, Y.: Legendrian and transversal knots in tight contact 3-manifolds. In: Topological Methods in Modern Mathematics, Stony Brook, NY, 1991, pp. 171–193. Publish or Perish, Houston (1993)
150. Eliashberg, Y., Gromov, M.: Embeddings of Stein manifolds of dimension n into the affine space of dimension $3n/2+1$. Ann. Math. (2) **136**(1), 123–135 (1992)
151. Eliashberg, Y., Kharlamov, V.: Some remarks on the number of complex points of a real surface in a complex one. In: Proc. Leningrad Int. Topology Conf., 1982. Nauka Leningrad Otdel., Leningrad (1983)
152. Eliashberg, Y., Mishachev, N.: Introduction to the h-Principle. Graduate Studies in Mathematics, vol. 48. Am. Math. Soc., Providence (2002)
153. Enoki, I.: On surfaces of class VII_0 with curves. Proc. Jpn. Acad., Ser. A, Math. Sci. **56**(6), 275–279 (1980)
154. Enoki, I.: Surfaces of class VII_0 with curves. Tohoku Math. J. (2) **33**(4), 453–492 (1981)
155. Eremenko, A.: Exceptional values in holomorphic families of entire functions. Mich. Math. J. **54**(3), 687–696 (2006)
156. Fatou, P.: Sur les équations fonctionnelles. Bull. Soc. Math. Fr. **47**, 161–271 (1919)
157. Fintushel, R., Stern, R.J.: Immersed spheres in 4-manifolds and the immersed Thom conjecture. Turk. J. Math. **19**(2), 145–157 (1995)
158. Fischer, G.: Complex Analytic Geometry. Lecture Notes in Math., vol. 538. Springer, Berlin (1976)
159. Fischer, W., Grauert, H.: Lokal-triviale Familien kompakter komplexer Mannigfaltigkeiten. Nachr. Akad. Wiss. Gött. Math.-Phys. Kl., 2B **1965**, 89–94 (1965)
160. Flenner, H., Kaliman, S., Zaidenberg, M.: A Gromov-Winkelmann type theorem for flexible varieties. J. Eur. Math. Soc. **18**(11), 2483–2510 (2016)
161. Flenner, H., Zaidenberg, M.: Locally nilpotent derivations on affine surfaces with a $\mathbb{C}^*$-action. Osaka J. Math. **42**(4), 931–974 (2005)
162. Fornæss, J.E.: Short $\mathbb{C}^k$. In: Complex Analysis in Several Variables—Memorial Conference of Kiyoshi Oka's Centennial Birthday. Adv. Stud. Pure Math., vol. 42, pp. 95–108. Math. Soc. Japan, Tokyo (2004)
163. Fornæss, J.E., Narasimhan, R.: The Levi problem on complex spaces with singularities. Math. Ann. **248**(1), 47–72 (1980)
164. Fornæss, J.E., Sibony, N.: Increasing sequences of complex manifolds. Math. Ann. **255**(3), 351–360 (1981)
165. Fornæss, J.E., Sibony, N.: Complex Hénon mappings in $\mathbb{C}^2$ and Fatou-Bieberbach domains. Duke Math. J. **65**(2), 345–380 (1992)
166. Fornæss, J.E., Sibony, N.: Complex dynamics in higher dimension. In: Several Complex Variables, Berkeley, CA, 1995–1996. Math. Sci. Res. Inst. Publ., vol. 37, pp. 273–296. Cambridge Univ. Press, Cambridge (1999)
167. Fornæss, J.E., Stensønes, B.: Stable manifolds of holomorphic hyperbolic maps. Int. J. Math. **15**(8), 749–758 (2004)
168. Fornæss, J.E., Wold, E.F.: Non-autonomous basins with uniform bounds are elliptic. Proc. Am. Math. Soc. **144**(11), 4709–4714 (2016)
169. Forster, O.: Some remarks on parallelizable Stein manifolds. Bull. Am. Math. Soc. **73**, 712–716 (1967)
170. Forster, O.: Zur Theorie der Steinschen Algebren und Moduln. Math. Z. **97**, 376–405 (1967)
171. Forster, O.: Plongements des variétés de Stein. Comment. Math. Helv. **45**, 170–184 (1970)
172. Forster, O.: Complete intersections in affine algebraic varieties and Stein spaces. In: Complete Intersections, Acireale, 1983. Lecture Notes in Math., vol. 1092, pp. 1–28. Springer, Berlin (1984)
173. Forster, O.: Lectures on Riemann Surfaces. Graduate Texts in Mathematics, vol. 81. Springer, New York (1991). Translated from the 1977 German original by Bruce Gilligan. Reprint of the 1981 English translation
174. Forster, O., Ramspott, K.J.: Singularitätenfreie analytische Raumkurven als vollständige Durchschnitte. Bayer. Akad. Wiss. Math.-Natur. Kl. S.-B. **1965**(II), 1–10 (1966)

175. Forster, O., Ramspott, K.J.: Analytische Modulgarben und Endromisbündel. Invent. Math. **2**, 145–170 (1966)
176. Forster, O., Ramspott, K.J.: Okasche Paare von Garben nicht-abelscher Gruppen. Invent. Math. **1**, 260–286 (1966)
177. Forster, O., Ramspott, K.J.: Homotopieklassen von Idealbasen in Steinschen Algebren. Invent. Math. **5**, 255–276 (1968)
178. Forster, O., Ramspott, K.J.: Über die Anzahl der Erzeugenden von projektiven Steinschen Moduln. Arch. Math. (Basel) **19**, 417–422 (1968)
179. Forstnerič, F.: Surjective holomorphic maps onto Oka manifolds. In: Proceedings of the INdAM Meeting Complex and Symplectic Geometry, Cortona, 2016. Springer INdAM Series. In press. http://www.springer.com/series/10283
180. Forstnerič, F.: On totally real embeddings into $\mathbb{C}^n$. Expo. Math. **4**(3), 243–255 (1986)
181. Forstnerič, F.: Some totally real embeddings of three-manifolds. Manuscr. Math. **55**(1), 1–7 (1986)
182. Forstnerič, F.: Complex tangents of real surfaces in complex surfaces. Duke Math. J. **67**(2), 353–376 (1992)
183. Forstnerič, F.: Approximation by automorphisms on smooth submanifolds of $\mathbb{C}^n$. Math. Ann. **300**(4), 719–738 (1994)
184. Forstnerič, F.: Complements of Runge domains and holomorphic hulls. Mich. Math. J. **41**(2), 297–308 (1994)
185. Forstnerič, F.: Equivalence of real submanifolds under volume-preserving holomorphic automorphisms of $\mathbb{C}^n$. Duke Math. J. **77**(2), 431–445 (1995)
186. Forstnerič, F.: A theorem in complex symplectic geometry. J. Geom. Anal. **5**(3), 379–393 (1995)
187. Forstnerič, F.: Actions of $(\mathbb{R}, +)$ and $(\mathbb{C}, +)$ on complex manifolds. Math. Z. **223**(1), 123–153 (1996)
188. Forstnerič, F.: Interpolation by holomorphic automorphisms and embeddings in $\mathbb{C}^n$. J. Geom. Anal. **9**(1), 93–117 (1999)
189. Forstnerič, F.: On complete intersections. Ann. Inst. Fourier (Grenoble) **51**(2), 497–512 (2001)
190. Forstnerič, F.: The Oka principle for sections of subelliptic submersions. Math. Z. **241**(3), 527–551 (2002)
191. Forstnerič, F.: Noncritical holomorphic functions on Stein manifolds. Acta Math. **191**(2), 143–189 (2003)
192. Forstnerič, F.: The Oka principle for multivalued sections of ramified mappings. Forum Math. **15**(2), 309–328 (2003)
193. Forstnerič, F.: Stein domains in complex surfaces. J. Geom. Anal. **13**(1), 77–94 (2003)
194. Forstnerič, F.: Holomorphic submersions from Stein manifolds. Ann. Inst. Fourier (Grenoble) **54**(6), 1913–1942 (2005). 2004
195. Forstnerič, F.: Extending holomorphic mappings from subvarieties in Stein manifolds. Ann. Inst. Fourier (Grenoble) **55**(3), 733–751 (2005)
196. Forstnerič, F.: Holomorphic flexibility properties of complex manifolds. Am. J. Math. **128**(1), 239–270 (2006)
197. Forstnerič, F.: Runge approximation on convex sets implies the Oka property. Ann. Math. (2) **163**(2), 689–707 (2006)
198. Forstnerič, F.: Manifolds of holomorphic mappings from strongly pseudoconvex domains. Asian J. Math. **11**(1), 113–126 (2007)
199. Forstnerič, F.: Oka manifolds. C. R. Math. Acad. Sci. Paris **347**(17–18), 1017–1020 (2009)
200. Forstnerič, F.: Invariance of the parametric Oka property. In: Complex Analysis. Trends Math., pp. 125–144. Birkhäuser/Springer, Basel AG/Basel (2010)
201. Forstnerič, F.: Oka maps. C. R. Math. Acad. Sci. Paris **348**(3–4), 145–148 (2010)
202. Forstnerič, F.: The Oka principle for sections of stratified fiber bundles. Pure Appl. Math. Q. **6**(3), 843–874 (2010). Special Issue: In honor of Joseph J. Kohn. Part 1

203. Forstnerič, F.: Holomorphic families of long $\mathbb{C}^2$'s. Proc. Am. Math. Soc. **140**(7), 2383–2389 (2012)
204. Forstnerič, F.: Oka manifolds: from Oka to Stein and back. Ann. Fac. Sci. Toulouse Math. (6) **22**(4), 747–809 (2013). With an appendix by Finnur Lárusson
205. Forstnerič, F.: A complex surface admitting a strongly plurisubharmonic function but no holomorphic functions. J. Geom. Anal. **25**(1), 329–335 (2015)
206. Forstnerič, F.: Noncritical holomorphic functions on Stein spaces. J. Eur. Math. Soc. **18**(11), 2511–2543 (2016)
207. Forstnerič, F., Globevnik, J., Rosay, J.-P.: Nonstraightenable complex lines in $\mathbb{C}^2$. Ark. Mat. **34**(1), 97–101 (1996)
208. Forstnerič, F., Ivarsson, B., Kutzschebauch, F., Prezelj, J.: An interpolation theorem for proper holomorphic embeddings. Math. Ann. **338**(3), 545–554 (2007)
209. Forstnerič, F., Kozak, J.: Strongly pseudoconvex handlebodies. J. Korean Math. Soc. **40**(4), 727–745 (2003)
210. Forstnerič, F., Lárusson, F.: The parametric h-principle for minimal surfaces in $\mathbb{R}^n$ and null curves in $\mathbb{C}^n$. Comm. Anal. Geom. In press
211. Forstnerič, F., Lárusson, F.: Survey of Oka theory. N.Y. J. Math. **17A**, 11–38 (2011)
212. Forstnerič, F., Lárusson, F.: Holomorphic flexibility properties of compact complex surfaces. Int. Math. Res. Not. **13**, 3714–3734 (2014)
213. Forstnerič, F., Lárusson, F.: Oka properties of groups of holomorphic and algebraic automorphisms of complex affine space. Math. Res. Lett. **21**(5), 1047–1067 (2014)
214. Forstnerič, F., Løw, E.: Global holomorphic equivalence of smooth submanifolds in $\mathbb{C}^n$. Indiana Univ. Math. J. **46**(1), 133–153 (1997)
215. Forstnerič, F., Løw, E., Øvrelid, N.: Solving the d- and $\overline{\partial}$-equations in thin tubes and applications to mappings. Mich. Math. J. **49**(2), 369–416 (2001)
216. Forstnerič, F., Ohsawa, T.: Gunning-Narasimhan's theorem with a growth condition. J. Geom. Anal. **23**(3), 1078–1084 (2013)
217. Forstnerič, F., Prezelj, J.: Oka's principle for holomorphic fiber bundles with sprays. Math. Ann. **317**(1), 117–154 (2000)
218. Forstnerič, F., Prezelj, J.: Extending holomorphic sections from complex subvarieties. Math. Z. **236**(1), 43–68 (2001)
219. Forstnerič, F., Prezelj, J.: Oka's principle for holomorphic submersions with sprays. Math. Ann. **322**(4), 633–666 (2002)
220. Forstnerič, F., Ritter, T.: Oka properties of ball complements. Math. Z. **277**(1–2), 325–338 (2014)
221. Forstnerič, F., Rosay, J.-P.: Approximation of biholomorphic mappings by automorphisms of $\mathbb{C}^n$. Invent. Math. **112**(2), 323–349 (1993). Erratum: Invent. Math. **118**(3), 573–574 (1994)
222. Forstnerič, F., Slapar, M.: Deformations of Stein structures and extensions of holomorphic mappings. Math. Res. Lett. **14**(2), 343–357 (2007)
223. Forstnerič, F., Slapar, M.: Stein structures and holomorphic mappings. Math. Z. **256**(3), 615–646 (2007)
224. Forstnerič, F., Smrekar, J., Sukhov, A.: On the Hodge conjecture for q-complete manifolds. Geom. Topol. **20**(1), 353–388 (2016)
225. Forstnerič, F., Stout, E.L.: A new class of polynomially convex sets. Ark. Mat. **29**(1), 51–62 (1991)
226. Forstnerič, F., Wold, E.F.: Bordered Riemann surfaces in $\mathbb{C}^2$. J. Math. Pures Appl. (9) **91**(1), 100–114 (2009)
227. Forstnerič, F., Wold, E.F.: Fibrations and Stein neighborhoods. Proc. Am. Math. Soc. **138**(6), 2037–2042 (2010)
228. Forstnerič, F., Wold, E.F.: Embeddings of infinitely connected planar domains into $\mathbb{C}^2$. Anal. PDE **6**(2), 499–514 (2013)
229. Forstnerič, F., Wold, E.F.: Fatou-Bieberbach domains in $\mathbb{C}^n \setminus \mathbb{R}^k$. Ark. Mat. **53**(2), 259–270 (2015)

230. Forstnerič, F.: Proper holomorphic immersions into Stein manifolds with the density property. ArXiv e-prints (2017). arXiv:1703.08594
231. Freedman, M.H., Quinn, F.: Topology of 4-Manifolds. Princeton Mathematical Series, vol. 39. Princeton University Press, Princeton (1990)
232. Frenkel, J.: Cohomologie non abélienne et espaces fibrés. Bull. Soc. Math. Fr. **85**, 135–220 (1957)
233. Friedland, S., Milnor, J.: Dynamical properties of plane polynomial automorphisms. Ergod. Theory Dyn. Syst. **9**(1), 67–99 (1989)
234. Friedman, R., Morgan, J.W.: Smooth Four-Manifolds and Complex Surfaces. Ergebnisse der Mathematik und ihrer Grenzgebiete, 3. Folge, vol. 27. Springer, Berlin (1994)
235. Fulton, W.: Intersection Theory, 2nd edn. Ergebnisse der Mathematik und ihrer Grenzgebiete, 3. Folge, vol. 2. Springer, Berlin (1998)
236. Givental', A.: Lagrangian imbeddings of surfaces and unfolded Whitney umbrella. Funct. Anal. Appl. **20**, 197–203 (1986)
237. Globevnik, J.: A bounded domain in $\mathbb{C}^N$ which embeds holomorphically into $\mathbb{C}^{N+1}$. Ark. Mat. **35**(2), 313–325 (1997)
238. Globevnik, J.: On Fatou-Bieberbach domains. Math. Z. **229**(1), 91–106 (1998)
239. Globevnik, J.: Interpolation by proper holomorphic embeddings of the disc into $\mathbb{C}^2$. Math. Res. Lett. **9**(4), 567–577 (2002)
240. Globevnik, J.: A complete complex hypersurface in the ball of $\mathbb{C}^N$. Ann. Math. (2) **182**(3), 1067–1091 (2015)
241. Globevnik, J.: Holomorphic functions unbounded on curves of finite length. Math. Ann. **364**(3–4), 1343–1359 (2016)
242. Globevnik, J., Stensønes, B.: Holomorphic embeddings of planar domains into $\mathbb{C}^2$. Math. Ann. **303**(4), 579–597 (1995)
243. Goerss, P.G., Jardine, J.F.: Simplicial Homotopy Theory. Progress in Mathematics, vol. 174. Birkhäuser, Basel (1999)
244. Golubitsky, M., Guillemin, V.: Stable Mappings and Their Singularities. Graduate Texts in Mathematics, vol. 14. Springer, New York (1973)
245. Gompf, R.E.: Minimal genera of open 4-manifolds. Geom. Topol. **21**(1), 107–155 (2017)
246. Gompf, R.E.: Handlebody construction of Stein surfaces. Ann. Math. (2) **148**(2), 619–693 (1998)
247. Gompf, R.E.: Stein surfaces as open subsets of $\mathbb{C}^2$. J. Symplectic Geom. **3**(4), 565–587 (2005)
248. Gompf, R.E., Stipsicz, A.I.: 4-Manifolds and Kirby Calculus. Graduate Studies in Mathematics, vol. 20. Am. Math. Soc., Providence (1999)
249. Gong, X.: On totally real spheres in complex space. Math. Ann. **309**(4), 611–623 (1997)
250. Goresky, M., MacPherson, R.: Stratified Morse Theory. Ergebnisse der Mathematik und ihrer Grenzgebiete, 3. Folge, vol. 14. Springer, Berlin (1988)
251. Grauert, H.: Charakterisierung der holomorph vollständigen komplexen Räume. Math. Ann. **129**, 233–259 (1955)
252. Grauert, H.: Approximationssätze für holomorphe Funktionen mit Werten in komplexen Rämen. Math. Ann. **133**, 139–159 (1957)
253. Grauert, H.: Holomorphe Funktionen mit Werten in komplexen Lieschen Gruppen. Math. Ann. **133**, 450–472 (1957)
254. Grauert, H.: Analytische Faserungen über holomorph-vollständigen Räumen. Math. Ann. **135**, 263–273 (1958)
255. Grauert, H.: On Levi's problem and the imbedding of real-analytic manifolds. Ann. Math. (2) **68**, 460–472 (1958)
256. Grauert, H.: Theory of q-convexity and q-concavity. In: Several Complex Variables, VII. Encyclopaedia Math. Sci., vol. 74, pp. 259–284. Springer, Berlin (1994)
257. Grauert, H., Kerner, H.: Approximation von holomorphen Schnittflächen in Faserbündeln mit homogener Faser. Arch. Math. (Basel) **14**, 328–333 (1963)
258. Grauert, H., Remmert, R.: Komplexe Räume. Math. Ann. **136**, 245–318 (1958)

259. Grauert, H., Remmert, R.: Analytische Stellenalgebren. Grundlehren Math. Wiss., vol. 176. Springer, Berlin (1971). Unter Mitarbeit von O. Riemenschneider
260. Grauert, H., Remmert, R.: Theory of Stein Spaces. Grundlehren Math. Wiss., vol. 236. Springer, Berlin (1979). Translated from the German by Alan Huckleberry
261. Grauert, H., Remmert, R.: Coherent Analytic Sheaves. Grundlehren Math. Wiss., vol. 265. Springer, Berlin (1984)
262. Green, M.L.: Holomorphic maps into complex projective space omitting hyperplanes. Trans. Am. Math. Soc. **169**, 89–103 (1972)
263. Greene, R.E., Wu, H.: Embedding of open Riemannian manifolds by harmonic functions. Ann. Inst. Fourier (Grenoble) **25**(1), 215–235 (1975)
264. Griffiths, P., Harris, J.: Principles of Algebraic Geometry. Wiley Classics Library. John Wiley & Sons, New York (1994). Reprint of the 1978 original
265. Gromov, M.: Partial Differential Relations. Ergebnisse der Mathematik und ihrer Grenzgebiete, 3. Folge, vol. 9. Springer, Berlin (1986)
266. Gromov, M.: Oka's principle for holomorphic sections of elliptic bundles. J. Am. Math. Soc. **2**(4), 851–897 (1989)
267. Gromov, M.L.: Stable mappings of foliations into manifolds. Izv. Akad. Nauk SSSR, Ser. Mat. **33**, 707–734 (1969)
268. Gromov, M.L.: Convex integration of differential relations. I. Izv. Akad. Nauk SSSR, Ser. Mat. **37**, 329–343 (1973)
269. Gromov, M.L., Eliashberg, J.M.: Nonsingular mappings of Stein manifolds. Funkc. Anal. Ego Prilož. **5**(2), 82–83 (1971)
270. Grothendieck, A.: On the de Rham cohomology of algebraic varieties. Inst. Hautes Études Sci. Publ. Math. **29**, 95–103 (1966)
271. Grunewald, F., Mennicke, J., Vaserstein, L.: On the groups $SL_2(\mathbf{Z}[x])$ and $SL_2(k[x, y])$. Isr. J. Math. **86**(1–3), 157–193 (1994)
272. Gunning, R.C.: Introduction to Holomorphic Functions of Several Variables, vol. III: Homological Theory. The Wadsworth & Brooks/Cole Mathematics Series. Wadsworth & Brooks/Cole Advanced Books & Software, Monterey (1990)
273. Gunning, R.C., Narasimhan, R.: Immersion of open Riemann surfaces. Math. Ann. **174**, 103–108 (1967)
274. Gunning, R.C., Rossi, H.: Analytic Functions of Several Complex Variables. AMS Chelsea Publishing, Providence (2009). Reprint of the 1965 original
275. Guralnick, R.M.: Similarity of holomorphic matrices. Linear Algebra Appl. **99**, 85–96 (1988)
276. Hamm, H.A.: Zur Homotopietyp Steinscher Räume. J. Reine Angew. Math. **338**, 121–135 (1983)
277. Hamm, H.A.: Zum Homotopietyp q-vollständiger Räume. J. Reine Angew. Math. **364**, 1–9 (1986)
278. Hanysz, A.: Oka properties of some hypersurface complements. Proc. Am. Math. Soc. **142**(2), 483–496 (2014)
279. Hanysz, A.: Holomorphic flexibility properties of the space of cubic rational maps. J. Geom. Anal. **25**(3), 1620–1649 (2015)
280. Hartogs, F.: Zur Theorie der analytischen Funktionen mehrerer unabhängiger Veränderlichen, insbesondere über die Darstellung derselben durch Reihen, welche nach Potenzen einer Veränderlichen fortschreiten. Math. Ann. **62**(1), 1–88 (1906)
281. Hartshorne, R.: Algebraic Geometry. Graduate Texts in Mathematics, vol. 52. Springer, New York (1977)
282. Harvey, F.R., Wells, R.O. Jr.: Zero sets of non-negative strictly plurisubharmonic functions. Math. Ann. **201**, 165–170 (1973)
283. Hatcher, A.E.: Notes on basic 3-manifold topology. http://www.math.cornell.edu/~hatcher/3M/3Mdownloads.html
284. He, Z.-X., Schramm, O.: Fixed points, Koebe uniformization and circle packings. Ann. Math. (2) **137**(2), 369–406 (1993)

285. He, Z.-X., Schramm, O.: Koebe uniformization for "almost circle domains". Am. J. Math. **117**(3), 653–667 (1995)
286. Heinzner, P., Kutzschebauch, F.: An equivariant version of Grauert's Oka principle. Invent. Math. **119**(2), 317–346 (1995)
287. Henkin, G., Leiterer, J.: The Oka-Grauert principle without induction over the base dimension. Math. Ann. **311**(1), 71–93 (1998)
288. Henkin, G.M.: Integral representation of functions which are holomorphic in strictly pseudoconvex regions, and some applications. Mat. Sb. (N.S.) **78**(120), 611–632 (1969)
289. Henkin, G.M., Leiterer, J.: Theory of functions on complex manifolds. Mathematische Lehrbücher und Monographien, II. Abteilung: Mathematische Monographien [Mathematical Textbooks and Monographs, Part II: Mathematical Monographs], vol. 60. Akademie-Verlag, Berlin (1984)
290. Hénon, M.: A two-dimensional mapping with a strange attractor. Commun. Math. Phys. **50**(1), 69–77 (1976)
291. Heunemann, D.: An approximation theorem and Oka's principle for holomorphic vector bundles which are continuous on the boundary of strictly pseudoconvex domains. Math. Nachr. **127**, 275–280 (1986)
292. Heunemann, D.: Extension of the complex structure from Stein manifolds with strictly pseudoconvex boundary. Math. Nachr. **128**, 57–64 (1986)
293. Heunemann, D.: Theorem B for Stein manifolds with strictly pseudoconvex boundary. Math. Nachr. **128**, 87–101 (1986)
294. Hill, C.D., Taiani, G.: Families of analytic discs in $\mathbb{C}^n$ with boundaries on a prescribed CR submanifold. Ann. Sc. Norm. Super. Pisa, Cl. Sci. (4) **5**(2), 327–380 (1978)
295. Hirsch, M.W.: Immersions of manifolds. Trans. Am. Math. Soc. **93**, 242–276 (1959)
296. Hirschhorn, P.S.: Model Categories and Their Localizations. Mathematical Surveys and Monographs, vol. 99. Am. Math. Soc., Providence (2003)
297. Hirzebruch, F., Hopf, H.: Felder von Flächenelementen in 4-dimensionalen Mannigfaltigkeiten. Math. Ann. **136**, 156–172 (1958)
298. Ho, P.T., Jacobowitz, H., Landweber, P.: Optimality for totally real immersions and independent mappings of manifolds into $\mathbb{C}^N$. N.Y. J. Math. **18**, 463–477 (2012)
299. Hörmander, L.: L^2 estimates and existence theorems for the $\bar{\partial}$ operator. Acta Math. **113**, 89–152 (1965)
300. Hörmander, L.: An Introduction to Complex Analysis in Several Variables, 3rd edn. North-Holland Mathematical Library, vol. 7. North-Holland Publishing Co., Amsterdam (1990)
301. Hörmander, L.: Notions of Convexity. Progress in Mathematics, vol. 127. Birkhäuser Boston, Boston (1994)
302. Hörmander, L., Wermer, J.: Uniform approximation on compact sets in $\mathbb{C}^n$. Math. Scand. **23**, 5–21 (1969)
303. Hovey, M.: Model Categories. Mathematical Surveys and Monographs, vol. 63. Am. Math. Soc., Providence (1999)
304. Hurewicz, W., Wallman, H.: Dimension Theory. Princeton Mathematical Series, vol. 4. Princeton University Press, Princeton (1941)
305. Husemoller, D.: Fibre Bundles, 3rd edn. Graduate Texts in Mathematics, vol. 20. Springer, New York (1994)
306. Ivarsson, B., Kutzschebauch, F.: Holomorphic factorization of mappings into $SL_n(\mathbb{C})$. Ann. Math. (2) **175**(1), 45–69 (2012)
307. Ivarsson, B., Kutzschebauch, F.: On the number of factors in the unipotent factorization of holomorphic mappings into $SL_2(\mathbb{C})$. Proc. Am. Math. Soc. **140**(3), 823–838 (2012)
308. Jacobson, N.: Basic Algebra. I. W. H. Freeman and Co., San Francisco (1974)
309. Jardine, J.F.: Intermediate model structures for simplicial presheaves. Can. Math. Bull. **49**(3), 407–413 (2006)
310. Jelonek, Z.: The Jacobian conjecture and the extensions of polynomial embeddings. Math. Ann. **294**(2), 289–293 (1992)

311. Jelonek, Z.: A hypersurface which has the Abhyankar-Moh property. Math. Ann. **308**(1), 73–84 (1997)
312. Jonsson, M., Varolin, D.: Stable manifolds of holomorphic diffeomorphisms. Invent. Math. **149**(2), 409–430 (2002)
313. Jöricke, B.: Local polynomial hulls of discs near isolated parabolic points. Indiana Univ. Math. J. **46**(3), 789–826 (1997)
314. Jöricke, B.: Envelopes of holomorphy and holomorphic discs. Invent. Math. **178**(1), 73–118 (2009)
315. Josefson, B.: On the equivalence between locally polar and globally polar sets for plurisubharmonic functions on $\mathbb{C}^n$. Ark. Mat. **16**(1), 109–115 (1978)
316. Jouanolou, J.P.: Une suite exacte de Mayer-Vietoris en K-théorie algébrique. In: Algebraic K-Theory, I: Higher K-Theories, Proc. Conf., Battelle Memorial Inst., Seattle, Wash., 1972. Lecture Notes in Math., vol. 341, pp. 293–316. Springer, Berlin (1973)
317. Jung, H.W.E.: Über ganze birationale Transformationen der Ebene. J. Reine Angew. Math. **184**, 161–174 (1942)
318. Kaliman, S.: Extensions of isomorphisms between affine algebraic subvarieties of k^n to automorphisms of k^n. Proc. Am. Math. Soc. **113**(2), 325–334 (1991)
319. Kaliman, S.: Isotopic embeddings of affine algebraic varieties into $\mathbb{C}^n$. In: The Madison Symposium on Complex Analysis, Madison, WI, 1991. Contemp. Math., vol. 137, pp. 291–295. Am. Math. Soc., Providence (1992)
320. Kaliman, S.: Some facts about Eisenman intrinsic measures. Complex Var. Theory Appl. **27**(2), 163–173 (1995)
321. Kaliman, S., Kutzschebauch, F.: Algebraic (volume) density property for affine homogeneous spaces. Math. Ann. **367**(3–4), 1311–1332 (2017). doi:10.1007/s00208-016-1451-9
322. Kaliman, S., Kutzschebauch, F.: Criteria for the density property of complex manifolds. Invent. Math. **172**(1), 71–87 (2008)
323. Kaliman, S., Kutzschebauch, F.: Density property for hypersurfaces $UV = P(\overline{X})$. Math. Z. **258**(1), 115–131 (2008)
324. Kaliman, S., Kutzschebauch, F.: Algebraic volume density property of affine algebraic manifolds. Invent. Math. **181**(3), 605–647 (2010)
325. Kaliman, S., Kutzschebauch, F.: On the density and the volume density property. In: Complex Analysis and Geometry. Springer Proc. Math. Stat., vol. 144, pp. 175–186. Springer, Tokyo (2015)
326. Kaliman, S., Kutzschebauch, F.: On algebraic volume density property. Transform. Groups **21**(2), 451–478 (2016)
327. Kaliman, S., Kutzschebauch, F., Truong, T.T.: On subelliptic manifolds. ArXiv e-prints (2016). arXiv:1611.01311
328. Kaliman, S., Zaĭdenberg, M.: A transversality theorem for holomorphic mappings and stability of Eisenman-Kobayashi measures. Trans. Am. Math. Soc. **348**(2), 661–672 (1996)
329. Kallin, E.: Polynomial convexity: the three spheres problem. In: Proc. Conf. Complex Analysis, Minneapolis, 1964, pp. 301–304. Springer, Berlin (1965)
330. Kasuya, N., Takase, M.: Knots and links of complex tangents. ArXiv e-prints (2016). arXiv:1606.03704
331. Kaup, L.: Eine topologische Eigenschaft Steinscher Räume. Nachr. Akad. Wiss. Göttingenamath.-Phys. Kl. II **1966**, 213–224 (1966)
332. Kelley, J.L.: General Topology. Graduate Texts in Mathematics, vol. 27. Springer, New York (1975). Reprint of the 1955 edition, Van Nostrand, Toronto, Ont.
333. Kenig, C.E., Webster, S.M.: The local hull of holomorphy of a surface in the space of two complex variables. Invent. Math. **67**(1), 1–21 (1982)
334. Kervaire, M.A.: Non-parallelizability of the n-sphere for $n > 7$. Proc. Natl. Acad. Sci. USA **44**(3), 280–283 (1958)
335. Kervaire, M.A.: Sur le fibré normal à une sphère immergée dans un espace euclidien. Comment. Math. Helv. **33**, 121–131 (1959)

336. Klein, M., Ramspott, K.J.: Ein Transformationssatz für Idealbasen holomorpher Funktionen. Bayer. Akad. Wiss. Math.-Natur. Kl. Sitzungsber., 93–100 (1988), 1987
337. Klimek, M.: Pluripotential Theory. London Mathematical Society Monographs. New Series, vol. 6. The Clarendon Press/Oxford University Press, New York (1991)
338. Kneser, H.: Die Randwerte einer analytischen Funktion zweier Veränderlichen. Monatshefte Math. Phys. **43**(1), 364–380 (1936)
339. Kobayashi, S.: Hyperbolic Manifolds and Holomorphic Mappings. Pure and Applied Mathematics, vol. 2. Marcel Dekker, New York (1970)
340. Kobayashi, S.: Intrinsic distances, measures and geometric function theory. Bull. Am. Math. Soc. **82**(3), 357–416 (1976)
341. Kobayashi, S.: Hyperbolic Complex Spaces. Grundlehren Math. Wiss., vol. 318. Springer, Berlin (1998)
342. Kobayashi, S., Ochiai, T.: Meromorphic mappings onto compact complex spaces of general type. Invent. Math. **31**(1), 7–16 (1975)
343. Kodaira, K., Spencer, D.C.: On deformations of complex analytic structures. I, II. Ann. Math. (2) **67**, 328–466 (1958)
344. Koebe, P.: Über die Uniformisierung beliebiger analytischer Kurven. Zweite Mitteilung. Nachr. Ges. Wiss. Göttingen, Math.-Phys. Kl. **1907**, 633–669 (1907)
345. Kohn, J.J.: Harmonic integrals on strongly pseudo-convex manifolds. I. Ann. Math. (2) **78**, 112–148 (1963)
346. Kohn, J.J.: Harmonic integrals on strongly pseudo-convex manifolds. II. Ann. Math. (2) **79**, 450–472 (1964)
347. Kolarič, D.: Approximation of holomorphic maps with a lower bound on the rank. Proc. Am. Math. Soc. **136**(4), 1273–1284 (2008)
348. Kolarič, D.: Tame sets in the complement of algebraic variety. J. Geom. Anal. **19**(4), 847–863 (2009)
349. Kolarič, D.: Parametric H-principle for holomorphic immersions with approximation. Differ. Geom. Appl. **29**(3), 292–298 (2011)
350. Kronecker, L.: Grundzüge einer arithmetischen Theorie der algebraische Grössen. J. Reine Angew. Math. **92**, 1–122 (1882)
351. Kronheimer, P.B., Mrowka, T.S.: The genus of embedded surfaces in the projective plane. Math. Res. Lett. **1**(6), 797–808 (1994)
352. Kronheimer, P.B., Mrowka, T.S.: Embedded surfaces and the structure of Donaldson's polynomial invariants. J. Differ. Geom. **41**(3), 573–734 (1995)
353. Kruzhilin, N.G.: Two-dimensional spheres on the boundaries of pseudoconvex domains in $\mathbb{C}^2$. Izv. Akad. Nauk SSSR, Ser. Mat. **55**(6), 1194–1237 (1991)
354. Kusakabe, Y.: Dense holomorphic curves in spaces of holomorphic maps and applications to universal maps. Int. J. Math. **28**(4), 1750028 (2017). 15 pp.
355. Kusunoki, Y., Sainouchi, Y.: Holomorphic differentials on open Riemann surfaces. J. Math. Kyoto Univ. **11**, 181–194 (1971)
356. Kutzschebauch, F.: Andersén-Lempert-theory with parameters: a representation theoretic point of view. J. Algebra Appl. **4**(3), 325–340 (2005)
357. Kutzschebauch, F., Lárusson, F., Schwarz, G.W.: Homotopy principles for equivariant isomorphisms. Trans. Am. Math. Soc. **369**, 7251–7300 (2017). doi:10.1090/tran/6797
358. Kutzschebauch, F., Lárusson, F., Schwarz, G.W.: Sufficient conditions for holomorphic linearisation. Transform. Groups **22**(2), 475–485 (2017). doi:10.1007/s00031-016-9376-7
359. Kutzschebauch, F., Lárusson, F., Schwarz, G.W.: An Oka principle for equivariant isomorphisms. J. Reine Angew. Math. **706**, 193–214 (2015)
360. Kutzschebauch, F., Lárusson, F., Schwarz, G.W.: An equivariant parametric Oka principle for bundles of homogeneous spaces. ArXiv e-prints (2016). arXiv:1612.07372
361. Kutzschebauch, F., Leuenberger, M.: The Lie algebra generated by locally nilpotent derivations on a Danielewski surface. Ann. Sc. Norm. Super. Pisa, Cl. Sci. **5**(15), 183–207 (2016)
362. Kutzschebauch, F., Leuenberger, M., Liendo, A.: The algebraic density property for affine toric varieties. J. Pure Appl. Algebra **219**(8), 3685–3700 (2015)

363. Kutzschebauch, F., Lind, A.: Holomorphic automorphisms of Danielewski surfaces I—density of the group of overshears. Proc. Am. Math. Soc. **139**(11), 3915–3927 (2011)
364. Kutzschebauch, F., Lodin, S.: Holomorphic families of nonequivalent embeddings and of holomorphic group actions on affine space. Duke Math. J. **162**(1), 49–94 (2013)
365. Kutzschebauch, F., Løw, E., Wold, E.F.: Embedding some Riemann surfaces into $\mathbb{C}^2$ with interpolation. Math. Z. **262**(3), 603–611 (2009)
366. Kutzschebauch, F., Ramos-Peon, A.: An Oka principle for a parametric infinite transitivity property. J. Geom. Anal. **27**(3), 2018–2043 (2017). doi:10.1007/s12220-016-9749-0
367. Kutzschebauch, F., Wold, E.F.: Carleman approximation by holomorphic automorphisms of $\mathbb{C}^n$. J. Reine Angew. Math. (2015). doi:10.1515/crelle-2015-0056
368. Lai, H.F.: Characteristic classes of real manifolds immersed in complex manifolds. Trans. Am. Math. Soc. **172**, 1–33 (1972)
369. Lärkäng, R., Lárusson, F.: Extending holomorphic maps from Stein manifolds into affine toric varieties. Proc. Am. Math. Soc. **144**(11), 4613–4626 (2016)
370. Lárusson, F.: Excision for simplicial sheaves on the Stein site and Gromov's Oka principle. Int. J. Math. **14**(2), 191–209 (2003)
371. Lárusson, F.: Model structures and the Oka principle. J. Pure Appl. Algebra **192**(1–3), 203–223 (2004)
372. Lárusson, F.: Mapping cylinders and the Oka principle. Indiana Univ. Math. J. **54**(4), 1145–1159 (2005)
373. Lárusson, F.: Affine simplices in Oka manifolds. Doc. Math. **14**, 691–697 (2009)
374. Lárusson, F.: Applications of a parametric Oka principle for liftings. In: Complex Analysis. Trends Math., pp. 205–211. Birkhäuser/Springer, Basel AG/Basel (2010)
375. Lárusson, F.: Deformations of Oka manifolds. Math. Z. **272**(3–4), 1051–1058 (2012)
376. Lárusson, F.: Absolute neighbourhood retracts and spaces of holomorphic maps from Stein manifolds to Oka manifolds. Proc. Am. Math. Soc. **143**(3), 1159–1167 (2015)
377. Lárusson, F., Ritter, T.: Proper holomorphic immersions in homotopy classes of maps from finitely connected planar domains into $\mathbb{C} \times \mathbb{C}^*$. Indiana Univ. Math. J. **63**(2), 367–383 (2014)
378. Lárusson, F., Truong, T.T.: Algebraic subellipticity and dominability of blow-ups of affine spaces. Doc. Math. **22**, 151–163 (2017)
379. Laufer, H.B.: Imbedding annuli in $\mathbb{C}^2$. J. Anal. Math. **26**, 187–215 (1973)
380. Laufer, H.B.: On the infinite dimensionality of the Dolbeault cohomology groups. Proc. Am. Math. Soc. **52**, 293–296 (1975)
381. Lees, J.A.: On the classification of Lagrange immersions. Duke Math. J. **43**(2), 217–224 (1976)
382. Leiterer, J.: On the similarity of holomorphic matrices. ArXiv e-prints (2017). arXiv:1703.09530
383. Leiterer, J.: Analytische Faserbündel mit stetigem Rand über streng pseudokonvexen Gebieten. I. Garbentheoretische Hilfsmittel. Math. Nachr. **71**, 329–344 (1976)
384. Leiterer, J.: Analytische Faserbündel mit stetigem Rand über streng pseudokonvexen Gebieten. II. Topologische Klassifizierung. Math. Nachr. **72**, 201–217 (1976)
385. Leiterer, J.: Theorem B für analytische Funktionen mit stetigen Randwerten. Beitr. Anal. **8**, 95–102 (1976)
386. Leiterer, J.: Holomorphic vector bundles and the Oka-Grauert principle. In: Several Complex Variables. IV. Algebraic Aspects of Complex Analysis. Encyclopaedia Math. Sci., vol. 10, 63–103 (1990); translation from Itogi Nauki Tekh., Ser. Sovrem. Probl. Mat., Fundam. Napravleniya **10**, 75–121 (1986)
387. Leiterer, J., Vâjâitu, V.: A relative Oka-Grauert principle on 1-convex spaces. J. Reine Angew. Math. **564**, 85–104 (2003)
388. Lelong, P.: Définition des fonctions plurisousharmoniques. C. R. Acad. Sci. Paris **215**, 398–400 (1942)
389. Lelong, P.: Fonctions plurisousharmoniques et formes différentielles positives. Gordon & Breach, Paris (1968)

390. Lempert, L.: Algebraic approximations in analytic geometry. Invent. Math. **121**(2), 335–353 (1995)
391. Lempert, L.: Holomorphic functions on (generalized) loop spaces. Math. Proc. R. Ir. Acad. **104A**(1), 35–46 (electronic) (2004)
392. Leuenberger, M.: (Volume) density property of a family of complex manifolds including the Koras-Russell cubic threefold. Proc. Am. Math. Soc. **144**(9), 3887–3902 (2016)
393. Levi, E.E.: Sulle ipersuperficie dello spazio a 4 dimensioni che possono essere frontiera del campo di esistenza di una funzione analitica di due variabili complesse. Ann. Mat. **3**(18), 69–79 (1911)
394. Lieb, I., Michel, J.: The Cauchy-Riemann Complex. Integral Formulae and Neumann Problem. Aspects of Mathematics, vol. E34. Friedr. Vieweg & Sohn, Braunschweig (2002)
395. Lieb, I., Range, R.M.: Lösungsoperatoren für den Cauchy-Riemann-Komplex mit $\mathcal{C}^k$-Abschätzungen. Math. Ann. **253**(2), 145–164 (1980)
396. Lieb, I., Range, R.M.: Estimates for a class of integral operators and applications to the $\overline{\partial}$-Neumann problem. Invent. Math. **85**(2), 415–438 (1986)
397. Lieb, I., Range, R.M.: Integral representations and estimates in the theory of the $\overline{\partial}$-Neumann problem. Ann. Math. (2) **123**(2), 265–301 (1986)
398. Lisca, P., Matić, G.: Tight contact structures and Seiberg-Witten invariants. Invent. Math. **129**(3), 509–525 (1997)
399. Loday, J.-L.: Applications algébriques du tore dans la sphère et de $S^p \times S^q$ dans S^{p+q}. In: Algebraic K-Theory, II: "Classical" Algebraic K-Theory and Connections with Arithmetic, Proc. Conf., Seattle Res. Center, Battelle Memorial Inst., 1972. Lecture Notes in Math., vol. 342, pp. 79–91. Springer, Berlin (1973)
400. Løw, E., Pereira, J.V., Peters, H., Wold, E.F.: Polynomial completion of symplectic jets and surfaces containing involutive lines. Math. Ann. **364**(1–2), 519–538 (2016)
401. Løw, E., Wold, E.F.: Polynomial convexity and totally real manifolds. Complex Var. Elliptic Equ. **54**(3–4), 265–281 (2009)
402. Luna, D.: Slices étales. In: Sur les groupes algébriques. Bull. Soc. Math. France, Paris, Mémoire, vol. 33, pp. 81–105. Soc. Math. France, Paris (1973)
403. Majcen, I.: Closed holomorphic 1-forms without zeros on Stein manifolds. Math. Z. **257**(4), 925–937 (2007)
404. Majcen, I.: Embedding certain infinitely connected subsets of bordered Riemann surfaces properly into $\mathbb{C}^2$. J. Geom. Anal. **19**(3), 695–707 (2009)
405. Majcen, I.: Proper holomorphic embeddings of finitely connected planar domains into $\mathbb{C}^n$. Ark. Mat. **51**(2), 329–343 (2013)
406. Makar-Limanov, L.: On the hypersurface $x + x^2y + z^2 + t^3 = 0$ in $\mathbb{C}^4$ or a $\mathbb{C}^3$-like threefold which is not $\mathbb{C}^3$. Isr. J. Math. **96**, 419–429 (1996). Part B
407. Manne, P.E.: Carleman approximation on totally real submanifolds of a complex manifold. In: Several Complex Variables, Stockholm, 1987/1988. Math. Notes, vol. 38, pp. 519–528. Princeton University Press, Princeton (1993)
408. Manne, P.E., Wold, E.F., Øvrelid, N.: Holomorphic convexity and Carleman approximation by entire functions on Stein manifolds. Math. Ann. **351**(3), 571–585 (2011)
409. Massey, W.S.: Proof of a conjecture of Whitney. Pac. J. Math. **31**, 143–156 (1969)
410. May, J.P.: Simplicial Objects in Algebraic Topology. Chicago Lectures in Mathematics. University of Chicago Press, Chicago (1992). Reprint of the 1967 original
411. May, J.P., Ponto, K.: More Concise Algebraic Topology. Localization, Completion, and Model Categories. Chicago Lectures in Mathematics. University of Chicago Press, Chicago (2012)
412. McDuff, D., Salamon, D.: J-Holomorphic Curves and Symplectic Topology, 2nd edn. Amer. Math. Soc. Colloquium Publications, vol. 52. Am. Math. Soc., Providence (2012)
413. Mergelyan, S.N.: On the representation of functions by series of polynomials on closed sets. Dokl. Akad. Nauk SSSR (N.S.) **78**, 405–408 (1951)
414. Merker, J., Porten, E.: Holomorphic extension of CR functions, envelopes of holomorphy, and removable singularities. IMRS Int. Math. Res. Surv., 287 (2006)

415. Michael, E.: Continuous selections. I. Ann. Math. (2) **63**, 361–382 (1956)
416. Michel, J., Perotti, A.: C^k regularity for the $\overline{\partial}$-equation on strictly pseudoconvex domains with piecewise smooth boundaries. Math. Z. **203**(3), 415–427 (1990)
417. Milnor, J.: Morse Theory. Annals of Mathematics Studies, vol. 51. Princeton University Press, Princeton (1963)
418. Milnor, J.W., Stasheff, J.D.: Characteristic Classes. Annals of Mathematics Studies, vol. 76. Princeton University Press/University of Tokyo Press, Princeton/Tokyo (1974)
419. Morgan, J.W.: The Seiberg-Witten Equations and Applications to the Topology of Smooth Four-Manifolds. Mathematical Notes, vol. 44. Princeton University Press, Princeton (1996)
420. Morgan, J.W., Szabó, Z., Taubes, C.H.: A product formula for the Seiberg-Witten invariants and the generalized Thom conjecture. J. Differ. Geom. **44**(4), 706–788 (1996)
421. Moser, J.K., Webster, S.M.: Normal forms for real surfaces in $\mathbb{C}^2$ near complex tangents and hyperbolic surface transformations. Acta Math. **150**(3–4), 255–296 (1983)
422. Murolo, C., Trotman, D.J.A., du Plessis, A.A.: Stratified transversality by isotopy. Trans. Am. Math. Soc. **355**(12), 4881–4900 (2003)
423. Narasimhan, R.: Holomorphic mappings of complex spaces. Proc. Am. Math. Soc. **11**, 800–804 (1960)
424. Narasimhan, R.: Imbedding of holomorphically complete complex spaces. Am. J. Math. **82**, 917–934 (1960)
425. Narasimhan, R.: The Levi problem for complex spaces. Math. Ann. **142**, 355–365 (1960/1961)
426. Narasimhan, R.: On the homology groups of Stein spaces. Invent. Math. **2**, 377–385 (1967)
427. Narasimhan, R.: Analysis on Real and Complex Manifolds. North-Holland Mathematical Library, vol. 35. North-Holland Publishing Co., Amsterdam (1985). Reprint of the 1973 edition
428. Nash, J.: Real algebraic manifolds. Ann. Math. (2) **56**, 405–421 (1952)
429. Nemirovski, S.: Adjunction inequality and coverings of Stein surfaces. Turk. J. Math. **27**(1), 161–172 (2003)
430. Nemirovski, S., Siegel, K.: Rationally convex domains and singular Lagrangian surfaces in $\mathbb{C}^2$. Invent. Math. **203**(1), 333–358 (2016)
431. Nemirovskiĭ, S.Y.: Complex analysis and differential topology on complex surfaces. Usp. Mat. Nauk **54**(4(328)), 47–74 (1999)
432. Newlander, A., Nirenberg, L.: Complex analytic coordinates in almost complex manifolds. Ann. Math. (2) **65**, 391–404 (1957)
433. Nishimura, Y.: Examples of analytic immersions of two-dimensional Stein manifolds into $\mathbb{C}^2$. Math. Jpn. **26**(1), 81–83 (1981)
434. Nishimura, Y.: Applications holomorphes injectives à jacobien constant de deux variables. J. Math. Kyoto Univ. **26**(4), 697–709 (1986)
435. Noguchi, J.: Analytic Function Theory of Several Variables. Elements of Oka's Coherence. Springer, Singapore (2016)
436. Nomura, R.: Kähler manifolds with negative holomorphic sectional curvature, Kähler-Ricci flow approach. ArXiv e-prints (2016). arXiv:1610.01976
437. Norguet, F.: Sur les domaines d'holomorphie des fonctions uniformes de plusieurs variables complexes. (Passage du local au global.) Bull. Soc. Math. Fr. **82**, 137–159 (1954)
438. O'Grady, K.G.: Higher-dimensional analogues of $K3$ surfaces. In: Current Developments in Algebraic Geometry. Math. Sci. Res. Inst. Publ., vol. 59, pp. 257–293. Cambridge Univ. Press, Cambridge (2012)
439. Ohsawa, T.: Completeness of noncompact analytic spaces. Publ. Res. Inst. Math. Sci. **20**(3), 683–692 (1984)
440. Ohsawa, T.: Holomorphic embedding of compact s.p.c. manifolds into complex manifolds as real hypersurfaces. In: Differential Geometry of Submanifolds, Kyoto, 1984. Lecture Notes in Math., vol. 1090, pp. 64–76. Springer, Berlin (1984)

441. Ohsawa, T.: Analysis of Several Complex Variables. Translations of Mathematical Monographs, vol. 211. Am. Math. Soc., Providence (2002). Translated from the Japanese by Shu Gilbert Nakamura
442. Ohsawa, T.: L^2 Approaches in Several Complex Variables. Development of Oka-Cartan Theory by L^2 Estimates for the $\overline{\partial}$ Operator. Springer Monographs in Mathematics. Springer, Tokyo (2015)
443. Oka, K.: Sur les fonctions analytiques de plusieurs variables. III. Deuxième problème de Cousin. J. Sci. Hiroshima Univ., Ser. A **9**, 7–19 (1939)
444. Oka, K.: Collected Papers. Springer, Berlin (1984). Translated from the French by R. Narasimhan. With commentaries by H. Cartan. Edited by R. Remmert
445. Orevkov, S.Y.: Proper analytic embedding of $\mathbb{CP}^1$ without a Cantor set into $\mathbb{C}^2$. Usp. Mat. Nauk **63**(1(379)), 155–156 (2008)
446. Osserman, R.: A Survey of Minimal Surfaces, 2nd edn. Dover Publications, New York (1986)
447. Ozsváth, P., Szabó, Z.: The symplectic Thom conjecture. Ann. Math. (2) **151**(1), 93–124 (2000)
448. Palais, R.S.: Foundations of Global Non-linear Analysis. W. A. Benjamin, New York (1968)
449. Palais, R.S.: Banach manifolds of fiber bundle sections. In: Actes du Congrès International des Mathématiciens (Nice, 1970), Tome 2, pp. 243–249. Gauthier-Villars, Paris (1971)
450. Pankov, A.A.: Analytic fibre bundles that are continuous on the boundary of a domain. Sakharth. SSR Mecn. Akad. Moambe **81**(2), 277–280 (1976)
451. Phillips, A.: Submersions of open manifolds. Topology **6**, 171–206 (1967)
452. Phillips, A.: Foliations on open manifolds. I. Comment. Math. Helv. **43**, 204–211 (1968)
453. Phillips, A.: Foliations on open manifolds. II. Comment. Math. Helv. **44**, 367–370 (1969)
454. Phillips, A.: Smooth maps transverse to a foliation. Bull. Am. Math. Soc. **76**, 792–797 (1970)
455. Phillips, A.: Smooth maps of constant rank. Bull. Am. Math. Soc. **80**, 513–517 (1974)
456. Poincaré, H.: Sur l'uniformisation des fonctions analytiques. Acta Math. **31**, 1–64 (1907)
457. Poletsky, E.A.: Stein neighborhoods of graphs of holomorphic mappings. J. Reine Angew. Math. **684**, 187–198 (2013)
458. Pommerenke, C.: Boundary Behaviour of Conformal Maps. Grundlehren Math. Wiss., vol. 299. Springer, Berlin (1992)
459. Prezelj, J.: Interpolation of embeddings of Stein manifolds on discrete sets. Math. Ann. **326**(2), 275–296 (2003)
460. Prezelj, J.: Weakly regular embeddings of Stein spaces with isolated singularities. Pac. J. Math. **220**(1), 141–152 (2005)
461. Prezelj, J.: A relative Oka-Grauert principle for holomorphic submersions over 1-convex spaces. Trans. Am. Math. Soc. **362**(8), 4213–4228 (2010)
462. Prezelj, J.: Positivity of metrics on conic neighborhoods of 1-convex submanifolds. Int. J. Math. **27**(5), 24 (2016)
463. Prezelj, J., Slapar, M.: The generalized Oka-Grauert principle for 1-convex manifolds. Mich. Math. J. **60**(3), 495–506 (2011)
464. Putinar, M.: A reciprocal to the Oka-Grauert principle. Math. Nachr. **159**, 149–152 (1992)
465. Quillen, D.G.: Homotopical Algebra. Lecture Notes in Math., vol. 43. Springer, Berlin (1967)
466. Ramspott, K.-J.: Stetige und holomorphe Schnitte in Bündeln mit homogener Faser. Math. Z. **89**, 234–246 (1965)
467. Range, R.M., Siu, Y.-T.: Uniform estimates for the $\bar{\partial}$-equation on domains with piecewise smooth strictly pseudoconvex boundaries. Math. Ann. **206**, 325–354 (1973)
468. Range, R.M., Siu, Y.T.: $\mathcal{C}^k$ approximation by holomorphic functions and $\overline{\partial}$-closed forms on $\mathcal{C}^k$ submanifolds of a complex manifold. Math. Ann. **210**, 105–122 (1974)
469. Reinhardt, K.: Über Abbildungen durch analytische Funktionen zweier Veränderlicher. Math. Ann. **83**(3–4), 211–255 (1921)
470. Remmert, R.: Sur les espaces analytiques holomorphiquement séparables et holomorphiquement convexes. C. R. Acad. Sci. Paris **243**, 118–121 (1956)

471. Remmert, R.: Holomorphe und meromorphe Abbildungen komplexer Räume. Math. Ann. **133**, 328–370 (1957)
472. Ritter, T.: Acyclic embeddings of open Riemann surfaces into new examples of elliptic manifolds. Proc. Am. Math. Soc. **141**(2), 597–603 (2013)
473. Ritter, T.: A strong Oka principle for embeddings of some planar domains into $\mathbb{C} \times \mathbb{C}^*$. J. Geom. Anal. **23**(2), 571–597 (2013)
474. Rosay, J.-P.: The polynomial hull of nonconnected tube domains, and an example of E. Kallin. Bull. Lond. Math. Soc. **21**(1), 73–78 (1989)
475. Rosay, J.P.: A counterexample related to Hartogs' phenomenon (a question by E. Chirka). Mich. Math. J. **45**(3), 529–535 (1998)
476. Rosay, J.-P.: Extension of holomorphic bundles to the disc (and Serre's problem on Stein bundles). Ann. Inst. Fourier (Grenoble) **57**(2), 517–523 (2007)
477. Rosay, J.-P., Rudin, W.: Holomorphic maps from $\mathbb{C}^n$ to $\mathbb{C}^n$. Trans. Am. Math. Soc. **310**(1), 47–86 (1988)
478. Rosenberg, J.: Algebraic K-Theory and Its Applications. Graduate Texts in Mathematics, vol. 147. Springer, New York (1994)
479. Rossi, H.: Vector fields on analytic spaces. Ann. Math. (2) **78**, 455–467 (1963)
480. Rudin, W.: Pairs of inner functions on finite Riemann surfaces. Trans. Am. Math. Soc. **140**, 423–434 (1969)
481. Rudin, W.: Totally real Klein bottles in $\mathbb{C}^2$. Proc. Am. Math. Soc. **82**(4), 653–654 (1981)
482. Rudin, W.: Injective polynomial maps are automorphisms. Am. Math. Mon. **102**(6), 540–543 (1995)
483. Rudin, W.: Function Theory in the Unit Ball of $\mathbb{C}^n$. Classics in Mathematics. Springer, Berlin (2008). Reprint of the 1980 edition
484. Rudolph, L.: Embeddings of the line in the plane. J. Reine Angew. Math. **337**, 113–118 (1982)
485. Sathaye, A.: On linear planes. Proc. Am. Math. Soc. **56**, 1–7 (1976)
486. Schneider, M.: Über eine Vermutung von Hartshorne. Math. Ann. **201**, 221–229 (1973)
487. Schneider, M.: Tubenumgebungen Steinscher Räume. Manuscr. Math. **18**(4), 391–397 (1976)
488. Schneider, M.: Vollständige, fast-vollständige und mengen-theoretisch-vollständige Durchschnitte in Steinschen Mannigfaltigkeiten. Math. Ann. **260**(2), 151–174 (1982)
489. Schneider, M.: On the number of equations needed to describe a variety. In: Complex Analysis of Several Variables, Madison, Wis., 1982. Proc. Sympos. Pure Math., vol. 41, pp. 163–180. Am. Math. Soc., Providence (1984)
490. Schumacher, G.: Über die Entwicklung der komplexen Analysis in Deutschland vom Ausgang des 19. Jahrhunderts bis zum Anfang der siebziger Jahre. Jahresber. Dtsch. Math.-Ver. **98**(2), 41–133 (1996)
491. Schur, I.: Ein Satz über quadratische Formen mit komplexen Koeffizienten. Am. J. Math. **67**, 472–480 (1945)
492. Schürmann, J.: Embeddings of Stein spaces into affine spaces of minimal dimension. Math. Ann. **307**(3), 381–399 (1997)
493. Schwarz, G.W.: An Oka principle for Stein G-manifolds. Indiana Univ. Math. J. To appear. ArXiv e-prints (2016). arXiv:1608.05156
494. Sebbar, A.: Principe d'Oka-Grauert dans A^∞. Math. Z. **201**(4), 561–581 (1989)
495. Seiberg, N., Witten, E.: Electric-magnetic duality, monopole condensation, and confinement in $N = 2$ supersymmetric Yang-Mills theory. Nucl. Phys. B **426**(1), 19–52 (1994)
496. Serre, J-P.: Quelques problèmes globaux relatifs aux variétés de Stein. In: Colloque sur les fonctions de plusieurs variables, tenu à Bruxelles, 1953, pp. 57–68. Georges Thone/Masson & Cie, Liège/Paris (1953)
497. Serre, J-P.: Faisceaux algébriques cohérents. Ann. Math. (2) **61**, 197–278 (1955)
498. Serre, J-P.: Géométrie algébrique et géométrie analytique. Ann. Inst. Fourier **6**, 1–42 (1955/1956)

499. Serre, J-P.: Trees. Springer Monographs in Mathematics. Springer, Berlin (2003). Translated from the French original by John Stillwell. Corrected 2nd printing of the 1980 English translation
500. Shestakov, I.P., Umirbaev, U.U.: The tame and the wild automorphisms of polynomial rings in three variables. J. Am. Math. Soc. **17**(1), 197–227 (electronic) (2004)
501. Shevchishin, V.V.: Lagrangian embeddings of the Klein bottle and the combinatorial properties of mapping class groups. Izv. Ross. Akad. Nauk Ser. Mat. **73**(4), 153–224 (2009)
502. Shub, M.: Global Stability of Dynamical Systems. Springer, New York (1987). With the collaboration of Albert Fathi and Rémi Langevin. Translated from the French by Joseph Christy
503. Siu, Y.T.: Techniques of Extension of Analytic Objects. Lecture Notes in Pure and Applied Mathematics
504. Siu, Y.T.: Every Stein subvariety admits a Stein neighborhood. Invent. Math. **38**(1), 89–100 (1976/1977)
505. Siu, Y.-T.: Recent techniques in hyperbolicity problems. In: Several Complex Variables, Berkeley, CA, 1995–1996. Math. Sci. Res. Inst. Publ., vol. 37, pp. 429–508. Cambridge Univ. Press, Cambridge (1999)
506. Siu, Y.-T.: Hyperbolicity of generic high-degree hypersurfaces in complex projective space. Invent. Math. **202**(3), 1069–1166 (2015)
507. Siu, Y.-T., Yeung, S.-K.: A generalized Bloch's theorem and the hyperbolicity of the complement of an ample divisor in an abelian variety. Math. Ann. **306**(4), 743–758 (1996). Erratum: Math. Ann. **326**(4), 205–207 (2003)
508. Siu, Y.-T., Yeung, S.-K.: Hyperbolicity of the complement of a generic smooth curve of high degree in the complex projective plane. Invent. Math. **124**(1–3), 573–618 (1996)
509. Siu, Y.-T., Yeung, S.-K.: Defects for ample divisors of abelian varieties, Schwarz lemma, and hyperbolic hypersurfaces of low degrees. Am. J. Math. **119**(5), 1139–1172 (1997). Addendum: Am. J. Math. **125**(3), 441–448 (2003)
510. Skoda, H.: Fibrés holomorphes à base et à fibre de Stein. C. R. Acad. Sci. Paris Sér. A-B **284**(19), A1199–A1202 (1977)
511. Skoda, H.: Fibrés holomorphes à base et à fibre de Stein. Invent. Math. **43**(2), 97–107 (1977)
512. Slapar, M.: On Stein neighborhood basis of real surfaces. Math. Z. **247**(4), 863–879 (2004)
513. Slapar, M.: Real surfaces in elliptic surfaces. Int. J. Math. **16**(4), 357–363 (2005)
514. Slapar, M.: Cancelling complex points in codimension two. Bull. Aust. Math. Soc. **88**(1), 64–69 (2013)
515. Slapar, M.: Modeling complex points up to isotopy. J. Geom. Anal. **23**(4), 1932–1943 (2013)
516. Slapar, M.: CR regular embeddings and immersions of compact orientable 4-manifolds into $\mathbb{C}^3$. Int. J. Math. **26**(5), 5 (2015)
517. Slapar, M.: On complex points of codimension 2 submanifolds. J. Geom. Anal. **26**(1), 206–219 (2016)
518. Smale, S.: The classification of immersions of spheres in Euclidean spaces. Ann. Math. (2) **69**, 327–344 (1959)
519. Snow, D.M.: Reductive group actions on Stein spaces. Math. Ann. **259**(1), 79–97 (1982)
520. Spanier, E.H.: Algebraic Topology. Springer, New York (1981). Corrected reprint of the 1966 original
521. Spring, D.: Convex Integration Theory. Solutions to the h-Principle in Geometry and Topology. Monographs in Mathematics, vol. 92. Birkhäuser, Basel (1998)
522. Steenrod, N.: The Topology of Fibre Bundles. Princeton Landmarks in Mathematics. Princeton University Press, Princeton (1999). Reprint of the 1957 edition, Princeton Paperbacks
523. Stehlé, J.-L.: Plongements du disque dans $\mathbb{C}^2$. In: Séminaire Pierre Lelong (Analyse), Année 1970–1971. Lecture Notes in Math., vol. 275, pp. 119–130. Springer, Berlin (1972)
524. Stein, K.: Analytische Funktionen mehrerer komplexer Veränderlichen zu vorgegebenen Periodizitätsmoduln und das zweite Cousinsche Problem. Math. Ann. **123**, 201–222 (1951)
525. Stein, K.: Überlagerungen holomorph-vollständiger komplexer Räume. Arch. Math. (Basel) **7**, 354–361 (1956)

526. Stensønes, B.: Fatou-Bieberbach domains with $\mathcal{C}^\infty$-smooth boundary. Ann. Math. (2) **145**(2), 365–377 (1997)
527. Sternberg, S.: Local contractions and a theorem of Poincaré. Am. J. Math. **79**, 809–821 (1957)
528. Stolzenberg, G.: Polynomially and rationally convex sets. Acta Math. **109**, 259–289 (1963)
529. Stolzenberg, G.: The analytic part of the Runge hull. Math. Ann. **164**, 286–290 (1966)
530. Stolzenberg, G.: Uniform approximation on smooth curves. Acta Math. **115**, 185–198 (1966)
531. Stopar, K.: Approximation of holomorphic mappings on 1-convex domains. Int. J. Math. **24**(14), 33 (2013)
532. Stout, E.L.: The Theory of Uniform Algebras. Bogden & Quigley, Tarrytown-on-Hudson (1971)
533. Stout, E.L.: Algebraic domains in Stein manifolds. In: Proceedings of the Conference on Banach Algebras and Several Complex Variables, New Haven, Conn., 1983. Contemp. Math., vol. 32, pp. 259–266. Am. Math. Soc., Providence (1984)
534. Stout, E.L.: Polynomial Convexity. Progress in Mathematics, vol. 261. Birkhäuser Boston, Boston (2007)
535. Stout, E.L.: Manifold-valued holomorphic approximation. Can. Math. Bull. **54**(2), 370–380 (2011)
536. Stout, E.L., Zame, W.R.: A Stein manifold topologically but not holomorphically equivalent to a domain in $\mathbb{C}^n$. Adv. Math. **60**(2), 154–160 (1986)
537. Sukhov, A., Tumanov, A.: Pseudoholomorphic discs near an elliptic point. Tr. Mat. Inst. Steklova **253**, 296–303 (2006). Kompleks. Anal. i Prilozh.
538. Sukhov, A., Tumanov, A.: Deformations and transversality of pseudo-holomorphic discs. J. Anal. Math. **116**, 1–16 (2012)
539. Suslin, A.A.: The structure of the special linear group over rings of polynomials. Izv. Akad. Nauk SSSR, Ser. Mat. **41**(2), 235–252 (1977). 477
540. Suzuki, M.: Propriétés topologiques des polynômes de deux variables complexes, et automorphismes algébriques de l'espace $\mathbf{C}^2$. J. Math. Soc. Jpn. **26**, 241–257 (1974)
541. Suzuki, M.: Sur les opérations holomorphes du groupe additif complexe sur l'espace de deux variables complexes. Ann. Sci. Éc. Norm. Supér. (4) **10**(4), 517–546 (1977)
542. Takagi, T.: On an algebraic problem related to an analytic theorem of Carathéodory and Fejér and on an allied theorem of Landau. Jpn. J. Math. **1**, 83–93 (1924)
543. Taubes, C.H.: The Seiberg-Witten invariants and symplectic forms. Math. Res. Lett. **1**(6), 809–822 (1994)
544. Taubes, C.H.: SW $\Rightarrow$ Gr: from the Seiberg-Witten equations to pseudo-holomorphic curves. J. Am. Math. Soc. **9**(3), 845–918 (1996)
545. Teleman, A.: Donaldson theory on non-Kählerian surfaces and class VII surfaces with $b_2 = 1$. Invent. Math. **162**(3), 493–521 (2005)
546. Thom, R.: Les singularités des applications différentiables. Ann. Inst. Fourier **6**, 43–87 (1955/1956)
547. Thom, R.: Un lemme sur les applications différentiables. Bol. Soc. Mat. Mexicana (2) **1**, 59–71 (1956)
548. Thurston, W.: The theory of foliations of codimension greater than one. Comment. Math. Helv. **49**, 214–231 (1974)
549. Thurston, W., Vaserstein, L.: On K_1-theory of the Euclidean space. Topol. Appl. **23**(2), 145–148 (1986)
550. Thurston, W.P.: Existence of codimension-one foliations. Ann. Math. (2) **104**(2), 249–268 (1976)
551. Toën, B., Vezzosi, G.: Homotopical algebraic geometry. I. Topos theory. Adv. Math. **193**(2), 257–372 (2005)
552. Tosatti, V., Yang, X.: An extension of a theorem of Wu-Yau. J. Differ. Geom. Accepted for publication. ArXiv e-prints (2015). arXiv:1506.01145
553. Tóth, Á., Varolin, D.: Holomorphic diffeomorphisms of complex semisimple Lie groups. Invent. Math. **139**(2), 351–369 (2000)

554. Tóth, Á., Varolin, D.: Holomorphic diffeomorphisms of semisimple homogeneous spaces. Compos. Math. **142**(5), 1308–1326 (2006)
555. Trotman, D.J.A.: Stability of transversality to a stratification implies Whitney (a)-regularity. Invent. Math. **50**(3), 273–277 (1978/1979)
556. Ugolini, R.: A parametric jet-interpolation theorem for holomorphic automorphisms of $\mathbb{C}^n$. J. Geom. Anal. (2017). doi:10.1007/s12220-017-9778-3
557. Vâjâitu, V.: Some convexity properties of morphisms of complex spaces. Math. Z. **217**(2), 215–245 (1994)
558. Varolin, D.: A general notion of shears, and applications. Mich. Math. J. **46**(3), 533–553 (1999)
559. Varolin, D.: The density property for complex manifolds and geometric structures. II. Int. J. Math. **11**(6), 837–847 (2000)
560. Varolin, D.: The density property for complex manifolds and geometric structures. J. Geom. Anal. **11**(1), 135–160 (2001)
561. Vaserstein, L.N.: Reduction of a matrix depending on parameters to a diagonal form by addition operations. Proc. Am. Math. Soc. **103**(3), 741–746 (1988)
562. Voevodsky, V.: $\mathbf{A}^1$-homotopy theory. In: Proc. Internat. Congress Math. (Berlin, 1998), Doc. Math., vol. I, pp. 579–604 (1998)
563. Warner, F.W.: Foundations of Differentiable Manifolds and Lie Groups. Graduate Texts in Mathematics, vol. 94. Springer, New York (1983). Corrected reprint of the 1971 edition
564. Webster, S.M.: Minimal surfaces in a Kähler surface. J. Differ. Geom. **20**(2), 463–470 (1984)
565. Weinstein, A.: Lectures on Symplectic Manifolds. Regional Conference Series in Mathematics, vol. 29. Am. Math. Soc., Providence (1977). Expository lectures from the CBMS Regional Conference held at the University of North Carolina, March 8–12, 1976
566. Weinstein, A.: Contact surgery and symplectic handlebodies. Hokkaido Math. J. **20**(2), 241–251 (1991)
567. Wells, R.O. Jr.: Compact real submanifolds of a complex manifold with nondegenerate holomorphic tangent bundles. Math. Ann. **179**, 123–129 (1969)
568. Wells, R.O. Jr.: Differential Analysis on Complex Manifolds, 3rd edn. Graduate Texts in Mathematics, vol. 65. Springer, New York (2008)
569. Whitehead, G.W.: Elements of Homotopy Theory. Graduate Texts in Mathematics, vol. 61. Springer, New York (1978)
570. Whitehead, J.H.C.: The immersion of an open 3-manifold in Euclidean 3-space. Proc. Lond. Math. Soc. **3**(11), 81–90 (1961)
571. Whitney, H.: Differentiable manifolds. Ann. Math. (2) **37**(3), 645–680 (1936)
572. Whitney, H.: The self-intersections of a smooth n-manifold in $2n$-space. Ann. Math. (2) **45**, 220–246 (1944)
573. Whitney, H.: Local properties of analytic varieties. In: Differential and Combinatorial Topology (A Symposium in Honor of Marston Morse), pp. 205–244. Princeton University Press, Princeton (1965)
574. Whitney, H.: Complex Analytic Varieties. Addison-Wesley Publishing Co., Reading (1972)
575. Whitney, H., Bruhat, F.: Quelques propriétés fondamentales des ensembles analytiques-réels. Comment. Math. Helv. **33**, 132–160 (1959)
576. Winkelmann, J.: On automorphisms of complements of analytic subsets in $\mathbf{C}^n$. Math. Z. **204**(1), 117–127 (1990)
577. Winkelmann, J.: The Oka-principle for mappings between Riemann surfaces. Enseign. Math. (2) **39**(1–2), 143–151 (1993)
578. Winkelmann, J.: On tameness and growth conditions. Doc. Math. **13**, 97–101 (2008)
579. Witten, E.: Monopoles and four-manifolds. Math. Res. Lett. **1**(6), 769–796 (1994)
580. Wold, E.F.: Fatou-Bieberbach domains. Int. J. Math. **16**(10), 1119–1130 (2005)
581. Wold, E.F.: Embedding Riemann surfaces properly into $\mathbb{C}^2$. Int. J. Math. **17**(8), 963–974 (2006)
582. Wold, E.F.: Proper holomorphic embeddings of finitely and some infinitely connected subsets of $\mathbb{C}$ into $\mathbb{C}^2$. Math. Z. **252**(1), 1–9 (2006)

583. Wold, E.F.: Embedding subsets of tori properly into $\mathbb{C}^2$. Ann. Inst. Fourier (Grenoble) **57**(5), 1537–1555 (2007)
584. Wold, E.F.: A Fatou-Bieberbach domain in $\mathbb{C}^2$ which is not Runge. Math. Ann. **340**(4), 775–780 (2008)
585. Wold, E.F.: A long $\mathbb{C}^2$ which is not Stein. Ark. Mat. **48**(1), 207–210 (2010)
586. Wold, E.F.: A Fatou-Bieberbach domain intersecting the plane in the unit disk. Proc. Am. Math. Soc. **140**(12), 4205–4208 (2012)
587. Wright, D.: The amalgamated free product structure of $GL_2(k[X_1, \cdots, X_n])$ and the weak Jacobian theorem for two variables. J. Pure Appl. Algebra **12**(3), 235–251 (1978)
588. Wright, D.: Abelian subgroups of $\mathrm{Aut}_k(k[X, Y])$ and applications to actions on the affine plane. Ill. J. Math. **23**(4), 579–634 (1979)
589. Wu, D., Yau, S.-T.: Negative holomorphic curvature and positive canonical bundle. Invent. Math. **204**(2), 595–604 (2016)
590. Wu, D., Yau, S.-T.: A remark on our paper "Negative holomorphic curvature and positive canonical bundle" [MR3489705]. Commun. Anal. Geom. **24**(4), 901–912 (2016)
591. Zaffran, D.: Serre problem and Inoue-Hirzebruch surfaces. Math. Ann. **319**(2), 395–420 (2001)
592. Zeron, E.S.: On the complement of sets with a system of Stein neighbourhoods. Bol. Soc. Mat. Mexicana (3) **13**(1), 87–93 (2007)

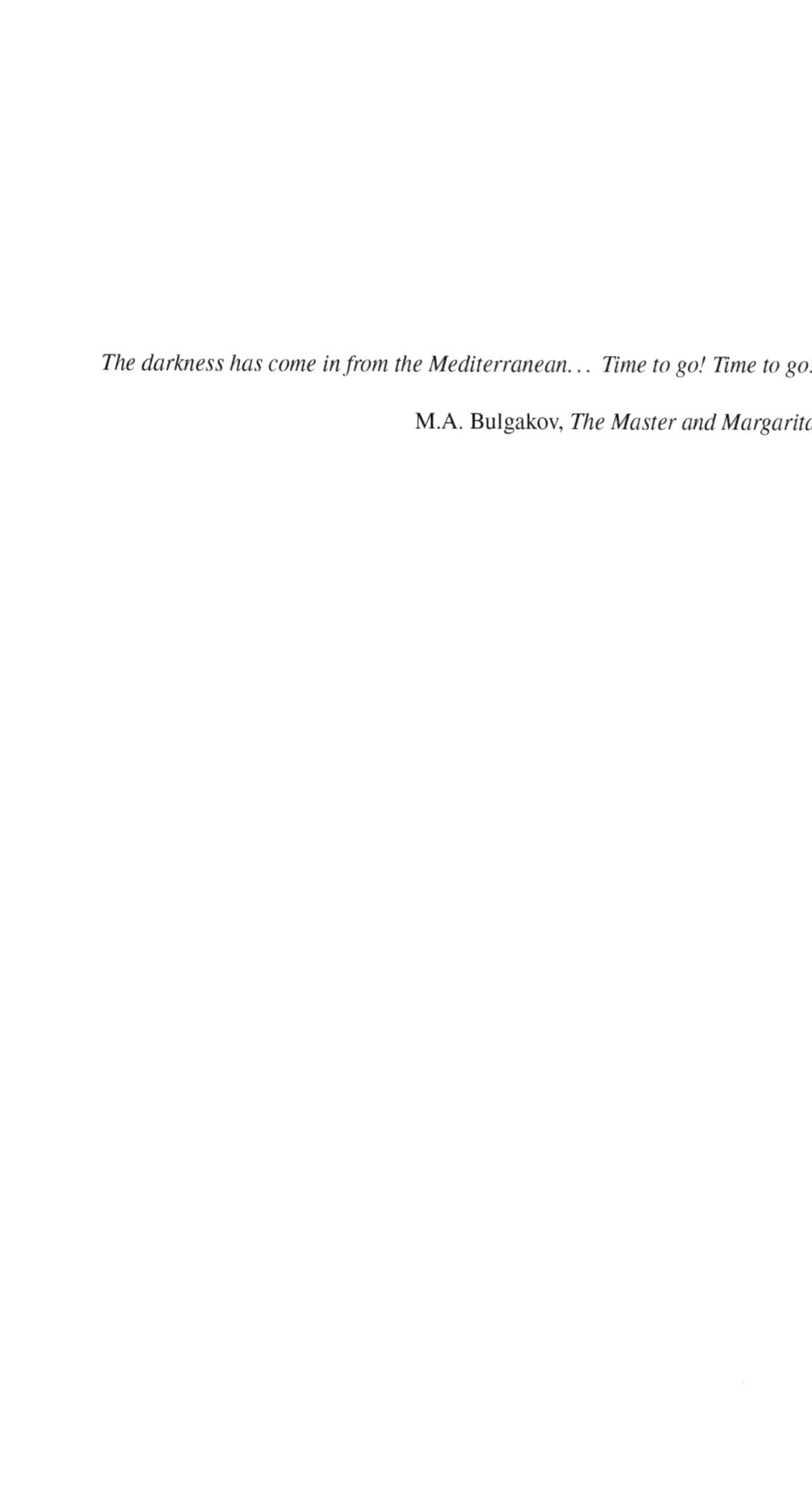

The darkness has come in from the Mediterranean... Time to go! Time to go!

M.A. Bulgakov, *The Master and Margarita*

Index

F. Forstnerič, *Stein Manifolds and Holomorphic Mappings*, Ergebnisse der Mathematik und ihrer Grenzgebiete. 3. Folge / A Series of Modern Surveys in Mathematics 56, DOI 10.1007/978-3-319-61058-0

GPSR Compliance
The European Union's (EU) General Product Safety Regulation (GPSR) is a set of rules that requires consumer products to be safe and our obligations to ensure this.

If you have any concerns about our products, you can contact us on

ProductSafety@springernature.com

In case Publisher is established outside the EU, the EU authorized representative is:

Springer Nature Customer Service Center GmbH
Europaplatz 3
69115 Heidelberg, Germany

www.ingramcontent.com/pod-product-compliance
Ingram Content Group UK Ltd.
Pitfield, Milton Keynes, MK11 3LW, UK
UKHW021919270726
14059UKWH00002B/94

* 9 7 8 3 3 1 9 6 1 0 5 7 3 *